1624
Drebble, Incubator

1728
Watt, Flyball governor

1868
Maxwell, Flyball stability analysis

1877
Routh, Stability

1890
Liapunov, Nonlinear stability

1910
Sperry, Gyroscope and autopilot

1927
Black, Feedback electronic amplifier: Bush, Differential analyzer

1932
Nyquist, Nyquist stability criterion

1938
Bode, Frequency response methods

1942
Wiener, Optimal filter design
Ziegler-Nichols PID tuning

1947
Hurewicz, Sampled data systems; Nichols, Nichols chart

1948
Evans, Root locus

1950
Kochenberger, Nonlinear analysis

1956
Pontryagin, Maximum principle

1957
Bellman, Dynamic programming

1960
Draper, Inertial navigation; Kalman, Optimal estimation

1969
Hoff, Microprocessor

Franklin

IIII

NO POSTAGE
NECESSARY IF
MAILED IN THE
UNITED STATES

BUSINESS REPLY MAIL

FIRST CLASS   PERMIT NO. 82   NATICK, MA

POSTAGE WILL BE PAID BY ADDRESSEE

THE MATHWORKS, INC.
24 Prime Park Way
Natick, MA 01760-9889

III∣∣∣∣∣∣IIIII∣∣I∣I∣II∣II∣∣I∣I∣I∣I∣I∣I∣I∣I∣I∣I∣∣∣∣II∣III

THIRD EDITION

◆ Feedback Control of ◆
Dynamic Systems

Gene F. Franklin
Stanford University

J. David Powell
Stanford University

Abbas Emami-Naeini
Integrated Systems, Inc.

Addison-Wesley Publishing Company

Reading, Massachusetts ◆ Menlo Park, California ◆ New York
Don Mills, Ontario ◆ Wokingham, England ◆ Amsterdam ◆ Bonn ◆ Sydney
Singapore ◆ Tokyo ◆ Madrid ◆ San Juan ◆ Milan ◆ Paris

To
Gertrude, David, Carole
Valerie, Daisy, Annika, Dave
Malahat, Sheila, Nima

This book is in the
Addison-Wesley Series in Electrical and Computer Engineering:
Control Engineering

Development Editor: Laurie McGuire *Manufacturing Manager:* Roy Logan
Sponsoring Editor: George J. Horesta *Illustrator:* Scientific Illustrators
Production Supervisor: Helen Wythe *Art Coordinator:* Alena Konecny
Production Coordinator: Amy Willcutt *Text Design/Layout:* Sandra Rigney
Cover Designer: Eileen Hoff *Cover photo:* T. Tracy/FPG International

Cover: Photograph of a robot soldering an element on to a p.c. board.

Many of the designations used by manufacturers and sellers to distinguish their products are claimed as trademarks. Where those designations appear in this book, and Addison-Wesley was aware of a trademark claim, the designations have been printed in initial caps or all caps.

MATLAB is a registered trademark of The MathWorks, Inc.,
24 Prime Park Way, Natick, MA 01760-1520.
Phone: (508) 653-1415, Fax: (508) 653-2997
Email: info@mathworks.com

Library of Congress Cataloging-in-Publication Data
Franklin, Gene F.
      Feedback control of dynamic systems / Gene F. Franklin, J. David
Powell. Abbas Emami-Naeini.—3rd ed.
          p.   cm.
      Includes index.
      ISBN 0-201-52747-2
      1. Feedback control systems.   I. Powell, J. David.
II. Emami-Naeini, Abbas.   III. Title.
TJ216.F723   1994
629.8'3—dc20   93-3414
                                                              CIP

Reprinted with corrections June, 1995

    5 6 7 8 9 10-MA-95

◆ **Preface** ◆

As we planned the writing and production of a third edition of *Feedback Control of Dynamic Systems* we were guided by three central objectives. First, we wanted to retain what was best about a basically good book. Second, we wanted to present the material in a way that would significantly improve its pedagogical effectiveness as a support for both students and teachers of feedback control. Finally, we wanted to enhance the book in ways that would reflect recent advances in the availability and use of computers in control system design.

The goals we have followed for each previous edition of this text serve as a basis for the third edition. These goals include developing insight into the problems of control and intuition about methods available to solve them, emphasizing design in parallel with analysis techniques, showing the unity among the several individual design techniques and synthesizing them into a "toolbox" of problem-solving methods, and presenting this interdisciplinary material in a way that is easily understood by students from any engineering background.

In order to meet these basic goals and to improve this edition's pedagogical effectiveness, we made a number of changes, including the addition of study aids to help students organize and review the large quantity of information in the book, as well as approximately three times as many problem-solving examples as the previous edition, ranging from basic drill to sophisticated design. We also provided expanded and clarified explanations of ideas and concepts that many students find difficult to understand. Finally we added computer commands for many operations and made available a MATLAB toolbox with files that will reproduce many of the figures of the book. These figures can be studied to see exactly how a given design solution works and to explore alternatives.

In formulating these goals and developing these changes, we were guided by our own experiences and benefited from extensive and very detailed reviews provided by numerous other instructors who have used the book. We made a serious effort to incorporate each of the reviewers' suggested changes, provided the change was consistent with the book's overall goals. As a result, we feel that this third edition presents the material with more pedagogical support, motivation, and accessibility than ever before while remaining a solid foundation for meeting the educational challenges of a study of feedback control.

Addressing the Educational Challenges

Some of the educational challenges facing students of feedback control are long-standing; others have emerged in recent years. Some of the challenges remain for students across their entire engineering education; others are unique to this relatively sophisticated course. Whether they are old or new, general or particular, the educational challenges we perceived were critical to the evolution of this text. Here we will state several educational challenges and describe our approaches to each of them.

CHALLENGE   *Students must master design as well as analysis techniques.*
Design is central to all of engineering and especially so to control systems. Students find that design issues, with their corresponding opportunities to tackle practical applications, particularly motivating. But students also find design problems difficult because of the poorly posed problem statements and the lack of unique solutions. Because of both its inherent importance for and its motivational effect on students, design is emphasized throughout this text so that confidence in solving design problems is developed from the start.

The emphasis on design begins in Chapter 4 following the development of modeling and dynamic response. The basic idea of feedback is introduced first, showing its influence on disturbance rejection and tracking accuracy and robustness to parameter changes. The design orientation continues with a uniform treatment of the root locus, frequency response, and state feedback techniques. All the treatments are aimed at providing the knowledge necessary to find a good feedback control design with no more complex mathematical development than is essential to clear understanding.

Throughout the text, examples are used to compare and contrast the design techniques afforded by the different design methods and, in the capstone case studies of Chapter 9, complex real-world design problems are attacked using all the methods in a unified way.

CHALLENGE   *New ideas continue to be introduced into control.*
Control is an active field of research and hence there is a steady influx of new concepts, ideas, and techniques. In time, some of these elements de-

velop to the point where they join the list of things every control engineer must know. This text is devoted to supporting students equally in their need to grasp both traditional and more modern topics.

Two of the modern aspects (1) balanced coverage of frequency response and state-space topics, and (2) an introduction to digital control are carried over from the earlier editions. The state-space methods are increasingly part of the well-prepared control engineer's toolbox, especially with the widespread availability of computer aids to carry out the computations associated with this method. Thus state-space ideas are introduced in the early chapters on models and state-space design methods are given equal emphasis (Chapter 7) with the s-plane (Chapter 5) and frequency-domain (Chapter 6) topics. Digital technology is used increasingly to implement controls and a preliminary description of digital design methods is presented in Chapter 8. However, the chapter is written in such a way that the basic ideas can be covered early in the course and can be applied throughout the book if desired.

New to the third edition is the integration of computer-aided methods. We recognize that this is a sensitive issue about which opinions vary from instructor to instructor. Some feel that time spent on computer assignments is time lost from learning basic concepts. Others feel that failure to teach computer-aided methods is a failure to prepare students properly for professional practice. We believe that computer aids are important and have included computer exercises and the ideas of Computer Aided Control System Design (CACSD) throughout the text. However, the inclusion of computer aids has been done in such a way that the main flow of the book does not depend on them and the book can be used by instructors who choose not to cover the computer-based methods at all. The computer aids are distinguished by a computer icon (🖥) for easy identification.

CHALLENGE   *Students need to manage a great deal of information.*

The vast array of systems to which feedback control is applied and the growing variety of techniques available for the solution of control problems means that today's student of feedback control must learn many new ideas. How do students keep their perspective as they plow through lengthy and complex textual passages? How to do they identify highlights and draw conclusions? How do they review for exams? Helping students with these tasks was a criterion in developing several new features for the third edition. We outline these features below.

FEATURE	REFERENCE EXAMPLE
Chapter openers offer perspective and overview. They place the specific chapter topic in the context of the discipline as a whole and they briefly overview the chapter sections.	Chapter 3 opener, pp. 85–86

FEATURE	REFERENCE EXAMPLE
Margin notes help students scan for chapter highlights. They point to important definitions, equations, and concepts.	pp. 44–45; 126–127; and 179–180
Boxed highlights identify key concepts within the running text. They also function to summarize important design procedures.	Advantage of feedback, p. 174; root-locus guidelines, p. 260
Bulleted chapter summaries help with student review and prioritization. These summaries briefly reiterate the key concepts and conclusions of the chapter.	Chapter 2 summary, p. 68–70
Synopsis of design aids. Relationships used in design and throughout the book are collected in one place for easy reference.	Inside back cover
The color blue is used (1) to highlight useful pedagogical features; (2) to highlight components under particular scrutiny within block diagrams; (3) to distinguish curves on graphs; and (4) to lend a more realistic look to figures of physical systems.	Figure 4.14, p. 188; Figure 3.22, p. 134; Figure 3.42, p. 158

CHALLENGE *Students of feedback control come from a wide range of disciplines.*

Feedback control is an interdisciplinary field in that control is applied to systems in every conceivable area of engineering. Consequently, some schools have separate introductory courses for control within the standard disciplines. However, to restrict the examples to one field is to miss much of the range and power of feedback. Covering the whole range of applications is overwhelming. In this book we aim to develop the interdisciplinary nature of the field and to provide review material for several of the most common technologies so that students from many disciplines will be comfortable with the presentation. For Electrical Engineering students who typically have a good background in transform analysis, we include in Chapter 2 an introduction to writing equations of motion for mechanical mechanisms. For mechanical engineers who typically have had a course in dynamics, we include in Chapter 3 a review of the Laplace Transform and dynamic response as needed in control. In addition, we introduce other technologies briefly and, from time to time, we present the equations of motion of a physical system without derivation but with enough physical description to be understood from a response point of view. Examples of some of the physical systems represented in the text include the read-write head for a computer disk drive, a satellite tracking system, the fuel-air ratio in an automobile engine, and an airplane automatic pilot system. A quadruple diamond icon (❖) appears next to examples containing such real-world systems.

Outline of the Book

The contents of the book are organized into nine chapters and five appendixes. The chapters include some sections of advanced or enrichment material marked with a box (■) that can be omitted without interfering with the flow of the material. The appendixes include background and reference material such as Laplace transform tables, a review of complex variables, and a review of matrix theory.

In Chapter 1, the essential ideas of feedback and some of the key design issues are introduced. It also contains a brief history of contol, from the ancient beginnings of process control to the contributions of flight control and electronic feedback amplifiers. This brief history intends to introduce the student to the origins of this field and the key figures who contributed to its development.

Chapter 2 is a short presentation of dynamic modeling and includes mechanical, electrical, electro-mechanical, fluid, and thermodynamic devices. It also discusses the state variable formulation of differential equations. This material can be omitted, used as the basis of review homework to smooth out the usual nonuniform preparation of students, or covered in depth.

Chapter 3 covers dynamic response. Again, much of this material may have been covered previously, especially by electrical engineering students. For many departments, the new material for a controls course concerns the correlation between pole locations and transient response and the effects of extra zeros and poles on dynamic response. This material needs to be covered carefully.

Chapter 4 introduces feedback in the most elementary context, permitting concentration on the essential effects of feedback on tracking accuracy, disturbance rejection, and sensitivity to model errors. Here, in the context of a first-order model for speed control, the concepts of proportional, derivative, and integral (PID) control are introduced. In this way, the student gets the idea of what control is all about *before* the tedious rules of root locus or the Nyquist Stability Criterion are developed. In this approach, the central issues of control design are brought forward and can remain in the foreground during the development of the necessary analysis that goes with construction of sophisticated design tools. The idea of steady-state tracking error, system type, and elementary stability ideas are also treated here.

Following the overview of feedback, the core of the book presents the design methods based on root locus, frequency response, and state variable feedback in Chapters 5, 6, and 7, respectively.

Chapter 8 develops the tools needed to design feedback control for implementation in a digital computer. The chapter is written so that the material can be used throughout the course. For example, a feedback control design found by using the root locus method in Chapter 5 could be implemented in a digital computer by covering Section 8.2. For a more in-depth treatment of the discrete case, Section 8.3 could be covered immediately after Chapter 3. Sections 8.4 and 8.5 might be covered after Chapter 5. However, for a complete treatment of feedback control using digital computers, the reader is referred to the companion text, *Digital Control of Dynamic Systems,* by Franklin, Powell, and Workman.

In Chapter 9 the three primary approaches are integrated in several case studies and a framework for design is described that includes a touch of the real-world context of practical control design.

Course Configurations

The material in this text can be covered flexibly. Most first-course students in controls will have some dynamics and Laplace transforms. Therefore, Chapter 2 and Section 3.2 would be a review for those students. In a ten-week quarter, it is possible to cover the remaining sections of Chapter 3, and all of Chapters 1, 4, 5, and 6. Most boxed sections should be omitted. In the second quarter, Chapters 7 and 9 can be covered comfortably including the boxed sections. Alternatively, some boxed sections could be omitted and selected portions of Chapter 8 included. A semester course should comfortably accommodate Chapters 1–7, including the review material of Chapters 2 and 3, if needed. If time remains after this core coverage, selected case studies from Chapter 9 or some introduction of digital control from Chapter 8 may be added.

Chapter 8 (digital control) may be included in several ways. The sequence shown in Fig. P.1 carries both continuous and discrete concepts along in parallel throughout the course. On the other hand, covering Section 8.2 after Chapter 5 simply gives an introduction on how to implement a controller in a digital computer without coverage of the z-transform. That material could also be used by the students for homework associated with Chapters 6 and 7. The entire book can also be used for a three-quarter sequence of courses consisting of modeling and dynamic response (Chapters 2 and 3), classical control (Chapters 4, 5, 6), and modern control (Chapters 7, 8, 9).

All the material in the book except Chapters 2 and 8 is now being used as the basic sequence of two ten-week quarter courses at Stanford University. This course sequence is taken by seniors and first-year graduate students, mostly in the departments of Aeronautics and Astronautics, Mechanical Engineering, and Electrical Engineering. The course sequence complements a graduate course in linear systems and is the prerequisite to

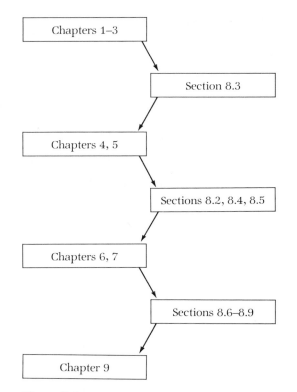

Figure P.1
A suggested outline for integrating digital controls

courses in digital control, optimal control, flight control, and smart product design. Prerequisites for the course sequence include dynamics or circuit analysis and Laplace transforms.

Prerequisites to This Feedback Control Course

This book is for a first course at the senior level for all engineering majors. For the core topics in Chapters 4–7, prerequisite understanding of modeling and dynamic response is necessary. Many students will come into the course with sufficient background in those concepts from previous courses in physics, circuits, and dynamic response. For those needing review, Chapters 2 and 3 should fill in the gaps.

An elementary understanding of matrix algebra is necessary to understand the state-space material. While all students will have much of this in prerequisite math courses, a review of the basic relations is given in Appendix C and a brief treatment of particular material needed in control is given at the start of Chapter 7. The emphasis is on the relations between linear dynamic systems and linear algebra.

A Note on Computer Integration

Some of the most common CACSD software packages are MATLAB, a product of The MathWorks, Inc., MATRIX$_x$ from Integrated Systems, Inc., CTRL-C, a product of Systems Control Technology, Inc., and CC from Systems Technology, Inc. We assume that any engineer doing control design today will have access to one or more of these tools or their equivalent. We firmly believe that to use these tools effectively the engineer must understand the basics of the method being used so that the results from the computer can be evaluated and checked for reasonableness by independent analysis. We have chosen to embed MATLAB statements throughout the text to illustrate the use of a CACSD tool to aid in particular aspects of control system design. For those students using one of the other CACSD software tools, we have included a cross reference table (Table F.1 and the inside back cover) that allows the name of a particular function for any of the software tools above to be determined easily.

As a further aid to students in quick and effective learning of CACSD tool use, all the graphical figures in the text were generated using MATLAB. These files are available at no cost by returning the card enclosed in the back of the book. The files are also available on the disk that is supplied with the instructor's manual. With these files, a user can duplicate the figure or edit the file to generate a similar figure with revised parameters or procedures. Our goal is to provide the student with sufficient resources to reap the rewards of computer calculations without the pain of programming or learning complex program procedures. We do not intend to replace the software manual by the statements in the text. Rather, our goal is to guide the student to the appropriate place in the manual and to expedite the process of making full use of the power of the computer.

Supplements

An instructor's manual with complete solutions to homework problems, categorization of problems by subtopic, and disk with MATLAB M-files is available to adopters of the third edition. The disk in the instructor's manual may be copied and distributed to students.

Acknowledgments

Finally, we wish to acknowledge our great debt to all those who have contributed to the development of feedback control into the exciting field it is today and specifically to the considerable help and education we have received from our students and our colleagues. In particular, we have benefited in this effort by many discussions with the following who taught introductory control at Stanford: A. E. Bryson, Jr., R. H. Cannon, Jr., D. B. Debra, and S. Rock. We also appreciate the comments and help of Profs. M. Anderson, J. Chiasson, S. Desa, and M. Rabins.

The thorough and thoughtful recommendations of the following reviewers were instrumental in improving the third edition: S. Centinkunt, University of Illinois at Chicago; R. I. Egbert, Wichita State University; T. Jordanides, California State University at Long Beach; J. Fleming, Texas A & M University; R. Langari, Texas A & M University; D. G. Meyer, University of Colorado at Boulder; C. P. Neuman, Carnegie-Mellon University; K. Passino, Ohio State University; and W. R. Perkins, University of Illinois at Urbana. Professors Egbert and Passino also contributed new homework problems. Special thanks go to A. L. Swindlehurst, Brigham Young University, for his thorough review, constructive suggestions, problem editing, and careful proofing of the third edition. We thank you all for your feedback and trust that our "loop compensation" has used it effectively!

The organization, editing, and guidance for addressing the needs of students provided to us by Laurie McGuire was an enormous help. Many thanks to her, Helen Wythe, and the rest of the Addison-Wesley staff.

G.F.F.
J.D.P.
A.E.-N.
Stanford, California

◆ Contents ◆

An Overview and Brief History of Feedback Control

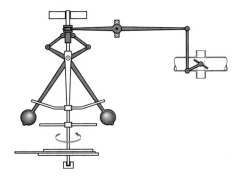

A Perspective on Feedback Control

Control is a very common concept. The term refers to specific kinds of human–machine interactions. For example, in driving an automobile, it is necessary to control the vehicle in order to arrive safely at a planned destination. Such systems require **manual control. Automatic control** involves machines only, as in room-temperature control, where the temperature is controlled by a furnace in winter and an air conditioner in summer: Both machines are in turn controlled (turned on and off) according to a thermostat reading. From these specific scenarios we can

generalize an engineering definition of control: **Control** is the process of causing a system variable to conform to some desired value, called a **reference value**. In the furnace-and-air-conditioner example, temperature is the system variable being controlled.

An extensive body of knowledge common to both manual and automatic control has evolved into the discipline of control systems design, the subject of this book. More specifically, we are concerned

with the special class of control systems that use feedback. **Feedback** is the process of measuring the controlled variable (for example, room temperature) and using that information to influence the value of the controlled variable.

1

The purposes and uses of feedback control have never been more exciting than they are today. Examples include airplane autopilots that land the airplane in dense fog, space telescopes that have a pointing accuracy of about a millionth of a degree, computer disk-drive read heads that are accurate to about a micron, and robots that deliver your meals in the hospital. The next century will no doubt see well-trained control engineers create even more imaginative applications of feedback control.

Chapter Overview

We open this chapter with an exploration of feedback control using a specific example: a household furnace controlled by a thermostat. The generic components of a control system are identified within the context of this example. In another example—an automobile cruise control—we assign actual numerical values to elements of the system model and perform a very simple mathematical analysis that shows the fundamental design issues in feedback control design. Section 1.3 provides a brief history of control theory and design. And finally, Section 1.4 provides a brief overview of the contents and organization of this text.

1.1 A Simple Feedback System

Block diagram

In feedback systems the variable being controlled—such as temperature or speed—is measured by a sensor, and the information is fed back to the process to influence the controlled variable. The principle is readily illustrated by a very common system, the household furnace controlled by a thermostat. The components of this system are shown in Fig. 1.1 as a **block diagram**. Such a picture identifies the major components as blocks in the system, omits details, and without equations shows the major directions of information and energy flow from one component to another.

FIGURE 1.1
Component block diagram of a room-temperature control system

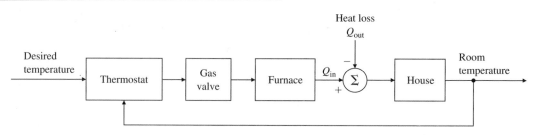

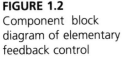

FIGURE 1.2

Component block diagram of elementary feedback control

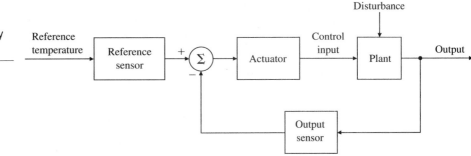

We can easily analyze the operation of this system qualitatively. Suppose both the temperature in the home where the thermostat is located and the outside temperature are significantly below the desired temperature when power is applied. The thermostat will be *on*, and the control logic will transmit power to the furnace gas valve, which will open. This will cause the furnace to fire, the blower to run, and heat to be supplied to the house. If the furnace is properly designed, the input heat Q_{in} will be much larger than the heat loss Q_{out}, and the room temperature will rise until it exceeds the thermostat setting by a small amount. At this time the furnace will be turned off, and room temperature will start to fall toward the outside value. When it falls a degree or so below the set point of the thermostat, the thermostat will come on again, and the cycle will repeat. From this example we can identify the generic components of the elementary feedback control system, shown in Fig. 1.2.

Plant

Disturbance

Actuator

Reference sensor, output sensor, comparator

The central component is the **process** or **plant,**[1] one of whose output variables is to be controlled. In our example the process is the house, the output is the house temperature, and the **disturbance** to the process is the flow of heat from the house due to conduction through the walls, roof, windows, and doors to the lower outside temperature. (The outward flow of heat also depends on other factors such as wind and open doors.) The **actuator** is the device that can influence the controlled variable of the process; in our case, the actuator is the gas furnace. Actually, the furnace has a pilot light, a gas valve, and a blower fan, which turns on or off depending on the air temperature in the furnace. These details illustrate the fact that many feedback systems contain components that themselves form other feedback systems.[2] The component labeled *thermostat* in Fig. 1.1 is divided into three parts in Fig. 1.2: the **reference** and **output sensors** and the **comparator** (the summation symbol). For purposes of feedback control we need to measure the output variable (house temperature), sense the reference

1. Since there are many control problems in factories, or "plants," the latter has become a generic term for the object to which control is applied.

2. Jonathan Swift (1733) said it this way: "So, Naturalists observe, a flea Hath smaller fleas that on him prey; And these have smaller still to bite 'em; And so proceed, *ad infinitum.*"

variable (desired temperature), and compare the two. As with this system, it is sometimes not possible for the controlled variable and the sensed variable to be the same. Although we wish to control the house temperature as a whole, the thermostat is in one particular room, which may or may not be the same temperature as the rest of the house. For example, if the thermostat is set to 75°F but is placed in the living room near a roaring fireplace, a person working in the study could still feel uncomfortably cold.

This text will present methods for analyzing feedback control systems and their components and describe the most important design techniques the engineer can use with confidence in applying feedback to solve control problems. We will also study the specific advantages of feedback that compensate for the additional complexity it demands.

1.2   A First Analysis of Feedback

The value of feedback can be readily demonstrated by quantitative analysis of a simplified model of a familiar system, the cruise control of an automobile (Fig. 1.3).

A mathematical model for cruise control

To study this situation analytically, we need a **model** of our system, that is, a set of mathematical relationships among the variables. For this example we ignore the dynamic response of the car and consider only the steady behavior. (Dynamics will play a major role in later chapters.) Furthermore, we assume that, for the range of speeds to be used by the system we may approximate the relations as linear. After measuring the speed of the vehicle on a level road at 55 mph, we find that a 1° change in the throttle angle (our control variable) causes a 10-mph change in speed. When the grade changes by 1%, we measure a speed change of 5 mph. The speedometer is found to be accurate to a fraction

FIGURE 1.3

Component block diagram of automobile cruise control

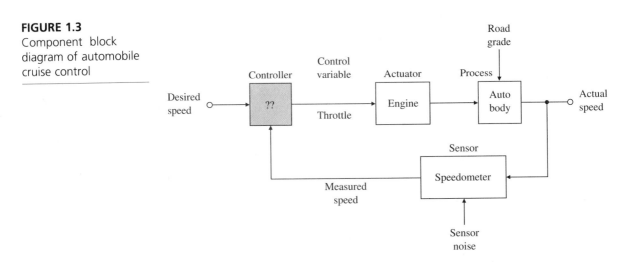

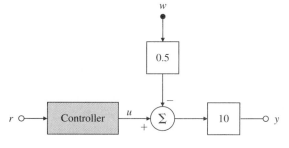

r = reference speed, mph
u = throttle angle, degrees
y = actual speed, mph
w = road grade, %

A functional-block-
diagram model of cruise
control

of 1 mph and thus can be considered exact. With these relations, we can draw
a **functional block diagram** (Fig. 1.4), which represents specific mathematical
relationships. In this kind of diagram the lines are like wires that carry signals
and a block is like an amplifier which multiplies the signal at its input by the
number or gain marked in the block to give the output signal. To sum two or
more signals, we show lines for the signals coming into a summer, a circle with
Σ inside. An algebraic sign (plus or minus) beside each arrow head indicates
whether the input adds to or subtracts from the total output of the summer.
In Fig. 1.4 no relation is given for the controller; we wish to compare the effects
of a 1% grade when the speed is set for 55 with and without feedback to the
controller.

Open-loop control

In the first case, called open-loop, the controller does not use the
speedometer reading and sets $u = r/10$ as shown in Fig. 1.5. This is an example
of an **open-loop control system**, in which there is no measurement of the output
and no subsequent use of that output to make the plant conform to the desired
output. The term *open loop* refers to the fact that there is no closed path or loop
around which the signals can go in the block diagram. In our simple example,

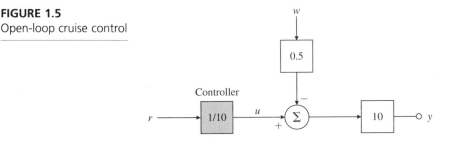

the output speed is given by

$$y_{ol} = 10(u - 0.5\,w)$$

$$= 10\left(\frac{r}{10} - 0.5\,w\right)$$

$$= r - 5w$$

If $w = 0$, a level road, and $r = 55$, then the speed will be 55 and there is no error. However, if $w = 1$ (a 1% grade), then the speed will be 50 and we have a 5 mph (or 10%) error in the speed. For a grade of 2%, the speed error would be 10 mph (or 20%), and so on. Although this example shows that there would be no error when $w = 0$, in fact, this result depends on the control gain (1/10) being the exact inverse of the plant gain of 10. In practice, this exact determination of the plant gain is virtually impossible to achieve and is typically subject to change in any case. If there is an error in the plant gain, the percent speed error would be the same as the percent plant-gain error.

Let us compare this with the feedback scheme shown in Fig. 1.6, where the feedback gain has a numerical value of 100. **Feedback gain** is the multiplication factor between the output error and the plant input. Note that the topology of this block diagram is a closed loop. A **closed-loop control system** is one that measures its output and adjusts its input accordingly by using feedback. In this case the equations are

$$y_{cl} = 10u - 5w,$$
$$u = 100(r - y_{cl}).$$

Combining them yields

$$y_{cl} = 1000r - 1000y_{cl} - 5w,$$
$$1001y_{cl} = 1000r - 5w,$$
$$y_{cl} = 0.999r - 0.005w.$$

Feedback gain

Closed-loop control

FIGURE 1.6
Closed-loop cruise
control

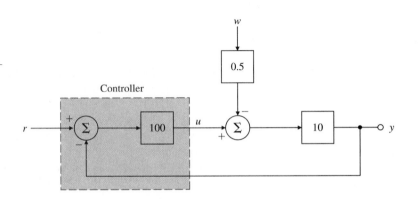

If the input set point $r = 55$ mph and the grade $w = 1\%$, then

$$y_{cl} = 0.999(55) - 0.005(1) = 54.95,$$

which is an error of only 0.05 mph, or 0.1%. If the grade $w = 10\%$, then

$$y_{cl} = 0.999(55) - 0.005(10) = 54.9,$$

an error of only 0.1 mph, or 0.2%. Thus the feedback in closed-loop control has drastically reduced the sensitivity of the speed error to the grade by a factor of 100 compared with the open-loop system. Note, however, that there is now a small speed error on level ground, because when $w = 0$,

$$y_{cl} = 0.999r = 54.945 \text{ mph.}$$

That error will be very small as long as the feedback gain $(= 100)$ is large. Unlike the open-loop case, an error in the plant gain causes little error. If the plant gain were 9 instead of the 10 used by the control designer, the feedback control equations would yield 54.939 mph whereas the open-loop control would yield 49.5 mph. The error that results when there is no disturbance is part of the **steady-state error** of a feedback system. The design of a feedback system is influenced substantially by a system's steady-state error, an issue that will come up repeatedly in the coming chapters.

Steady-state error

The large reduction of the speed sensitivity to grade disturbances and plant gain in our example is due to the relatively large gain of 100 in the feedback. Unfortunately, there are limits to how high this gain can be made; when dynamics are introduced, the feedback can make the response worse than before, even unstable. The dilemma is illustrated by another familiar situation where it is easy to change a feedback gain. If one tries to raise the gain of a public-address amplifier too much, the sound system will squeal in a most unpleasant way. This is a situation where the gain in the feedback loop—from the speakers to the microphone through the amplifier back to the speakers—is too much. The issue of how to get the gain as large as possible to reduce the errors without making the system become unstable and squeal is what much of feedback control design is all about.

The design tradeoff

1.3   A Brief History

An interesting history of early work on feedback control has been written by O. Mayr (1970), who traces the control of mechanisms to antiquity. Two of the earliest examples are the control of flow rate to regulate a water clock and the control of liquid level in a wine vessel, which is thereby kept full regardless of how many cups are dipped from it. The control of fluid flow rate is reduced to the control of fluid level, since a small orifice will produce constant flow if the pressure is constant, which is the case if the level of the liquid above the orifice is constant. The mechanism of the liquid-level control invented in antiquity and

Liquid-level control

FIGURE 1.7
Early historical control of liquid level and flow

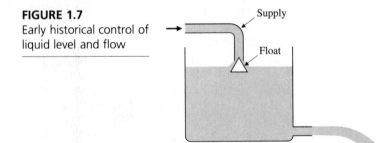

still used today (for example, in the water tank of the ordinary flush toilet) is the **float valve**. As the liquid level falls, so does the float, allowing the flow into the tank to increase; as the level rises, the flow is reduced and if necessary cut off. Figure 1.7 shows how a float valve operates. Notice here that sensor and actuator are not separate devices but contained in the carefully shaped float-and-supply-tube combination.

Drebbel's incubator

A more recent invention described by Mayr (1970) is a system, designed by Cornelis Drebbel in about 1620, to control the temperature of a furnace used to heat an incubator (Fig. 1.8). The furnace consists of a box to contain the fire, with a flue at the top fitted with a damper. Inside the fire box is the double-walled incubator box, the hollow walls of which are filled with water to transfer the heat evenly to the incubator. The temperature sensor is a glass vessel filled with alcohol and mercury and placed in the water jacket around the incubator box. As the fire heats the box and water, the alcohol expands and the riser floats up, lowering the damper on the flue. If the box is too cold, the alcohol contracts, the damper is opened, and the fire burns hotter. The desired temperature is set

FIGURE 1.8
Drebbel's incubator for hatching chicken eggs. *(Adapted from Mayr, 1970)*

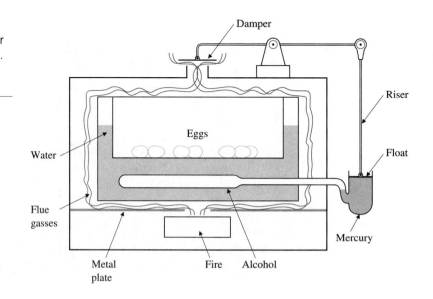

by the length of the riser, which sets the opening of the damper for a given expansion of the alcohol.

A famous problem in the chronicles of control systems was the search for a means to control the rotation speed of a shaft. Much early work (Fuller, 1976) seems to have been motivated by the desire to automatically control the speed of the grinding stone in a wind-driven flour mill. Of various methods attempted, the one with the most promise used a conical pendulum, or **fly-ball governor**, to measure the speed of the mill. The sails of the driving windmill were rolled up or let out with ropes and pulleys, much like a window shade, to maintain fixed speed. However, it was adaptation of these principles to the steam engine in the laboratories of James Watt around 1788 that made the fly-ball governor famous. An early version is shown in Fig. 1.9, while Figs. 1.10 and 1.11 show a close-up of a fly-ball governor and a sketch of its components.

The action of the fly-ball governor (also called a centrifugal governor) is simple to describe. Suppose the engine is operating in equilibrium. Two weighted balls spinning around a central shaft can be seen to describe a cone of a given angle with the shaft. When a load is suddenly applied to the engine, engine speed will slow, and the balls of the governor will drop to a smaller cone. Thus the ball angle is used to sense the output speed. This action, through the levers, will open the main valve to the steam chest (which is the actuator) and admit more steam to the engine, restoring most of the lost speed. To hold the steam valve at a new position it is necessary for the fly balls to rotate at a different angle, implying that the speed under load is not exactly the same as before. We saw this effect earlier with cruise control, where feedback control gave a very small error. To recover the exact same speed in the system, it would require resetting the desired speed setting by changing the length of the rod from the lever to the valve. Subsequent inventors introduced mechanisms that integrated

Fly-ball governor

FIGURE 1.9
A steam engine from the shop of James Watt. *(British Crown Copyright, Science Museum, London)*

FIGURE 1.10
Watts steam engine (1789–1800;) with fly-ball governor. *(British Crown Copyright, Science Museum, London)*

the speed error to provide automatic reset. In Chapter 4 we will analyze these systems to show that such integration can result in feedback systems with zero steady-state error to constant disturbances.

Because Watt was a practical man, like the millwrights before him, he did not engage in theoretical analysis of the governor. Fuller (1976) has traced the

FIGURE 1.11

Operating parts of a fly-ball governor

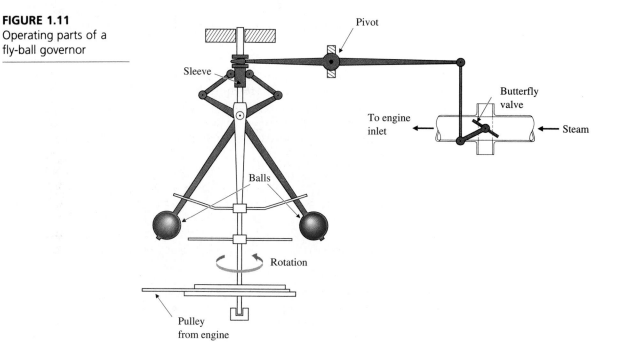

Beginnings of Control
Theory

early development of control theory to a period of studies from Christian Huygens in 1673 to James Clerk Maxwell in 1868. Fuller gives particular credit to the contributions of G. B. Airy, professor of mathematics and astronomy at Cambridge University from 1826 to 1835 and Astronomer Royal at Greenwich Observatory from 1835 to 1881. Airy was concerned with speed control; if his telescopes could be rotated counter to the rotation of the earth, a fixed star could be observed for extended periods. Using the centrifugal-pendulum governor he discovered that it was capable of unstable motion—"and the machine (if I may so express myself) became perfectly wild" (Airy, 1840; quoted in Fuller, 1976). According to Fuller, Airy was the first worker to discuss instability in a feedback control system and the first to analyze such a system using differential equations. These attributes signal the beginnings of the study of feedback control dynamics.

Stability analysis

The first systematic study of the stability of feedback control was apparently given in the paper "On Governors" by J. C. Maxwell (1868).[3] In this paper, Maxwell developed the differential equations of the governor, linearized them about equilibrium, and stated that stability depends on the roots of a certain (characteristic) equation having negative real parts. Maxwell attempted to derive conditions on the coefficients of a polynomial that would hold if all the roots had negative real parts. He was successful only for second- and third-order cases. Determining criteria for stability was the problem for the Adams Prize of 1877, which was won by E. J. Routh.[4] His criterion, developed in his essay, remains of sufficient interest that control engineers are still learning how to apply his simple technique. Analysis of the characteristic equation remained the foundation of control theory until the invention of the electronic feedback amplifier by H. S. Black in 1927 at Bell Telephone Laboratories.

Shortly after publication of Routh's work, the Russian mathematician A. M. Lyapunov (1893) began studying the question of stability of motion. His studies were based on the nonlinear differential equations of motion and also included results for linear equations that are equivalent to Routh's criterion. His work was fundamental to what is now called the state-variable approach to control theory but was not introduced into the control literature until about 1958.

Frequency response

The development of the feedback amplifier is briefly described in an interesting article based on a talk by H. W. Bode (1960) reproduced in Bellman and Kalaba (1964). With the introduction of electronic amplifiers, long-distance telephoning became possible in the decades following World War I. However, as distances increased, so did the loss of electrical energy; in spite of using larger-diameter wire, increasing numbers of amplifiers were needed to replace the lost energy. Unfortunately, large numbers of amplifiers resulted in much distortion since the small nonlinearities of the vacuum tubes then used in electronic

3. An exposition of Maxwell's contribution is given in Fuller (1976).

4. E. J. Routh was first academically in his class at Cambridge University in 1854, while J. C. Maxwell was second. In 1877 Maxwell was on the Adams Prize Committee that chose the problem of stability as the topic for the year.

amplifiers were multiplied many times. To solve the problem of reducing distortion, Black proposed the feedback amplifier. As mentioned earlier in connection with the automobile cruise control, the more we wish to reduce errors (or distortion), the more feedback we need to apply. The loop gain from actuator to plant to sensor to actuator must be made very large. With high gain the feedback loop begins to squeal and is unstable. Here was Maxwell's and Routh's stability problem again, except that in this technology the dynamics were so complex (with differential equations of order 50 being common) that Routh's criterion was not very helpful. So the communications engineers at Bell Telephone Laboratories, familiar with the concept of frequency response and the mathematics of complex variables, turned to complex analysis. In 1932 H. Nyquist published a paper describing how to determine stability from a graphical plot of the loop-frequency response. From this theory there developed an extensive methodology of feedback-amplifier design described by Bode (1945) and extensively used still in the design of feedback controls. Nyquist and Bode plots are discussed in more detail in Chapter 6.

Simultaneous with the development of the feedback amplifier, feedback control of industrial processes was becoming standard. This field, characterized by processes that are not only highly complex but also nonlinear and subject to relatively long time delays between actuator and sensor, developed **proportional-integral-derivative (PID) control**. The PID controller was first described by Callender et al. (1936). This technology was based on extensive experimental work and simple linearized approximations to the system dynamics. It led to standard experiments suitable to application in the field and eventually to satisfactory "tuning" of the coefficients of the PID controller. (PID controllers are covered in Chapter 4.) Also under development at this time were devices for guiding and controlling aircraft; especially important was the development of sensors for measuring aircraft altitude and speed. An interesting account of this branch of control theory is given in McRuer (1973).

PID control

An enormous impulse was given to the field of feedback control during World War II. In the United States engineers and mathematicians at the MIT Radiation Laboratory combined their knowledge to bring together not only Bode's feedback amplifier theory and the PID control of processes but also the theory of stochastic processes developed by N. Wiener (1930). The result was the development of a comprehensive set of techniques for the design of **servomechanisms**, as control mechanisms came to be called. Much of this work was collected and published in the records of the Radiation Laboratory by James et al. (1947).

Another approach to control systems design was introduced in 1948 by W. R. Evans, who was working in the field of guidance and control of aircraft. Many of his problems involved unstable or neutrally stable dynamics, which made the frequency methods difficult, so he suggested returning to the study of the characteristic equation that had been the basis of the work of Maxwell and Routh nearly 70 years earlier. However, Evans developed techniques and rules allowing one to follow graphically the paths of the roots of the characteristic

Root locus

State-variable design

Modern control
Classical control

equation as a parameter was changed. His method, the **root locus**, is suitable for design as well as for stability analysis and remains an important technique today. The root-locus method developed by Evans is covered in Chapter 5.

During the 1950s several authors, including R. Bellman and R. E. Kalman in the United States and L. S. Pontryagin in the U.S.S.R., began again to consider the ordinary differential equation (ODE) as a model for control systems. Much of this work was stimulated by the new field of control of artificial earth satellites, in which the ODE is a natural form for writing the model. Supporting this endeavor were digital computers, which could be used to carry out calculations unthinkable 10 years before. (Now, of course these calculations can be done by any engineering student with a desktop computer). The work of Lyapunov was translated into the language of control at about this time, and the study of optimal controls, begun by Wiener and Phillips during World War II, was extended to optimizing trajectories of nonlinear systems based on the calculus of variations. Much of this work was presented at the first conference of the newly formed International Federation of Automatic Control held in Moscow in 1960.[5] This work did not use the frequency response or the characteristic equation but worked directly with the ODE in "normal" or "state" form and typically called for extensive use of computers. Even though the foundations of the study of ODEs were laid in the late 19th century, this approach is now often called **modern control** to distinguish it from **classical control**, which uses the complex variable methods of Bode and others. In the period from the 1970s continuing through the present, we find a growing body of work that seeks to use the best features of each technique.

Thus we come to the current state of affairs. We feel that the well-prepared engineer should understand many techniques, should be able to choose the method best suited to the problem at hand, and should be able to use that method to guide and verify automatic computer calculations.[6]

1.4   An Overview of the Book

The central purpose of this book is to introduce the most important techniques for control systems design. Chapter 2 will review the techniques necessary to obtain models of the dynamic systems that we wish to control: These include model making for mechanical, electric, electromechanical, and a few other physical systems. Also described in Chapter 2 is the linearization of nonlinear models.

---

5. Optimal control gained a large boost when Bryson and Denham (1962) showed that the path of a supersonic aircraft should actually dive at one point in order to reach a given altitude in minimum time. This nonintuitive result was later demonstrated to skeptical fighter pilots in flight tests.

6. For more background on the history of control, see the set of survey papers appearing in the *IEEE Control Systems Magazine* (November 1984).

In Chapter 3 we will discuss the analysis of dynamic response using Laplace transforms, along with the relationship between time response and the poles and zeros of the transfer function. The chapter will also include a brief discussion of numerical simulation and how to construct models from experimental data.

In Chapter 4 we will cover the basic features of feedback. An analysis of the effects of feedback on disturbance rejection, on tracking accuracy, and on dynamic response will be given. In the most elementary of situations we will present the essential problems of control and how they can be addressed.

In Chapters 5 through 7 we will introduce the techniques for realizing, in more complex dynamic systems, the objectives identified in Chapter 4. These methods include root locus, frequency response, and state-variable pole placement. They are alternative means to the same end and have different advantages and disadvantages. The methods are fundamentally complementary, and each needs to be understood to achieve the most effective control systems design.

In Chapter 8 we will introduce briefly the ideas associated with implementing control in a digital computer. The chapter addresses how one "digitizes" the control equations developed in Chapters 4 through 7, how the analysis of sampled systems requires another analysis tool—the z-transform—and how the sampling introduces a delay that tends to destabilize the system.

Application of all the techniques to problems of substantial complexity will be discussed in Chapter 9. There you will find all the design methods brought to bear simultaneously on specific case studies.

Computer aids

Most control designers today make extensive use of computer-aided control systems design (CACSD) software tools that are commercially available. These tools are capable of performing many of the calculations that are discussed in this book. Furthermore, most instructional programs in control systems design will also make CACSD tools available to the students. The most widely used software are MATLAB, MATRIX$_x$, CTRL-C, and CC.[7] You will find references to MATLAB routines throughout the text to help illustrate this method of solution. They are not meant to be a substitute for the software manual; rather, they will help guide you to the appropriate place in the manual for specific calculations. Appendix F contains a table that cross references the routine in MATLAB with those from the other software products. Furthermore, many of the figures in the book were created using MATLAB, and the files for their creation are in a disk that is available free of charge from the Mathworks, Inc., by returning the card at the back of the book. They should be helpful to you in learning how to use CACSD methods and in solving similar problems.

Needless to say, due to space limitations many topics are not treated in the book. We do not extend the methods to systems with more than one input or output, so-called multivariable control. Nor is optimal control treated in more than a very introductory manner. We have had to omit any real consideration of nonlinear control, despite the fact that many real design problems are for

---

7. MATLAB is a product of The Mathworks, Inc.; MATRIX$_x$ is from Integrated Systems, Inc.; Ctrl-C is put out by Systems Control Technology; and CC is a product of Systems Technology, Inc.

nonlinear plants. Also beyond the scope of this text is a detailed treatment of the experimental testing and modeling of real hardware, which is the ultimate test of whether any design really works. The book concentrates on analysis and design of linear control for linear plant models—not because we think that is the final test of a design, but because that is the best way to grasp the basic ideas of feedback and to be able to quickly answer questions of error characteristics and stability. We believe that mastery of the material here will provide a foundation of understanding on which you can build knowledge of these more advanced and realistic topics—a foundation strong enough to allow you to build your own design method in the tradition of all those who worked to give us the knowledge we present here.

Summary

- **Control** is the process of making a system variable adhere to a particular value, called the **reference value**. Two kinds of control were defined and illustrated: In **open-loop control** the system does not measure the output, and there is no compensation of that output to make it conform to the desired output. In **closed-loop control** the system uses **feedback**, which is the process of measuring a control variable and returning the output to influence the value of the variable.

- A simple feedback system consists of the **actuator**, the **comparator**, the **process** or **plant**, and **reference** and **output sensors**.

- Block diagrams are helpful for visualizing what is happening in control systems. A **component block diagram** represents in block form the major components in a control system and shows the major directions of information and energy flow from one component to another. A **functional block diagram** illustrates the mathematical relationships between the components in a control system.

- There are two present-day approaches to control: **Classical control** methods use complex variables, while **modern control** methods also rely on ordinary differential equations (ODEs).

Problems

1.1 Draw the component block diagram for each of the following common feedback control systems:

    a) the manual steering system of an automobile
    b) Drebbel's incubator
    c) the water level controlled by a float and valve
    d) Watt's steam engine with fly-ball governor
    e) the automatic steering of an ocean-going ship
    f) a public address system

FIGURE 1.12

A papermaking machine. *(From Åström, 1970, p. 192. Reprinted with permission)*

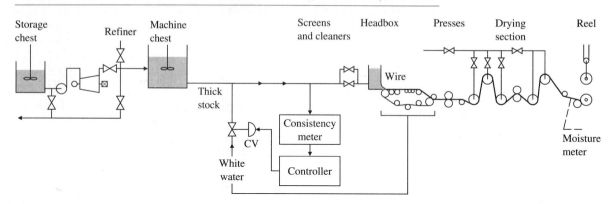

Identify the following components in each diagram:

- process
- actuator
- sensor
- reference input
- controlled output
- actuator output
- sensor output

1.2 The thermostat in your home or office might use one of the following controls: a tube filled with liquid, a bimetallic strip, spring bellows, or an electronic controller. Describe how each works.

1.3 A machine for making paper is diagrammed in Fig. 1.12. There are two main parameters under feedback control: the density of fibers as controlled by the consistency of the thick stock that flows from the headbox onto the wire, and the moisture content of the final product that comes out of the dryers. Stock from the machine chest is diluted by white water returning from under the wire as controlled by a control valve (CV). A meter supplies a reading of the consistency. At the "dry end" of the machine, there is a moisture sensor. Draw a component block diagram, and identify the seven components listed in Problem 1.1 for

a) control of consistency

b) control of moisture.

1.4 Many variables in the human body are under feedback control. For each of the following controlled variables, draw the component block diagram showing the process being controlled, the sensor that measures the variable, the actuator that causes it to increase and/or decrease, the information path that completes the feedback path, and the disturbances that upset the variable. You may need to consult an encyclopedia or textbook on human physiology for information.

a) blood pressure

b) blood sugar concentration

c) heart rate

d) eye-pointing angle

e) eye-pupil diameter

f) blood calcium level

1.5 Draw the component block diagram for temperature control in a refrigerator.

1.6 Draw the component block diagram for an elevator-position control. Indicate how you would measure the position of the elevator car. Consider a combined coarse and fine measurement system. What accuracies do you suggest for each sensor?

1.7 Feedback control requires being able to sense the variable being controlled. Because electrical signals can be transmitted, amplified, and processed easily, often we want to use this fact by having the sensor deliver as output a voltage proportional to the variable being measured. Devise the principles of operation, and sketch a diagram suitable to explain the operation of the different kinds of sensors that would measure the following:

a) temperature

b) pressure

c) liquid level

d) flow of liquid along a pipe (or blood along an artery)

e) linear position

f) rotational position

g) linear velocity

h) rotational speed

i) translational acceleration

j) force

k) torque

In each case comment on the expected dynamic range, the linearity, and the signal–to–noise ratio of your device.

1.8 Each of the variables listed in Problem 1.7 can be brought under feedback control. Describe alternative actuators that can influence each variable. Comment on the range of power levels available in each case.

Dynamic Models

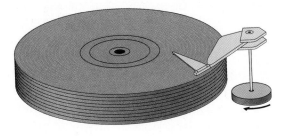

A Perspective on Dynamic Models

The overall goal of feedback control is to use the principle of feedback to cause the output variable of a dynamic process to follow a desired reference variable accurately regardless of the reference variable's path and of any external disturbances or any changes in the dynamics of the process. This complex goal is met as the result of a number of simple, distinct steps. The first of these is to develop a mathematical description (called a **dynamic model**) of the process to be controlled. The term **model**, as it is used and understood by control engineers, means a set of differential equations that describe the dynamic behavior of the system.

In many cases the modeling of complex systems is difficult and expensive, especially when the important steps of building and testing prototypes are included. However, in this introductory text, we will focus on the most basic principles of modeling for the most common physical systems. More comprehensive sources and specialized texts will be referenced throughout the text where appropriate for those wishing more detail.

In later chapters we will explore a variety of analysis methods for dealing with the equations of motion, including the state-variable method, which is introduced briefly in this chapter. Because it lends itself to computer analysis, this method is of particular interest to contemporary students of control.

Chapter Overview

The fundamental step in building a dynamic model is writing the equations of motion for the system. Through discussion and a variety of examples, Section 2.1 demonstrates how to write the equations of motion for a variety of mechanical systems. Once such equations are written, they can be expressed in the state-variable form, as defined and shown in Section 2.2. Models in state-variable form allow us to apply the computational efficiency of analysis tools such as MATLAB; an example using this tool is shown in Section 2.2.

In addition to the mechanical systems discussed, electric circuits, electromechanical systems, and heat and fluid-flow systems are modeled in Sections 2.3 to 2.5, respectively. Finally, Section 2.6 concludes with a discussion of linearization and scaling. The differential equations developed in modeling are often nonlinear. Because they are significantly more challenging to solve than linear ones and because linear models are usually adequate, this concluding explanation of how to linearize nonlinear equations is intended to make possible more efficient analysis and design.

In order to focus on the important first step of developing mathematical models, we will defer explanation of the computational methods used to solve the problems in this chapter until Chapter 3.

2.1 Dynamics of Mechanical Systems

Newton's law for translational motion

The cornerstone for obtaining a mathematical model, or the **equations of motion**, for any mechanical system is Newton's law,

$$\mathbf{F} = m\mathbf{a}, \tag{2.1}$$

where

$\mathbf{F}$ = the vector sum of all forces applied to each body in a system, Newtons (N) or pounds (lb),

$\mathbf{a}$ = the vector acceleration of each body with respect to an inertial reference frame (that is, one that is neither accelerating nor rotating with respect to the stars); often called **inertial acceleration**, m/sec² or ft/sec²,

m = mass of the body, kg or slug.

Note that here in Eq. (2.1), as throughout the text, we use the convention of boldfacing the type to indicate that the quantity is a matrix or vector, possibly a vector function.

In SI units a force of one Newton will impart an acceleration of $1 \, \text{m/sec}^2$ to a mass of 1 kilogram. In English units a force of 1 lb will impart an acceleration of $1 \, \text{ft/sec}^2$ to a mass of 1 slug. The "weight" of an object is mg, where g is the acceleration of gravity ($=9.81 \, \text{m/sec}^2 = 32.2 \, \text{ft/sec}^2$). In English units it is common usage to refer to the mass of an object in terms of its weight in pounds, which is the quantity measured on scales. To obtain the mass in slugs, divide the weight by g. Therefore, an object weighing 1 lb has a mass of $1/32.2$ slugs. A slug has dimensions $\text{lb} \cdot \text{sec}^2/\text{ft}$. In SI units, scales are typically calibrated in kilograms, which is a direct measure of mass.

Application of this law typically involves defining convenient coordinates to account for the body's motion (position, velocity, and acceleration), determining the forces on the body using a free-body diagram, and then writing the equations of motion from Eq. (2.1). The procedure is simplest when the coordinates chosen express the position with respect to an inertial frame because in this case the accelerations needed for Newton's law are simply the second derivatives of the position coordinates.

Use of free-body
diagram in applying
Newton's law

♦ **EXAMPLE 2.1**   *A Simple System; Cruise-control Model*

   Write the equations of motion for the speed and forward motion of the car shown in Fig. 2.1 assuming that the engine imparts a force u as shown.

Solution.   For simplicity we assume that the rotational inertia of the wheels is negligible and that there is friction retarding the motion of the car, friction that is proportional to the car's speed. The car can then be approximated for modeling purposes using the free-body diagram seen in Fig. 2.2, which defines coordinates, shows all forces acting on the body (heavy lines), and indicates the acceleration (dashed lines). The coordinate of the car's position, x, is the distance from the reference line shown and chosen so that positive is to the right. Note that in this case the inertial acceleration is simply the second derivative of x (that is, $\mathbf{a} = \ddot{x}$) because the car position is measured with respect to an inertial reference. The equation of motion is found using Eq. (2.1). The friction force acts opposite to the direction of motion; therefore it is drawn opposite the direction of positive

FIGURE 2.1
Cruise-control model

FIGURE 2.2
Free-body diagram for
cruise control

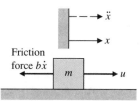

motion and entered as a negative force in Eq. (2.1). The result is

$$u - b\dot{x} = m\ddot{x},\tag{2.2}$$

or

$$\ddot{x} + \frac{b}{m}\dot{x} = \frac{u}{m}.\tag{2.3}$$

For the case of the automotive cruise control where the variable of interest is the speed, $v\ (=\dot{x})$, the equation of motion becomes

$$\dot{v} + \frac{b}{m}v = \frac{u}{m}.\tag{2.4}$$

◆

Newton's law also can be applied to systems with more than one mass. In this case it is particularly important to draw the free-body diagram of each mass showing the applied external forces as well as the equal and opposite internal forces that act from each mass on the other.

◆ **EXAMPLE 2.2**   *A Two-mass System: Suspension Model*

Fig. 2.3 shows an automobile suspension system. Write the equations of motion for the automobile and wheel motion assuming one-dimensional vertical motion of a mass

FIGURE 2.3
Automobile suspension

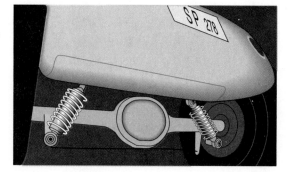

FIGURE 2.4
The quarter-car model

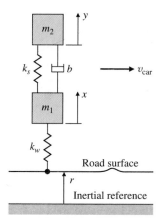

above one wheel. A system comprised of one of the four wheel suspensions is usually referred to as a quarter-car model.

Solution.   The system can be approximated by the simplified system shown in Fig. 2.4. The coordinates of the two masses, x and y, are the displacements of the masses from their equilibrium conditions. The equilibrium positions are offset from the springs' unstretched positions because of the force of gravity. The shock absorber is represented in the schematic diagram by a dashpot symbol with friction constant b. The magnitude of the force from the shock absorber is assumed to be proportional to the rate of change of the relative displacement of the two masses, that is, to $b(\dot{y} - \dot{x})$.

The force from the spring acts on both masses in proportion to their relative displacement with spring constant k_s. Figure 2.5 shows the free-body diagram of each mass. Note that the forces from the spring on the two masses are equal in magnitude but act in opposite directions, which is also the case for the damper. A positive displacement y of mass m_2 will result in a force from the spring on m_2 in the direction shown and a force from the spring on m_1 in the direction shown. However, a positive displacement x of mass m_1 will result in a force from the spring k_s on m_1 in the opposite direction to that drawn in Fig. 2.5 as indicated by the *minus x*-term for the spring force.

The lower spring k_w represents the tire compressibility, for which there is insufficient damping (velocity-dependent force) to warrant including a dashpot in the model. The force from this spring is proportional to the distance the wheel axle rides above the road, $x - r$, and a force will result if either the road surface has a bump (r changes) or the wheel

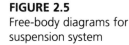

FIGURE 2.5
Free-body diagrams for suspension system

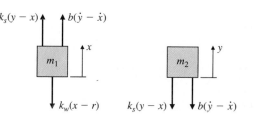

bounces (x changes). The motion of the simplified car over a bumpy road will result in a value of $r(t)$ that is not constant.

There is a constant force of gravity acting on each mass; however, this force has been omitted, as have the equal and opposite forces from the springs. Gravitational forces can always be omitted from vertical-spring mass systems (1) if the position coordinates are defined from the equilibrium position that results when gravity is acting, and (2) if the spring forces used in the analysis are actually the perturbation in spring forces from those forces acting at equilibrium.

Applying Eq. (2.1) to each mass and noting that some forces on each mass are in the negative (down) direction yields the system of equations

$$b(\dot{y} - \dot{x}) + k_s(y - x) - k_w(x - r) = m_1\ddot{x},$$
$$-k_s(y - x) - b(\dot{y} - \dot{x}) = m_2\ddot{y}. \tag{2.5}$$

Some rearranging results in

$$\ddot{x} + \frac{b}{m_1}(\dot{x} - \dot{y}) + \frac{k_s}{m_1}(x - y) + \frac{k_w}{m_1}x = \frac{k_w}{m_1}r, \tag{2.6a}$$

$$\ddot{y} + \frac{b}{m_2}(\dot{y} - \dot{x}) + \frac{k_s}{m_2}(y - x) = 0. \tag{2.6b}$$

The most common source of error in writing equations for systems like these are sign errors. The method for keeping the signs straight in the development above involved mentally picturing the displacement of the masses and drawing the resulting force in the direction that the displacement would produce. Once you have obtained the equations for a system, a check on the signs for systems that are obviously stable from physical reasoning can be quickly carried out. You will see in Chapter 4 that a stable system always has the same signs on each variable. For this system Eq. (2.6a) shows that the signs on the $\ddot{x}$-, $\dot{x}$-, and x-terms are all positive, as they must be for stability. Likewise, the signs on the $\ddot{y}$-, $\dot{y}$-, and y-terms are all positive in the second equation.

Check for sign errors.

◆

Application of Newton's law to one-dimensional rotational systems requires that Eq. (2.1) be modified to

Newton's law for rotational motion

$$M = I\alpha, \tag{2.7}$$

where

M = the sum of all moments about the center of mass of a body, N·m or lb·ft,

I = the body's moment of inertia about its center of mass, kg·m² or slug·ft²,

α = the angular acceleration of the body, rad/sec².

◆ **EXAMPLE 2.3** *Rotational Motion: Satellite Control Model*

Satellites, as shown in Fig. 2.6, usually require attitude control so that antennas, sensors, and solar panels are properly oriented. Antennas are usually pointed toward a particular location on earth, while solar panels need to be oriented toward the sun for maximum power generation. To gain insight into the full three-axis attitude-control system, it is helpful to consider one axis at a time. Write the equations of motion for one axis of this system.

FIGURE 2.6
Communications satellite
*(Courtesy Space
Systems/Loral)*

FIGURE 2.7
Satellite control
schematic

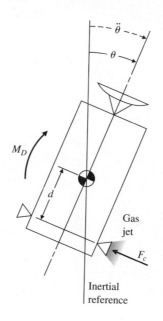

Solution. Figure 2.7 depicts this case, where motion is only allowed about the axis perpendicular to the page. The angle θ that describes the satellite orientation must be measured with respect to an inertial reference, that is, a reference that has no angular acceleration. The control force comes from reaction jets that produce a moment of $F_c d$ about the mass center. There may also be small disturbance moments on the satellite, M_D, which arise primarily from solar pressure acting on any asymmetry in the solar panels. Applying Eq. (2.7) yields the equation of motion

$$F_c d + M_D = I\ddot{\theta}. \tag{2.8}$$

Double-integrator plant

The output of this system, θ, results from integrating the sum of the input torques twice; hence this type of system is often referred to as the **double-integrator plant**.

In many cases a system, such as the disk-drive read/write head shown in Fig. 2.8, in reality has some flexibility, which can cause problems in the design of

FIGURE 2.8
Disk read/write mechanism *(Photo courtesy of Hewlett-Packard Company.)*

a control system. Particular difficulty arises when there is flexibility, as in this case, between the sensor and actuator locations. Therefore, it is often important to include this flexibility in the model even when the system seems to be quite rigid.

◆ **EXAMPLE 2.4**   *Flexibility: Flexible Read/write for a Disk Drive*

❖   Assume there is some flexibility between the read head and the drive motor in Fig. 2.8. Find the equations of motion relating the motion of the read head to a torque applied to the base.

Solution.   Schematically the dynamic model for this situation is as shown in Fig. 2.9. This model is dynamically similar to the resonant system shown in Fig. 2.4 and results in equations of motion that are similar in form to Eqs. (2.6). The moments on each body are shown in the free-body diagrams in Fig. 2.10. The discussion of the moments on each body is essentially the same as the discussion for Example 2.2, except the springs and damper in that case produced forces, instead of moments that act on each inertia, as in

FIGURE 2.9
Disk read/write head

FIGURE 2.10
Free-body diagrams of
the disk read/write head

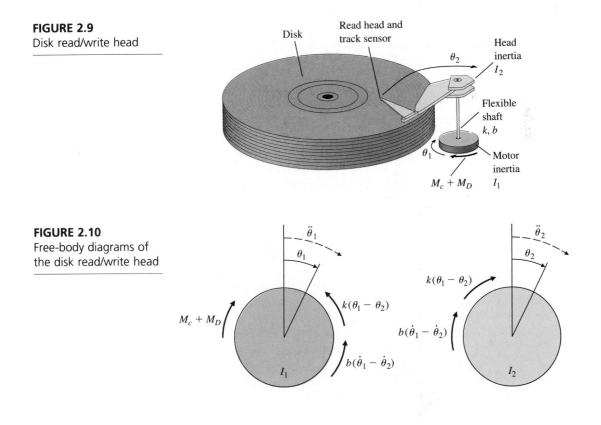

this case. When the moments are summed, equated to the accelerations according to Eq. (2.7), and rearranged, the result is

$$I_1\ddot{\theta}_1 + b(\dot{\theta}_1 - \dot{\theta}_2) + k(\theta_1 - \theta_2) = M_c + M_D$$
$$I_2\ddot{\theta}_2 + b(\dot{\theta}_2 - \dot{\theta}_1) + k(\theta_2 - \theta_1) = 0.$$

(2.9)

In the special case where a point in a rotating body is fixed with respect to an inertial reference, as is the case with a pendulum, Eq. (2.7) can be applied such that M is the sum of all moments about the *fixed* point and I is the moment of inertia about the fixed point.

◆ **EXAMPLE 2.5**   *Rotational Motion: Pendulum*

Write the equations of motion for the simple pendulum shown in Fig. 2.11.

FIGURE 2.11
Pendulum

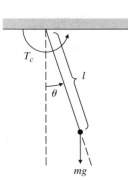

Solution.   The moment of inertia about the pivot point is $I = ml^2$. The sum of moments about the pivot point contains a term from gravity as well as the applied torque T_c. The equation of motion, obtained from Eq. (2.7), is

$$T_c - mgl\sin\theta = I\ddot{\theta},$$

(2.10)

which is usually written in the form

$$\ddot{\theta} + \frac{g}{l}\sin\theta = \frac{T_c}{ml^2}.$$

(2.11)

This equation is nonlinear due to the $\sin\theta$ term. If the motion is assumed to be small enough that we can let $\sin\theta \cong \theta$, Eq. (2.11) then becomes the linear equation

$$\ddot{\theta} + \frac{g}{l}\theta = \frac{T_c}{ml^2}.$$

(2.12)

If there is no applied torque, this represents the motion of a harmonic oscillator with a natural frequency of

$$\omega_n = \sqrt{\frac{g}{l}}.$$

(2.13)

In some cases, mechanical systems contain both translational and rotational portions. The procedure is the same; sketch the free-body diagrams, define coordinates and positive directions, determine all forces and moments acting, and apply Eqs. (2.1) and/or (2.7).

Typically, the resulting equations of motion will be nonlinear. Such equations are much more difficult to solve than linear ones, and the kinds of possible motions resulting from a nonlinear model are much more difficult to categorize than those resulting from a linear model. It is therefore useful to linearize models in order to gain access to linear analysis methods. It may be that the linear models and linear analysis are used only for the design of the control system (whose function may be to maintain the system in the linear region). Once a control system is synthesized and shown to have desirable performance based on linear analysis, it is then prudent to carry out an accurate numerical simulation of the system with all the nonlinearities in order to validate that performance.

◆ **EXAMPLE 2.6**   *Rotational and Translational Motion: Hanging Crane*

Write the equations of motion for the hanging crane pictured in Fig. 2.12 and shown schematically in Fig. 2.13. Linearize the equations about $\theta = 0$, which would typically be valid for the hanging crane. Also linearize the equations for $\theta = \pi$, which represents the situation for the inverted pendulum shown in Fig. 2.14.

FIGURE 2.12
Crane with a hanging
load *(Photo courtesy of
Harnischfeger
Corporation, Milwaukee,
Wisconsin, 1992.)*

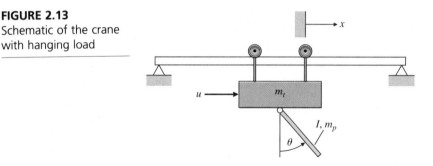

Solution.   A schematic diagram of the hanging crane is shown in Fig. 2.13 while the free-body diagrams are shown in Fig. 2.15. In the case of the pendulum, the forces are shown with bold lines, while the components of the inertial acceleration of its center of mass are shown with dashed lines. Because the pivot point of the pendulum is *not* fixed with respect to an inertial reference, the rotation of the pendulum and the motion of its mass center must be considered. The inertial acceleration needs to be determined because the vector **a** in Eq. (2.1) is given with respect to an inertial reference. The total inertial acceleration of the pendulum's mass center is the vector sum of the three dashed arrows shown in Fig. 2.15(b). The derivation of the components of an object's acceleration is called **kinematics** and is usually studied as a prelude to the application of Newton's laws. Alternatively, one could express the center of mass of the pendulum as a vector from an inertial reference and then differentiate that vector twice to obtain an inertial accelera-tion. The results of a kinematic study are shown in Fig. 2.15(b). The component of acceleration along the pendulum is $l\dot{\theta}^2$ and is called the centripetal acceleration. It is present for any object whose velocity is changing direction. The $\ddot{x}$-component of acceleration is a consequence of the pendulum pivot point accelerating at the trolley's acceleration and will always have the same direction and magnitude as those of the trolley's. The $l\ddot{\theta}$ component is a result of angular acceleration of the pendulum and is always perpendicular to the pendulum.

To understand the $l\ddot{\theta}$ and $l\dot{\theta}^2$ terms better, consider the situation in Fig. 2.15(c) where the $\hat{\mathbf{i}}$ and $\hat{\mathbf{j}}$ axes are inertially fixed. A vector **r** describing the position of the pendulum center of mass can be expressed as

$$\mathbf{r} = x\hat{\mathbf{i}} + l(\hat{\mathbf{i}} \sin \theta - \hat{\mathbf{j}} \cos \theta).$$

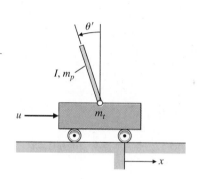

FIGURE 2.15
Hanging crane: (a) free-body diagram of the trolley; (b) free-body diagram of the pendulum; (c) position vector of the pendulum

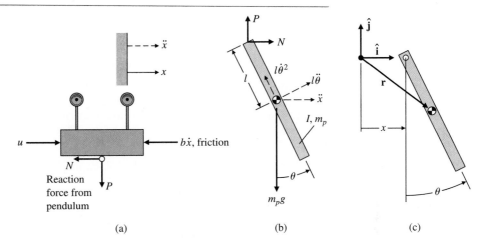

| | | |
| (a) | (b) | (c) |

The first derivative of **r** is

$$\dot{\mathbf{r}} = \dot{x}\hat{\mathbf{i}} + l\dot{\theta}(\hat{\mathbf{i}}\cos\theta + \hat{\mathbf{j}}\sin\theta).$$

Likewise, the second derivative of **r** is

$$\ddot{\mathbf{r}} = \ddot{x}\hat{\mathbf{i}} + l\ddot{\theta}(\hat{\mathbf{i}}\cos\theta + \hat{\mathbf{j}}\sin\theta) - l\dot{\theta}^2(\hat{\mathbf{i}}\sin\theta - \hat{\mathbf{j}}\cos\theta).$$

Note that the equation for $\ddot{\mathbf{r}}$ confirms the acceleration components shown in Fig. 2.15(b). The $l\dot{\theta}^2$ term is aligned along the pendulum pointing toward the axis of rotation and the $l\ddot{\theta}$ term is aligned perpendicular to the pendulum pointing in the direction of a positive rotation.

Having all the forces and accelerations for the two bodies, we now proceed to apply Eq. (2.1). In the case of the trolley, Fig. 2.15(a), we see that it is constrained by the tracks to move only in the x-direction; therefore, application of Eq. (2.1) in this direction yields

$$m_t\ddot{x} + b\dot{x} = u - N. \tag{2.14}$$

Conceptually, Eq. (2.1) can be applied to the pendulum of Fig. 2.15(b) in the vertical and horizontal directions, and Eq. (2.7) can be applied for rotational motion to yield three equations in the three unknowns: N, P, and θ. These three equations can then be manipulated to eliminate the reaction forces N and P so that a single equation results describing the motion of the pendulum, that is, a single equation in θ. For example, application of Eq. (2.1) in the x-direction yields

$$N = m_p\ddot{x} + m_p l\ddot{\theta}\cos\theta - m_p l\dot{\theta}^2\sin\theta. \tag{2.15}$$

However, considerable algebra will be avoided if Eq. (2.1) is applied perpendicular to the pendulum to yield

$$P\sin\theta + N\cos\theta - m_p g\sin\theta = m_p l\ddot{\theta} + m_p\ddot{x}\cos\theta. \tag{2.16}$$

Application of Eq. (2.7) for the rotational pendulum motion where the moments are summed about the center of mass yields

$$-Pl\sin\theta - Nl\cos\theta = I\ddot\theta, \tag{2.17}$$

where I is the moment of inertia about the pendulum's mass center. The reaction forces N and P can now be eliminated by combining Eqs. (2.16) and (2.17). This yields the equation

$$(I + m_p l^2)\ddot\theta + m_p gl\sin\theta = -m_p l\ddot{x}\cos\theta. \tag{2.18}$$

It is identical to a pendulum equation of motion, except that it contains a forcing function that is proportional to the trolley's acceleration.

An equation describing the trolley motion was found in Eq. (2.14), but it contains the unknown reaction force N. By combining Eqs. (2.15) and (2.14) N can be eliminated to yield,

$$(m_t + m_p)\ddot{x} + b\dot{x} + m_p l\ddot\theta\cos\theta - m_p l\dot\theta^2\sin\theta = u. \tag{2.19}$$

Equations (2.18) and (2.19) are the nonlinear differential equations that describe the motion of the crane with its hanging load. For an accurate calculation of the motion of the system, these nonlinear equations need to be solved.

To linearize the equations for small motions about $\theta = 0$, let $\cos\theta \cong 1$, $\sin\theta \cong \theta$, and $\dot\theta^2 \cong 0$; thus the equations are approximated by

$$(I + m_p l^2)\ddot\theta + m_p gl\theta = -m_p l\ddot{x}$$
$$(m_t + m_p)\ddot{x} + b\dot{x} + m_p l\ddot\theta = u. \tag{2.20}$$

For the inverted pendulum in Fig. 2.14 where $\theta \cong \pi$, assume that $\theta = \pi + \theta'$, where θ' represents motion from the vertical *upward* direction. In this case, $\cos\theta \cong -1$, $\sin\theta \cong -\theta'$ in Eqs. (2.18) and (2.19), and Eqs. (2.20) become[1]

Inverted pendulum equations

$$(I + m_p l^2)\ddot\theta' - m_p gl\theta' = m_p l\ddot{x}$$
$$(m_t + m_p)\ddot{x} + b\dot{x} - m_p l\ddot\theta' = u. \tag{2.21}$$

◆

Summary: Developing Equations of Motion for Rigid Bodies

The physics necessary to write the equations of motion of a rigid body is entirely given by Newton's laws of motion. The method is as follows:

1. Assign variables such as x and θ that are both necessary and sufficient to describe an *arbitrary* position of the object.
2. Draw a free-body diagram of each component, and indicate *all*

---

1. The inverted pendulum is often described with the angle of the pendulum being positive for *clockwise* motion. If defined that way, then reverse the sign on all terms in Eqs. (2.21) in θ' or $\ddot\theta'$.

forces acting on each body and the accelerations of the center of mass with respect to an inertial reference.

3. Apply Newton's law in translation (Eq. 2.1) and rotation (Eq. 2.7) form.

4. Combine the equations to eliminate internal forces.

The combination of the equations to eliminate internal forces is sometimes expedited by an intelligent choice of directions for the application of Eq. (2.1), as illustrated by Example 2.6. It is often useful to try alternate directions and then evaluate the algebra required to eliminate the internal reaction forces. Applying Newton's laws in different directions sometimes yields what at first appear to be quite different sets of equations; however, after manipulation, you can always show the equations to be equivalent.

■ Distributed Parameter Systems

All the preceding examples contained one or more rigid bodies, although some were connected to others by springs. Actual structures—for example, satellite solar panels, airplane wings, or robot arms—usually bend; Fig. 2.16(a) shows a very simple flexible beam. The equation describing its motion is a fourth-order *partial* differential equation that arises because the mass elements are continuously distributed along the beam with a small amount of flexibility between each element. This type of system is called a **distributed parameter system**. The dynamic analysis methods presented in this section are not sufficient to analyze this case; however, more advanced texts (Thomson, 1981) show that the result is

$$EI \frac{\partial^4 w}{\partial x^4} + \rho \frac{\partial^2 w}{\partial t^2} = 0, \tag{2.22}$$

where

E = Young's modulus,

I = beam cross-sectional moment of inertia,

ρ = beam density,

w = beam deflection.

The exact solution to Eq. (2.22) is too cumbersome to use in designing control systems, but it is often important to account for the gross effects of bending in control systems design.

The continuous beam in Fig. 2.16(b) has an infinite number of vibration-mode shapes, all with different frequencies. Typically, the lowest-frequency modes have the largest amplitude and are the most important to approximate well. The simplified model in Fig. 2.16(c) can be made to duplicate the essential

FIGURE 2.16
(a) Flexible robot arm used for research at Stanford University; (b) model for a continuous flexible beam; (c) simplified model for the first bending mode; (d) model for the first and second bending modes. *(Photo courtesy of E. Schmitz.)*

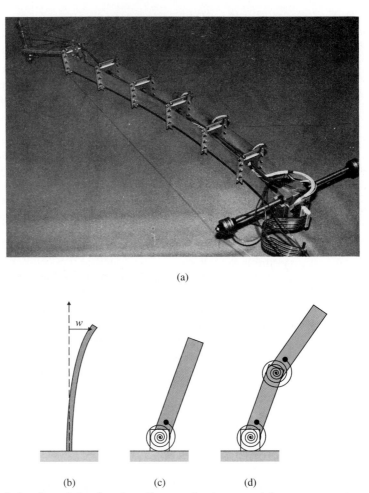

(a)

(b)   (c)   (d)

behavior of the first bending mode shape and frequency and would usually be adequate for controller design. If frequencies higher than the first bending mode are anticipated in the control system operation, it may be necessary to model the beam as shown in Fig. 2.16(d), which can be made to approximate the first two bending modes and frequencies. Likewise, higher-order models can be used if such accuracy and complexity are deemed necessary (Thomson, 1981; Schmitz, 1985). When a continuously bending object is approximated as two or more rigid bodies connected by springs, the resulting model is sometimes referred to as a **lumped parameter model**.

A flexible structure can be approximated by a lumped parameter model

2.2   Differential Equations in State-variable Form

Use of Newton's law and the free-body diagram in Section 2.1 typically leads to second-order differential equations, that is, equations that contain the second derivative such as $\ddot{x}$ in Eq. (2.3). Differential equations can be expressed as a set of simultaneous first-order differential equations. These are represented in the **state-variable form** as the vector equation

$$\dot{\mathbf{x}} = \mathbf{f}(\mathbf{x}, u), \tag{2.23}$$

where the input is u and the output is

$$y = h(\mathbf{x}, u). \tag{2.24}$$

The column vector $\mathbf{x}$ is called the **state of the system** and contains n elements for an nth-order system. For mechanical systems, the state vector elements usually consist of the positions and velocities of the separate bodies. The vector function $\mathbf{f}$ relates the first derivative of the state to the state itself and the input u. Parameters in the function $\mathbf{f}$ could be dependent on time as well as position and velocity, but we will not explicitly account for that in this text. For much of the analysis and design we will approximate a nonlinear system with a linear model.

For the linear case, Eqs. (2.23) and (2.24) can be rewritten as

Standard form of linear differential equations

$$\dot{\mathbf{x}} = \mathbf{F}\mathbf{x} + \mathbf{G}u, \tag{2.25a}$$

$$y = \mathbf{H}\mathbf{x} + Ju, \tag{2.25b}$$

where, for an nth-order system, $\mathbf{F}$ is an $n \times n$ **system matrix**, $\mathbf{G}$ is an $n \times 1$ **input matrix**, $\mathbf{H}$ is a $1 \times n$ row matrix referred to as the **output matrix**, and J is a scalar called the **direct transmission term**.[2] To save space we will sometimes refer to a state vector by its **transpose**, $\mathbf{x} = [x_1 \ x_2]^T$, which is equivalent to

$$\mathbf{x} = \begin{bmatrix} x_1 \\ x_2 \end{bmatrix}$$

In Chapter 7 we will consider control systems design using the state-variable form. Some aspects of matrix theory that aid in the analysis of systems described in this manner are contained in Section 7.1 and Appendix C.

The state-variable method of specifying differential equations is used by computer-aided control systems design (CACSD) software packages. Therefore, in order to specify linear differential equations to the computer, you need to know the values of the matrices $\mathbf{F}$, $\mathbf{G}$, $\mathbf{H}$, and the constant J. Another option available for specifying a dynamic system to CACSD software is to describe the system in terms of its **transfer function**, a concept that will be explained in Chapter 3.

MATLAB, MATRIXx, and CtrlC are popular CACSD tools

---

2. It is also common, particularly on the East Coast, to use the notation $\mathbf{A}$, $\mathbf{B}$, $\mathbf{C}$, and D in place of $\mathbf{F}$, $\mathbf{G}$, $\mathbf{H}$, and J. This is sometimes referred to as the "Rocky Mountain transformation."

◆ **EXAMPLE 2.7** *Satellite Model in State-variable Form*

Write equations of motion in state-variable form for the satellite-attitude control model in Example 2.3 with $M_D = 0$.

Solution. Define the attitude and the angular velocity of the satellite as the state-variables x_1 and x_2, so that $x_1 \triangleq \theta$ and $x_2 \triangleq \dot{\theta} = \dot{x}_1$.[3] The single second-order equation (2.8) can then be written in an equivalent way as two first-order equations:

$$\dot{x}_1 = x_2,$$

$$\dot{x}_2 = \frac{d}{I} F_c.$$

These equations are expressed using Eq. (2.25a), $\dot{\mathbf{x}} = \mathbf{Fx} + \mathbf{Gu}$, as

$$\begin{bmatrix} \dot{x}_1 \\ \dot{x}_2 \end{bmatrix} = \begin{bmatrix} 0 & 1 \\ 0 & 0 \end{bmatrix} \begin{bmatrix} x_1 \\ x_2 \end{bmatrix} + \begin{bmatrix} 0 \\ d/I \end{bmatrix} F_c.$$

The output of the system is the satellite attitude, $y = x_1 = \theta$. Using Eq. (2.25b), $y = \mathbf{Hx} + Ju$, this relation is expressed as

$$y = \begin{bmatrix} 1 & 0 \end{bmatrix} \begin{bmatrix} x_1 \\ x_2 \end{bmatrix}.$$

The matrices for the state-variable form are then

$$\mathbf{F} = \begin{bmatrix} 0 & 1 \\ 0 & 0 \end{bmatrix}, \qquad \mathbf{G} = \begin{bmatrix} 0 \\ d/I \end{bmatrix}, \qquad \mathbf{H} = \begin{bmatrix} 1 & 0 \end{bmatrix}, \qquad J = 0,$$

and the input $u \triangleq F_c$.

For this example the state-variable form is a more cumbersome way of writing the differential equation than the second-order version in Eq. (2.8). However, the method is not more cumbersome for most systems, and the advantages of having a standard form have led to its widespread use.

◆

The following example has more complexity and shows how to use MATLAB to find the solution of linear differential equations. A common solution method for evaluating the performance of a control system is to assume that the initial conditions are zero and the input $u(t)$ is a step with a magnitude of 1; that is, $u(t) = 0$ for $t < 0$ and $u(t) = 1$ for $t \geq 0$. This is called the **unit step response**. For steps in u of some magnitude other than 1, the solution is called simply the **step response**. The methods used in MATLAB to calculate the response will be discussed in Section 3.6 so, for now, you are not expected to understand how this calculation is made.

Step Response

---

3. The symbol $\triangleq$ means "is defined to be."

◆ **EXAMPLE 2.8**   *Cruise-control Step Response*

❖ (a) Rewrite the equation of motion from Example 2.1 in state-variable form where the output is the car position x.

(b) Use MATLAB to find the response of the velocity of the car for the case where the input jumps from being $u = 0$ at time $t = 0$ to a constant $u = 500$ N thereafter. Assume that the car mass m is 1000 kg and $b = 50$ N·sec/m.

Solution. (a) First we need to express the differential equation describing the plant, Eq. (2.3), as a set of simultaneous first-order equations. To do so, we define the position and the velocity of the car as the state-variables x_1 and x_2, so that $x_1 = x$ and $x_2 = \dot{x} = \dot{x}_1$. The single second-order equation, Eq. (2.3), can then be rewritten as a set of two first-order equations:

$$\dot{x}_1 = x_2$$

$$\dot{x}_2 = -\frac{b}{m}x_2 + \frac{1}{m}u$$

Next, we use the standard form of Eq. (2.25a), $\dot{\mathbf{x}} = \mathbf{F}\mathbf{x} + \mathbf{G}u$, to express these equations:

$$\begin{bmatrix} \dot{x}_1 \\ \dot{x}_2 \end{bmatrix} = \begin{bmatrix} 0 & 1 \\ 0 & -b/m \end{bmatrix} \begin{bmatrix} x_1 \\ x_2 \end{bmatrix} + \begin{bmatrix} 0 \\ 1/m \end{bmatrix} u. \qquad (2.26)$$

The output of the system is the car position $y = x_1 = x$, which is expressed in matrix form as

$$y = \begin{bmatrix} 1 & 0 \end{bmatrix} \begin{bmatrix} x_1 \\ x_2 \end{bmatrix},$$

or

$$y = \mathbf{H}\mathbf{x}.$$

So the state-variable-form matrices defining this example are

$$\mathbf{F} = \begin{bmatrix} 0 & 1 \\ 0 & -b/m \end{bmatrix}, \qquad \mathbf{G} = \begin{bmatrix} 0 \\ 1/m \end{bmatrix}, \qquad \mathbf{H} = \begin{bmatrix} 1 & 0 \end{bmatrix}, \qquad J = 0.$$

(b) The equations of motion are those given in part (a) except that now the output is $v = x_2$. Therefore, the output matrix is

$$\mathbf{H} = \begin{bmatrix} 0 & 1 \end{bmatrix}.$$

The coefficients required are $b/m = 0.05$ and $1/m = 0.001$. The numerical values of the matrices defining the system are thus

$$\mathbf{F} = \begin{bmatrix} 0 & 1 \\ 0 & -0.05 \end{bmatrix}, \qquad \mathbf{G} = \begin{bmatrix} 0 \\ 0.001 \end{bmatrix}, \qquad \mathbf{H} = \begin{bmatrix} 0 & 1 \end{bmatrix}, \quad J = 0.$$

The step function in MATLAB calculates the time response of a linear system to a unit step input. Because the system is linear, the output for this case can be multiplied by the magnitude of the input step to derive any step response. Equivalently, the $\mathbf{G}$ matrix can be multiplied by the magnitude of the input step.

FIGURE 2.17
Response of the car
velocity to a step in u

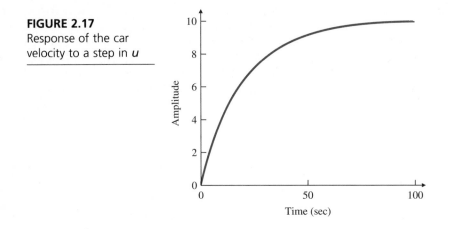

The statement

$$\text{step(F,500*G,H,J)}$$

calculates and plots the time response for an input step with a 500-N magnitude. The
result is shown in Fig. 2.17.

◆

The state-variable form can be applied to a system of any order. Example
2.9 illustrates the method for a fourth-order system.

◆ **EXAMPLE 2.9** *Disk-Drive in State-variable Form*

Find the state-variable form of the differential equations for Example 2.4 where the
output is θ_2.

Solution. Define the state vector to be

$$\mathbf{x} = [\theta_1 \quad \dot{\theta}_1 \quad \theta_2 \quad \dot{\theta}_2]^T.$$

Then solve Eqs. (2.9) for $\ddot{\theta}_1$ and $\ddot{\theta}_2$ so that the state-variable form is more apparent. The
resulting matrices are

$$\mathbf{F} = \begin{bmatrix} 0 & 1 & 0 & 0 \\ -\dfrac{k}{I_1} & -\dfrac{b}{I_1} & \dfrac{k}{I_1} & \dfrac{b}{I_1} \\ 0 & 0 & 0 & 1 \\ \dfrac{k}{I_2} & \dfrac{b}{I_2} & -\dfrac{k}{I_2} & -\dfrac{b}{I_2} \end{bmatrix}, \qquad \mathbf{G} = \begin{bmatrix} 0 \\ \dfrac{1}{I_1} \\ 0 \\ 0 \end{bmatrix}, \qquad \mathbf{H} = [0 \quad 0 \quad 1 \quad 0], J = 0.$$

◆

In some cases, such as in Eqs. (2.20), more than one second derivative will appear in a differential equation. This makes transformation to state-variable form more difficult. Problem 2.8 illustrates both the difficulty and the solution. Another difficulty arises if the differential equation contains derivatives of the input u. Techniques to handle this situation will be discussed in Section 7.1.

2.3   Models of Electric Circuits

Electric circuits consist of interconnections between sources of electric voltage and current, and other electronic elements such as resistors, capacitors, and transistors. Electric circuits are frequent components of feedback-control systems because they give the designer enormous flexibility in modifying and processing signals. Operational amplifiers are themselves examples of complex feedback systems. Some of the most important methods of feedback system design were developed by the designers of high-gain feedback amplifiers, mainly at Bell Telephone Laboratories between 1925 and 1940. Electric and electronic components also play a central role in electromechanical devices such as electric motors and generators and electrical sensors. In this brief survey we cannot derive the physics of electricity or give a comprehensive review of all the important analytical techniques. We will define the variables, describe the relationships imposed on them by typical elements and circuits, and describe a few of the most effective methods available for solving the resulting equations.

Symbols for some linear circuit elements and their current-voltage relationships are given in Fig. 2.18. Passive circuits consist of interconnections among resistors, capacitors, and inductors.

Operational amplifier

With electronics we increase the set of electrical elements by adding active devices, including diodes, transistors, and amplifiers. In signal processing and in circuits for control systems, the most common amplification device is the operational amplifier, or op amp.[4] The symbol of the op amp is shown in Fig. 2.19(a), and its equivalent circuit is shown in Fig. 2.19(b). In typical amplifiers the zero frequency or DC gain A_o varies from 10^5 to 10^7. We can take A_o to be infinity with satisfactory accuracy for most control uses.

Kirchhoff's laws

The basic equations of electric circuits are given by **Kirchhoff's laws:**

Kirchhoff's current law (KCL): The algebraic sum of currents leaving a junction or node equals the algebraic sum of the currents entering that node.

---

4. Oliver Heaviside introduced the mathematical operation p to signify differentiation so that $pv = dv/dt$. The Laplace transform incorporates this idea, using the complex variable s. Ragazzini *et al.* (1947) demonstrated that an ideal, high-gain electronic amplifier permitted one to realize arbitrary "operations" in the Laplace transform variable s, so they named it the operational amplifier.

FIGURE 2.18
Elements of electric
circuits

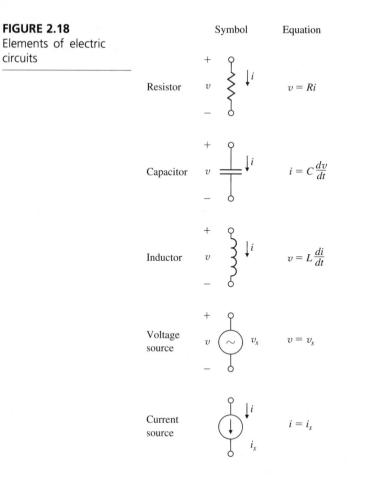

	Symbol	Equation
Resistor		$v = Ri$
Capacitor		$i = C\dfrac{dv}{dt}$
Inductor		$v = L\dfrac{di}{dt}$
Voltage source		$v = v_s$
Current source		$i = i_s$

Kirchhoff's voltage law (KVL): **The algebraic sum of all voltages taken around a closed path in a circuit is zero.**

With complex circuits of many elements it is essential to write the equations in a careful, well-organized way. Of the numerous methods for doing this we choose for description and illustration the popular and powerful scheme known as **node analysis**. One node is selected as a reference, and we assume the voltages of all other nodes to be unknowns. The choice of reference is arbitrary in theory, but in actual electronic circuits the common, or ground, terminal is the obvious and standard choice. Next, we write equations for the selected unknowns using the current law (KCL) at each node. We express these currents in terms of the selected unknowns by using the element equations in Fig. 2.18. If the circuit contains voltage sources, we must substitute a voltage law (KVL) for such sources. Example 2.10 illustrates how node analysis works.

FIGURE 2.19
The operational
amplifier: (a) the symbol;
(b) the equivalent circuit

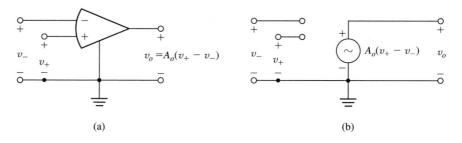

(a) (b)

◆ **EXAMPLE 2.10** *Equations for the Bridged Tee Circuit*

Determine the differential equations for the circuit shown in Fig. 2.20.

Solution. We select node ④ as the reference and the voltages $v_①$, $v_②$, and $v_③$ as the unknowns.

We start with the degenerate KVL relationship

$$v_① = v_i. \tag{2.27}$$

At node ② the KCL is

$$-\frac{v_① - v_②}{R_1} + \frac{v_② - v_③}{R_2} + C_1 \frac{dv_②}{dt} = 0, \tag{2.28}$$

and at node ③ the KCL is

$$\frac{v_③ - v_②}{R_2} + C_2 \frac{d(v_③ - v_①)}{dt} = 0. \tag{2.29}$$

These three equations describe the circuit.

In order to write the equations in the state-variable form,—that is, as a set of simultaneous first-order differential equations—we select the capacitor voltages v_{C_1} and v_{C_2} as the state elements, so that $\mathbf{x} \triangleq [v_{C_1} \ v_{C_2}]^T$, and v_i as the input, so that $u \triangleq v_i$. Here $v_{C_1} = v_②$, $v_{C_2} = v_① - v_③$, and still $v_① = v_i$. Thus $v_① = v_i$, $v_② = v_{C_1}$, and

FIGURE 2.20
Bridged tee circuit

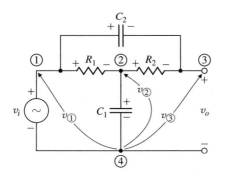
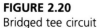

$v_{③} = v_i - v_{C_2}$. In terms of v_{C_1} and v_{C_2}, Eq. (2.28) becomes

$$\frac{v_{C_1} - v_i}{R_1} + \frac{v_{C_1} - (v_i - v_{C_2})}{R_2} + C_1 \frac{dv_{C_1}}{dt} = 0.$$

Rearranging this equation into standard form, we get

$$\frac{dv_{C_1}}{dt} = -\frac{1}{C_1}\left(\frac{1}{R_1} + \frac{1}{R_2}\right) v_{C_1} - \frac{1}{C_1}\left(\frac{1}{R_2}\right) v_{C_2} + \frac{1}{C_1}\left(\frac{1}{R_1} + \frac{1}{R_2}\right) v_i. \tag{2.30}$$

In terms of v_{C_1} and v_{C_2}, Eq. (2.29) becomes

$$\frac{v_i - v_{C_2} - v_{C_1}}{R_2} + C_2 \frac{d}{dt}(v_i - v_{C_2} - v_i) = 0.$$

In standard form the equation is

$$\frac{dv_{C_2}}{dt} = -\frac{v_{C_1}}{C_2 R_2} - \frac{v_{C_2}}{C_2 R_2} + \frac{v_i}{C_2 R_2}. \tag{2.31}$$

Equations (2.27) to (2.29) are entirely equivalent to the state-variable form, Eqs. (2.30) and (2.31), in describing the circuit. The standard matrix definitions are

$$\mathbf{F} = \begin{bmatrix} -\dfrac{1}{C_1}\left(\dfrac{1}{R_1} + \dfrac{1}{R_2}\right) & -\dfrac{1}{C_1 R_2} \\ -\dfrac{1}{C_2 R_2} & -\dfrac{1}{C_2 R_2} \end{bmatrix}, \qquad \mathbf{G} = \begin{bmatrix} \dfrac{1}{C_1}\left(\dfrac{1}{R_1} + \dfrac{1}{R_2}\right) \\ \dfrac{1}{C_2 R_2} \end{bmatrix}.$$

Kirchhoff's laws can also be applied to circuits that contain an operational amplifier, as the next example illustrates.

◆ **EXAMPLE 2.11**   *Equations for a Circuit with an Op-amp*

Determine the differential equations for the standard op amp circuit shown in Fig. 2.21(a).

Solution.   Node analysis gives us the following equation from KCL at node ①:

$$\frac{v_① - v_i}{R_1} + \frac{v_① - v_o}{R_2} + C \frac{d}{dt}(v_① - v_o) = 0. \tag{2.32}$$

The law of the amplifier as given on the right side of Fig. 2.19(a) is in this case

$$v_o = -A_o v_①.$$

If we let $A_0 \to \infty$, then $v_{-①} \to 0$ for fixed values of v_o, and Eq. (2.32) reduces to

$$-\frac{v_i}{R_1} - \frac{v_o}{R_2} - \frac{C \, dv_o}{dt} = 0.$$

FIGURE 2.21
(a) A simple op-amp
circuit; (b) the electronic
integrator

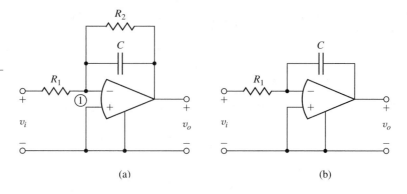

(a)   (b)

In state-variable standard form, with $\mathbf{x} \triangleq v_o$, the above equation becomes

$$\frac{dv_o}{dt} = -\frac{v_o}{CR_2} - \frac{v_i}{CR_1}.$$

Finally, if we let $R_2 \to \infty$ (an open circuit), then we have the circuit of Fig. 2.21(b) and the equation

$$-\frac{C\,dv_o}{dt} = \frac{v_i}{R_1},$$

or

$$v_o = -\frac{1}{R_1 C}\int_0^t v_i(\tau)\,d\tau + v_o(0). \tag{2.33}$$

OP amp as integrator

Thus the ideal op amp in this circuit performs the operation of integration and the device is usually referred to as an **integrator**.

The equations of much more complicated circuits can be established by nodal analysis. However, when the equations are linear, it is more common to use the Laplace transform directly on the element equations (Fig. 2.18) reducing them to algebraic form and to solve the resulting linear algebraic equations. We will develop this technique in Chapter 3.

2.4 Models of Electromechanical Systems

Electric current and magnetic fields interact in many ways, three of which are important to an understanding of the operation of most electromechanical-energy conversion devices such as linear and rotary motors and electric sensors of motion. In this section we will state these relationships and sketch a few typical configurations used in control devices. For greater detail respecting electromagnetic principles as discovered by researchers such as Ampere,

FIGURE 2.22

Toroid wrapped with N turns of wire

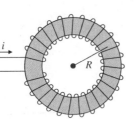

Faraday, Weber, Tesla, and others, consult a book on physics, such as Halliday and Resnick (1986). For a general introduction to the application of these principles to electromechanical energy conversion devices, see R. J. Smith (1987). For a thorough treatment of direct-current (DC) motors, see Reliance Motion Control Corp. (1980).

The first important relationship between current and magnetism to note is that an electric current establishes a magnetic field. The strength of the field at a particular point depends on the intensity of the current, the materials, and the geometry of the situation. A typical idealized case is sketched in Fig. 2.22; here a toroid of radius R and made of a material with permeability μ is wrapped with N turns of wire carrying i amperes. The field strength at the center of the toroid is given by

Magnetic field is proportional to current.

$$B = \frac{\mu}{2\pi R}\ Ni. \tag{2.34}$$

Thus the field strength is proportional to the current strength and the number of turns of wire around the toroid. The permeability μ is a property of the material and represents the extent to which the material aids the current in establishing the field. For air, $\mu_o = 4\pi \times 10^{-7}$ Wb/A·m. For a ferromagnetic material such as iron under typical nonsaturated conditions, the permeability can be several thousand times μ_o. The net effect of this relationship between current and magnetic field is that a strong field can be established by sending current through many turns of wire wrapped around a ferromagnetic material. If a small gap is cut in the toroid of the scheme in Fig. 2.22, then the field will be slightly reduced, but it will be accessible in the gap to interact with current-carrying conductors, which can be placed there. Such a device is an electromagnet, so called because the magnetic field is formed and controlled by the electric current in the wire. Some materials, known since prehistoric times, can establish a magnetic field because of the electrical properties of their constituent molecules. These are called **permanent magnets**, since the field does not come and go with an external current. In many cases, such as in small motors and generators used in many control applications, a permanent magnetic may be entirely satisfactory for establishing the required magnetic field.

Force is proportional to current and magnetic field.

The second electromagnetic effect of interest is the fact that a charge q moving with vector velocity $\mathbf{v}$ in a magnetic field of intensity $\mathbf{B}$ experiences a force $\mathbf{F}$ given by the vector cross-product $\mathbf{F} = q\mathbf{v} \times \mathbf{B}$. If the moving charge is composed of a current of i amperes in a conductor of length l meters arranged at right angles to the field strength of B tesla, then the force is at right angles to the plane of i and B, and has the magnitude

Law of motors

$$F = Bli \text{ Newtons.} \qquad (2.35)$$

This equation is the basis of using a magnetic field to convert electric energy from the source of current i into mechanical energy by arranging for the force F to do work on some mechanical object. Such devices are called **motors**, and Eq. (2.35) is the **law of motors**.

◆ **EXAMPLE 2.12**   *The Law of Motors: Modeling a Loudspeaker*

A loudspeaker for reproducing sound from electric signals contains such a motor. A typical geometry is sketched in Fig. 2.23. The permanent magnet establishes a radial field in the cylindrical gap between the poles of the magnet. The force of a current flowing in the wire wound on the bobbin causes the bobbin to move left or right and causes the cone to move against the air. As the cone moves back and forth, the fluctuations of air pressure propagate and are heard as sound. Approximate the effects of the air as if the cone had equivalent mass M and viscous friction coefficient b. Then write the equations of motion of the device. Assume that the magnet establishes a uniform field B of 0.5 tesla and the bobbin has 20 turns at a 2-cm diameter.

Solution.   The current is at right angles to the field, and the force of interest is at right angles to the plane of i and B, so Eq. (2.35) applies. In this case the parameter values are $B = 0.5$ tesla and

$$l = 20 \times \frac{2\pi}{100} m = 1.26\,m.$$

Thus

$$F = 0.5 \times 1.26 \times i = 0.63i \text{ newtons.}$$

FIGURE 2.23
Geometry of a loudspeaker: (a) overall configuration; (b) the electromagnet

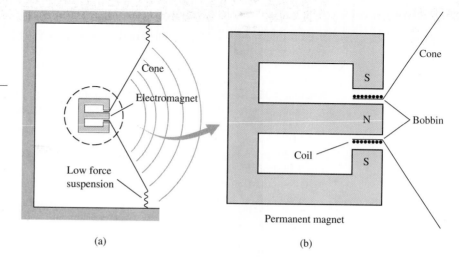

(a)   (b)

The mechanical equation follows from Eq. (2.1) and is very similar to Eq. (2.2). For a mass M and friction coefficient b, the equation is

$$M\ddot{x} + b\dot{x} = 0.63i. \tag{2.36}$$

This differential equation describes the motion of the loudspeaker cone as a function of the input current i driving the system. To put the model in state-variable form, a logical state vector for this system would be $\mathbf{x} \triangleq [x \ \dot{x}]^T$, which leads to the standard matrices

$$\mathbf{F} = \begin{bmatrix} 0 & 1 \\ 0 & -b/M \end{bmatrix} \quad \text{and} \quad \mathbf{G} = \begin{bmatrix} 0 \\ 0.63/M \end{bmatrix}.$$

◆

Equation (2.35) can also be used to find the force on the rotor of a motor and in many other useful applications.

The third important electromechanical relationship is the effect of mechanical motion on electricity. It is the opposite of Eq. (2.35), which expresses force (a mechanical variable) in terms of current (an electrical variable). If a charge is moving in a magnetic field and is forced along a conductor, an electric voltage is established between the ends of the conductor. If a conductor (which is filled with charged particles) of length l meters is moved at a velocity v meters per second through a constant field of B teslas at right angles to the direction of the field, then the voltage between the ends of the conductor is given by

$$e(t) = Blv \text{ volts.} \tag{2.37}$$

Law of the generator

This expression is called the **law of the generator**, because it is the basis for generating electric power by moving a conductor in a field and letting e cause current to flow in an external circuit.

◆ **EXAMPLE 2.13**   *Law of the Generator: Loudspeaker with Circuit*

For the loudspeaker in Fig. 2.23 and the circuit driving it in Fig. 2.24, find the differential equations relating the input voltage v_a to the output cone displacement x. Assume the amplifier output resistance is R and the speaker equivalent inductance is L.

Solution.   The loudspeaker motion acts according to Eq. (2.36), and this motion results in a voltage across the coil as given by Eq. (2.37), where the velocity is $\dot{x}$. The result is

$$e_{\text{coil}} = Bl\dot{x} = 0.63\,\dot{x}. \tag{2.38}$$

This induced voltage is important and needs to be added to the analysis of the circuit in Fig. 2.24. For the mechanical parts the equation of motion is Eq. (2.36). For the electrical parts the equation of motion is obtained using KVL loop analysis to get

$$L\frac{di}{dt} + Ri = v_a - 0.63\,\dot{x}. \tag{2.39}$$

FIGURE 2.24
A loudspeaker showing
the electric circuit

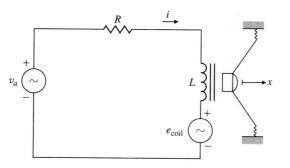

These two coupled equations comprise the dynamic model for the loudspeaker. A logical state vector for this third-order system would be $\mathbf{x} \triangleq [x \; \dot{x} \; i]^T$, which leads to the standard matrices

$$\mathbf{F} = \begin{bmatrix} 0 & 1 & 0 \\ 0 & -b/M & 0.63/M \\ 0 & -0.63/L & -R/L \end{bmatrix} \quad \text{and} \quad \mathbf{G} = \begin{bmatrix} 0 \\ 0 \\ 1/L \end{bmatrix},$$

where now the input $u \triangleq v_a$.

DC motor actuators

     A common actuator in control systems is the DC motor. It directly provides rotary motion and, coupled with wheels or drums with cables, can provide translational motion. A sketch of the basic components of a DC motor is given in Fig. 2.25. In addition to housing and bearings, the nonturning part (stator) has magnets, which establish a field across the turning part (rotor). The magnets may be electromagnets or, for small motors, permanent magnets. The brushes force current through the wire wound around the rotor. The (rotating) commutator causes the current always to be sent through the armature, a collection of conductor windings, so that it will produce the maximum torque in

FIGURE 2.25
Sketch of a DC motor

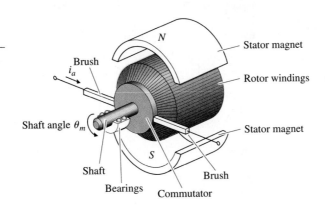

the desired direction. If the direction of the current is reversed, the direction of the torque is reversed.

The underlying principles of a DC motor are given by the motor and generator laws (Eqs. 2.35 and 2.37). However, rather than express these relationships in terms of the field strength and winding geometry, it is standard to relate the torque T developed in the rotor in terms of the armature current i_a and a torque constant K_t, and to express the voltage generated as a result of rotation[5] in terms of the shaft's rotational velocity $\dot{\theta}_m$ and an electric or "emf" (electromotive force) constant K_e. Thus

Torque
$$T = K_t i_a, \tag{2.40}$$

Back EMF
$$e = K_e \dot{\theta}_m. \tag{2.41}$$

In consistent units it is true that K_t equals K_e, but in some cases a torque constant will be given in other units such as ounce-inches per ampere, and the electric constant may be expressed in units of volts per 1000 rpm. In such cases the engineer must make the necessary translations to be certain the equations are correct.

◆ **EXAMPLE 2.14**   *Modeling a DC Motor*

Find a set of equations for the DC motor with the armature driven by the electric circuit shown in Fig. 2.26(a). Assume the rotor has inertia J_m and friction coefficient b.

Solution.   In our analysis we need to include the back emf for the electrical circuit. For the mechanical part of the system we need to include the motor torque in analyzing the rotor. The free-body diagram for the rotor, shown in Fig. 2.26(b), defines the positive direction and shows the two applied torques, T and $b\dot{\theta}_m$. Application of Eq. (2.7) yields

$$J_m \ddot{\theta}_m + b\dot{\theta}_m = K_t i_a. \tag{2.42}$$

KVL loop analysis shows the electrical equation to be

$$L_a \frac{di_a}{dt} + R_a i_a = v_a - K_e \dot{\theta}_m. \tag{2.43}$$

A state vector for this system is $\mathbf{x} \triangleq [\theta_m \ \dot{\theta}_m \ i_a]^T$, which leads to the standard matrices

$$\mathbf{F} = \begin{bmatrix} 0 & 1 & 0 \\ 0 & -\dfrac{b}{J_m} & \dfrac{K_t}{J_m} \\ 0 & -\dfrac{K_e}{L_a} & -\dfrac{R_a}{L_a} \end{bmatrix} \quad \text{and} \quad \mathbf{G} = \begin{bmatrix} 0 \\ 0 \\ \dfrac{1}{L_a} \end{bmatrix},$$

where the input $u \triangleq v_a$.

5. Because the generated electromotive force (emf) works against the applied armature voltage, we call it the **back emf**.

FIGURE 2.26
DC motor: (a) electric circuit of the armature; (b) free-body diagram of the rotor

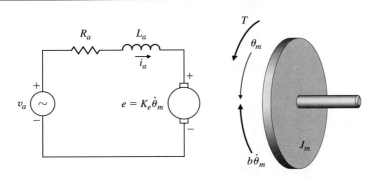

In most cases the electrical circuit response is much faster than the rotor motion; thus an applied voltage results in essentially an instantaneous change in the current flow. Therefore, it is often possible to neglect the existence of the inductance L_a in Eq. (2.43). In this case we combine Eqs. (2.42) and (2.43) into one equation to get

$$J_m \ddot{\theta}_m + \left(b + \frac{K_t K_e}{R_a} \right) \dot{\theta}_m = \frac{K_t}{R_a} v_a. \tag{2.44}$$

From Eq. (2.44) it is clear that the effect of the back emf is indistinguishable from the friction, in that they both act identically to damp the motion of the motor.

◆

AC motor
actuators

Another device used for electromechanical energy conversion is the alternating-current (AC) induction motor invented by N. Tesla and based on the following concept. Suppose we construct a rotor of conductors with no commutator or brushes. Now suppose the stator magnetic field can be made to rotate at high speed (perhaps 1800 rpm). The moving field will induce currents (the principle of induction) in the rotor, which will in turn be subject to forces according to the motor law. These forces will cause the rotor to turn so that it chases the rotating magnetic field.

Elementary analysis of the AC motor is more complex than that of the DC motor. Rather than give the details in terms of the motor geometry, we will present the experimental facts in the form of a set of curves of speed versus torque for fixed frequency and varying amplitude of applied voltage. A typical set appears in Fig. 2.27(a) for a low-rotor-resistance machine used for power applications and in Fig. 2.27(b) for a high-resistance machine typical of servo-motor applications. Although the data in the figure are for a constant engine speed, they can be used to extract the motor constants that will provide a dynamic model for the motor.

FIGURE 2.27
Torque-speed curves for two typical induction motors: (a) low-rotor-resistance machine; (b) high-rotor-resistance machine showing four values of armature voltage v_a.

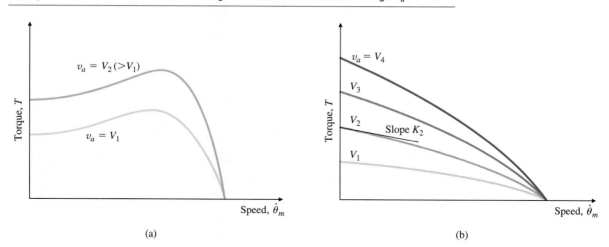

(a)   (b)

For analysis of a control problem involving an AC motor such as that described by Fig. 2.27(b), we make a linear approximation to the curves for speed near zero and at a midrange voltage to obtain the expression

$$T = K_1 v_a - K_2 \dot{\theta}_m. \tag{2.45}$$

The constant K_1 represents the ratio of a change in torque to a change in voltage at zero speed and is proportional to the distance between the curves at zero speed. The constant K_2 represents the ratio of a change in torque to a change in speed at zero speed and a midrange voltage; therefore, it is the slope of a curve at zero speed as shown by the line at V_2. For the electrical portion, values for the armature resistance R_a and inductance L_a are determined by experiment. Once we have values for K_1, K_2, R_a, and L_a, the analysis proceeds like the analysis in Example 2.14 for the DC motor. For the case where the circuit is so fast that the inductor can be neglected, we can substitute K_1 and K_2 into Eq. (2.44) in place of K_t/R_a and $K_t Ke/R_a$, respectively.

In addition to the DC and AC motors mentioned here, control systems use brushless DC motors (Reliance Motion Control Corp., 1980) and stepping motors (Kuo, 1982). Models for these machines, developed in the works just cited, do not differ in principle from the motors considered in this section. In general, the analysis, supported by experiment, develops a function of the torque in terms of voltage and speed similar to the AC-motor torque-speed curves given in Fig. 2.27. From these curves we can obtain a linearized formula such as Eq. (2.45) to use in the mechanical part of the system, and an equivalent circuit consisting of a resistance and an inductance to use in the electrical part.

2.5 Heat and Fluid-flow Models

Thermodynamics, heat transfer, and fluid dynamics are each the subject of complete textbooks. For purposes of generating dynamic models for use in control systems, the most important aspect of the physics is to represent the dynamic interaction between the variables. Experiments are usually required to determine the actual values of the parameters and thus to complete the dynamic model for purposes of control systems design.

2.5.1 Heat Flow

Some control systems involve regulation of temperature for portions of the system. The dynamic models of temperature control systems involve the flow and storage of heat energy. Heat energy flows through substances at a rate proportional to the temperature difference across the substance; that is,

$$q = \frac{1}{R}(T_1 - T_2),$$ (2.46)

where

q = heat energy flow, joules per second (J/sec),

R = thermal resistance, $°C/J \cdot sec$,

T = temperature, $°C$.

The net heat-energy flow into a substance affects the temperature of the substance according to the relation

$$\dot{T} = \frac{1}{C}q,$$ (2.47)

where C is the thermal capacity in J/°C. Typically, there are several paths for heat to flow into or out of a substance, and q in Eq. (2.47) is the sum of heat flows obeying Eq. (2.46).

◆ **EXAMPLE 2.15** *Equations for Heat Flow*

❖❖ A room with all but two sides insulated ($1/R = 0$) is shown in Fig. 2.28. Find the differential equations that determine the temperature in the room.

Solution. Application of Eqs. (2.46) and (2.47) yields

$$\dot{T}_I = \frac{1}{C_I}\left(\frac{1}{R_1} + \frac{1}{R_2}\right)(T_O - T_I),$$

FIGURE 2.28
Dynamic model for room
temperature

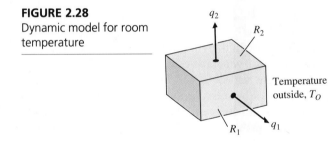

where

$\qquad C_I$ = thermal capacity of air within the room,

$\qquad T_O$ = temperature outside,

$\qquad T_I$ = temperature inside,

$\qquad R_2$ = thermal resistance of the room ceiling,

$\qquad R_1$ = thermal resistance of the room wall.

Normally the material properties are given in tables as follows:

1. The specific heat at constant volume c_v, which is converted to heat capacity by

Specific heat

$$C = mc_v, \tag{2.48}$$

where m is the mass of the substance;

2. The thermal conductivity[6] k, which is related to thermal resistance R by

Thermal conductivity

$$\frac{1}{R} = \frac{kA}{l},$$

where A is the cross-sectional area and l is the length of the heat-flow path.

In addition to flow due to transfer as expressed by Eq. (2.46), heat can also flow when a warmer mass flows into a cooler mass, or vice versa. In this case,

$$q = \dot{m}c_v(T_1 - T_2), \tag{2.49}$$

where $\dot{m}$, often referred to as w, is the mass flow rate of the fluid at T_1 flowing into the reservoir at T_2. For a more complete discussion of dynamic models for temperature control systems, see Cannon (1967) or textbooks on heat transfer.

6. In the case of insulation for houses, resistance is quoted as R-values; for example, R-11 refers to a substance that has a resistance to heat flow equivalent to that given by 11 in. of solid wood.

◆ **EXAMPLE 2.16**   *Equations for Modeling a Heat Exchanger*

A heat exchanger is shown in Fig. 2.29. Steam enters the chamber through the controllable valve at the top, and cooler steam leaves at the bottom. There is a constant flow of water through the pipe that winds through the middle of the chamber so that it picks up heat from the steam. Find the differential equations that describe the dynamics of the measured water outflow temperature as a function of the area, As, of the steam-inlet control valve when open. The sensor that measures the water outflow temperature, being downstream from the exit temperature in the pipe, lags the temperature by t_d seconds.

Solution.   The temperature of the water in the pipe will vary continuously along the pipe as the heat flows from the steam to the water. The temperature of the steam will also reduce in the chamber as it passes over the maze of pipes. An accurate thermal model of this process is therefore quite involved because the actual heat transfer from the steam to the water will be proportional to the local temperatures of each fluid. For many control applications it is not necessary to have great accuracy because the feedback will correct for a considerable amount of error in the model. Therefore, it makes sense to combine the spatially varying temperatures into the single temperatures T_s and T_w for the outflow steam and water temperatures, respectively. We then assume that the heat transfer from steam to water is proportional to the difference in these temperatures as given by Eq. (2.46). There is also a flow of heat into the chamber from the inlet steam that depends on the steam flow rate and its temperature according to Eq. (2.49):

$$q_{in} = w_s c_{vs}(T_{si} - T_s),$$

where

$w_s = K_s A_s$, mass flow rate of the steam,

A_s = area of the steam inlet valve,

K_s = flow coefficient of the inlet valve,

c_{vs} = specific heat of the steam,

T_{si} = temperature of the inflow steam,

T_s = temperature of the outflow steam.

The net heat flow into the chamber is the difference between the heat from the hot incoming steam and the heat flowing out to the water. This net flow determines the rate of temperature change of the steam according to Eq. (2.47):

$$C_s \dot{T}_s = A_s K_s c_{vs}(T_{si} - T_s) - \frac{1}{R}(T_s - T_w), \qquad (2.50)$$

where

$C_s = m_s c_{vs}$ is the thermal capacity of the steam in the chamber with mass m_s,

R = the thermal resistance of the heat flow averaged over the entire exchanger.

FIGURE 2.29

Heat exchanger

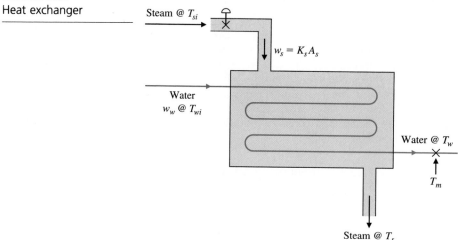

Likewise, the differential equation describing the water temperature is

$$C_w \dot{T}_w = w_w c_{vw}(T_{wi} - T_w) + \frac{1}{R}(T_s - T_w), \tag{2.51}$$

where

w_w = mass flow rate of the water,

c_{vw} = specific heat of the water,

T_{wi} = temperature of the incoming water,

T_w = temperature of the outflowing water.

To complete the dynamics, the time delay between the measurement and the exit flow is described by the relation

$$T_m = T_w(t - t_d),$$

where T_m is the measured downstream temperature of the water and t_d is the time delay. Equation (2.50) is nonlinear because the state variable T_s is multiplied by the control input A_s. The equation can be linearized about T_{so} (a specific value of T_s) so that $T_{si} - T_s$ becomes a constant, which we will define as ΔT_s. In order to eliminate the T_{wi} term in Eq. (2.51), it is convenient to measure all temperatures in terms of deviation in degrees from T_{wi}. The resulting equations are then

$$C_s \dot{T}_s = -\frac{1}{R} T_s + \frac{1}{R} T_w + K_s c_{vs} \Delta T_s A_s$$

$$C_w \dot{T}_w = -\left(\frac{1}{R} + w_w c_{vw}\right) T_w + \frac{1}{R} T_s$$

$$T_m = T_w(t - t_d).$$

2.5.2 Incompressible Fluid Flow

Fluid flows are common in many control systems components, one of the most common being the hydraulic actuators used to move the control surfaces on airplanes.

The physical relations governing fluid flow are continuity, force equilibrium, and resistance. The continuity relation is simply a statement of the conservation of matter:

The continuity relation

$$\dot{m} = w_{in} - w_{out}, \tag{2.52}$$

where

$m =$ fluid mass within a prescribed portion of the system,

$w_{in} =$ mass flow rate into the prescribed portion of the system,

$w_{out} =$ mass flow rate out of the prescribed portion of the system.

◆ **EXAMPLE 2.17**   *Equations for Describing Water Tank Height*

Determine the differential equation describing the height of the water in the tank in Fig. 2.30.

Solution.   Application of Eq. (2.52) yields

$$\dot{h} = \frac{1}{A\rho} (w_{in} - w_{out}), \tag{2.53}$$

where

$A =$ area of the tank,

$\rho =$ density of water,

$h = m/A\rho =$ height of water,

$m =$ mass of water in the tank.

FIGURE 2.30
Water-tank example

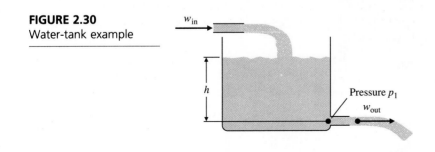

w_{in}

h

Pressure p_1

w_{out}

Force equilibrium must apply exactly as described by Eq. (2.1) for mechanical systems. However, for fluid flow systems some forces may result from fluid pressure acting on a piston. In this case the force from the fluid is

$$f = pA, \tag{2.54}$$

where

f = force,

p = pressure in the fluid,

A = area on which the fluid acts.

◆ **EXAMPLE 2.18** *Modeling a Hydraulic Piston*

 Determine the differential equation describing the motion of the piston actuator shown in Fig. 2.31 given that there is a force F_D acting on it and a pressure p in the chamber.

Solution. Equations (2.1) and (2.54) apply directly, where the forces include the fluid pressure as well as the applied force. The result is

$$M\ddot{x} = Ap - F_D,$$

where

A = area of the piston,

p = pressure in the chamber,

M = mass of the piston,

x = position of the piston.

FIGURE 2.31
Hydraulic piston actuator

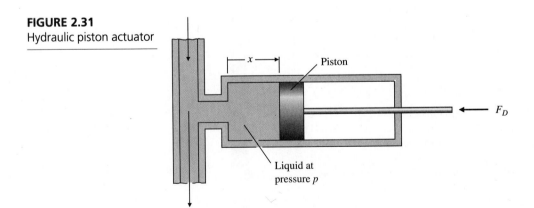

In many cases of fluid-flow problems the flow is restricted either by a constriction in the path or by friction. The general form of the effect of resistance is given by

$$w = \frac{1}{R} (p_1 - p_2)^{1/\alpha},$$ (2.55)

where

w = mass flow rate,

p_1, p_2 = pressures at ends of the path through which flow is occurring,

R, α = constants whose values depend on the type of restriction.

The constant α takes on values between 1 and 2. The most common value is approximately 2 for high flow rates (those having a Reynolds number $Re > 10^5$) through pipes or through short constrictions or nozzles. For very slow flows through long pipes or porous plugs where the flow remains laminar ($Re < 1100$), $\alpha = 1$. Flow rates in between these extremes can yield intermediate values of α. The Reynolds number indicates the relative importance of inertial forces and viscous forces in the flow. It is proportional to a material's velocity and density and to the size of the restriction, and it is inversely proportional to the viscosity. When Re is small, the viscous forces predominate and the flow is laminar. When Re is large, the inertial forces predominate and the flow is turbulent.

Note that a value of $\alpha = 2$ indicates that the flow is proportional to the square root of the pressure difference and therefore will produce a nonlinear differential equation. For the initial stages of control systems analysis and design, it is typically very useful to linearize these equations so that the design techniques described in this book can be applied. Linearization involves selecting an operating point and expanding the nonlinear term to be a small perturbation from that point.

Linearization of a
fluid-flow problem

◆ **EXAMPLE 2.19**   *Linearization of Water Tank Height and Outflow*

Find the nonlinear differential equation describing the height of the water in the tank in Fig. 2.30. Assume there is a relatively short restriction at the outlet and that $\alpha = 2$. Also linearize your equation about the operating point h_0.

Solution.   Applying Eq. (2.55) yields the flow out of the tank as a function of the height of the water in the tank:

$$w_{out} = \frac{1}{R} (p_1 - p_a)^{1/2},$$ (2.56)

where

$p_1 = \rho g h$, the hydrostatic pressure,

p_a = ambient pressure outside the restriction.

Substituting Eq. (2.56) into Eq. (2.53) yields the nonlinear differential equation for the height:

$$\dot{h} = \frac{1}{A\rho}\left(w_{in} - \frac{1}{R}\sqrt{\rho g h - p_a}\right).$$

Linearization involves selecting the operating point $p_0 = \rho g h_0$ and substituting $p_1 = p_0 + \Delta p$ into Eq. (2.56). Then we expand the nonlinear term according to the relation

$$(1 + \varepsilon)^\beta \cong 1 + \beta\varepsilon, \tag{2.57}$$

where $\varepsilon \ll 1$. Equation (2.56) can thus be written as

$$w_{out} = \frac{\sqrt{p_0 - p_a}}{R}\left(1 + \frac{\Delta p}{p_0 - p_a}\right)^{1/2}$$

$$\cong \frac{\sqrt{p_0 - p_a}}{R}\left(1 + \frac{1}{2}\frac{\Delta p}{p_0 - p_a}\right). \tag{2.58}$$

The linearizing approximation made in Eq. (2.58) is valid as long as $\Delta p \ll p_0 - p_a$; that is, as long as the deviations of the system pressure from the chosen operating point are relatively small.

Combining Eqs. (2.53) and (2.58) yields the following linearized equation of motion for the water tank level:

$$\Delta\dot{h} = \frac{1}{A\rho}\left[w_{in} - \frac{\sqrt{p_0 - p_a}}{R}\left(1 + \frac{1}{2}\frac{\Delta p}{p_0 - p_a}\right)\right].$$

Since $\Delta p = \rho g\,\Delta h$, this equation reduces to

$$\Delta\dot{h} = -\frac{g}{2AR\sqrt{p_0 - p_a}}\,\Delta h + \frac{w_{in}}{A\rho} - \frac{\sqrt{p_0 - p_a}}{\rho A R},$$

which is a linear differential equation for Δh. Because the operating point is not an equilibrium point that is requiring some control input to maintain, the linearized equation does not fit the standard form of Eqs. (2.25) due to the extra constant on the right. The implication of this is that when the system is at the operating point ($\Delta h = 0$) with no input ($w_{in} = 0$), it will move from that point ($\Delta\dot{h} \neq 0$).

◆

Hydraulic actuators

Hydraulic actuators are used extensively in control systems because they can supply large forces with low inertia and low weight. They are often used to move the aerodynamic control surfaces of airplanes, to gimbal rocket nozzles, to move the linkages in earth-moving equipment, and to move robot arms. They obey the same fundamental relationships we saw in the water tank: continuity (Eq. 2.52), force balance (Eq. 2.54), and flow resistance (Eq. 2.55).

◆ **EXAMPLE 2.20**   *Modeling a Hydraulic Actuator*

(a) Find the nonlinear differential equations relating the movement of the control surface, θ, to the input displacement x of the pilot valve for the hydraulic actuator shown in Fig. 2.32. (b) Find the linear approximation to the equations of motion when $\dot{y}$ = constant, with and without an applied load; that is, when $F \neq 0$ and when $F = 0$. Assume that θ motion is small.

Solution.   (a) When the pilot valve is at $x = 0$, both passages are closed and no motion results. When $x > 0$, as shown in Figure 2.32, the oil flows clockwise as shown and the piston is forced to the left. When $x < 0$, the fluid flows counter clockwise: The oil supply at high pressure p_s enters the *left* side of the large piston chamber, forcing the piston to the right. This causes the oil to flow out of the pilot valve chamber from the right-most channel instead of the left.

We assume the flow through the orifice formed by the pilot valve is proportional to x; that is,

$$w_1 = \frac{1}{R_1}(p_s - p_1)^{1/2}x. \tag{2.59}$$

FIGURE 2.32
Hydraulic actuator with pilot valve

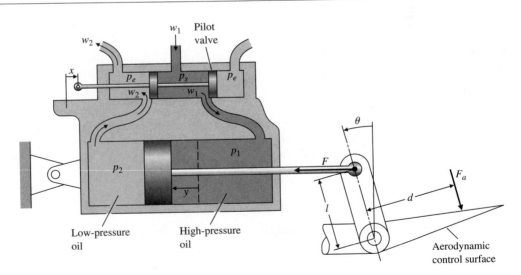

Similarly,

$$w_2 = \frac{1}{R_2}(p_2 - p_e)^{1/2}x. \tag{2.60}$$

The continuity relation yields

$$\rho A \dot{y} = w_1 = w_2, \tag{2.61}$$

where

ρ = fluid density,

A = piston area.

The force balance on the piston yields

$$A(p_1 - p_2) - F = m\ddot{y}, \tag{2.62}$$

where

m = mass of the piston and the attached rod

F = force applied by the piston rod to the control surface attachment point.

Furthermore, the moment balance of the control surface using Eq. (2.7) yields

$$I\ddot{\theta} = Fl\cos\theta - F_a d, \tag{2.63}$$

where

I = moment of inertia of the control surface and attachment about the hinge,

F_a = applied aerodynamic load.

To solve this set of five equations, we require the additional kinematic relationship between θ and y,

$$y = l\sin\theta. \tag{2.64}$$

The actuator is usually constructed so that the pilot valve exposes the two passages equally; therefore, $R_1 = R_2$, and infer from Eqs. (2.59) to (2.61) that

$$p_s - p_1 = p_2 - p_e. \tag{2.65}$$

These relations complete the nonlinear differential equations of motion; they are formidable and difficult to solve.

(b) For the case where $\dot{y} = $ a constant ($\ddot{y} = 0$) and there is no applied load ($F = 0$), Eqs. (2.62) and (2.65) indicate that

$$p_1 = p_2 = \frac{p_s + p_e}{2}. \tag{2.66}$$

Therefore, using Eq. (2.61) and letting $\sin\theta = \theta$ (since θ is assumed to be small), we get

$$\dot{\theta} = \frac{\sqrt{p_s - p_e}}{\sqrt{2A\rho Rl}}x. \tag{2.67}$$

This represents a single integration between the input x and the output θ where the proportionality constant is a function only of the supply pressure and the fixed parameters of the actuator. The state-variable form of this equation is

$$\mathbf{x} \triangleq \theta, \qquad \mathbf{F} = 0, \qquad \mathbf{G} = \frac{\sqrt{p_s - p_e}}{\sqrt{2A\rho R l}}, \qquad u \triangleq x.$$

For the case $\dot{y} = $ constant but $F \neq 0$, Eqs. (2.62) and (2.65) indicate that

$$p_1 = \frac{p_s + p_e + F/A}{2}$$

and

$$\dot{\theta} = \frac{\sqrt{p_s - p_e - F/A}}{\sqrt{2A\rho R l}} \, x. \tag{2.68}$$

This result is also a single integration between the input x and the output θ, but the proportionality constant now depends on the applied load F.

As long as the commanded values of x produce θ motion that has a sufficiently small value of $\ddot{\theta}$, the approximation given by Eqs. (2.67) or (2.68) is valid and no other linearized dynamic relationships are necessary. However, as soon as the commanded values of x produce accelerations where the inertial forces ($m\ddot{y}$ and the reaction to $I\ddot{\theta}$) are a significant fraction of $p_s - p_e$, the approximations are no longer valid. We must then incorporate these forces into the equations, thus obtaining a dynamic relationship between x and θ that is much more involved than the pure integration implied by Eqs. (2.67) or (2.68). Typically, for initial control system designs, hydraulic actuators are assumed to obey the simple relationship of Eqs. (2.67) or (2.68). ◆

2.6   Linearization and Scaling

The differential equations of motion for almost all processes selected for control are nonlinear. On the other hand, as will be evident in the next chapter, both analysis and control design are far easier for linear than for nonlinear models. **Linearization** is the process of finding a linear model that approximates a nonlinear one. Fortunately, as Lyapunov proved over 100 years ago, if a small-signal linear model is valid near an equilibrium and is stable, then there is a region (which may be small, of course) containing the equilibrium within which the nonlinear system is stable.[7] So we can safely make a linear model and design a linear control for it such that, at least in the neighborhood of the equilibrium, our design will be stable. Since a very important role of feedback

---

7. In 1949 the Russian scientist Aizerman conjectured that if a certain class of systems were stable with any linear gain between two limits, then the nonlinear system with a nonlinear gain characteristic that was kept between the same limits would also be stable. Unfortunately, this conjecture is not true.

control is to maintain the process variables near equilibrium, such small-signal linear models are a frequent starting point for control models. Small-signal linearization is discussed in Section 2.6.1.

An alternative approach to obtain a linear model for use as the basis of control system design is to use part of the control effort to cancel the nonlinear terms and to design the remainder of the control based on linear theory. This approach—linearization by feedback—is popular in the field of robotics, where it is called the **method of computed torque**. It is also a research topic for control of aircraft. Section 2.6.2 takes a brief look at this method.

The magnitude of the values of the variables in a problem is often very different, sometimes so much so that numerical difficulties arise. This was a serious problem years ago when equations were solved using analog computers and it was routine to *scale* the variables so that all had similar magnitudes. Today's widespread use of digital computers for solving differential equations has largely eliminated the need to scale a problem unless the number of variables is very large because computers are now capable of accurately handling numbers with wide variations in magnitude. Nevertheless, it is wise to understand the principle of scaling for the few cases where extreme variations in magnitude exist and scaling is necessary. Sections 2.6.3 and 2.6.4 discuss two kinds of scaling.

2.6.1   Small-signal Linearization

A nonlinear differential equation is one where the derivatives of the state have a nonlinear relationship to the state itself and/or the control. In other words, the differential equations *cannot* be written in the form

$$\dot{\mathbf{x}} = \mathbf{F}\mathbf{x} + \mathbf{G}u$$

but must be left in the form

$$\dot{\mathbf{x}} = \mathbf{f}(\mathbf{x}, u).$$

For small-signal linearization we first determine equilibrium values of $\mathbf{x}_0$, $\mathbf{u}_0$, that is, values where $\dot{\mathbf{x}}_0 = \mathbf{0} = \mathbf{f}(\mathbf{x}_0, u_0)$. We then expand the nonlinear equation in terms of perturbations from these equilibrium values; that is, we let $\mathbf{x} = \mathbf{x}_0 + \delta\mathbf{x}$ and $u = u_0 + \delta u$, so that

$$\dot{\mathbf{x}}_0 + \delta\dot{\mathbf{x}} \cong \mathbf{f}(\mathbf{x}_0, u_0) + \mathbf{F}\delta\mathbf{x} + \mathbf{G}\delta u$$

where $\mathbf{F}$ and $\mathbf{G}$ are the best linear fits to the nonlinear function $\mathbf{f}(\mathbf{x}, u)$ at $\mathbf{x}_0$ and u_0. Subtracting out the equilibrium solution, this reduces to

$$\delta\dot{\mathbf{x}} = \mathbf{F}\delta\mathbf{x} + \mathbf{G}\delta u, \tag{2.69}$$

which is a linear differential equation approximating the dynamics of the motion *about* the equilibrium point.

In developing the models discussed so far in this chapter, we have encountered nonlinear equations on several occasions: the pendulum in

Example 2.5, the hanging crane in Example 2.6, the AC induction motor in Section 2.4, the tank flow in Example 2.19, and the hydraulic actuator in Example 2.20. In each case, we assumed either that the motion was small or that motion from some operating point was small, so that nonlinear functions became linear functions. The steps followed in those examples essentially involved finding **F** and **G** in order to linearize the differential equations to the form of Eq. (2.69).

◆ **EXAMPLE 2.21**   *Linearization of Motion in a Ball Levitator*

Figure 2.33 shows a magnetic bearing used in large turbo machinery. The magnetics are energized using feedback control methods so that the axle is always in the center and never touches the magnets, thus keeping friction to an almost nonexistent level. A simplified version of a magnetic bearing that can be built in a laboratory is shown in Fig. 2.34, where one electromagnet is used to levitate a ball bearing. The physical arrangement of the levitator is depicted in Fig. 2.35. The equation for motion of the ball, derived from Newton's equation (2.1), is

$$m\ddot{x} = f_m(x, i) - mg, \tag{2.70}$$

where the force $f_m(x, i)$ is caused by the field of the electromagnet. Theoretically, the force from an electromagnet falls off with an inverse square relationship to the distance from the magnet, but the exact relationship for the laboratory levitator is difficult to derive from physical principles because its magnetic field is so complex. However, the forces can

FIGURE 2.33
A magnetic bearing
*(Photo courtesy of
Magnetic Bearings, Inc.)*

be measured with a scale. Figure 2.36 shows the experimental curves for a ball with a 1-cm diameter and a mass of 8.4×10^{-3} kg. At the current value of $i_2 = 600$ mA and the displacement x_1, the magnetic force f_m just cancels the gravity force $mg = 82 \times 10^{-3}$ N. (The mass of the ball is 8.4×10^{-3} kg, and the acceleration of gravity is 9.8 m/sec².) Therefore the point (x_1, i_2) represents an equilibrium. Using the data, find the linearized equations of motion about the equilibrium point.

Solution. First we write in expansion form the force in terms of deviations from the equilibrium values x_1 and i_2:

$$f_m(x_1 + \delta x, i_2 + \delta i) \cong f_m(x_1, i_2) + K_x \delta x + K_i \delta i. \tag{2.71}$$

FIGURE 2.35
Model for ball levitation

The linear gains are found as follows: K_x is the slope of the force versus x along the curve $i = i_2$, as shown in Fig. 2.36, and is found to be about 14 N/m. K_i is the change of force with current for the value of fixed $x = x_1$. We find that for $i = i_1 = 700$ mA at $x = x_1$, the force is about 122×10^{-3} N, and at $i = i_3 = 500$ mA at $x = x_1$, it is about 42×10^{-3} N. Thus

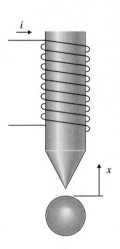

$$K_i \cong \frac{122 \times 10^{-3} - 42 \times 10^{-3}}{700 - 500} = \frac{80 \times 10^{-3} \text{ N}}{200 \text{ mA}}$$

$$\cong 400 \times 10^{-3} \text{ N/A}$$

$$\cong 0.4 \text{ N/A}.$$

Substituting these values into Eq. (2.71) leads to the following linear approximation for the force in the neighborhood of equilibrium:

$$f_m \cong 82 \times 10^{-3} + 14\delta x + 0.4\delta i.$$

Substituting this expression into Eq. (2.70) and using the numerical values for mass and gravity force, we get for the linearized model

$$(8.4 \times 10^{-3})\ddot{x} = 82 \times 10^{-3} + 14\delta x + 0.4\delta i - 82 \times 10^{-3}.$$

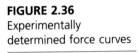

FIGURE 2.36
Experimentally
determined force curves

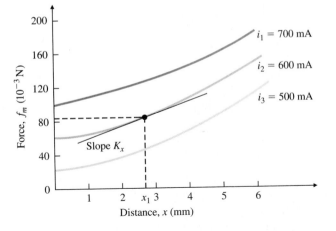

Since $x = x_1 + \delta x$, then $\ddot{x} = \delta\ddot{x}$. The equation in terms of δx is thus

$$(8.4 \times 10^{-3})\delta\ddot{x} = 14\delta x + 0.4\,\delta i$$

$$\delta\ddot{x} = 1667\delta x + 47.6\,\delta i, \tag{2.72}$$

which is the desired linearized equation of motion about the equilibrium point. A logical state vector is $\mathbf{x} = [\delta x\ \ \delta\dot{x}]^T$, which leads to the standard matrices

$$\mathbf{F} = \begin{bmatrix} 0 & 1 \\ 1667 & 0 \end{bmatrix} \quad \text{and} \quad \mathbf{G} = \begin{bmatrix} 0 \\ 47.6 \end{bmatrix},$$

and the control $u = \delta i$.

2.6.2 Linearization by Feedback

Linearization by feedback is accomplished by subtracting the nonlinear terms out of the equations of motion and adding them to the control. The result is a linear system, provided the computer implementing the control has enough capability to compute the nonlinear terms fast enough. A more detailed understanding of the method is best achieved through example.

An example of
the method of
computed torque

To illustrate linearization by feedback, we consider the equation of a simple pendulum developed in Example 2.5, (Eq. 2.10):

$$ml^2\ddot{\theta} + mgl \sin\theta = T_c. \tag{2.73}$$

If we compute the torque T_c to be

$$T_c = mgl \sin\theta + u, \tag{2.74}$$

then the motion is described by

$$ml^2\ddot{\theta} = u. \tag{2.75}$$

Eq. (2.75) is a linear equation *no matter how large the angle θ becomes.* We use it as the model for purposes of control design because it enables us to use linear analysis techniques. The resulting linear control will provide the value of u based on measurements of θ; however, the value of the torque actually sent to the equipment would derive from Eq. (2.74). For robots with two or three rigid links, this computed-torque approach has led to effective control. It is also being researched for the control of aircraft, where the linear models change considerably in character with the flight regime.

2.6.3 Amplitude Scaling

There are two types of scaling that are sometimes carried out: amplitude scaling and time scaling. **Amplitude scaling** is usually performed unwittingly by simply picking units that make sense for the problem at hand. For the ball levitator, expressing the motion in millimeters and the current in milliamps would keep the numbers within a range that is easy to work with. Equation (2.72) was developed in the standard SI units of meters, kilograms, and amperes, but in computing the motion of a rocket going into orbit, using kilometers makes more sense. The equations of motion are usually solved using CACSD software, which is often capable of working in any units. For higher-order systems it becomes important to scale the problem so that the elements of the state vector have similar numerical variations. A method for accomplishing the best scaling for a complex system is first to estimate the maximum values for each state element and then to scale the system so that each element varies between -1 and 1.

In general, we can perform amplitude scaling by defining the scaled variables for each state element: If

$$x' = S_x x, \tag{2.76}$$

then

$$\dot{x}' = S_x \dot{x} \qquad \text{and} \qquad \ddot{x}' = S_x \ddot{x}. \tag{2.77}$$

We then pick S_x to result in the appropriate scale change, substitute Eqs. (2.76) and (2.77) into the equations of motion, and recompute the coefficients.

◆ **EXAMPLE 2.22**   *Scaling for the Ball Levitator*

Scale the variables for the ball levitator in Example 2.21 to result in units of millimeters and milliamps instead of meters and amps.

Solution.   Referring to Eq. (2.76), we define

$$\delta x' = S_x \delta x \qquad \text{and} \qquad \delta i' = S_i \delta i$$

such that both S_x and S_i have a value of 1000 in order to convert δx and δi in meters and amps to $\delta x'$ and $\delta i'$ in millimeters and milliamps. Substituting these relations into Eq. (2.72) and taking note of Eq. (2.77) yields

$$\delta \ddot{x}' = 1667 \delta x' + 47.6 \frac{S_x}{S_i} \delta i'.$$

In this case $S_x = S_i$, so Eq. (2.72) remains unchanged. Had we scaled the two quantities by different amounts, there would have been a change in the last coefficient in the equation.

$\blacklozenge$

2.6.4 Time Scaling

The unit of time when using SI units or English units is seconds. CACSD software is *usually* able to compute results accurately no matter how fast or slow the particular problem at hand. However, if a dynamic system responds in a few milliseconds, or if there are characteristic frequencies in the system on the order of several kHz; the problem may become ill-conditioned, so that the numerical routines produce errors. This can be particularly troublesome for high-order systems. The same holds true for an extremely slow system. It is therefore important to know how to change the units of time.

We define the new scaled time to be

$$\tau = \omega_0 t \tag{2.78}$$

such that, if t is measured in seconds and $\omega_0 = 1000$, then τ will be measured in milliseconds. The effect of the time scaling is to change the differentiation so that

$$\dot{x} = \frac{dx}{dt} = \frac{dx}{d(\tau/\omega_0)} = \omega_0 \frac{dx}{d\tau}, \tag{2.79}$$

and

$$\ddot{x} = \frac{d^2 x}{dt^2} = \omega_0^2 \frac{d^2 x}{d\tau^2}. \tag{2.80}$$

Putting the equation into state-variable form allows a more concise way of stating time scaling. For the system described by

$$\dot{\mathbf{x}} = \mathbf{F}\mathbf{x} + \mathbf{G}u, \tag{2.81}$$

we say it is time-scaled, using $\tau = \omega_0 t$, by the system

$$\dot{\mathbf{x}} = \frac{1}{\omega_0} \mathbf{F}\mathbf{x} + \frac{1}{\omega_0} \mathbf{G}u. \tag{2.82}$$

◆ **EXAMPLE 2.23**   *Time Scaling an Oscillator*

The equation for an oscillator was derived in Example 2.5. For a case with a very fast natural frequency $\omega_n = 15,000$ rad/sec (about 2 kHz), Eq. (2.12) can be rewritten as

$$\ddot{\theta} + 15,000^2 \cdot \theta = 10^6 \cdot T_c$$

Determine the time-scaled equation so that the unit of time is milliseconds.

Solution.   The value of ω_0 in Eq. (2.78) is 1000. Equation (2.80) shows that

$$\frac{d^2\theta}{d\tau^2} = 10^{-6} \cdot \ddot{\theta},$$

and the time-scaled equation becomes

$$\frac{d^2\theta}{d\tau^2} + 15^2 \cdot \theta = T_c$$

In practice, we would then solve the equation

$$\ddot{\theta} + 15^2 \cdot \theta = T_c \tag{2.83}$$

and label the plots in milliseconds instead of seconds.

In state-variable form with a state vector $\mathbf{x} = [\theta \ \dot{\theta}]^T$, the unscaled matrices are

$$\mathbf{F} = \begin{bmatrix} 0 & 1 \\ -15,000^2 & 0 \end{bmatrix} \quad \text{and}$$

$$\mathbf{G} = \begin{bmatrix} 0 \\ 10^6 \end{bmatrix}.$$

Applying Eq. (2.82) results in

$$\mathbf{F} = \begin{bmatrix} 0 & \dfrac{1}{1000} \\ -\dfrac{15,000^2}{1000} & 0 \end{bmatrix} \quad \text{and}$$

$$\mathbf{G} = \begin{bmatrix} 0 \\ 10^3 \end{bmatrix},$$

which are equivalent to Eq. (2.83).

◆

Summary

● Mathematical modeling of the system to be controlled is the first step in analyzing and designing the required system controls. In this chapter we developed models for representative systems. Important equations for each category of system are summarized as follows:

SUMMARY TABLE 2.1   Key Equations for Dynamic Models

System	Important Laws or Relationships	Associated Equations	Page
Mechanical	Translational motion (Newton's law)	$F = ma$	20
	Rotational motion	$M = I\alpha$	24
	Motion of nonrigid bodies		33
Electrical	Kirchhoff's current law (KCL)		39
	Kirchhoff's voltage law (KVL)		40
Electromechanical	Electromagnetic field strength	$B \propto i$	44
	Law of motors	$F = Bli$	45
	Law of the generator	$e(t) = Blv$	46
	Torque developed in a rotor	$T = K_t i_a:$	48
	Voltage generated as a result of rotation of a rotor	$e = K_e \dot{\theta}_m$	48
Heat flow	Heat-energy flow	$q = \dfrac{1}{R}(T_1 - T_2)$	51
	Temperature as a function of heat-energy flow	$\dot{T} = \dfrac{1}{C}q$	51
	Specific heat	$C = mc,$	52
	Thermal conductivity	$\dfrac{1}{R} = \dfrac{kA}{L}$	52
Fluid Flow	Continuity relation (conservation of matter)	$\dot{m} = w_{in} - w_{out}$	55
	Force of a fluid acting on a piston	$f = pA$	56
	Effect of resistance to fluid flow	$w = \dfrac{1}{R}(p_1 - p_2)^{1/\gamma}$	57

- An alternative way of expressing the second-order differential equations that characterize the model of a system is the **state-variable form**,

$$\dot{\mathbf{x}} = \mathbf{f}(\mathbf{x}, u),$$

$$y = h(\mathbf{x}, u).$$

For a linear system the state-variable form can be expressed in matrix notation of the form

$$\dot{\mathbf{x}} = \mathbf{F}\mathbf{x} + \mathbf{G}u,$$

$$y = \mathbf{H}\mathbf{x} + Ju.$$

Equations in state-variable form are conducive to solution by computer packages that were developed especially for matrix equations (such as MATLAB, MATRIX$_x$, and Ctrl-C). In Section 2.2 we briefly introduced the state-variable form; it will be explored in more depth in Chapter 7.

- **Linearization** and **scaling** (Section 2.6) are methods by which certain complications of dealing with differential equations can be minimized. In linearization, nonlinear differential equations are simplified to linear ones by either (1) considering a small-signal linear model that is acurate near an equilibrium, or (2) linearization by feedback. Scaling of variables results in numerical values that fall within a narow-enough range of magnitude to minimize errors and allow for ease of computation.

Problems

2.1   Write the differential equations for the mechanical systems shown in Fig. 2.37.

FIGURE 2.37
Mechanical systems

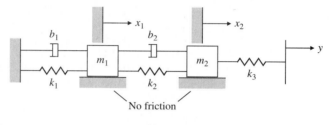

(a)

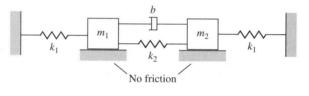

(b)

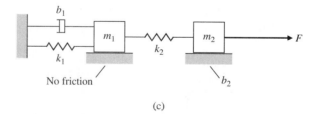

(c)

2.2   Write the equations of motion of a pendulum consisting of a thin, 2-kg stick of length l suspended from a pivot. How long should the rod be in order for the period to be exactly 1 sec? (The inertia I of a thin stick about an endpoint is $\frac{1}{3}ml^2$. Assume θ is small enough that $\sin\theta \cong \theta$.)

2.3   Write the equations of motion for the double-pendulum system shown in Fig. 2.38. Assume the displacement angles of the pendulums are small enough to ensure that

FIGURE 2.38
Double pendulum

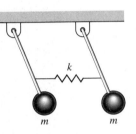

the spring is always horizontal. The pendulum rods are taken to be massless, of length l, and the springs are attached 2/3 of the way down.

2.4 Write the equations of motion for a body of mass M suspended from a fixed point by a spring with a constant k. Carefully define where the body's displacement is zero.

2.5 a) For the car suspension discussed in Example 2.2, write the equations of motion (Eqs. 2.6) in state-variable form. Use the state vector $\mathbf{x} = [x\ \dot{x}\ y\ \dot{y}]^T$.

   b) Plot the position of the car and the wheel after the car hits a "unit bump" (i.e., r is a unit step) using MATLAB or another CACSD tool. Assume that $m_1 = 10\,\text{kg}$, $m_2 = 250\,\text{kg}$, $K_w = 500{,}000\,\text{N/m}$, $K_s = 10{,}000\,\text{N/m}$. Find the value of b that you would prefer if you were a passenger in the car.

2.6 Automobile manufacturers are contemplating building active suspension systems that have the ability to supply an equal force in opposite directions on the wheel axle and the car body. Modify the equations of motion in Example 2.2 to include such a force input.

2.7 In many mechanical positioning systems the movement of a large unwieldy object is controlled by manipulating a much smaller object that is mechanically coupled with it. Figure 2.39 depicts such a situation, where a force u is applied to a small mass m in order to position a larger mass M. The coupling between the objects is modeled by a spring constant k with a damping coefficient b. Write the equations of motion governing this system, identify appropriate state variables, and express these equations in state-variable form.

FIGURE 2.39
Schematic of mechanical positioning system

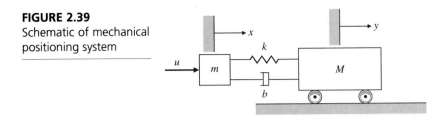

2.8 a) Try to put the equations of motion for the inverted pendulum, Eqs. (2.21), into state-variable form using the state vector $\mathbf{x} = [\theta\ \dot{\theta}\ x\ \dot{x}]^T$. Why is it not possible?

   b) Write the equations of the inverted pendulum in the "descriptor" form

$$\mathbf{E}\dot{\mathbf{x}} = \mathbf{F}'\mathbf{x} + \mathbf{G}'u,$$

and define values for $\mathbf{E}$, $\mathbf{F}'$, and $\mathbf{G}'$ (note that $\mathbf{E}$ is a 4×4 matrix). Then show how you would compute **F** and **G** for the standard state-variable description of the equations of motion.

2.9 Use node analysis to write the dynamic equations for the circuits shown in Fig. 2.40 and listed below:

a) lead network
b) lag network
c) notch network

FIGURE 2.40

Lead (a), lag (b) and notch (c) circuits

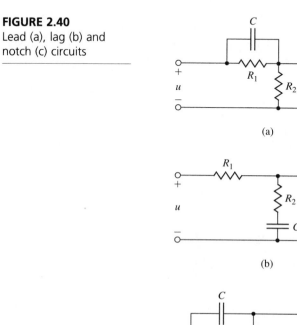

(a)

(b)

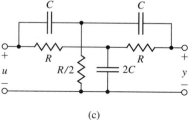

(c)

2.10 Using the step command in MATLAB, compare the response of each of the circuits in Problem 2.9 to a unit-step input. Assume $R = R_1 = R_2 = 1\,\Omega$ and $C = 1\,\text{F}$.

2.11 Use node analysis to write the dynamic equations for the op-amp circuits in Fig. 2.41 and listed below. Assume ideal operational amplifiers in every case.

a) first-order op-amp lead network
b) second-order op-amp circuit
c) Sallen-Key circuit

2.12 Write the state equations for the electrical network shown in Fig. 2.42. Choose as state variables the capacitor voltage v_C and inductor current i_L.

FIGURE 2.41
Op-amp circuits

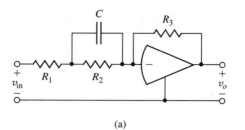

(a)

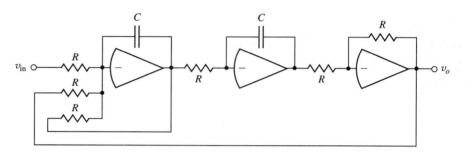

(b)

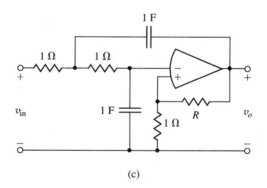

(c)

FIGURE 2.42
Simple electrical network
with voltage source

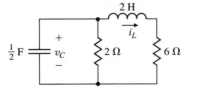

2.13 Write the state equations for the electric circuit shown in Fig. 2.43. Choose as state variables the capacitor voltages and inductor current.

FIGURE 2.43
Simple electric circuit
wth current source

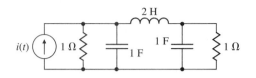

2.14 Find the differential equations for the circuit shown in Fig. 2.44, and put them in state-variable form.

FIGURE 2.44
Circuit for Problem 2.14

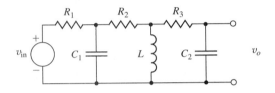

2.15 Find the differential equations for the circuit shown in Fig. 2.45, and put them in state-variable form.

FIGURE 2.45
Circuit for Problem 2.15

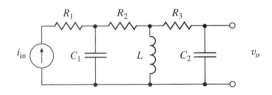

2.16 The torque constant of a motor is the ratio of torque to current and is often given in ounce-inches per ampere. The electric constant of a motor is the ratio of back emf to speed and is often given in volts per 1000 rpm. In consistent units the two constants are the same for a given motor.

a) Show that the units ounce-inches per ampere are proportional to volts per 1000 rpm by reducing both to MKS (SI) units.

b) A certain motor has a back emf of 25 V at 1000 rpm. What is its torque constant in ounce-inches per ampere?

2.17 A simplified sketch of a computer tape drive is given in Fig. 2.46.

a) Write the equations of motion in terms of the parameters listed below. K and B represent the spring constant and the damping of tape stretch, respectively, and ω_1 and ω_2 are angular velocities.

$J_1 = 4 \times 10^{-5} \, \text{kg} \cdot \text{m}^2$, motor and capstan

$B_1 = 1 \times 10^{-2} \, \text{N} \cdot \text{m} \cdot \text{sec}$, motor damping

$r_1 = 2 \times 10^{-2} \, \text{m}$

$K_t = 3 \times 10^{-2}\,\text{N} \cdot \text{m/A}$, motor-torque constant

$K = 2 \times 10^4\,\text{N/m}$

$B = 20\,\text{N/m} \cdot \text{sec}$

$r_2 = 2 \times 10^{-2}\,\text{m}$

$J_2 = 1 \times 10^{-5}\,\text{kg} \cdot \text{m}^2$

$B_2 = 1 \times 10^{-2}\,\text{N} \cdot \text{m} \cdot \text{sec}$, viscous damping, idler

$F = 6\,\text{N}$, constant force

$\dot{x}_1$ = tape velocity (variable to be controlled)

b) Use the values in part (a) to write the equations in state-variable form as a set of first-order differential equations. Use the variables $(x_1, \omega_1, x_2, \omega_2, i_a)$.

FIGURE 2.46

Tape drive

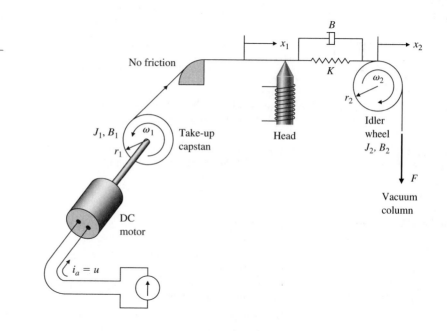

2.18 Assume the driving force on the hanging crane of Fig. 2.13 is provided by a motor mounted on the cab with one of the support wheels connected directly to the armature shaft. The motor constants are K_e, K_t, and the circuit driving the motor has a resistance R_a and no inductance. The wheel has a radius r. Write the equations of motion relating the applied motor voltage to the cab position and load angle.

2.19 The electromechanical system shown in Fig. 2.47 represents a simplified model of a capacitor microphone. The system consists in part of a parallel plate capacitor connected into an electric circuit. Capacitor plate a is rigidly fastened to the microphone frame. Sound waves pass through the mouthpiece and exert a force $f_s(t)$ on plate b, which has mass M and is connected to the frame by a set of springs and dampers. The capacitance C is a function of the distance x between the plates, as follows:

$$C(x) = \frac{\varepsilon A}{x},$$

where

ε = dielectric constant of the material between the plates,

A = surface area of the plates.

The charge q and the voltage e across the plates are related by

$$q = C(x)e.$$

The electric field in turn produces the following force f_e on the movable plate that opposes its motion:

$$f_e = \frac{q^2}{2\varepsilon A}.$$

Write differential equations in state-variable form that describe the operation of this system. Use the state vector $\mathbf{x} = [q \; \dot{q} \; x \; \dot{x}]^T$.

2.20 A very typical problem of electromechanical position control is an electric motor driving a load that has one dominant vibration mode. The problem arises in computer-disk-head control, reel-to-reel tape drives, and many other applications. A schematic diagram is sketched in Fig. 2.48. The motor has an electrical constant K_e, a torque constant K_t, an armature inductance L_a, and a resistance R_a. The rotor has an inertia J_1 and a viscous friction B. The load has an inertia J_2. The two inertias are connected by a shaft with a spring constant k and an equivalent viscous damping b.

FIGURE 2.47
Simplified model for
capacitor microphone

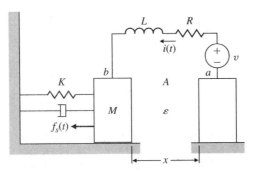

a) Write the equations of motion.

b) Write the equation as a set of simultaneous first-order equations in state-variable form. Use the state vector $\mathbf{x} = [\theta_2 \; \dot{\theta}_2 \; \theta_1 \; \dot{\theta}_1 \; i_a]^T$.

2.21 A precision-table leveling scheme shown in Fig. 2.49 relies on thermal expansion of the two actuators under the table corners to raise and lower the table. Find the differential equations relating the height of the actuator d versus the applied voltage

FIGURE 2.48

Motor with a flexible load

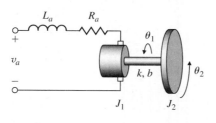

v_i, where

T_{act} = actuator temperature,

T_{amb} = ambient air temperature,

R_f = heat-flow coefficient between the actuator and the air,

C = thermal capacity of the actuator,

R = resistance of the heater.

FIGURE 2.49

(a) Precision table kept level by actuators; (b) side view of one actuator

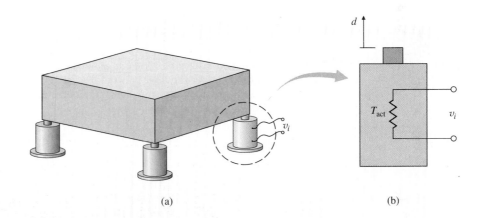

(a)   (b)

Assume that (1) the actuator acts as a pure electric resistance, (2) the heat flow into the actuator is proportional to the electric power input, and (3) d is proportional to the difference between T_{act} and T_{amb}.

2.22 An air conditioner supplies cold air at the same temperature to each room on the fourth floor of the high-rise building shown in Fig. 2.50(a). The floor plan is shown in Fig. 2.50(b). The cold air flow produces an equal amount of heat flow q out of each room. Write a set of differential equations governing the temperature in each room, where

T_o = temperature outside the building,

R_o = resistance to heat flow through the outer walls,

R_i = resistance to heat flow through the inner walls.

Assume that (1) all rooms are perfect squares, (2) there is no heat flow through the

FIGURE 2.50
Building air conditioning:
(a) high-rise building;
(b) floor plan of the
fourth floor

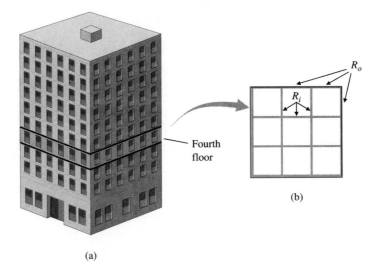

Fourth
floor

(b)

(a)

floors or ceilings, and (3) the temperature in each room is constant throughout the room. Take advantage of symmetry to reduce the number of differential equations to three.

2.23 For the two-tank fluid-flow system shown in Fig. 2.51, find the differential equations relating the flow into the first tank to the flow out of the second tank. Put the equations in state-variable form, and define whatever intermediate quantities are required.

FIGURE 2.51
Two-tank fluid-flow
system for Problem 2.23

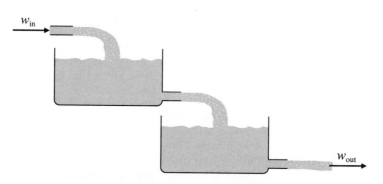

2.24 A laboratory experiment in the flow of water through two tanks is sketched in Fig. 2.52. Assume that Eq. (2.56) describes flow through the equal-sized holes at points A, B, or C.

a) With holes at A and C but none at B, write the equations of motion for this system in terms of h_1 and h_2. Assume that $h_3 = 20$ cm and $h_2 < 20$ cm. When $h_2 = 10$ cm, the outflow is 200 g/min.

b) At $h_1 = 30$ cm and $h_2 = 10$ cm, compute a linearized model and the transfer function from pump flow (in cubic centimeters per minute) to h_2.

c) Repeat parts (a) and (b) assuming hole A is closed and hole B is open.

FIGURE 2.52
Two-tank fluid-flow
system for Problem 2.24

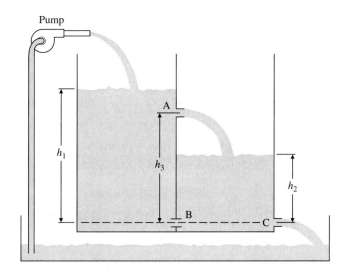

2.25 Figure 2.53 shows a simple pendulum system in which a cord is wrapped around a fixed cylinder. The motion of the system that results is described by the differential equation

$$(l + R\theta)\ddot{\theta} + g\sin\theta + R\dot{\theta}^2 = 0,$$

where

l = length of the cord in the vertical (down) position,

R = radius of the cylinder.

a) Write the state-variable equations for this system.
b) Linearize the equation around the point $\theta = 0$, and show that for small values of θ the system equation reduces to an equation for a simple pendulum, that is,

$$\ddot{\theta} + (g/l)\theta = 0.$$

FIGURE 2.53
Motion of cord wrapped
around a fixed cylinder

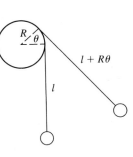

2.26 The flow of traffic in a single lane can be described by the following equation:

$$\frac{dy}{dt} = V - Ae^{-\alpha/y(t)},$$

where

$y(t)$ = relative distance between two cars,

V = constant velocity of the lead car,

A, α = real positive constants.

a) Obtain the equilibrium value Y that results in $\dot{y}(t) = 0$.
b) Obtain the range of V/A in $0 \le V/A \le \infty$ for which $Y > 0$.
c) Linearize the equation around Y, and write the resulting linearized equation.

2.27 Consider the circuit shown in Fig. 2.54 where u_1 and u_2 are voltage and current sources, respectively, and R_1 and R_2 are nonlinear resistors with the following characteristics:

$$\text{Resistor 1:} \quad i_1 = G(v_1) = v_1^3,$$

$$\text{Resistor 2:} \quad v_2 = r(i_2),$$

where the function r is shown in Fig. 2.55.

FIGURE 2.54
A nonlinear circuit for
Problem 2.27

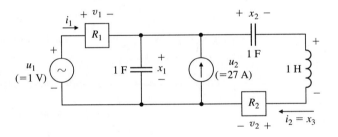

FIGURE 2.55
Nonlinear resistance

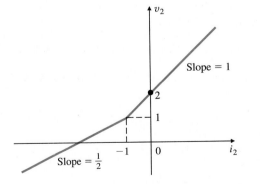

a) Show that the circuit equations can be written as

$$\dot{x}_1 = G(u_1 - x_1) + u_2 - x_3,$$

$$\dot{x}_2 = x_3,$$

$$\dot{x}_3 = x_1 - x_2 - r(x_3).$$

Suppose we have a constant voltage source of 1 V at u_1 and a constant current source of 27 A; that is, $u_1^0 = 1$, $u_2^0 = 27$. Find the *equilibrium state* $\mathbf{x}^0 = [x_1^0, x_2^0, x_3^0]^T$ for the network. For a particular input $\mathbf{u}^0$, an equilibrium state of the system is defined to be any constant state vector whose elements satisfy the relation

$$\dot{x}_1 = \dot{x}_2 = \dot{x}_3 = 0.$$

Consequently, any system started in one of its equilibrium states will remain there indefinitely until a different input is applied.

b) Due to disturbances, the initial state (capacitance, voltages, and inductor current) is slightly different from the equilibrium and so are the independent sources; that is,

$$u(t) = u^0 + \delta u(t),$$

$$x(t_0) = x^0(t_0) + \delta x(t_0).$$

Do a small-signal analysis of the network about the equilibrium found in part (a), displaying the equations in the form

$$\delta \dot{x}_1 = f_{11}\delta x_1 + f_{12}\delta x_2 + f_{13}\delta x_3 + g_1\delta u_1 + g_2\delta u_2.$$

c) Draw the circuit diagram that corresponds to the linearized model, and give the values of the elements.

2.28 The circuit shown in Fig. 2.56 has a nonlinear conductance G such that $i_G = g(v_G) = v_G(v_G - 1)(v_G - 4)$. The state differential equations are

$$\frac{di}{dt} = -i + v,$$

$$\frac{dv}{dt} = -i + g(u - v),$$

where i and v are the states and u is the input.

a) One equilibrium state occurs when $u = 1$ yielding $i_1 = v_1 = 0$. Find the other two pairs of v and i that will produce equilibrium.

b) Find the linearized model of the system about the equilibrium point $u = 1$, $i = v_1 = 0$.

c) Find the linearized models about the other two equilibrium points.

FIGURE 2.56

Nonlinear circuit for Problem 2.28

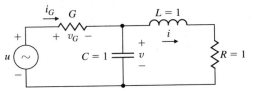

2.29 Consider the hydraulic system with an accumulator shown in Fig. 2.57.

a) Find the differential equations relating the outflow w_{out} to the resistance of the valve, R_c, which will vary with time. Put them in state-variable form, and define whatever intermediate quantities are required. Note that the supply pressure is a fixed value, P_o, as is the pressure in the accumulator above the bladder separating the gas from the fluid. Assume there is no restriction to the flow of liquid out of the accumulator.

b) Use a numerical simulation of the system to compare the outflow as a function of time with and without the accumulator for the case when the valve repeatedly opens for 20 msec and then shuts for 40 sec. Assume that $P_o = 300\,\text{kPa}$, $\alpha = 2$, $R = 500\,\text{kPa}^{1/2} \cdot \text{sec/cm}^3$, and the resistance of the valve, when open, is $R_c = 200\,\text{kPa}^{1/2} \cdot \text{sec/cm}^3$.

c) Linearize the equations using the average pressure levels from the simulation in part (b), and compare the linearized results with the exact simulation.

FIGURE 2.57

Hydraulic system for Problem 2.29

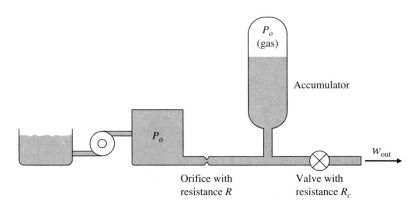

Orifice with resistance R — Valve with resistance R_c

 2.30 Modify the equation of motion for the cruise control in Example 2.1 (Eq. 2.4) so that it has a control law; that is, let

$$u = K(v_r - v),$$

where

v_r = reference speed,

K = constant.

This is a "proportional" control law where the difference between v_r and the actual speed v is used as a signal to speed the engine up or slow it down. Put the equations in the standard state-variable form with v_r as the input and v as the state. Assume that $m = 1000\,\text{kg}$ and $b = 50\,\text{N} \cdot \text{sec/m}$, and find the response for a unit step in v_r using MATLAB or other CACSD tool. Using trial and error, find a value of K that you think would result in a control system in which the actual speed converges as quickly as possible to the reference speed with no objectionable behavior.

2.31 The linearized equations of motion of a Boeing-747 are given in Eq. (9.18). Using MATLAB or other computer aid:

a) Determine the response of the altitude h for a 2-sec pulse of the elevator with a magnitude of $2°$. Note that, since Eq. (9.18) represents a set of linearized equations, the state variables actually represent the deviation of the state from the nominal operating point. For example, h represents the amount the altitude of the aircraft differs from 20,000 ft.

b) Consider using the feedback law

$$\delta_e = K_h h + \delta_{e,ext},$$

where the elevator input angle is the sum of a term proportional to the error in altitude h plus an external input (a disturbance or command input). Note from part (a) that a positive change in the elevator causes a negative change in altitude, so that the proposed proportional feedback law has the logical sign to anticipate a stable system provided $K_h > 0$. By trial and error, try to find a value for the feedback gain K_h such that a $2°$ pulse of 2 sec on $\delta_{e,ext}$ yields a more stable altitude response.

c) If you have trouble finding a value of K_h that produces a stable response, try modifying the feedback law to include information on pitch rate q:

$$\delta_e = K_h h + K_q q + \delta_{e,ext}.$$

Use trial and error to pick appropriate values for both K_h and K_q. Assume the same type of pulse input for $\delta_{e,ext}$ as in part (b).

d) Show that the further introduction of pitch-angle feedback θ, such that

$$\delta_e = K_h h + K_q q + K_\theta \theta + \delta_{e,ext},$$

allows you to decrease the time it takes for the altitude to settle back to its nominal value, as well as to decrease the value of K_q required for a stable response. Note that, although $K_h = 0$ produces stable altitude behavior, we require $K_h > 0$ in order to guarantee that $h \to 0$ (so there will be no steady-state error).

Dynamic Response

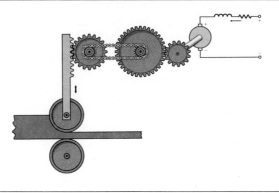

A Perspective on System Response

We saw in Chapter 2 how to obtain the dynamic model of a system. In designing a control system it is important to see how well a trial design matches the desired performance. We do this by solving the equations of the system model.

There are two ways to approach solving the dynamic equations. For a quick, *approximate* analysis we use linear analysis techniques. The resulting approximations of system response provide insight into why the solution has certain features and how the system might be changed to modify the response in a desired direction. In contrast, a *precise* picture of the system response typically calls for numerical techniques to solve the system equations. Although this chapter focuses primarily on linear analysis, we will also touch briefly on the numerical methods and computer tools that can be used to solve for the time response of both nonlinear and linear systems.

There are three domains within which to study dynamic response: the **s-plane**, the **frequency response**, and the **state space** (analysis using the state-variable description). The well-prepared control engineer needs to be fluent in all of them, so they will be treated in depth in Chapters 5 to 7, respectively. The purpose of this chapter is to discuss some of the fundamental mathematical tools needed before studying analysis in the *s*-plane and frequency response.

Chapter Overview

The Laplace transform, reviewed in Section 3.1, is the mathematical tool for transforming differential equations into an easier-to-manipulate algebraic form. In addition to the mathematical tools at our disposal, there are two visual tools that can help us to model a system and identify the pertinent mathematical relationships between elements of the system. One is the block diagram, which was introduced in Chapter 1; the other is the signal-flow graph. Both of these are discussed in Section 3.2.

Once the transfer function has been determined, we can identify its poles and zeros, which tell us something about system characteristics. Sections 3.3 to 3.5 focus on poles and zeros and some of the ways for manipulating them to steer system characteristics in a desired way. Instead of analyzing system dynamics through the transfer function, we may in certain circumstances prefer to use numerical solution techniques, as discussed in Section 3.6. Finally, a method for developing a model based on experimental time-response data is discussed in Section 3.7.

3.1 The Laplace Transform

Two attributes of linear constant systems form the basis for almost all analytical techniques applied to these systems:

1. A linear system response obeys the principle of superposition.
2. The response of a linear *constant* system can be expressed as the convolution of the input with the unit impulse response of the system.

The concepts of superposition, convolution, and impulse response will be defined shortly.

From the second property (as we will show), it follows immediately that the response of a linear constant system to an exponential input is also exponential. This result is the principle reason for the usefulness of Fourier and Laplace transforms in the study of linear constant systems.

3.1.1 Response by Convolution

Superposition

The **principle of superposition** states that if the system has an input that can be expressed as a sum of signals, then the response of the system can be expressed as the same sum of the individual responses to the respective signals. We can express superposition mathematically: Consider the system to have input u and output y; suppose further that, with the system at rest, we apply the input $u_1(t)$ and observe the output $y_1(t)$. After restoring the system to rest, we apply a second input $u_2(t)$ and again observe the output, which we call $y_2(t)$. Then, we form the composite input $u(t) = \alpha_1 u_1(t) + \alpha_2 u_2(t)$. Finally, if superposition applies, then the response will be $y(t) = \alpha_1 y_1(t) + \alpha_2 y_2(t)$. Superposition will apply if and only if the system is linear.

◆ **EXAMPLE 3.1**   *Superposition*

Show that superposition holds for the system modeled by the first-order linear differential equation

$$\dot{y} + ky = u.$$

Solution.   We let $u = \alpha_1 u_1 + \alpha_2 u_2$ and assume that $y = \alpha_1 y_1 + \alpha_2 y_2$. Then $\dot{y} = \alpha_1 \dot{y}_1 + \alpha_2 \dot{y}_2$. If we substitute these expressions into the system equation, we get

$$\alpha_1 \dot{y}_1 + \alpha_2 \dot{y}_2 + k(\alpha_1 y_1 + \alpha_2 y_2) = \alpha_1 u_1 + \alpha_2 u_2.$$

From this it follows that

$$\alpha_1(\dot{y}_1 + ky_1 - u_1) + \alpha_2(\dot{y}_2 + ky_2 - u_2) = 0. \tag{3.1}$$

If y_1 is the solution with input u_1 and y_2 is the solution with input u_2, then Eq. (3.1) is satisfied, the response is the sum of the individual responses, and superposition holds.

◆

Using the principle of superposition, we can solve for the system responses to a set of elementary signals. We are then able to solve for the response to a general signal simply by decomposing the given signal into a sum of the elementary components and, by superposition, concluding that the response to the general signal is the sum of the responses to the elementary signals. In order for this process to work, the elementary signals need to be sufficiently "rich" that any reasonable signal can be expressed as a sum of them, and their responses have to be easy to find. The most common candidates for elementary signals for use in linear systems are the impulse and the exponential.

The idea for the impulse comes from dynamics. Suppose we wish to study the motion of a baseball hit by a bat. The details of the collision between the bat and ball can be very complex as the ball deforms and the bat bends; however, for purposes of computing the path of the ball, we can summarize the effect of the collision as the net velocity change of the ball over a very short time period. We assume that the ball is subjected to an **impulse**, a very intense force for a very short time. The physicist Paul Dirac suggested that such forces could be represented by the mathematical concept of an impulse, $\delta(t)$, which has the property that if $f(t)$ is continuous at $t = \tau$, then

Impulse response

$$\int_{-\infty}^{\infty} f(\tau)\delta(t - \tau)\,d\tau = f(t). \tag{3.2}$$

In other words, the impulse is so short and so intense that no value of f matters except over the short range the δ occurs. Since integration is a limit of a summation process, Eq. (3.2) can be viewed as representing the function f as a sum of impulses. If we replace f by u, then Eq. (3.2) represents an input $u(t)$ as a sum of impulses of intensity $u(t-\tau)$. To find the response to an arbitrary input, the principle of superposition tells us we need only to find the response to a unit impulse. For a general linear system we can express the impulse response as $h(t, \tau)$, the response at t to an impulse applied at τ. The total response is then the sum (integral) of these with intensity u:

$$y(t) = \int_{-\infty}^{\infty} u(\tau)h(t, \tau)\, d\tau.$$

The superposition integral

This is the **superposition integral**. If the system is not only linear but also constant, then the impulse response is given by $h(t-\tau)$ since the response at t to an input applied at τ depends only on the difference between the time the impulse is applied and the time we are observing the response. Constant systems are called shift-invariant for this reason. For constant systems the superposition integral then takes the special form

$$y(t) = \int_{-\infty}^{\infty} u(\tau)h(t-\tau)\, d\tau,$$

or

$$y(t) = \int_{-\infty}^{\infty} u(t-\tau)h(\tau)\, d\tau. \tag{3.3}$$

The convolution integral is a special form of the superposition integral.

Equation (3.3) is called the **convolution integral**. The integral occurs so often that the notation $y = u*h$ is often used for Eq. (3.3).

◆ **EXAMPLE 3.2**   *Convolution*

We can illustrate convolution with a simple system. Consider the impulse response for the system described by the differential equation

$$\dot{y} + ky = u = \delta(t),$$

with an initial condition of $y(0)=0$ before the impulse. Because $\delta(t)$ has an effect only near $t=0$, we can integrate this equation from just before zero to just after zero with the result

$$\int_{0^-}^{0^+} \dot{y}\, dt + k \int_{0^-}^{0^+} y\, dt = \int_{0^-}^{0^+} \delta(t)\, dt.$$

The integral of $\dot{y}$ is simply y, the integral of y over so small a range is zero, and the integral of the impulse over the same range is unity. Therefore,

$$y(0^+) - y(0^-) = 1.$$

Since the system was at rest before application of the impulse, $y(0^-)=0$. Thus the effect of the impulse is that $y(0^+) = 1$. For positive time we have the differential equation

$$\dot{y} + ky = 0, \qquad y(0^+) = 1.$$

If we assume a solution $y = Ae^{st}$, then $\dot{y} = Ase^{st}$. The above equation then becomes

$$Ase^{st} + kAe^{st} = 0$$
$$s + k = 0$$
$$s = -k.$$

Since $y(0^+) = 1$, it is necessary that $A = 1$. Thus the solution for the impulse response is $y(t) = h(t) = e^{-kt}$ for $t > 0$. To take care of the fact that $h(t) = 0$ for negative time, we define the unit step function

$$1(t) = \begin{cases} 0, & t < 0, \\ 1, & t \geqslant 0. \end{cases}$$

With this definition the impulse response of the first-order system becomes

$$h(t) = e^{-kt} 1(t).$$

The response to a general input is given by the convolution of this impulse response and the input:

$$y(t) = \int_{-\infty}^{\infty} h(\tau) u(t - \tau) \, d\tau$$

$$= \int_{-\infty}^{\infty} e^{-k\tau} 1(\tau) u(t - \tau) \, d\tau$$

$$= \int_{0}^{\infty} e^{-k\tau} u(t - \tau) \, d\tau.$$

3.1.2   Transfer Functions

An immediate consequence of convolution is that an input of the form e^{st} results in an output $H(s)e^{st}$. Note that both input and output are exponential time functions, and that the output differs from the input only in the amplitude $H(s)$. $H(s)$ is the **transfer function** of the system. The constant s may be complex, expressed as $s = \sigma + j\omega$. Thus both the input and the output may be complex. If we let $u(t) = e^{st}$ in Eq. (3.3), then

Transfer function

$$y(t) = \int_{-\infty}^{\infty} h(\tau) u(t - \tau) \, d\tau$$

$$= \int_{-\infty}^{\infty} h(\tau) e^{s(t - \tau)} \, d\tau$$

$$= \int_{-\infty}^{\infty} h(\tau) e^{st} e^{-s\tau} \, d\tau$$

$$= \int_{-\infty}^{\infty} h(\tau) e^{-s\tau} \, d\tau \, e^{st}$$

$$= H(s) e^{st} \tag{3.4}$$

where

$$H(s) = \int_{-\infty}^{\infty} h(\tau)e^{-s\tau}\, d\tau. \tag{3.5}$$

The integral in Eq. (3.5) does not need to be computed to find the transfer function of a system. Instead, one can assume a solution of the form of Eq. (3.4), substitute that into the differential equation of the system, then solve for the transfer function $H(s)$.

◆ **EXAMPLE 3.3**   *Transfer Function*

Compute the transfer function for the system of Example 3.2, and find the output y for the input $u = e^{st}$.

Solution.   The system equation from Example 3.2 is

$$\dot{y}(t) + ky(t) = u(t) = e^{st}. \tag{3.6}$$

We assume that we can express $y(t)$ as $H(s)e^{st}$. With this form, we have $\dot{y} = sH(s)e^{st}$, and Eq. (3.6) reduces to

$$sH(s)e^{st} + kH(s)e^{st} = e^{st}. \tag{3.7}$$

Solving for the transfer function $H(s)$, we get

$$H(s) = \frac{1}{s+k}.$$

Substituting this back into Eq. (3.4) yields the output

$$y = \frac{e^{st}}{s+k}.$$

◆

A very common way to use the exponential response of linear constant systems is in finding the **frequency response**, or response to a sinusoid. First we express the sinusoid as a sum of two exponential expressions (Euler's relation):

Frequency response

$$A\cos(\omega t) = \frac{A}{2}\left(e^{j\omega t} + e^{-j\omega t}\right).$$

If we let $s = j\omega$ in the basic response formula (Eq. 3.4), then the response to $u(t) = e^{j\omega t}$ is $y(t) = H(j\omega)e^{j\omega t}$; similarly, the response to $u(t) = e^{-j\omega t}$ is $H(-j\omega)e^{-j\omega t}$. By superposition, the response to the sum of these two exponentials, which make up the cosine signal, is the sum of the responses:

$$y(t) = \frac{A}{2}\left[H(j\omega)e^{j\omega t} + H(-j\omega)e^{-j\omega t}\right]. \tag{3.8}$$

The transfer function $H(j\omega)$ is a complex number that can be represented in polar form or in magnitude-and-phase form as $H(j\omega) = M(\omega)e^{j\varphi(\omega)}$ or simply $H = Me^{j\varphi}$. With this substitution, Eq. (3.8) becomes

$$y(t) = \frac{A}{2} M(e^{j(\omega t + \varphi)} + e^{-j(\omega t + \varphi)})$$
$$= AM \cos(\omega t + \varphi). \tag{3.9}$$

This means that if a system represented by the transfer function $H(s)$ has a sinusoidal input with magnitude A, the output will be sinusoidal at the same frequency with magnitude AM and will be shifted in phase by the angle φ.

◆ **EXAMPLE 3.4** *Frequency Response*

For the system in Example 3.2, find the response to the sinusoidal input $u = A \cos(\omega t)$. That is, find the frequency response.

Solution. In Example 3.3 we found the transfer function. To find the frequency response, we let $s = j\omega$ and find that

$$H = \frac{1}{j\omega + k}.$$

From this we get

$$M = \frac{1}{\sqrt{\omega^2 + k^2}} \quad \text{and} \quad \varphi = \tan^{-1}\left(-\frac{\omega}{k}\right).$$

Therefore, the response of this system to a sinusoid will be

$$y(t) = AM \cos(\omega t + \varphi).$$

M is usually referred to as the **amplitude ratio** and ϕ is referred to as the **phase**. ◆

This example illustrates that the response of a linear constant system to a sinusoid of frequency ω is a sinusoid with the *same* frequency and with an amplitude ratio equal to the magnitude of the transfer function evaluated at the input frequency. Furthermore, the phase difference between input and output signals is given by the phase of the transfer function evaluated at the input frequency. The magnitude ratio and phase difference can be computed from the transfer function as just discussed; it can also be measured experimentally quite easily in the laboratory by driving the system with a known sinusoidal input and measuring the amplitude and phase of the system's output.

We can generalize the frequency response by defining the **Laplace trans-**

form of a signal $f(t)$ as

$$F(s) = \int_{-\infty}^{\infty} f(t)e^{-st}\,dt. \tag{3.10}$$

If we apply this definition to both $u(t)$ and $y(t)$ and use the convolution integral (Eq. 3.3), we find that

$$Y(s) = H(s)U(s), \tag{3.11}$$

The key property of
Laplace transforms

where $Y(s)$ and $U(s)$ are the Laplace transforms of $y(t)$ and $u(t)$, respectively. We will prove this result in Section 3.1.4 starting with Eq. (3.20).

Laplace transforms such as Eq. (3.10) can be used to study the complete response characteristics of feedback systems, including the transient response; that is, the time response to an initial condition. This is in contrast to the use of Fourier transforms, where the steady-state response is the main concern. With Eq. (3.10) we have a means for computing the response of linear constant systems to quite general inputs. Given any input into a system, we compute the transform of the input and the transfer function for the system. The transform of the output is then given by Eq. (3.11) as the product of these two. If we wanted the time function of the output, we would need to "invert" $Y(s)$ to get what is called the **inverse transform**; this step is typically not carried out explicitly. Nevertheless, understanding the process necessary for deriving $y(t)$ from $Y(s)$ is important because it leads to insight into the behavior of linear systems.

While it is possible to determine the transient response properties of the system using Eq. (3.10), it is generally more useful to use a simpler version of the Laplace transform based on the input beginning at time zero.

3.1.3   The $\mathscr{L}_-$ Laplace Transform

In many applications it is useful to define a **one-sided** (or **unilateral**) **Laplace transform**, which uses 0^- (that is, a value just before $t=0$) as the lower limit of integration in Eq. (3.10). The $\mathscr{L}_-$ Laplace transform of $f(t)$, denoted by $\mathscr{L}_-\{f(t)\} = F(s)$, is a function of the complex variable $s = \sigma + j\omega$, where

$$F(s) \triangleq \int_{0^-}^{\infty} f(t)e^{-st}\,dt. \tag{3.12}$$

The decaying exponential term in the integrand in effect provides a built-in convergence factor. This means that even if $f(t)$ does not vanish as $t\to\infty$, the integrand will vanish for sufficiently large values of σ if f does not grow at a faster than exponential rate. The fact that the lower limit of integration is at 0^- allows the use of an impulse function at $t=0$ as illustrated in Example 3.2; however, this distinction between $t=0^-$ and $t=0$ does not usually come up in practice. We will therefore for the most part drop the minus superscript on $t=0$; however, we will return to using the notation $t=0^-$ when an impulse at $t=0$ is involved and the distinction is of practical value.

If Eq. (3.12) is a one-sided transform, then by extension, Eq. (3.10) is a **two-sided Laplace transform.**[1] We will use the $\mathscr{L}$ symbol from here on to mean $\mathscr{L}_-$.

Based on the formal definition in Eq. (3.12), we can ascertain the properties of Laplace transforms and compute the transforms of common time functions. The analysis of linear systems by means of Laplace transforms usually involves using tables of common properties and time functions, so we have provided this information in Appendix A. The tables of time functions and their Laplace transforms, together with the table of properties, permit us to find transforms of complex signals from simpler ones. For a thorough study of Laplace transforms and extensive tables, see Churchill (1972) and Campbell and Foster (1948). For more study of the two-sided transform, see Van der Pol and Bremmer (1955). These authors show that the time function can be obtained from the Laplace transform by the relation

$$f(t) = \frac{1}{2\pi j} \int_{\sigma_c - j\infty}^{\sigma_c + j\infty} F(s)e^{st} \, ds, \tag{3.13}$$

where σ_c is a selected value to the right of all the singularities of $F(s)$ in the s-plane. In practice this relation is seldom used. Instead, complex Laplace transforms are broken down into simpler ones that are listed in the tables along with their corresponding time responses.

Let us compute a few Laplace transforms of some typical time functions.

◆ **EXAMPLE 3.5**   *Step and Ramp Transforms*

Find the Laplace transform of the step, $a1(t)$, and ramp, $bt1(t)$, functions.

Solution.   For a step of size a, from Eq. (3.12) we have,

$$F(s) = \int_0^\infty ae^{-st} \, dt = \frac{-ae^{-st}}{s} \Big|_0^\infty = 0 - \frac{-a}{s} = \frac{a}{s}.$$

For the ramp signal $f(t) = bt\,1(t)$, again from Eq. (3.12), we have

$$F(s) = \int_0^\infty bte^{-st} \, dt = \left[-\frac{bte^{-st}}{s} - \frac{be^{-st}}{s^2} \right]_0^\infty = \frac{b}{s^2},$$

where we employed the technique of integration by parts,

$$\int u \, dv = uv - \int v \, du,$$

with $u = bt$ and $dv = e^{-st} \, dt$.

◆

---

1. The other possible one-sided transform is, of course, $\mathscr{L}_+$, in which the lower limit of the integral is 0^+. It is sometimes used in other applications.

A more subtle example is that of the impulse function.

◆ **EXAMPLE 3.6**   *Impulse Function Transform*

Find the Laplace transform of the unit impulse function.

Solution.   From Eq. (3.12) we get

$$F(s) = \int_{0^-}^{\infty} \delta(t)e^{-st}\, dt = \int_{0^-}^{0^+} \delta(t)\, dt = 1. \tag{3.14}$$

◆

It is the transform of the unit impulse function that led us to choose the $\mathscr{L}_-$ transform rather than the $\mathscr{L}_+$ transform.

◆ **EXAMPLE 3.7**   *Sinusoid Transform*

Find the Laplace transform of the sinusoid function.

Solution.   Again we use Eq. (3.12) to get

$$\mathscr{L}\{\sin \omega t\} = \int_{0}^{\infty} (\sin \omega t)e^{-st}\, dt. \tag{3.15}$$

If we substitute the relation

$$\sin \omega t = \frac{e^{j\omega t} - e^{-j\omega t}}{2j}$$

into Eq. (3.15), we find that

$$\mathscr{L}\{\sin \omega t)\} = \int_{0}^{\infty} \left(\frac{e^{j\omega t} - e^{-j\omega t}}{2j} \right) e^{-st}\, dt$$

$$= \frac{1}{2j} \int_{0}^{\infty} (e^{(j\omega - s)t} - e^{-(j\omega + s)t})\, dt$$

$$= \frac{\omega}{s^2 + \omega^2}.$$

◆

Table A.2 in the Appendix lists Laplace transforms for elementary time functions. Each entry in the table follows from direct application of the transform definition as demonstrated by Examples 3.5 to 3.7.

3.1.4   Properties of Laplace Transforms

In this section we will address each of the significant properties of the Laplace transform listed in Table A.1.

Superposition

One of the more important properties of the Laplace transform is that it is linear. We can prove this as follows,

$$\mathcal{L}\{\alpha f_1(t) + \beta f_2(t)\} = \int_0^\infty [\alpha f_1(t) + \beta f_2(t)]e^{-st}\, dt$$

$$= \alpha \int_0^\infty f_1(t)e^{-st}\, dt + \beta \int_0^\infty f_2(t)e^{-st}\, dt$$

$$= \alpha F_1(s) + \beta F_2(s).$$

The scaling property is a special case of this; that is,

$$\mathcal{L}\{\alpha f(t)\} = \alpha F(s).$$

Time Delay

Suppose a function $f(t)$ is delayed by $\lambda > 0$ units of time. Its Laplace transform is

$$F_1(s) = \int_0^\infty f(t - \lambda)e^{-st}\, dt.$$

Let us define $t' = t - \lambda$. Then $dt' = dt$ since λ is a constant and $f(t) = 0$ for $t < 0$. Thus

$$F_1(s) = \int_{-\lambda}^\infty f(t')e^{-s(t' + \lambda)}dt' = \int_0^\infty f(t')e^{-s(t' + \lambda)}dt'.$$

Since $e^{-s\lambda}$ is independent of time, it can be taken out of the integrand, so

$$F_1(s) = e^{-s\lambda} \int_0^\infty f(t')e^{-st'}\, dt' = e^{-s\lambda}F(s).$$

From this result we see that a time delay of λ corresponds to multiplication of the transform by $e^{-s\lambda}$.

Time Scaling

It is sometimes useful to time-scale equations of motion. For example, in the control system of a tape drive it is meaningful to measure time in milliseconds (see also Chapters 2 and 9). If the time t is scaled by a factor a, then the Laplace transform of the time-scaled signal is

$$F_1(s) = \int_0^\infty f(at)e^{-st}\,dt.$$

Again we define $t' = at$. As before, $dt' = a\,dt$ and

$$F_1(s) = \int_0^\infty f(t')\frac{e^{-st'/a}}{|a|}\,dt' = \frac{1}{|a|}F\left(\frac{s}{a}\right).$$

Shift in Frequency

Multiplication (modulation) of $f(t)$ by an exponential expression in the time domain corresponds to a shift in frequency:

$$F_1(s) = \int_0^\infty e^{-at}f(t)e^{-st}\,dt = \int_0^\infty f(t)e^{-(s+a)t}\,dt = F(s+a). \qquad (3.16)$$

Differentiation

The transform of the derivative of a signal is related to its Laplace transform and its initial condition as follows:

$$\mathcal{L}\left\{\frac{df}{dt}\right\} = \int_{0^-}^\infty \left(\frac{df}{dt}\right)e^{-st}dt = e^{-st}f(t)\big|_{0^-}^\infty + s\int_{0^-}^\infty f(t)e^{-st}\,dt. \qquad (3.17)$$

Since $f(t)$ is assumed to have a Laplace transform, $e^{-st}f(t) \to 0$ as $t \to \infty$. Thus

$$\mathcal{L}[\dot{f}] = -f(0^-) + sF(s). \qquad (3.18)$$

Another application of Eq. (3.18) leads to

$$\mathcal{L}\{\ddot{f}\} = s^2F(s) - sf(0^-) - \dot{f}(0^-). \qquad (3.19)$$

Repeated application of Eq. (3.18) leads to

$$\mathcal{L}\{f^m(t)\} = s^m F(s) - s^{m-1}f(0^-) - s^{m-2}\dot{f}(0^-) - \cdots - f^{(m-1)}(0^-),$$

where $f^m(t)$ denotes the mth derivative of $f(t)$ with respect to time.

Integration

Let us assume we wish to determine the Laplace transform of the integral of a time function, that is, to find

$$F_1(s) = \mathcal{L}\left\{\int_0^t f(\xi)d\xi\right\} = \int_0^\infty \left[\int_0^t f(\xi)d\xi\right]e^{-st}dt.$$

Employing integration by parts, where

$$u = \int_0^t f(\xi)d\xi \qquad \text{and} \qquad dv = e^{-st}dt,$$

we get

$$F_1(s) = \left[-\frac{1}{s} e^{-st} \left(\int_0^t f(\xi)d\xi \right) \right]_0^\infty - \int_0^\infty -\frac{1}{s} e^{-st} f(t)dt = \frac{1}{s} F(s).$$

Convolution

We have seen previously that the response of a system is determined by convolving the input with the impulse response of the system, or by forming the product of the transfer function and the Laplace transform of the input. The following discussion extends this concept to various time functions.

Convolution in the time domain corresponds to multiplication in the frequency domain. Assume that $\mathcal{L}\{f_1(t)\} = F_1(s)$ and $\mathcal{L}\{f_2(t)\} = F_2(s)$. Then

$$\mathcal{L}\{f_1(t)*f_2(t)\} = \int_0^\infty f_1(t)*f_2(t)e^{-st}dt = \int_0^\infty \left[\int_0^t f_1(\tau)f_2(t-\tau)d\tau \right] e^{-st}dt \qquad (3.20)$$

Reversing the order of integration and changing the limits of integration yield

$$\mathcal{L}\{f_1(t)*f_2(t)\} = \int_0^\infty \int_\tau^\infty f_1(\tau)f_2(t-\tau)e^{-st}dtd\tau.$$

Multiplying by $e^{-s\tau}e^{s\tau}$ results in

$$\mathcal{L}\{f_1(t)*f_2(t)\} = \int_0^\infty f_1(\tau)e^{-s\tau} \left[\int_\tau^\infty f_2(t-\tau)e^{-s(t-\tau)}dt \right] d\tau.$$

If $t' \triangleq t - \tau$, then

$$\mathcal{L}\{f_1(t)*f_2(t)\} = \int_0^\infty f_1(\tau)e^{-s\tau}d\tau \int_0^\infty f_2(t')e^{-st'}dt'$$

$$\mathcal{L}\{f_1(t)*f_2(t)\} = F_1(s)F_2(s).$$

A similar, or dual, of this result is discussed next.

Time Product

Multiplication in the time domain corresponds to convolution in the frequency domain:

$$\mathcal{L}\{f_1(t)f_2(t)\} = \frac{1}{2\pi j} \int_{\sigma_c - j\infty}^{\sigma_c + j\infty} F_1(\xi)F_2(s - \xi)d\xi.$$

To see this, consider the relation

$$\mathcal{L}\{f_1(t)f_2(t)\} = \int_0^\infty f_1(t)f_2(t)e^{-st}dt.$$

Substituting the expression for $f_1(t)$ given by Eq. (3.13) yields

$$\mathcal{L}\{f_1(t)f_2(t)\} = \int_0^\infty \left[\frac{1}{2\pi j} \int_{\sigma_c - j\infty}^{\sigma_c + j\infty} F_1(\xi)e^{\xi t}\,d\xi \right] f_2(t)e^{-st}\,dt.$$

Changing the order of integration results in

$$\mathcal{L}\{f_1(t)f_2(t)\} = \frac{1}{2\pi j} \int_{\sigma_c - j\infty}^{\sigma_c + j\infty} F_1(\xi) \int_0^\infty f_2(t)e^{-(s - \xi)t}\,dt\,d\xi.$$

Using Eq. (3.16), we get

$$\mathcal{L}\{f_1(t)f_2(t)\} = \frac{1}{2\pi j} \int_{\sigma_c - j\infty}^{\sigma_c + j\infty} F_1(\xi)F_2(s - \xi)d\xi.$$

Multiplication by Time

Multiplication by time corresponds to differentiation in the frequency domain. Let us consider

$$\frac{d}{ds} F(s) = \frac{d}{ds} \int_0^\infty e^{-st}f(t)\,dt$$

$$= \int_0^\infty -te^{-st}f(t)dt$$

$$= -\int_0^\infty e^{-st}[tf(t)]dt$$

$$= -\mathcal{L}\{tf(t)\}.$$

Then

$$\mathcal{L}\{tf(t)\} = -\frac{d}{ds} F(s),$$

which is the desired result.

3.1.5   Partial-fraction Expansion

The easiest way to find $f(t)$ from its Laplace transform $F(s)$, if $F(s)$ is rational, is to expand $F(s)$ as a sum of simpler terms that can be found in the tables. The basic tool for performing this operation is called **partial-fraction expansion**. Consider the general form for the rational function $F(s)$ consisting of the ratio of two polynomials:

$$F(s) = \frac{b_1 s^m + b_2 s^{m-1} + \cdots + b_{m+1}}{s^n + a_1 s^{n-1} + \cdots + a_n}.$$

By factoring the polynomials this same function could also be expressed in terms of the product of factors as

$$F(s) = K \frac{\prod_{i=1}^{m}(s - z_i)}{\prod_{i=1}^{n}(s - p_i)}.$$

Zeros and poles

For a transform $F(s)$ representing the response of any physical system, $m \leq n$. When $s = z_i$, s is referred to as a **zero** of the function, and when $s = p_i$, s is referred to as a **pole** of the function. Assuming for now that the poles $\{p_i\}$ are real or complex but distinct, we rewrite $F(s)$ as the partial fraction

$$F(s) = \frac{C_1}{s - p_1} + \frac{C_2}{s - p_2} + \cdots + \frac{C_n}{s - p_n}. \tag{3.21}$$

Next we determine the set of constants $\{C_i\}$. We multiply both sides of Eq. (3.21) by the factor $s - p_1$ to get

$$(s - p_1)F(s) = C_1 + \frac{s - p_1}{s - p_2}C_2 + \cdots + \frac{(s - p_1)C_n}{s - p_n}. \tag{3.22}$$

If we let $s = p_1$ on both sides of Eq. (3.22), then all the C_i terms will equal zero except for the first one. For this term,

$$C_1 = (s - p_1)F(s)|_{s = p_1}. \tag{3.23}$$

The other coefficients can be expressed in a similar form:

$$C_i = (s - p_i)F(s)|_{s = p_i}.$$

The cover-up method of determining coefficients

This process is called the **cover-up method** because in the factored form of $F(s)$ (Eq. 3.21), we can cover up the individual denominator terms, evaluate the rest of the expression with $s = p_i$, and determine the coefficients C_i. Once this has been completed, the time function becomes

$$f(t) = \sum_{i=1}^{n} C_i e^{p_i t} 1(t)$$

because, as entry 7 in Table A.2 shows, if

$$F(s) = \frac{1}{s - p_i}$$

then

$$f(t) = e^{p_i t} 1(t).$$

◆ **EXAMPLE 3.8** *Partial-fraction Expansion: Distinct Real Roots*

Suppose you have computed $Y(s)$ and found that

$$Y(s) = \frac{(s + 2)(s + 4)}{s(s + 1)(s + 3)}.$$

Find $y(t)$.

Solution. We may write $Y(s)$ in terms of its partial-fraction expansion:

$$Y(s) = \frac{C_1}{s} + \frac{C_2}{s+1} + \frac{C_3}{s+3}.$$

Using the cover-up method, we get

$$C_1 = \frac{(s+2)(s+4)}{(s+1)(s+3)}\bigg|_{s=0} = \frac{8}{3}.$$

In a similar fashion,

$$C_2 = \frac{(s+2)(s+4)}{s(s+3)}\bigg|_{s=-1} = -\frac{3}{2}$$

and

$$C_3 = \frac{(s+2)(s+4)}{s(s+1)}\bigg|_{s=-3} = -\frac{1}{6}.$$

We can check the correctness of the result by adding the components again to verify that the original function has been recovered. With the partial fraction the solution can be looked up in the tables at once to be

$$y(t) = \frac{8}{3}\,1(t) - \frac{3}{2}\,e^{-t}1(t) - \frac{1}{6}\,e^{-3t}1(t).$$

In the case of quadratic factors in the denominator, the numerator of the quadratic factor is chosen to be first-order as shown in Example 3.9.

◆ **EXAMPLE 3.9**   *Partial-fraction Expansion: Distinct Complex Roots*

Find the function $f(t)$ for which the Laplace transform is

$$F(s) = \frac{1}{s(s^2 + s + 1)}.$$

Solution. We rewrite $F(s)$ as

$$F(s) = \frac{C_1}{s} + \frac{C_2 s + C_3}{s^2 + s + 1}.$$

Using the cover-up method, we find C_1 to be

$$C_1 = sF(s)|_{s=0} = 1.$$

We equate the numerators,

$$(s^2 + s + 1) + (C_2 s + C_3)s = 1.$$

After equating like powers of s on the two sides of this equation, we find that $C_2 = -1$ and $C_3 = -1$. To make it more suitable for using the tables, we rewrite the partial fraction as

$$F(s) = \frac{1}{s} - \frac{s + \frac{1}{2} + \frac{1}{2}}{(s + \frac{1}{2})^2 + \frac{3}{4}}.$$

From the tables we have,

$$f(t) = \left(1 - e^{-t/2} \cos \sqrt{\tfrac{3}{4}} t - \frac{1}{\sqrt{3}} e^{-t/2} \sin \sqrt{\tfrac{3}{4}} t\right) 1(t).$$

◆

Whenever there exists a complex conjugate pair of poles such as

$$F(s) = \frac{C_1}{s - p_1} + \frac{C_2}{s - p_1^*},$$

we can show that

$$C_2 = C_1^*,$$

(see Problem 3.1), and that

$$f(t) = C_1 e^{p_1 t} + C_1^* e^{p_1^* t} = 2\mathrm{Re}(C_1 e^{p_1 t}).$$

Assuming that $p_1 = \alpha + j\beta$, we may rewrite $f(t)$ as

$$f(t) = 2\mathrm{Re}\{C_1 e^{p_1 t}\} = 2\mathrm{Re}\{|C_1| e^{j\arg(C_1)} e^{(\alpha + j\beta)t}\}$$
$$= 2|C_1| e^{\alpha t} \cos[\beta t + \arg(C_1)].$$

For the case where $F(s)$ has repeated roots, the procedure to compute the partial-fraction expansion must be modified. If p_1 is repeated three times, we write the partial fraction as

$$F(s) = \frac{C_1}{s - p_1} + \frac{C_2}{(s - p_1)^2} + \frac{C_3}{(s - p_1)^3} + \frac{C_4}{s - p_4} + \cdots + \frac{C_n}{s - p_n}.$$

We determine the constants C_4 through C_n as discussed previously. If we multiply both sides of the above equation by $(s - p_1)^3$,

$$(s - p_1)^3 F(s) = C_1(s - p_1)^2 + C_2(s - p_1) + C_3 + \cdots + \frac{C_n(s - p_1)^3}{s - p_n}, \quad (3.24)$$

and set $s = p_1$, then all the factors on the right side of Eq. (3.24) will go to zero except C_3, which is

$$C_3 = (s - p_1)^3 F(s)|_{s = p_1}$$

as before. To determine the other factors, we differentiate Eq. (3.24) with respect to the Laplace variable s:

$$\frac{d}{ds}[(s - p_1)^3 F(s)] = 2C_1(s - p_1) + C_2 + \cdots + \frac{d}{ds}\left[\frac{C_n(s - p_1)^3}{s - p_n}\right]. \quad (3.25)$$

Again, if we set $s = p_1$, we have

$$C_2 = \frac{d}{ds}[(s - p_1)^3 F(s)]_{s=p_1}.$$

Similarly, if we differentiate Eq. (3.25) again and set $s = p_1$ a second time, we get

$$C_1 = \frac{1}{2}\frac{d^2}{ds^2}[(s - p_1)^3 F(s)]_{s=p_1}.$$

In general, we may compute C_i for a factor with multiplicity k as

$$C_{k-i} = \frac{1}{i!}\left[\frac{d^i}{ds^i}[(s - p_1)^k F(s)]\right]_{s=p_1}, \qquad i = 0, \ldots, k-1.$$

◆ **EXAMPLE 3.10**   *Partial-fraction Expansion: Repeated Real Roots*

Find the function $f(t)$ that has the Laplace transform

$$F(s) = \frac{s+3}{(s+1)(s+2)^2}.$$

Solution.   We write the partial fraction as

$$F(s) = \frac{C_1}{s+1} + \frac{C_2}{s+2} + \frac{C_3}{(s+2)^2}.$$

Then

$$C_1 = (s+1)F(s)|_{s=-1} = \frac{s+3}{(s+2)^2}\bigg|_{s=-1} = 2,$$

$$C_2 = \frac{d}{ds}(s+2)^2 F(s)|_{s=-2} = -2,$$

$$C_3 = (s+2)^2 F(s)|_{s=-2} = \frac{s+3}{s+1}\bigg|_{s=-2} = -1.$$

The function $f(t)$ is

$$f(t) = (2e^{-t} - 2e^{-2t} - te^{-2t})1(t).$$

3.1.6   Laplace Transform Theorems

A useful property of the Laplace transform known as the **Final Value Theorem** allows us to compute the constant steady-state value of a time function given its Laplace transform. The theorem follows from the development of partial-fraction expansion. Suppose we have a transform $Y(s)$ of a signal $y(t)$ and wish to know the final value $y(t)$ from $Y(s)$. There are three possibilities for the limit.

It can be constant, undefined, or unbounded. If $Y(s)$ has any poles in the right half of the s-plane—that is, if the real part of any $p_i > 0$—then $y(t)$ will grow and the limit will be unbounded. If $Y(s)$ has a pair of poles on the imaginary axis of the s-plane, i.e., $p_i = \pm j\omega$, then $y(t)$ will contain a sinusoid that persists forever and the final value will not be defined. Only one case can provide a nonzero constant final value: If all poles of $Y(s)$ are in the left half of the s-plane except for one at $s = 0$, then all terms of $y(t)$ will decay to zero except the term corresponding to the pole at $s = 0$, and that term corresponds to a constant in time. Thus the final value is given by the coefficient associated with the pole at $s = 0$. Therefore, the Final Value Theorem is:

The Final Value Theorem

If all poles of $sY(s)$ are in the left half of the s-plane, then

$$\lim_{t \to \infty} y(t) = \lim_{s \to 0} s Y(s). \qquad (3.26)$$

We may prove this result as follows. The derivative relationship developed in Eq. (3.18) is

$$\mathscr{L} \left\{ \frac{df}{dt} \right\} = sF(s) - f(0^-) = \int_{0^-}^{\infty} e^{-st} \frac{df}{dt} \, dt.$$

We assume we are interested in the case where $s \to 0$. Then

$$\lim_{s \to 0} [sF(s) - f(0)] = \lim_{s \to 0} \left(\int_{0}^{\infty} e^{-st} \frac{df}{dt} \, dt \right) = \lim_{t \to \infty} [f(t) - f(0)]$$

and we have,

$$\lim_{t \to \infty} f(t) = \lim_{s \to 0} sF(s).$$

Another way to see this same result is to note the partial-fraction expansion of $F(s)$ (Eq. 3.21):

$$F(s) = \frac{C_1}{s - p_1} + \frac{C_2}{s - p_2} + \cdots + \frac{C_n}{s - p_n}.$$

Let us say that $p_1 = 0$ and all other p_i are in the left half-plane so that C_1 is the steady-state value of $f(t)$. Using Eq. (3.23) we see that

$$C_1 = \lim_{t \to \infty} f(t) = sF(s)|_{s=0},$$

which is the same result as above.

◆ **EXAMPLE 3.11**   *Final Value Theorem*

Find the steady-state value of the system corresponding to:

$$Y(s) = \frac{3(s+2)}{s(s^2 + 2s + 10)}.$$

Solution.   Applying the Final Value Theorem, we obtain

$$y(\infty) = s Y(s)|_{s=0} = \frac{3 \cdot 2}{10} = 0.6.$$

Thus, after the transients have decayed to zero, $y(t)$ will settle to a constant value of 0.6.

◆

Use the Final Value
Theorem on stable
systems only.

Care must be taken to apply the Final Value Theorem only to stable systems. While one could use Eq. (3.26) on any $Y(s)$, doing so could result in erroneous results, as shown in the next example.

◆ **EXAMPLE 3.12**   *Incorrect Use of the Final Value Theorem*

Find the final value of the signal corresponding to:

$$Y(s) = \frac{3}{s(s-2)}.$$

Solution.   If we blindly apply Eq. (3.26), we obtain

$$y(\infty) = s Y(s)|_{s=0} = -\tfrac{3}{2}.$$

However,

$$y(t) = \left(-\tfrac{3}{2} + \tfrac{3}{2} e^{2t} \right) 1(t),$$

and Eq. (3.26) yields the constant term only. Of course, the true final value is unbounded.

◆

Calculating DC gain by
the Final Value Theorem

The theorem can also be used to find the DC gain of a system. The **DC gain** is the ratio of the output of a system to its input (presumed constant) after all transients have decayed. To find the DC gain, we assume there is a unit-step input $[U(s) = 1/s]$ and use the Final Value Theorem to compute the steady-state value of the output. Therefore, for a system transfer function $G(s)$,

$$\text{DC gain} = \lim_{s \to 0} s G(s) \frac{1}{s} = \lim_{s \to 0} G(s). \qquad (3.27)$$

◆ **EXAMPLE 3.13**   *DC Gain*

Find the DC gain of the system whose transfer function is

$$G(s) = \frac{3(s+2)}{(s^2 + 2s + 10)}.$$

Solution.   Applying Eq. (3.27), we get

$$\text{DC gain} = G(s)|_{s=0} = \frac{3 \cdot 2}{10} = 0.6.$$

◆

A second valuable Laplace transform theorem is the **Initial Value Theorem,** which states that it is always possible to determine the initial value of the time function $f(t)$ from its Laplace transform. We may also state the theorem in this way:

The Initial Value Theorem

For any Laplace transform pair,

$$\lim_{s \to \infty} sF(s) = f(0^+) \qquad (3.28)$$

We may show this as follows. Using Eq. (3.18),

$$\mathcal{L}\left\{\frac{df}{dt}\right\} = sF(s) - f(0^-) = \int_{0^-}^{\infty} e^{-st} \frac{df}{dt} \, dt \qquad (3.29)$$

Let us consider the case where $s \to \infty$ and rewrite the integral as

$$\int_{0^-}^{\infty} e^{-st} \frac{df(t)}{dt} \, dt = \int_{0^+}^{\infty} e^{-st} \frac{df(t)}{dt} \, dt + \int_{0^-}^{0^+} e^{-st} \frac{df(t)}{dt} \, dt.$$

Taking the limit of Eq. (3.29) as $s \to \infty$, we get

$$\lim_{s \to \infty} [sF(s) - f(0^-)] = \lim_{s \to \infty} \left[\int_{0^-}^{0^+} e^0 \frac{df(t)}{dt} \, dt + \int_{0^+}^{\infty} e^{-st} \frac{df(t)}{dt} \, dt \right]$$

The second term on the right side of the above equation approaches zero as $s \to \infty$ because $e^{-st} \to 0$. Hence

$$\lim_{s \to \infty} [sF(s) - f(0^-)] = \lim_{s \to \infty} [f(0^+) - f(0^-)] = f(0^+) - f(0^-)$$

or,

$$\lim_{s \to \infty} sF(s) = f(0^+).$$

In contrast with the Final Value Theorem, the Initial Value Theorem can be applied to any function $F(s)$.

◆ **EXAMPLE 3.14**   *Initial Value Theorem*

Find the initial value of the signal in Example 3.12.

Solution.   From the Initial Value Theorem, we get

$$y(0^+)=\lim_{s\to\infty} sY(s)=\lim_{s\to\infty} s\frac{3}{s(s-2)}=0,$$

which checks with the expression for $y(t)$ computed in Example 3.12.

3.1.7   Using Laplace Transforms to Solve Problems

Laplace transforms can be used to solve differential equations using the properties already developed. First, we find the Laplace transform of the differential equation using the differentiation properties in Eqs. (3.18) and (3.19). Then we find the Laplace transform of the output; using partial-fraction expansion and Table A.2, this can be converted to a time response function. We will illustrate this with three examples.

◆ **EXAMPLE 3.15**   *Homogeneous Differential Equation Solution*

Find the solution to the differential equation

$$\ddot{y}(t)+y(t)=0 \qquad \text{where } y(0)=\alpha,\ \dot{y}(0)=\beta.$$

Solution.   Using Eq. (3.19), the Laplace transform of the differential equation is

$$s^2Y(s)-\alpha s-\beta+Y(s)=0,$$
$$(s^2+1)Y(s)=\alpha s+\beta,$$
$$Y(s)=\frac{\alpha s}{s^2+1}+\frac{\beta}{s^2+1}.$$

After looking up in the transform tables the two terms on the right side of the above equation, we get

$$y(t)=[\alpha\cos t+\beta\sin t]1(t).$$

We can verify that this solution is correct by substituting it back into the differential equation.

Another example will illustrate the solution when the equations are not homogeneous, that is, when the system is forced.

◆ **EXAMPLE 3.16**   *Forced Differential Equation Solution*

Find the solution to the differential equation,

$$\ddot{y}(t) + 5\dot{y}(t) + 4y(t) = 3 \qquad \text{where } y(0) = \alpha, \ \dot{y}(0) = \beta.$$

Solution.   Taking the Laplace transform of both sides using Eqs. (3.18) and (3.19), we get

$$s^2 Y(s) - s\alpha - \beta + 5[s Y(s) - \alpha] + 4Y(s) = \frac{3}{s}.$$

Solving for $Y(s)$ yields

$$Y(s) = \frac{s(s\alpha + \beta + 5\alpha) + 3}{s(s+1)(s+4)}.$$

The partial-fraction expansion is

$$Y(s) = \frac{\frac{3}{4}}{s} - \frac{\frac{3 - \beta - 4\alpha}{3}}{s+1} + \frac{\frac{3 - 4\alpha - 4\beta}{12}}{s+4}.$$

Therefore, the time function is given by

$$y(t) = \left(\frac{3}{4} + \frac{-3 + \beta + 4\alpha}{3} e^{-t} + \frac{3 - 4\alpha - 4\beta}{12} e^{-4t} \right) 1(t).$$

By differentiating this solution twice and substituting the result in the original differential equation, we can verify that this solution satisfies the differential equation.

◆

The solution is especially simple if the initial conditions are all zero.

◆ **EXAMPLE 3.17**   *Forced Equation Solution with Zero Initial Conditions*

Find the solution to

$$\ddot{y}(t) + 5\dot{y}(t) + 4y(t) = u(t), \qquad y(0) = 0, \ \dot{y}(0) = 0, \ u(t) = 2e^{-2t} 1(t).$$

Solution.   Taking the Laplace transform of both sides, we get

$$s^2 Y(s) + 5s Y(s) + 4 Y(s) = \frac{2}{s+2}.$$

Solving for $Y(s)$ yields

$$Y(s) = \frac{2}{(s+2)(s+1)(s+4)}.$$

The partial- fraction expansion is

$$Y(s) = -\frac{1}{s+2} + \frac{\frac{2}{3}}{s+1} + \frac{\frac{1}{3}}{s+4}.$$

Therefore, the time function is given by,

$$y(t) = (-1e^{-2t} + \tfrac{2}{3}e^{-t} + \tfrac{1}{3}e^{-4t})1(t).$$

◆

The primary value of using the method of solving differential equations through the Laplace transform is that it provides information concerning the qualitative characteristic behavior of the response. Once we know the values of the poles of $Y(s)$, we know what kind of characteristic terms will appear in the response. In the last example the pole at $s = -1$ produced a decaying $y = Ce^{-t}$ term in the response. The pole at $s = -4$ produced a $y = Ce^{-4t}$ term in the response, which decays faster. If there had been a pole at $s = +1$, there would have been a growing $y = Ce^{+t}$ term in the response. Using the pole locations to understand in essence how the system will respond is a powerful tool and will be developed further in Section 3.3. Control systems designers often manipulate design parameters so that the poles have values that would give acceptable responses, and they skip the steps associated with converting those poles to actual time responses until the final stages of the design. They use trial-and-error design methods (as described in Chapter 5) that graphically present how changes in design parameters affect the pole locations. Once a design has been obtained with pole locations predicted to give acceptable responses, the control designer determines a time response to verify that the design is satisfactory. However, because finding a time response through the Laplace transform is often tedious, the designer solves the differential equations directly by using numerical computer methods. This bypasses the need for performing a partial-fraction expansion and converting the results to time responses.

Poles indicate response character

3.1.8   Laplace Transforms Using CACSD Software

In Chapter 2 we saw that a linear dynamic system may be represented by its differential equations, possibly in state-variable form,

$$\dot{\mathbf{x}} = \mathbf{F}\mathbf{x} + \mathbf{G}u$$

$$y = \mathbf{H}\mathbf{x} + Ju.$$

Three ways of representing linear systems

Earlier in the chapter we saw that a linear dynamic system may be represented by the Laplace transform of its differential equation, that is, its transfer function. The transfer function can be described either as a ratio of two polynomials in s,

$$H(s)=\frac{b_1 s^m + b_2 s^{m-1} + \cdots + b_{m+1}}{s^n + a_1 s^{n-1} + \cdots + a_n},$$

or in factored zero-pole form

$$H(s)=K\frac{\prod_{i=1}^{m}(s-z_i)}{\prod_{i=1}^{n}(s-p_i)}.$$

CACSD packages typically accept the specification of a system in any of these three forms. MATLAB refers to the descriptions as ss, tf, and zp (for state space, transfer function, and zero-pole, respectively) and can transform the system description from any one form to another. Therefore, by specifying a system by its state-space description to MATLAB and then transforming the description to the polynomial transfer-function form, we get the Laplace transform of the system. The relationship between the state-space description (F, G, H, and J) and the transfer-function description $H(s)$ is derived in Section 7.2 and shown by Eq. (7.36).

◆ **EXAMPLE 3.18**   *Cruise-control Transfer Function Using MATLAB*

❖ 🖳   Find the transfer function between the input u and the position of the car x in the cruise-control system in Example 2.8.

Solution.   In Example 2.8 we found the matrices F, G, and H and the value of J to be

$$\mathbf{F}=\begin{bmatrix} 0 & 1 \\ 0 & -0.05 \end{bmatrix}, \qquad \mathbf{G}=\begin{bmatrix} 0 \\ 0.001 \end{bmatrix}, \qquad \mathbf{H}=[1 \quad 0], \qquad J=0.$$

The MATLAB statement

    [num,den]=ss2tf(F,G,H,J)

will compute and display the coefficients of the numerator polynomial as the row vector num and the denominator as den. The results for this example are

num $=[0 \quad 0 \quad 0.001]$ and   den $=[1 \quad 0.05 \quad 0]$.

This means that the transfer function of the system is

$$H(s)=\frac{0.001}{s^2+0.05s+0}=\frac{0.001}{s(s+0.05)}.$$

◆

When a dynamic system is represented by a single differential equation of any order, finding the polynomial form of the transfer function from that differential equation is usually easier than finding the state-variable description from that differential equation. Therefore, you will find it best in these cases to specify a system directly in terms of its transfer function.

◆ **EXAMPLE 3.19**   *Transformations Using* MATLAB

Find the transfer function and state-variable description of the system whose differential equation is

$$\ddot{y} + 6\dot{y} + 25y = 9u + 3\dot{u}.$$

Solution.   Using the differentiation rules given by Eqs. (3.18) and (3.19), we see by inspection that

$$\frac{Y(s)}{U(s)} = \frac{3s+9}{s^2+6s+25}.$$

The MATLAB statements

```
num=[3 9]

den=[1 6 25]

[F,G,H,J]=tf2ss(num,den)
```

will provide values of the state-variable description of the system:

$$\mathbf{F} = \begin{bmatrix} -6 & -25 \\ 1 & 0 \end{bmatrix}, \qquad \mathbf{G} = \begin{bmatrix} 1 \\ 0 \end{bmatrix}, \qquad \mathbf{H} = [3 \quad 9], \qquad J = 0.$$

Note that because of the derivatives of u in the differential equations, it is not as straightforward to write the state-variable description of this system as it was for Example 2.8. Note also that the system output y is not one of the state elements, as evidenced by the fact that **H** contains two nonzero elements.

If the transfer function were desired in factored form, it could be obtained by transforming either the ss or tf description. Therefore, either of the following MATLAB statements

```
[z,p,k]=ss2zp(F,G,H,J)

[z,p,k]=tf2zp(num,den)
```

would result in

$$z = -3, \qquad p = \begin{bmatrix} -3+4j \\ -3-4j \end{bmatrix}, \qquad k = 3.$$

This means the transfer function could also be written as

$$\frac{Y(s)}{U(s)} = \frac{3(s+3)}{(s+3-4j)(s+3+4j)}.$$

◆

3.2 System Modeling Diagrams

3.2.1 The Block Diagram

To obtain the transfer function, we need to find the Laplace transform of the equations of motion and solve the resulting algebraic equations for the relationship between the input and the output. In many control systems the system equations can be written so that their components do not interact except by having the input of one part be the output of another part. In these cases it is easy to draw a block diagram that represents the mathematical relationships in a similar manner to that used for the component block diagram in Chapter 1. The transfer function of each component is placed in a box, and the input-output relationships between components are indicated by lines and arrows. We can then solve the equations by graphical simplification, which is often easier and more informative than algebraic manipulation, even though the methods are in every way equivalent. Drawings of three elementary block diagrams are seen in Fig. 3.1. It is convenient to think of each block as representing an electronic amplifier with the transfer function printed inside. The interconnections of blocks include summing points, where any number of signals may be added together. These are represented by a circle with the symbol Σ inside. In Fig. 3.1(a) the block with transfer function $G_1(s)$ is in series with the block with transfer function $G_2(s)$, and the overall transfer function is given by the product $G_2 G_1$. In Fig. 3.1(b) two systems are in parallel with their outputs added, and the overall transfer function is given by the sum $G_1 + G_2$. These diagrams derive simply from the equations that describe them.

Figure 3.1(c) shows a more complicated case. Here the two blocks are connected in a feedback arrangement so that each feeds into the other. When the feedback $Y_2(s)$ is *subtracted*, as shown in the figure, we call it **negative feedback**. As you will see in Chapter 4, negative feedback is usually required for system stability. For now we will simply solve the equations and then relate them back to the diagram. The equations are

$$U_1(s) = R(s) - Y_2(s),$$
$$Y_2(s) = G_2(s)G_1(s)U_1(s),$$
$$Y_1(s) = G_1(s)U_1(s),$$

Negative feedback

FIGURE 3.1
Three examples of elementary block diagrams

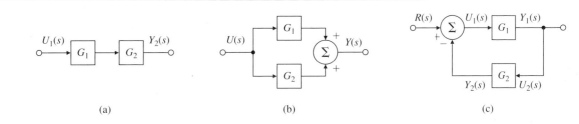

(a) (b) (c)

and their solution is

$$Y_1(s) = \frac{G_1(s)}{1 + G_1(s)G_2(s)} R(s).$$ (3.30)

We can express the solution by the rule:

The gain of a single-loop negative feedback system is given by the forward gain divided by the sum of 1 plus the loop gain.

Positive feedback

When the feedback is added instead of subtracted, we call it **positive feedback**. In this case the gain is given by the forward gain divided by the sum of 1 minus the loop gain.

The three elementary cases given in Fig. 3.1 can be used in combination to solve by repeated reduction any transfer function defined by a block diagram. However, the manipulations can be tedious and subject to error when the topology of the diagram is complicated. Figure 3.2 shows examples of block-diagram algebra that complement those shown in Fig. 3.1. Figures 3.2(a) and (b) show how the interconnections of a block diagram can be manipulated without affecting the mathematical relationships. Figure 3.2(c) shows how the manipulations can be used to convert a general system (on the left) to a system

FIGURE 3.2

Examples of block-diagram algebra

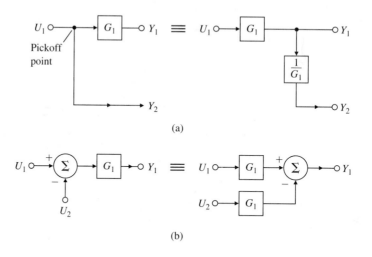

(a)

(b)

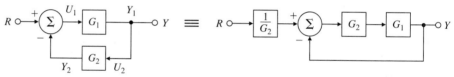

(c)

Unity feedback system

without a component in the feedback path, usually referred to as a **unity feedback system**.

   In all cases the basic principle is to simplify the topology while maintaining exactly the same relationships among the remaining variables of the block diagram. In relation to the algebra of the underlying linear equations, block-diagram reduction is a pictorial way to solve equations by eliminating variables.

◆ **EXAMPLE 3.20**   *Transfer Function from the Block Diagram*

Find the transfer function of the system shown in Fig. 3.3(a).

Solution.   First we simplify the block diagram. Using the principles of Eq. (3.30) we replace the feedback loop involving G_1 and G_3 by its equivalent transfer function,

FIGURE 3.3
Example for block-diagram simplification

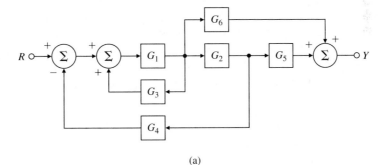

(a)

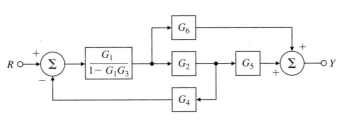

(b)

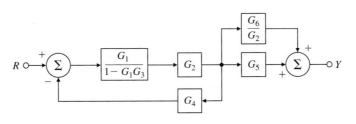

(c)

noting that it is a positive feedback loop. The result is Fig. 3.3(b). The next step is to move the pick-off point preceding G_2 to its output (see Fig. 3.2a), as shown in Fig. 3.3(c). The negative feedback loop on the left is in series with the subsystem on the right, which is comprised of the two parallel blocks G_5 and G_6/G_2. The overall transfer function can be written using all three rules for reduction given by Fig. 3.1:

$$\frac{Y(s)}{R(s)} = \frac{\dfrac{G_1 G_2}{1 - G_1 G_3}}{1 + \dfrac{G_1 G_2 G_4}{1 - G_1 G_3}} \left(G_5 + \frac{G_6}{G_2} \right)$$

$$= \frac{G_1 G_2 G_5 + G_1 G_6}{1 - G_1 G_3 + G_1 G_2 G_4}.$$

As we have seen, a system of algebraic equations may be represented by a block diagram that represents individual transfer functions by blocks and has interconnections that correspond to the system equations. A block diagram is a convenient tool to visualize the system as a collection of interrelated subsystems that emphasize the relationships among the system variables.

■ 3.2.2   Mason's Rule and the Signal-flow Graph

Signal-flow graph

An alternative notation to the block diagram is given by the **signal-flow graph** introduced by S. J. Mason (1953, 1956). As with the block diagram, the signal-flow graph offers a visual tool for representing the causal relationships between the components of the system. The method consists of characterizing the system by a network of directed branches and associated gains (transfer functions) connected at nodes. Several block diagrams and their corresponding signal-flow graphs are shown in Fig. 3.4. The two ways of depicting a system are equivalent, and you can use either diagram to apply Mason's rule (to be defined shortly).

In a signal-flow graph the internal signals in the diagram, such as the common input to several blocks or the output of a summing junction, are called **nodes**. The system input point and the system output point are also nodes; the input node has outgoing branches only, and the output node has incoming branches only. Mason defined a **path** through a block diagram as a sequence of connected blocks, the route passing from one variable to another *in the direction of signal flow of the blocks* without including any variable more than once. A **forward path** is a path from the input to output such that no node is included more than once. If the nodes are numbered in a convenient order, then a forward path can be identified by the numbers that are included. Any closed path that returns to its starting node is a **loop**, and a path that leads from a given variable back to the same variable is a **loop path**. The **path gain** is the product of component gains (transfer functions) making up the path. Similarly, the **loop gain** is the path gain associated with a loop, that is, the

FIGURE 3.4
Block diagrams and
corresponding signal
flow graphs

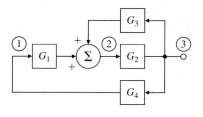 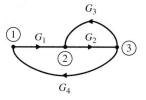

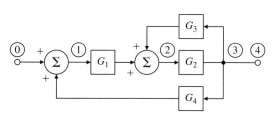 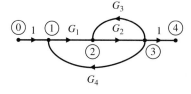

(a)

(b)

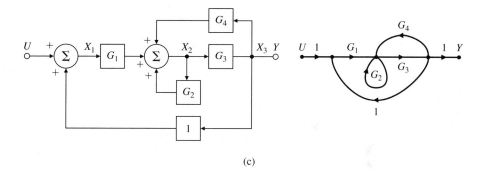

(c)

product of gains in a loop. If two paths have a common component, they are said to touch. Notice particularly in this connection that the input and the output of a summing junction are not the same and that the summing junction is a one-way device from its inputs to its output.

Mason's rule relates the graph to the algebra of the simultaneous equations they represent.[2] Consider Fig. 3.4 (c), where the signal at each node has been given a name and the gains are marked. Then the block diagram (or the signal-flow graph) represents the following system of equations:

$$X_1(s) = X_3(s) + U(s),$$
$$X_2(s) = G_1(s)X_1(s) + G_2(s)X_2(s) + G_4(s)X_3(s),$$
$$Y(s) = 1X_3(s).$$

2. The derivation is based on Cramer's rule for solving linear equations by determinants and is described in Mason's papers.

Mason's rule states that the input-output transfer function associated with a signal-flow graph is given by

$$G(s) = \frac{Y(s)}{U(s)} = \frac{1}{\Delta} \sum_i G_i \Delta_i$$

where

G_i = path gain of the ith forward path.

Δ = the system determinant = $1 - \Sigma$ (all individual loop gains) + Σ (gain products of all possible two loops that do not touch) $- \Sigma$ (gain products of all possible three loops that do not touch) $+ \cdots$,

Δ_i = ith forward path determinant

= value of Δ for that part of the block diagram that does *not* touch the ith forward path.

◆ **EXAMPLE 3.21**   *Mason's Rule in a Simple System*

Find the transfer function for the block diagram in Fig. 3.5.

Solution.   From the block diagram shown in Fig. 3.5 we have

forward path	path gain
1236	$G_1 = 1 \left(\dfrac{1}{s}\right) (b_1)(1)$
12346	$G_2 = 1 \left(\dfrac{1}{s}\right)\left(\dfrac{1}{s}\right) (b_2)(1)$
123456	$G_3 = 1 \left(\dfrac{1}{s}\right)\left(\dfrac{1}{s}\right)\left(\dfrac{1}{s}\right) (b_3)(1)$

	loop path gain
232	$\ell_1 = -a_1/s$
2342	$\ell_2 = -a_2/s^2$
23452	$\ell_3 = -a_3/s^3$

and the determinants are

$$\Delta = 1 - \left(-\frac{a_1}{s} - \frac{a_2}{s^2} - \frac{a_3}{s^3}\right) + 0$$

$$\Delta_1 = 1 - 0$$
$$\Delta_2 = 1 - 0$$
$$\Delta_3 = 1 - 0.$$

FIGURE 3.5
Block diagram for
Example 3.21

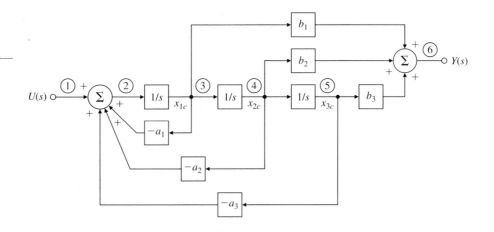

Applying Mason's rule, we find the transfer function to be

$$\frac{Y(s)}{U(s)} = \frac{(b_1/s) + (b_2/s^2) + (b_3/s^3)}{1 + (a_1/s) + (a_2/s^2) + (a_3/s^3)}$$

$$= \frac{b_1 s^2 + b_2 s + b_3}{s^3 + a_1 s^2 + a_2 s + a_3}.$$

Mason's rule is particularly useful for more complex systems where there are several loops, some of which do not sum into the same point.

◆ **EXAMPLE 3.22** *Mason's Rule in a Complex System*

Find the transfer function for the system shown in Fig. 3.6.

FIGURE 3.6
Block diagram for
Example 3.22

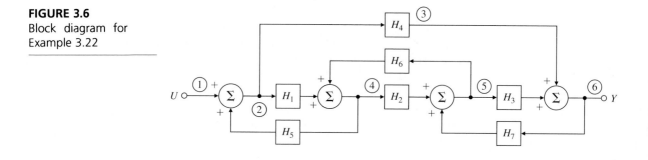

Solution.   From the block diagram, we find that

$$
\begin{array}{cc}
\textit{forward path} & \textit{path gain} \\
12456 & G_1 = H_1 H_2 H_3 \\
1236 & G_2 = H_4 \\
\end{array}
$$

$$
\begin{array}{cc}
& \textit{loop path gain} \\
242 & \ell_1 = H_1 H_5 \text{ (does not touch } \ell_3) \\
454 & \ell_2 = H_2 H_6 \\
565 & \ell_3 = H_3 H_7 \text{ (does not touch } \ell_1) \\
236542 & \ell_4 = H_4 H_7 H_6 H_5 \\
\end{array}
$$

and the determinants are

$$\Delta = 1 - (H_1 H_5 + H_2 H_6 + H_3 H_7 + H_4 H_7 H_6 H_5) + (H_1 H_5 H_3 H_7)$$
$$\Delta_1 = 1 - 0$$
$$\Delta_2 = 1 - H_2 H_6.$$

Therefore,

$$\frac{Y(s)}{U(s)} = \frac{H_1 H_2 H_3 + H_4 - H_4 H_2 H_6}{1 - H_1 H_5 - H_2 H_6 - H_3 H_7 - H_4 H_7 H_6 H_5 + H_1 H_5 H_3 H_7}.$$

◆

Mason's rule is useful for solving relatively complicated block diagrams by hand. It yields the solution in the sense that it provides an explicit input-output relationship for the system represented by the diagram. The advantage compared with path-by-path block-diagram reduction is that it is systematic and algorithmic rather than problem-dependent. Many CACSD packages allow you to specify a system in terms of individual blocks in an overall system, and the software algorithms perform the required block-diagram reduction; therefore, Mason's rule is less important today than in the past. However, there are some derivations that rely on the concepts embodied by the rule, so it still has a role in the control designer's toolbox.

3.3   Response versus Pole Locations

Once the transfer function has been determined by any of the available methods, we can start to analyze the response of the system it represents. When the system equations are simultaneous ordinary differential equations (ODEs), the transfer function that results will be a ratio of polynomials; that is,

$$H(s) = b(s)/a(s).$$

If we assume that b and a have no common factors (as is usually the case), then values of s such that $a(s) = 0$ will represent points where $H(s)$ is infinity. As we

Poles

Zeros

The impulse response is
the natural response.

Stability

Time constant τ

saw in Section 3.1.5, these *s*-values are called poles of $H(s)$. Values of *s* such that $b(s)=0$ are points where $H(s)=0$, and the corresponding *s*-locations are called zeros. The effect of zeros on the response will be discussed in Section 3.5. These poles and zeros completely describe $H(s)$ except for a constant multiplier. Since the impulse response is given by the time function corresponding to the transfer function, we call the impulse response the **natural response** of the system. We can use the poles and zeros to compute the corresponding time response and thus identify time histories with pole locations in the *s*-plane. For example, the poles identify the classes of signals contained in the impulse response, as may be seen by a partial-fraction expansion of $H(s)$. For a first-order pole,

$$H(s) = \frac{1}{s + \sigma},$$

Table A.2, entry 7, indicates that the impulse response will be an exponential function; that is,

$$h(t) = e^{-\sigma t} 1(t).$$

When $\sigma > 0$, the pole is located at $s < 0$, the exponential expression decays, and we say the impulse response is **stable**. If $\sigma < 0$, the pole is to the right of the origin. Because the exponential expression here grows with time, the impulse response is referred to as **unstable**. Figure 3.7 shows a typical response and defines the **time constant**

$$\tau = 1/\sigma \qquad (3.31)$$

as the time when the response is $1/e$ times the initial value. The straight line is tangent to the exponential curve at $t=0$ and terminates at $t=\tau$. This characteristic of an exponential expression is useful in sketching a time plot or checking computer results.

FIGURE 3.7
First-order system
response

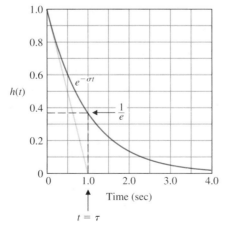

◆ **EXAMPLE 3.23**   *Response versus Pole Locations, Real Roots*

Compare the time response with the pole locations for the system with a transfer function between input and output given by

$$H(s) = \frac{2s+1}{s^2+3s+2}. \tag{3.32}$$

Solution.   The numerator is

$$b(s) = 2(s+\tfrac{1}{2}),$$

and the denominator is

$$a(s) = s^2 + 3s + 2 = (s+1)(s+2).$$

The poles of $H(s)$ are therefore at $s = -1$ and $s = -2$, and the one zero is at $s = -\tfrac{1}{2}$. A complete description of this transfer function is shown by the plot of the locations of the poles and the zeros in the s-plane (Fig. 3.8).

A partial-fraction expansion of $H(s)$ results in

$$H(s) = -\frac{1}{s+1} + \frac{3}{s+2}.$$

From Table A.2 we can look up the inverse of each term in $H(s)$, which will give us the time function $h(t)$ that would result if the system input were an impulse. In this case,

$$h(t) = \begin{cases} -e^{-t} + 3e^{-2t} & t \geq 0, \\ 0 & t < 0. \end{cases} \tag{3.33}$$

We see that the shape of the component parts of $h(t)$, which are e^{-t} and e^{-2t}, are determined by the poles at $s = -1$ and -2. This is true of more complicated cases as well: In general, the shape of the natural response is determined by the location of the poles of the transfer function.

A sketch of these pole locations and corresponding natural responses is given in Fig. 3.9, along with other pole locations including complex ones, which will be discussed shortly. The role of the numerator in the process of partial-fraction expansion is to influence the size of the coefficient that multiplies each component. Because e^{-2t} decays faster than e^{-t}, the signal corresponding to the pole at -2 decays faster than the signal corresponding to the pole at -1. For brevity we simply say that the pole at -2 is faster than the pole at -1. In general, poles farther to the left in the s-plane are associated

> "Fast poles" and "slow poles" refer to relative rate of signal decay.

FIGURE 3.8
Sketch of s-plane showing poles as crosses and zeros as circles

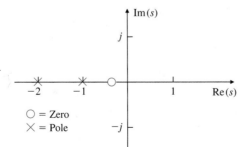

○ = Zero
✕ = Pole

FIGURE 3.9
Time functions
associated with points in
the *s*-plane (LHP, left
half-plane; RHP, right
half-plane)

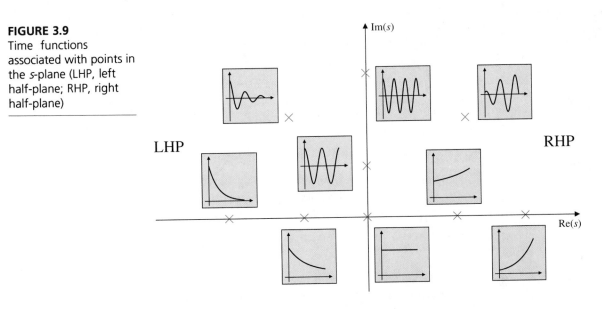

LHP   RHP

with natural signals that decay faster than those associated with poles closer to the imaginary axis. If the poles had been located with positive values of *s* (in the right half of the *s*-plane), the response would have been a growing exponential function and thus unstable. Figure 3.10 shows that the fast $3e^{-2t}$ term dominates the early part of the time history and that the $-e^{-t}$ term is the primary contributor later on.

The purpose of this example is to illustrate the relationship between the poles and the character of the response, which can only be done by finding the inverse Laplace transform and examining each term as above. However, if we simply wanted to plot the impulse response for this example, the expedient way would be to use the MATLAB sequence

Impulse response using
MATLAB

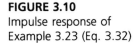

```
num=[2  1]

den=[1  3  2]

impulse(num,den)
```

FIGURE 3.10
Impulse response of
Example 3.23 (Eq. 3.32)

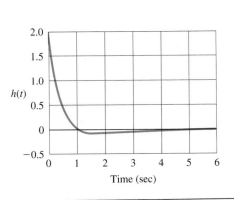

Complex poles can be defined in terms of their real and imaginary parts, traditionally referred to as

$$s = -\sigma \pm j\omega_d.$$

This means that a pole has a negative real part if σ is positive. Since complex poles always come in complex conjugate pairs, the denominator corresponding to a complex pair will be

$$a(s) = (s + \sigma - j\omega_d)(s + \sigma + j\omega_d) = (s + \sigma)^2 + \omega_d^2. \tag{3.34}$$

When finding the transfer function from differential equations, we typically write the result in the polynomial form

$$H(s) = \frac{\omega_n^2}{s^2 + 2\zeta\omega_n s + \omega_n^2}. \tag{3.35}$$

By multiplying out the form given by Eq. (3.34) and comparing it with the coefficients of the denominator of $H(s)$ in Eq. (3.35), we find the correspondence between the parameters to be

$$\sigma = \zeta\omega_n \quad \text{and} \quad \omega_d = \omega_n\sqrt{1 - \zeta^2} \tag{3.36}$$

Damping ratio; damped and undamped natural frequency

where the parameter ζ is the **damping ratio**[3] and ω_n is the **undamped natural frequency**. The poles of this transfer function are located at a radius ω_n in the s-plane and at an angle $\theta = \sin^{-1}\zeta$, as shown in Fig. 3.11. Therefore, the damping ratio reflects the level of damping as a fraction of the critical damping value where the poles become real. In rectangular coordinates the poles are at $s = -\sigma \pm j\omega_d$. When $\zeta = 0$, we have no damping, $\theta = 0$, and the damped natural frequency $\omega_d = \omega_n$, the undamped natural frequency.

For purposes of finding the time response from Table A.2 corresponding to a complex transfer function, it is easiest to manipulate the $H(s)$ so that the complex poles fit the form of Eq. (3.34), because then the time response can be found directly from the table. Equation (3.35) can be rewritten as

$$H(s) = \frac{\omega_n^2}{(s + \zeta\omega_n)^2 + \omega_n^2(1 - \zeta^2)}.$$

Therefore, from entry number 20 in Table A.2 and the definitions in Eq. (3.36), we see that the impulse response is

$$h(t) = \frac{\omega_n}{\sqrt{1 - \zeta^2}} e^{-\sigma t}(\sin \omega_d t)1(t).$$

Figure 3.12(a) plots $h(t)$ for several values of ζ such that time has been normalized to the undamped natural frequency ω_n. Note that the actual

FIGURE 3.11

s-plane plot for a pair of complex poles

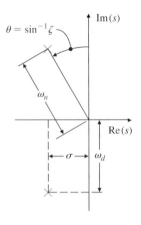

3. In communications and filter engineering, the standard second-order transfer function is written $H = 1/[1 + Q(s/\omega_n + \omega_n/s)]$. Here ω_n is called the **band center** and Q is the **quality factor**. Comparison with Eq. (3.35) shows that $Q = 1/2\zeta$.

FIGURE 3.12
Responses of second-
order systems versus ζ:
(a) impulse responses; (b)
step responses

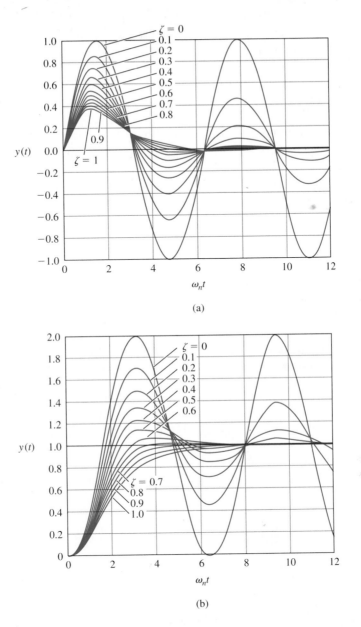

(a)

(b)

frequency ω_d decreases slightly as the damping ratio increases. Also note that
for very low damping the response is oscillatory, while for large damping
(ζ near 1) the response shows no oscillation. A few of these responses are
sketched in Fig. 3.9 to show qualitatively how changing pole locations in the
s-plane affect impulse responses. You will find it useful as a control designer to
commit the image of Fig. 3.9 to memory so that you can understand instantly
how changes in pole locations influence the time response.

FIGURE 3.13
Pole locations
corresponding to three
values of ζ

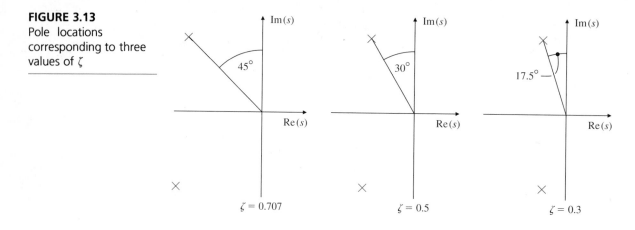

$\zeta = 0.707$ $\zeta = 0.5$ $\zeta = 0.3$

Three pole locations are shown in Fig. 3.13 for comparison with the corresponding impulse responses in Fig. 3.12(a). The negative real part of the pole, σ, determines the decay rate of an exponential envelope that multiplies the sinusoid, as shown in Fig. 3.14. Note that if $\sigma < 0$ (and the pole is in the RHP, or right half-plane), then the natural response will grow with time so, as defined earlier the system is said to be unstable. If $\sigma = 0$, the natural response neither grows nor decays, so stability is open to debate. If $\sigma > 0$, the natural response decays, so the system is stable.

Stability depends on whether natural response grows or decays.

It is also interesting to examine the step response of $H(s)$, that is, the response of the system $H(s)$ to a unit step input $u = 1(t)$ where $U(s) = 1/s$. The step-response transform is given by $Y(s) = H(s)U(s)$, which is found in Table A.2, entry 21. Figure 3.12(b), which plots $y(t)$ for several values of ζ, shows that the basic transient response characteristics from the impulse response carries over quite well to the step response; the difference between the two responses is that the step response's final value is the commanded unit step.

Step response

FIGURE 3.14
Second-order system
response with an
exponential envelope

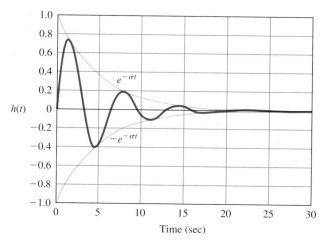

◆ **EXAMPLE 3.24**   *Oscillatory Time Response*

Discuss the correlation between the poles of

$$H(s) = \frac{2s+1}{s^2+2s+5} \tag{3.37}$$

and the impulse response of the system, and find the exact impulse response.

Solution.   From the form of $H(s)$ given by Eq. (3.35), we see that

$$\omega_n^2 = 5 \Rightarrow \omega_n = \sqrt{5} = 2.24 \text{ rad/sec}$$

and

$$2\zeta\omega_n = 2 \Rightarrow \zeta = \frac{1}{\sqrt{5}} = 0.447.$$

This indicates we should expect a frequency around 2 rad/sec with very little oscillatory motion. In order to obtain the exact response, we manipulate $H(s)$ until the denominator is in the form of Eq. (3.34):

$$H(s) = \frac{2s+1}{s^2+2s+5} = \frac{2s+1}{(s+1)^2+2^2}.$$

From this equation we see that the poles of the transfer function are complex, with real part -1 and imaginary parts $\pm 2j$. Table A.2 has two entries, numbers 19 and 20, that match the denominator. The right side of the above equation needs to be broken into two parts so that they match the numerators of the entries in the table:

$$H(s) = \frac{2s+1}{(s+1)^2+2^2} = 2\frac{s+1}{(s+1)^2+2^2} - \frac{1}{2}\frac{2}{(s+1)^2+2^2}$$

thus, the impulse response is

$$h(t) = (2e^{-t}\cos 2t - \tfrac{1}{2}e^{-t}\sin 2t)1(t).$$

Figure 3.15 is a plot of the response and shows how the envelope attenuates the

FIGURE 3.15
System response for
Example 3.24

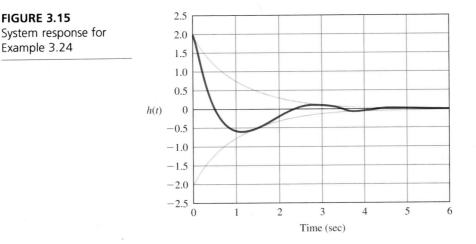

Time (sec)

sinusoid, the domination of the $2 \cos 2t$ term, and the small phase shift caused by the $-\frac{1}{2} \sin 2t$ term.

Impulse response by
MATLAB

As in the previous example, the expedient way of determining the impulse response would be to use the MATLAB sequence

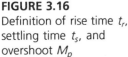

```
num = [2  1]

den = [1  2  5]

impulse(num,den)
```

◆

3.4   Time-domain Specifications

Specifications for a control system design often involve certain requirements associated with the time response of the system. The requirements for a step response are expressed in terms of the standard quantities illustrated in Fig. 3.16:

Definitions of rise time, settling time, overshoot, and peak time

The **rise time** t_r is the time it takes the system to reach the vicinity of its new set point.

The **settling time** t_s is the time it takes the system transients to decay.

The **overshoot** M_p is the maximum amount the system overshoots its final value divided by its final value (and often expressed as a percentage).

The **peak time** t_p is the time it takes the system to reach the maximum overshoot point.

For a second-order system the time responses shown in Fig. 3.12(b) yield information about the specifications that is too complex to be remembered unless converted to a simpler form. By examining these curves in light of the

FIGURE 3.16
Definition of rise time t_r, settling time t_s, and overshoot M_p

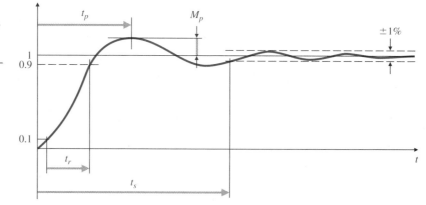

definitions given in Fig. 3.16, we can relate the curves to the pole-location parameters ζ and ω_n. For example, all the curves rise in roughly the same time. If we consider the curve for $\zeta = 0.5$ to be an average, the rise time from $y = 0.1$ to 0.9 is approximately $\omega_n t_r = 1.8$. Thus we can say that

Rise time t_r

$$t_r \cong \frac{1.8}{\omega_n}. \tag{3.38}$$

Although this relationship could be embellished by including the effect of the damping ratio, it is important to keep in mind how Eq. (3.38) is typically used. It is only accurate for a second-order system with no zeros; for all other systems it is a rough approximation to the relationship between t_r and ω_n. Most systems being analyzed for control systems design are more complicated than the pure second-order system, so designers use Eq. (3.38) with the knowledge that it is a rough approximation only.

For the overshoot M_p we can be more analytical. This value occurs when the derivative is zero, which can be found from calculus. The time history of the curves in Fig. 3.12(b), found from the inverse Laplace transform of $H(s)/s$, is

$$y(t) = 1 - e^{-\sigma t}\left(\cos \omega_d t + \frac{\sigma}{\omega_d}\sin \omega_d t\right),$$

where $\omega_d = \omega_n\sqrt{1-\zeta^2}$ and $\sigma = \zeta\omega_n$.

The derivative of this function is set to zero:

$$\dot{y}(t) = \sigma e^{-\sigma t}\left(\cos \omega_d t + \frac{\sigma}{\omega_d}\sin \omega_d t\right) - e^{-\sigma t}(-\omega_d \sin \omega_d t + \sigma \cos \omega_d t) = 0$$

$$= e^{-\sigma t}\left(\frac{\sigma^2}{\omega_d}\sin \omega_d t + \omega_d \sin \omega_d t\right) = 0.$$

This occurs when $\sin \omega_d t = 0$, so

$$\omega_d t_p = \pi$$

and thus

Peak time t_p

$$t_p = \frac{\pi}{\omega_d}. \tag{3.39}$$

Substituting Eq. (3.39) into the expression for $y(t)$, we compute

$$y(t_p) \triangleq 1 + M_p = 1 - e^{-\sigma\pi/\omega_d}\left(\cos \pi + \frac{\sigma}{\omega_d}\sin \pi\right).$$

$$= 1 + e^{-\sigma\pi/\omega_d}$$

Thus we have the formula

Overshoot M_p

$$M_p = e^{-\pi\zeta/\sqrt{1-\zeta^2}}, \qquad 0 \leqslant \zeta < 1, \tag{3.40}$$

FIGURE 3.17
Overshoot M_p versus
damping ratio ζ for the
second-order system

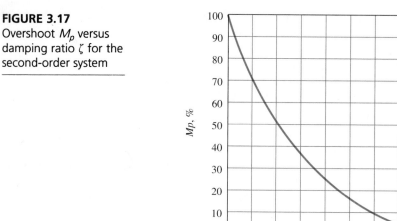

whicn is plotted in Fig. 3.17. Two frequently used values from this curve are $M_p = 0.16$ for $\zeta = 0.5$ and $M_p = 0.05$ for $\zeta = 0.7$.

The final parameter of interest from the transient response is the settling time t_s. This is the time required for the transient to decay to a small value so that $y(t)$ is almost in the steady state. Various measures of smallness are possible. For illustration we will use 1% as a reasonable measure; in other cases 2% or 5% are used. As an analytic computation we notice that the deviation of y from 1 is the product of the decaying exponential $e^{-\sigma t}$ and the circular functions sine and cosine. The duration of this error is essentially decided by the transient exponential, so we can define the settling time as that value of t_s when the decaying exponential reaches 1%:

$$e^{-\zeta \omega_n t_s} = 0.01.$$

Therefore,

$$\zeta \omega_n t_s = 4.6$$

or

Settling time t_s

$$t_s = \frac{4.6}{\zeta \omega_n} = \frac{4.6}{\sigma}, \tag{3.41}$$

where σ is the negative real part of the pole, as may be seen from Fig. 3.11.

Equations (3.38), (3.40), and (3.41) characterize the transient response of a system having no finite zeros and two complex poles with undamped natural frequency ω_n, damping ratio ζ, and negative real part σ. In analysis and design, they are used to estimate rise time, overshoot, and settling time, respectively,

Design synthesis

for just about any system. In design synthesis we wish to specify t_r, M_p, and t_s and to ask where the poles need to be so that the actual responses are less than or equal to these specifications. For specified values of t_r, M_p, and t_s, the synthesis form of the equation is then

$$\omega_n \geqslant \frac{1.8}{t_r}, \tag{3.42a}$$

$$\zeta \geqslant \zeta(M_p) \qquad \text{(from Fig. 3.17),} \tag{3.42b}$$

$$\sigma \geqslant \frac{4.6}{t_s}. \tag{3.42c}$$

These equations, which can be graphed in the s-plane as shown in Fig. 3.18(a–c), will be used in later chapters to guide the selection of pole and zero locations to meet control system specifications for dynamic response.

It is important to keep in mind that Eqs. (3.42a–c) are qualitative guides and not precise design formulas. They are meant to provide only a starting point for the design iteration. After the control design is complete, the time response should always be checked by an exact calculation, usually by numerical simulation, to verify whether the time specifications have actually been met. If not, another iteration of the design is required.

For a first-order system,

$$H(s) = \frac{\sigma}{s + \sigma},$$

the transform of the step response is

$$Y(s) = \frac{\sigma}{s(s + \sigma)}.$$

We see from entry 11 in Table A.2 that $Y(s)$ corresponds to

$$y(t) = (1 - e^{-\sigma t})1(t). \tag{3.43}$$

FIGURE 3.18
Graphs of regions in the s-plane delineated by certain transient requirements: (a) rise time; (b) overshoot; (c) settling time; and (d) composite of all three requirements

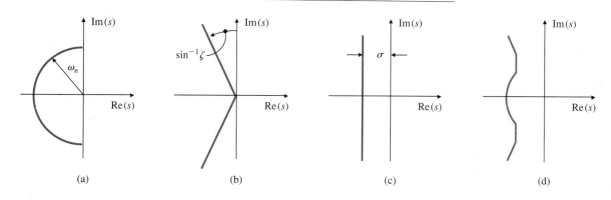

(a)   (b)   (c)   (d)

Comparison with the development for Eq. (3.41) shows that the value of t_s for a first-order system is the same:

$$t_s = \frac{4.6}{\sigma}.$$

No overshoot is possible, so $M_p = 0$. The rise time from $y = 0.1$ to $y = 0.9$ can be seen from Fig. 3.7 to be

$$t_r = \frac{\ln\ 0.9 - \ln\ 0.1}{\sigma} = \frac{2.2}{\sigma}.$$

However, it is more typical to describe a first-order system in terms of its time constant, which was defined in Fig. 3.7 to be $\tau = 1/\sigma$.

Time constant τ

◆ **EXAMPLE 3.25**   *Transformation of the Specifications to the s-Plane*

Find the allowable regions in the s-plane for the poles of a transfer function of a system if the system response requirements are $t_r \leqslant 0.6$ sec, $M_p \leqslant 10\%$, and $t_s \leqslant 3$ sec.

Solution.   Without knowing whether or not the system is second-order with no zeros, it is impossible to find the allowable region accurately. Regardless of the system, we can obtain a first approximation using the relationships for a second-order system. Equation (3.42a) indicates that

$$\omega_n \geqslant \frac{1.8}{t_r} = 3.0 \text{ rad/sec},$$

Eq. (3.42b) and Fig. 3.17 indicate that

$$\zeta \geqslant 0.6,$$

FIGURE 3.19
Allowable region in
s-plane for Example 3.25

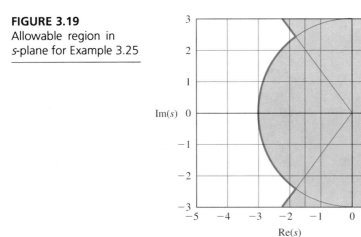

and Eq. (3.42c) indicates that

$$\sigma \geqslant \frac{4.6}{3} = 1.5 \text{ sec.}$$

The allowable region is anywhere to the left of the solid line in Fig. 3.19. Note that any pole meeting the ζ and ω_n restrictions will automatically meet the σ restriction. ◆

3.5 Effects of Zeros and Additional Poles

Relationships such as those shown in Fig. 3.18 are correct for the simple second-order system; for more complicated systems they can be used only as guidelines: If a certain design has an inadequate rise time (is too slow), we must raise the natural frequency; if the transient has too much overshoot, then the damping needs to be increased; and if the transient persists too long, the poles need to be moved to the left in the s-plane.

Thus far only the poles of $H(s)$ have entered into the discussion. Because $H(s) = b(s)/a(s)$, there may be finite values of s for which $b(s) = 0$; as mentioned earlier (Section 3.4), these s-values are called the zeros of $H(s)$.[4] At the level of transient analysis, the zeros exert their influence by modifying the coefficients of the exponential terms whose shape is decided by the poles. To illustrate this, consider the following two transfer functions, which have the same poles but different zeros:

Effect of zeros

$$H_1(s) = \frac{2}{(s+1)(s+2)}$$

$$= \frac{2}{s+1} - \frac{2}{s+2}, \tag{3.44}$$

$$H_2(s) = \frac{2(s+1.1)}{1.1(s+1)(s+2)}$$

$$= \frac{2}{1.1}\left(\frac{0.1}{s+1} + \frac{0.9}{s+2}\right)$$

$$= \frac{0.18}{s+1} + \frac{1.64}{s+2}. \tag{3.45}$$

The effect of zeros near poles

They are normalized to have the same DC gain, i.e., gain at $s = 0$. Notice that the coefficient of the $(s+1)$ term has been modified from 2 in $H_1(s)$ to 0.18 in $H_2(s)$. This dramatic reduction is brought about by the zero at $s = -1.1$ in

---

4. We assume that $b(s)$ and $a(s)$ have no common factors. If this is not so, it is possible for $b(s)$ and $a(s)$ to be zero at the same location and for $H(s)$ to not equal zero there. The implications of this case will be discussed in Chapter 7, when we have a state-space description.

$H_2(s)$, which almost cancels the pole at $s = -1$. If we put the zero exactly at $s = -1$, this term will vanish completely. In general, a zero near a pole reduces the amount of that term in the total response. From the equation for the coefficients in a partial-fraction expansion, Eq. (3.23),

$$C_1 = (s - p_1)F(s)|_{s=p_1}$$

we can see that, if $F(s)$ has a zero near the pole at $s = p_1$, the value of $F(s)$ will be small because the value of s is near the zero. Therefore, the coefficient C_1, which reflects how much of that term appears in the response, will be small.

In order to take into account how zeros affect the transient response when designing a control system, we consider transfer functions with two complex poles and one zero. To expedite the plotting for a wide range of cases, we write the transform in a form with normalized time and zero locations:

$$H(s) = \frac{(s/\alpha\zeta\omega_n) + 1}{(s/\omega_n)^2 + 2\zeta(s/\omega_n) + 1}. \tag{3.46}$$

The zero is located at $s = -\alpha\zeta\omega_n = -\alpha\sigma$. If α is large, the zero will be far removed from the poles; because the coefficient on the s term is so small, the zero will have little effect on the response. If $\alpha \cong 1$, the value of the zero will be close to that of the real part of the poles and can be expected to have a substantial influence on the response. The step-response curves for $\zeta = 0.5$ and for several values of α are plotted in Fig. 3.20. We see that the major effect of the zero is to increase the overshoot M_p, whereas it has very little influence on the settling time. A plot of M_p versus α is given in Fig. 3.21. The plot shows that the zero has very little effect on M_p if $\alpha > 3$, but as α decreases below 3, it has an increasing effect, especially when $\alpha < 1$.

FIGURE 3.20

Plots of the step response of a second-order system with a zero ($\zeta = 0.5$)

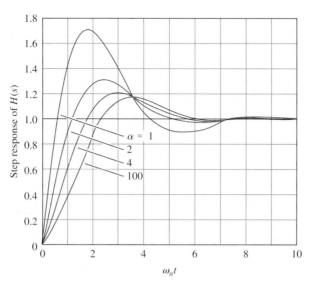

FIGURE 3.21
Plot of overshoot M_p as a function of normalized zero location α. At $\alpha = 1$, the real part of the zero equals the real part of the poles

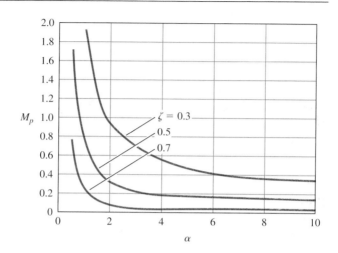

Figure 3.20 can be explained in terms of Laplace-transform analysis. First we replace s/ω_n with s:

$$H(s) = \frac{s/\alpha\zeta + 1}{s^2 + 2\zeta s + 1}.$$

This has the effect of normalizing frequency in the transfer function and normalizing time in the corresponding step responses; thus $\tau = \omega_n t$. We then rewrite the transfer function as the sum of two terms:

$$H(s) = \frac{1}{s^2 + 2\zeta s + 1} + \frac{1}{\alpha\zeta}\frac{s}{s^2 + 2\zeta s + 1}. \tag{3.47}$$

The first term, which we will call $H_o(s)$, is the original term (having no finite zero), and the second term $H_d(s)$, which is introduced by the zero, is a product of a constant $(1/\alpha\zeta)$ times s times the original term. The Laplace transform of df/dt is $sF(s)$, so $H_d(s)$ corresponds to a product of a constant times the *derivative* of the original term. The step responses of $H_o(s)$ and $H_d(s)$ are plotted in Fig. 3.22. Looking at these curves, we can see why the zero increased the overshoot: The derivative has a large hump in the early part of the curve, and adding this to the $H_o(s)$ response lifts up the total response of $H(s)$ to produce the overshoot. This analysis is also very informative for the case when $\alpha < 0$ and the zero is in the RHP where $s > 0$. (This is typically called a **RHP zero** and sometimes referred to as a **nonminimum-phase zero**, a topic to be discussed

RHP or nonminimum-phase zero

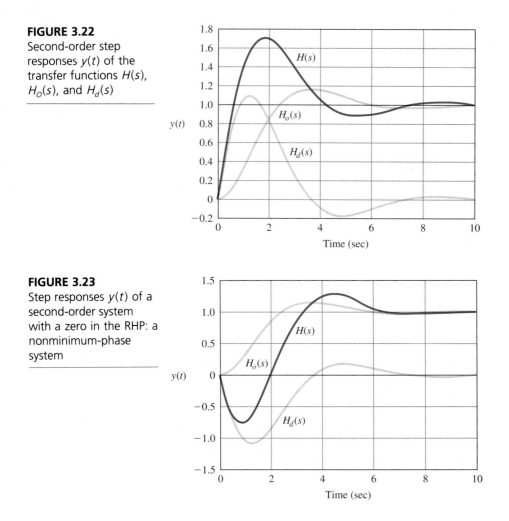

FIGURE 3.22
Second-order step
responses $y(t)$ of the
transfer functions $H(s)$,
$H_o(s)$, and $H_d(s)$

FIGURE 3.23
Step responses $y(t)$ of a
second-order system
with a zero in the RHP: a
nonminimum-phase
system

in more detail in Section 6.1.1.) In this case the derivative term is subtracted rather than added. A typical case is sketched in Fig. 3.23.

◆ **EXAMPLE 3.26** *Aircraft Response Using* MATLAB

The transfer function between the elevator and altitude of the Boeing-747 aircraft described in Section 9.3.2 can be approximated as

$$\frac{h(s)}{\delta_e(s)} = \frac{30(s-6)}{s(s^2+4s+13)}.$$

(a) Use MATLAB to plot the altitude time history for a $-1°$ impulsive elevator input. Explain the response, noting the physical reasons for the nonminimum phase nature of the response.

(b) Examine the accuracy of the approximations for t_r, t_s, and M_p (Eqs. (3.38) and (3.41) and Fig. 3.17).

Solution.   (a) The MATLAB statements to create the impulse response for this case are

u = −1

num = 30*[1 − 6]

den = [1 4 13 0]

impulse(u*num,den)

The result is the plot shown in Fig. 3.24. Notice how the altitude drops initially and then rises to a new final value. The final value is predicted by the Final Value Theorem:

$$h(\infty) = s \left. \frac{30(s-6)(-1)}{s(s^2+4s+13)} \right|_{s=0} = \frac{30(-6)(-1)}{13} = +13.8.$$

Response of a nonminimum-phase system

The fact that the response has a finite final value for an impulsive input is due to the s-term in the denominator. This represents a pure integration, and the integral of an impulse function is a finite value. If the input had been a step, the altitude would have continued to increase with time; in other words the integral of a step function is a ramp function.

The initial drop is predicted by the RHP zero in the transfer function. The negative elevator deflection is defined to be upward by convention (see Fig. 9.30). The upward deflection of the elevators drives the tail down, which rotates the craft nose up and produces the climb. The deflection at the initial instant causes a downward force before the craft has rotated; therefore, the initial altitude response is down. After rotation the increased lift resulting from the increased angle of attack of the wings causes the airplane to climb.

FIGURE 3.24

Response of an airplane's altitude to an impulsive elevator input

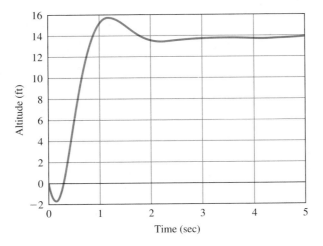

(b) The rise time from Eq. (3.38) is

$$t_r = \frac{1.8}{\omega_n} = \frac{1.8}{\sqrt{13}} = 0.5 \text{ sec.}$$

We find the damping ratio ζ from the relation

$$2\zeta\omega_n = 4 \Rightarrow \zeta = \frac{2}{\sqrt{13}} = 0.55.$$

From Fig. 3.17 we find the overshoot M_p to be 0.14. Since $2\zeta\omega_n = 2\sigma = 4$, (Eq. 3.41) shows that

$$t_s = \frac{4.6}{\sigma} = \frac{4.6}{2} = 2.3 \text{ sec.}$$

Detailed examination of the time history $h(t)$ from MATLAB output shows that $t_r \cong 0.43$ sec, $M_p \cong 0.14$, and $t_s \cong 2.6$ sec, which are reasonably close to the estimates. The only significant effect of the nonminimum phase zero was to cause the initial response to go the "wrong direction" and make the response be somewhat sluggish.

◆

Effect of extra pole

In addition to studying the effects of zeros, it is useful to consider the effects of an extra pole on the standard second-order step response. In this case we take the transfer function to be

$$H(s) = \frac{1}{(s/\alpha\zeta\omega_n + 1)[(s/\omega_n)^2 + 2\zeta(s/\omega_n) + 1]}. \tag{3.48}$$

Plots of the step response for this case are shown in Fig. 3.25 for $\zeta = 0.5$ and

FIGURE 3.25
Step responses for several third-order systems with $\zeta = 0.5$

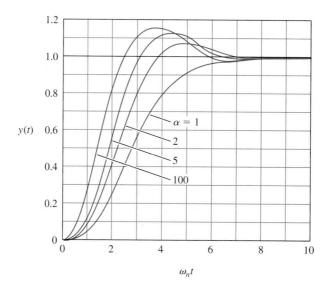

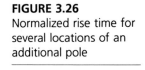

FIGURE 3.26
Normalized rise time for
several locations of an
additional pole

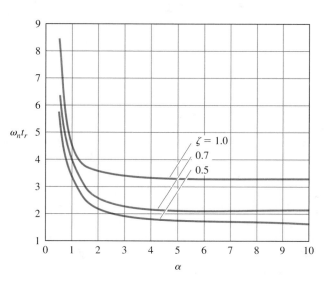

several values of α. In this case the major effect is to increase the rise time. A plot of the rise time versus α is shown in Fig. 3.26 for several values of ζ.

From this discussion we can draw several conclusions about the dynamic response of a simple system as revealed by its pole-zero patterns:

Effects of Pole-zero Patterns on Dynamic Response

1. For a second-order system with no finite zeros, the transient response parameters are approximated by

$$\text{Rise time: } t_r \cong \frac{1.8}{\omega_n}$$

$$\text{Overshoot: } M_p \cong \begin{cases} 5\%, & \zeta = 0.7 \\ 16\%, & \zeta = 0.5 \\ 35\%, & \zeta = 0.3 \end{cases} \quad \text{(See Fig. 3.17)}$$

$$\text{Settling time: } t_s \cong \frac{4.6}{\sigma}$$

2. An additional zero in the left half-plane will increase the overshoot if the zero is within a factor of 4 of the real part of the complex poles. A plot is given in Fig. 3.21.

3. An additional zero in the right half-plane will depress the overshoot (and may cause the step response to start out in the wrong direction).

4. An additional pole in the left half-plane will increase the rise time significantly if the extra pole is within a factor of 4 of the real part of the complex poles. A plot is given in Fig. 3.26.

3.6 Numerical Simulation

Numerical solution of differential equations has become common with the widespread access to computers. Examination of the Laplace transform in Section 3.1 led us to insight into the character of solutions that we exploited in Sections 3.3 to 3.5. The Laplace transform also leads to a method of solving linear ODEs; however, this method is not used by CACSD software to compute the solution of differential equations and is almost never used to compute an actual time history by hand. Numerical solution methods are capable of solving linear as well as nonlinear differential equations. However, some efficiencies are possible if the equations are linear; and typically, CACSD software exploits these. Often a control design is carried out using linear analysis to arrive at a candidate feedback control scheme based solely on examination of pole locations in the s-plane. Then the design engineer examines a nonlinear simulation of the time response to acquire information that either confirms the acceptability of the design or indicates the need for more iteration.

3.6.1 Solution of Nonlinear Differential equations

In Chapter 2 we gave the state-variable form of the differential equations for the nonlinear case (Eq. 2.23):

$$\dot{\mathbf{x}} = \mathbf{f}(\mathbf{x}, u). \tag{3.49}$$

The task at hand is to solve for $\mathbf{x}(t)$ given the initial condition $\mathbf{x}(0)$ and the forcing function $u(t)$.

Euler's method

The fundamental ideas behind numerical integration are illustrated by deriving a particularly simple scheme called **Euler's method**. It follows from the definition of a derivative that

$$\frac{dx}{dt} = \lim_{\Delta t \to 0} \frac{\Delta x}{\Delta t}. \tag{3.50}$$

Even if Δt is not quite equal to zero, this relationship will be approximately true. Let us examine the differential equation for two values of time, t_i and t_{i+1}, where Δt is sufficiently close to zero that Eq. (3.50) is approximately correct. In this case, Eq. (3.49) becomes

$$\frac{\mathbf{x}_{i+1} - \mathbf{x}_i}{\Delta t} = \mathbf{f}(\mathbf{x}_i, u_i), \tag{3.51}$$

which can be solved to obtain

$$\mathbf{x}_{i+1} = \mathbf{x}_i + \Delta t\, \mathbf{f}(\mathbf{x}_i, u_i). \tag{3.52}$$

Many repeated evaluations of Eq. (3.52) lead to the desired solution. At the initial point $\mathbf{x}_i = \mathbf{x}(0)$ and for subsequent values of the index $i, \mathbf{x}_i$ takes on the value from the previous calculation of $\mathbf{x}_{i+1}$. Any arbitrary time history of the input u_i can be used: steps, ramps, sinusoids, random sequences, or stock market indices. As the step size Δt decreases, the accuracy of the method improves and the required computation time increases.

It is instructive to examine Euler's method from a graphical viewpoint. Figure 3.27 shows a time history of $\dot{x}$. Since

$$x(t) = x(0) + \int_0^t \dot{x}\, dt, \tag{3.53}$$

the value of $x(t)$ at time t_i is equal to $x(0)$ plus the shaded area under the $\dot{x}$ curve up until t_i. Euler's method approximates the area under the curve by assuming that $\dot{x}$ remains constant through the interval:

$$\int_{t_i}^{t_{i+1}} \dot{x}\, dt \cong \Delta t \dot{x}(t_i). \tag{3.54}$$

Although the graph can only be drawn for a scalar variable x, the ideas are the same for each element of a state vector $\mathbf{x}$, which is the case for the integration formula in Eq. (3.52).

Figure 3.27 suggests that a more accurate formula is to use the average of the values of the derivative at t_i and t_{i+1}. Because this is essentially a straight-line approximation to the $\dot{x}$ curve between t_i and t_{i+1}, it is called the **trapezoidal rule**. Unfortunately, it is impossible to implement for numerical integration because the derivative at t_{i+1} depends on $x(t_{i+1})$, which is not known yet. Many different numerical integration schemes have been devised to approximate the area under the curve; however, they are all iterative because the derivative at the endpoint is not initially known.

Trapezoidal rule

Runge–Kutta methods comprise one popular set of integration schemes. The second-order Runge–Kutta method obtains an approximate value of the

Runge–Kutta methods

FIGURE 3.27
Graphical interpretation of numerical integration

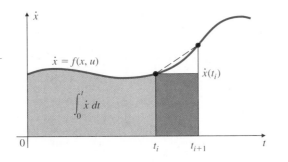

endpoint $\mathbf{x}_{i+1}$ using the Euler method, estimates the derivative at the endpoint using the approximate $\mathbf{x}_{i+1}$, and then arrives at the final value for $\mathbf{x}_{i+1}$ using an average of the two derivatives:

$$\mathbf{k}_1 = \mathbf{f}(\mathbf{x}_i, u_i)$$

$$\mathbf{k}_2 = \mathbf{f}(\mathbf{x}_i + \mathbf{k}_1 \Delta t, u_{i+1}) \tag{3.55}$$

$$\mathbf{x}_{i+1} = \mathbf{x}_i + \frac{\Delta t}{2}(\mathbf{k}_1 + \mathbf{k}_2).$$

Third- and higher-order Runge–Kutta methods use this same basic idea; they differ from the second-order formula by using estimates of derivatives at midpoints as well as endpoints and including them in a weighted average to arrive at the final estimate of $\mathbf{x}_{i+1}$.

Integration errors will decrease as step size decreases.

Most CACSD software includes some form of numerical integration capability such as the Runge–Kutta method and most will include some sort of automatic Δt step size determination. Any method will become more accurate as the step size decreases; however, initially, neither the computer algorithms nor the user knows what step size is the best compromise between accuracy and speed. A commonly used scheme is to integrate using two different methods (perhaps a second- and third-order Runge–Kutta formula), compare the difference, and then cut the step size in half if the error exceeds a certain tolerance. The step size will continue to be cut in half until the error tolerance is met. Numerous books devoted entirely to the subject of numerical integration can provide more detail; see, for example, Shampine and Gordon (1975).

3.6.2   Solution of Linear Differential Equations

When the differential equations are linear, they can be written in the form

$$\dot{\mathbf{x}} = \mathbf{F}\mathbf{x} + \mathbf{G}u. \tag{3.56}$$

The solution to the homogeneous portion,

$$\dot{\mathbf{x}} = \mathbf{F}\mathbf{x}, \tag{3.57}$$

is found by assuming a solution of the form

$$\mathbf{x}(t) = e^{\mathbf{F}t}\mathbf{x}(0) \qquad t \geqslant 0, \tag{3.58}$$

where the matrix exponential $e^{\mathbf{F}t}$ is a power series exactly like a scalar exponential, i.e.,

$$e^{\mathbf{F}t} = \mathbf{I} + \mathbf{F}t + \frac{\mathbf{F}^2 t^2}{2!} + \frac{\mathbf{F}^3 t^3}{3!} + \cdots.$$

By differentiating this expression with respect to time, we find that, just like the scalar exponential,

$$\frac{d}{dt}\left(e^{\mathbf{F}t}\right) = \mathbf{F}e^{\mathbf{F}t}.$$

Substituting this back into Eq. (3.57) along with Eq. (3.58) shows that the assumed solution was correct. This solution is also the impulse response of the system in Eq. (3.56); therefore, the convolution integral in Eq. (3.3) can be used to write the total solution as

$$\mathbf{x}(t) = e^{\mathbf{F}(t-t_0)}\mathbf{x}(t_0) + \int_{t_0}^{t} e^{\mathbf{F}(t-\tau)}\mathbf{G}u(\tau)\,d\tau. \tag{3.59}$$

This equation can be specialized for solving between two time points t_i and t_{i+1}:

$$\mathbf{x}(t_{i+1}) = e^{\mathbf{F}(t_{i+1}-t_i)}\mathbf{x}(t_i) + \int_{t_i}^{t_{i+1}} e^{\mathbf{F}(t_{i+1}-\tau)}\mathbf{G}u(\tau)\,d\tau. \tag{3.60}$$

By assuming that $\Delta t = t_{i+1} - t_i$ is small and that $u(t)$ does not change over that time interval, the solution simplifies to

$$\mathbf{x}(t_{i+1}) = e^{\mathbf{F}\Delta t}\mathbf{x}(t_i) + \left[\int_0^{\Delta t} e^{\mathbf{F}\eta}d\eta\right]\mathbf{G}u(t_i). \tag{3.61}$$

Difference equation

Eq. (3.61) is called a **difference equation** and usually written

$$\mathbf{x}(t_{i+1}) = \boldsymbol{\Phi}\mathbf{x}(t_i) + \boldsymbol{\Gamma}u(t_i), \tag{3.62}$$

where

Discrete standard form

$$\boldsymbol{\Phi} = e^{\mathbf{F}\Delta t} \quad \text{and} \quad \boldsymbol{\Gamma} = \left[\int_0^{\Delta t} e^{\mathbf{F}\eta}\,d\eta\right]\mathbf{G}. \tag{3.63}$$

This is the standard form for discrete systems and corresponds to the standard form for continuous systems given by Eqs. (2.25).

Efficiency of linear vs. nonlinear numerical simulation

The $\boldsymbol{\Phi}$ and $\boldsymbol{\Gamma}$ matrices are constant for *all* time steps. Therefore, to compute the response of a linear system at times that are spaced Δt seconds apart, the only quantities that require extensive computations, as in the Runge–Kutta method described in Section 3.6.1, are the $\boldsymbol{\Phi}$ and $\boldsymbol{\Gamma}$ matrices. The numerical computation of the solution of $\mathbf{x}(t_i)$ is obtained by repeated evaluation of Eq. (3.62). The entire process is much more efficient than the nonlinear case because $\boldsymbol{\Phi}$ and $\boldsymbol{\Gamma}$ are only calculated once and the time step Δt can be larger than for nonlinear equations.

◆ **EXAMPLE 3.27** *Numerical Simulation of a Pendulum Using* MATLAB

Find and plot the first 10 sec of the pendulum time history using the equations from Example 2.5 using MATLAB. Assume the control torque is a pulse for the first second; that is, $T_c = 1$ N for $0 < t \leqslant 1$ sec and zero thereafter. The length $l = 1$ m, and its mass $m = 0.2$ kg.

(a) Use the linearized equation of motion (Eq. 2.12). (b) Use the nonlinear equation of motion (Eq. 2.11). (c) Repeat parts (a) and (b) for a smaller pulse of $T_c = 0.3$ N.

Solution. (a) The most expedient method for entering this system into MATLAB for linear analysis is to write the equation of motion as a transfer function:

$$\frac{\theta(s)}{T_c(s)} = \frac{1/ml^2}{s^2 + g/l},$$

where $(1/ml^2) = 5$ and $g/l = 9.82$. The lsim routine is required rather than the step routine used in Example 2.8 because the input time history is constant for 1 sec and then becomes zero thereafter. lsim will accept any input time history. Therefore, the statements in MATLAB

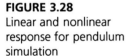

Linear simulation using MATLAB

```
num = 5
den = [1  0  9.82]
t = 0:.05:10
u = [ones(21,1);zeros(180,1)]
lsim(num,den,u,t)
```

will produce the desired time history shown in Fig. 3.28. The index of the input time history vector u corresponds to the time history vector t and produces the square pulse of control for the first second.

FIGURE 3.28
Linear and nonlinear response for pendulum simulation

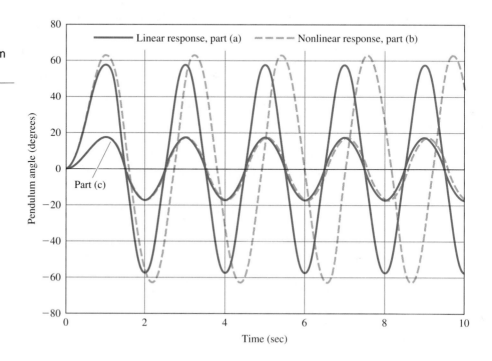

Nonlinear simulation
using MATLAB

(b) For the nonlinear solution, the differential equations need to be put into the standard first-order form of Eq. (3.49). We accomplish this by defining the state $\mathbf{x}=[\theta\ \dot\theta]^T$ and rewriting Eq. (2.11) as

$$\dot x_1 = x_2$$

$$\dot x_2 = -\frac{g}{l}\sin x_1 + \frac{T_c}{ml^2}.$$

A Runge–Kutta integration routine in MATLAB that performs both second- and third-order calculations and compares the two to determine the step size is called ode23. This routine requires that a separate function be provided that computes the derivatives. The following MATLAB statements

```
global Wn2 Uo
Wn2 = 9.82
Uo = 5
to = 0
tf = 10
Xo = [0;0]
[t,x] = ode23('pendot',to,tf,Xo)
```

call the integration routine, which in turn calls the derivative function, labeled pendot for this case. The derivative function is created by the following MATLAB statements:

```
function xdot = pendot(t,x)
if t <= 1,
U = Uo
else U = 0
end
xdot(1) = x(2)
xdot(2) = -Wn2*sin(x(1)) + U
```

This produces the large-amplitude (dashed) time histories shown in Fig. 3.28. For this level of the control input, the nonlinear response exhibits a slightly longer period and larger response amplitude than the linear case. This is because $\sin\theta$ is less than θ at the high angles excited by the 1-N control pulse, thus lowering the restoring gravity torque.

(c) To repeat part (a) for $T_c=0.3$ N, we need to change the num = 5 statement to num = 1.5 in order to reduce the magnitude of the input impulse. To repeat part (b) for $T_c=0.3$ N, we need to change Uo = 5 to Uo = 1.5. In this case, because the angles never exceed 20°, the linear approximation in Fig. 3.28 is almost indistinguishable from the nonlinear one.

3.7   Obtaining Models from Experimental Data

There are several reasons for using experimental data to obtain a model of the dynamic system to be controlled. In the first place, the best theoretical model built from equations of motion is still only an approximation of reality. Sometimes, as in the case of a very rigid spacecraft, the theoretical model is extremely good. Other times, as with many chemical processes such as papermaking or metalworking, the theoretical model is very approximate. In every case, before the final control design is done, it is important and prudent to verify the theoretical model with experimental data. Secondly, in situations where the theoretical model is especially complicated or the physics of the process is poorly understood, the only reliable information on which to base the control design is the experimental data. Finally, the system is sometimes subject to on-line changes, which occur when the environment of the system changes. Examples include when an aircraft changes altitude or speed, a paper machine is given a different composition of fiber, or a nonlinear system moves to a new operating point. On these occasions we need to "retune" the controller by changing the control parameters. This requires a model for the new conditions, and experimental data are often the most effective, if not the only, information available for the new model.

There are four kinds of experimental data for generating a model:

Four sources of experimental data

1. **transient response**, such as comes from an impulse or a step;
2. **frequency response data**, which result from exciting the system with sinusoidal inputs at many frequencies;
3. **stochastic steady-state information**, as might come from flying an aircraft through turbulent weather or from some other natural source of randomness;
4. **pseudorandom-noise data**, as may be generated in a digital computer.

Each class of experimental data has its properties, advantages, and disadvantages.

Transient response

Transient response data are quick and relatively easy to obtain. They are also often representative of the natural signals to which the system is subjected. Thus a model derived from such data can be reliable for designing the control system. On the other hand, in order for the signal-to-noise ratio to be sufficiently high, the transient response must be highly noticeable. Thus the method is rarely suitable for normal operations, so the data must be collected as part of special tests. A second disadvantage is that the data do not come in a form suitable for standard control systems designs, and some parts of the model, such as poles and zeros, must be computed from the data.[5] This computation can be simple in special cases or complex in the general case.

5. Ziegler and Nichols (1943), building on the earlier work of Callender *et al.* (1936), use the step response directly in designing the controls for certain classes of processes. See Chapter 4 for details.

Frequency response

Frequency-response data (see Chapter 6) is simple to obtain but substantially more time-consuming than transient-response information. This is especially so if the time constants of the process are large, as often occurs in chemical processing industries. As with the transient-response data, it is important to have a good signal-to-noise ratio, so obtaining frequency-response data can be very expensive. On the other hand, as we will see in Chapter 6, frequency-response data are exactly in the right form for frequency-response design methods, so once the data have been obtained, the control design can proceed immediately.

Stochastic steady state

Normal operating records from a natural stochastic environment at first appear to be an attractive basis for modeling systems since such records are by definition nondisruptive and inexpensive to obtain. Unfortunately, the quality of such data is inconsistent, tending to be worst just when the control is best, because then the upsets are minimal and the signals are smooth. At such times some or even most of the system dynamics are hardly excited. Since they contribute little to the system output, they will not be found in the model constructed to explain the signals. The result is a model that represents only part of the system and is sometimes unsuitable for control. In some instances, as occurs when trying to model the dynamics of the electroencephalogram (brain waves) of a sleeping or anesthetized person to locate the frequency and intensity of alpha waves, normal records are the only possibility. Usually they are the last choice for control purposes.

Pseudorandom noise (PRBS)

Finally, the pseudorandom signals that can be constructed using digital logic have much appeal. Especially interesting for model making is the pseudorandom binary signal (PRBS). The PRBS takes on the value $+A$ or $-A$ according to the output (1 or 0) of a feedback shift register. The feedback to the register is a binary sum of various states of the register that have been selected to make the output period (which must repeat itself in finite time) as long as possible. For example, with a register of 20 bits, $2^{20} - 1$ (over a million) pulses are produced before the pattern repeats. Analysis beyond the scope of this text has revealed that the resulting signal is almost like a broad-band random signal. Yet this signal is entirely under the control of the engineer who can set the level (A) and the length (bits in the register) of the signal. The data obtained from tests with a PRBS must be analyzed by computer, and both special-purpose hardware and programs for general-purpose computers have been developed to perform this analysis.

3.7.1 Models from Transient-response Data

To obtain a model from transient data we assume that a step response is available. If the transient is a simple combination of elementary transients, then a reasonable low-order model can be estimated using hand calculations. For example, consider the step response shown in Fig. 3.29. The response is monotonic and smooth. If we assume that it is given by a sum of exponentials,

FIGURE 3.29
A step response characteristic of many chemical processes

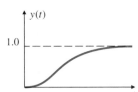

we can write

$$y(t) = y(\infty) + Ae^{-\alpha t} + Be^{-\beta t} + Ce^{-\gamma t} + \cdots. \tag{3.64}$$

Subtracting off the final value and assuming that $-\alpha$ is the slowest pole, we write

$$y - y(\infty) \cong Ae^{-\alpha t}$$
$$\log_{10}[y - y(\infty)] \cong \log_{10} A - \alpha t \log_{10} e,$$
$$\cong \log_{10} A - 0.4343 \alpha t. \tag{3.65}$$

This is the equation of a line whose slope determines α and intercept determines A. If we fit a line to the plot of $\log_{10}[y - y(\infty)]$ (or $\log_{10}[y(\infty) - y]$ if A is negative), then we can estimate A and α. Once these are estimated, we plot $y - [y(\infty) + Ae^{-\alpha t}]$, which as a curve approximates $Be^{-\beta t}$ and on the log plot is equivalent to $\log_{10} B - 0.4345 \beta t$. We repeat the process, each time removing the slowest remaining term, until the data stop being accurate. Then we plot the final model step response and compare it with data so we can assess the quality of the computed model. It is possible to get a good fit to the step response and yet be far off from the true time constants (poles) of the system. However, the method gives a good approximation for control of processes whose step responses look like Fig. 3.29.

◆ **EXAMPLE 3.28**   *Determining the Model from Time-response Data*

Find the transfer function that generates the data given in Table 3.1 and which are plotted in Fig. 3.30.

Solution.   Table 3.1 shows and Fig. 3.30 implies that the final value of the data is $y(\infty) = 1$. We know that A is negative because $y(\infty)$ is greater than $y(t)$. Therefore, the first step in the process is to plot $\log_{10}[y(\infty) - y]$, which is shown in Fig. 3.31. From

TABLE 3.1   Step-response Data[a]

t	y(t)	t	y(t)
0.1	0.000	1.0	0.510
0.1	0.005	1.5	0.700
0.2	0.034	2.0	0.817
0.3	0.085	2.5	0.890
0.4	0.140	3.0	0.932
0.5	0.215	4.0	0.975
		∞	1.000

[a]Sinha and Kuszta (1983).

FIGURE 3.30
Step response data in
Table 3.1

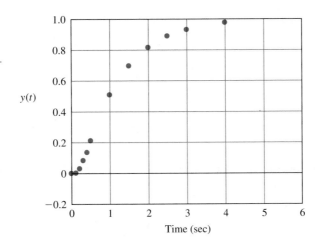

the line (fitted by eye) the values are

$$\log_{10}|A| = 0.125,$$

$$0.4343\alpha = \frac{1.602 - 1.167}{\Delta t} = \frac{0.435}{1} \Rightarrow \alpha \cong 1.$$

Thus

$$A = -1.33,$$

$$\alpha = 1.0.$$

FIGURE 3.31
$\log_{10}[y(\infty) - y]$ versus t

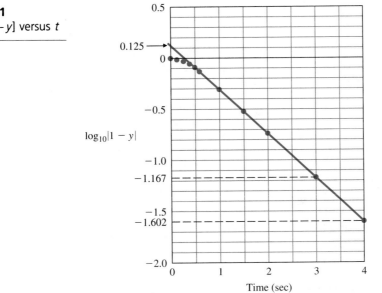

FIGURE 3.32
$\log_{10}[y-(1+Ae^{-\alpha t})]$
versus t

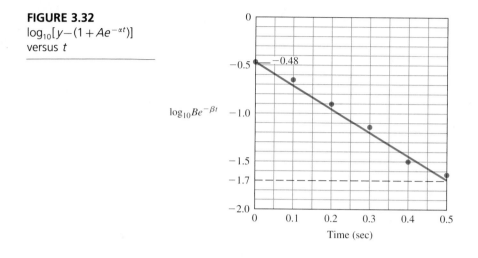

If we now subtract $1 + Ae^{\alpha t}$ from the data and plot the log of the result, we find the plot of Fig. 3.32. Here we estimate

$$\log_{10}B = -0.48,$$

$$0.4343\beta = \frac{-0.48 - (-1.7)}{0.5} = 2.5,$$

$$\beta \cong 5.8,$$

$$B = 0.33.$$

Combining these results, we arrive at the y estimate

$$\hat{y}(t) \cong 1 - 1.33e^{-t} + 0.33e^{-5.8t}. \tag{3.66}$$

Equation (3.66) is plotted as the colored line in Fig. 3.33 and shows a reasonable fit to the data, although some error is noticeable near $t = 0$.

From $\hat{y}(t)$ we compute

$$\hat{Y}(s) = \frac{1}{s} - \frac{1.33}{s+1} + \frac{0.33}{s+5.8}$$

$$= \frac{(s+1)(s+5.8) - 1.33s(s+5.8) + 0.33s(s+1)}{s(s+1)(s+5.8)}$$

$$= \frac{-0.58s + 5.8}{s(s+1)(s+5.8)}.$$

The resulting transfer function is

$$G(s) = \frac{-.58(s-10)}{(s+1)(s+5.8)}.$$

Notice that this method has given us a system with a zero in the RHP, even though the data showed no values of y that were negative. Very small differences in the

FIGURE 3.33
Model fits to the
experimental data

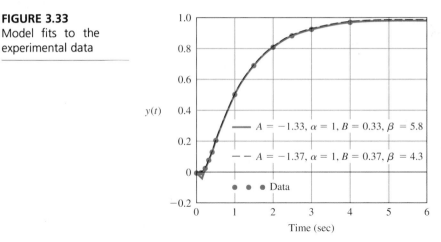

estimated value for A, all of which approximately fit the data, can cause values of β to range from 4 to 6. This illustrates the sensitivity of pole locations to the quality of the data and emphasizes the need for a good signal-to-noise ratio.

By using a computer to perform the plotting, we are better able to iterate the four parameters to achieve the best overall fit. The data presentation in Figs. 3.31 and 3.32 can be obtained directly by using a semilog plot. This eliminates having to calculate $\log_{10}$ and the exponential expression to find the values of the parameters. The equations of the lines to be fit to the data are $y(t) = Ae^{\alpha t}$ and $y(t) = Be^{\beta t}$, which are straight lines on a semilog plot. The parameters A and α, or B and β, are iteratively selected so that the straight line comes as close as possible to passing through the data. This process produces the improved fit shown by the dashed black line in Fig. 3.33. The revised parameters, $A = -1.37$, $B = 0.37$, and $\beta = 4.3$ result in the transfer function

$$G(s) = \frac{-0.22s + 4.3}{(s+1)(s+4.3)}.$$

The RHP zero is still present, but it is now located at $s \cong +20$ and has no noticeable effect on the time response.

This set of data was fitted quite well by a second-order model. In many cases a higher-order model is required to explain the data and the modes may not be as well separated.

◆

If the transient response has oscillatory modes, then these can sometimes be estimated by comparing them with the standard plots of Fig. 3.12. The period will give the frequency ω_d, and the decay from one period to the next will afford an estimate of the damping ratio. If the response has a mixture of modes not well separated in frequency, then more sophisticated methods need to be used. One such is **least-squares identification**, in which a numerical optimization routine selects the best combination of parameters so as to

Least-squares
identification

minimize the fit error. The fit error is defined to be a scalar **cost function**

$$J = \sum_i (y_{data} - y_{model})^2, \qquad i = 1, 2, 3, \ldots \text{ for each data point,}$$

so that fit errors at all data points are taken into account in determining the best value for the parameters.

3.7.2   Models from Other Data

As mentioned early in Section 3.7, we can also generate a model using frequency-response data, which are obtained by exciting the system with a set of sinusoids and plotting $H(j\omega)$. In Chapter 6 we will show how such plots can be used directly for design. Alternatively, we can use the frequency response to estimate the poles and zeros of a transfer function using straight-line asymptotes on a logarithmic plot.

The construction of dynamic models from normal stochastic operating records or from the response to a PRBS can be based either on the concept of cross-correlation or on the least-squares fit of a discrete equivalent model, both topics in the field of **system identification**. They require substantial presentation and background that are beyond the scope of this text. An introduction to system identification can be found in Chapter 8 of Franklin et al. (1990), and a comprehensive treatment is given in Ljüng (1987).

Summary

- The Laplace transform is the primary tool used to determine the behavior of linear systems. The Laplace transform of a time function $f(t)$ is given by

$$\mathscr{L}[f(t)] = F(s) = \int_0^\infty f(t)e^{-st}dt. \tag{3.10}$$

This relationship leads to the key property of Laplace transforms, namely,

$$\mathscr{L}[\dot{f}(t)] = sF(s) - f(0) \tag{3.18}$$

This property allows us to find the transfer function of a linear ODE. Given the transfer function $G(s)$ of a system and the input $u(t)$, with transform $U(s)$, the system output transform is

$$Y(s) = G(s)U(s).$$

- Normally, inverse transforms are found by referring to Table A.2 or by computer. Properties of Laplace transforms and their inverses are summarized in Table A.1.

- The Final Value Theorem is useful in finding steady-state errors for stable signals:

$$\lim_{t \to \infty} y(t) = \lim_{s \to 0} s Y(s). \tag{3.26}$$

- Block diagrams are a convenient way to show the relationships between the components of a system. They can usually be simplified using the relations in Fig. 3.2 and Eq. (3.30), that is

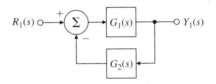

is equivalent to

$$Y_1(s) = \frac{G_1(s)}{1 + G_1(s)G_2(s)} R_1(s). \tag{3.30}$$

- The locations of poles in the s-plane determine the character of the response, as shown in Figure 3.9.
- The location of a pole in the s-plane is defined by the parameters shown in Fig. 3.11. These parameters are related to the time-domain quantities of rise time t_r, settling time t_s, and overshoot M_p, which are defined in Fig. 3.16. The correspondences between them, for a second-order system with no zeros, are given by:

$$t_r \cong \frac{1.8}{\omega_n} \tag{3.38}$$

$$M_p = e^{-\pi\zeta/\sqrt{1-\zeta^2}} \tag{3.40}$$

$$t_s = \frac{4.6}{\zeta\omega_n}. \tag{3.41}$$

- When a zero is present, the overshoot increases. This effect is summarized in Figs. 3.20 and 3.21.
- When an additional pole is present, the system response is more sluggish. This effect is summarized in Figs. 3.25 and 3.26.
- Numerical simulation is usually used to find the time history of a solution to a linear or nonlinear ODE. This is carried out using CACSD software. In MATLAB the step, lsim, or ODE23 functions accomplish the task.
- Determining a model from experimental data, or verifying an analytically based model by experiment, is an important step in system design.

Problems

3.1 Show that, in a partial-fraction expansion, complex conjugate poles have coefficients that are also complex conjugates. (The result of this relationship is that whenever complex conjugate pairs of poles are present, only one of the coefficients needs to be computed.)

3.2   Find the Laplace transform of the following time functions:

a) $f(t) = \alpha + \beta t$
b) $f(t) = \alpha + \beta t + \gamma t^2 + \delta(t)$
c) $f(t) = \alpha \sin \omega t + \beta \cos \omega t + e^{-at} \sin \omega t$
d) $f(t) = \alpha e^{-at} + \beta e^{-bt} + te^{-ct}$
e) $f(t) = \alpha \sin^2 t + \beta \cos^2 t$
f) $f(t) = te^{-at} + 2t \cos t$

3.3   Find the Laplace transform of the following time functions:

a) $f(t) = t \sin 3t - 2t \cos t$
b) $f(t) = \sin t * \sin t$
c) $f(t) = \sin t/t$

3.4   Find the Laplace transform of the following time functions:

a) $f(t) = t \sin t$
b) $f(t) = t \cos at$
c) $f(t) = t^2 + e^{-at} \sin bt$
d) $f(t) = a1(t) + bt \cos at$

3.5   Find the Laplace transform of the following time functions:

a) $f(t) = \sin t \sin 3t$
b) $f(t) = (t+1)^2$
c) $f(t) = 3 \cos 6t$
d) $f(t) = \int_0^t \cos(t-\tau) \sin \tau \, d\tau$
e) $f(t) = \sinh t$

3.6   Find the time function corresponding to each of the following Laplace transforms:

a) $F(s) = \dfrac{1}{s^6}$

b) $F(s) = \dfrac{4}{s^4 + 4}$

c) $F(s) = \dfrac{e^{-\tau s}}{s^2}$

d) $F(s) = \dfrac{s+1}{s^2}$

e) $F(s) = \dfrac{(s^2 - 1)}{(s^2 + 1)^2}$

f) $F(s) = \arctan \dfrac{1}{s}$

3.7   Using partial-fraction expansions, find the function $f(t)$ of the following:

a) $F(s) = \dfrac{2}{s(s+2)}$

b) $F(s) = \dfrac{10}{s(s+1)(s+10)}$

c) $F(s) = \dfrac{1}{s(s+2)^2}$

d) $F(s) = \dfrac{2(s+2)}{(s+1)(s^2+4)}$

e) $F(s) = \dfrac{2(s+2)(s+5)^2}{(s+1)(s^2+4)^2}$

f) $F(s) = \dfrac{2(s^2+s+1)}{s(s+1)^2}$

g) $F(s) = \dfrac{1}{s^2+4}$

h) $F(s) = \dfrac{3s^2+9s+12}{(s+2)(s^2+5s+11)}$

i) $F(s) = \dfrac{2s^2+s+1}{s^3-1}$

j) $F(s) = \dfrac{3s+2}{s^2+4s+20}$

k) $F(s) = \dfrac{s^3+2s+4}{s^4-16}$

3.8 Solve the following ordinary differential equations using Laplace transforms:

a) $\ddot{y}(t) + \dot{y}(t) + 3y(t) = 0$; $y(0) = \alpha$, $\dot{y}(0) = \beta$

b) $\ddot{y}(t) - 2\dot{y}(t) + 4y(t) = 0$; $y(0) = \alpha$, $\dot{y}(0) = \beta$

c) $\ddot{y}(t) + \dot{y}(t) = \sin t$; $y(0) = \alpha$, $\dot{y}(0) = \beta$

d) $\ddot{y}(t) + 3y(t) = \sin t$; $y(0) = \alpha$, $\dot{y}(0) = \beta$

e) $\ddot{y}(t) + 2\dot{y}(t) = e^t$; $y(0) = \alpha$, $\dot{y}(0) = \beta$

f) $\ddot{y}(t) + y(t) = t$; $y(0) = 1$, $\dot{y}(0) = -1$

3.9 Given that the Laplace transform of $f(t)$ is $F(s)$, find the Laplace transform of the following:

a) $g(t) = f(t) \cos \omega t$

b) $g(t) = \displaystyle\int_0^t \int_0^{t_1} f(\tau) d\tau \, dt_1$

3.10 Write the dynamic equations describing the circuit in Fig. 3.34. Write the equations in both state-variable form and as a second-order differential equation in $y(t)$. Assuming a zero input, solve the differential equation for $y(t)$ using Laplace-transform methods for the parameter values and initial conditions shown in the figure. Verify your answer using the initial command in MATLAB.

3.11 Consider the standard second-order system

$$G(s) = \frac{\omega_n^2}{s^2 + 2\zeta\omega_n s + \omega_n^2}.$$

Find the output of the system for the input shown in Fig. 3.35.

FIGURE 3.34
Circuit for Problem 3.10

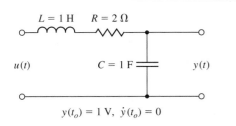

$y(t_o) = 1$ V, $\dot{y}(t_o) = 0$

FIGURE 3.35
Plot of input for Problem 3.11

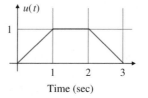

Time (sec)

3.12 For the electric circuit shown in Fig. 3.36, find the following:

    a) the time-domain equation relating $i(t)$ and $v_1(t)$;

    b) the time-domain equation relating $i(t)$ and $v_2(t)$;

    c) assuming all initial conditions are zero, the transfer function $V_2(s)/V_1(s)$ and the damping ratio ζ and undamped natural frequency ω_n of the system;

    d) the values of R that will result in $v_2(t)$ having an overshoot of no more than 25%, assuming $v_1(t)$ is a unit step, L $= 10\,\text{mH}$, and $C = 4\,\mu\text{F}$.

FIGURE 3.36
Circuit for Problem 3.12

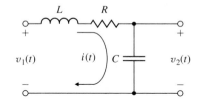

3.13 Compute the transfer functions for the circuits in the following figures:

    a) Fig. 2.40(a)   **c)** Fig. 2.40(c)   **e)** Fig. 2.41(b)

    b) Fig. 2.40(b)   **d)** Fig. 2.41(a)   **f)** Fig. 2.41(c)

3.14 A rotating load is connected to a field-controlled DC motor with negligible field inductance. A test results in the output load reaching a speed of 1 rad/sec within 1/2 sec when a constant input of 100 V is applied to the motor terminals. The output steady-state speed from the same test is found to be 2 rad/sec. Determine the transfer function $\theta(s)/V_f(s)$ of the motor.

3.15 For the tape drive shown in Fig. 2.46, compute the following, using the numbers given in Problem 2.17(a):

    a) the transfer function from the motor current to the tape position;

    b) the poles and zeros for the transfer function in part (a).

3.16 For the tape drive shown in Fig. 2.46, compute the following, using the numbers given in Problem 2.17(a):

    a) the transfer function from the motor current to the tape tension;

    b) the poles and zeros for the transfer function in part (a).

3.17 For the system in Fig. 2.48, compute the transfer function from the motor voltage to position θ_2.

3.18 Compute the transfer function for the two-tank system in Fig. 2.52 with holes at A and C.

3.19 For a second-order system with transfer function

$$G(s) = \frac{3}{s^2 + 2s - 3},$$

determine the following:

    a) DC gain;

    b) the final value to a step input.

3.20 a) Compute the transfer function for the block diagram shown in Fig. 3.37. Note that a_i and b_i are constants.

    b) Write the third-order differential equation that relates y and u. (*Hint*: Consider the transfer function.)

    c) Write three simultaneous first-order (state-variable) differential equations using variables x_1, x_2, and x_3, as defined on the block diagram in Fig. 3.37. Notice

FIGURE 3.37
Block diagram for
Problem 3.20

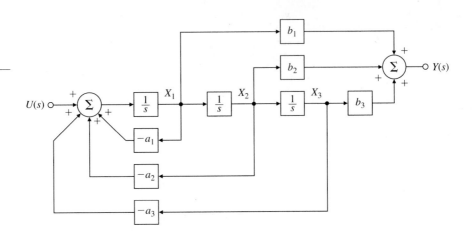

how the same constant parameters enter the transfer function, the differential
equations, and the matrices of the state-variable form. (This special structure is
called the control canonical form and will be discussed further in Chapter 7.)

3.21 Find the transfer functions for the block diagrams in Fig. 3.38.

FIGURE 3.38
Block diagrams for
Problem 3.21

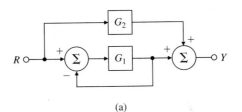

(a)

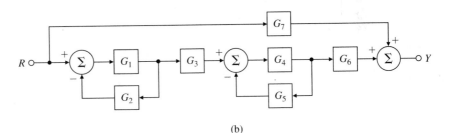

(b)

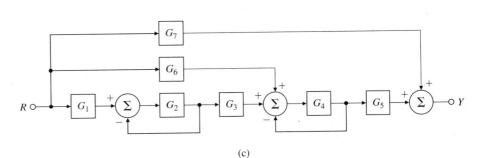

(c)

FIGURE 3.39
Block diagrams for Problem 3.22

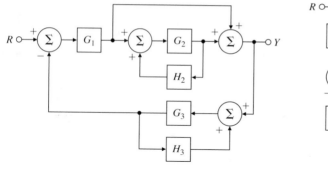

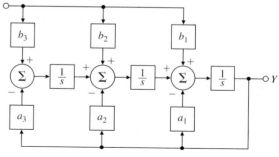

(a)

(b)

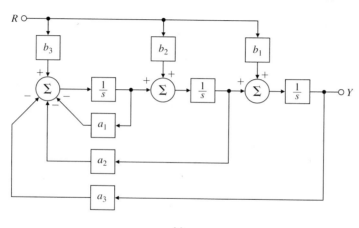

(c)

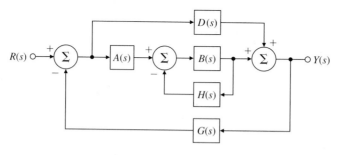

(d)

3.22 Find the transfer functions for the block diagrams in Fig. 3.39, using the following:

a) the ideas of Figs. 3.1 and 3.2;

b) Mason's rule.

3.23 Use block-diagram algebra or Mason's rule to determine the transfer function between $R(s)$ and $Y(s)$ in Fig. 3.40.

FIGURE 3.40
Block diagram for
Problem 3.23

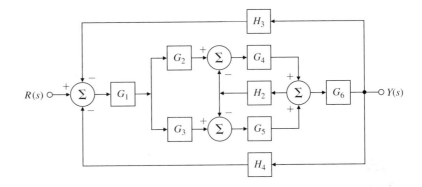

3.24 Use block-diagram algebra to determine the transfer function between $R(s)$ and $Y(s)$ in Fig. 3.41.

FIGURE 3.41
Block diagram for
Problem 3.24

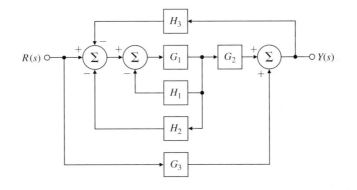

3.25 Consider the continuous rolling mill depicted in Fig. 3.42. Suppose that the motion of the adjustable roller has a damping coefficient b, and that the force exerted by the rolled material on the adjustable roller is proportional to the material's change in thickness: $F_s = c(T - x)$. Suppose further that the DC motor has a torque constant K_t and a back-emf constant K_e, and that the rack-and-pinion has an effective radius of R.

a) What are the inputs to this system? The output?

b) Without neglecting the effects of gravity on the adjustable roller, draw a block diagram of the system that explicitly shows the following quantities: $V_s(s)$, $I_a(s)$, $F_m(s)$ (the force the motor exerts on the adjustable roller), and $X(s)$.

c) Simplify your block diagram as much as possible while still identifying the output and each input separately.

FIGURE 3.42
Continuous rolling mill

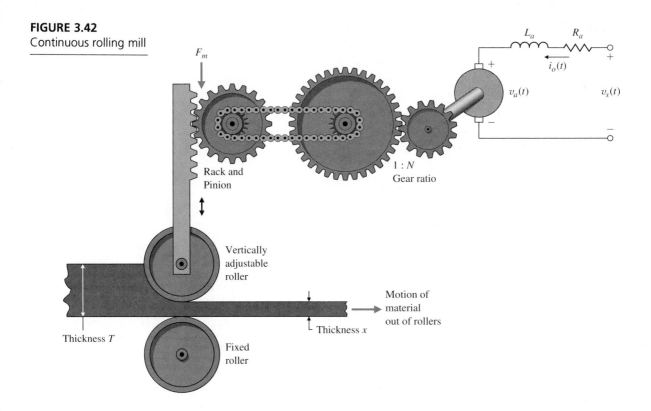

3.26 In aircraft control systems, an ideal pitch response (q_o) versus a pitch command (q_c) is described by the transfer function

$$\frac{Q_o(s)}{Q_c(s)} = \frac{\tau\omega_n^2(s + 1/\tau)}{s^2 + 2\zeta\omega_n s + \omega_n^2}.$$

The actual aircraft response is more complicated than this ideal transfer function; nevertheless, the ideal model is used as a guide for autopilot design. Assume that t_r is the desired rise time, and that

$$\omega_n = \frac{1.789}{t_r}$$

$$\frac{1}{\tau} = \frac{1.6}{t_r}$$

$$\zeta = 0.89.$$

Show that this ideal response possesses a fast settling time and minimal overshoot by plotting the step response for $t_r = 0.8$, 1.0, 1.2, and 1.5 sec.

3.27 For the unity feedback system shown in Fig. 3.43, specify the gain K of the proportional controller so that the output $y(t)$ has an overshoot of no more than 10% in response to a unit step.

FIGURE 3.43
Unity feedback system
for Problem 3.27

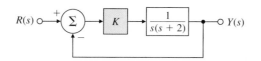

$R(s)$ with $+$, $-$ summing junction Σ, K, $\dfrac{1}{s(s+2)}$, $Y(s)$

3.28 For the unity feedback system shown in Fig. 3.44, specify the gain and pole location of the compensator so that the overall closed-loop response to a unit-step input has an overshoot of no more than 25%, and a 2% settling time of no more than 0.1 sec. Verify your design using MATLAB.

FIGURE 3.44
Unity feedback system
for Problem 3.28

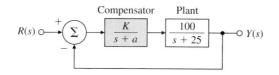

Compensator   Plant

$R(s)$ with Σ, $\dfrac{K}{s+a}$, $\dfrac{100}{s+25}$, $Y(s)$

3.29 Consider the system shown in Fig. 3.45, where

$$G(s)=\frac{1}{s(s+3)} \quad \text{and} \quad D(s)=\frac{K(s+z)}{s+p}.$$

Find K, z, and p so that the closed-loop system has a 10% overshoot to a step input and a settling time of 1.5 sec (1% criterion).

FIGURE 3.45
Unity feedback system
for Problem 3.29

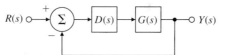

$R(s)$ with Σ, $D(s)$, $G(s)$, $Y(s)$

3.30 Sketch the step response of a system with the transfer function

$$G(s)=\frac{s/2+1}{(s/40+1)[(s/4)^2+s/4+1]}.$$

Justify your answer based on the locations of the poles and zeros (do not find the inverse Laplace transform). Then compare your answer with the step response computed using MATLAB.

3.31 Suppose you desire the peak time of a given second-order system to be less than t'_p. Draw the region in the s-plane that corresponds to values of the poles that meet the specification $t_p<t'_p$.

3.32 Suppose you are to design a unity feedback controller for a first-order plant, as depicted in Fig. 3.46. (As you will learn in Chapter 4, the configuration shown here

FIGURE 3.46
Unity feedback system
for Problem 3.32

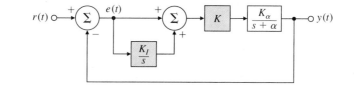

$r(t)$ with Σ, $e(t)$, Σ, K, $\dfrac{K_\alpha}{s+\alpha}$, $y(t)$, and $\dfrac{K_I}{s}$ feedback

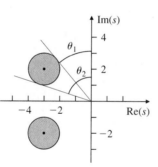

is referred to as a proportional-integral controller). You are to design the controller so that the closed-loop poles lie within the shaded regions shown in Fig. 3.47.

a) What values of ω_n and ζ correspond to the shaded regions in Fig. 3.47? (A simple estimate from the figure is sufficient.)

b) Let $K_\alpha = \alpha = 2$. Find values for K and K_I so that the poles of the closed-loop system lie within the shaded regions.

c) Prove that no matter what the values of K_α and α are, the controller provides enough flexibility to place the poles anywhere in the complex plane.

3.33 The open-loop transfer function of a unity feedback system is

$$G(s) = \frac{K}{s(s+2)}.$$

The desired system response to a step input is specified as peak time $t_p = 1$ sec and overshoot $M_p = 5\%$.

a) Determine whether both specifications can be met simultaneously by selecting the right value of K.

b) Sketch the associated region in the s-plane where both specifications are met, and indicate what root locations are possible for some likely values of K.

c) Pick a suitable value for K, and use MATLAB to verify that the specifications are satisfied.

3.34 a) Show that the second-order system

$$\ddot{y} + 2\zeta\omega_n\dot{y} + \omega_n^2 y = 0, \qquad y(0) = y_o, \quad \dot{y}(0) = 0,$$

has the response

$$y(t) = y_o \frac{e^{-\sigma t}}{\sqrt{1-\zeta^2}} \sin(\omega_d t + \cos^{-1}\zeta).$$

b) Prove that, for the underdamped case ($\zeta < 1$), the response oscillations decay at a predictable rate (see Fig. 3.48) called the **logarithmic decrement** δ, where

$$\delta = \ln\frac{y_o}{y_1} = \sigma\tau_d$$

$$= \ln\frac{\Delta y_1}{y_1} \cong \ln\frac{\Delta y_i}{y_i},$$

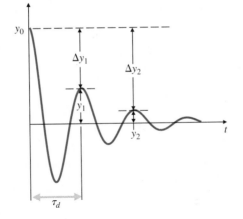

and τ_d is the damped natural period of vibration

$$\tau_d = \frac{2\pi}{\omega_d}.$$

3.35 Consider the two nonminimum phase systems,

$$G_1(s) = -\frac{2(s-1)}{(s+1)(s+2)}; \tag{i}$$

$$G_2(s) = \frac{3(s-1)(s-2)}{(s+1)(s+2)(s+3)}. \tag{ii}$$

a) Sketch the unit step responses for $G_1(s)$ and $G_2(s)$, paying close attention to the transient part of the response.

b) Explain the difference in the behavior of the two responses as it relates to the zero locations.

c) Consider a stable, strictly proper system (that is, m zeros and n poles, where $m < n$). Let $y(t)$ denote the step response of the system. The step response is said to have an undershoot if it initially starts off in the "wrong" direction. Prove that a stable, strictly proper system has an undershoot if and only if its transfer function has an *odd* number of *real* RHP zeros.

3.36 Consider the following second-order system with an extra pole:

$$H(s) = \frac{\omega_n^2 p}{(s+p)(s^2 + 2\zeta\omega_n s + \omega_n^2)}.$$

Show that the unit step response is

$$y(t) = 1 + Ae^{-pt} + Be^{-\sigma t}\sin(\omega_d t - \theta),$$

where

$$A = \frac{-\omega_n^2}{\omega_n^2 - 2\zeta\omega_n p + p^2}$$

$$B = \frac{p}{\sqrt{(p^2 - 2\zeta\omega_n p + \omega_n^2)(1 - \zeta^2)}}$$

$$\theta = \tan^{-1}\frac{\sqrt{1-\zeta^2}}{-\zeta} + \tan^{-1}\frac{\sqrt{1-\zeta^2}}{p - \zeta\omega_n}.$$

a) Which term dominates $y(t)$ as p gets large?
b) Give approximate values for A and B for small values of p.
c) Which term dominates as p gets small? (Small with respect to what?)
d) Using the explicit expression for $y(t)$ above or the step command in MATLAB, and assuming $\omega_n = 1$ and $\zeta = 0.7$, plot the step response of the system above for several values of p ranging from very small to very large. At what point does the extra pole cease to have much effect on the system response?

3.37 A measure of the degree of instability in an unstable aircraft response is the amount of time it takes for the *amplitude* of the time response to double (see Fig. 3.49) given some nonzero initial condition.

a) For a first-order system, show that the **time to double** τ_2 is

$$\tau_2 = \frac{\ln 2}{p},$$

where p is the pole location in the RHP.

b) For a second-order system (with two complex poles in the RHP), show that

$$\tau_2 = \frac{\ln 2}{-\zeta\omega_n}.$$

FIGURE 3.49
Time to double

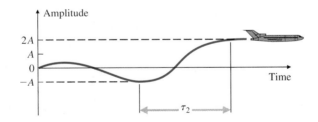

3.38 Samples from a step response are given in Table 3.2. Plot this data on a linear scale [$y(t)$ vs. t] and semilog scale [$\log(y - y_\infty)$ vs. t], and obtain an estimate of the transfer function.

3.39 The equations of motion for the DC motor shown in Fig. 2.26 were given in Eq. (2.44) as

$$J_m\ddot{\theta}_m + \left(b + \frac{K_t K_e}{R_a}\right)\dot{\theta}_m = \frac{K_t}{R_a}v_a.$$

TABLE 3.2   Step-response Data for Problem 3.38

t	y(t)	t	y(t)	t	y(t)
0	0	0.20	0.0138	0.90	0.4409
0.02	0.0001	0.22	0.0395	1.00	0.4924
0.04	0.0005	0.24	0.0480	1.50	0.6904
0.06	0.0014	0.26	0.0571	2.00	0.8121
0.08	0.0031	0.28	0.0668	2.50	0.8860
0.10	0.0057	0.30	0.0771	3.00	0.9309
0.12	0.0091	0.50	0.1979	3.50	0.9581
0.14	0.0135	0.60	0.2624	4.00	0.9746
0.16	0.0187	0.70	0.3253	5.00	0.9907
0.18	0.0248	0.80	0.3851		

Assume that

$$J_m = 0.01 \ \text{kg·m}^2,$$

$$b = 0.001 \ \text{N·m·sec},$$

$$K_e = 0.02 \ \text{V·sec},$$

$$K_t = 1 \ \text{N·m/A},$$

$$R_a = 10 \ \Omega.$$

a) Find the transfer function between the applied voltage v_a and the motor speed $\dot{\theta}_m$.

b) What is the steady-state speed of the motor after a voltage $v_a = 10$ V has been applied?

c) Find the transfer function between the applied voltage v_a and the shaft angle θ_m.

d) Suppose feedback is added to the system in part (c) so that it becomes a position servo device such that the applied voltage is given by

$$v_a = K(\theta_r - \theta_m),$$

where K is the feedback gain. Find the transfer function between θ_r and θ_m.

e) What is the maximum value of K that can be used if an overshoot $M_p < 20\%$ is desired?

f) What values of K will provide a rise time of less than 4 sec? (Ignore the M_p constraint.)

g) Use MATLAB to plot the step response of the position servo system for values of the gain $K = 0.01$, 0.02, and 0.04. Find the overshoot and rise time of the three step responses by examining your plots. Are the plots consistent with your calculations in parts (e) and (f)?

3.40 You wish to control the elevation of the satellite-tracking antenna shown in Figs. 3.50 and 3.51. The antenna and drive parts have a moment of inertia J and a damping B; these arise to some extent from bearing and aerodynamic friction but mostly from the back emf of the DC drive motor. The equations of motion are

$$J\ddot{\theta} + B\dot{\theta} = T_c,$$

where T_c is the torque from the drive motor. Assume that

$$J = 600{,}000 \ \text{kg·m}^{2\cdot} \qquad B = 20{,}000 \ \text{N·m·sec}$$

FIGURE 3.50
Satellite-tracking
antenna. *(Courtesy
Space Systems/Loral)*

a) Find the transfer function between the applied torque T_c and the antenna angle θ.

b) Suppose the applied torque is computed so that θ tracks a reference command θ_r according to the feedback law

$$T_c = K(\theta_r - \theta),$$

where K is the feedback gain. Find the transfer function between θ_r and θ.

c) What is the maximum value of K that can be used if you wish to have an overshoot $M_p < 10\%$?

d) What values of K will provide a rise time of less than 80 sec? (Ignore the M_p constraint.)

FIGURE 3.51
Schematic of antenna
for Problem 3.40

e) Use MATLAB to plot the step response of the antenna system for $K = 200, 400,$ 1000, and 2000. Find the overshoot and rise time of the four step responses by examining your plots. Do the plots confirm your calculations in parts (c) and (d)?

3.41 The block diagram of an autopilot designed to maintain the pitch attitude θ of an aircraft is shown in Fig. 3.52. The transfer function relating the elevator angle δ_e and the pitch attitude θ is

$$\frac{\theta(s)}{\delta_e(s)} = G(s) = \frac{50(s+1)(s+2)}{(s^2 + 5s + 40)(s^2 + 0.03s + 0.06)},$$

where θ is the pitch attitude in degrees and δ_e is the elevator angle in degrees. The autopilot controller uses the pitch attitude error ε to adjust the elevator according to the following transfer function:

$$\frac{\delta_e(s)}{\varepsilon(s)} = D(s) = \frac{K(s+3)}{s+10}.$$

Using MATLAB, find a value of K that will provide an overshoot of less than 10% and a rise time faster than 0.5 sec for a unit step change in θ_r. After examining the step response of the system for various values of K, comment on the difficulty associated with making rise-time and overshoot measurements for complicated systems.

FIGURE 3.52
Block diagram of autopilot

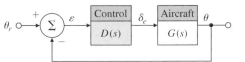

· 4 ·

Basic Properties of Feedback

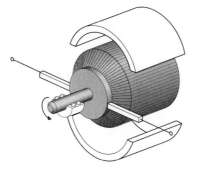

A Perspective on Properties of Feedback

Prepared with an understanding of models and the transfer function, we now begin to consider some simple control systems, the advantages of feedback, and the types of feedback available. As mentioned in Chapter 1 there are two basic system structures for achieving control of dynamic systems: open-loop control (Fig. 4.1), and feedback control, also known as closed-loop control (Fig. 4.2). The difference between the two control structures is that the closed-loop control requires the use of an output sensor, which usually introduces noise as well as a measure of the output variable. In this chapter we will compare the effectiveness of open- and closed-loop control strategies. We will explore the consequences that control has on the dynamic response of the system.

FIGURE 4.1
Open-loop control system

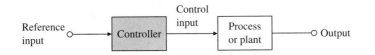

FIGURE 4.2
Feedback control system

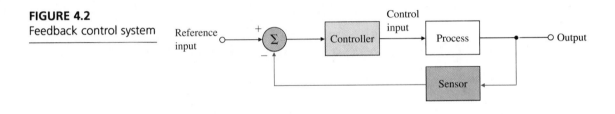

Chapter Overview

The chapter begins with a case study using speed control to illustrate the properties and advantages of feedback control compared with open-loop control. In Section 4.2 we consider several common types of feedback: constant-gain control, which in the process industries is called proportional control; integral control, which, when combined with proportional control, improves steady-state error properties; and derivative control, which improves transient properties. Together these three kinds of control—called proportional-integral-derivative (PID) control—constitute the heuristic approach to controller design that has found wide acceptance in the process industries. We also discuss parameter sensitivity, PID tuning, and integrator anti-windup, topics of much practical significance. In Section 4.3 we consider the general case of steady-state tracking of systems that do not have constant inputs but rather inputs that can be approximated as a polynomial in time. Finally, in Section 4.4 we consider the definition of system stability and Routh's test, which can determine stability by examining the coefficients of the system's characteristic equation.

4.1   A Case Study of Speed Control

Equations (2.42) and (2.43), which describe the dynamics of a DC motor are duplicated here as

Modeling a DC motor or
speed control using
equations

$$J_m\ddot{\theta}_m + b\dot{\theta}_m = K_t i_a + T_\ell,$$ (4.1)

$$K_e\dot{\theta}_m + L_a\frac{di_a}{dt} + R_a i_a = v_a,$$ (4.2)

with the difference that Eq. (4.1) includes a new term, T_ℓ, for load-torque disturbance. For simplicity let us define the output to be $y \triangleq \dot{\theta}_m$ = velocity and rename the disturbance $w \triangleq T_\ell$ to obtain

$$J_m \dot{y} + by = K_t i_a + w, \tag{4.3}$$

$$K_e y + L_a \frac{di_a}{dt} + R_a i_a = v_a. \tag{4.4}$$

Taking the Laplace transform of Eqs. (4.3) and (4.4) we obtain

$$s J_m Y(s) + b Y(s) = K_t I_a(s) + W(s), \tag{4.5}$$

$$K_e Y(s) + s L_a I_a(s) + R_a I_a(s) = V_a(s). \tag{4.6}$$

After substituting I_a from Eq. (4.5) into Eq. (4.6), the equation for motor speed becomes

$$(J_m L_a s^2 + b L_a s + J_m R_a s + b R_a + K_t K_e) Y(s) = K_t V_a(s) + (Ra + L_a s) W(s), \tag{4.7}$$

or

$$\left(\frac{J_m L_a}{b R_a + K_t K_e} s^2 + \frac{J_m R_a + b L_a}{b R_a + K_t K_e} s + 1 \right) Y(s)$$

$$= \frac{K_t}{b R_a + K_t K_e} V_a(s) + \frac{(R_a + L_a s)}{(b R_a + K_t K_e)} W(s). \tag{4.8}$$

We may rewrite this equation as

$$(\tau_1 s + 1)(\tau_2 s + 1) Y(s) = A V_a(s) + B W(s), \tag{4.9}$$

where the time constants and gains are given by[1]

$$\tau_{1,2}^{-1} = \frac{(J_m R_a + b L_a) \pm \sqrt{(J_m R_a + b L_a)^2 - 4 J_m L_a (b R_a + K_t K_e)}}{2 J_m L_a}$$

$$A = \frac{K_t}{b R_a + K_t K_e},$$

$$B = \frac{1}{b R_a + K_t K_e}.$$

We may also rewrite Eq. (4.9) in transfer-function form:

$$Y(s) = \frac{A}{(\tau_1 s + 1)(\tau_2 s + 1)} V_a(s) + \frac{B}{(\tau_1 s + 1)(\tau_2 s + 1)} W(s). \tag{4.10}$$

If w and v_a are constants, then the steady-state solution of these equations is

$$y_{ss} = A v_a + B w. \tag{4.11}$$

---

1. If $b = 0$ and L_a is small, then $\tau_2 = L_a/R_a$ and is called the **electrical time constant** and $\tau_1 = R_a J_m / K_t K_e$ is called the **mechanical time constant**.

FIGURE 4.3
Open-loop speed-control
system

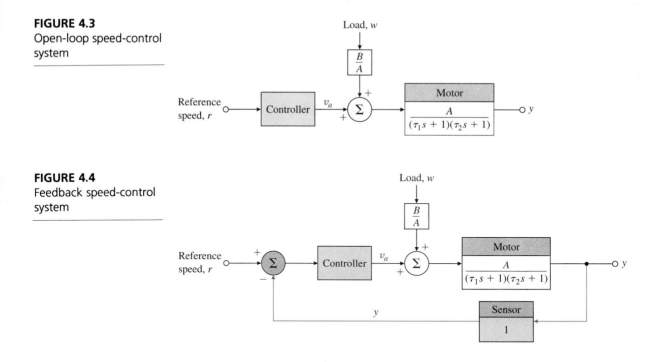

FIGURE 4.4
Feedback speed-control
system

The transfer functions in Eq. (4.10) are labeled "motor" and "load" in Fig. 4.3, which shows open-loop control. In contrast, the block diagram in Fig. 4.4 shows a closed-loop system because it includes an output sensor. In this case the sensor is a tachometer, which is usually a small permanent-magnet DC machine that produces a voltage proportional to the shaft speed $y(=\dot{\theta}_m)$. The controller in Fig. 4.4 is an electronic amplifier of gain K that produces a voltage proportional to the difference between the voltages that represent the reference speed r and motor speed y.

Modeling a DC motor or speed control by block diagram

A similar block diagram would result for many different types of speed-control systems. For example, in the case of a steam engine controlled by a fly-ball governor the centrifugal weights together comprise the speed sensor, and their motion directly operates a valve (the controller) that changes the speed of the steam engine.

4.1.1   Disturbance Rejection

Now let us compare feedback control and open-loop control with respect to how well in the steady-state each system maintains a constant reference speed in the face of load or disturbance torques. For the open-loop controller (Fig. 4.3), the amplifier voltage is taken to be

$$v_a = Kr. \tag{4.12}$$

The gain K is determined so that output speed y equals reference speed r when the load torque w is zero. From Eq. (4.11) this value is found to be

$$K = \frac{1}{A}.$$

In the steady state ($\dot{y} = 0$) and with no load torque ($w = 0$), Eq. (4.11) gives the output speed as

$$y_{ss} = Av_a = A\,\frac{1}{A}\,r = r.$$

With the feedback controller (Fig. 4.4) the amplifier output voltage is taken to be

$$v_a = K(r - y). \tag{4.13}$$

Because the control effort is proportional to the speed signals, this is called proportional control. (We will study proportional control in more detail in Section 4.2.) Combining the motor model given by Eq. (4.9) with the feedback controller added according to Eq. (4.13), the formula for the closed-loop system is

$$[(\tau_1 s + 1)(\tau_2 s + 1) + AK]Y(s) = AKR(s) + BW(s). \tag{4.14}$$

For no load torque ($w = 0$) and in the steady-state ($\dot{y} = 0$), the speed is given by

$$y_{ss} = \frac{AK}{1 + AK}\,r. \tag{4.15}$$

If the gain K is selected so that $AK \gg 1$, then $y_{ss} \cong r$.

With these controls both approaches provide the desired result, $y_{ss} \cong r$. However, the gains have been set with zero load torque; now let us compute what effect a nonzero torque will have on the steady-state speed in the two cases. Equation (4.11) shows that in the open-loop system the steady-state speed is

$$y_{ss} = AKr + Bw. \tag{4.16}$$

Here, with $K = 1/A$, the speed with load torque for the open-loop case is

$$y_{ss} = r + Bw,$$

and the variation in speed caused by the load torque for the open-loop case is $y_{ss} - r = \delta y$, where

$$\delta y = Bw.$$

So we see that the speed error is proportional to the disturbing load torque, and the control designer has no influence over the parameter B that determines the size of the error.

For the feedback case (Eq. 4.14), the steady-state speed for constant values of r and w becomes

$$y_{ss} = \frac{AK}{1 + AK} r + \frac{B}{1 + AK} w. \tag{4.17}$$

If it is possible for the designer to pick values of K so that $AK \gg 1$ and $AK \gg B$, no significant error will result with or without a disturbing load torque. In fact, as long as $AK > 0$, comparison of Eqs. (4.16) and (4.17) reveals that the feedback-controller speed errors due to a load torque will be less than the open-loop errors by exactly $1 + AK$. This is our first conclusion about feedback:

Advantage of feedback

System errors can be made less sensitive to disturbances with feedback than they are in open-loop systems (by a factor of $1 + AK$).

◆ **EXAMPLE 4.1**   *Open-loop Speed Control: Disturbance Response*

For a certain motor and load, it is determined that $\tau_1 = \frac{1}{60}$, $\tau_2 = \frac{1}{600}$, $A = 10$, and $B = 50$. The reference speed is 100 rad/sec. Find the steady-state open-loop armature voltage needed to get this speed with zero load torque. What is the steady-state speed with this load voltage if the load torque is $w = -0.1\,\text{N} \cdot \text{m}$?

Solution.   For the open-loop system with an armature voltage input v_a and a load torque w, the steady-state output is given by Eq. (4.11) with $K = \frac{1}{10}$

$$y_{ss} = 10v_a + 50w.$$

Assuming $w = 0$, the output is

$$y_{ss} = 100 \text{ rad/sec.}$$

Now suppose a constant load torque of $-0.1\,\text{N} \cdot \text{m}$ is applied. The steady-state value of the output with the disturbance is

$$y_{ss} = 100 + 50(-0.1) = 95 \text{ rad/sec.}$$

◆

Now consider the same example using proportional feedback of the output.

◆ **EXAMPLE 4.2**   *Closed-loop Proportional Speed Control: Disturbance Response*

Consider the feedback control structure shown in Fig. 4.4, where tachometer feedback and proportional control are employed. You are to improve the ability of the system to reject steady-state disturbances by a factor of at least 100 compared with the open-loop system. The parameter values are the same as those given in Example 4.1.

Solution.   We assume first that the reference speed $r = 0$. The steady-state output due to the input of the unit-step disturbance torque is (Eq. 4.17)

$$y_{ss} = \frac{B}{1 + KA} = \frac{50}{1 + 10K}.$$

To improve on this by a factor of 100 relative to the open-loop case, let $1 + 10K = 100$, so $K = 9.9$. With this value of K and assuming that $w = 0$, the steady-state value of the output is

$$y_{ss} = \frac{AK}{1 + AK} r_{ss} = \frac{99}{100} 100$$

$$= 99 \text{ rad/sec.}$$

With the disturbance, $y_{ss} = 99 - 0.05 = 98.95$ rad/sec.

◆

From Examples 4.1 and 4.2 we see that while the open-loop speed changed by 5%, due to the disturbance, the closed-loop speed changed by only 0.05%.

4.1.2   Sensitivity: Effects of Gain Changes

As another comparison of open- and closed-loop control, let us suppose that a change in operating temperature causes the motor gain to drift from its original value of A to $A + \delta A$. In the open-loop case the controller gain remains $K = 1/A$, and the new overall system gain would be

$$T_{ol} + \delta T_{ol} = K(A + \delta A) = \frac{1}{A}(A + \delta A) = 1 + \frac{\delta A}{A},$$

where T_{ol} is the open-loop torque. The gain error is

$$\frac{\delta A}{A}.$$

In terms of percent changes, defined as $\delta T/T$, we find that

$$\frac{\delta T_{ol}}{T_{ol}} = \frac{\delta A}{A}. \tag{4.18}$$

Sensitivity

This means that a 10% error in A would yield a 10% error in T_{ol}. H. W. Bode called the ratio of $\delta T/T$ to $\delta A/A$ the **sensitivity** of the gain from r to y_{ss} with respect to A.

Applying the same change in A to the feedback case (Eq. 4.15) yields the new steady-state feedback gain

$$T_{cl} + \delta T_{cl} = \frac{(A + \delta A)K}{1 + (A + \delta A)K},$$

where T_{cl} is the (closed-loop) torque. We see from Eq. (4.18) that for the open-loop case the sensitivity is 1.0. We can compute the sensitivity directly if we use differential calculus. The (closed-loop) steady-state gain is

$$T_{cl} = \frac{AK}{1 + AK}.$$

The first-order variation is proportional to the derivative and is given by

$$\delta T = \frac{dT}{dA} \delta A.$$

For the closed-loop case,

$$\frac{\delta T}{T} = \left(\frac{A}{T} \frac{dT}{dA} \right) \frac{\delta A}{A}$$

$$= \text{(sensitivity)} \frac{\delta A}{A}.$$

From this formula the sensitivity is[2]

$$S_A^{T_{cl}} \triangleq \text{sensitivity of } T_{cl} \text{ with respect to } A$$

$$\triangleq \frac{A}{T_{cl}} \frac{dT_{cl}}{dA},$$

so

$$S_A^{T_{cl}} = \frac{A}{AK/(1 + AK)} \frac{(1 + AK)K - K(AK)}{(1 + AK)^2}$$

$$= \frac{1}{1 + AK}. \qquad (4.19)$$

This result exhibits another major advantage of feedback:

Advantage of feedback

In feedback control, the error in the controlled quantity is less sensitive to variations in the plant gain A by a factor of $1 + AK$ than when open-loop control is used.

Therefore, when the system gain is subject to change, it is still possible to achieve precise control using feedback. This is often very important when the system being controlled—for example, an electric motor or steam engine—is a large, high-powered device whose gain is not only difficult to compute but naturally subject to substantial variation.

2. A factor such as $1 + AK$, for this example, occurs so frequently in feedback design that Bode named it the **return difference** of the feedback path (see Chapter 6).

The superiority of feedback control is equally apparent when we consider errors in the control gain K. Because K always appears as a product with A, the analysis for an error in A applies directly to an error in K.

◆ **EXAMPLE 4.3**   *Open- and Closed-Loop Sensitivity to Parameter Changes*

Compare the sensitivity of the open-loop and closed-loop transfer function to changes in the parameter A for the DC motor of Example 4.1.

Solution.   For the open-loop case, Eq. (4.18) shows that $S_A^{T_{ol}} = 1$.
   For the closed-loop situation we use Eq. (4.19) to find that

$$S_A^{T_{cl}} = \frac{1}{1 + 10(9.9)} = 0.01.$$

Thus, for the closed-loop case a 10% change in motor gain A will cause only a 0.1% change in the steady-state speed. The open-loop controller is 100 times more sensitive to gain changes.

◆

The results in this section so far have been computed for the steady-state in the presence of constant inputs, reference, and disturbance. Very similar results can be obtained for the steady-state behavior in the presence of sinusoidal reference and disturbance signals. This is important because there are times when such signals naturally occur: a disturbance of 60 Hz due to power-line interference in an electronic system, for example. Otherwise the concept is important because more complex time-varying signals can be described as containing signals over a band of frequencies. For example, it is well known that human hearing is restricted to signals in the frequency range of about 60 to 15,000 Hz. A feedback amplifier and loudspeaker system designed for high-fidelity sound must accurately track signals over this range. If we take the controller in Fig. 4.5 to have the transfer function $D(s)$ and we take the process to have the transfer function $G(s)$, then the steady-state open-loop gain at the sinusoidal signal of frequency ω_o will be $|D(j\omega_o)G(j\omega_o)|$, and the error of the

FIGURE 4.5
Simple feedback control block diagram where w = disturbance, v = sensor noise

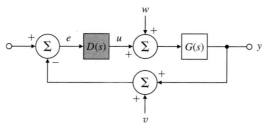

feedback system will be

$$|E(j\omega_o)| = |R(j\omega_o)| \left| \frac{1}{1 + D(j\omega_o)G(j\omega_o)} \right|.$$

Thus, to reduce errors to 1% of the input, we must make $|1 + DG| \geqslant 100$ or $|D(j\omega_o)G(j\omega_o)| \geqslant 100$. In a similar fashion, we can determine the sensitivity of a closed-loop gain $T_{cl}(j\omega_o)$ at frequency ω_o to a parameter such as the amplifier gain K.

4.1.3   Dynamic Tracking

Thus far we have looked at steady-state properties in the presence of a constant reference and a constant disturbance. However, the systems we are interested in are dynamic, and tracking time-varying inputs is also an important role for control. While a constant-gain open-loop controller has no effect on the dynamics of the system for either reference or disturbance inputs, feedback does change the dynamics of the system. If the open-loop controller includes a filter to change the response to the reference, the plant dynamics will still determine the system's response to disturbances. In the case of open-loop speed control, Eq. (4.10) shows that the simple second-order plant dynamics are described by the two time constants τ_1 and τ_2. The dynamics with proportional feedback control are described by Eq. (4.14), and the roots of the characteristic equation of this system are at

$$s_{1,2} = \frac{-(\tau_1 + \tau_2) \pm \sqrt{(\tau_1 + \tau_2)^2 - 4\tau_1\tau_2(1 + AK)}}{2\tau_1\tau_2}.$$

Therefore, the time constants $1/|s_{1,2}|$ change as the feedback gain K is increased. It is often true that closed-loop systems have a faster response as the feedback gain is increased, and if there are no other effects, this is generally desirable. As we will see, however, systems typically also become less well damped and even unstable as the gain increases. Thus a definite limit exists on how high we can

Design tradeoffs between gain and disturbance

make the gain in our efforts to reduce the effects of disturbances and changes in the plant parameters. Attempts to resolve the conflict between small steady-state errors and good transient or dynamic responses have in the past led to introducing proportional (P), integral (I), and derivative (D) terms in feedback controllers, as we will see in Section 4.2.

◆ **EXAMPLE 4.4**   *Open-loop Speed Control: Transient Disturbance Response*
Again consider the servomechanism in Example 4.1 with $\tau_1 = \frac{1}{60}$, $\tau_2 = \frac{1}{600}$, $A = 10$, and $B = 50$, and with the gain set at 0.1. Determine the output for a step load input of $w = -0.1\,\mathrm{N \cdot m}$.

FIGURE 4.6
Transient response to a
disturbance input

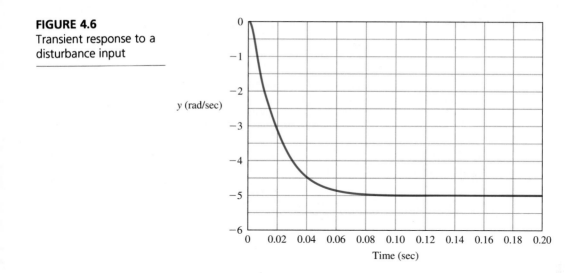

Solution.   For the open-loop system with an armature voltage input v_a and a load torque w, the output is given by Eq. (4.10), with $K = 1/A$, as

$$Y(s) = \frac{10}{(\frac{1}{60}s + 1)(\frac{1}{600}s + 1)} V_a(s) + \frac{50}{(\frac{1}{60}s + 1)(\frac{1}{600}s + 1)} W(s).$$

Now we suppose that the load torque is a step input of $w = -0.1\,\mathrm{N\cdot m}$. Then

$$W(s) = -\frac{0.1}{s},$$

and assuming $r = 0$, we find the output due to the disturbance to be

$$Y(s) = \frac{-5}{(\frac{1}{60}s + 1)(\frac{1}{600}s + 1)s}.$$

The response of the system is shown in Fig. 4.6. Notice that the dynamic response is determined by the two open-loop time constants and that there is a constant offset due to the disturbance input, as we saw earlier.

◆ **EXAMPLE 4.5**   *Closed-loop Proportional Speed Control: Dynamic Response*

For the system described in Example 4.2, set the controller gain for an improvement by a factor of 100 in the system's ability to reject steady-state disturbances relative to that in open-loop control. Determine the transient-response properties of the system.

Solution. The transform of the output due to the disturbance input is

$$Y(s) = \frac{50}{\tau_1 \tau_2 s^2 + (\tau_1 + \tau_2)s + (1 + AK)} W(s)$$

For the required error reduction, we again let $1 + AK = 100$, so $K = 9.9$.
So that,

$$Y(s) = \frac{1800000}{s + 330 \pm 1868.4j} W(s),$$

which corresponds to $\zeta = 0.174$ and $\omega_n = 1897.4\,\text{rad/sec}$. Using the Final Value Theorem for $w = -0.1$, we find that

$$y_{ss} = \lim_{s \to 0} \frac{50s}{\tau_1 \tau_2 s^2 + (\tau_1 + \tau_2)s + (1 + AK)} \frac{(-0.1)}{s}$$

$$= -\frac{5.0}{1 + AK}.$$

With the given value of K, the error is

$$y_{ss} = -0.050.$$

The disturbance response of the system is shown in Fig. 4.7(a).
The transfer function from the reference to the output is[3]

$$T(s) = \frac{Y(s)}{R(s)} = \frac{10K}{\tau_1 \tau_2 s^2 + (\tau_1 + \tau_2)s + 1 + AK} = \frac{3564000}{s + 330 \pm 1868.4j}.$$

3. $T(s)$ will denote the closed-loop transfer function from here on.

FIGURE 4.7
Transient responses to (a) disturbance input and (b) reference input

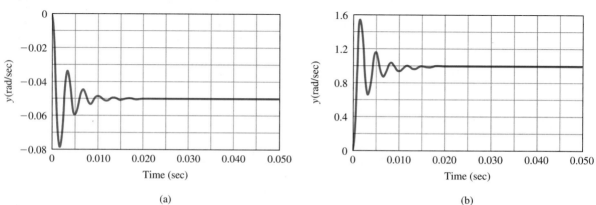

(a)   (b)

Figure 4.7(b) shows the response of the system to a step reference input. Now we notice that dynamic response here is much different from the open-loop case—faster and more poorly damped.

◆

4.2   The Classical Three-term PID Controller

We have seen in Section 4.1 that proportional feedback control can reduce error responses to disturbances but that it still allows a non-zero steady-state error. In addition, proportional feedback increases the speed of response but has a much larger transient overshoot. As we will see in this section, when the controller includes a term proportional to the integral of the error, then the steady-state error can be eliminated, although this comes at the expense of further deterioration in the dynamic response. Finally, addition of a term proportional to the derivative of the error can damp the dynamic response. Combined, these three kinds of control form the classical PID controller, which is widely used in the process industries and whose tuning rules have an interesting history. We will study the buildup of the three-term controller part by part.

The PID (proportional-integral-derivative) controller

4.2.1   Proportional Feedback Control

As we saw in the last section, when the feedback control signal is made to be linearly proportional to the error in the measured output, we call the result **proportional feedback**. This is the case for the feedback (Eq. 4.13) in the controller in Section 4.1. The general form of proportional control is

General form of proportional control

$$u = Ke.$$

Therefore, the controller transfer function $D(s)$ in Fig. 4.5 is

$$D(s) = K. \tag{4.20}$$

We can view the proportional controller as an amplifier with a "knob" to adjust the gain up or down. The system with proportional control may have a steady-state offset (or droop) in response to a constant reference input and may not be entirely capable of rejecting a constant disturbance.

◆ **EXAMPLE 4.6**   *Proportional Speed Control Revisited: Dynamic Closed-loop Poles*

Again consider the DC motor (Eqs. 4.1 and 4.2) with the second-order transfer function,

$$G(s) = \frac{A}{(\tau_1 s + 1)(\tau_2 s + 1)}.$$

Determine the behavior of the closed-loop poles for proportional feedback.

FIGURE 4.8
Locus of roots of Eq.
(4.21) for values of K

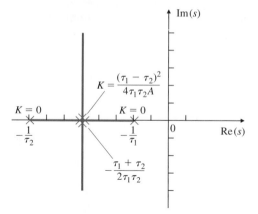

Solution. With proportional feedback the closed-loop characteristic equation in the form $1 + KG$ is

$$\tau_1\tau_2 s^2 + (\tau_1 + \tau_2)s + 1 + AK = 0.$$

Without feedback, $K = 0$, and so the open-loop poles are at

$$s_1 = -\frac{1}{\tau_1}, \qquad s_2 = -\frac{1}{\tau_2}.$$

With feedback the closed-loop poles are a function of the feedback gain K,

$$s_{1,2} = \frac{-(\tau_1 + \tau_2) \pm \sqrt{(\tau_1 + \tau_2)^2 - 4\tau_1\tau_2(1 + AK)}}{2\tau_1\tau_2}, \qquad (4.21)$$

and are plotted in Fig. 4.8 for varying values of K. The figure shows that the roots are at $-1/\tau_1$, $-1/\tau_2$ for the open-loop case ($K = 0$) and move increasingly toward each other for increasing values of K over the range $0 < K < (\tau_1 - \tau_2)^2/4\tau_1\tau_2 A$, which implies a decreasing time constant for the system. For values of $K > (\tau_1 - \tau_2)^2/4\tau_1\tau_2 A$, the real part of the roots does not change with K but the imaginary part increases, and thus according to the analysis of Chapter 3, the rise time of the system decreases but a decreasing damping ratio (and therefore more overshoot) accompanies the faster response.

◆

Limits of proportional feedback

For higher-order systems, large values of the proportional feedback gain will typically lead to instability. For most systems there is an upper limit on the proportional feedback gain in order to achieve a well-damped stable response, and this limit may still have an unacceptable steady-state error.

Increasing the value of K increases the magnitude of DG, which we saw in Section 4.1 to be effective in reducing the errors of the system. Example 4.6 showed that the dynamic response often places a limit on how high K can be made; therefore, there is a limit on how much the errors can be reduced by using proportional feedback only. One way to improve the steady-state accuracy of control without adding extremely high proportional gains is to introduce integral control, which we discuss in the following section.

4.2.2 Proportional-Integral (PI) Feedback Control

The primary reason for integral control is to reduce or eliminate constant steady-state errors, but this benefit typically comes at the cost of worse transient response. Integral feedback has the form

Integral control

$$u(t) = \frac{K}{T_I} \int_{t_0}^{t} e \, d\eta; \tag{4.22}$$

therefore, the $D(s)$ in Fig. 4.5 becomes

$$\frac{U(s)}{E(s)} = D(s) = \frac{K}{T_I s}, \tag{4.23}$$

where T_I is called the **integral**, or **reset time**, and $1/T_I$ is a measure of the speed of response and is referred to as the **reset rate**. T_I is the time for the integrator output to reach $1 \cdot K$ with an input of unity. This feedback has the primary virtue that it can provide a finite value of control signal with no error-signal input e. This comes about because u is a function of all past values of e rather than just the current value, as in the proportional case. Therefore, past errors e "charge up" the integrator to some value that will remain, even if the error becomes zero and stays there. This feature means that disturbances (see Fig. 4.5) can be canceled with zero error because e no longer has to be finite to produce a control that will counteract the (constant) disturbance w. Another way of seeing this is to inspect Fig. 4.5 and note that it is now possible for $D(s)$ to have an infinite gain at $s = 0$, which would yield a zero y response to a constant disturbance w.

Consider again the speed controller discussed in Section 4.1. To see the effects of integral control, let the voltage v_a in Eq. (4.4) be the integral of the error between y and r:

$$v_a = \frac{K}{T_I} \int_{0}^{t} (r - y) \, d\eta. \tag{4.24}$$

The equation of motion with this controller is

$$\tau_1 \tau_2 \ddot{y} + (\tau_1 + \tau_2) \dot{y} + y = A \left[\frac{K}{T_I} \int_{0}^{t} (r - y) \, d\eta \right] + Bw. \tag{4.25}$$

To remove the integral in (4.25), we differentiate once to obtain

$$(\tau_1 \tau_2) \dddot{y} + (\tau_1 + \tau_2) \ddot{y} + \dot{y} = A \frac{K}{T_I} (r - y) + B\dot{w},$$

$$(\tau_1 \tau_2) \dddot{y} + (\tau_1 + \tau_2) \ddot{y} + \dot{y} + A \frac{K}{T_I} y = A \frac{K}{T_I} r + B\dot{w}.$$

We can see from this equation that several limitations of proportional control have been removed. If the load torque is a constant, then $\dot{w} = 0$ and the steady-state response to this class of load disturbance is completely gone. Furthermore, if the reference speed is a constant, then in the steady state we have

$$A \frac{K}{T_I} y = A \frac{K}{T_I} r$$

or

$$y_{ss} = r.$$

Thus, as long as the system remains stable, the motor speed equals the reference speed *regardless* of the value of K. The final concern is with the dynamic response. For this we need to look at the characteristic equation, which is now

$$\tau_1 \tau_2 s^3 + (\tau_1 + \tau_2)s^2 + s + A \frac{K}{T_I} = 0. \tag{4.26}$$

Increasing the gain K/T_I of this system will ultimately result in lightly damped roots for large values of K. The roots of Eq. (4.26) vary as the control gain K is changed and are plotted in Fig. 4.9 as a function of K.[4] Notice that two of the roots move toward the right half-plane (RHP) when K is increased. If the designer wishes to increase the dynamic speed of response with large integral

---

4. This kind of plot, a plot of the roots of the characteristic equation as one parameter is changed, is extremely useful and was developed into a major design tool by W. Evans, who called the plot a **root locus**. We will study Evans' technique in Chapter 5.

FIGURE 4.9
Locus of roots of Eq. (4.26) for values of K

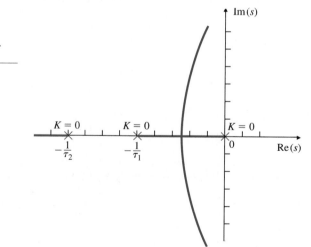

gain, then the response bcomes very oscillatory. A way to avoid this behavior in some cases is to use both proportional and integral control at the same time.

In general, even though integral control improves the steady-state tracking response, it has the effect of slowing down the response if we keep the overshoot unchanged. This characteristic will be discussed in more depth in the following chapters.

With both proportional and integral control, the control voltage for our speed-control problem becomes

Proportional-integral (PI) control for the speed-control example

$$v_a = K \left[r - y + \frac{1}{T_I} \int_0^t (r - y) \, d\eta \right], \tag{4.27}$$

where in this form K is called the **proportional gain**. If we substitute this equation for the voltage into the time-domain version of Eq. (4.9) and differentiate once to remove the integral, the equation of motion becomes

$$\tau_1 \tau_2 \dddot{y} + (\tau_1 + \tau_2)\ddot{y} + \dot{y} = A \left[K(\dot{r} - \dot{y}) + \left(\frac{K}{T_I}\right)(r - y) \right] + B\dot{w}. \tag{4.28}$$

As with integral control, when both the disturbance torque and the reference speed are constant, the steady-state speed is exactly equal to the reference speed regardless of the plant gain A. The characteristic equation corresponding to Eq. (4.28) is

$$\tau_1 \tau_2 s^3 + (\tau_1 + \tau_2)s^2 + (1 + AK)s + \frac{AK}{T_I} = 0.$$

Now it can be seen that by choosing K and T_I the designer can independently set values for the coefficient of s and the constant term and thus has independent control over two of the three terms in the characteristic equation and can provide better transient response than can be done with integral control alone.

4.2.3   Derivative Feedback Control

Derivative feedback (also called **rate feedback**) has the form

Derivative control

$$u(t) = K T_D \dot{e}. \tag{4.29}$$

Therefore the $D(s)$ in Fig. 4.5 becomes

$$D(s) = K T_D s, \tag{4.30}$$

and T_D is called the **derivative time**. It is used in conjunction with proportional and/or integral feedback to increase the damping and generally improve the stability of a system. In practice, pure derivative feedback is not practical to implement for reasons that were discussed in Chapter 2; however, approximations to derivative feedback can be implemented and will be discussed in Chapters 5 and 6. Another reason derivative feedback is not used by itself is

FIGURE 4.10
Anticipatory nature of
derivative control

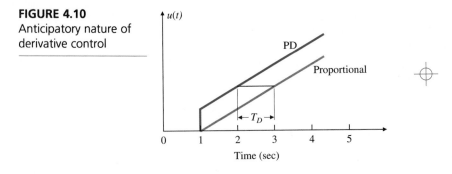

FIGURE 4.11
Alternative ways of configuring rate feedback

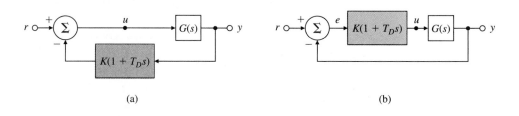

(a)   (b)

that, if $e(t)$ remained constant, the output of a derivative controller would be
zero and a proportional or integral term would be needed to provide a control
signal at this time.

In the derivative control the correction depends on the rate of change of the
error. As a result, a controller with derivative control exhibits an anticipatory
response: Proportional-derivative behavior leads the proportional-only action
by T_D seconds, as shown in Fig. 4.10 for a ramp error $e(t)$.

Derivative control may be introduced into the feedback loop in two ways:
as shown in Fig. 4.11(a), which applies to the tachometer in a DC motor, or as
part of a dynamic compensator in the forward loop, as shown in Fig. 4.11(b). In
both cases the closed-loop characteristic equation is the same but the zeros from
r to y are, of course, different; also, with the derivative in the feedback, the
reference is not differentiated, which may be a desirable result. In Fig. 4.11(b)
there is a closed-loop zero at $-1/T_D$; a step change in the reference input will
cause an initial pulse in the control signal and will result in a faster response (and
higher overshoot) than the system in Fig. 4.11(a).

◆ **EXAMPLE 4.7**   *Derivative Control: DC Motor Revisited*

For the DC motor system of Example 4.6, show the effect of derivative control on the
system roots.

Solution.   The characteristic equation is $1 + DG$, and with $D(s) = KT_D s$ it becomes

$$\tau_1 \tau_2 s^2 + (\tau_1 + \tau_2 + AKT_D)s + 1 = 0.$$

Comparing the left side of this equation to $(s/\omega_n)^2 + 2\zeta(s/\omega_n) + 1$, we see that the derivative term enters with the damping-ratio term and thus has the potential to make ζ larger, indicating the stabilizing effect of derivative control.

◆

4.2.4   Proportional-Integral-Derivative (PID) Control

For control over steady-state and transient errors we can combine all three control strategies we have discussed so far to get **proportional-integral-derivative (PID) control**. Here the control signal is a linear combination of the error, the time integral of the error, and the time rate of change of the error. All three gain constants are adjustable. The PID combination is sometimes able to provide an aceptable degree of error reduction simultaneously with acceptable stability and damping. PID controllers are so effective that PID control is standard in such processing industries as petroleum refining, papermaking, and metalworking. In some cases, just proportional and derivative control are combined to yield **PD** control.

In the generic unity-sensor-gain topology shown in Fig. 4.5 the controller transfer function is given by

The general form for PID control

$$D(s) = K\left(1 + \frac{1}{T_I s} + T_D s\right). \tag{4.31}$$

To design a particular control loop the engineer merely has to adjust the constants K, T_I, and T_D in Eq. (4.31) to arrive at acceptable performance. This adjustment process is called **tuning the controller**. Criteria for tuning are based on the ideas presented in Section 4.1. Increasing K and $1/T_I$ tends to reduce system errors but may not be capable of also producing adequate stability, while increasing T_D tends to improve stability.

For the speed-control problem we have

$$v_a = K\left[r - y + \frac{1}{T_I}\int_0^t (r - y)\,d\eta + T_D(\dot{r} - \dot{y})\right]. \tag{4.32}$$

Taking the derivative of the closed-loop system equation yields

$$\tau_1 \tau_2 \dddot{y} + (\tau_1 + \tau_2)\ddot{y} + \dot{y} = A\left[K(\dot{r} - \dot{y}) + \frac{K}{T_I}(r - y) + KT_D(\ddot{r} - \ddot{y})\right] + B\dot{w}.$$

The characteristic equation is

$$\tau_1 \tau_2 T_I s^3 + T_I[(\tau_1 + \tau_2) + AKT_D]s^2 + T_I(1 + AK)s + AK = 0. \tag{4.33}$$

If we divide through by $\tau_1 \tau_2 T_I$, we have three coefficients and three parameters

(K, T_I, and T_D). Thus, in theory we can set the poles anywhere we like. The addition of derivative control in this system has given us complete control over its dynamics.

◆ **EXAMPLE 4.8**   *PID Control: DC Motor Revisited*

Consider the DC motor discussed in Example 4.1 and let $K = 5$, $T_D = 0.0004$, and $T_I = 0.01$. Discuss the effect of proportional, PI-, and PID control on the response of the system.

Solution.   Figure 4.12(a) illustrates the effects of proportional, PI, and PID feedback on the step disturbance response of the system. Note that adding the integral term, increases the oscillatory behavior but lowers the error, and that adding the derivative term reduces the oscillation while maintaining a low error.

The response of the system to a reference step input is shown in Fig. 4.12(b). It shows the presence of a steady-state offset for proportional control and no steady-state errors for PI or PID control. Note the reduction of the oscillatory behavior due to the addition of the derivative term.

These responses were computed using MATLAB. As an example, for the PI controller the transfer function from the disturbance input to the output is

$$\frac{Y(s)}{W(s)} = \frac{T_I B s}{T_I \tau_1 \tau_2 s^3 + T_I(\tau_1 + \tau_2)s^2 + T_I(1 + AK)s + AK},$$

and from the reference to the output is

$$\frac{Y(s)}{R(s)} = \frac{AK(T_I s + 1)}{T_I \tau_1 \tau_2 s^3 + T_I(\tau_1 + \tau_2)s^2 + T_I(1 + AK)s + AK}.$$

FIGURE 4.12

Transient responses to (a) step disturbance input and (b) step reference input.

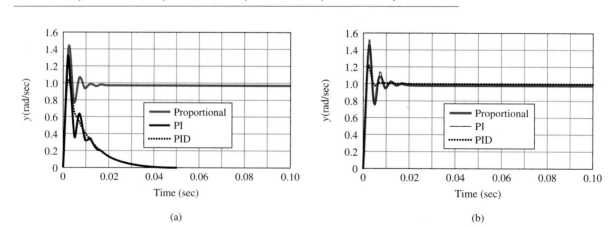

The step responses can be computed by forming the numerator and denominator coefficient vectors (in descending powers of s) and using the step function in MATLAB. For example, the sequence

Step response via MATLAB

```
numG = [TI * B  0];
denG = [TI * TAU1 * TAU2  TI * (TAU1 + TAU2)  TI * (1 + A * K)  A * K],
y = step(numG, denG).
```

produces a plot of y versus time for a disturbance input w.

■ 4.2.5 Time-response Sensitivity to Parameters

In earlier sections we considered the effects of errors on the steady-state performance of a dynamic system and showed how feedback control, especially integral control, can reduce these errors. Since the actual response of the system is a time signal, the sensitivity of the time response to parameter changes can be very useful to explore. Computing and solving an expression for time-response sensitivity is quite straightforward using a tool for computer-aided control systems design (CACSD).

Consider the output response $y(t, \theta)$ that shows the explicit dependence on a parameter θ, which may be a plant or controller parameter. The effect of a perturbation in the parameter, $\delta\theta$, on the nominal response can be determined using the Taylor's series expansion

$$y(t, \theta + \delta\theta) = y(t, \theta) + \frac{\partial y}{\partial \theta} \delta\theta + \cdots. \tag{4.34}$$

For a first-order approximation the parameter perturbation effect is determined by the sensitivity function

$$\delta y(t) = \frac{\partial y}{\partial \theta} \delta\theta. \tag{4.35}$$

An example of generating the sensitivity function

The sensitivity function can be generated from the system itself by using superposition to manipulate the block diagram (see Perkins *et al.*, 1991). As an example, let us consider the feedback system with the input $R(s)$ and the output $Y(s)$ shown in Fig. 4.13. We will assume that the reciprocal of the integral time, $1/T_I$, is the design parameter. We redraw the block diagram for the system as in

FIGURE 4.13
System with a PI controller

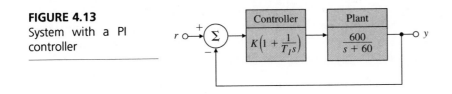

FIGURE 4.14
Manipulating the block diagram of the system in Fig. 4.13: (a) PI controller block
diagram; (b) redrawing of (a); (c) isolation of the parameter θ

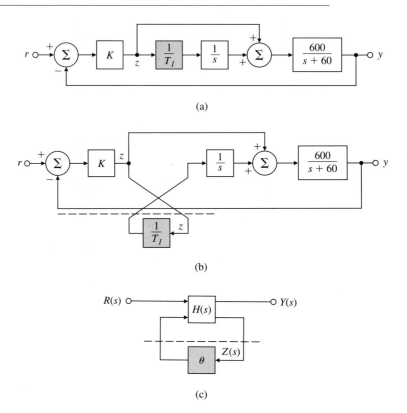

Fig. 4.14(a) and then isolate $\theta = 1/T_I$ as shown in Fig. 4.14(b, c). Next we assume
that the nominal parameter θ is perturbed by the amount $\delta\theta$. This is reflected in
the redrawing of the system in Fig. 4.15(a), where $z(s) + \delta z(s)$ is defined as an
input. This system is equivalent to the one shown in Fig. 4.15(b). The principle of
superposition tells us that if the input $R(s)$ is set to zero, the output will contain
only a component due to the other "input," as shown in Fig. 4.15(c). If we scale
the output by $\delta\theta$ and let $\delta\theta \to 0$, then the output of the diagram is the desired
sensitivity function (Fig. 4.15d). Furthermore, the "input" z can be generated
directly from the nominal plant and connected to the sensitivity model, as
shown in Fig. 4.15(e). Hence the complete sensitivity model (Fig. 4.15e) contains
two copies of the system, and the sensitivity model is driven by the plant signal
taken from the same point as the parameter of interest. The copy of the system
generates the partial derivative of the sensitivity function.

FIGURE 4.15
Further block-diagram manipulation

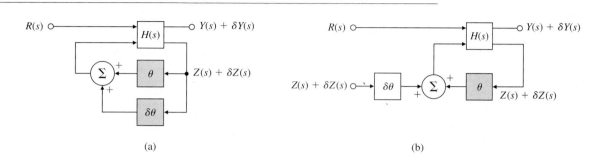

(a)

(b)

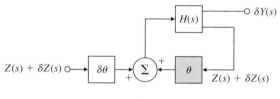

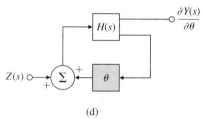

(c)

(d)

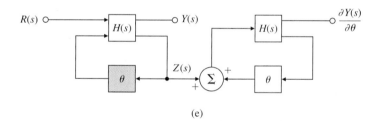

(e)

Parametric changes are normally expressed as percentages of the nominal value, which allows us to scale the sensitivity function in logarithmic form:

Logarithmic sensitivity
function

$$\frac{\partial y(t, \theta)}{\partial \ln \theta} = \frac{\dfrac{\partial y(t, \theta)}{\partial \theta}}{\dfrac{\partial \ln \theta}{\partial \theta}} = \theta \frac{\partial y(t, \theta)}{\partial \theta}. \tag{4.36}$$

This logarithmic sensitivity function represents the change in the output with respect to percentage change in the parameter θ and can be generated as shown in Fig. 4.16. We can use this concept to tune the parameters in a PID controller to achieve improved response, as shown by the following example.

◆ **EXAMPLE 4.9** *PI Control: Time-domain Sensitivity*

Consider the first-order system with the PI controller shown in Fig. 4.16 where the reciprocal of the integral time $1/T_I$ is the design parameter. Find the output sensitivity function with respect to $1/T_I$. Assume the proportional gain $K = 0.18$, and the nominal value of $1/T_I$ is 220.

Solution. The sensitivity function is shown in Fig. 4.17(a), where $H(s)$ (see Fig. 4.16) is the closed-loop transfer function. The logarithmic sensitivity function indicates that an increase in the parameter $1/T_I$ will cause a decrease in the rise time and an increase in the overshoot. Figure 4.17(b) shows the step response when $1/T_I$ is increased from the nominal value to 300. As expected, the overshoot increases.

FIGURE 4.16
Block diagram for sensitivity-function generation

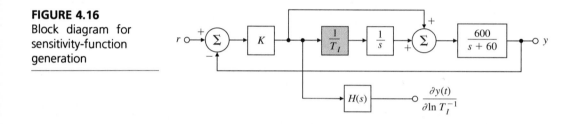

FIGURE 4.17
(a) Sensitivity function; (b) step responses

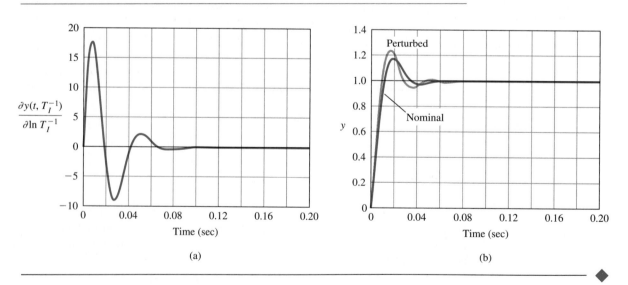

4.2.6   Ziegler–Nichols Tuning of PID Regulators

As we will see in later chapters, sophisticated methods are available to develop a controller that will meet steady-state and transient specifications for both tracking input references and rejecting disturbances. These methods require that control of the process use complete dynamic models in the form of equations of motion or transfer functions.

Callender *et al.* (1936) attempted to select a class of common systems dynamics and widely used PID controllers, and to specify satisfactory values for the controller settings based on estimates of the plant parameters that an operating engineer could make from experiments on the process itself. J. G. Ziegler and N. B. Nichols (1942, 1943) recognized that the step responses of a large number of process control systems exhibit a **process reaction curve** like that shown in Fig. 4.18. This curve can be generated from either experimental data or dynamic simulation of the plant. The *S*-shape of the curve is characteristic of many high-order systems, and such plant transfer functions may be approximated by

Transfer function for a high-order system with a characteristic process reaction curve

$$\frac{Y(s)}{U(s)} = \frac{Ke^{-t_d s}}{\tau s + 1},\qquad (4.37)$$

which is simply a first-order system plus a time delay of t_d seconds. The constants in Eq. (4.37) can be determined from the unit step response of the process. If a tangent is drawn at the inflection point of the reaction curve, then the slope of the line is $R = K/\tau$ and the intersection of the tangent line with the time axis identifies the time delay $L = t_d$.

Ziegler and Nichols gave two methods for tuning the controller for such a model. In the first method the choice of controller parameters is based on a decay ratio of approximately 0.25. This means that a dominant transient decays to a quarter of its value after one period of oscillation, as shown in Fig. 4.19. A

Tuning by decay ratio of 0.25

FIGURE 4.18
Process reaction curve

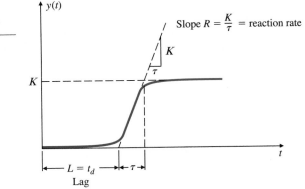

FIGURE 4.19
Quarter decay ratio

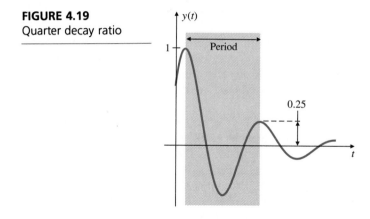

TABLE 4.1

Ziegler–Nichols Tuning for the Regulator $D(s) = K(1 + 1/T_I s + T_D s)$, for a Decay Ratio of 0.25

Type of Controller	Optimum Gain
Proportional	$K = 1/RL$
PI	$\begin{cases} K = 0.9/RL, \\ T_I = L/0.3 \end{cases}$
PID	$\begin{cases} K = 1.2/RL, \\ T_I = 2L, \\ T_D = 0.5L \end{cases}$

quarter decay corresponds to $\zeta = 0.21$ and is a good compromise between quick response and adequate stability margins. The authors simulated the equations for the system on an analog computer and adjusted the controller parameters until the transients showed a decay of 25% in one period. The regulator parameters suggested by Ziegler and Nichols for the common controller terms are shown in Table 4.1.

In the second method the criteria for adjusting the parameters are based on evaluating the system at the limit of stability rather than on taking a step response. We increase the proportional gain until we observe continuous oscillations, that is, until the system becomes marginally stable. The correspond-

Tuning by evaluation at limit of stability (ultimate sensitivity method)

FIGURE 4.20
Determination of the ultimate gain and period

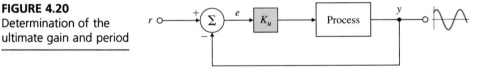

FIGURE 4.21
Marginally stable system

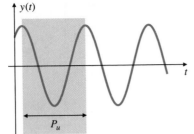

TABLE 4.2

Ziegler–Nichols Tuning for the Regulator $D(s) = K(1 + 1/T_I s + T_D s)$, Based on a Stability Boundary

Type of Controller	Optimum Gain
Proportional	$K = 0.5K_u$
PI	$\begin{cases} K = 0.45K_u \\ T_I = 1/1.2P_u \end{cases}$
PID	$\begin{cases} K = 0.6K_u \\ T_I = \frac{1}{2}P_u \\ T_D = \frac{1}{8}P_u \end{cases}$

ing gain K_u (also called the **ultimate gain**) and the period of oscillation P_u (also called the **ultimate period**) are determined as shown in Figs. 4.20 and 4.21. P_u should be measured when the amplitude of oscillation is quite small. Then we "back off" from this gain, as shown in Table 4.2.

Experience has shown that the controller settings according to Ziegler–Nichols rules provide a good closed-loop response for many systems. The process operator can do the final tuning of the controller iteratively to yield satisfactory control.[5]

5. For a recent revisit, see Hang *et al.* (1990).

◆ **EXAMPLE 4.10**   *Tuning of a Heat Exchanger: Quarter Decay Ratio*

❖   Consider the heat exchanger of Example 2.16. The process reaction curve of this system is shown in Fig. 4.22. Determine proportional and PI regulator gains for the system using the Zeigler–Nichols rules to achieve a quarter decay ratio. Plot the corresponding step responses.

FIGURE 4.22
A measured process
reacton curve

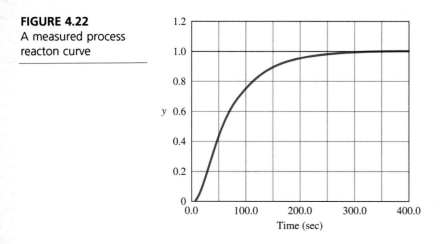

Solution.   From the process reaction curve we measure the maximum slope to be $R \cong \frac{1}{90}$ and the time delay to be $L \cong 13$ sec. According to the Zeigler–Nichols rules the gains are

Proportional: $\qquad\qquad K = \dfrac{1}{RL} = \dfrac{90}{13} = 6.92,$

PI: $\qquad\qquad\qquad K = \dfrac{0.9}{RL} = 6.22 \qquad \text{and} \qquad T_I = \dfrac{L}{0.3} = \dfrac{13}{0.3} = 43.3.$

FIGURE 4.23
Closed-loop step responses

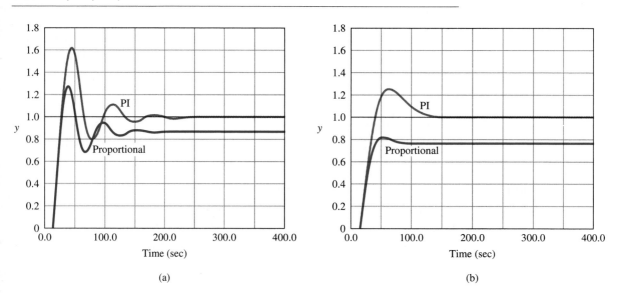

(a)   (b)

Figure 4.23(a) shows the step responses of the closed-loop system to these two regulators. Note that the proportional regulator results in a steady-state offset, while the PI regulator tracks the step exactly in the steady state. Both regulators are rather oscillatory and have considerable overshoot. If we arbitrarily reduce the gain K by a factor of 2 in each case, the overshoot and oscillatory behaviors are substantially reduced, as shown in Fig. 4.23(b). These simulations were carried out using the **SystemBuild** graphical simulation software package.

◆

◆ **EXAMPLE 4.11**   *Tuning of a Heat Exchanger: Oscillatory Behavior*

❖   Proportional feedback was applied to the heat exchanger in the previous example until the system showed nondecaying oscillations in response to a short pulse (impulse) input, as shown in Fig. 4.24. The ultimate gain was $K_u = 15.3$, and the period was measured at $P_u = 42$ sec. Determine the proportional and PI regulators according to the Zeigler–Nichols rules based on oscillatory behavior. Plot the corresponding step responses.

Solution.   The regulators from Table 4.2 are

Proportional: $K = 0.5K_u = 7.65,$

PI: $K = 0.45K_u = 6.885$ and $T_I = \dfrac{1}{1.2}P_u = 35.$

The step responses of the closed-loop system are shown in Fig. 4.25(a). Note that the

FIGURE 4.24
Ultimate period

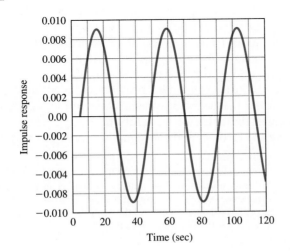

FIGURE 4.25
Closed-loop step response

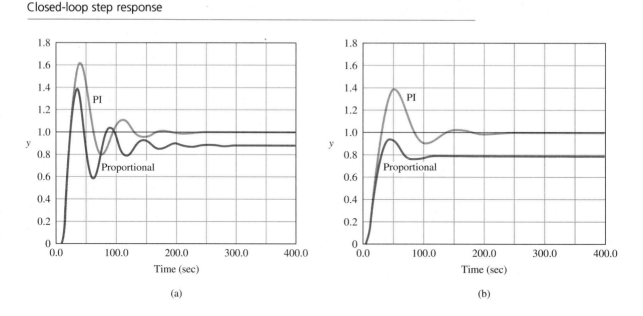

(a)

(b)

responses are similar to those in Example 4.10. If we reduce K by 50%, then the overshoot is substantially reduced, as shown in Fig. 4.25(b). Again, the simulations were carried out using SystemBuild.

4.2.7 Integrator Antiwindup

In almost all systems, actuators saturate because the dynamic range of practical actuators is usually limited; for example, a valve saturates when it is fully open or closed, and the control surfaces in an aircraft cannot be deflected beyond a certain angle from their nominal positions. Whenever control saturation happens, we have to stop integrating with the integral control law; otherwise the integrator will keep integrating, and this charge must be removed later, resulting in substantial overshoot. This problem is called integrator windup.[6]

Consider the feedback system shown in Fig. 4.26. Suppose a large reference step causes the actuator to saturate at u_{max}. Then the integrator keeps integrating the error e, and the signal u_c keeps growing. However, the input to

---

6. In process control, integral control is usually called **reset control** and so integrator windup is referred to as **reset windup**. Without integral control, a given setpoint of, say, 10 results in a response of less value, say 9.9. The operator must then reset to 10.1 to bring the output to the desired value of 10. With integral control the controller automatically brings the output to 10 with a setpoint of 10; hence the integrator does the reset.

FIGURE 4.26
Feedback system with actuator saturation

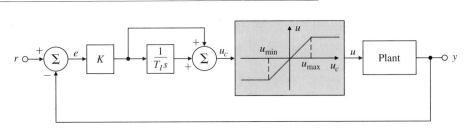

the plant is still at its maximum value, namely, $u = u_{max}$, so the error remains large. The increase in u_c is not helpful since the input to the plant is not changing. The integrator output may become quite large if saturation lasts a long time. It will take a considerable negative error e to bring the integrator output u_c back within the proportional band where the control is not saturated and is ready for subsequent operations.

Example of an antiwindup method

As an example, consider the first-order plant with the PI regulator

$$u = -K \left[e + \frac{1}{T_I} \int_0^t e(t)\, dt \right], \qquad (4.38)$$

and a control that saturates at ± 1. With the system at rest, an input of $r = 1$ will cause e to equal 1, the control to saturate, and the integral to ramp up. The integrator output u_c will grow until the plant output exceeds the reference input so that $e < 0$.

The solution to this problem is **integrator antiwindup**, which "turns off" the integral action as soon as the actuator saturates. (This can be done quite easily if the controller is implemented digitally; see Chapter 8.) Two equivalent anti-windup schemes are shown in Fig. 4.27(a, b) using nonlinearities with a PI regulator. The method in Fig. 4.27(a) is easier to understand, whereas the one in Fig. 4.27(b) is easier to implement. In these schemes, as soon as the actuator saturates, the feedback loop around the integrator moves rapidly to keep the input to the integrator e_1 at zero. During this time the integrator essentially becomes a fast first-order lag. To see this, note that we can redraw the portion of the block diagram in Fig. 4.27(a) from e to u_c as shown in Fig. 4.27(c). The integrator part then becomes the first-order lag shown in Fig. 4.27(d). The slope of the dead-zone nonlinearity K_a should be chosen to be large enough that the antiwindup circuit is capable of following e to keep the output from saturating.

Practical applications of integral control require corresponding use of antiwindup methods.

The effect of the antiwindup is to reduce both the overshoot and the control effort in the feedback system. Implementation of such antiwindup schemes is a necessity in any practical applications of integral control. Omission of this technique may lead to deterioration of response and even instability.

FIGURE 4.27
Integrator antiwindup techniques: (a) PI controller with antiwindup; (b) implementation
of antiwindup with a single nonlinearity; (c) equivalent block diagram during saturation;
(d) first-order lag equivalent of an antiwindup integrator during saturation

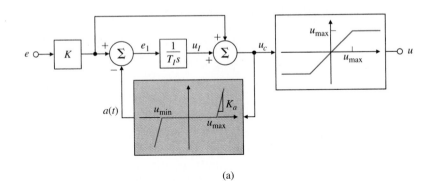

(a)

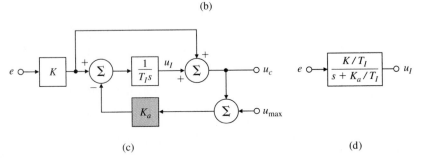

(b)

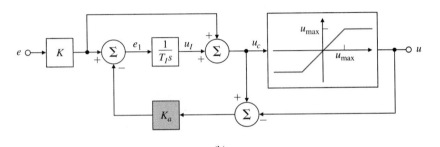

(c)   (d)

◆ **EXAMPLE 4.12**   *Antiwindup Compensation for a PI Controller*

Consider a plant with the following transfer function for small signals,

$$G(s) = \frac{1}{s},$$

and a PI controller,

$$D(s) = K + \frac{K_I}{s} = 2 + \frac{4}{s},$$

in the unity feedback configuration. The input to the plant is limited to ± 1.0. Study the effect of antiwindup on the response of the system.

Solution.   Suppose we use an antiwindup element (dead-zone nonlinearity) with a slope of 10, as shown in Fig. 4.28. Figure 4.29(a) shows the step response of the system with and

FIGURE 4.28
Antiwindup example

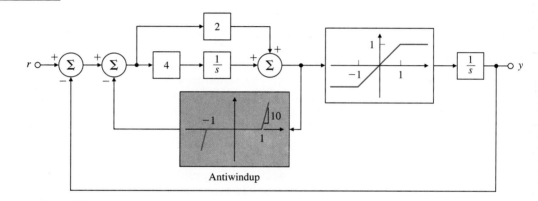

FIGURE 4.29
Integrator antiwindup: (a) step response; (b) control effort

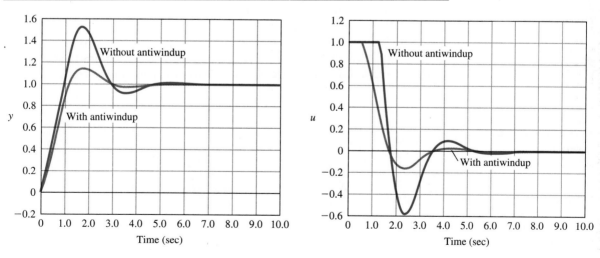

(a)   (b)

without the antiwindup element. Figure 4.29(b) shows the corresponding control effort. Note that the system with antiwindup has substantially less overshoot and less control effort.

◆

4.3   Steady-state Tracking and System Type

In the speed-control case study in Section 4.1 we considered constant reference inputs and constant disturbances. We found in Section 4.2 that for a system with such signals, integral control could keep the steady-state error at zero even when the motor gain differed from the one used in the design. In a number of important cases the reference input will not be constant but can be approximated as a linear function of time for a time span long enough for the system to reach steady-state; we would like to know what the steady-state error is in this case. For example, when an antenna is tracking the elevation angle to a satellite, the time history as the satellite approaches overhead is the S-shaped curve shown in Fig. 4.30. This signal may be approximated by a ramp function for a large portion of the signal relative to the speed of response of the servomechanism. The change in set points in many process control problems is the change from one constant reference value to another and may be represented by step inputs. In the position control of elevators, a ramp function with a constant velocity allows the elevator to go from one floor to another.

The general method is to approximate the input as a polynomial in time and then consider the steady-state tracking errors that result for polynomials of different degrees. As we will see, the error is zero for input polynomials up to a certain degree, a nonzero constant for that degree of polynomial, and then unbounded for higher degrees. Stable systems can be classified into **system types** according to the polynomial degree for which the error is constant. System types can be defined with regard to reference inputs or disturbance inputs, and in this section we will consider both classifications. The system type is very dependent on the point at which the reference or disturbance signals enter into the feedback system. Determining the system type involves calculating a certain transfer

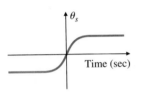

System type is classified by the degree of input polynomial for which the steady-state tracking error is constant.

FIGURE 4.31
Typical single-loop system

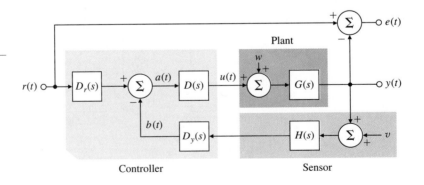

function (for example, from reference input to system error or from disturbance input to system error) and then applying the Final Value Theorem. As we will see, determination of system type is easiest for the special (and most common) case of unity feedback.

To set the problem in a general context, we consider the single-loop feedback system shown in Fig. 4.31, where

$r(t)$ = reference input,

$u(t)$ = plant input (control signal),

$y(t)$ = plant output,

$e(t)$ = system error = $r(t) - y(t)$,

$a(t)$ = the actuating signal,

$b(t)$ = output of feedback part of dynamic compensation,

$G(s)$ = plant transfer function including actuator,

$H(s), D_y(s)$ = feedback components of sensor transfer function and dynamic compensation, respectively,

$D(s)$ = controller transfer function,

$D_r(s)$ = input signal processing or prefiltering;

v, w = sensor and disturbance noise, respectively.

The feedback transfer function $H(s)D_y(s)$ typically represents the sensor action to convert the output $y(t)$ to an electrical output signal $b(t)$. Likewise, we often require a transfer function $D_r(s)$ to convert the reference input (such as speed or temperature) into an electrical signal that combines with $b(t)$ in the controller to generate the actuating signal $a(t)$. The controller with transfer function $D(s)$ converts the electrical actuating signal into the control signal $u(t)$. HD_y and D_r usually have the same dimensions (such as volts per 1000 rpm or volts per degree Celsius), and we can assume that D_r is not zero. Using this information we can simplify Fig. 4.31 to obtain the structure shown in Fig. 4.32.

Now let us assume that the reference input is a polynomial of degree k and compute the steady-state error for this general system in terms of the given

FIGURE 4.32
Simplified block diagram of a typical single-loop system

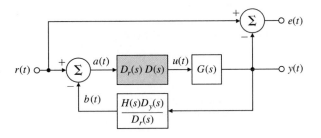

transfer functions. If the reference input is

$$r(t) = \frac{t^k}{k!} 1(t),$$

then the transform of the input is

$$R(s) = \frac{1}{s^{k+1}}.$$

If $k = 0$, the input is a step function of unit amplitude; if $k = 1$, the input is a ramp function with a unit slope; if $k = 2$, the input is a parabola with a unit second derivative, and so on. These inputs are called position, velocity, and acceleration inputs, respectively. In order to compute the steady-state errors, we need the transfer function from input to system error and then we can apply the Final Value Theorem. From Fig. 4.32 we see that the transfer function $T(s)$ from R to Y is

$$\frac{Y(s)}{R(s)} = T(s) = \frac{D_r DG}{1 + HD_y DG}, \tag{4.39}$$

and the error is

$$E(s) = R(s) - Y(s) = R(s) - T(s)R(s).$$

The reference-to-error transfer function is thus

$$\frac{E(s)}{R(s)} = 1 - T(s),$$

and the system error transform is

$$E(s) = [1 - T(s)]R(s).$$

We assume that the conditions of the Final Value Theorem are satisfied, namely that all poles of $sE(s)$ are in the left half-plane. In that case the steady-state error is given by applying the Final Value Theorem to get

$$e_{ss} = \lim_{t \to \infty} e(t) = \lim_{s \to 0} sE(s) = \lim_{s \to 0} s[1 - T(s)]R(s). \tag{4.40}$$

With the test input the error transform is

$$E(s) = \frac{1}{s^{k+1}} [1 - T(s)],$$

and the steady-state error is given again by the Final Value Theorem:

$$e_{ss} = \lim_{s \to 0} s \frac{1 - T(s)}{s^{k+1}} = \lim_{s \to 0} \frac{1 - T(s)}{s^k}. \tag{4.41}$$

The result of evaluating the limit in Eq. (4.41) can be zero, a nonzero constant, or infinite. If the solution to Eq. (4.41) is a nonzero constant, the system is referred

Type 0 and type I systems to as **type** k. For example, if $k = 0$, and the solution to Eq. (4.41) is a constant then the system is type 0. In that case, the system has a steady-state offset for a step reference input. Similarly, if $k = 1$ and the solution to Eq. (4.41) is a constant, then the system is type I. This means that it has a zero steady-state error to a step input and a constant steady-state error due to a ramp reference input. Type I systems are by far the most common in practice.

It is important to note that a system of type I or higher has a closed-loop DC gain of 1, which means that $T(0) = 1$. Also apparent from Eq. (4.41) is that, for the steady-state error to be zero, with a constant input, $T(s)$ should approach unity as s approaches zero.

4.3.1   A Special Case of System Type: Unity Feedback

A further simplification of Fig. 4.32 is possible if $HD_y = 1$, a case that is quite common. We can then model the system as the unity feedback system shown in Fig. 4.33 and take advantage of the fact that now the actuating signal is the error, that is, $a(t) = e(t)$.[7] In Fig. 4.33 we have defined $G_o = D_r DG$ as the overall feedforward or open-loop transfer function. The analysis that follows applies only to the special case of unity feedback. (Recall from Chapter 3 that it is possible to transform any nonunity feedback system into an equivalent unity feedback system using Fig. 3.2c).

If our system is described by Fig. 4.33, then (with $D_r(s) = 1$)

$$1 - T(s) = \frac{1}{1 + G_o(s)},$$

and the system error is

$$E(s) = \frac{1}{1 + G_o(s)} R(s). \tag{4.42}$$

Using the Final Value Theorem we find the system error for the test input to be

$$e_{ss} = \lim_{s \to 0} s \frac{1}{[1 + G_o(s)]s^{k+1}} = \lim_{s \to 0} \frac{1}{[1 + G_o(s)]s^k}. \tag{4.43}$$

If the system is type 0 (having a constant steady-state error for a step-input

---

7. The authors of some textbooks treat the actuating signal $a(t)$ as if it were the system error even in the general (nonunity) feedback case. Take care to distinguish unity from nonunity feedback regarding this point.

FIGURE 4.33
Unity feedback system

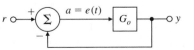

reference signal), then we set $k = 0$ and the error is

$$e_{ss} = \frac{1}{1 + G_o(0)} = \frac{1}{1 + K_p}. \qquad (4.44)$$

Position-error constant

The constant K_p in Eq. (4.44) is called the **position-error constant** because when the signal being controlled is mechanical motion and $k = 0$, the reference is a position and so the error is a position error. The form of Eq. (4.44) leads to the simple computation

$$K_p = \lim_{s \to 0} G_o(s). \qquad (4.45)$$

In other words, if the loop is unity feedback and the system is type 0, then the position-error constant is the zero-frequency or DC gain of the open-loop transfer function.

For ramp inputs, Eq. (4.43) with $k = 1$ results in

$$e_{ss} = \lim_{s \to 0} \frac{1}{[1 + G_o(s)]s} = \lim_{s \to 0} \frac{1}{sG_o(s)}.$$

For this error to exist, $G_o(s)$ must have at least one pole at $s = 0$. Suppose it has exactly one; then the system is type I and the error is, by definition,

$$e_{ss} = \frac{1}{K_v},$$

where the error constant K_v is the **velocity constant** and with unity feedback is given by

Velocity constant

$$K_v = \lim_{s \to 0} sG_o(s). \qquad (4.46)$$

Type II systems

The relationship between K_v and the steady-state error to a ramp input is shown in Fig. 4.34. Similarly, for type II systems in the unity feedback case, $G_o(s)$ has

FIGURE 4.34
Relationship between
ramp response and K_v

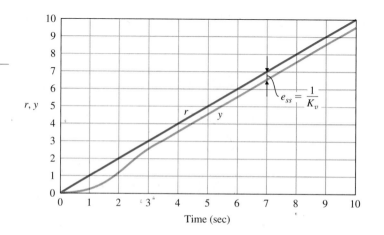

TABLE 4.3   Errors According to System Type for Unity Feedback

	Step	Ramp	Parabola
Type 0	$1/(1 + K_p)$	∞	∞
Type I	0	$1/K_v$	∞
Type II	0	0	$1/K_a$

two poles at $s = 0$. Here the error is defined as

$$e_{ss} = \frac{1}{K_a},$$

and the **acceleration constant** K_a can be computed in the unity-feedback-gain case as

$$K_a = \lim_{s \to 0} s^2 G_o(s) \tag{4.47}$$

Table 4.3 summarizes the three system types. In general, the error constants must be computed from Eq. (4.41), but in the unity-feedback-gain case they are given by

$$K_p = \lim_{s \to 0} G_o(s).$$

$$K_v = \lim_{s \to 0} s G_o(s).$$

$$K_a = \lim_{s \to 0} s^2 G_o(s).$$

The definition of system type helps us to identify quickly the ability of a system to track polynomials. In the unity feedback structure, if the process gain changes in a type I system, the velocity-error constant changes, but the system will still have zero steady-state error in response to a constant input and will still be type I. The same is true for type II or higher. Thus we can say that system type is a **robust property** in the unity feedback structure. Robustness is the major reason for preferring unity feedback over other kinds of control. System type can also be defined with respect to disturbance inputs, as we shall investigate next.

Robustness of system type

◆ **EXAMPLE 4.13**   *System Type for a Satellite Antenna*

Consider the satellite-antenna tracking problem represented by a nonunity feedback system with the plant transfer function

$$G(s) = \frac{1}{s(1 + s\tau)},$$

where in the feedback loop $H(s) = h$, with $h > 0$. Determine the system type with respect to reference inputs.

Solution.   The system error is

$$E(s) = R(s) - Y(s)$$

$$= R(s) - T(s)R(s)$$

$$= R(s) - \frac{G(s)}{1 + hG(s)} R(s)$$

$$= \frac{1 + (h-1)G(s)}{1 + hG(s)} R(s).$$

The steady-state system error from Eq. (4.41) is

$$e_{ss} = \lim_{s \to 0} sR(s)[1 - T(s)].$$

For a step reference input, $R(s) = 1/s$ and hence

$$e_{ss} = \lim_{s \to 0} [1 - T(s)] = \lim_{s \to 0} \frac{s(1 + s\tau) + h - 1}{s(1 + s\tau) + h}$$

$$= \frac{h - 1}{h}.$$

Therefore, if the system is not unity feedback, the system is type 0 in spite of the fact that the plant has a pure integrator. If the system is unity feedback, then $h = 1$, in which case $e_{ss} = 0$ and the system is type I.

◆

4.3.2   System Type with Respect to Disturbance Inputs

In most control systems, disturbances of one type or another exist. In practice, these disturbances can sometimes be usefully approximated by polynomial time functions such as steps or ramps. This would suggest that system type can also be defined with respect to the system's ability to reject disturbance inputs in a way analogous to the classification scheme based on reference inputs. System type with regard to disturbance inputs specifies the degree of the polynomial expressing those input disturbances that the system can reject in the steady-state. Knowing the system type, we can then determine the response of the system to disturbance inputs such as step or ramp signals. How the system will respond depends on where the disturbance enters in the control system.

To determine the system type, we form the transfer function from the disturbance input $w(t)$ to the output $y(t)$:

$$\frac{Y(s)}{W(s)} = T_w(s). \tag{4.48}$$

Error constants

Equation (4.48) determines the system error resulting from a disturbance input. The system is type 0 if a step input disturbance results in a constant steady-state output error. The system is type I if, for a ramp input disturbance, the steady-state value of the output is a constant:

$$y_{ss} = \lim_{s \to 0} \left[s T_w(s) \frac{1}{s^2} \right] = \text{constant}.$$

This means that the steady-state value of the output for a step input disturbance is zero:

$$y_{ss} = \lim_{s \to 0} \left[s T_w(s) \frac{1}{s} \right] = 0.$$

Higher-order system types (such as type II) may be defined in a similar fashion. For the special case of unity feedback, any integrator(s) must precede the point at which the disturbance enters the system to reject the disturbance in the steady-state. This will make the system type I (or higher) with respect to the disturbances. In a system that is type I (or higher), the transfer function from w to y will have DC gain of zero; that is,

$$T_w(0) = 0,$$

a fact that is also apparent from earlier equations.

◆ **EXAMPLE 4.14**   *System Type for a DC Motor*

Consider the simplified model of the DC motor employing proportional feedback as shown in Fig. 4.35(a). Using the parameter values

$$\tau = 1, \qquad A = 1, \qquad K_I = 1,$$

determine the system type and properties with respect to (a) step reference, (b) step disturbance, and (c) ramp reference inputs.

FIGURE 4.35
DC motor with (a) proportional feedback and (b) unity feedback

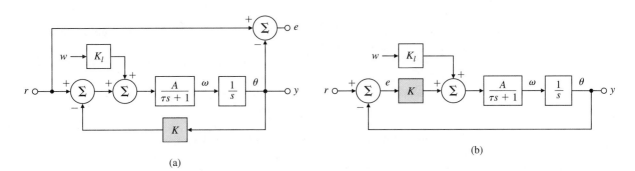

Solution.   (a) The closed-loop transfer function from r to y (where $w = 0$) is

$$T(s) = \frac{1}{s(s + 1) + K} .$$

The system error is

$$E(s) = R(s)\left[1 - \frac{Y(s)}{R(s)}\right] = [1 - T(s)]R(s)$$

$$= \frac{s(s + 1) + K - 1}{s(s + 1) + K} R(s).$$

The system error due to a unit-step input $R(s) = 1/s$ is given by

$$E(s) = \frac{1 - T(s)}{s} .$$

Using the Final Value Theorem, we get

$$e_{ss} = \lim_{s \to 0} s \frac{1 - T(s)}{s} = \frac{K - 1}{K} .$$

This equation tells us that the feedback system is not capable of following a step reterence input with zero steady-state error unless $K = 1$. As we saw in Example 4.13, this happens because nonunity feedback prevents the integrator from achieving zero steady-state error to a constant input. We can clarify this situation by implementing the unity-feedback configuration shown in Fig. 4.35(b). By inspection of this figure we see that the system is type I with respect to the reference input. Now the transfer function and system error are

$$T(s) = \frac{K}{s(s + 1) + K} ,$$

$$E(s) = \frac{s(s + 1)}{s(s + 1) + K} R(s).$$

For a step reference input, $R(s) = 1/s$, so

$$e_{ss} = 0,$$

and hence

$$y_{ss} = r_{ss}.$$

This means that unity feedback provides zero error to a step reference input without help from a dynamic compensator.

(b) For both setups in Fig. 4.35, the transfer function from the disturbance to the output is

$$\frac{Y(s)}{W(s)} = T_w(s) = \frac{1}{s(s + 1) + K} .$$

For a unit step disturbance input the steady-state output will be

$$y_{ss} = \frac{1}{K} .$$

Thus the system is incapable of rejecting the disturbance completely, so there is a steady-state offset. The integrator in the plant cannot counter the offset because the disturbance enters the system before the integrator. The system is type 0 with respect to this disturbance input.

We may probe a little deeper into the reason for the tracking properties of the system. Note that the transfer function from r to e in Fig. 4.35(a) (for $K = 1$) and in Fig. 4.35(b) (regardless of the value of K) is

$$\frac{E(s)}{R(s)} = \frac{s(s + 1)}{s(s + 1) + K},$$

Blocking zero

which has a zero at the origin. This **blocking zero** is what determines the tracking properties. Note that no such blocking zero exists in the transfer function from the disturbance to the output. Hence the system is not capable of rejecting the constant disturbances exactly.

(c) Let us now investigate the tracking properties of the system with respect to a ramp reference input signal. For the system in Fig. 4.35(a),

$$e_{ss} = \lim_{s \to 0} s \, \frac{s(s + 1) + K - 1}{s(s + 1) + K} \, \frac{1}{s^2},$$

which will be unbounded unless $K = 1$, that is, when there is unity feedback. In that case

$$e_{ss} = \frac{1}{K} = 1.$$

The system in Fig. 4.35(b) will have the same steady-state error to a ramp input.

◆

◆ **EXAMPLE 4.15**   *Satellite Attitude Control*

Consider the system shown in Fig. 4.36(a) where

J = moment of inertia,

w = disturbance torque,

K = sensor gain,

$D(s)$ = the compensator.

If the compensator is proportional-derivative (PD) feedback of the form $D(s) = 1 + T_D s$, then the system can be redrawn as Fig. 4.36(b). Note that the rule for conversion to unity feedback (Fig. 3.2c) allows this. Figure 4.36(c) shows the same system but with PID rather than PD control. Determine the ability of the systems in Fig. 4.36(b, c) to respond to reference inputs and to reject disturbance.

Solution.   We see from inspection of Fig. 4.36(b) that the system is type II with respect to reference inputs. The tracking error for a ramp reference input is

$$e_{ss} = \lim_{s \to 0} s \left[\frac{J s^2}{J s^2 + K(1 + T_D s)} \right],$$

FIGURE 4.36
System type for satellite attitude control: (a) satellite attitude control system with
dynamic compensator; (b) PD control; (c) PID control

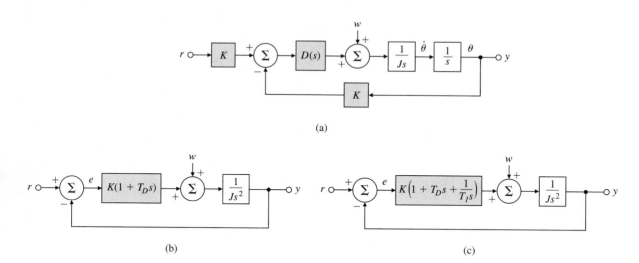

(a)

(b)   (c)

which is zero. For a parabolic reference input, where $R(s) = 1/s^3$, there will be the
constant error

$$e_{ss} = \frac{J}{K}.$$

Note the presence of the two blocking zeros at the origin.

For PD control, the system has a finite error for a constant disturbance torque.

$$y_{ss} = \lim_{s \to 0} \left[s \frac{1}{Js^2 + K(1 + T_D s)} \frac{1}{s} \right] = \frac{1}{K},$$

and is thus type 0 for disturbances.

If the compensator were a PID controller, then

$$D(s) = K\left(1 + T_D s + \frac{1}{T_I s}\right).$$

Now the system has zero steady-state error for reference step, ramp, and parabolic inputs.
Furthermore, this system has no error to a constant disturbance because

$$y_{ss} = \lim_{s \to 0} \left[s \frac{T_I s}{JT_I s^3 + K(T_D s^2 + T_I s + 1)} \frac{1}{s} \right] = 0,$$

and the integrator from the compensator creates a blocking zero from the disturbance to
the output. Therefore, the system is type III for reference inputs and type I for disturbance
inputs.

4.3.3   Truxal's Formula

In this chapter we have derived formulas for the error constants in terms of the system transfer function. The most common case is the type I system whose error constant is K_v, the velocity error constant. Truxal (1955) derived a formula for the velocity constant in terms of the closed-loop poles and zeros, a formula that connects the steady-state error to the dynamic response. Since control design often requires a tradeoff between these two characteristics, Truxal's formula can be useful to know. Its derivation is quite direct. Suppose the closed-loop transfer function $T(s)$ of a type I system is

$$T(s) = K \frac{(s - z_1)(s - z_2)\cdots(s - z_m)}{(s - p_1)(s - p_2)\cdots(s - p_n)}. \tag{4.49}$$

Since the steady-state error in response to a step input in a type I system is zero, the DC gain is unity; thus

$$T(0) = 1. \tag{4.50}$$

The system error is given by

$$E(s) \triangleq R(s) - Y(s) = R(s)\left[1 - \frac{Y(s)}{R(s)}\right] = R(s)[1 - T(s)]. \tag{4.51}$$

The system error due to a unit ramp input is given by

$$E(s) = \frac{1 - T(s)}{s^2}. \tag{4.52}$$

Using the Final Value Theorem, we get

$$e_{ss} = \lim_{s \to 0} \frac{1 - T(s)}{s}. \tag{4.53}$$

Using L'Hôpital's rule we rewrite Eq. (4.53) as

$$e_{ss} = -\lim_{s \to 0} \frac{dT}{ds} \tag{4.54}$$

or

$$e_{ss} = -\lim_{s \to 0} \frac{dT}{ds} = \frac{1}{K_v}. \tag{4.55}$$

Equation (4.55) implies that $1/K_v$ is related to the slope of the transfer function at the origin. Using Eq. (4.50) we can rewrite Eq. (4.55) as

$$e_{ss} = -\lim_{s \to 0} \frac{dT}{ds} \frac{1}{T} \tag{4.56}$$

or

$$e_{ss} = -\lim_{s \to 0} \frac{d}{ds}[\ln T(s)]. \tag{4.57}$$

Substituting Eq. (4.49) into Eq. (4.57), we get

$$e_{ss} = -\lim_{s \to 0} \frac{d}{ds} \left\{ \ln \left[K \frac{\prod\limits_{i=1}^{m} (s - z_i)}{\prod\limits_{i=1}^{n} (s - p_i)} \right] \right\} \tag{4.58}$$

$$= -\lim_{s \to 0} \frac{d}{ds} \left[K + \sum_{i=1}^{m} \ln(s - z_i) - \sum_{i=1}^{n} \ln(s - p_i) \right], \tag{4.59}$$

or

Truxal's formula

$$\frac{1}{K_v} = -\left. \frac{d \ln T}{ds} \right|_{s=0} = \sum_{i=1}^{n} -\frac{1}{p_i} + \sum_{i=1}^{m} \frac{1}{z_i}. \tag{4.60}$$

We observe from Eq. (4.60) that K_v increases as the closed-loop poles move away from the origin. We will use Truxal's formula in Chapter 7 in our discussion of dynamic compensator design. Similar relationships exist for other error coefficients, and these are explored in the problems.

◆ **EXAMPLE 4.16**   *Truxal's Formula*

A third order type I system has closed-loop poles at $-2 \pm 2j$ and -0.1. The system has only one closed-loop zero. Where should the zero be if a $K_v = 10$ is desired?

Solution.   From Truxal's formula we have,

$$\frac{1}{K_v} = -\frac{1}{-2 + 2j} - \frac{1}{-2 - 2j} - \frac{1}{-0.1} + \frac{1}{z}$$

or

$$0.1 = 0.5 + 10 + \frac{1}{z}.$$

Therefore, the closed-loop zero should be at $z = -0.1$.

◆

4.4   Stability

A system that always gives responses appropriate to the stimulus is considered to be stable. In testing the stability of a dynamic system we must carefully define the words *response*, *stimulus*, and *appropriate*. For example, it is possible for a system to have a well-behaved output while an internal variable is growing without bound. If we consider only the output as the response, then we might

Definitions of stability can be varying and subjective.

call such a system stable; however, if we consider the internal variable to be the response, then we might just as reasonably call this system unstable. The stimulus might be a signal persisting for all time, or it might be only a set of initial conditions. Finally, we might require that the response not grow or that it go to zero if the stimulus is removed. For systems that are nonlinear or time-varying, the number of definitions of stability is large, and some of the definitions can be quite complicated in order to account for the varieties of responses the systems can display. Here we will describe a few elementary results that are useful in the study of linear constant systems. Later in the text we will cover other aspects of stability, such as Nyquist's frequency-response stability test (Chapter 6) and Lyapunov stability (Chapter 7).

4.4.1 Bounded Input–Bounded Output Stability

A system is said to have **bounded input–bounded output (BIBO) stability** if every bounded input results in a bounded output (regardless of what goes on inside the system). A test for this property is readily available when the system response is given by convolution. If the system has input $u(t)$, output $y(t)$, and impulse response $h(t)$, then

$$y(t) = \int_{-\infty}^{\infty} h(\tau)u(t - \tau)\, d\tau. \tag{4.61}$$

If $u(t)$ is bounded, then there is a constant M such that $|u| \leqslant M < \infty$, and the output is bounded by

$$|y| = \left| \int hu\, d\tau \right|$$

$$\leqslant \int |h|\,|u|\, d\tau$$

$$\leqslant M \int_{-\infty}^{\infty} |h(\tau)|\, d\tau.$$

Thus the output will be bounded if $\int_{-\infty}^{\infty} |h|\, d\tau$ is bounded.

On the other hand, suppose the integral is not bounded, and the bounded input $u(t - \tau) = +1$ if $h(\tau) > 0$ and $u(t - \tau) = -1$ if $h(\tau) < 0$. In this case,

$$y(t) = \int_{-\infty}^{\infty} |h(\tau)|\, d\tau, \tag{4.62}$$

and the output is not bounded. We conclude that

Mathematical definition of BIBO stability

The system with impulse response $h(t)$ is **BIBO**-stable if and only if the integral

$$\int_{-\infty}^{\infty} |h(\tau)|\, d\tau < \infty.$$

FIGURE 4.37

Capacitor driven by
current source

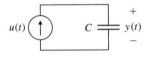

As an example, consider the capacitor driven by a current source sketched in Fig. 4.37. The capacitor voltage is the output, and the current is the input. The impulse response of this setup is $h(t) = 1(t)$, the unit step. Now for this response,

$$\int_{-\infty}^{\infty} |h(\tau)| \, d\tau = \int_{0}^{\infty} d\tau \qquad (4.63)$$

is not bounded. The capacitor is not BIBO-stable. Notice that the transfer function of the system is $1/s$ and has a pole on the imaginary axis. Physically we can see that constant input current will cause the voltage to grow, and thus the system response is neither bounded nor stable. In general, if a constant system has any pole on the imaginary axis or in the right half-plane (RHP), the response will not be BIBO-stable; if every pole is inside the left half-plane (LHP), then the response will be BIBO-stable. Thus for these systems, pole locations of the transfer function can be used to check for stability.

*Determination of BIBO
stability by pole location*

An alternative to computing the integral of the impulse response or even to locating the roots of the characteristic equation is given by Routh's stability criterion, which we will discuss in Section 4.4.3.

4.4.2   Stability of Constant Systems

A concept of stability due to Lyapunov is that the output and all the internal variables never become unbounded and go to zero as time goes to infinity for sufficiently small initial conditions. Consider the linear constant system whose characteristic equation is

$$s^n + a_1 s^{n-1} + a_2 s^{n-2} + \cdots + a_n = 0. \qquad (4.64)$$

Assume that the roots $\{p_i\}$ of the characteristic equation are real or complex but are distinct. Note that Eq. (4.64) shows up as the denominator in the transfer function for the system

$$T(s) = \frac{Y(s)}{R(s)} = \frac{b_0 s^m + b_1 s^{m-1} + \cdots + b_m}{s^n + a_1 s^{n-1} + \cdots + a_n}$$

$$= \frac{K \prod_{i=1}^{m} (s - z_i)}{\prod_{i=1}^{n} (s - p_i)}, \qquad m \leqslant n. \qquad (4.65)$$

The solution to the differential equation whose characteristic equation is given by Eq. (4.64) may be written as

$$y(t) = \sum_{i=1}^{n} K_i e^{p_i t} \qquad (4.66)$$

where $\{p_i\}$ are the roots to Eq. (4.64) and $\{K_i\}$ depend on the initial conditions.

The system is stable if and only if every term in Eq. (4.66) goes to zero as $t \to \infty$:

$$e^{p_i t} \to 0 \qquad \text{for all } p_i.$$

This will happen if all the poles of the system are strictly in the LHP, where

$$\text{Re}\{p_i\} < 0. \tag{4.67}$$

Asymptotic internal stability occurs when all poles are strictly in the LHP

This is called **asymptotic internal stability**. Therefore, the stability of a system can be determined by computing the location of the roots to the characteristic equation and determining whether they are all in the LHP. CACSD software makes this computation easy.

4.4.3 Routh's Stability Criterion

There are several methods of obtaining information about the locations of the roots of a polynomial without actually solving for the roots. These methods were especially useful before the availability of CACSD software. They are still useful for determining the ranges of coefficients of polynomials for stability, especially when the coefficients are in symbolic (nonnumerical) form. Consider the characteristic equation of an nth-order system:[8]

$$a(s) = s^n + a_1 s^{n-1} + a_2 s^{n-2} + \cdots + a_{n-1} s + a_n. \tag{4.68}$$

It is possible to make certain statements about the stability of the system without actually solving for the roots of the polynomial. This is a classical problem, and several methods exist for the solution.

A necessary condition for Routh stability

A *necessary condition for stability* of the system is that all of the roots of Eq. (4.68) have negative real parts, which in turn requires that all the $\{a_i\}$ be positive.[9] If any of the coefficients are missing (are zero) or are negative, then the system will have poles located outside the LHP. This condition can be checked by inspection. Once the elementary necessary conditions have been satisfied, we need a more powerful test. Equivalent tests were independently proposed by Routh in 1874 and Hurwitz in 1895; we will discuss the former version. Routh's formulation requires the computation of a triangular array that is a function of the $\{a_i\}$. He showed that a *necessary and sufficient condition for stability* is that all of the elements in the first column of this array be positive.

A necessary and sufficient condition for stability

To determine the Routh array, we first arrange the coefficients of the characteristic polynomial in two rows, beginning with the first and second coefficients and followed by the even-numbered and odd-numbered coefficients:

Routh array

$$
\begin{array}{llll}
s^n & : & 1 & a_2 & a_4 & \cdots \\
s^{n-1} & : & a_1 & a_3 & a_5 & \cdots
\end{array}
$$

8. Without loss of generality we can assume the polynomial to be monic (that is, the coefficient of the highest power of s is 1).

9. This is easy to see if we construct the polynomial as a product of first- and second-order factors.

We then add subsequent rows to complete the **Routh array**:

Row n	s^n:	1	a_2	a_4	$\cdots$
Row $n-1$	s^{n-1}:	a_1	a_3	a_5	$\cdots$
Row $n-2$	s^{n-2}:	b_1	b_2	b_3	$\cdots$
Row $n-3$	s^{n-3}:	c_1	c_2	c_3	$\cdots$
$\vdots$	$\vdots$	$\vdots$	$\vdots$	$\vdots$	
Row 2	s^2:	$*$	$*$		
Row 1	s:	$*$			
Row 0	s^0:	$*$			

We compute the elements from the $(n-2)$th and $(n-3)$th rows as follows:

$$b_1 = -\frac{\det\begin{bmatrix} 1 & a_2 \\ a_1 & a_3 \end{bmatrix}}{a_1} = \frac{a_1 a_2 - a_3}{a_1},$$

$$b_2 = -\frac{\det\begin{bmatrix} 1 & a_4 \\ a_1 & a_5 \end{bmatrix}}{a_1} = \frac{a_1 a_4 - a_5}{a_1},$$

$$b_3 = -\frac{\det\begin{bmatrix} 1 & a_6 \\ a_1 & a_7 \end{bmatrix}}{a_1} = \frac{a_1 a_6 - a_7}{a_1},$$

$$c_1 = -\frac{\det\begin{bmatrix} a_1 & a_3 \\ b_1 & b_2 \end{bmatrix}}{b_1} = \frac{b_1 a_3 - a_1 b_2}{b_1},$$

$$c_2 = -\frac{\det\begin{bmatrix} a_1 & a_5 \\ b_1 & b_3 \end{bmatrix}}{b_1} = \frac{b_1 a_5 - a_1 b_3}{b_1},$$

$$c_3 = -\frac{\det\begin{bmatrix} a_1 & a_7 \\ b_1 & b_4 \end{bmatrix}}{b_1} = \frac{b_1 a_7 - a_1 b_4}{b_1}.$$

Note that the elements of the $(n - 2)$th row and the rows beneath it are formed from the two previous rows using determinants, with the two elements in the first column and other elements from successive columns. Normally these are $n + 1$ elements in the first column when the array terminates. If these are all positive, then all the roots of the characteristic polynomial are in the LHP. However, if the elements of the first column are not all positive, then the number of roots in the RHP equals the number of sign changes in the column. A pattern of $+$, $-$, $+$ is counted as *two* sign changes: one change from $+$ to $-$ and another from $-$ to $+$.

◆ **EXAMPLE 4.17**   *Routh's Test*

The polynomial

$$a(s) = s^6 + 4s^5 + 3s^4 + 2s^3 + s^2 + 4s + 4$$

satisfies the necessary condition for stability since all the $\{a_i\}$ are positive and nonzero. Determine whether any of the roots of the polynomial are in the RHP.

Solution.   The Routh array for this polynomial is

s^6:	1	3	1	4
s^5:	4	2	4	0
s^4:	$\frac{5}{2} = \dfrac{4 \cdot 3 - 1 \cdot 2}{4}$	$0 = \dfrac{4 \cdot 1 - 4 \cdot 1}{4}$	$4 = \dfrac{4 \cdot 4 - 1 \cdot 0}{4}$	
s^3:	$2 = \dfrac{\frac{5}{2} \cdot 2 - 4 \cdot 0}{\frac{5}{2}}$	$-\frac{12}{5} = \dfrac{\frac{5}{2} \cdot 4 - 4 \cdot 4}{\frac{5}{2}}$	0	
s^2:	$3 = \dfrac{2 \cdot 0 - \frac{5}{2}(-\frac{12}{5})}{2}$	$4 = \dfrac{2 \cdot 4 - (\frac{5}{2} \cdot 0)}{2}$		
s:	$-\frac{76}{15} = \dfrac{3(-\frac{12}{5}) - 8}{3}$	0		
s^0:	$4 = \dfrac{-\frac{76}{15} \cdot 4 - 0}{-\frac{76}{15}}.$			

We conclude that the polynomial has RHP roots, since the elements of the first column are not all positive. In fact, there are two poles in the RHP since there are two sign changes.[10]

◆

Note that in computing the Routh array we can simplify the rest of the calculations by multiplying or dividing a row by a positive constant. Also note that the last two rows each have one nonzero element.

If the first term in one of the rows is zero or if an entire row is zero, then the standard Routh array cannot be formed, so we have to use one of the special techniques described next.

Special Cases

Special case I

If only the first element in one of the rows is zero, then we can replace the zero with a small positive constant $\epsilon > 0$ and proceed as before. We then apply the stability criterion by taking the limit as $\epsilon \to 0$.

10. The actual roots of the polynomial are -3.2644, $0.7797 \pm 0.7488j$, $-0.6046 \pm 0.9935j$, and -0.8858, which, of course, agree with our conclusion.

◆ **EXAMPLE 4.18**   *Routh's Test for Special Case I*

Consider the polynomial

$$a(s) = s^5 + 3s^4 + 2s^3 + 6s^2 + 6s + 9.$$

Determine whether any of the roots are in the RHP.

Solution.   The Routh array is

s^5:	1	2	6
s^4:	3	6	9
s^3:	0	3	0
New s^3:	ϵ	3	0	← Replace zero by ϵ
s^2:	$\dfrac{2\epsilon - 3}{\epsilon}$	3	0
s:	$3 - \dfrac{3\epsilon^2}{2\epsilon - 3}$	0	0
s^0:	3	0	

There are two sign changes in the first column of the array, which means there are two poles not in the LHP.[11]

An alternative procedure is to define the auxiliary variable

$$z \triangleq \frac{1}{s}, \tag{4.69}$$

and express the characteristic equation in terms of z. This usually produces a Routh array whose first-column elements are all nonzero. We can then deduce the stability properties of the system from this new array.

Special case II   Another special case occurs when an entire row of the Routh array is zero. This indicates that there are complex conjugate pairs of roots that are mirror images of each other with respect to the imaginary axis. If the ith row is zero, we form the following auxiliary equation from the previous (nonzero) row:

$$a_1(s) = \beta_1 s^{i+1} + \beta_2 s^{i-1} + \beta_3 s^{i-3} + \cdots, \tag{4.70}$$

where $\{\beta_i\}$ are the coefficients of the $(i + 1)$th row in the array. We then replace the ith row by the coefficients of the *derivative* of the auxiliary polynomial, and complete the array. However, the roots of the auxiliary polynomial in Eq. (4.70) are also roots of the characteristic equation, and these must be tested separately.

11. The actual roots are at $-2.9043, 0.6567 \pm 1.2881j, -0.7046 \pm 0.9929j$.

◆ **EXAMPLE 4.19** *Routh Test for Special Case II*

For the polynomial

$$a(s) = s^5 + 5s^4 + 11s^3 + 23s^2 + 28s + 12,$$

determine whether there are any roots on the $j\omega$ axis or in the RHP.

Solution. The Routh array is

s^5:	1	11	28
s^4:	5	23	12
s^3:	6.4	25.6	0
s^2:	3	12	
s:	0	0	$\leftarrow a_1(s) = 3s^2 + 12$
New s:	6	0	$\leftarrow \dfrac{da_1(s)}{ds} = 6s$
s^0:	12		

There are no sign changes in the first column. Hence all the roots have negative real parts except for a pair on the imaginary axis. We may deduce this as follows. When we replace the zero in the first column by $\epsilon > 0$, there are no sign changes. If we let $\epsilon < 0$, then there are two sign changes. Thus, if $\epsilon = 0$, there are two poles on the imaginary axis, which are the roots of

$$a_1(s) = s^2 + 4 = 0,$$

or

$$s = \pm j2.$$

This agrees with the fact that the actual roots are at -3, $\pm 2j$, -1, and -1.

◆

Routh's method is also useful in determining the range of parameters for which a feedback system remains stable.

◆ **EXAMPLE 4.20** *Stability Versus Parameter Range*

Consider the system shown in Fig. 4.38. The stability properties of the system are a function of the proportional feedback gain K. Determine the range of K over which the system is asymptotically stable.

FIGURE 4.38
A feedback system for testing asymptotic stability

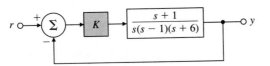

Solution.   The characteristic equation for the system is given by

$$1 + K \frac{s+1}{s(s-1)(s+6)} = 0,$$

or

$$s^3 + 5s^2 + (K-6)s + K = 0.$$

The corresponding Routh array is

$$
\begin{array}{lll}
s^3: & 1 & K-6 \\
s^2: & 5 & K \\
s: & (4K-30)/5 & \\
s^0: & K &
\end{array}
$$

For the system to remain stable, it is necessary that

$$\frac{4K-30}{5} > 0 \quad \text{and} \quad K > 0,$$

or

$$K > 7.5 \quad \text{and} \quad K > 0.$$

Thus Routh's method provides an analytical answer to the stability question. Although any gain satisfying this inequality stabilizes the system, the dynamic response could be quite different depending on the specific value of K. Given a specific value of the gain, we may compute the closed-loop poles by finding the roots of the characteristic polynomial. The characteristic polynomial has the coefficients represented by the row vector (in descending powers of s)

$$\text{denG} = [1 \; 5 \; K-6 \; K],$$

Computing roots
by MATLAB

and we may compute the roots using the **MATLAB** function

$$\text{roots(denG).}$$

For $K = 7.5$ the roots are at -5 and $\pm 1.22j$, and the system is marginally stable. Note that Routh's method predicts the presence of poles on the $j\omega$ axis for $K = 7.5$. If we set $K = 13$, the closed-loop poles are at -4.06 and $-0.47 \pm 1.7j$, and for $K = 25$ they are at -1.90 and $-1.54 \pm 3.27j$. In both these cases the system is stable as predicted by Routh's method. Figure 4.39 shows the transient responses for the three gain values. To obtain these transient responses, we compute the closed-loop transfer function

$$\frac{Y(s)}{R(s)} = \frac{K(s+1)}{s^3 + 5s^2 + (K-6)s + K},$$

so that the numerator polynomial is expressed as

$$\text{numG} = [K \; K]$$

FIGURE 4.39
Transient responses for
the system in Fig. 4.38

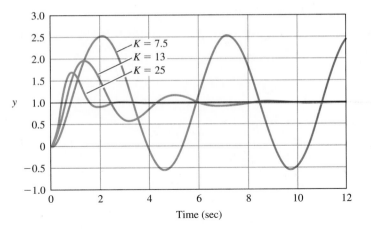

and denG is as before. The MATLAB function

$$\text{step(numG, denG)}$$

produces a plot of the (unit) step response.

◆ **EXAMPLE 4.21** *Stability Versus Two Parameter Ranges*

Find the range of the controller gains (K, K_I) so that the PI feedback system in Fig. 4.40 is asymptotically stable.

Solution. The characteristic equation of the closed-loop system is

$$1 + \left(K + \frac{K_I}{s} \right) \frac{1}{(s+1)(s+2)} = 0,$$

which we may rewrite as

$$s^3 + 3s^2 + (2 + K)s + K_I = 0.$$

The corresponding Routh array is

$$
\begin{array}{lll}
s^3: & 1 & 2 + K \\
s^2: & 3 & K_I \\
s: & (6 + 3K - K_I)/3 & \\
s^0: & K_I &
\end{array}
$$

FIGURE 4.40
System with PI control

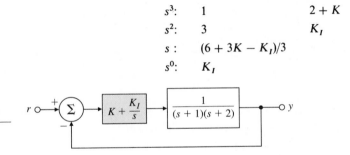

FIGURE 4.41
Allowable region for stability

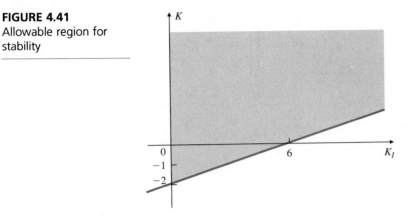

For asymptotic stability we must have

$$K_I > 0 \quad \text{and} \quad K > \tfrac{1}{3}K_I - 2.$$

The allowable region is the shaded area in the (K_I, K) plane shown in Fig. 4.41, which represents an analytical solution to the stability question. The closed-loop transfer function is

$$\frac{Y(s)}{R(s)} = \frac{Ks + K_I}{s^3 + 3s^2 + (2 + K)s + K_I}.$$

As in Example 4.20, we may compute the closed-loop poles for different values of the dynamic compensator gains by using the MATLAB function roots on the denominator polynomial

Computing roots
by MATLAB

$$\text{denG} = [1 \ 3 \ 2 + K \ \text{KI}].$$

FIGURE 4.42
Transient response for the system in Fig. 4.40

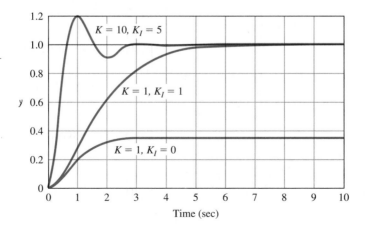

Similarly, we may find the zero by finding the root of the numerator polynomial

$$numG = [K \ KI].$$

The closed-loop zero of the system is at $-K_I/K$. Figure 4.42 shows the transient response for three sets of feedback gains. For $K = 1$ and $K_I = 0$ the closed-loop poles are at 0 and $-1.5 \pm 0.86j$, and there is a zero at the origin. For $K = K_I = 1$ the poles and zeros are all at -1. For $K = 10$ and $K_I = 5$ the closed-loop poles are at -0.46 and $-1.26 \pm 3.03j$, and the zero is at -0.5. The step responses were again obtained using the MATLAB function

$$step(numG, denG).$$

There is a large steady-state error in this case when $K_I = 0$.

Summary

- Feedback can be used to stabilize systems, speed up transient response, improve steady-state characteristics, provide disturbance rejection, and decrease sensitivity to parameter variations.

- Proportional feedback reduces errors, but high gains may destabilize the system. Integral control improves the steady-state error and provides robustness with respect to parameter variations, but it also reduces stability. Derivative control increases damping and improves stability. These three kinds of control can be combined within the same system. Consider the following unity feedback system:

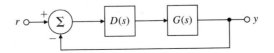

then,

Proportional: $u(t) = Ke(t)$

PI: $u(t) = K\left[e(t) + \dfrac{1}{T_I} \displaystyle\int_0^t e(\eta)\,d\eta \right]$

PID: $u(t) = K\left[e(t) + \dfrac{1}{T_I} \displaystyle\int_0^t e(\eta)\,d\eta + T_D\dot{e}(t) \right]$

- The PID controller

$$U(s) = K\left(1 + \frac{1}{T_I s} + T_D s \right) E(s) = D(s)E(s)$$

is ubiquitous in the process-control industry and is the basic ingredient in many control systems. Useful guidelines for tuning PID controllers were presented in Tables 4.1 and 4.2.

- System type indicates the ability of a system to follow or reject particular time functions with zero steady-state error. The unity feedback system shown in the above figure is type k with respect to the reference input if

$$D(s)G(s) = \frac{(s + z_1)(s + z_2)\cdots}{s^k(s + p_1)(s + p_2)\cdots}.$$

For a list of steady-state errors to a reference input of such a system, see Table 4.3.

- For a stable system, all the closed-loop poles must be in the LHP.
- For a stable system all the elements in the first column of the Routh array must be positive. To determine the Routh array, refer to the formulas in Section 4.4.3.

Problems

4.1 Consider a system with the configuration of Fig. 4.5 where D is the gain of the controller and G is the process. The nominal values of these gains are $D = 5$ and $G = 7$. Suppose a constant disturbance w is added to the control input u before the signal goes to the process.

a) Compute the gain from w to y in terms of D and G.

b) Suppose the system designer knows that an increase by a factor of 6 in the loop gain DG can be tolerated before the system goes out of specification. Where should the designer place the extra gain if the objective is to minimize the system error $r - y$ due to the disturbance? For example, either D or G could be increased by a factor of 6, or D could be doubled and G tripled, and so on. Which choice is the best?

4.2 Bode defined the sensitivity function relating a transfer function G to one of its parameters k as the ratio of percent change in k to percent change in G. We define the reciprocal of Bode's function as

$$S_k^G = \frac{dG/G}{dk/k} = \frac{d \ln G}{d \ln k} = \frac{k}{G} \frac{dG}{dk}.$$

Thus, when the parameter k changes by a certain percentage, S tells us what percent change to expect in G. In control systems design we are almost always interested in the sensitivity at zero frequency, or when $s = 0$. The purpose of this exercise is to examine the effect of feedback on sensitivity. In particular, we would like to compare the topologies shown in Fig. 4.43 for connecting three amplifier stages with a gain of $-K$ into a single amplifier with a gain of -10.

a) For each topology in Fig. 4.43, compute β_i so that if $K = 10$, $y = -10r$.

b) For each topology, compute S_k^G when $G = y/r$. [Use the respective β_i values found in part (a).] Which case is the *least* sensitive?

c) Compute the sensitivities of the systems in Fig. 4.43(b, c) to β_2 and β_3. Using your results, comment on the relative need for precision in sensors and actuators.

FIGURE 4.43
Three-amplifier
topologies for Problem
4.2

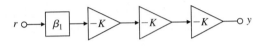

(a)

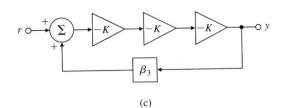

(b)

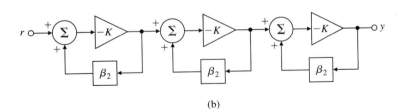

(c)

FIGURE 4.44
Block diagrams for Problem 4.3

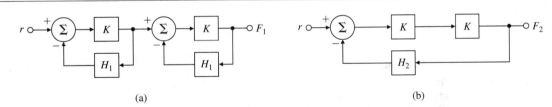

(a)   (b)

4.3 Compare the two structures shown in Fig. 4.44 with respect to sensitivity to changes in the overall gain due to changes in the amplifier gain. Use the relation

$$S = \frac{d \ln F}{d \ln K} = \frac{K}{F} \frac{dF}{dK}$$

as the measure. Select H_1 and H_2 so that the nominal system outputs satisfy $F_1 = F_2$, and assume $KH_1 > 0$.

4.4 A DC-motor speed control is described by the differential equation

$$\dot{y} + 60y = 600v_a - 1500w,$$

where y is the motor speed, v_a is the armature voltage, and w is the load torque. Assume the armature voltage is computed using the PI control law

$$v_a = -\left(K_1 y + K_I \int_0^t y\, dt\right).$$

a) Compute the transfer function from w to y as a function of K_1 and K_I.
b) Compute values for K_1 and K_I so that the characteristic equation of the closed-loop system will have roots at $-60 \pm 60j$.

4.5 Consider the system shown in Fig. 4.45, which consists of a prefilter and a unity feedback system.

a) Determine the transfer function from r to y.

b) Determine the steady-state error due to a step input.

c) Discuss the effect of different values of (K_r, a) on the system's response.

d) For each of the following three cases,

$$(1) \ K = 1, \ T = 1, \qquad (2) \ K = 10, \ T = 1, \qquad (3) \ K = 1, \ T = 2,$$

use MATLAB or a similar CACSD tool to find values for K_r and a so that (if possible) the rise time is less than 1.5 sec, the settling time is less than 10 sec, the overshoot is less than 20%, and the steady-state error is less than 5%. In cases where the specifications are easily met, try to make the rise time as small as possible. If the specifications cannot be met, find the best design possible.

FIGURE 4.45

Unity feedback system with prefilter

 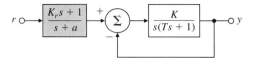

4.6 A unity feedback control system has the open-loop transfer function

$$G(s) = \frac{A}{s(s + a)}.$$

a) Compute the sensitivity of the closed-loop transfer function to changes in the parameter A.

b) Compute the sensitivity of the closed-loop transfer function to changes in the parameter a.

c) If the unity gain in the feedback changes to a value of $\beta \neq 1$, compute the sensitivity of the closed-loop transfer function with respect to β.

d) Assuming $A = 1$ and $a = 1$, plot the magnitude of each of the above sensitivity functions for $s = j\omega$ using the loglog or semilogx commands in MATLAB. Comment on the relative effect of parameter variations in A, a, and β at different frequencies ω, paying particular attention to DC (when $\omega = 0$).

4.7 For the second-order DC motor modeled by Eqs. (4.1) and (4.2), compute $y(t)$ and $i_a(t)$ explicitly in terms of the parameters (J_m, R_a, K_t, K_e, b).

4.8 Consider the satellite-attitude control problem shown in Fig. 4.46 where

J = spacecraft inertia,

θ_r = reference satellite attitude,

θ = actual satellite attitude,

K_s = sensor gain,

K_r = reference scaling gain,

w = disturbance torque.

a) Use proportional control, $D(s) = K$, and study the properties of the controller with respect to providing additional damping.

b) Use PD control and let $D(s) = K(1 + T_D s)$. Determine the tracking and disturbance-rejection properties for step inputs on θ_r and w.

c) Let $D(s) = K(1 + 1/T_I s)$, and discuss the effect of this controller on damping and tracking and disturbance-rejection properties.

d) Let $D(s) = K(1 + 1/T_I s + T_D s)$. Discuss the effect of PID control on the stability and performance properties of the system.

FIGURE 4.46

Satellite attitude control

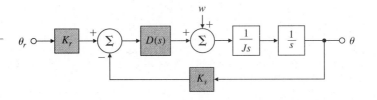

4.9 Consider the second-order plant

$$G(s) = \frac{1}{(s + 1)(5s + 1)}.$$

a) Determine the tracking response properties of the system for proportional, PD, and PID regulators (as configured in Fig. 4.5) with $K = 19$, $T_I = 2$, and $T_D = \frac{4}{19}$. Assume there is a unit step reference input $r(t)$.

b) Determine the disturbance-rejection properties of the system for each of the three regulators in part (a) with respect to a unit step disturbance $w(t)$ at the input to the plant.

c) Qualitatively compare the responses obtained in parts (a) and (b).

d) Verify your results for parts (a) and (b) using MATLAB.

4.10 Consider a system with the plant transfer function $G(s) = 1/s(s+1)$. You wish to add a dynamic controller so that $\omega_n = 2$ rad/sec and $\zeta \geqslant 0.5$. Several dynamic controllers have been proposed:

(1) $D(s) = s + 2$,

(2) $D(s) = \dfrac{s + 2}{s + 4}$,

(3) $D(s) = \dfrac{s + z)}{s + p}$, with zero $z = 2$ and poles $p = 10$ or 20,

(4) $D(s) = \dfrac{(s+z)(s + 0.1)}{(s + p)(s + 0.01)}$, with zero $z = 2$ and poles $p = 10$ or 20.

Using MATLAB, you are to simulate the closed-loop system response (see Fig. 4.5) for each controller, and fully investigate the resulting transient and steady-state responses.

a) Experiment with the location of the controller zero z and pole p, and explore the possibility of achieving improved system performance using a modified dynamic controller.

b) How sensitive is the system response to perturbations in the location of z and p?

c) For which values of z and p does the system response resemble a standard second-order system (with a "dominant" pair of poles)?

FIGURE 4.47
Paper-machine response data for Problem 4.11

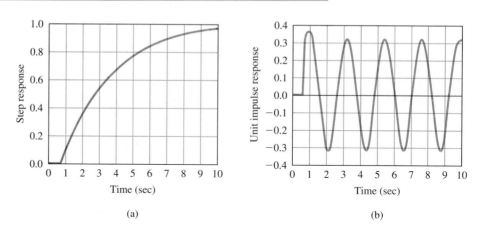

(a)

(b)

d) Compare each system for control effort, that is, magnitude of plant input $u(t)$.

e) Based on your results from parts (a) to (d), recommend a dynamic controller for the system from the four candidate designs.

4.11 The unit-step response of a paper machine is shown in Fig. 4.47(a) where the input into the system is stock flow onto the wire and the output is basis weight (thickness). The time delay and slope of the transient response may be determined from the figure.

a) Find the proportional, PI, and PID-controller parameters using the Zeigler–Nichols transient-response method.

b) Using proportional feedback control, control designers have obtained a closed-loop system with the unit impulse response shown in Fig. 4.47(b). When the gain $K_u = 8.556$, the system is on the verge of instability. Determine the proportional-, PI-, and PID-controller parameters according to the Zeigler–Nichols ultimate sensitivity method.

4.12 A paper machine has the transfer function

$$G(s) = \frac{e^{-2s}}{3s + 1},$$

where the input is stock flow onto the wire and the output is basis weight or thickness.

a) Find the PID-controller parameters using the Zeigler–Nichols tuning rules.

b) The system becomes marginally stable for a proportional gain of $K_u = 3.044$ as shown by the unit impulse response in Fig. 4.48. Find the optimal PID-controller parameters according to the Zeigler–Nichols tuning rules.

4.13 Consider the system with the plant transfer function

$$G(s) = \frac{1}{s^2 + 1}.$$

FIGURE 4.48
Unit impulse response
for paper-machine in
Problem 4.12

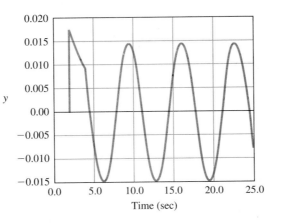

We would like to use integral control on this system. It is known that the system's actuator is a saturation nonlinearity with a slope of 1 and $|u| \leq 10$. Design an integral controller with antiwindup using the two techniques discussed in this chapter. Using computer simulations, experiment with different slopes for the deadband nonlinearity, and select the "best" value for the slope. Plot both the step tracking response and the control effort. Qualitatively describe the effect of the antiwindup on both the output response and the control effort.

4.14 A certain control system has the following specifications: rise time $t_r \leq 0.010$ sec, overshoot $M_p \leq 16\%$, and steady-state error to unit ramp $e_{ss} \leq 0.005$.

   a) Sketch the allowable region in the s-plane for the dominant second-order poles of an acceptable system.
   b) If $Y/R = G/(1 + G)$, what condition must $G(s)$ satisfy near $s = 0$ for the closed-loop system to meet specifications; that is, what is the required low-frequency behavior of $G(s)$?

4.15 For the system in Problem 4.4(b), compute the following steady-state errors:
   a) to a unit-step reference input;
   b) to a unit-ramp reference input;
   c) to a unit-step disturbance input;
   d) to a unit-ramp disturbance input.
   e) Verify your answers to parts (a) to (d) using MATLAB.

4.16 Consider the system shown in Fig. 4.49.

   a) Discuss the effect of different values of (a, b, K) on the shape of the step response of the system.
   b) Determine the system type with respect to r.

FIGURE 4.49
Control system for
Problem 4.16

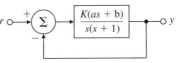

FIGURE 4.50
DC control system for Problem 4.17

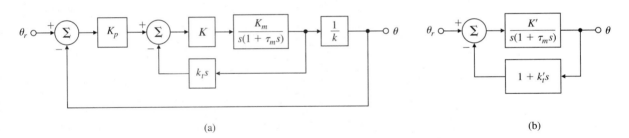

(a)   (b)

4.17 Consider the DC-motor control system with rate feedback shown in Fig. 4.50(a).

a) Show that the system may be redrawn as Fig. 4.50(b).
b) Determine the system type with respect to θ_r.

4.18 Consider the system shown in Fig. 4.51, where

$$D(s) = K \frac{s + \alpha}{s^2 + \omega^2}.$$

a) Prove that the system is capable of tracking a sinusoidal reference input $r = \sin \omega t$ with zero steady-state error.
b) Find the range of K and α such that the closed-loop system remains stable. For the special case where $\alpha = 1$ is fixed, determine whether the closed-loop system would be asymptotically stable for some set of values for K.

FIGURE 4.51
Control system for
Problem 4.18

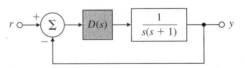

4.19 Consider the system shown in Fig. 4.52.

a) What condition must $D(s)$ satisfy so that the system can track a ramp reference input with constant steady-state error?
b) For a transfer function $D(s)$ that satisfies the condition in part (a), find the class of disturbances $w(t)$ that the system can reject.
c) Show that although a PI controller satisfies the condition derived in part (a), it will not yield a stable closed-loop system. Will a PID controller work; that is, satisfy part (a) *and* stabilize the system? If so, what constraints must K, T_I, and T_D satisfy?
d) With the aid of MATLAB, use the Ziegler–Nichols ultimate sensitivity method to design a suitable PID controller for the system. Check the disturbance-rejection properties of your closed-loop design. Also, discuss the effects of varying the controller parameters K, T_I, and T_D on the system's step response (in terms of rise time, overshoot, transient behavior, and so forth).

FIGURE 4.52
Control system for
Problem 4.19

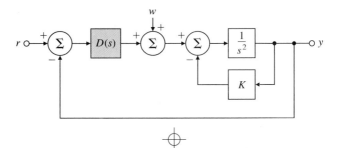

4.20 For the second-order system

$$G(s) = \frac{\omega_n^2}{s^2 + 2\zeta\omega_n s + \omega_n^2},$$

compute K_v and K_a in terms of (ζ, ω_n). Can K_v and K_a be selected independently?

4.21 Consider the second-order system

$$G(s) = \frac{\omega_n^2}{s^2 + 2\zeta\omega_n s + \omega_n^2}.$$

We would like to add a transfer function of the form $D(s) = K(s + a)/(s + b)$ in series with $G(s)$ so that K_v for the system becomes infinite. What are suitable values for $K, a,$ and b?

4.22 For a unity feedback system with the open-loop transfer function

$$G(s) = \frac{A}{s(s + a)},$$

find values for (A, a) so that $\zeta = 0.707$, $K_v = 20 \text{ sec}^{-1}$, and $K_a = 30$.

4.23 Consider the system shown in Fig. 4.53(a).

    a) What is the system type? Compute the steady-state tracking error due to a ramp input $r(t) = At$ where $A > 0$.

    b) Show that by modifying the system to match Fig. 4.53(b), you can make the steady-state tracking error to the ramp input in part (a) to be zero.

FIGURE 4.53
Control system for Problem 4.23

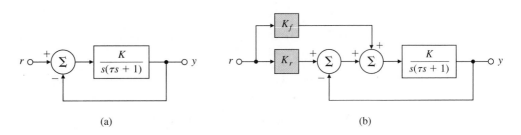

(a)   (b)

4.24 A unity feedback system has the open-loop transfer function

$$G(s) = \frac{s+1}{s^3 + s^2}.$$

Find the error constants K_p, K_v, and K_a for this system.

4.25 Consider the system shown in Fig. 4.54. Assume that the sensor dynamics are $H(s) = 1$.

a) Can the system follow a step reference input r with zero steady-state error? Explain.

b) Can the system reject a step disturbance w with zero steady-state error? Explain.

c) Compute the sensitivity of the closed-loop transfer function to changes in the plant pole at -2.

d) In some instances there are dynamics in the sensor. Repeat parts (a) to (c) for $H(s) = 5/(s + 20)$.

FIGURE 4.54
Control system for
Problem 4.25

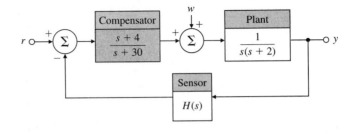

4.26 Consider the system shown in Fig. 4.55.

a) Determine the transfer function from r to y.

b) Determine the transfer function from w to y.

c) Find the range of (K_1, K_2) for which the system is stable.

d) What is the system type with respect to r and w?

FIGURE 4.55
Control system for
Problem 4.26

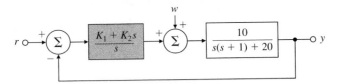

4.27 The unity feedback system shown in Fig. 4.56 has disturbance inputs w_1, w_2, and w_3 and is asymptotically stable. Also,

$$G_1(s) = \frac{K_1 \prod_{i=1}^{m_1} (s + z_{1i})}{s^{l_1} \prod_{i=1}^{m_1} (s + p_{1i})}, \quad G_2(s) = \frac{K_2 \prod_{i=1}^{m_1} (s + z_{2i})}{s^{l_2} \prod_{i=1}^{m_1} (s + p_{2i})}.$$

FIGURE 4.56
Single input-single output unity feedback system with disturbance inputs

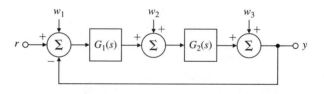

a) Show that the system is of type 0, type l_1, and type $(l_1 + l_2)$ with respect to disturbance inputs w_1, w_2, and w_3.
b) Consider the multivariable system shown in Fig. 4.57. Find y_1 and y_2 for constant disturbances. What is the system type with respect to disturbanes at w_1? At w_2?

FIGURE 4.57
Multivariable system

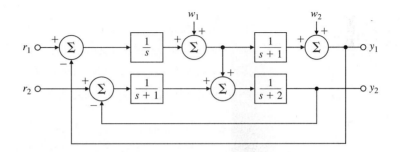

4.28 One possible representation of an automobile speed-control system with integral control is shown in Fig. 4.58.

a) With a zero reference input ($v_c = 0$), find the transfer function relating the output speed v to the wind disturbance w.
b) What is the steady-state response of v if w is a unit ramp function?
c) What type is this system in relation to reference inputs? What is the value of the static error constants K_p and K_v?
d) What type is this system in relation to the disturbance w? What is the error constant due to the disturbance?

FIGURE 4.58
System using integral control

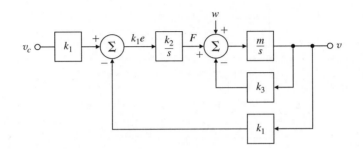

FIGURE 4.59
Control system for
Problem 4.29

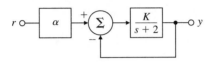

4.29 For the feedback system shown in Fig. 4.59, find the value of α that will make the system type I for $K = 5$. Show that the system is not robust by using this value of α and computing the tracking error $e = r - y$ to a step reference for $K = 4$ and $K = 6$.

4.30 A position control system has the overall transfer function (meter/meter) given by

$$\frac{Y(s)}{R(s)} = \frac{b_0 s + b_1}{s^2 + a_1 s + a_2}.$$

Choose the parameters (a_1, a_2, b_0, b_1) so that the following specifications are satisfied simultaneously:

- The rise time $t_r \leqslant 0.1$ sec.
- The overshoot $M_p \leqslant 20\%$.
- The settling time $t_s \leqslant 0.5$ sec.
- The steady-state error to a step reference is zero.
- The steady-state error to a ramp input of 0.1 m/sec is not more than 1 mm.

Verify your answer via computer simulation.

4.31 Suppose you are given the system depicted in Fig. 4.60(a), where the plant parameter a is subject to variations.

a) Find $G(s)$ so that the system shown in Fig. 4.60(b) has the same transfer function from r to y as the system in Fig. 4.60(a).

b) Assume that $a = 1$ is the nominal value of the plant parameter. Find the error constants K_p, K_v, and K_a for the nominal system. What is the system type in this case?

c) Now assume that $a = 1 + \delta a$, where δa is some perturbation to the plant parameter. What are the error constants and system type for the perturbed system?

FIGURE 4.60
Control system for Problem 4.31

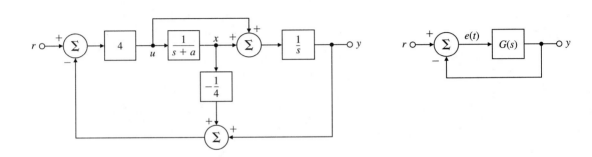

FIGURE 4.61
Two feedback systems for Problem 4.32

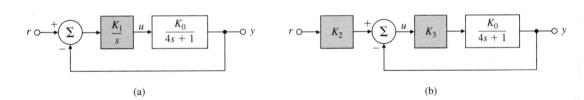

(a)   (b)

4.32 Two feedback systems are shown in Fig. 4.61.

a) Determine values for K_1, K_2, and K_3 so that both systems

- exhibit zero steady-state error to step inputs (that is, both are type I), and
- whose static velocity-error constant $K_v = 1$ when $K_0 = 1$.

b) Suppose K_0 undergoes a small perturbation: $K_0 \rightarrow K_0 + \delta K_0$. What effect does this have on the system type in each case? Which system do you think would be preferred?

c) Estimate the transient response of both systems to a step reference input, and compute t_s, t_r, and M_p. In your opinion, which system has a better transient response at the nominal parameter values?

4.33 You are given the system shown in Fig. 4.62, where the feedback gain β is subject to variations. You are to design a controller for this system so that the output $y(t)$ accurately tracks the reference input $r(t)$.

a) Let $\beta = 1$. You are given the following three options for the controller $D_i(s)$:

$$D_1(s) = K, \qquad D_2(s) = \frac{K}{s}, \qquad D_3(s) = \frac{K}{s^2},$$

where K is a constant. Choose the controller (including a particular value for K) that will result in a type I system with a steady-state error of less than $\frac{1}{10}$.

b) Next, suppose that due to harsh conditions there is some attenuation in the feedback path that is best modeled by $\beta = 0.9$. Find the steady-state error due to a ramp input for your choice of $D_i(s)$ in part (a).

c) If $\beta = 0.9$, what is the system type for part (b)? What are the values of the appropriate error constants?

FIGURE 4.62
Control system for Problem 4.33

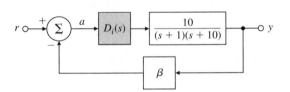

4.34 Consider the system shown in Fig. 4.63.

a) Find the transfer function from the input to the tracking error.

b) For this system to respond to inputs of the form $r(t) = t^n$ (where $n \ll q$) with zero steady-state error, what must be true of the open-loop poles $p_1, p_2, \ldots, p_q$?

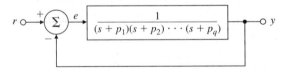

4.35 The feedback control system shown in Fig. 4.64 is to be designed to satisfy the following specifications: (1) steady-state error of less than 10% to a ramp input, (2) maximum overshoot for a unit step input of less than 5%, and (3) 1% settling time of less than 3 sec.

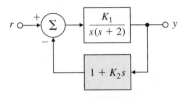

a) Compute the closed-loop transfer function.

b) Find the error due to a unit ramp input in terms of K_1 and K_2.

c) What does specification 1 imply about the possible values of K_1?

d) What does specification 3 imply about the closed-loop poles?

e) Sketch the region in the complex plane where the closed-loop poles may lie.

f) Suppose $K_1 = 32$. Find the value of K_2 that yields closed-loop poles on the right-hand boundary of the feasible region. Use MATLAB to check whether this choice for K_2 satisfies the desired specifications. If not, adjust K_2 until it does.

g) Using $K_1 = 32$ and the value for K_2 computed in part (f), estimate the settling time of the system. Use MATLAB to check your answer.

4.36 The transfer functions of a magnetic-tape-drive speed-control system are shown in Fig. 4.65. The speed sensor is fast enough that its dynamics can be neglected.

a) Assuming $\omega_r = 0$, what is the steady-state error due to a step disturbance torque of 1 N·m? What must the amplifier gain K be in order to make the steady-state error $e_{ss} \leq 0.01$ rad/sec?

b) Plot the roots of the closed-loop system in the complex plane, and accurately sketch the time response $\omega(t)$ for a step input ω_r using the gain K computed in part (a). Are these roots desirable? Why or why not?

c) Plot the region in the complex plane corresponding to the specifications of a 1% settling time $t_s \leq 0.1$ sec and an overshoot $M_p \leq 5\%$.

d) Suggest a simple control scheme that can be added to this proportional control system to meet the specifications in part (c), and show that the new scheme meets the specifications.

e) How would the disturbance-induced steady-state error change with the new control scheme in part (d)? How could the steady-state error be eliminated entirely?

FIGURE 4.65
Speed-control system for
a magnetic tape drive

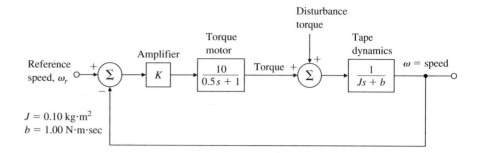

4.37 A linear ODE model of the DC motor with negligible armature inductance ($L_a = 0$) and disturbance torque w was given earlier in the chapter; it is restated here, in slightly different form, as

$$\frac{JR_a}{K_t}\ddot{\theta} + K_e\dot{\theta} = v_a + \frac{R_a}{K_t}w, \qquad (4.3)$$

where θ is measured in radians. Dividing through by the coefficient of $\ddot{\theta}$, we obtain

$$\ddot{\theta} + a_1\dot{\theta} = b_0 v_a + c_0 w,$$

where

$$a_1 = \frac{K_t K_e}{JR_a}, \qquad b_0 = \frac{K_t}{JR_a}, \qquad c_0 = \frac{1}{J}.$$

With rotating potentiometers, it is possible to measure the positioning error between θ and the reference angle θ_r or $e = \theta_r - \theta$. With a tachometer we can measure the motor speed $\dot{\theta}$. Consider using feedback of the error e and the motor speed $\dot{\theta}$ in the form

$$v_a = K(e - T_D\dot{\theta}),$$

where K and T_D are controller gains to be determined.

a) Draw a block diagram of the resulting feedback system, using θ and $\dot{\theta}$ as state variables to represent the motor.

b) Suppose the numbers work out so that $a_1 = 65$, $b_0 = 200$, and $c_0 = 10$. If there is *no* load torque ($w = 0$), what speed (in rpm) results from $v_a = 100$ V?

c) Using the parameter values given in part (b), find K and T_D so that a step change in θ_r with zero load torque results in a transient that has an approximately 17% overshoot and that settles to within 5% of steady-state in less than 0.05 sec.

d) Derive an expression for the steady-state error to an arbitrary reference-angle input, and compute its value for your design in part (c) assuming $\theta_r = 1$ rad.

e) Derive an expression for the steady-state error to an arbitrary disturbance torque when $\theta_r = 0$, and compute its value for your design in part (c) assuming $w = 1.0$.

4.38 We wish to design an automatic speed control for an automobile. Assume that (1) the car has a mass m of 1000 kg, (2) the accelerator is the control U and supplies a force on the automobile of 10 N per degree of accelerator motion, and (3) air drag provides a friction force (proportional to velocity) of 10 N · sec/m.

a) Obtain the transfer function from U to the velocity of the automobile.

b) Assume the velocity changes are given by

$$V(s) = \frac{1}{s + 0.02} U(s) + \frac{0.05}{s + 0.02} G(s),$$

where V is given in meters per second and G is the percent grade of the road. Design a proportional control law that will maintain a velocity error of less than 1 m/sec in the presence of a constant 2% grade.

c) Discuss what advantage (if any) integral control would have for this problem.

d) Assuming that pure integral control (that is, no proportional term) is advantageous, select the feedback gain so that the roots have critical damping ($\zeta = 1$).

4.39 Consider the automobile speed-control system depicted in Fig. 4.66.

a) Find the transfer function from $W(s)$ and $R(s)$ to $Y(s)$.

b) Assume that the desired speed is a constant reference r, so that $R(s) = r/s$. Assume that the road is level, so $w(t) = 0$. Discuss how the gains K, K_r, and K_f should be chosen to guarantee that

$$\lim_{t \to \infty} y(t) = r.$$

Include both the feedforward (assuming $K_f = 0$) and feedback cases ($K_f \neq 0$) in your discussion.

c) Repeat part (b) assuming that a constant grade disturbance $W(s) = w/s$ is present in addition to the reference input. In particular, find the variation in speed due to the grade change for both the feedforward and feedback cases. Use your results to explain (1) why feedback control is necessary and (2) how the gain K should be chosen to reduce steady state error.

d) Assume that $w(t) = 0$ and that the gain K_a undergoes the perturbation $K_a + \delta K_a$. Determine the error in speed due to the gain change for both the feedforward and feedback cases. How should the gains be chosen in this case to reduce the effects of δK_a?

FIGURE 4.66
Automobile speed-control system

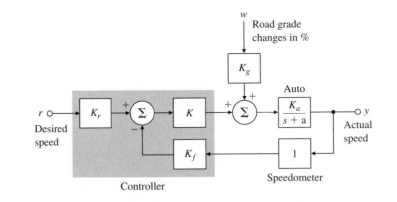

Controller

4.40 Prove that for a type II system,

$$\frac{1}{K_a} = \frac{1}{2}\left(\sum_{i=1}^{m} \frac{1}{z_i^2} - \sum_{i=1}^{n} \frac{1}{p_i^2} \right),$$

where z_i and p_i are the closed-loop zeros and poles of the system.

4.41 For a system with impulse response $h(t)$, prove that

$$\frac{1}{K_v} = \int_0^\infty t h(t)\, dt,$$

and

$$\frac{1}{K_a} = -\frac{1}{2}\int_0^\infty t^2 h(t)\, dt.$$

4.42 For a system with no overshoot, prove that

$$\frac{1}{2\pi} t_r^2 = -\frac{2}{K_a} - \frac{1}{K_v^2}$$

and

$$T_d \cong \frac{1}{K_v},$$

where T_d = time to 50% of final value.

4.43 Suppose that unity feedback is to be applied around the following open-loop systems. Use Routh's stability criterion to determine whether the resulting closed-loop systems will be stable.

a) $KG(s) = \dfrac{4(s+2)}{s(s^3 + 2s^2 + 3s + 4)}$ **c)** $KG(s) = \dfrac{4(s^3 + 2s^2 + s + 1)}{s^2(s^3 + 2s^2 - s - 1)}$

b) $KG(s) = \dfrac{2(s+4)}{s^2(s+1)}$

4.44 Use Routh's stability criterion to determine how many roots with positive real parts the following equations have.

a) $s^4 + 8s^3 + 32s^2 + 80s + 100 = 0.$
b) $s^5 + 10s^4 + 30s^3 + 80s^2 + 344s + 480 = 0.$
c) $s^4 + 2s^3 + 7s^2 - 2s + 8 = 0.$
d) $s^3 + s^2 + 20s + 78 = 0.$
e) $s^4 + 6s^2 + 25 = 0.$

4.45 Find the range of K for which all the roots of the following polynomial are in the LHP.

$$s^5 + 5s^4 + 10s^3 + 10s^2 + 5s + K = 0.$$

Use MATLAB to verify your answer by plotting the roots of the polynomial in the s-plane for various values of K.

4.46 The transfer function of a typical tape-drive system is given by

$$G(s) = \frac{K(s+4)}{s[(s+0.5)(s+1)(s^2 + 0.4s + 4)]},$$

where time is measured in milliseconds. Using Routh's stability criterion, determine the range of K for which this system is stable when the characteristic equation is $1 + G(s) = 0.$

4.47 Automatic ship steering is particularly useful in heavy seas when it is important to maintain the ship along an accurate path. Such a control system for a large tanker is shown in Fig. 4.67, with the plant transfer function relating heading changes to rudder deflection in radians.

a) Write the differential equation that relates the heading angle to rudder angle for the ship *without* feedback.

b) This control system uses simple proportional control and has a gain of 1. Is it stable as shown? (*Hint:* Use Routh's criterion.)

c) Is it possible to stabilize the system by changing the proportional gain from 1 to some lower value?

d) Use MATLAB to design a controller for the ship steering system so that the resulting closed-loop system is stable, and in response to a step heading command has zero steady-state error and less than 10% overshoot. Check the rise and settling times of your resulting closed-loop design. Are these reasonable values for a large tanker?

FIGURE 4.67
Ship-steering control system

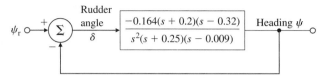

4.48 Consider the system shown in Fig. 4.68.

a) Compute the closed-loop characteristic polynomial.

b) For what values of (T, A) is the system stable? *Hint:* An approximate answer may be found using

$$e^{-Ts} \cong 1 - Ts$$

or

$$e^{-Ts} \cong \frac{1}{\left(1 + \dfrac{T}{3}s\right)^3}$$

for the pure delay. As an alternative, you could use the computer to simulate the system or to find the roots of the system's characteristic polynomial for various values of T and A.

FIGURE 4.68
Control system for Problem 4.48

4.49 Modify the Routh criterion so that it applies to the case where all the poles are to be to the left of $-\alpha$ when $\alpha > 0$. Apply the modified test to the polynomial

$$s^3 + (6 + K)s^2 + (5 + 6K)s + 5K = 0,$$

finding those values of K for which all poles have a real part less than -1.

4.50 Suppose the characteristic polynomial of a given closed-loop system is computed to be

$$s^4 + (11 + K_2)s^3 + (121 + K_1)s^2 +$$
$$(K_1 + K_1K_2 + 110K_2 + 210)s + 11K_1 + 100 = 0.$$

Find constraints on the two gains K_1 and K_2 that guarantee a stable closed-loop system, and plot the allowable region(s) in the (K_1, K_2) plane. You may wish to use the computer to help solve this problem.

4.51 Overhead electric power lines sometimes experience a low-frequency, high-amplitude vertical oscillation, or **gallop**, during winter storms when the line conductors become covered with ice. In the presence of wind, this ice can assume aerodynamic lift and drag forces that result in a gallop up to several meters in amplitude. Large-amplitude gallop can cause clashing conductors and structural damage to the line support structures caused by the large dynamic loads. These effects in turn can lead to power outages. Assume that the line conductor is a rigid rod, constrained to vertical motion only, and suspended by springs and dampers as shown in Fig. 4.69. A simple model of this conductor galloping is

$$m\ddot{y} + \frac{D(\alpha)\dot{y} - L(\alpha)v}{(\dot{y}^2 + v^2)^{1/2}} + T\left(\frac{n\pi}{\ell}\right)y = 0,$$

where

   m = mass of conductor,

   y = conductor's vertical displacement,

   D = aerodynamic drag force,

   L = aerodynamic lift force,

   v = wind velocity,

   α = aerodynamic angle of attack = $-\tan^{-1}(\dot{y}/v)$,

   T = conductor tension,

   n = number of harmonic frequencies,

   ℓ = length of conductor.

Assume that $L(0) = 0$ and $D(0) = D_0$ (a constant), and linearize the equation around the value $y = \dot{y} = 0$. Use Routh's stability criterion to show that galloping can occur whenever

$$\frac{\partial L}{\partial \alpha} + D_0 < 0.$$

FIGURE 4.69
Electric power-line conductor

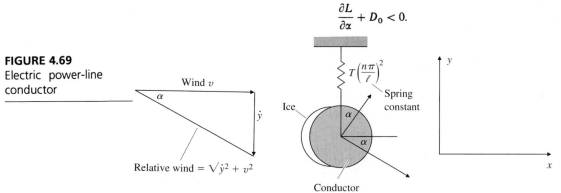

·5·

The Root-locus Design Method

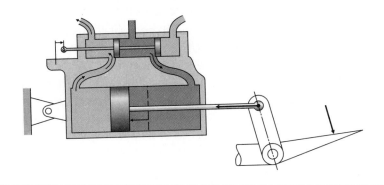

A Perspective on the Root-locus Design Method

In Chapter 3 we related the features of a step response, such as rise time, overshoot, and settling time, to pole locations in the s-plane characterized by the natural frequency ω_n, the damping ratio ζ, and the real part σ. We also examined the changes in these transient-response features when a pole or a zero is added to a simple transfer function. In Chapter 4 we saw how feedback can influence stability and dynamic response by changing the system's pole locations.

In this chapter we present a specific technique, the root-locus method, which shows how changes in the system's feedback character-istics and other parameters influence the pole locations. Using this technique we can plot the locus of the closed-loop pole locations in the s-plane as a parameter varies; this produces a root locus (hence the name of the method).

In Fig. 4.8 we used a root locus (without identifying it as such) to show the effect of changing the feedback gain in a proportional control system. Root locus is most commonly used to study the effect of open-loop gain variations; however, because the method is general, it can be used to plot the roots of any polynomial expressed in root-locus form. For example, the root-locus method can be used to plot the roots of the characteristic equation that results from the analysis of digitally controlled systems using the z-transform, a topic we will discuss in Chapter 8.

Chapter Overview

We open in Section 5.1 by showing, through discussion and examples, the root-locus calculation for some basic feedback systems. Section 5.2 contains a step-by-step approach to sketching a root locus, and Section 5.3 applies this approach to a number of root-locus problems. Section 5.4 demonstrates the use of the root-locus method to gauge how the changes in parameters other than loop gain affect system design.

In Section 5.5 we discuss how to select gain based on analysis of the root locus of a system. When gain adjustment alone cannot produce a satisfactory design, lead and lag compensation can be used to modify system response, as described in Section 5.6. Section 5.7 extends the use of root-locus analysis to such problems as time delays and nonlinear systems. Finally, because drawing a root locus by hand is a somewhat tedious (though useful) step in system design, the chapter concludes in Section 5.8 with a discussion and demonstration of computer-aided graphing of the root locus.

5.1   Root Locus of a Basic Feedback System

We begin with the basic feedback system shown in Fig. 5.1. For this system the closed-loop transfer function is

$$\frac{Y(s)}{R(s)} = \frac{K_A K_G G(s)}{1 + K_A K_G G(s)}, \tag{5.1}$$

and the characteristic equation, whose roots are the poles of this transfer function, is

$$1 + K_A K_G G(s) = 0. \tag{5.2}$$

The closed-loop roots depend on the amplifier gain K_A, and we will have some influence over the closed-loop dynamic response through our choice of K_A. W. R. Evans suggested that we plot the locus of all possible roots of Eq. (5.2) as K_A varies from zero to infinity and then use the resulting plot to aid us in

FIGURE 5.1

Basic closed-loop block diagram for developing the root-locus method

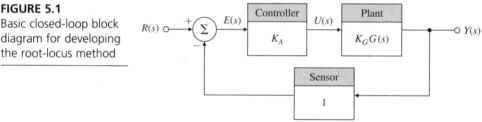

selecting the best value of K_A. Furthermore, by studying the effects of additional poles and zeros on this graph, we can determine the consequences of additional dynamics in the loop. We thus have a tool not only for selecting the gain but for designing the dynamic compensation as well. Similarly, we can extend this technique to examine the effect of other plant-parameter changes in order to achieve the best overall control design.

The graph of all possible roots of Eq. (5.2) relative to some particular variable (whether system gain or some other parameter) is called the **root locus,** and the design technique based on this graph is called the root-locus **method of Evans**. We begin our discussion of the method with the mechanics of constructing a root locus, using the amplifier gain as the variable parameter.

To set the notation for our study, we assume here that the system's open-loop transfer function $K_G G(s)$ is a rational function whose numerator is $K_G b(s)$, where $b(s)$ is a monic[1] polynomial of degree m, and whose denominator is a monic polynomial $a(s)$ of degree n such that $n \geq m$.[2] Furthermore,

Root locus
Evans's method

$$b(s) = s^m + b_1 s^{m-1} + \cdots + b_m \tag{5.3}$$

$$= (s - z_1)(s - z_2) \cdots (s - z_m)$$

$$= \prod_{i=1}^{m} (s - z_i),$$

$$a(s) = s^n + a_1 s^{n-1} + \cdots + a_n \tag{5.4}$$

$$= \prod_{i=1}^{n} (s - p_i).$$

For our initial study, we assume plant gain K_G is positive and define the root-locus parameter K as

$$K \triangleq K_A K_G.$$

The roots of $b(s) = 0$ are the zeros of $G(s)$ and are labeled z_i; similarly, the roots of $a(s) = 0$ are the poles of $G(s)$ and are labeled p_i.

We may now state the root-locus problem expressed in Eq. (5.2) in several equivalent but useful ways. Each of the following equations has the same roots:

$$1 + KG(s) = 0, \tag{5.5a}$$

$$1 + K\frac{b(s)}{a(s)} = 0, \tag{5.5b}$$

Root-locus form

$$a(s) + Kb(s) = 0, \tag{5.5c}$$

$$G(s) = -\frac{1}{K}. \tag{5.5d}$$

---

1. Monic means the coefficient of the highest power of s is 1.

2. When $G(s)$ represents a physical system, it is necessary that $n \geq m$ or else the system would have an infinite response to a finite input.

Eqs. (5.5) are sometimes referred to as the **root-locus form** of a characteristic equation. The root locus is the set of values of s for which Eqs. (5.5) holds for some positive real value of K (and K_A). Because the solutions to Eqs. (5.5) are the roots of the closed-loop system, the root-locus method can be thought of as a method for inferring properties of the closed-loop system given the open-loop transfer function $KG(s)$.

For purposes of the root locus development, we have defined $G(s)$ here to be the ratio of two monic polynomials, $b(s)$ and $a(s)$. In previous and future chapters, $G(s)$ is used as the plant transfer function and includes the constant K_G in Fig. 5.1. No matter how $G(s)$ is defined in a particular case, the root locus gain, K, should always be determined based on the definitions here.

◆ **EXAMPLE 5.1**   *Root Locus for a DC Motor*

A normalized transfer function of the DC motor from Eq. (2.44) is

$$\frac{\theta_m(s)}{v_a(s)} = K_G G(s) = \frac{1}{s(s + 1)}.$$

Solve for the locus of roots with respect to K_A of the closed-loop system created by feeding back the output θ_m as shown in Fig. 5.1. Solve by using direct calculations of the root locations.

Solution.   In terms of our notation

$$m = 0, \qquad K_G = 1, \qquad b(s) = 1, \qquad K_A = K$$

$$n = 2, \qquad a(s) = s^2 + s, \qquad p_i = 0, \ -1.$$

From Eq. (5.5c) the root locus is a graph of the roots of the quadratic equation

$$s^2 + s + K = 0. \tag{5.6}$$

Using the quadratic formula we can immediately express the roots of Eq. (5.6) as

$$r_1, r_2 = -\frac{1}{2} \pm \frac{\sqrt{1 - 4K}}{2}. \tag{5.7}$$

A plot of the corresponding root locus is shown in Fig. 5.2. For $0 \leqslant K \leqslant \frac{1}{4}$, the roots are real between -1 and 0. At $K = \frac{1}{4}$ there are two roots at $-\frac{1}{2}$, and for $K > \frac{1}{4}$ the roots become complex, with a real part of $-\frac{1}{2}$ and an imaginary part that increases essentially in proportion to the square root of K. The dashed lines in Fig. 5.2 correspond to roots with a damping ratio $\zeta = 0.5$. The poles of $G(s)$ at $s = 0$ and $s = -1$ are marked by the symbol ×, and the points where the locus crosses the 0.5 damping-ratio lines are marked with dots (●). We can compute K at the point where the locus crosses $\zeta = 0.5$ because we know that, if $\zeta = 0.5$, then $\theta = 30°$ and the magnitude of the imaginary part of the root is $\sqrt{3}$ times the magnitude of the real part. Since the size of the real part is $\frac{1}{2}$, then from Eq.

FIGURE 5.2
Root locus for $G(s) = 1/$
$[s(s + 1)]$ as a function
of open-loop gain K

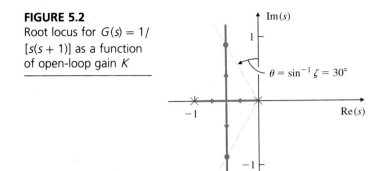

(5.7), we have

$$\frac{\sqrt{4K - 1}}{2} = \frac{\sqrt{3}}{2},$$

and, therefore, $K = 1$.

We can observe several features of this simple locus by looking at Eqs. (5.6) and (5.7) and Fig. 5.2. First, there are two roots and thus two branches to the root locus. At $K = 0$ these branches begin at the poles of $G(s)$ (which are at 0 and -1) as they should, since for $K = 0$ the system is open-loop. As K is increased, the roots move toward each other, coming together at $s = -\frac{1}{2}$. At that point they break away from the real axis. After the **breakaway point** the roots move off to infinity with equal real parts, so the sum of the two roots is always -1. From the viewpoint of design, we see that, by altering the value of the gain K, we can cause the closed-loop poles to fall at any point along the cross that makes up the root locus in Fig. 5.2. If some points along this locus correspond to a satisfactory transient response, then we can complete the design by choosing the corresponding value of K; otherwise, we are forced to consider a more complex controller.

> Breakaway points are where roots move away from the real axis.

◆

As we pointed out earlier, Evans's technique is not limited to focusing on the system gain (K in Example 5.1); the same ideas are applicable for finding the locus with respect to *any* parameter in the characteristic equation.

◆ **EXAMPLE 5.2**   *Root Locus with Respect to Plant Parameter*

Consider the $G(s)$ function in Example 5.1, except that now K is fixed ($= 1$), and the motor time constant $\tau = 1/c$ is the parameter of interest:

$$G(s) = \frac{1}{s(s + c)}. \tag{5.8}$$

Find the root locus with respect to c.

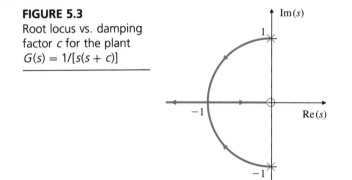

FIGURE 5.3
Root locus vs. damping
factor c for the plant
$G(s) = 1/[s(s + c)]$

Solution.   The corresponding closed-loop characteristic equation is

$$1 + G(s) = 0$$

or

$$s^2 + cs + 1 = 0. \tag{5.9}$$

The same forms of Eq. (5.5), with the associated definitions of poles and zeros, will apply if
we let

$$a(s) = s^2 + 1,$$
$$b(s) = s,$$
$$K = c.$$

Thus, a root-locus form of the characteristic equation is

$$1 + c\,\frac{s}{s^2 + 1} = 0.$$

The solutions to Eq. (5.9) are easily computed as

$$r_1, r_2 = -\frac{c}{2} \pm \frac{\sqrt{c^2 - 4}}{2}. \tag{5.10}$$

The locus of solutions versus the parameter c is shown in Fig. 5.3, with the poles [roots of
$a(s)$] again indicated by $\times$ s and the zero [root of $b(s)$] by the $\bigcirc$ symbol. Note that
when $c = 0$, the roots are at the $\times$ s and are oscillatory, while the damping ratio ζ grows
as c grows. At $s = -1$ the two locus segments from the poles abruptly change direction
Break-in point   and move in opposite directions along the real axis; this point is called the **break-in point**.

Of course, computing the root locus for a quadratic equation is easy to do
since we can solve the characteristic equation (5.6 or 5.9) for the roots, as is done
in Eq. (5.7) and (5.10), and directly plot these as a function of the parameter K or
c. To be useful the method must be suitable for much higher-order systems
where explicit solutions are impossible to obtain. General properties of the root

locus and rules for the construction of complex graphs of roots were developed by Evans, and together they constitute the mechanics of drawing a root locus, the core of the method. In the next section we will present, as step-by-step guidelines, the general method for sketching any root locus. The techniques will be developed using the loop gain K as the parameter of interest. In Section 5.4 we will modify the technique to apply to other parameters.

5.2   Guidelines for Sketching a Root Locus

As we saw in the previous section, the purpose of the root locus is to show in a graphical form the general trend of the roots of the closed-loop system as we vary some parameter. There are two reasons for learning how to generate a root locus by hand: (1) to use the manual method for designing fairly simple systems and (2) to verify and understand computer-generated root loci. Our goal in this section is to extract as much information as possible from Eqs. (5.5) to aid us in sketching a reasonably accurate picture of the root locus for the given $G(s)$.

When using the method to verify and understand the computer results, we are interested more in the general shape of the locus than in specific values, so some of the following rules may be omitted. We can judge important features from a computer-generated locus as correct or not by understanding how loci must behave. For a manual control design to be feasible, we need to be able to sketch a locus quickly and with reasonable accuracy so that several attempts at the trial-and-error process are possible. Experience allows us to do this and, in addition, to make good estimates of how an additional pole or zero will change a locus and permit a better control system design.

In Eq. (5.5a) we defined the root locus this way:

> *Definition I:* The root locus is the set of values of s for which $1 + KG(s) = 0$ is satisfied as the real parameter K varies from 0 to $+\infty$. Often $G(s)$ is the open-loop transfer function of a system; in this case, roots on the locus are **closed-loop poles** of that system.

Now suppose we look at Eq. (5.5d). If K is to be real and positive, $G(s)$ must be real and negative. In other words, if we arrange $G(s)$ in polar form as magnitude and phase, then $G(s)$ must have the opposite phase of K in order to satisfy Eq. (5.5d). We can thus define the root locus in terms of this **phase condition** as follows.

The basic root-locus rule; the phase of $G(s) = 180°$.

> *Definition II:* The root locus of $G(s)$ is the set of points in the s-plane where the phase of $G(s)$ is 180°.

Since the phase remains unchanged if an integral multiple of 360° is added, we can express Definition II as $\angle G = 180° + 360° l$, where l is any integer. The immense merit of Definition II is that, while it is very difficult to solve a high-order polynomial, computing the phase is relatively easy. The usual case is when K is real and positive; we call this case the **positive or 180° locus**. When K is real

FIGURE 5.4

Measuring the phase of
Eq. (5.11)

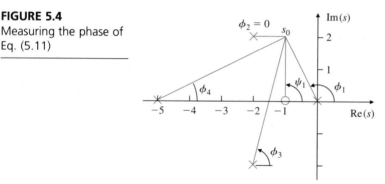

and negative, $G(s)$ must be real and positive for s to be on the locus. Therefore, the phase of $G(s)$ must be $0°$. This special case is called a **$0°$ or negative locus**. See Section 5.4.2.

From Definition II we can in principle sketch a root locus for a complex transfer function by measuring the phase and marking those places where we find $180°$. This direct approach can be illustrated by considering the example

$$G(s) = \frac{s + 1}{s[(s + 2)^2 + 4](s + 5)}. \tag{5.11}$$

In Fig. 5.4 the poles of this $G(s)$ are marked $\times$ and the zero is marked $\bigcirc$. Suppose we select the test point $s_0 = -1 + 2j$. We would like to test whether or not s_0 lies on the root locus for some value of K. For this point to be on the locus, we must have $\angle G(s_0) = 180° + 360°l$ for some integer l, or equivalently, from Eq. (5.11)

$$\angle(s_0 + 1) - \angle s_0 - \angle[(s_0 + 2)^2 + 4] - \angle(s_0 + 5) = 180° + 360°l \tag{5.12}$$

The angle from the zero term $s_0 + 1$ can be computed[3] by drawing a line from the location of the zero at -1 to the test point s_0. In this case the line is vertical and has a phase angle, marked ψ_1 on Fig. 5.4. In similar fashion, the vector s_0 is shown with angle ϕ_1, and the angles of the two vectors from the complex poles at $-2 \pm 2j$ to s_0 are shown with angles ϕ_2 and ϕ_3. The phase of the vector $s_0 + 5$ is shown with angle ϕ_4. From Eq. (5.12) we find the total phase of $G(s)$ at $s = s_0$ to be the sum of the phase of the zero term minus the phases of the denominator terms corresponding to the poles:

$$\angle G = \psi_1 - \phi_1 - \phi_2 - \phi_3 - \phi_4$$
$$= 90° - 116.6° - 0° - 76° - 26.6°$$
$$= -129.2°.$$

---

3. The graphical evaluation of the magnitude and phase of a complex number is reviewed in Appendix B, Section B.3.

FIGURE 5.5

Step 1: Mark the poles and zeros

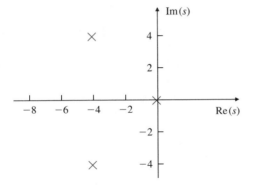

Since the phase of G is not $180°$, we conclude that s_0 is *not* on the root locus, so we must select another point and try again. Although measuring phase is easy, measuring phase at every point in the s-plane is hardly practical. Therefore, we need some general guidelines for determining where the root locus is. For simple reference later, these are presented here as a list of steps, with the following transfer function for illustration:

$$G(s) = \frac{1}{s[(s+4)^2 + 16]}. \qquad (5.13)$$

STEP 1.   *Draw the axes of the s-plane to a suitable scale and enter an $\times$ on this plane for each pole of $G(s)$ and a $\bigcirc$ for each zero of $G(s)$.* See Fig. 5.5.

STEP 2.   *Find the real axis portions of the locus.* If we take a test point on the real axis such as s_0 in Fig. 5.6, we find that the angles ϕ_1 and ϕ_2 of the two factors in $a(s)$ due to the complex poles satisfy $\phi_1 = -\phi_2$ and they cancel each other. The same would be true, of course, for the angles from complex zeros. The result is that the angle of $G(s_0)$ for s_0 on the real axis is given by the angles from poles and zeros *on the real axis only.* But these angles are $0°$ if the test point is to the right and $180°$ if the test point is to the *left* of a given pole or zero. For the

FIGURE 5.6

Step 2: Find the real-axis part of the locus.

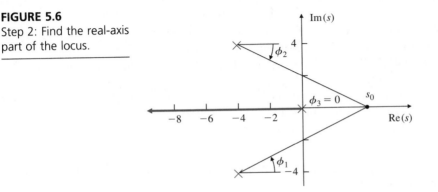

angle to add to $180° + 360°l$, the test point must be to the left of an *odd* number of real-axis poles plus zeros. In our test case there is one real-axis pole at the origin, so the locus is on the real axis for all s_0 to the left of the origin. Accordingly, we mark this segment with a heavy line in Fig. 5.6.

STEP 3.   *Draw the asymptotes for large values of K.* As $K \to \infty$, the equation $1 + KG(s) = 0$ or

$$G(s) = -\frac{1}{K}$$

can be satisfied only if $G(s) = 0$. This can occur in two apparently different ways. In the first instance, since $G = b(s)/a(s)$, we have $G = 0$ if $b(s) = 0$. Therefore, for large values of K a root of $1 + KG = 0$ will be found near the zeros of G, which is where $b(s) = 0$.

   To see the second manner in which $G(s)$ may go to zero, we substitute Eqs. (5.3) and (5.4) into Eq. (5.5b),

$$1 + K\frac{b(s)}{a(s)} = 0,$$

to get

$$1 + K\frac{s^m + b_1 s^{m-1} + \cdots + b_m}{s^n + a_1 s^{n-1} + \cdots + a_n} = 0. \tag{5.14}$$

We are interested in what happens for large values of K. Since $n > m$ when $G(s)$ represents a physical system, it is clear that $G(s)$ goes to zero as $s \to \infty$. In fact, for very large values of s, the highest-order power of s in Eq. (5.14) predominates, allowing it to be approximated[4] by

$$1 + K\frac{1}{(s - \alpha)^{n-m}} = 0. \tag{5.15}$$

This is the equation for a system where there are $n - m$ poles, all clustered at $s = \alpha$. Another way to see this same result is to consider the picture we would see if we could observe the locations of poles and zeros from a vantage point near infinity: They would appear to cluster near the s-plane origin. Thus m zeros would cancel the effects of m poles, and the other $n - m$ poles would appear to be in the same place. We say that the locus of Eq. (5.14) is asymptotic to the locus of Eq. (5.15) for large values of K and s. We need to find the locus for the asymptotic system and to compute α.

   To find the locus, we choose our search point s_0 such that $s_0 = Re^{j\phi}$ for some large fixed value of R and some variable ϕ. Since all poles of this simple system are in the same place, the angle of its transfer function is $180°$ if all $n - m$

---

4. This approximation can be obtained by dividing $a(s)$ by $b(s)$ and matching the dominant two terms (highest powers in s) to the expansion $(s - \alpha)^{n-m}$.

angles, each equal to ϕ_l, sum to 180°. Therefore, ϕ_l is given by

$$(n - m)\phi_l = 180° + 360°l$$

for some integer l and for any value of R. Thus, the asymptotic root locus consists of radial lines at the $n - m$ distinct angles given by

There are $n - m$ asymptotes.

$$\phi_l = \frac{180° + 360°(l - 1)}{n - m}, \qquad l = 1, 2, \ldots, n - m. \tag{5.16}$$

For our example system, $n - m = 3$ and $\phi_{1,2,3} = 60°$, 180°, and 300°. The lines of the asymptotic locus come from $s_0 = \alpha$ on the real axis.

To determine α we make use of a simple property of polynomials. Suppose we have a monic polynomial with coefficients a_i and roots p_i, as in Eq. (5.4), and we equate the polynomial form with the factored form

$$s^n + a_1 s^{n-1} + a_2 s^{n-2} + \cdots + a_n = (s - p_1)(s - p_2) \cdots (s - p_n).$$

If we multiply out the factors on the right side of this equation, we see that the coefficient of s^{n-1} is $-p_1 - p_2 - \cdots - p_n$. On the left side of the equation we see that this term is a_1. Thus $a_1 = -\Sigma p_i$; in other words, the coefficient of the *second*-highest term in a monic polynomial is the negative sum of its roots—in this case, the poles of $G(s)$. Applying this result to the polynomial $b(s)$ we find the negative sum of the zeros to be b_1. These results can be written as

$$\begin{aligned} -b_1 &= \sum z_i, \\ -a_1 &= \sum p_i. \end{aligned} \tag{5.17}$$

Finally, we apply this result to the closed-loop characteristic polynomial obtained from Eq. (5.14):

$$s^n + a_1 s^{n-1} + \cdots + a_n + K(s^m + b_1 s^{m-1} + \cdots + b_m) = 0.$$

Note that the sum of the poles is the negative of the coefficient of s^{n-1}, and is *independent of K* if $m < n - 1$. But since this is the closed-loop characteristic equation, this coefficient is the negative of the sum of the roots of the *closed-loop system* Σr_i. We have thus shown that the center point of the roots does not change with K if $m < n - 1$ and that the common open-loop and closed-loop sum is $-a_1$, which can be expressed as

$$-\sum r_i = -\sum p_i. \tag{5.18}$$

For large values of K, m of the roots r_i are approximately equal to the zeros z_i and $n - m$ of the roots are from the asymptotic $1/(s - \alpha)^{n-m}$ system, whose poles add up to $(n - m)\alpha$. Combining these results we conclude that the sum of all the roots equals the sum of those roots that go to infinity plus the sum of those roots that go to the zeros of $G(s)$:

$$+\sum r_i = +(n - m)\alpha + \sum z_i = +\sum p_i.$$

FIGURE 5.7
Step 3: Draw the
asymptotes.

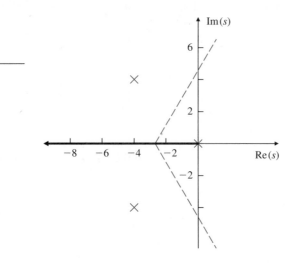

Solving for α we get

The asymptotes are
centered at α.

$$\alpha = \frac{\sum p_i - \sum z_i}{n - m}. \tag{5.19}$$

Notice that in the sums $\sum p_i$ and $\sum z_i$ the imaginary parts *always* add to zero since complex poles and zeros always occur in complex conjugate pairs. Thus Eq. (5.19) requires information about the real parts only. For our example system,

$$\alpha = \frac{-4 - 4 + 0}{3 - 0}$$

$$= -\tfrac{8}{3} = -2.67.$$

The asymptotes at $\pm 60°$ are shown dashed in Fig. 5.7. Notice that they cross the imaginary axis at $\pm(2.67)j\sqrt{3} = \pm 4.62j$. The asymptote at $180°$ was already found in step 2.

STEP 4. *Compute the departure and arrival angles.* At this point we know that the locus begins at the points marked $\times$ [the poles of $G(s)$] and that it goes either to the points marked $\bigcirc$ [the zeros of $G(s)$] or to infinity along the radial asymptotic lines. We next compute the angle by which a branch of the locus departs from one of the poles. For this purpose we take a test point s_0 very near pole 2 at $-4 + 4j$ and compute the angle of $G(s_0)$. The situation is sketched in Fig. 5.8. We select the test point close enough to pole 2 that the angles ϕ_1 and ϕ_3 to the test point can be considered the same as those angles to pole 2. Thus, $\phi_1 = 90°$ and $\phi_3 = 135°$, and ϕ_2 can be calculated from the angle condition

$$-90° - \phi_2 - 135° = +180° + 360°l \tag{5.20}$$

FIGURE 5.8
Step 4: Compute the
departure and arrival
angles.

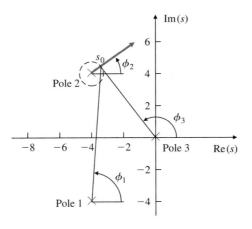

where the integer l is selected so that $-180° < \phi_2 < +180°$. If we let $l = -1$, then

$$-\phi_2 = 90° + 135° + 180° - 360°,$$

$$\phi_2 = -45°.$$

By the complex conjugate symmetry of the plots, the angle of departure of the locus near pole 1 at $-4 - 4j$ will be $+45°$. Because of the angle condition on the real axis discussed in Step 2, the angle of departure from pole 3 at the origin is $180°$, so the locus departs along the real axis.

If there had been some zeros in $G(s)$, the angles from pole 2 to the zeros would have been added to the left side of Eq. (5.20). For the general case, we can see from Eq. (5.20) that the angle of departure from a single pole is

$$\phi_{\text{dep}} = \sum \psi_i - \sum \phi_i - 180° - 360°l \tag{5.21}$$

where $\sum \phi_i$ is the sum of the angles to the remaining poles and $\sum \psi_i$ is the sum of the angles to all the zeros. For a multiple pole of order q, we must count the angle from the pole q times. This alters Eq. (5.21) to

Rule for departure angles

$$q\phi_{\text{dep}} = \sum \psi_i - \sum \phi_i - 180° - 360°l, \tag{5.22}$$

where l takes on q values because there are q branches of the locus that depart from that multiple pole.

The process of calculating a departure angle for small values of K, as shown in Fig. 5.8, is also valid for computing the angle by which a root locus arrives at a zero of $G(s)$ for large values of K. The general formula that results is

Rule for arrival angles

$$q\psi_{\text{arr}} = \sum \phi_i - \sum \psi_i + 180° + 360°l \tag{5.23}$$

where $\sum \phi_i$ is the sum of the angles to all the poles, $\sum \psi_i$ is the sum of the angles to the remaining zeros, and l is determined as it was for Eq. (5.22).

Routh's stability criterion can be used for Step 5.

STEP 5. *Estimate (or compute) the points where the locus crosses the imaginary axis.*[5] A root of the characteristic equation in the right half-plane (RHP) implies that the closed-loop system is unstable, a fact that can be tested by Routh's stability criterion (see Chapter 4). If we compute the Routh array using K as a parameter, we can locate those values of K for which an incremental change will cause the number of roots in the RHP to change. Such values must correspond to a root locus crossing the imaginary axis. Given the facts that K is known and that a root exists for $s_0 = j\omega_0$, it is often not hard to solve for the frequency ω_0.

For the third-order example we are using, the characteristic equation is

$$1 + \frac{K}{s[(s+4)^2 + 16]} = 0,$$

which is equivalent to

$$s^3 + 8s^2 + 32s + K = 0. \tag{5.24}$$

The Routh array for this polynomial is

$$
\begin{array}{lll}
s^3: & 1 & 32 \\
s^2: & 8 & K \\
s^1: & \dfrac{8 \cdot 32 - K}{8} & 0 \\
s^0: & K &
\end{array}
$$

In this case we see that the equation has no roots in the RHP for $0 < K < 8 \cdot 32 = 256$. Since we are interested in the locus for positive values of K, we focus on the upper bound. For $K < 256$ there are no roots in the RHP, and if $K > 256$, the Routh test indicates that there are two roots in the RHP. Thus $K = 256$ must correspond to a solution at $s = j\omega_0$ for some ω_0. Substituting this data into Eq. (5.24) we find

$$(j\omega_0)^3 + 8(j\omega_0)^2 + 32(j\omega_0) + 256 = 0. \tag{5.25}$$

For Eq. (5.25) to be true, both real and imaginary parts must equal zero.[6] This gives us

$$-8\omega_0^2 + 256 = 0,$$

and

$$-\omega_0^3 + 32\omega_0 = 0. \tag{5.26}$$

The solution to Eq. (5.26) is $\omega_0 = \pm\sqrt{32} = \pm 5.66$, which is plotted in Fig. 5.9.

---

5. This step is tedious, especially for fourth- or higher-order systems, and is usually omitted.

6. Notice that we could have substituted an unknown value of K into Eq. (5.25) and computed $K = 256$ without using the Routh array in this case.

FIGURE 5.9

Step 5: Find the Im(s)-axis crossing points.

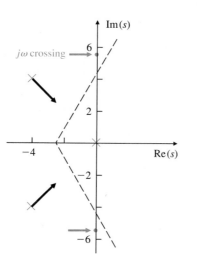

We note that the asymptotic locus crosses the imaginary axis at $s = 4.62j$, which is below the actual locus crossing.

STEP 6.   *Estimate locations of multiple roots, especially on the real axis, and determine the arrival and departure angles at these locations.*[7] As with any polynomial, it is possible for a characteristic polynomial of a degree greater than 1 to have multiple roots. In the second-order locus in Fig. 5.2, there are two roots at $s = -\frac{1}{2}$ when $K = \frac{1}{4}$. Here the horizontal branches of the locus come together and the vertical branches break away from the real axis, becoming complex for $K > \frac{1}{4}$. The locus arrives at $0°$ and $180°$ and departs at $+90°$ and $-90°$. It is possible to use a formula for points of multiple roots, but the formula is only slightly easier to solve than the original root locus, so it is usually only used for simple systems.

Let us look again at the second-order root locus of Fig. 5.2 and consider what happens to the gain K between -1 and 0 along the real axis. Notice that $K = 0$ at $s = -1$, that K increases steadily to $K = \frac{1}{4}$ at $s = -\frac{1}{2}$, and that, *staying on the real axis*, K decreases back to zero at $s = 0$. The gain has a maximum at the point of multiple roots, $s = -\frac{1}{2}$. Since $G(s)$ is a smooth function except at a pole and since $K = -1/G$, the change in K must also be smooth. Therefore, if K is to have a maximum at this point, then $dK/ds = 0$ at the locus breakaway or multiple-root point. Using the relation $K = -1/G$, we thus obtain the formula

$$\frac{d}{ds}\left(-\frac{1}{G}\right)_{s=s_0} = 0. \tag{5.27}$$

---

7. Like Step 5, this step is also not often done by hand, but understanding the principles aids in the interpretation of computer-generated loci.

Since $G = b/a$, Eq. (5.27) requires that

$$\frac{d}{ds}\left(-\frac{a}{b}\right) = -\frac{1}{b^2}\left(b\frac{da}{ds} - a\frac{db}{ds}\right) = 0. \qquad (5.28)$$

For the second-order case in Example 5.1, we have

$$b = 1, \qquad \frac{db}{ds} = 0;$$

$$a = s^2 + s, \qquad \frac{da}{ds} = 2s + 1.$$

Substituting these expressions into Eq. (5.28), we find

$$2s_0 + 1 = 0$$

$$s_0 = -\tfrac{1}{2}, \qquad (5.29)$$

which confirms our previous observation of the breakaway point in Fig. 5.2. For the third-order case given by the plots in Figs. 5.5 to 5.10, the polynomials are

$$b = 1, \qquad \frac{db}{ds} = 0;$$

$$a = s^3 + 8s^2 + 32s, \qquad \frac{da}{ds} = 3s^2 + 16s + 32.$$

The points of possible multiple roots are given by

$$3s^2 + 16s + 32 = 0,$$

or

$$s_0 = -2.67 \pm 1.89j. \qquad (5.30)$$

However Fig. 5.10 shows that these points are *not* on the root locus and are therefore extraneous. This emphasizes that Eq. (5.28) does not indicate whether s_0 is a multiple root on the locus. The derivative condition is necessary but not sufficient to indicate a breakaway (multiple-root) situation.

In order to compute (estimate) the angles of arrival and departure from a point of multiple roots, it is useful to introduce two additional features of the root locus. The first is the concept of the **continuation locus**. We can imagine plotting a root locus for an initial range of K, perhaps for $0 \leqslant K \leqslant K_0$. If we let $K = K_0 + K_1$, we can then plot a new locus with parameter K_1, a locus whose starting poles are the roots of the original system at $K = K_0$. To see how this works, we return to the second-order root locus of Eq. (5.6) and let K_0 be the value corresponding to the breakaway point $K_0 = \tfrac{1}{4}$. If we let $K = \tfrac{1}{4} + K_1$, we have the locus equation $s^2 + s + \tfrac{1}{4} + K_1 = 0$, or

$$(s + \tfrac{1}{2})^2 + K_1 = 0. \qquad (5.31)$$

Continuation locus

The steps for plotting this locus are, of course, the same as for any other, except that now the initial departure of the K_1 locus of Eq. (5.31) corresponds to the breakaway point of the K locus of Eq. (5.6). Applying the rule for departure angles (Eq. (5.22) from the double pole at $s = -\frac{1}{2}$, we find that

$$2\phi_{\text{dep}} = -180° - 360°l,$$
$$\phi_{\text{dep}} = -90° - 180°l,$$
$$\phi_{\text{dep}} = \pm 90° \text{ (departure angles at breakaway).} \tag{5.32}$$

In this case the arrival angles at $s = -\frac{1}{2}$ are, from the original root locus, along the real axis and are clearly $0°$ and $180°$.

This same development can be used to show the second new feature of root loci. Two locus segments coming toward each other on the real axis will *always* break away at $\pm 90°$. Furthermore, two locus segments coming together at any point in the s-plane will always approach at a relative angle of $180°$ and then break away with a $\pm 90°$ change in direction. Three locus segments coming together will always approach at $120°$ angles with respect to each other and then depart at the $120°$ angles that are rotated $60°$ relative to the arrival angles.

STEP 7.   *Complete the sketch.* The complete locus for our third-order example is drawn in Fig. 5.10. It combines all the results found so far—that is, the real-axis segment, the number of asymptotes and their angles, the angles of departure from the poles, and the imaginary-axis crossing points. It is usually sufficient to sketch the locus by using only Steps 1 to 4. Step 5 can be used to refine a hand-drawn locus if no CACSD (computer-aided control systems design) software is available to provide the detail. Step 6 is almost never done today, but the information is useful to help interpret plots that come from the computer, especially when limited values of K have been plotted.

FIGURE 5.10
Step 7: Complete the sketch.

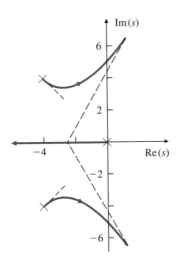

**Summary: Guidelines for
Plotting a 180° Root Locus**

1. Mark poles with an $\times$ and zeros with a $\bigcirc$.
2. Draw the locus on the real axis to the left of an odd number of real poles plus zeros.
3. Draw the asymptotes, centered at α and leaving at angles ϕ_l, where

$$n - m = \text{number of asymptotes}$$

$$\alpha = \frac{\sum p_i - \sum z_i}{n - m} = \frac{-a_1 + b_1}{n - m},$$

$$\phi_l = \frac{180° + 360°(l - 1)}{n - m}, \qquad l = 1, 2, \ldots, n - m.$$

4. Compute locus departure angles from the poles and arrival angles at the zeros:

$$q\phi_{\text{dep}} = \sum \psi_i - \sum \phi_i - 180° - 360°l,$$

$$q\psi_{\text{arr}} = \sum \phi_i - \sum \psi_i + 180° + 360°l,$$

where q is the order of the pole or zero and l takes on q integer values so that the angles are between $\pm 180°$.
5. If further refinement is required at the stability boundary: Assume $s_0 = j\omega_0$, and compute the point(s) where the locus crosses the imaginary axis for positive values of K, and/or use Routh's stability criterion. (This step may not be necessary.)
6. Use the results from the study of multiple roots to help in sketching how locus segments come together and break away: Two segments come together at 180° and break away at $\pm 90°$. Three locus segments approach each other at relative angles of 120° and depart at angles rotated by 60°.
7. Complete the locus, using the facts developed in the previous steps and making reference to the illustrative loci for guidance. The locus branches start at poles and end at zeros or infinity.

5.3   Selected Illustrative Root Loci

In order to gain facility with the root-locus method, it is helpful to sketch a number of locus plots and visualize the various alternative shapes a given locus can take. Using the seven-step procedure described in the previous section, we will now consider several cases that illustrate important features of root loci. All the loci in this section could be generated quite quickly with CACSD software; however, you will not have as good an understanding of why loci behave as they do if you opt for computer solutions at this point in the learning process.

For the first four cases we consider the locus of roots of $1 + KG(s) = 0$, where

$$G(s) = \frac{s + 1}{s^2(s + p)}, \tag{5.33}$$

for various values of p.

◆ **EXAMPLE 5.3**   *Illustrative Locus: Large Values of p*

Sketch the root locus for very large values of p in Eq. (5.33)—namely, the locus for

$$G(s) = (s + 1)/s^2.$$

Go through each of the seven steps of the guidelines.

Solution

STEP 1.   We mark the poles and zeros on the *s*-plane:

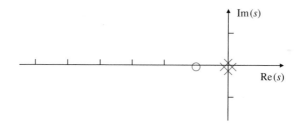

STEP 2.   We draw the locus on the real axis to the left of an odd number of poles plus zeros:

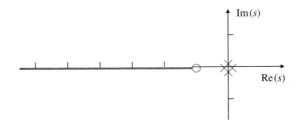

STEP 3.   We find the number of asymptotes and draw them. Here there is

$n - m = 2 - 1 = 1$ asymptote:

$$\phi_1 = \frac{180°}{2 - 1}$$

$$= 180°.$$

α does not matter, since there is only one asymptote. In this case we find that the asymptote was created by Step 2.

STEP 4.   We compute the departure and arrival angles. For the two poles at $s = 0$, we draw a small circle around them. The angles from the two poles are equal, and the angle from the zero is essentially zero, so the angle condition reduces to

$$-2\phi_l + 0° = 180° + 360°l,$$

$$\phi_1 = -90°, \quad \phi_2 = +90°.$$

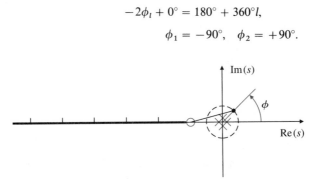

The locus leaves vertically, one branch up and the other down, as indicated by arrows:

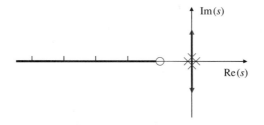

STEP 5.   We compute the points where the locus crosses the imaginary axis. The characteristic equation is

$$s^2 + Ks + K = 0,$$

and the Routh array is

s^2:	1	K
s^1:	K	
s^0:	K	

Since the entries in the first column are all positive if $K > 0$, the equation's roots are all in the LHP on the root locus and do *not* cross the imaginary axis.

FIGURE 5.11
Root locus for
$G(s) = (s + 1)/s^2$

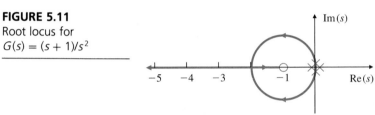

STEP 6.   We locate the possible multiple roots according to Eq. (5.28). In this case,

$$b(s) = s + 1, \qquad a(s) = s^2;$$

$$\frac{db}{ds} = 1, \qquad \frac{da}{ds} = 2s.$$

The root-locus condition is

$$-s^2(1) + (s + 1)(2s) = 0,$$
$$s^2 + 2s = 0,$$
$$s = 0, \; -2.$$

The root at $s = 0$ corresponds to the two poles of $G(s)$ that are on the locus at $K = 0$. The other case at $s = -2$ corresponds to a multiple root, since in Step 2 we found that $s = -2$ is on the locus. Therefore, we know that the locus break-in point is at $s = -2$.

STEP 7.   Finally, we sketch the locus shown in Fig. 5.11. In this case, part of the locus is a circle. (See also Problem 5.10).

◆

◆ **EXAMPLE 5.4**   *Illustrative Locus: p = 4*

Let $p = 4$ in Eq. (5.33), so

$$G(s) = \frac{s + 1}{s^2(s + 4)},$$

and find the root locus of $1 + KG(s) = 0$.

Solution

STEP 1.   We mark the poles and zeros on the s-plane:

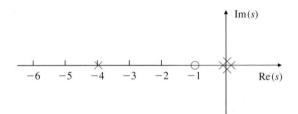

STEP 2. We draw the locus on the real axis to the left of an odd number of poles plus zeros:

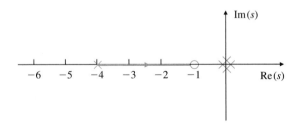

STEP 3. We draw the $n - m = 3 - 1 = 2$ asymptotes:

$$\phi_l = \frac{180° + 360°(l - 1)}{n - m}, \qquad\qquad \alpha = \frac{\sum p_i - \sum z_i}{n - m}$$

$$= \frac{180° + 360°(l - 1)}{3 - 1} \qquad\qquad = \frac{-4 - (-1)}{3 - 1}$$

$$= 90° + 180°(l - 1), \qquad\qquad = -\tfrac{3}{2}.$$

$$\phi_1 = +90°, \quad \phi_2 = -90°;$$

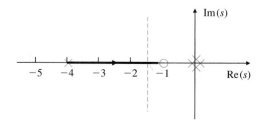

STEP 4. We compute the departure angles from the poles and the arrival angles toward the zeros. We draw a small circle around the two poles at $s = 0$. The angles from the zero at -1 and from the pole at -4 are both zero, and the angles from the two poles at the origin are the same. Therefore, the root-locus condition is

$$-2\phi_l - 0° + 0° = 180° + 360°l,$$

$$\phi_l = -90° - 180°l$$

$$= \pm 90°.$$

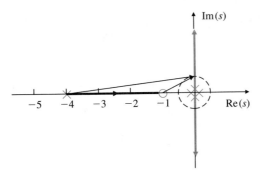

STEP 5.   We compute the point(s) where the locus crosses the imaginary axis:

$$1 + K \frac{s + 1}{s^2(s + 4)} = 0,$$

$$s^3 + 4s^2 + Ks + K = 0.$$

We let $s = j\omega$, to get

$$-j\omega^3 - 4\omega^2 + Kj\omega + K = 0,$$

and set real and imaginary parts separately to zero:

$$-4\omega^2 + K = 0, \qquad -\omega^3 + K\omega = 0,$$

$$K = 4\omega^2; \qquad K = \omega^2.$$

The two equations have no solution except $K = \omega = 0$, so we conclude that the branches of the locus do not cross the imaginary axis for $K > 0$. We can confirm this by performing the Routh test:

$$
\begin{array}{lll}
s^3: & 1 & K \\
s^2: & 4 & K \\
s^1: & \dfrac{4K - K}{4} & \\
s^0: & K. &
\end{array}
$$

The entries in the first column are all positive if $K > 0$, so the equation has no roots in the RHP for positive values of K.

STEP 6.   We locate the points of multiple roots, which will include breakaway and break-in points:

$$b(s) = s + 1, \qquad a(s) = s^3 + 4s^2;$$

$$\frac{db}{ds} = 1, \qquad \frac{da}{ds} = 3s^2 + 8s.$$

Substituting these expressions into Eq. (5.28), we get

$$(s + 1)(3s^2 + 8s) - (s^3 + 4s^2)(1) = 0,$$

$$-s^3 - 4s^2 + 3s^3 + 11s^2 + 8s = 0,$$

$$2s^3 + 7s^2 + 8s = 0,$$

$$s(2s^2 + 7s + 8) = 0.$$

The possible locations are

$$s = 0 \quad \text{and} \quad s = -1.75 \pm 0.97j.$$

Testing whether the mathematical solution is a multiple root

The root at $s = 0$ is on the locus and corresponds to the two poles of $G(s)$ at $s = 0$, which are on the locus for $K = 0$. The two other roots are not on the root locus and therefore do not correspond to multiple roots. To confirm this, we could compute the phase of $G(-1.75 \pm 0.97j)$ and determine whether it was $\cong 180°$. However, it does not make sense that two branches of the locus would come together at this location because there is only one branch in the upper half of the s-plane.

STEP 7.   The complete root locus is drawn in Fig. 5.12.

FIGURE 5.12
Root Locus for
$G(s) = (s + 1)/$
$[s^2(s + 4)]$

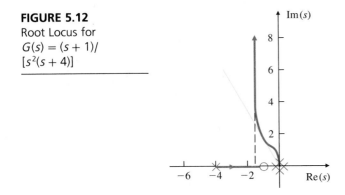

◆ **EXAMPLE 5.5**   *Illustrative Locus: p = 12*

Plot the root locus for $1 + KG(s) = 0$, where $p = 12$ in Eq. (5.33), so that

$$G(s) = \frac{s + 1}{s^2(s + 12)}.$$

Solution

STEP 1. We mark the poles and zero:

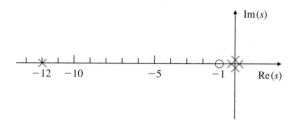

STEP 2. We draw the real-axis branch of the locus:

STEP 3. We draw the asymptotes:

$$\phi_l = \frac{180° + 360°(l - 1)}{3 - 1}, \qquad \alpha = \frac{-12 - 0 - (-1)}{3 - 1}$$

$$= \pm 90°, \qquad = -\frac{11}{2}.$$

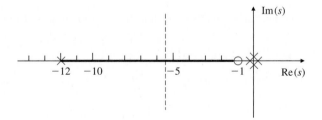

STEP 4. We compute the departure and arrival angles at the poles at $s = 0$: $\phi_{1,2} = \pm 90°$:

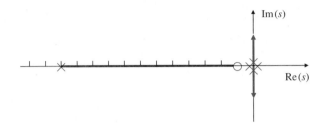

Step 5 may be skipped based on accumulated knowledge of this series of loci.

STEP 5. Based on what we have already learned about how this series of loci behaves and the fact that the asymptotes are at $\alpha = -\frac{11}{2}$, we would not expect this locus to cross the imaginary axis. Therefore, we can skip this step.

If we wish to be absolutely sure of our conclusion, we can use the Routh test. From the characteristic polynomial,

$$s^3 + 12s^2 + Ks + K = 0,$$

we create the corresponding Routh array:

$$
\begin{array}{c c c}
s^3: & 1 & K \\
s^2: & 12 & K \\
s^1: & \dfrac{12K - K}{12} & \\
s^0: & K. &
\end{array}
$$

Again we find that the signs of the terms in the first column are all positive if $K > 0$, so there are no crossings of the imaginary axis.

STEP 6. We locate the multiple roots:

$$b(s) = s + 1, \qquad a(s) = s^3 + 12s^2;$$

$$\frac{db}{ds} = 1, \qquad \frac{da}{ds} = 3s^2 + 24s.$$

Using Eq. (5.28) again, we find the possible points of multiple roots at the solution of

$$-(s^3 + 12s^2)(1) + (s + 1)(3s^2 + 24s) = 0,$$

$$-s^3 - 12s^2 + 3s^3 + 27s^2 + 24s = 0,$$

$$2s^3 + 15s^2 + 24s = 0.$$

The roots of this equation are

$$s = 0, -5.18, -2.31,$$

which are all on the locus and thus points of multiple roots. This means that the locus either breaks away or breaks in to the real axis at these points. The locus branches cannot cross away from the real axis because in that case there would be other multiple solutions. Therefore, the only possible configuration is for the locus to break away at $s = 0$, to break in at $s = -2.31$, and to break away again at $s = -5.18$.

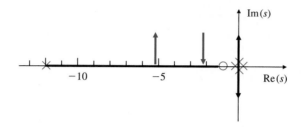

STEP 7.   The complete locus is drawn in Fig. 5.13. Note that after the breakaway from the real axis at -5.18, the roots approach the vertical asymptotes as the value of K gets large. Also recall that the locus both breaks in and breaks away at right angles.

FIGURE 5.13
Root locus for
$G(s) = (s + 1)/$
$[s^2(s + 12)]$

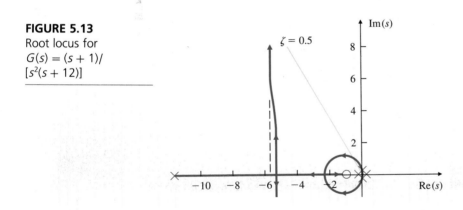

◆ **EXAMPLE 5.6**   *Illustrative Locus: p = 9*

Find the root locus for

$$G(s) = \frac{s + 1}{s^2(s + 9)}.$$

Solution

STEPS 1, 2.   We mark the axes and the real-axis segments.

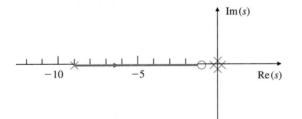

STEP 3.   We draw the asymptotes:

$$\phi_l = \frac{180° + 360°(l - 1)}{3 - 1}, \qquad \alpha = \frac{-9 - 0 - (-1)}{3 - 1}$$

$$= \pm 90°, \qquad\qquad = -4.$$

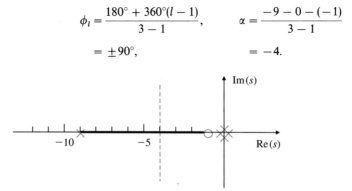

STEPS 4, 5.   As in the previous three cases, the locus segments at the two poles depart $s = 0$ at angles of $\pm 90°$, and there is no crossing of the imaginary axis.

STEP 6.   We estimate the possible multiple roots:

$$b(s) = s + 1, \qquad a(s) = s^3 + 9s^2,$$

$$\frac{db}{ds} = 1, \qquad \frac{da}{ds} = 3s^2 + 18s.$$

The possible multiple roots are at

$$-(s^3 + 9s^2)(1) + (3s^2 + 18s)(s + 1) = 0,$$
$$-s^3 - 9s^2 + 3s^3 + 21s^2 + 18s = 0,$$
$$2s^3 + 12s^2 + 18s = 0,$$
$$s = 0, -3, -3.$$

Again the points of multiple roots are on the locus, but in this case we have *repeated* roots in the derivative, which indicates that we have *three* roots at the same place. Applying the rule of departure angles ($K_1 > 0$) to the triple root at $s = -3$, we find

$$-3\phi_l + 180° = +180° + 360°l,$$

$$\phi_{\text{dep}} = 0°, \pm 120°.$$

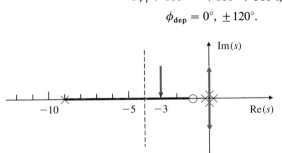

The arrival angles are rotated $60°$ relative to the departure angles:

$$\phi_{\text{arr}} = \pm 60°, 180°.$$

STEP 7.   The complete locus is sketched in Fig. 5.14.

FIGURE 5.14

Root locus for
$G(s) = (s + 1)/$
$[s^2(s + 9)]$

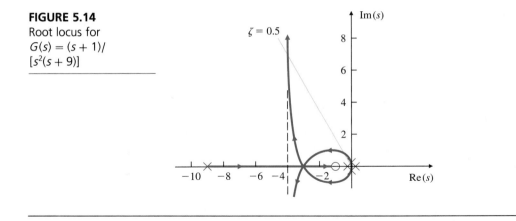

From Figs. 5.12 to 5.14, it is evident that when the third pole is near the zero, there is a modest distortion of the locus that would result for $G(s) = 1/s^2$, which consists of two straight-line locus segments departing at $\pm 90°$ from the two poles at $s = 0$. Then, as we increase p, the distortion becomes more extreme until at $p = 9$ the locus breaks in at -3 in a triple multiple root. As the pole p is moved beyond -9, the locus exhibits distinct break-in and break-away points, approaching, as p gets very large, the circular locus of one zero and two poles. Figure 5.14, when $p = -9$, is thus a transition locus between the two second-order extremes, which occur at $p = 1$ (where the zero is canceled) and $p = \infty$ (where the extra pole has no effect).

The next example illustrates multiple poles off the real axis.

◆ **EXAMPLE 5.7**   *Illustrative Locus: Multiple Poles*

Find the root locus of $1 + KG(s)$ where

$$G(s) = \frac{1}{s(s + 2)[(s + 1)^2 + 4]}.$$

Solution

STEPS 1,2.   We mark the s-plane and draw in the real-axis segments.

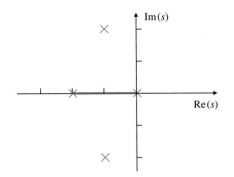

STEP 3.   We draw the asymptotes:

$$\phi_l = \frac{180° + 360°(l - 1)}{4 - 0}, \qquad \alpha = \frac{-2 - 1 - 1 - 0 + 0}{4 - 0}$$

$$= 45° + 90°(l - 1) \qquad\qquad = -1.$$

$$= 45°, 135°, -45°, -135°.$$

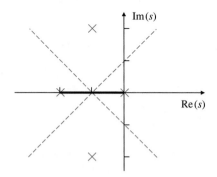

STEP 4.   The departure angle at the complex pole at $-1 + 2j$ is

$$\phi_{\text{dep}} = \phi_3 = -\phi_1 - \phi_2 - \phi_4 - 180° - 360°l,$$

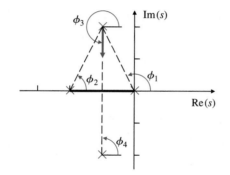

where

$$\phi_1 = \tan^{-1}\left(\frac{2}{-1}\right) = 116.6°,$$

$$\phi_2 = \tan^{-1}\left(\frac{2}{1}\right) = 63.4°,$$

$$\phi_4 = 90°.$$

Thus

$$\phi_3 = -116.6° - 63.4° - 90° - 180° - 360°l,$$
$$\phi_3 = -450° - 360°l,$$
$$\phi_3 = -90°.$$

We can observe at once that, along the line $s = -1 + j\omega$, ϕ_2, and ϕ_1 are angles of an isosceles triangle and always add to $180°$. Hence, the entire line from one complex pole to the other is on the locus *in this special case.*

STEP 5.   We compute the crossings of the imaginary axis. The characteristic equation is

$$s^4 + 4s^3 + 9s^2 + 10s + K = 0.$$

If we try a solution for $s = j\omega_0$, we find that ω_0 and K must satisfy the equation

$$\omega_0^4 - 4j\omega_0^3 - 9\omega_0^2 + 10j\omega_0 + K = 0.$$

Equating the real and imaginary parts to zero, we get the equations

$$\omega_0^4 - 9\omega_0^2 + K = 0, \tag{5.34a}$$
$$-4\omega_0^3 + 10\omega_0 = 0. \tag{5.34b}$$

From Eq. (5.34b) we get $\omega_0^2 = \frac{5}{2}$ and thus $\omega_0 = 1.58$. From both Eqs. (5.34), it follows that

$$K = 9\left(\frac{5}{2}\right) - \frac{25}{4}$$

$$= \frac{90 - 25}{4} = 16.25.$$

This value of K can also be verified by the Routh array.

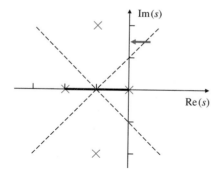

STEP 6.   We locate possible multiple roots:

$$b = 1, \qquad a = s^4 + 4s^3 + 9s^2 + 10s,$$

$$\frac{db}{ds} = 0, \qquad \frac{da}{ds} = 4s^3 + 12s^2 + 18s + 10.$$

The condition reduces to $da/ds = 0$. We could find solutions to this cubic, but there is a shorter way: From Step 4 we notice that the line at $s = -1 + j\omega$ is on the locus, so there must be a breakaway point at $s = -1$, which can be divided out. That is, we can easily show that

$$4s^3 + 12s^2 + 18s + 10 = (s + 1)(4s^2 + 8s + 10).$$

The quadratic has roots $-1 \pm j\sqrt{\frac{3}{2}} = -1 \pm 1.22j$. Since these points are on the line between the complex poles, they are true points of multiple roots on the locus.

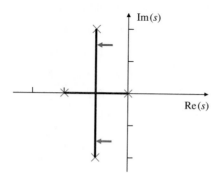

STEP 7.   Using this data, we sketch the locus in Fig. 5.15. Notice that here we have complex multiple roots. Branches of the locus come together at $-1 \pm 1.22j$ and break away at $0°$ and $180°$.

FIGURE 5.15
Root locus for $G(s) = 1/\{s(s+2)[(s+1)^2 + 4]\}$

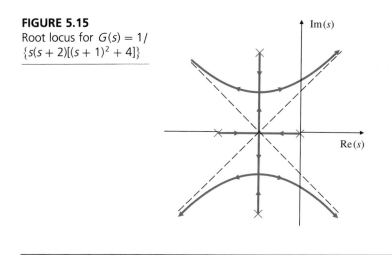

Like the locus in Example 5.6, the locus in Example 5.8 is a transition between two types of loci. In Problem 5.57, you are asked to examine this transition in more detail.

In the next two examples we examine the importance of the departure and arrival angles.

◆ **EXAMPLE 5.8**   *Illustrative Locus: Arrival and Departure Angles I*

This root locus is typical of servomechanisms with a flexible load like the disk-head mechanism in Example 2.4. In this case the position sensor is on the motor shaft; that is, the measured quantity is θ_1 in Fig. 2.9. This control system is referred to as having a **collocated sensor and actuator**.

Find the locus of roots of $1 + KG(s)$ where

$$G(s) = \frac{(s+0.1)^2 + 16}{s[(s+0.1)^2 + 25]}.$$

Solution.   Again we follow the steps.

STEPS 1, 2.   We mark the s-plane and draw in the real-axis segments:

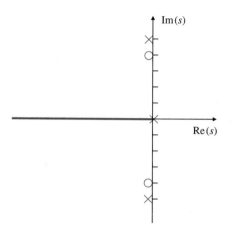

STEP 3.   We draw the asymptotes:

$$\phi_l = \frac{180° + 360°(l - 1)}{3 - 2} = 180°,$$

$$\alpha = \text{not relevant.}$$

STEP 4.   We compute the arrival and departure angles. For the arrival at the zero, the angle condition is

$$-\phi_1 - \phi_2 - \phi_3 + \psi_1 + \psi_2 = 180° + 360°l,$$

where

$$\phi_1 \cong 90°,$$

$$\phi_2 \cong -90°,$$

$$\phi_3 \cong 90°,$$

$$\psi_2 \cong 90°.$$

Thus $\psi_1 = 180°$, and the root arrives from the left. At the pole the departure angle is found from the same formula, except that now ϕ_2 is unknown and $\phi_1 \cong \phi_3 \cong \psi_1 \cong \psi_2 = 90°$. Hence

$$-90° - \phi_2 - 90° + 90° + 90° = 180° + 360°l.$$

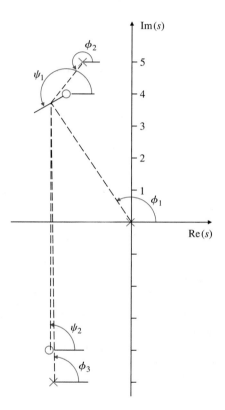

Thus $\phi_2 \cong 180°$, and departure is also toward the left. We pause here to sketch in our best guess at the locus:

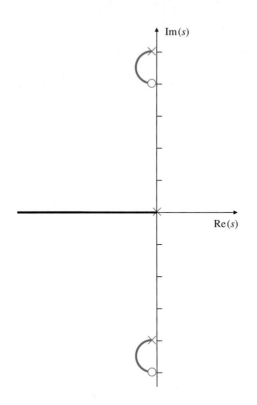

Information derived from the locus sketch in Step 4 is sufficient to allow us to skip Steps 5 and 6.

At this point the graph already shows that we cannot expect to find any root crossing of the imaginary axis, nor any multiple root points. Therefore, we will skip Steps 5 and 6.

STEP 7. The complete locus is sketched in Fig. 5.16. It shows that, when the sensor and actuator are collocated, proportional feedback has only a small effect on resonant-mode damping, but that small effect tends to be an improvement.

FIGURE 5.16
Root locus for
$G(s) = [(s + 0.1)^2 + 16]/\{s[(s + 0.1)^2 + 25]\}$

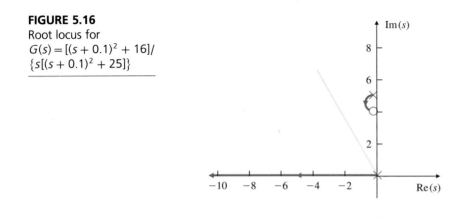

◆ **EXAMPLE 5.9**   *Illustrative Locus: Arrival and Departure Angles II*

For this example we use the same basic configuration discussed in Example 5.8, except that we interchange the complex poles and zeros. Thus

$$G(s) = \frac{(s + 0.1)^2 + 25}{s[(s + 0.1)^2 + 16]}.$$

Solution.   Since this example is very much like Example 5.8, we will show here only the steps resulting in different calculations.

STEP 4.   We compute the angles of arrival and departure. In this case we have the angle condition for departure at pole 2 at $-0.1 + 4j$ as $-\phi_1 - \phi_2 - \phi_3 + \psi_1 + \psi_2 = 180° + 360°l$, where $\phi_1 \cong \phi_3 \cong \psi_3 = 90°$ and $\psi_1 \approx -90°$. Thus

$$-90° - \phi_2 - 90° + 90° - 90° = 180° + 360°l; j \ \phi_2 = 0°.$$

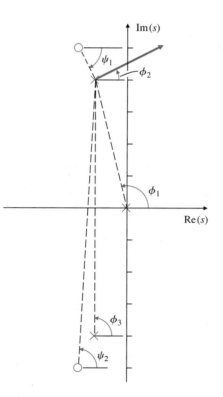

For arrival at the zero $-0.1 + 5j$, the angles are $\phi_1 \cong \phi_2 \cong \phi_3 \cong \psi_2 = 90°$, and the angle condition for the root locus is

$$-90° - 90° - 90° + 90° + \psi_1 = 180° + 360°l, \ \psi_1 = 0°.$$

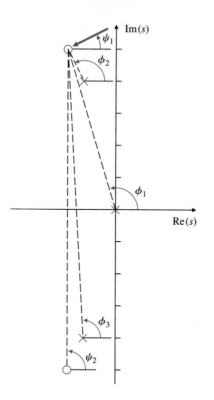

STEP 5. From Step 4 we see that both departure and arrival are *toward* the RHP; we would thus expect the locus to loop from the pole to the zero, with most of the loop in the RHP. Knowledge that the system is unstable for a wide range of K is usually sufficient to warrant stopping at this point and trying a new design. However, if it were important to pin down exactly where the locus crosses, we could use the Routh stability criterion. Let us suppose we do want to know exactly what values of K are on the stable part of the locus. Then the characteristic equation is

$$s^3 + (K + 0.2)s^2 + (0.2K + 16.01)s + 25.01K = 0,$$

and the Routh array is

s^3:	1	$0.2K + 16.01$
s^2:	$K + 0.2$	$25.01K$
s^1:	$\dfrac{(0.2K + 16.01)(K + 0.2) - 25.01K}{K + 0.2}$	
s^0:	$25.01K.$	

In this case, stability is decided (for $K > 0$) by the term in the s^1 row. Multiplying this out, we find that the imaginary axis will be crossed for real positive solutions to the equation

$$0.2K^2 - 8.96K + 3.20 = 0.$$

FIGURE 5.17
Root locus for
$G(s) = [(s + 0.1)^2 + 25]$
$/\{s[(s + 0.1)^2 + 16]\}$

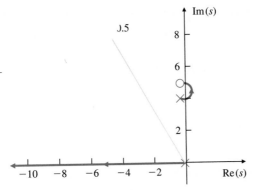

These are $K = 44.4$ and 0.362. With these values we may find the frequencies from the real part of the characteristic equation:

$$(K + 0.2)\omega^2 = 25.01 K,$$

$$\omega = \sqrt{\frac{25.01 K}{K + 0.2}}$$

$$= 4.99, 4.01.$$

Notice that the crossings are very close to the (open-loop) poles and zeros and that the system is unstable for $0.362 < K < 44.4$.

STEPS 6, 7.   Again we omit the calculation of possible multiple roots and plot the root locus in Fig. 5.17.

◆

Shifting the relative positions of poles and zeros can dramatically influence stability.

The major conclusion to draw from Examples 5.8 and 5.9 is that if the system has poles or zeros near the unstable boundary, then the departure angles will quickly show if, in fact, these will cause closed-loop stability problems. Furthermore, if we compare Figs. 5.16 and 5.17, it is obvious that shifting the relative position of poles and zeros can have a dramatic influence on stability. In Fig. 5.16 the zero is at a lower frequency than the pole and the locus stays in the LHP, whereas in Fig. 5.17 their positions are reversed and that portion of the locus represents an unstable system for almost any value of K; thus the system is unstable for almost any value of K. These kinds of pole-zero configurations arise in systems that contain lightly damped flexible modes.

5.4   Other Root-locus Usage

The root-locus technique is a graphical scheme to show locations of possible roots of an equation as a real parameter varies. Thus far we have considered only cases where the parameter is the loop gain in a negative feedback control system and where the gain can only be positive. Each of these assumptions can be altered to allow a far wider range of applications of the root-locus technique.

5.4.1   Loci versus Other Parameters

An idea mentioned early in Section 5.1, but beyond the developments detailed so far, is to consider parameters other than the loop gain K. In general, the root locus is a plot of solutions to the relation

$$1 + KG(s) = 0 \tag{5.35}$$

for some real value of K and a given transfer function $G(s)$. To consider a parameter other than K, it is necessary to manipulate the characteristic equation into the form of Eq. (5.35).

◆ **EXAMPLE 5.10**   *Root locus versus Damping Factor*

Repeat Example 5.2, using the guidelines for sketching a root locus; that is, find the locus versus c for a unity feedback loop around

$$G(s) = \frac{1}{s(s + c)}.$$

Solution.   The characteristic equation for the closed-loop case in Example 5.2 was given by Eq. (5.9) and is

$$s^2 + cs + 1 = 0.$$

It will be easier to draw the root locus if we manipulate the equation into the form

$$1 + \frac{cs}{s^2 + 1} = 0.$$

It is now apparent that, for purposes of sketching a root locus, the zero is at $s = 0$ and the poles are at $s = \pm 1j$, as shown in Fig. 5.18. There is a locus segment to the left of the zero at $s = 0$, and there is one asymptote at $180°$ because there is one more pole than zero. The angle of departure from the pole at $s = +j$ is

$$\phi_{\text{dep}} = 90° - 90° - 180° = -180°$$

and $\phi_{\text{dep}} = -180°$ for the pole at $s = -j$ as well.

To find the point of multiple roots where the locus segments from each pole break-in to the real axis, we use Eq. (5.28). In this case,

$$b(s) = s \quad \text{and} \quad a(s) = s^2 + 1,$$

so the condition for a multiple root is

$$s^2 - 1 = 0 \quad \text{or} \quad s = \pm 1.$$

FIGURE 5.18
Locus vs. damping c

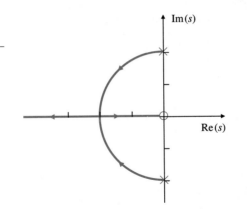

Only $s = -1$ is on the locus, so the locus segments join the real axis at that point, and the locus in Fig. 5.18 results.

◆

An example of using the root locus to guide tachometer and amplifier gain.

A block diagram of a relatively common servomechanism structure is shown in Fig. 5.19. Here a speed-measuring device (a tachometer) has been added, and the problem is to use the root locus to guide the selection of the tachometer gain K_T as well as the amplifier gain K_A. A root locus with respect to K_A is the standard situation we dealt with in previous sections. We take the view that a principal function of K_T is to influence the closed-loop system poles, and thus we will consider the locus of these roots as K_T is varied. The characteristic equation of the system in Fig. 5.19 is

$$1 + \frac{K_A}{s(s+1)} + \frac{K_T}{s+1} = 0,$$

which is not in the standard $1 + KG(s)$ form. However, after clearing fractions,

FIGURE 5.19
Block diagram of a servomechanism structure including velocity, or tachometer feedback

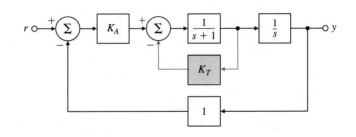

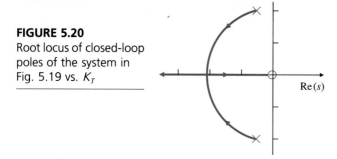

the characteristic polynomial becomes

$$s^2 + s + K_A + K_T s = 0. \tag{5.36}$$

Let us assume that K_A has a nominal value of 4. Now Eq. (5.36) is in suitable form for a root-locus study with respect to K_T. Next we identify $G(s)$ by rewriting Eq. (5.36) as

$$1 + K_T \frac{s}{s^2 + s + 4} = 0. \tag{5.37}$$

Thus, for root-locus purposes, the "zero" is at $s = 0$, and the "poles" are at the roots of $s^2 + s + 4 = 0$, or $s = -\frac{1}{2} \pm 1.94j$. A sketch of the locus is shown in Fig. 5.20.

We are now free to select K_T for a specific damping ratio or whatever else might guide the selection of K_T. Let us take $K_T = 1$. Having selected a trial value of K_T, we can consider the effects of changing K_A from 4 to another value. The locus, with respect to K_A is governed by Eq. (5.36). The characteristic polynomial in this case can be written as

$$s^2 + (1 + K_T)s + 4 + K = 0,$$

where we have assumed $K + 4 = K_A$. With respect to K, the locus (for $K_T = 1$) is governed by

$$s^2 + 2s + 4 + K = 0. \tag{5.38}$$

Note that the "poles" of the locus corresponding to Eq. (5.38) are the roots of the previous locus, which was drawn versus K_T. The situation is sketched in Fig. 5.21, with the previous locus versus K_T left dashed. We could draw a locus with respect to K for a while, stop, resolve the equation, and continue the locus with respect to K_T; in other words, we could go back and forth between these parameters at will.

To summarize, the technique for handling any single parameter is to write out the characteristic equation, collect terms that do *not* multiply the parameter

A summary of the technique for handling any parameter.

FIGURE 5.21

Root locus vs. K after choosing $K_T = 1$

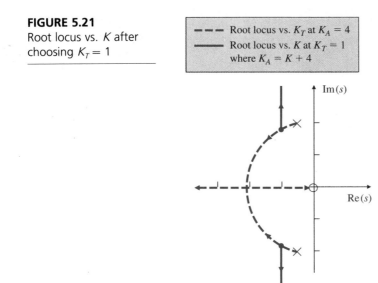

- – – – Root locus vs. K_T at $K_A = 4$
- ——— Root locus vs. K at $K_T = 1$
  where $K_A = K + 4$

as $a(s)$, and collect terms that *do* multiply the parameter as $b(s)$. After this formulation stage, the process for plotting the root locus proceeds as before.

5.4.2   Zero-degree Loci for Negative Parameters

We now consider modifying the root-locus procedure to permit analysis of negative values of the parameter. In the case of the servomechanism considered in Fig. 5.19, we sketched a root locus with respect to K where $K_A = 4 + K$. It would seem reasonable to consider values of K_A less than 4 as well as those greater than 4. In order to make K_A less than 4 we need to make K negative. What are the rules for a negative locus (a root locus relative to a negative parameter)? First of all, Eq. (5.35) must be satisfied for negative values of K, which implies that $G(s)$ is real and *positive*. In other words, for the negative locus the phase condition is:

The angle of $G(s)$ is $0° + 360°l$ for s on the negative locus.

The steps for plotting a negative locus are essentially the same as for the positive locus except that we search for $0° + 360°l$ instead of $180° + 360°l$. For this reason a negative locus is also referred to as a $0°$ root locus. This time we find that the locus is to the left of an *even* number of real poles plus zeros (zero being even). Computation of the asymptotes for large values of s is, as before, given by

$$\alpha = \frac{\sum p_i - \sum z_i}{n - m},$$

(5.39)

but we modify the angles to be

$$\phi_l = \frac{360°(l - 1)}{n - m}, \qquad l = 1, 2, 3, \ldots, n - m,$$

(shifted by 180° from the 180° locus). Following are the guidelines for plotting a 0° locus.

Summary: Guidelines for Plotting a 0° Root Locus

1. Mark the n poles as $\times$ and the m zeros as $\bigcirc$.
2. Draw the locus on the real axis to the left of an *even* number of real poles plus zeros.
3. Draw $n - m$ radial asymptotes centered at α and with angles ϕ_l where

$$\alpha = \frac{\sum p_i - \sum z_i}{n - m} = \frac{-a_1 + b_1}{n - m},$$

$$\phi_l = \frac{360°(l - 1)}{n - m}, \qquad l = 1, 2, 3, \ldots, n - m.$$

4. Compute departure angles from poles and arrival angles to zeros by searching for points *around* the pole or zero where the phase of $G(s)$ is 0°, so that

$$q\phi_{\text{dep}} = \sum \psi_i - \sum \phi_i - 360°l,$$

$$q\psi_{\text{arr}} = \sum \phi_i - \sum \psi_i + 360°l,$$

where q is the order of the pole or zero, and l takes on q integer values such that the angles are between $\pm 180°$.
5. Assume $s = j\omega_0$, and compute the points where the locus crosses the imaginary axis for negative values of K.
6. The equation has multiple roots at points *on the locus* [for which the angle of $G(s)$ is $0° + 360°l$] where $dG/ds = 0$, or

$$b\frac{da}{ds} - a\frac{db}{ds} = 0.$$

7. Fill in the locus using these calculations as guides, and the locus is complete.

The result of extending the guidelines for constructing root loci to include negative parameters is that we can visualize the root locus as a set of continuous curves showing the location of possible solutions to the equation $1 + KG(s)$ for *all real values of K*, both positive and negative. One branch of the locus departs from every pole in one direction for positive values of K, and another branch departs from the same pole in another direction for negative K. Likewise, all zeros will have two branches arriving, one with positive and the other with negative values of K. For $n - m$ excess poles, there will be $2(n - m)$ branches of the locus asymptotically approaching infinity. For a single pole or zero the angles of departure or arrival for the two locus branches will be 180° apart. For a double pole or zero the two positive branches will be 180° apart, and the two negative branches will be at 90° to the positive branches.

◆ **EXAMPLE 5.11**   *Zero-degree Locus*

Sketch the zero-degree locus for the $G(s)$ in Example 5.3, $G(s) = (s + 1)/s^2$, whose 180° locus is shown in Fig. 5.11.

Solution

STEPS 1, 2.   We mark the poles and zeros on the s-plane and draw the locus on the real axis to the left of an even number of poles plus zeros:

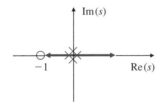

STEP 3.   We find the number of asymptotes and draw them:

$$n - m = 2 - 1 = 1 \text{ asymptote:}$$

$$\phi_1 = \frac{0°}{2 - 1} = 0°.$$

α does not matter, since there is only the one asymptote, which goes from the double pole at $s = 0$ to $+\infty$.

STEP 4.   We compute the departure and arrival angles. For the two poles at $s = 0$,

$$2\phi_{\text{dep}} = 360°l = 0°, \ 180°,$$

which verifies the locus that has already been determined by the real-axis rule. Likewise, the angle of arrival to the zero has already been found to be zero.

STEP 5.   There are no points where the locus crosses the imaginary axis, but two branches depart from the poles at $s = 0$.

FIGURE 5.22
0° root locus for
$G(s) = (s + 1)/s^2$

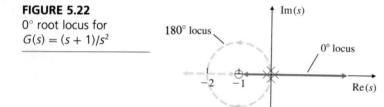

STEP 6.   We located the possible multiple roots according to Eq. (5.28) in Example 5.3 and found multiple roots at $s = 0$ and -2. Only the $s = 0$ point is on the 0° locus at $K = 0$, as we have already discovered.

STEP 7.   We sketch the locus as shown in Fig. 5.22. The 180° locus from Fig. 5.7, repeated here as a dashed line, illustrates how the 0° locus is a continuation of the 180° locus.

◆

The continuity of these loci encourages us to imagine a root starting at a pole of $G(s)$ and moving along a curve (a branch of the root locus) until the parameter has some value K_0. If we now let $K = K_0 + \bar{K}$, a new locus with respect to $\bar{K}$ is just the continuation of the old locus with the same zeros, but with the roots of $1 + K_0 G(s) = 0$ as the new set of poles. Thus we can visualize the roots moving along the locus branches; the **sensitivity** of the closed-loop poles to the parameter K at $K = K_0$ is given by how the roots move for both positive and negative values of $\bar{K}$ for $K = K_0 + \bar{K}$.

◆ **EXAMPLE 5.12**   *Zero-degree Locus for Eq. (5.38)*

Consider again Eq. (5.38), for which we sketched the locus of roots for positive variations of K_A from the value 4 and found the locus in Fig. 5.21. Now sketch the locus for negative variations of K_A from the value 4, and add to that the locus for positive variations.

Solution.   The governing equation is Eq. (5.38),

$$s^2 + 2s + 4 + K = 0,$$

where $K > 0$ means that $K_A > 4$ and $K < 0$ means that $K_A < 4$. Therefore, the poles of the root locus are at the roots of $s^2 + 2s + 4 = 0$, which are at $s = -1 \pm 1.73j$. These are shown in Fig. 5.23. The real-axis segment extends along the entire real axis because there is an even number (zero) of poles and zeros on the real axis. There are two asymptotes at 0° and 180°. The angle of departure from the complex root at $-1 + 1.73j$ is

$$\phi_{\text{dep}} = -90° - 0° = -90°.$$

Likewise, $\phi_{\text{dep}} = +90°$ from the pole at $-1 - 1.73j$. Figure 5.23 shows both the complete negative locus (solid line) and the positive locus taken from the solid line in Fig. 5.21 (shown dashed here).

FIGURE 5.23
0° root locus
corresponding to
$G(s) = 1/(s^2 + 2s + 4)$

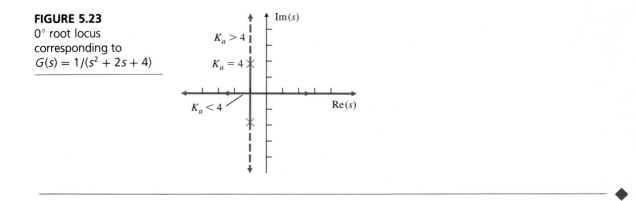

5.5   Selecting Gain from the Root Locus

The root locus is a plot of *all possible locations* for roots to the equation $1 + KG(s) = 0$ for some real positive value of K. The purpose of design is to select a particular value of K that will meet the specifications for static and dynamic response. We now turn to the issue of selecting K so that the roots are at specific places. Using Definition II of the locus, we developed seven steps to sketch a root locus from the phase of $G(s)$ alone. If the equation is actually to have a root at a particular place when the phase of $G(s)$ is 180°, then a **magnitude condition** must also be satisfied. This condition is given by Eq. (5.5d), rearranged as

$$K = -\frac{1}{G(s)}.$$

For values of s on the root locus the phase of G is 180°, so we can write the magnitude condition as

$$K = \frac{1}{|G|}. \tag{5.40}$$

Equation (5.40) has both an algebraic and a graphical interpretation. To see the latter, consider the locus of $1 + KG(s)$ where

$$G(s) = \frac{1}{s[(s + 4)^2 + 16]}. \tag{5.41}$$

For this transfer function, the locus was plotted in Fig. 5.10 and is repeated here showing all parts in Fig. 5.24. On Fig. 5.24 the lines corresponding to a damping ratio of $\zeta = 0.5$ are sketched, and the points where the locus crosses these lines are marked with dots ($\bullet$). Suppose we wish to set the gain so that the poles are located at the dots. This corresponds to selecting the gain so that two of the closed-loop system poles have a damping ratio of $\zeta = 0.5$. (We will find the third

FIGURE 5.24
Root locus for
$G(s) = 1/\{s[(s+4)^2 + 16]\}$
showing calculations of
gain K

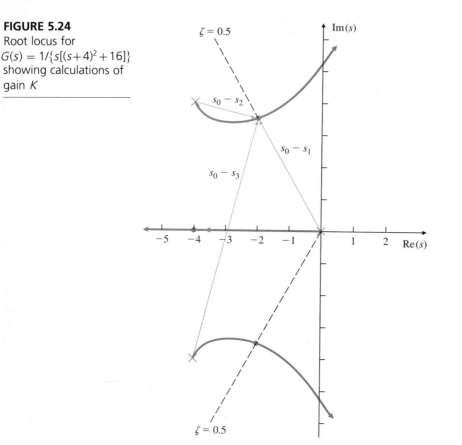

pole shortly.) What is the value of K when a root is at the dot? From Eq. (5.40), the value of K is given by 1 over the magnitude of $G(s_0)$, where s_0 is the coordinate of the dot. On the figure we have plotted three vectors marked $s_0 - s_1$, $s_0 - s_2$, and $s_0 - s_3$, which are the vectors from the poles of $G(s)$ to the point s_0. (Since $s_1 = 0$, the first vector equals s_0.) Algebraically, we have

$$G(s_0) = \frac{1}{s_0(s_0 - s_2)(s_0 - s_3)}. \tag{5.42}$$

Using Eq. (5.40), this becomes

$$K = \frac{1}{|G(s_0)|} = |s_0|\,|s_0 - s_2|\,|s_0 - s_3|. \tag{5.43}$$

Graphical calculation of
the desired gain

The graphical interpretation of Eq. (5.43) shows that its three magnitudes are the lengths of the corresponding vectors drawn on Fig. 5.24 (see Appendix

B). Hence we can compute the gain to place the roots at the dot ($s = s_0$) by measuring the lengths of these vectors and multiplying the lengths together, provided that the scale of the imaginary and real axes is identical. Using the scale of the figure we estimate that

$$|s_0| \cong 4.0,$$

$$|s_0 - s_2| \cong 2.1,$$

$$|s_0 - s_3| \cong 7.7.$$

Thus the gain is estimated to be

$$K = 4.0(2.1)(7.7) \cong 65.$$

We conclude that if K is set to the value 65, then a root of $1 + KG$ will be at s_0, which has a damping ratio of 0.5. Another root is at the conjugate of s_0. Where is the third root? The third branch of the locus lies along the negative real axis. Ordinarily, we would need to take a test point, compute a trial gain, and repeat this process until we found the point where $K = 65$. In this case it is more convenient to use the property of the root locus expressed in Eq. (5.18)—that the sum of the roots is constant (does not change as K changes) if $m < n - 1$. Thus, the unknown root must be moved far enough to the left to keep the sum fixed. From Fig. 5.24 we estimate that $s_0 = -2 + 3.4j$. Since the starting point was at $s = -4 + 4j$, this root has moved two units to the right. The conjugate has moved an equal distance, so the third root must have moved $2 + 2$ units to the left of where it began at $s = 0$. We have marked the new location at -4 with the third dot.

   A process with the transfer function given by Eq. (5.41) has one integrator and, in a unity feedback configuration, will be a type I control system. In this case the steady-state error in tracking a ramp input is given by the velocity constant:

$$K_v = \lim_{s \to 0} sKG(s)$$

$$= \lim_{s \to 0} s \, \frac{K}{s[(s + 4)^2 + 16]}$$

$$= \frac{K}{32}. \tag{5.44}$$

With the gain set for complex roots at a damping $\zeta = 0.5$, the root-locus gain $K = 65$, so from Eq. (5.44) we get $K_v = 65/32 \cong 2$. If the closed-loop dynamic response as determined by the root locations is satisfactory and the steady-state accuracy as measured by K_v is good enough, then the design can be completed by gain selection alone. However if no value of K satisfies all the constraints, as is typically the case, then additional modifications are necessary to meet the system specifications.

5.6   Dynamic Compensation

Lead and lag
compensation

If the process dynamics are of such a nature that a satisfactory design cannot be obtained by a gain adjustment alone, then some modification or compensation of the process dynamics is indicated. While the variety of compensation schemes is great, two categories have been found to be particularly simple and effective. These are lead and lag compensation.[8] **Lead compensation** acts mainly to lower rise time and decrease the transient overshoot; in a crude way, lead compensation approximates derivative control. **Lag compensation** is usually used to improve the steady-state accuracy of the system; it approximates integral control. In this section we will examine these two schemes in some detail.

Lead and lag compensation have historically been implemented using analog electronics and hence were often referred to as networks. Today, however, most new control system designs use digital computer technology, in which the compensation is implemented in the software. The dynamic form of compensation is the same in both digital and analog technologies, so the theory and design methodology are unchanged. An extra step is required, however, to transform the continuous dynamics represented by the lead or lag transfer functions into equations that can be coded in the control computer. This transformation process is described in Chapter 8 and in Franklin *et al.* (1990).

Compensation with a transfer function of the form

$$D(s) = \frac{s + z}{s + p} \tag{5.45}$$

is called lead compensation if $z < p$ and lag compensation if $z > p$. Compensation is typically placed in series with the plant in the feedforward path, as shown in Fig. 5.25. It can also be placed in the feedback path and in that location has the same effect on the overall system poles. The characteristic equation of the system in Fig. 5.25 is

$$1 + KD(s)G(s) = 0$$

where $K = K_A K_G$ as before. All the root locus rules apply to this case if the $G(s)$ from Sections 5.1 through 5.5 is replaced with $D(s)G(s)$.

---

8. The names of these compensation schemes derived from their frequency (sinusoidal) responses, where the output leads the input in one case (a positive phase shift) and lags the input in the other (a negative phase shift). See Chapter 6.

FIGURE 5.25
Feedback system with
compensation

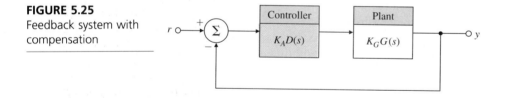

5.6.1 Lead Compensation

Lead compensation
approximates PD control.

To explain the basic stabilizing effect of lead compensation on a system, we first consider a simplified $D(s) = s + z$. This is the same as the proportional-derivative (PD) control discussed in Section 4.2. We will apply this compensation to the case of a second-order system with transfer function

$$KG(s) = \frac{K}{s(s + 1)},$$

which was analyzed in Example 5.1. $G(s)$ has the root locus shown in Fig. 5.2, repeated now as the solid-line portion of the locus in Fig. 5.26. Also shown in Fig. 5.26 is the locus produced by $D(s)G(s)$, where $D(s) = s + 2$; note that this adds a zero at $s = -2$. The modified locus is the circle sketched with dashed lines. Notice that the effect of the zero is to move the locus to the left, toward the more stable part of the s-plane. Clearly, if our speed-of-response specification calls for $\omega_n \cong 2$, then gain alone can produce only a very low value of damping ratio ζ; hence at the required gain the transient overshoot will be substantial. However, by adding the zero we can move the locus to a position having closed-loop roots at $\omega_n = 2$ and damping ratio $\zeta \geqslant 0.5$. We have "compensated" the given dynamics by using $D(s) = s + 2$.

The trouble with choosing $D(s)$ based on only a zero is that the physical realization would contain a differentiator that would greatly amplify the inevitable high-frequency noise present from the sensor signal. Furthermore, for reasons to be discussed in Chapter 6, it is impossible to build a pure differentiator. However, since the effect of the zero is most important near $\omega_n = 2$, the action of the compensation will not be greatly reduced if we add a

FIGURE 5.26
Root loci for $1 + KD(s)G(s) = 0$, $G(s) = 1/[s(s + 1)]$: without compensation (solid lines) and with the compensation $D(s) = s + 2$ (dashed lines)

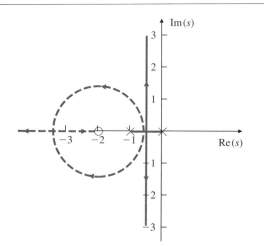

FIGURE 5.27
Root loci for three cases with $G(s) = 1/[s(s + 1)]$: (a) $D(s) = (s + 2)/(s + 20)$;
(b) $D(s) = (s + 2)/(s + 10)$; (c) $D(s) = s + 2$ (solid lines)

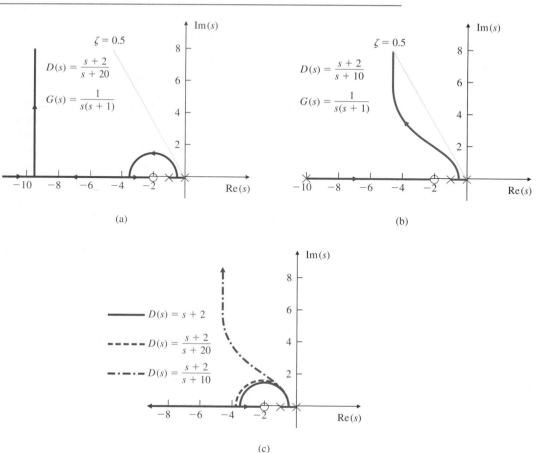

higher-frequency pole, perhaps at $s = -20$ to give

$$D(s) = \frac{s + 2}{s + 20}.$$

The resulting transfer function is thus lead compensation.

To see the effect of the pole on the compensation, consider the root loci for two cases shown in Fig. 5.27(a, b) with different pole locations. The important fact about these loci is that for small gains, before the real part of the root reaches -2, they are almost identical with the locus of Fig. 5.26, where $D(s) = s + 2$. To emphasize this point, in Fig. 5.27(c) all three loci are superimposed on the same plot. Note that the effect of the pole is to press the locus toward the right, but for the early part of the locus, the effect of the pole is not great.

Zero and pole selection

Selecting exact values of z and p in Eq. (5.45) for particular cases is done by trial and error, which can be minimized with experience. In general, the zero is placed in the neighborhood of the closed-loop ω_n, as determined by rise-time or settling-time requirements, and the pole is located at a distance 3 to 20 times the value of the zero location. The choice of pole location is a compromise between the conflicting effects of noise suppression and compensation effectiveness. In general, if the pole is too close to the zero, then, as seen in Fig. 5.27(c), the root locus moves back too far toward its uncompensated shape and the zero is not successful in doing its job. On the other hand, for reasons that will become clear in Chapter 6, when the pole is too far to the left, the magnification of noise at the output of $D(s)$ is too great, and the motor or other actuator of the process will be overheated by noise energy.

◆ **EXAMPLE 5.13**   *Design of Lead Compensation*

Find a compensation for $G(s) = 1/[s(s + 1)]$ that will provide a closed-loop damping $\zeta > 0.5$ and natural frequency $\omega_n > 7 \, \text{rad/sec}$.

Solution.   The trials of different compensation poles and zeros in Fig. 5.27(c) show that

$$D(s) = \frac{s + 2}{s + 10}$$

can meet the requirements with a sufficiently high value of K. Figure 5.28 shows

FIGURE 5.28
Root locus for
Example 5.13

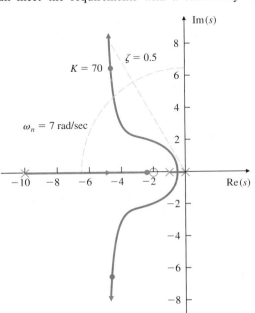

FIGURE 5.29
Step response for
Example 5.13

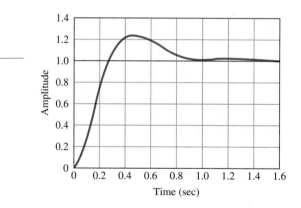

that $K = 70$ will comfortably meet both specifications, resulting in $\zeta = 0.56$ and $\omega_n = 7.7$ rad/sec. Figure 5.29 shows the step response of the system. The overshoot is more than the 14% that we would expect for a second-order system with $\zeta = 0.56$ according to Fig. 3.17; however, the fact that an extra zero and pole are present explains the additional overshoot. Typically, lead compensation in the feedforward path will increase the step-response overshoot because the compensation is differentiating the signal, as discussed in Section 3.5 and illustrated in Fig. 3.22.

◆

We can make the method for selecting the pole and zero of lead compensation more analytical if we select the desired closed-loop pole location first. Then we arbitrarily pick one lead-compensation parameter and use the root-locus angle condition to select the other. The method can be illustrated with the same plant. Suppose we wish to force the root locus to pass through the point $-1 + j\sqrt{3}$ corresponding to $\omega_n = 2$ and $\zeta = 0.5$ so that, as discussed in Section 3.5 and shown in Fig. 3.20, we should expect a rise time $t_r \cong 1$ sec and an overshoot $M_p \cong 20\%$. The corresponding s-plane is sketched in Fig. 5.30.

FIGURE 5.30
Construction for placing
a specific point on the
root locus

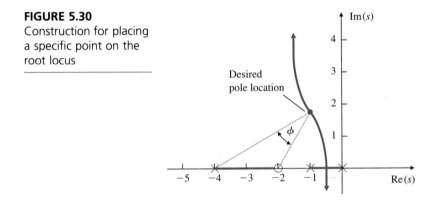

FIGURE 5.31
Circuit of a lead network

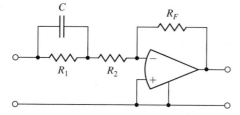

Because $\angle\, G(s) = -210°$ at the desired point for the closed-loop pole, the root-locus angle condition will be satisfied if the angle ϕ from $D(s)$ (the net contribution of the lead pole and zero) is $+30°$. Thus if we arbitrarily place the zero at -2, then the angle ϕ will be $30°$ if the lead compensation pole is placed at -4; thus the point $-1 + j\sqrt{3}$ will be on the locus, as shown in Fig. 5.30. Other combinations can also be selected that yield the required $30°$ from $D(s)$; the angle criterion for a specific closed-loop pole location specifies only the relationship between the pole and zero.

As stated earlier, the name *lead compensation* is a reflection of the fact that to sinusoidal signals these transfer functions impart phase lead. For example, the phase of Eq. (5.45) at $s = j\omega$ is given by

$$\phi = \tan^{-1}\left(\frac{\omega}{z}\right) - \tan^{-1}\left(\frac{\omega}{p}\right). \tag{5.46}$$

If $z < p$, then ϕ is positive, which by definition indicates phase lead. The details of this aspect of phase lead compensation will be treated in Chapter 6.

Lead compensation can be physically realized in many ways. In analog electronics a common method is to use an operational amplifier. An example of this is shown in Fig. 5.31. The transfer function of the circuit in Fig. 5.31 is readily found by the methods of Chapter 3 to be

Lead compensation can be implemented using analog electronics, but digital computers are preferred.

$$D(s) = -K_D \frac{T_1 s + 1}{\alpha T_1 s + 1}, \tag{5.47}$$

where

$$K_D = \frac{R_F}{R_1 + R_2} = 1 \quad \text{if} \quad R_F = R_1 + R_2,$$

$$T_1 = R_1 C,$$

$$\alpha = \frac{R_2}{R_1 + R_2}.$$

Notice that the zero is located at $z = -1/T_1$ and the pole is located at $p = -1/\alpha T_1$, so the parameter α sets the separation distance between pole and zero, typically a factor of 3 to 20.

5.6.2   Lag Compensation

Once satisfactory dynamic response has been obtained, perhaps by using one or more lead compensations, we may discover that the low-frequency gain—the value of the relevant steady-state error constant such as K_v—is still too low. In order to increase this constant, the equivalent of another integration at near-zero frequency is indicated. The improvement is thus made by a pole near $s = 0$, but usually we include a zero nearby so that the pole-zero pair (called a **dipole**) does not significantly interfere with the dynamic response of the overall system as determined by the lead compensation(s). Thus, we want an expression for $D(s)$ that will yield a significant gain at $s = 0$ to raise K_v (or another steady-state error constant) and that is nearly unity (no effect) at the higher frequencies where dynamic response is determined. The result is

Effect of lag compensation on steady-state errors

$$D(s) = \frac{s + z}{s + p}, \qquad z > p, \tag{5.48}$$

where the values of z and p are small (perhaps $z = 0.1$ and $p = 0.01$) yet $D(0) = z/p = 3$ to 10 (the value depending on the extent to which the steady-state gain requires boosting). Because $z > p$, the phase ϕ given by Eq. (5.46) is negative, corresponding to phase lag. Hence a device with this transfer function is called lag compensation.

An example of lag compensation

The effects of lag compensation on dynamic response can be studied by looking at the corresponding root locus. Again we take $G(s) = 1/[s(s + 1)]$, include the lead compensation $D_1(s) = (s + 2)/(s + 20)$ that produced the locus in Fig. 5.27(a), and raise the gain until the closed-loop roots correspond to a damping ratio of $\zeta = 0.707$. At this point, the root-locus gain of about 31 can be found using the methods described in Section 5.5. Thus the velocity constant is

$$K_v = \lim_{s \to 0} sKDG$$

$$= \lim_{s \to 0} s(31) \frac{s + 2}{s + 20} \frac{1}{s(s + 1)}$$

$$= \tfrac{31}{10} = 3.1.$$

Suppose we now add a lag compensation of $D_2(s) = (s + 0.1)/(s + 0.01)$. We selected this in order to increase the velocity constant by about 10 (that is, $z/p = 10$) and yet keep the values of both z and p very small so that $D_2(s)$ would have little effect on the portions of the locus representing the faster dynamics ($|s| > 0.2$). The resulting root locus is plotted in Fig. 5.32.

In Fig. 5.32(a) the locus is plotted showing the dominant roots at $-1.35 \pm 1.35j$. Notice the very small circle near the origin, however. When this region is expanded to get Fig. 5.32(b), we can see that the circle is a result of the lag compensation. Notice that a closed-loop root remains very near the lag-compensation zero at $-0.1 + 0j$. This root will correspond to a very slowly decaying transient, which has a small magnitude because the zero will almost

FIGURE 5.32
Root locus with both lead and lag compensations: (a) whole locus; (b) portion of part (a) expanded to show the root due to lag compensation

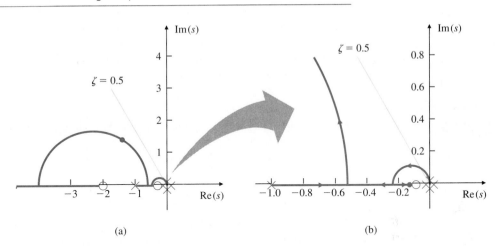

(a)   (b)

cancel the pole in the transfer function. Still, the decay is so slow that this term may seriously influence the settling time. Because of this effect it is important to place the lag pole-zero combination at as high a frequency as possible without causing major shifts in the dominant root locations. Also important to notice is the fact that the transfer function from a plant disturbance to the system error will not have the zero, and thus disturbance transients can be very long in duration in a system with lag compensation.

As with lead compensation, lag compensation is usually implemented using a digital computer as described in Chapter 8. However, it, too, can be implemented using analog electronics. A circuit diagram of a lag network is given in Fig. 5.33. The transfer function of this circuit can be shown to be

$$D(s) = -\frac{R_2}{R_i} \frac{1}{\beta} \frac{\beta Ts + 1}{Ts + 1},$$

FIGURE 5.33
Circuit diagram of a lag network

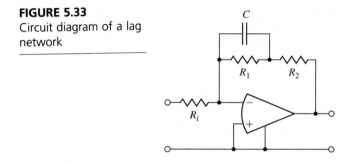

where

$$T = R_1 C,$$

$$\beta = \frac{R_2}{R_1 + R_2}.$$

Usually $R_i = R_2$, so the high-frequency gain is unity, but some gain adjustment can be made by selecting other values for R_i.

5.7   Extensions of the Root-locus Method

As we have seen in this chapter, the root-locus technique is a graphical scheme to show locations of possible roots of an equation as a real parameter varies. Thus far we have considered only polynomial equations for linear systems. However, these restrictions can be altered to allow us to apply this technique to a far wider range of cases. We examine two additional types of systems in this section: those with time delay and with nonlinear elements.

5.7.1   Time Delay

Time delays often arise in control systems, both from delays in the process itself and from delays in the processing of sensed signals. Chemical plants often have processes with a time delay representing the time material takes to flow through pipes. In measuring the attitude of a spacecraft en route to Mars, there is a significant time delay for the sensed quantity to arrive back on Earth due to the speed of light. There is also a small time delay in any digital control system due to the cycle time of the computer and the fact that data is processed at discrete intervals. Time delay *always* reduces the stability of a system; therefore, it is important to be able to analyze its effect. In this section we discuss how to use the root locus for such analysis. Although an easier method of analyzing time delay is to use the frequency-response methods to be described in Chapter 6, knowing several different ways to analyze a design allows the control designer to do a better job.

Time delays always reduce the stability of a system.

Consider the problem of designing a control system for the temperature of the heat exchanger in Example 2.16. The transfer function between the control A_s and the measured output temperature T_m is described by two first-order terms plus a time delay T_d of 5 sec. The time delay results because the temperature sensor is physically located far enough downstream from the exchanger that there is a delay in its reading. The transfer function is

An example of a root locus with time delay

$$G(s) = \frac{e^{-5s}}{(10s + 1)(60s + 1)}, \tag{5.49}$$

where the e^{-5s} term arises from the time delay, as shown in Table A.1 for entry number 2.

The corresponding root-locus equations are

$$1 + KG(s) = 0,$$

$$1 + K\frac{e^{-5s}}{(10s + 1)(60s + 1)} = 0,$$

$$600s^2 + 70s + 1 + Ke^{-5s} = 0. \tag{5.50}$$

How would we plot the root locus corresponding to Eq. (5.50)? Since it is not a polynomial, we cannot proceed with the methods used in previous examples. Instead, there are two basic approaches we will describe: approximation and direct application of the phase condition.

In the first approach we reduce the given problem to one we have previously solved by approximating the nonrational function e^{-5s} with a rational function. Since we are concerned with control systems and hence typically with low frequencies, we want an approximation that will be good for $s = 0$ and values nearby.[9] The most common means for finding such an approximation is attributed to H. Padé. It consists of matching the series expansion of the transcendental function e^{-5s} with the series expansion of a rational function whose numerator is a polynomial of degree p and whose denominator is a polynomial of degree q. The result is called a **(p, q) Padé approximant** to e^{-5s}. We will initially compute the approximants to e^{-s}, and in the final result we will substitute $T_d s$ for s to allow for any desired delay.

To illustrate the process, we begin with the $(1, 1)$ approximant. In this case we wish to select b_0, b_1, and a_0 so that the error

$$e^{-s} - \frac{b_0 s + b_1}{a_0 s + 1} = \varepsilon \tag{5.51}$$

is small. For the Padé approximant we expand both e^{-s} and the rational function into a McLauren series and match as many of the initial terms as possible. The two series are

$$e^{-s} = 1 - s + \frac{s^2}{2} - \frac{s^3}{3!} + \frac{s^4}{4!} - \cdots,$$

$$\frac{b_0 s + b_1}{a_0 s + 1} = b_1 + (b_0 - a_0 b_1)s - a_0(b_0 - a_0 b_1)s^2 + a_0^2(b_0 - a_0 b_1)s^3 + \cdots.$$

Matching coefficients of the first four terms, we get the following equations to

---

9. The nonrational function e^{-5s} is analytic for finite values of s and so may be approximated by a rational function. If nonanalytic functions such as $\sqrt{s}$ were involved, great caution would be needed in selecting the approximation.

solve:

$$b_1 = 1,$$
$$b_0 - a_0 b_1 = -1,$$
$$-a_0(b_0 - a_0 b_1) = \tfrac{1}{2},$$
$$a_0^2(b_0 - a_0 b_1) = -\tfrac{1}{6}.$$

Now we notice that there are an infinite number of equations but only three parameters. We substitute $T_d s$ for s and then match the first three coefficients. The resulting Padé approximant is

$$e^{-T_d s} \cong \frac{1 - (T_d s/2)}{1 + (T_d s/2)}. \tag{5.52}$$

If we assume $p = q = 2$, we have five parameters, and a better match is possible. In this case we have the $(2, 2)$ approximant

$$e^{-T_d s} \cong \frac{1 - T_d s/2 + (T_d s)^2/12}{1 + T_d s/2 + (T_d s)^2/12}. \tag{5.53}$$

The comparison of these approximants is best seen from their pole-zero configuration in Fig. 5.34. The locations of the poles are shown in the LHP; the zeros are in the RHP at the reflections of the poles.

In some cases a very crude approximation is acceptable, and the $(0, 1)$ approximant can be used, which is a first order lag given by

$$e^{-T_d s} \cong \frac{1}{1 + T_d s}. \tag{5.54}$$

FIGURE 5.34
Poles and zeros of the Padé approximants to e^{-s}, with superscripts identifying the corresponding approximants; for example, x^1 represents the $(1, 1)$ approximant

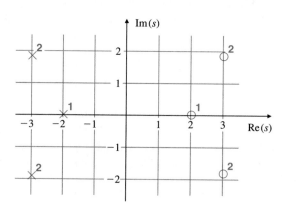

FIGURE 5.35
Root loci for the heat exchanger with and without time delay

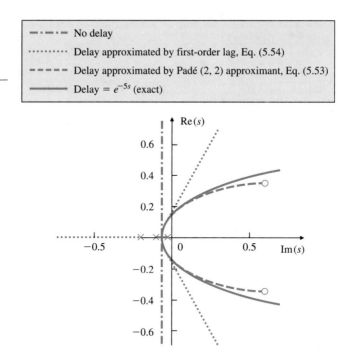

- ─ ·─ ·─ No delay
- ·········· Delay approximated by first-order lag, Eq. (5.54)
- ─ ─ ─ ─ Delay approximated by Padé (2, 2) approximant, Eq. (5.53)
- ──────── Delay = e^{-5s} (exact)

Contrasting methods of approximating delay

To illustrate the effect of a delay and the accuracy of the different approximations, root loci for the heat exchanger are drawn in Fig. 5.35 for four cases. Notice that for low gains and up to the point where the loci cross the imaginary axis, the approximate curves are very close to the exact. However, the Padé curve follows the exact curve much further than does the first-order lag, and its increased accuracy would be useful if the delay were larger. All analyses of the delay show its destabilizing effect and how it limits the achievable response time of the system.

While the Padé approximation leads to a rational transfer function, it is not necessary for plotting a root locus. A direct application of the phase condition can be used to plot an exact locus of a system with time delay. The phase-angle condition does not change if the transfer function of the process is nonrational, so we still must search for values of s for which the phase is $180° + 360°l$. If we write the transfer function as

$$G(s) = e^{-\lambda s}\bar{G}(s),$$

the phase of $G(s)$ is the phase of $\bar{G}(s)$ minus $\lambda\omega$ for $s = \sigma + j\omega$. Thus we can formulate a root-locus problem as searching for locations where the phase of $\bar{G}(s)$ is $180° + \lambda\omega + 360°l$. To plot such a locus, we would fix ω and search along a horizontal line in the s-plane until we found a point, then raise the value of ω, change the target angle, and repeat. Similarly, the departure angles are modified by $\lambda\omega$, where ω is the imaginary part of the pole from which the departure being computed. This kind of search method was used to obtain the exact (e^{-5s}) curve in Fig. 5.35.

An exact analysis of delay is much easier using frequency response; therefore, if a simple approximation for the delay such as Eq. (5.52) or the first-order lag in Eq. (5.54) is not sufficient, the expedient approach is to use the frequency-response design methods described in Chapter 6.

5.7.2   Nonlinear Systems

As we have tried to make clear, every real control system is nonlinear, and the linear analysis and design methods we have described use linear approximations to the real models. There is one important category of nonlinear systems for which some significant analysis (and design) can be done. This comprises the systems in which the nonlinearity has no dynamics and is well approximated as a gain that varies as the size of its input signal varies. Sketches of a few such nonlinear system elements and their common names are shown in Fig. 5.36.

The behavior of systems containing any one of these nonlinearities can be qualitatively described by considering the nonlinear element as a varying,

FIGURE 5.36
Nonlinear elements with no dynamics: (a) saturation, (b) relay, (c) relay with dead zone, (d) gain with dead zone, (e) preloaded spring, or coulomb plus viscous friction, and (f) quantization

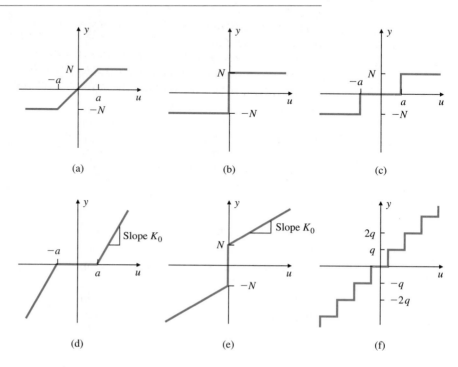

FIGURE 5.37
General shape of the
effective gain of
saturation

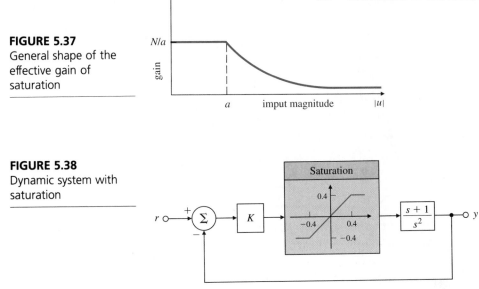

FIGURE 5.38
Dynamic system with
saturation

signal-dependent gain. For example, with the saturation element (Fig. 5.36a), it is clear that for input signals with magnitudes of less than a, the nonlinearity is linear with the gain N/a. However, for signals larger than a, the output size is bounded by N, while the input size can get much larger than a, so once the input exceeds a, the ratio of output to input goes down. Thus, saturation has the gain characteristics shown in Fig. 5.37. *All* actuators saturate at some level. If they did not, their output would increase to infinity, which is physically impossible.

Sizing the actuator

An important aspect of control system design is **sizing the actuator**, which means picking the size, weight, power required, cost, and saturation level of the device. Generally, higher saturation levels require bigger, heavier, and more costly actuators. The key factor that enters into the sizing is the effect of the saturation on the control system's performance.

A nonlinear example:
saturation

To illustrate saturation with an example, consider the system shown in Fig. 5.38. The root locus of this system versus K with the saturation removed is given by Fig. 5.39. At $K = 1$ the damping ratio is $\zeta = 0.5$. As the gain is reduced, the locus shows that the roots move toward the origin of the s-plane with less and

FIGURE 5.39
Root locus of $(s + 1)/s^2$,
the system in Fig. 5.38
with the saturation
removed

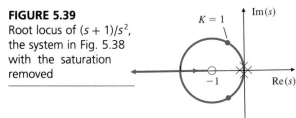

FIGURE 5.40

Step responses of system
in Fig. 5.38 for various
input step sizes

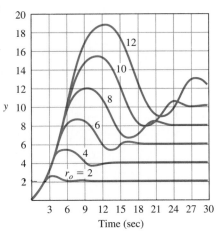

less damping. Plots of the step responses of this system were obtained using numerical simulation. A series of different step inputs r were introduced to the system with magnitudes $r_o = 2, 4, 6, 8, 10,$ and 12, and the results are shown in Fig. 5.40. As long as the signal entering the saturation remains less than 0.4, the system will be linear and should behave according to the roots at $\zeta = 0.5$. However, notice that as the input gets larger, the response has more and more overshoot and slower and slower recovery. This can be explained by noting that larger and larger input signals correspond to smaller and smaller effective gain K, as seen in Fig. 5.37. From the root-locus plot of Fig. 5.39, we see that as K decreases, the closed-loop poles move closer to the origin and have a smaller damping ζ. This results in the longer rise and settling times, increased overshoot, and greater oscillatory response.

A nonlinear example: stability depends on input magnitude.

As a second example of a nonlinear response described by signal-dependent gain, consider the system whose block diagram is drawn in Fig. 5.41 and whose root locus, excluding the saturation, is plotted in Fig. 5.42. From this locus we can readily calculate that the imaginary axis crossing occurs at $\omega_0 = 1$ and $K = \frac{1}{2}$. Systems such as this, which are stable for (relatively) large gains but unstable for smaller gains, are called **conditionally stable systems**. If $K = 2$,

Conditional stability

which corresponds to $\zeta = 0.5$ on the locus, the system should show responses

FIGURE 5.41

Block diagram of a
conditionally stable
system

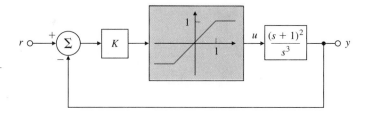

FIGURE 5.42
Root locus for $G(s) = (s + 1)^2/s^3$ from system in Fig. 5.41

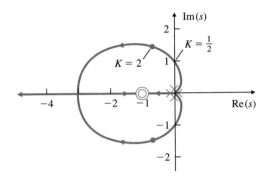

FIGURE 5.43
Step responses of system in Fig. 5.41

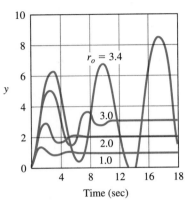

A nonlinear example: an oscillatory system with saturation

consistent with $\zeta = 0.5$ for small signals, become less well damped as the signal strength is increased, and become unstable for larger signals. Step responses from nonlinear simulation of the system with $K = 2$ for input steps of size $r_o = 1.0, 2.0, 3.0$, and 3.4 are shown in Fig. 5.43. These responses confirm our prediction. Furthermore, the unstable case shows oscillations near 1 rad/sec, which is predicted by the frequency at which the root locus crosses into the RHP.

The final illustration of the use of the root locus to give a qualitative description of the response of a nonlinear system is based on the block diagram in Fig. 5.44. This system is typical of electromechanical control problems where the designer perhaps at first is not aware of the resonant mode corresponding to the denominator term $s^2 + 0.2\,s + 1$, ($\omega = 1$, $\zeta = 0.1$). The root locus for this system versus K, excluding the saturation, is sketched in Fig. 5.45. The imaginary-axis crossing can be verified to be at $\omega_0 = 1$, $K = 0.2$; thus a gain of $K = 0.5$ is enough to force the roots of the resonant mode into the RHP, as

FIGURE 5.44
Block diagram of a system with an oscillatory mode

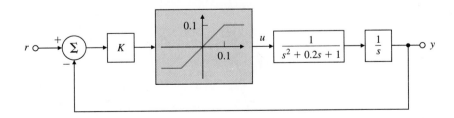

FIGURE 5.45
Root locus for the system
in Fig. 5.44, excluding
saturation

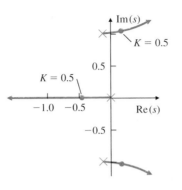

shown by the dots. If the system gain is set at $K = 0.5$, our analysis predicts a system that is initially unstable but becomes stable as the gain decreases. Thus we would expect the response of the system with the saturation to build up due to the instability until the magnitude is sufficiently large that the effective gain is lowered to $K = 0.2$ *and then stop growing*!

Plots of the step responses with $K = 0.5$ for three steps of size $r_o = 1, 4$, and 8 are shown in Fig. 5.46, and again our heuristic analysis is exactly correct: The error builds up to a fixed amplitude and then starts to oscillate. The oscillations have a frequency of 1 rad/sec and hold constant amplitude at any DC equilibrium value (for step sizes of $r_o = 1, 4$, or 8). In this case the response always approaches a periodic solution of fixed amplitude known as a **limit cycle**, so-called because the response is cyclic and is approached in the limit as time grows large. A nonlinear analysis method known as **describing functions** can be used to predict stability and limit-cycle amplitude for systems like these (Truxal, 1955, p. 566).

FIGURE 5.46
Step response of system
in Fig. 5.44

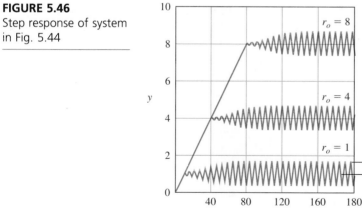

FIGURE 5.47
Root locus including
compensation

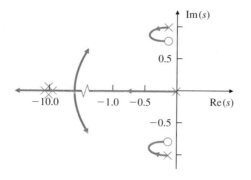

In order to prevent the limit cycle, the locus has to be modified by compensation so that no branches cross into the RHP. One common method to do this for a lightly damped oscillatory mode is to place compensation zeros near the poles, but at a slightly lower frequency. Example 5.8 demonstrated that a pole-zero pair located in this manner will cause a locus segment to go from the pole to the zero, looping to the left, and thus staying away from the RHP. Figure 5.47 shows the root locus for the system, $1/[s(s^2 + 0.2s + 1.0)]$, including a compensation with zeros located as just discussed. In addition, the compensation also includes two poles. They were both placed at $s = -10$, fast enough to not cause stability problems with the system, yet slow enough that high frequency noise would not be amplified too much. Thus the compensation used for the root locus is

$$D(s) = 123 \, \frac{s^2 + 0.18s + 0.81}{(s + 10)^2},$$

where the gain of 123 has been selected to make the compensation's DC gain equal to unity. This kind of compensation is called a **notch filter** because its response to sinusoidal inputs (that is, its frequency response) is such that it attenuates inputs in the vicinity of a specific frequency. In this case the notch is at $\omega_n^2 = 0.81$ or $\omega_n = 0.9$ rad/sec, so that any input from the plant resonance is attenuated and is thus prevented from detracting from the stability of the system. Figure 5.48 shows the system including the notch filter, and Fig. 5.49

FIGURE 5.48
Block diagram of the
system with a notch filter

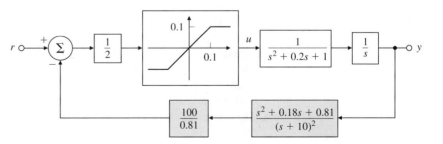

FIGURE 5.49
Step responses of the system in Fig. 5.48

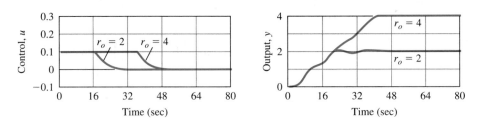

shows the time response for two step inputs. Both inputs $r_o = 2$ and 4 are sufficiently large so that the nonlinearity is saturated initially; however, because the system is stable, the saturation results only in lowering the gain, so the response is slower than predicted by linear analysis but still stable, as also predicted by linear analysis. In both cases the nonlinearity eventually becomes unsaturated, and the system stabilizes to its new commanded value of r.

5.8   Computer-aided Determination of the Root Locus

In order to use the root locus as a design tool and to verify computer-generated loci, it is very important, as we emphasized earlier, to be able to sketch root loci. The control engineer can then quickly predict, for design purposes, the effect of an added zero or pole, or even several of them, or can quickly confirm computer output in a qualitative sense. For this reason, you need to understand the guidelines for sketching loci and should be able to plot numerous example loci by hand. Once you have the aid of a computer, it can be used to determine accurate loci and to establish exact values for the parameters. It is especially useful in computing the closed-loop pole sensitivity to those parameters because their values may be known only to a certain tolerance at the time of the design and may be subject to drift over the life of the system.

There are two basic approaches to machine computation of the root locus. In the first approach we formulate the root-locus problem as a polynomial in the form $a(s) + Kb(s) = 0$. For a sequence of values of K varying from near zero to a large value, we solve the polynomial for its n roots by any of many available numerical techniques. An advantage of this method is that it computes all roots for each value of K, guaranteeing that a specific value—for example, that value of the loop gain that gives the design-velocity constant—is also included.

Polynomial factoring by computer

One of the disadvantages of the method is that the resulting root locations are very unevenly distributed in the s-plane. For example, near a point of multiple roots the sensitivity of the root locations to the K-value is very great, and the roots just "fly through" such points; the plots appear to be coarse, so it is easy to miss important features. (On the other hand, the method has the advantage that near a zero the root moves very slowly, since it takes an infinite value of the parameter to push the root all the way into the zero.)

A second disadvantage of this method is that the equation must be a polynomial. As we saw in Section 5.7.1, pure time delay involves a transcendental equation, so we must use an approximation such as the Padé method to reduce the given problem to an equivalent polynomial. Such approximations limit the range of values of the parameter for which the results are accurate, and checking the accuracy is difficult unless a means is available to solve the true equation at critical points.

A final disadvantage is that not many algorithms are able to solve polynomials easily at points of multiple roots. This problem is related to the great sensitivity of the roots to the parameter at these points, as mentioned earlier. A method related to polynomial factoring is possible when a state-variable formulation of the equations of motion is available. As we will see in Chapter 7, the closed-loop poles are eigenvalues of the system matrix $\mathbf{F}$; an algorithm called QR can be used with great accuracy to solve for the eigenvalues of systems of substantial complexity. For systems described by high-order ODEs, this is the most powerful method available and used by most CACSD software. This method will be examined further in Chapter 7.

Curve tracing by computer

The alternative to polynomial factoring is a method based on curve tracing. It hinges on the fact that a point on the positive root locus is a point where the phase of $G(s)$ is $180°$. Thus, given a point on the locus at s_0 with gain K_0, we can draw a circle of radius δ around s_0 and search on the circle for a new point where the angle condition is met and the new gain is larger than K_0. This method can be easily arranged to include a delay term such as $e^{-\lambda s}$; the resulting points will be spaced δ radians apart, a value that the designer can specify.

A disadvantage of this method is that only one branch of the locus is generated at a time (although computer logic can be easily set up to step through each of the open-loop poles to produce a complete graph). A second disadvantage is that the designer needs to monitor the selection of δ in order to ensure that the search for $180°$ converges on some points and to avoid wasting too much time with a small value of δ at less critical points.

With polynomial factoring or curve tracing, it is important for the software to provide a graph of the results, since the essential information in the locus is graphical. Many of the root loci plotted in this chapter were originally plotted by a computer program developed at Stanford and implemented on the HP-85. Subsequently, all loci have been recreated using MATLAB so you can see yourself exactly how to do it. As mentioned in the Preface, these files are available to all. Other commercially available CACSD software—Ctrl-C, MATRIX_x, and CC—also include root-locus capability.

◆ **EXAMPLE 5.14**   *Autopilot Design*

For the Piper Dakota shown in Fig. 5.50, the transfer function between the elevator input and the pitch attitude is

$$G(s) = \frac{\theta(s)}{\delta_e(s)} = \frac{160(s + 2.5)(s + 0.7)}{(s^2 + 5s + 40)(s^2 + 0.03s + 0.06)}, \tag{5.55}$$

where

θ = pitch attitude, degrees (see Fig. 9.29),

δ_e = elevator angle, degrees.

(For a more detailed discussion of longitudinal aircraft motion, refer to Section 9.3.2.)

(a) Design an auto pilot so that the response to a step elevator input has a rise time of 1 sec or less and an overshoot less than 10%.

(b) When there is a constant disturbing moment acting on the aircraft so that the pilot must supply a constant force on the controls for steady flight, it is said to be out of trim. The transfer function between the disturbing moment and the attitude is the same as

FIGURE 5.50
Auto-pilot design in the Piper Dakota, showing elevator and trim tab. *(Photo courtesy of Denise Freeman)*

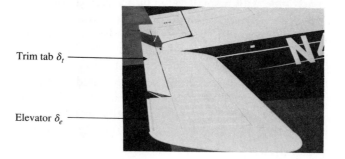

Trim tab δ_t

Elevator δ_e

FIGURE 5.51

Block diagrams for auto-pilot design: (a) open loop; (b) feedback scheme excluding trim control

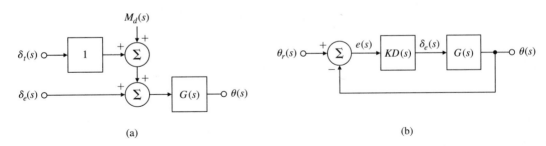

(a)   (b)

that due to the elevator; that is,

$$\frac{\theta(s)}{M_d(s)} = \frac{160(s + 2.5)(s + 0.7)}{(s^2 + 5s + 40)(s^2 + 0.03s + 0.06)}, \tag{5.56}$$

where M_d is the moment acting on the aircraft. There is a separate aerodynamic surface for trimming, δ_t, that can be actuated and will change the moment on the aircraft. It is shown in the close-up of the tail in Fig. 5.50. Its influence is depicted in the block diagram shown in Fig. 5.51(a). For both manual and auto-pilot flight it is desirable to adjust the trim so that there is no steady-state control effort required from the elevator (that is, so $\delta_e = 0$). In manual flight this means no force is required by the pilot to keep the aircraft at a constant attitude, whereas in auto-pilot control it means reducing the amount of electrical power required and saving wear and tear on the servomotor that drives the elevator. Design an auto pilot that will command the trim δ_t so as to drive the steady-state value of δ_e to zero for an arbitrary moment M_d as well as meet the specifications in part (a).

Solution. (a) To satisfy the requirement that the rise time $t_r \leqslant 1$ sec, Eq. 3.42 indicates that, for the ideal second-order case, ω_n must be greater than 1.8 rad/sec. And to provide an overshoot of less than 10%, Fig. 3.17 indicates that ζ should be greater than 0.6, again, for the ideal second-order case. In the design process we can examine a root locus for a candidate for feedback compensation and then look at the resulting time response when the roots appear to satisfy the design guidelines. However, since this is a fourth-order system, the design guidelines might not be sufficient, or they might be overly restrictive.

To initiate the design process, it is often instructive to look at the system charac-teristics with proportional feedback, that is, where $D(s) = 1$ in Fig. 5.51(b). The statements in MATLAB to create a root locus with respect to K and a time response for the proportional feedback case with $K = 0.3$ are:

Root locus and time
response for auto pilot
via MATLAB

```
numG = 160*conv([1  2.5],[1  0.7])
denG = conv([1  5  40],[1 .03 .06])
rlocus(numG,denG)

K = 0.3
[numCL,denCL] = feedback(K*numG,denG,1,1)
step(numCL,denCL)
```

FIGURE 5.52
Root loci for auto-pilot design

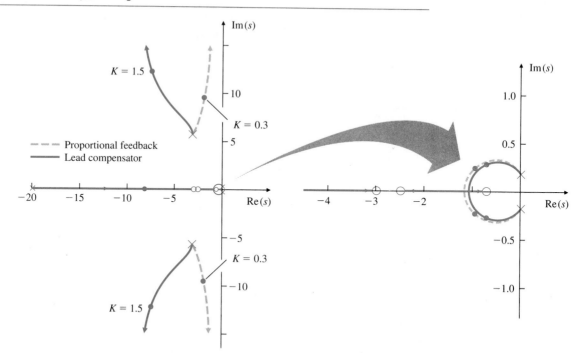

The resulting root locus and time response are shown with dashed lines in Figs. 5.52 and 5.53. Notice from Fig. 5.52 that the two faster roots will always have a damping ratio ζ that is less than about 0.2; therefore, proportional feedback probably will not be acceptable. Also, the slower roots have some effect on the time response shown in Fig. 5.53 (dashed curve) with $K = 0.3$ in that they cause a long-term settling. However, the dominating characteristic of the response that determines whether or not the compensation meets the specifications is the behavior in the first few seconds, which is dictated by the fast roots. The low damping of the fast roots causes the time response to be oscillatory, which leads to excess overshoot and a longer settling time than desired.

We saw in Section 5.6.1 that lead compensation causes the locus to shift to the left, a change needed here to increase the damping. Some trial and error will be required to arrive at a suitable pole and zero location. Values of $z = 3$ and $p = 20$ in Eq. (5.45) have a substantial effect in moving the fast branches of the locus to the left; thus

$$D(s) = \frac{s + 3}{s + 20}.$$

Trial and error is also required to arrive at a value of K that meets the specifications. The statements in **MATLAB** to add this compensation are

FIGURE 5.53
Time-response plots for
auto-pilot design

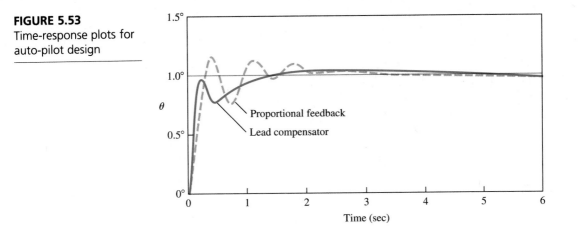

Lead compensation via
MATLAB

```
numD=[1  3]
denD=[1  20]
num=conv(numD,numG)
den=conv(denD,denG)
rlocus(num,den)
K=1.5
[numCL,denCL]=feedback(K*num,den,1,1)
step(numCL,denCL)
```

The root locus for this case and the corresponding time response are also shown in Figs. 5.52 and 5.53 by the solid lines. Note that the damping of the fast roots that corresponds to $K = 1.5$ is $\zeta = 0.52$, which is slightly lower than we would like; also, the natural frequency is $\omega_n = 15$ rad/sec, much faster than we need. However, these values are close enough to meeting the guidelines to suggest a look at the time response. In fact, the time response shows that $t_r \cong 0.9$ sec and $M_p \cong 8\%$, both within the specifications, although by a very slim margin.

In summary, the primary design path consisted of adjusting the compensation to influence the fast roots, examining their effect on the time response, and continuing the iteration until the time specifications were satisfied.

(b) The purpose of the trim is to provide a moment that will eliminate a steady-state nonzero value of the elevator. Therefore, if we integrate the elevator command δ_e and feed this integral to the trim device, the trim should eventually provide the moment required to hold an arbitrary attitude, thus eliminating the need for a steady-state δ_e. This idea is shown in Fig. 5.54(a). If the gain on the integral term K_I is small enough, the destabilizing effect of adding the integral should be small and the system should behave approximately as before since that feedback loop has been left intact. The block diagram in Fig. 5.54(a) can be reduced to that in Fig. 5.54(b) for analysis purposes by defining the compensation to be

$$D_I(s) = KD(s)\left(1 + \frac{K_I}{s}\right).$$

FIGURE 5.54

Block diagram showing the trim-command loop

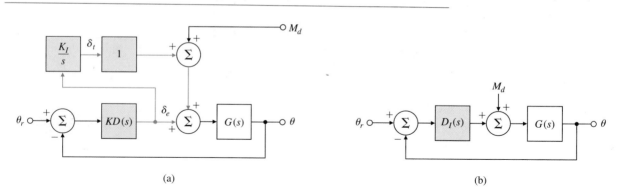

(a)   (b)

However, it is important to keep in mind that, physically, there will be two outputs from the compensation: δ_e (used by the elevator servomotor) and δ_t (used by the trim servomotor).

The characteristic equation of the system with the integral term is

$$1 + KDG + \frac{K_I}{s}KDG = 0.$$

To aid in the design process, it is desirable to find the locus of roots with respect to K_I, but the characteristic equation is not in any of the root-locus forms given by Eqs. (5.5). Therefore, some manipulation yields

$$1 + \frac{(K_I/s)KDG}{1 + KDG} = 0,$$

which reduces to[10]

$$1 + \frac{K_I \, \text{num}(KDG)}{s[\text{roots}(1 + KDG)]} = 0$$

because

$$1 + KDG = 1 + \frac{\text{num}(KDG)}{\text{den}(KDG)} = \frac{\text{den}(KDG) + \text{num}(KDG)}{\text{den}(KDG)} = \frac{\text{roots}(1 + KDG)}{\text{den}(KDG)}.$$

Therefore, the MATLAB statements to produce a root locus with respect to K_I are

Root locus versus K_I the integral gain

```
denI=[denCL 0]
rlocus(K*num,denI)
```

10. Here num, den, and roots correspond to the MATLAB quantities num, den, and the roots function.

The statements to obtain a time response with $K_I = 0.15$ are:

```
KI=0.15
numDI=K*conv(numD,[1 KI])
denDI=[denD 0]
numI=conv(numDI,numG)
denI=conv(denDI,denG)
[numCLI,denCLI]=feedback(numI,denI,1,1)
step(numCLI,denCLI)
```

It can be seen from the locus in Fig. 5.55 that the damping of the fast roots decreases as K_I increases. This shows the necessity for keeping the value of K_I as low as possible. After some trial and error we select $K_I = 0.15$. This value has little effect on the roots—note the roots are virtually on top of the previous roots obtained without the integral term—and little effect on the short-term behavior of the step response, as shown in Fig. 5.56(a), so the specifications are still met. $K_I = 0.15$ does cause the longer-term attitude behavior to approach the commanded value with no error, as we would expect with integral control. It also causes δ_e to approach zero (Fig. 5.56b shows it settling in approximately

FIGURE 5.55

Root locus versus K_I: assumes an added integral term and lead compensation with a gain $K = 1.5$; roots for $K_I = 0.15$ marked with •

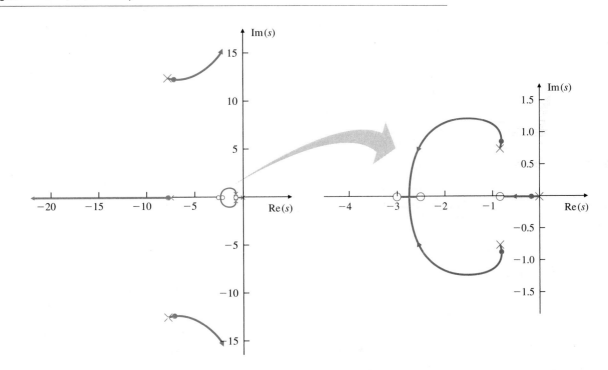

FIGURE 5.56
Step response for the
case with an integral
term and 5° command

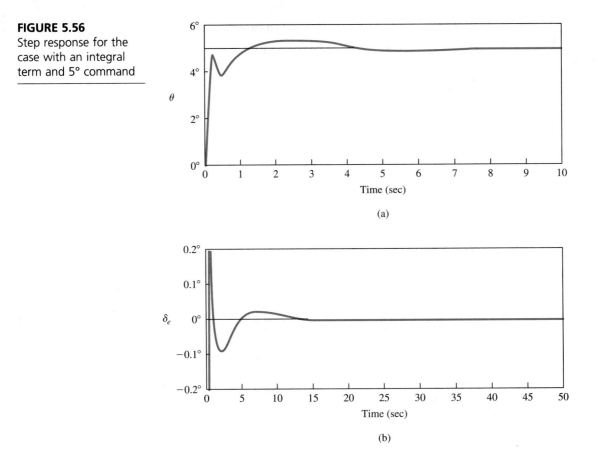

(a)

(b)

30 sec), which is good because this is the reason for choosing integral control in the first place. The time for the integral to reach the correct value is predicted by the new, slow real root that is added by the integral term at $s = -0.14$. The time constant associated with this root is $\tau = 1/0.14 \cong 7\,\text{sec}$. The settling time to 1% for a root with $\sigma = 0.14$ is shown by Eq. (3.41) to be $t_s = 33\,\text{sec}$, which agrees with the behavior in Fig. 5.56(b).

◆

Summary

- The root locus with respect to K is a graph of the values of s that are solutions to the equation

$$1 + KG(s) = 0. \qquad (5.5a)$$

When $K > 0$, s is on the locus if $\angle G(s) = 180°$, producing a 180° locus. When $K < 0$, s is on the locus if $\angle G(s) = 0°$, producing a 0° locus.

- If $G(s)$ is the open-loop transfer function of a system with negative feedback,

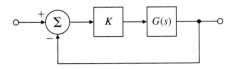

then the characteristic equation of the closed-loop system is

$$1 + KG(s) = 0,$$

and the root-locus method displays the effect of changing the gain K on the closed-loop system roots.

- The key steps in plotting a $180°$ locus are:

1. Mark the poles of $G(s)$ by $\times$ and zeros of $G(s)$ by $\circ$.
2. Draw the locus on the real axis to the left of an odd number of poles plus zeros.
3. Draw the asymptotes, centered at α and leaving at angles ϕ_l:

$$n = \text{number of poles},$$

$$m = \text{number of zeros},$$

$$n - m = \text{number of asymptotes},$$

$$\alpha = \frac{\Sigma\, p_i - \Sigma\, z_i}{n - m},$$

$$\phi_l = \frac{180° + 360°(l - 1)}{n - m}, \qquad l = 1, 2, \ldots, n - m.$$

4. Compute the locus departure angles from the poles and arrival angles at the zeros:

$$\phi_{\text{dep}} = \frac{1}{q}\left(\Sigma\, \psi_i - \Sigma\, \phi_i - 180° - 360°l\right),$$

$$\psi_{\text{arr}} = \frac{1}{q}\left(\Sigma\, \phi_i - \Sigma\, \psi_i + 180° + 360°l\right),$$

where

$$q = \text{order of the pole or zero},$$
$$\psi_i = \text{angles from the zeros},$$
$$\phi_i = \text{angles from the poles}.$$

- The gain K corresponding to a root at a particular point s_o on the locus can be found from

$$K = \frac{1}{|G(s_o)|}, \tag{5.40}$$

where $|G(s_o)|$ can be found graphically by measuring the distances from s_o to each of the poles and zeros.

- Lead compensation, given by

$$D(s) = \frac{s+z}{s+p}, \qquad z < p, \tag{5.45}$$

approximates proportional-derivative (PD) control. It moves the locus to the left and typically improves the system damping.

- Lag compensation, given by

$$D(s) = \frac{s+z}{s+p}, \qquad z > p, \tag{5.48}$$

approximates proportional-integral (PI) control. It improves the steady-state error by increasing the low-frequency gain, but usually it degrades stability.

- The root locus can be used to analyze the effect of time delay.

- Nonlinearities with no dynamics, such as saturation, can be analyzed using the root locus by considering the nonlinearity to be a variable gain.

- Root loci are usually obtained using computers; in MATLAB the rlocus function creates a root locus with respect to K given the numerator and denominator of $G(s)$.

- A working knowledge of how to draw a root locus by hand is useful for verifying the accuracy of computer results and for expediting the design iterations.

Problems

5.1 Set up the following characteristic equations in the form suited to Evans's root-locus method:

a) $s + (1/\tau) = 0$ versus parameter τ
b) $s^2 + bs + b + 1 = 0$ versus parameter b
c) $(s + b)^3 + A(Ts + 1) = 0$
  (1) versus parameter A,
  (2) versus parameter T,
  (3) versus the parameter b, if possible. Say why you can or can not. Can a locus be drawn versus b for given constant values of A and T?

FIGURE 5.57
Pole-zero maps

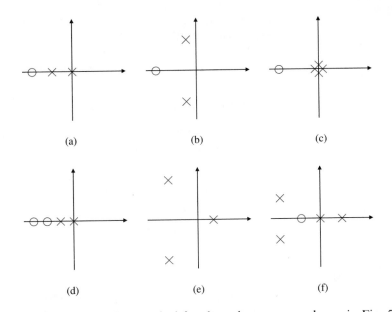

(a)   (b)   (c)

(d)   (e)   (f)

5.2 Roughly sketch the root loci for the pole-zero maps shown in Fig. 5.57. Show asymptotes, centroids, a rough evaluation of arrival and departure angles, and the loci for positive values of the parameter K. Each pole-zero map is from a characteristic equation of the form

$$1 + K\frac{b(s)}{a(s)} = 0,$$

where the roots of the numerator $b(s)$ are shown as small circles and the roots of the denominator $a(s)$ are shown as x's on the s-plane. Note that in Fig. 5.57(c), there are two poles at the origin.

5.3 For the characteristic equation

$$1 + \frac{K}{s(s + 1)(s + 5)} = 0:$$

a) Draw the real-axis segments of the corresponding root locus.
b) Sketch the asymptotes of the locus for $K \to \infty$.
c) For what value of K are the roots on the imaginary axis?

5.4 Sketch the root locus with respect to K for the following systems. Be sure to give the asymptotes, arrival and departure angles, and imaginary-axis crossings, if any. After completing the hand sketch verify your results using MATLAB.

a) $KG(s) = \dfrac{K}{s(s^2 + 2s + 10)}$ d) $KG(s) = \dfrac{K(s + 3)}{s(s + 1)(s^2 + 4s + 5)}$

b) $KG(s) = \dfrac{K(s^2 + 2s + 8)}{s(s^2 + 2s + 10)}$ e) $KG(s) = \dfrac{K(s + 2)}{s^4}$

c) $KG(s) = \dfrac{K(s^2 + 2s + 12)}{s(s^2 + 2s + 10)}$

5.5 Sketch the root locus with respect to K for the following systems, and verify your results using MATLAB:

a) $KG(s) = \dfrac{10K}{s^2 + 3s + 7}$

c) $KG(s) = \dfrac{K(s + 1)}{s(s + 2)}$

e) $KG(s) = \dfrac{K(s^2 + 1)}{s(s^2 + 4)}$

b) $KG(s) = \dfrac{10K}{s(s^2 + 3s + 7)^2}$

d) $KG(s) = \dfrac{K(s^2 + 1)}{s(s^2 + 100)}$

f) $KG(s) = \dfrac{K(s^2 + 4)}{s(s^2 + 1)}$

5.6 Sketch the root locus with respect to K for the following systems, and verify your results using MATLAB:

a) $KG(s) = \dfrac{K[(s+1.5)(s+4)]}{s[(s+1)(s+2.5)]}$

c) $KG(s) = \dfrac{K(s + 1)}{s(s - 1)(s^2 + 2s + 5)}$

b) $KG(s) = \dfrac{K[(s+1.5)(s+5)]}{(s+0.5)(s+2)(s+3)}$

5.7 Sketch the complete root locus with respect to K for the following systems, and verify your results using MATLAB:

a) $KG(s) = \dfrac{K(s + 2)}{s(s + 0.1)(s^2 + 2s + 2)}$

d) $KG(s) = \dfrac{K(s + 0.5)(s + 1.5)}{s(s^2 + 2s + 2)(s + 5)(s + 15)}$

b) $KG(s) = \dfrac{2K}{s(s^2 + 6s + 13)}$

e) $KG(s) = \dfrac{K(s + 1)(s - 0.2)}{s(s + 1)(s + 3)(s^2 + 5)}$

c) $KG(s) = \dfrac{2K(s^2 + 2s + 2)}{s(s^2 + 6s + 13)(s^2 + s + 2)}$

5.8 **a)** For the system given in Fig. 5.58, plot the root locus as the parameter K_1 is varied from 0 to ∞ with $\lambda = 2$.
  b) Repeat part (a) with $\lambda = 5$. What is special about this value?
  c) Repeat part (a) for $K_1 = 2$ with λ varying from 0 to ∞.

FIGURE 5.58
Control system for
Problem 5.8

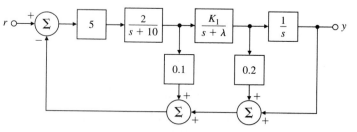

5.9 For the system shown in Fig. 5.59, sketch the root locus with respect to the parameter b. Be sure to show the direction in which b increases on the locus.

FIGURE 5.59
Control system for
Problem 5.9

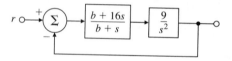

5.10 Suppose you are given a system with the transfer function

$$KG(s) = \frac{K(s-z)}{(s-p_1)(s-p_2)},$$

where z, p_1, p_2 are real, $z < p_1$, and $z < p_2$. Show that the root-locus for this system is a circle centered at z with radius given by

$$r = \sqrt{|p_1 - z||p_2 - z|}.$$

5.11 The loop transmission of a system has two poles at $s = -1$ and a zero at $s = -2$. There is a third real-axis pole located somewhere to the left of the zero. Several different root loci are possible, depending on the exact location of the third pole. The extreme cases occur when the pole is located at infinity or when it is located at $s = -2$. Sketch several different possible loci.

5.12 Sketch the root locus with respect to α for the system shown in Fig. 5.60. From your root-locus plot, estimate the closed-loop pole locations, and sketch the corresponding step responses when $\alpha = 0, 0.5$, and 2. Use MATLAB to check the accuracy of your approximate step responses.

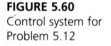

FIGURE 5.60
Control system for
Problem 5.12

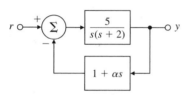

5.13 Consider the system with

$$KG(s) = \frac{K[(s+1)^2 + 1]}{s^2(s+2)(s+3)}.$$

We wish to investigate the root locus with respect to K.

a) Plot the location of poles and zeros of $G(s)$ showing the segment of the root locus on the real axis.
b) What are the arrival angles at the complex zeros?
c) What is the breakaway point?
d) What are the asymptotes?
e) Sketch the root locus.
f) Compare your sketch with the root locus generated by MATLAB.

5.14 Suppose you are given the plant

$$G(s) = \frac{1}{s^2 + (1+\alpha)s + (1+\alpha)},$$

where α is a system parameter that is subject to variations. Use root-locus methods to determine what variations in α can be tolerated before instability occurs. (Be sure to consider the case where $\alpha < 0$.)

5.15 Draw the root locus for the system

$$KG(s) = \frac{K(s+2)}{s(s-1)(s+6)^2},$$

and verify the imaginary-axis crossings using the Routh array.

5.16 Using root-locus techniques, determine the values of $\alpha < 0$ for which the polynomial $s^3 + 2s^2 + \alpha = 0$ has repeated roots.

5.17 Sketch the root locus for the system

$$KG(s) = \frac{K}{(s+1)(s+2)(s^2+4s+8)}.$$

5.18 Use MATLAB to study the behavior of the root locus for

$$KG(s) = \frac{K(s+a)}{s^4+9s^3+60s^2+52s}$$

as the parameter a is varied from 0 to 10, paying particular attention to the region between 2.5 and 3.5.

5.19 Sketch the root locus for the system

$$KG(s) = -\frac{8K}{(s-1)[(s+2)^2+3]}.$$

5.20 A simplified model of the longitudinal motion of a helicopter near hover has the transfer function

$$G(s) = \frac{9.8(s-0.25 \pm 2.4975j)}{(s+0.6565)(s-0.1183 \pm 0.3678j)}.$$

Sketch the root locus for this system, and verify your answer using MATLAB.

5.21 Let

$$G(s) = \frac{1}{(s+2)(s+3)} \quad \text{and} \quad D(s) = K\frac{s+a}{s+b}.$$

Using root-locus techniques, find values for the parameters a, b, and K of the compensation that will produce closed-loop poles at $s = -1 \pm j$ for the system shown in Fig. 5.61.

FIGURE 5.61
Unity feedback system for Problems 5.23 to 5.27 and 5.42

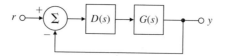

5.22 Suppose that in Fig. 5.61,

$$G(s) = \frac{1}{s(s^2 + 2s + 2)} \qquad \text{and} \qquad D(s) = \frac{K}{s + 2}.$$

Sketch the root-locus with respect to K for the closed-loop system, paying particular attention to points that generate multiple roots.

5.23 Suppose the unity feedback system of Fig. 5.61 has an open-loop plant given by $G(s) = 1/s^2$. Design a lead compensation to be added in series with the plant so that the dominant poles of the closed-loop system are located at $s = -2 \pm 2j$.

5.24 Assume that the unity feedback system of Fig. 5.61 has the open-loop plant

$$G(s) = \frac{K}{s(s + 3)(s + 6)}.$$

Design a lag compensation to meet the following specifications:

- The steady-state error to a unit ramp input must not exceed 10%.
- The damping ratio of the dominant closed-loop poles must be 0.5.
- The dominant poles must have a real part that is less than -1.

5.25 A numerically controlled machine tool has a transfer function given by

$$G(s) = \frac{1}{s(s + 1)}.$$

Performance specifications require that, in the unity feedback configuration of Fig. 5.6.1, the closed-loop poles be located at $s = -1 \pm j\sqrt{3}$.

a) Show that this specification cannot be achieved by choosing $D(s) = K$.
b) Design a lead compensator that will meet the specification.

5.26 An armature-controlled DC motor with negligible armature resistance has a transfer function

$$G(s) = \frac{4}{s(s + 0.5)}.$$

Design a series compensation $D(s)$ in the unity feedback configuration of Fig. 5.61 to meet the following closed-loop specifications:

- The steady-state error to a unit ramp input must be less than 0.02.
- The damping ratio of the dominant closed-loop poles must be equal to 0.5.
- The undamped natural frequency of the dominant closed-loop poles must be 5 rad/sec.

5.27 Assume the closed-loop system of Fig. 5.61 has a feedforward transfer function $G(s)$ given by

$$G(s) = \frac{K}{s(s + 2)}.$$

Design a lag compensation so that the dominant poles of the closed-loop system are located at $s = -1 \pm j$ and the steady-state error to a unit ramp input is less than 0.2.

FIGURE 5.62
Feedback system for Problem 5.28

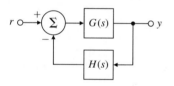

FIGURE 5.63
Feedback system for Problem 5.29

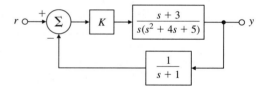

5.28 For the feedback configuration of Fig. 5.62, use asymptotes, center of asymptotes, angles of departure and arrival, and the Routh array to sketch root loci for the following feedback control systems. Use MATLAB to verify your results.

a) $G(s) = \dfrac{K}{s(s + 1 + 3j)(s + 1 - 3j)}$, $\quad H(s) = \dfrac{s + 2}{s + 8}$

b) $G(s) = \dfrac{K(s+1)}{s^2(s+3)}$, $\quad H(s) = 1$

c) $G(s) = \dfrac{K(s + 5)(s + 7)}{(s + 1)(s + 3)}$, $\quad H(s) = 1$

d) $G(s) = \dfrac{K(s + 3 + 4j)(s + 3 - 4j)}{s(s + 1 + 2j)(s + 1 - 2j)}$, $\quad H(s) = 1$

5.29 Consider the system in Fig. 5.63.

a) Using Routh's stability criterion, determine all values of K for which the system is stable.

b) Sketch the root locus. Include angles of departure and arrival, and find the values for K and s at all breakaway and break-in points and imaginary-axis crossings.

5.30 Consider the two systems shown in Fig. 5.64. Note that each system has an unstable plant and the same loop transfer function. Using root-locus methods, determine whether or not either system can be made stable for some positive value of K. Repeat your answer assuming that the transfer function in the feedback path is $s - 2 + \delta$ for Fig. 5.64(a) and $1/(s - 2 + \delta)$ for Fig. 5.64(b).

FIGURE 5.64
Feedback systems for Problem 5.30

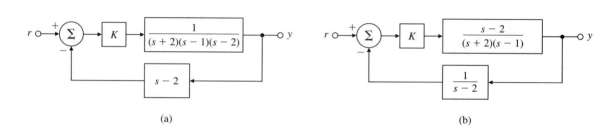

(a)   (b)

FIGURE 5.65
Feedback systems for Problem 5.31

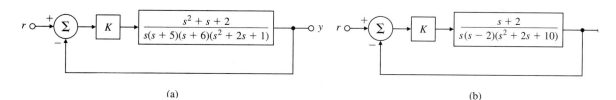

(a)   (b)

5.31 Using root-locus methods, find the range of the gain K for which the systems in Fig. 5.65 are stable.

5.32 Consider the root-locus problem where

$$G(s) = \frac{(s^2 + s + 9.25)(s + 2)}{s(s + 1)(s + 20)(s^2 + s + 16.25)}.$$

a) Sketch the portions of the real axis corresponding to the positive root locus.
b) Sketch the asymptotes for $K \to \infty$, explicitly indicating their angles and the point where they meet on the real axis.
c) Estimate the departure angle at the pole at $s = -\frac{1}{2} + 4j$.
d) Estimate the arrival angle at the zero at $s = -\frac{1}{2} + 3j$.
e) For each point on the locus that has multiple roots (that is, break-in and breakaway points), give the corresponding characteristic polynomial of the closed-loop system for that value of K.
f) Sketch the root locus, and compare it with the root locus generated by MATLAB.

5.33 Consider the system in Fig. 5.66.

a) Use Routh's criterion to determine the regions in the (K_1, K_2) plane for which the system is stable.
b) Use root-locus methods to verify your answer to part (a).

FIGURE 5.66
Feedback system for
Problem 5.33

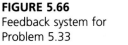 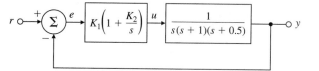

5.34 An elementary magnetic suspension scheme is depicted in Fig. 5.67. For small motions near the reference position, the voltage e on the photo detector is related to the ball displacement x (in meters) by $e = 100x$. The upward force (in newtons) on the ball caused by the current i (in amperes) may be approximated by $f = 0.5i + 20x$. The mass of the ball is 20 g, and the gravitational force is 9.8 N/kg. The power amplifier is a voltage-to-current device with an output (in amperes) of $i = u + V_0$.

a) Write the equations of motion for this setup.
b) Give the value of the bias V_0 that results in the ball being in equilibrium at $x = 0$.

FIGURE 5.67
Elementary magnetic
suspension

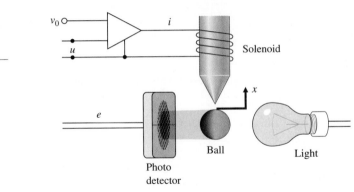

c) What is the transfer function from u to e?

d) Suppose the control input u is given by $u = -Ke$. Sketch the root locus of the closed-loop system as a function of K.

e) Suggest a different control law that yields improved performance over the one proposed in part (d).

5.35 Sketch the root locus for the system

$$G(s) = \frac{K(s + 1)}{s(s + 1)(s + 2)},$$

and determine the value of the root-locus gain for which the complex conjugate poles have a damping ratio of 0.5.

5.36 For the system in Fig. 5.68:

a) Find the locus of closed-loop roots with respect to K.

b) Is there a value of K that will cause all roots to have a damping ratio greater than 0.5?

c) Find the values of K that yield closed-loop poles with the damping ratio $\zeta = 0.707$.

FIGURE 5.68
Feedback system for
Problem 5.36

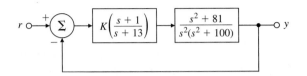

5.37 A unity (negative) feedback system has the transfer function

$$G(s) = \frac{2s - 4}{s^2 + Ks + 9},$$

where K is a positive constant. Use root-locus techniques to determine a value for K for which the closed-loop system has a damping ratio $\zeta = 0.707$.

FIGURE 5.69
Feedback system for
Problem 5.38

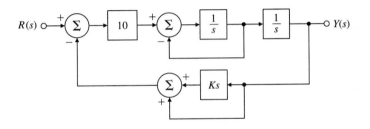

5.38 For the feedback system shown in Fig. 5.69, find the value of the gain K that results in dominant closed-loop poles with a damping ratio $\zeta = 0.5$.

5.39 Consider the rocket-positioning system shown in Fig. 5.70.

a) Show that if the sensor that measures x has a unity transfer function, the lead compensator

$$H(s) = K \frac{s+2}{s+4}$$

stabilizes the system.

b) Assume that the sensor transfer function is modeled by a single pole with a 0.1-sec time constant and unity DC gain. Using the root-locus procedure, find a value for the gain K that will provide the maximum damping ratio.

FIGURE 5.70
Block diagram for
rocket-positioning
control system

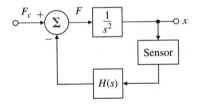

5.40 A unity feedback system has the plant transfer function

$$KG(s) = \frac{K}{s(s^2 + 6s + 12)}.$$

We wish to investigate the root-locus versus K.

a) Plot the location of poles and zeros of $G(s)$ showing segments of the root locus on the real axis.

b) What are the departure angles from the complex poles?

c) Where are the breakaway and break-in points?

d) Sketch the root locus, and compare your answer with that obtained from MATLAB.

e) What is the value of K at the point where the closed-loop complex roots have a damping ratio $\zeta = 0.5$?

FIGURE 5.71
Control system for Problem 5.41

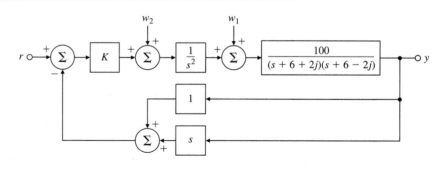

5.41 For the system in Fig. 5.71:

a) Sketch the locus of closed-loop roots with respect to K.

b) Find the maximum value of K for which the system is stable. Assume $K = 2$ for the remaining parts of this problem.

c) What is the steady-state error ($e = r - y$) for a step change in r?

d) What is the steady-state error in y for a constant disturbance w_1?

e) What is the steady-state error in y for a constant disturbance w_2?

f) If you wished to have more damping, what changes would you make to the system?

5.42 Consider the open-loop system

$$G(s) = \frac{bs + k}{s^2[mMs^2 + (M + m)bs + (M + m)k]}$$

in the unity feedback loop of Fig. 5.61. This is the transfer function relating the input force $u(t)$ and the position $y(t)$ of mass M in Problem 2.7. In this problem we will use root-locus techniques to design a controller $D(s)$ so that the closed-loop step response has a rise time of less than 0.1 sec and an overshoot of less than 10%. You may use MATLAB for any of the following questions.

a) Approximate $G(s)$ by assuming that $m \cong 0$, and let $M = 1$, $k = 1$, $b = 0.1$, and $D(s) = K$. Can K be chosen to satisfy the performance specifications? Why or why not?

b) Repeat part (a) assuming $D(s) = K(s + z)$, and show that K and z can be chosen to meet the specifications.

c) Repeat part (b) assuming

$$D(s) = \frac{Kp(s + z)}{s + p},$$

and pick p so that the values for K and z computed in part (b) remain more or less valid.

d) Now suppose that m is not negligible, and is given by $m = M/10$. Check to see if the controller you designed in part (c) still meets the given specifications. If not, adjust the controller parameters so that the specifications are met.

FIGURE 5.72
Positioning servomechanism (*Reprinted from Clark, 1962, with permission*)

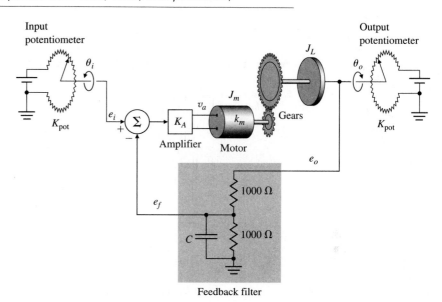

5.43 Consider the positioning servomechanism system shown Fig. 5.72, where

$$e_i = K_{pot}\theta_i, \qquad e_o = K_{pot}\theta_o, \qquad K_{pot} = 10 \text{ V/rad},$$

$$T = \text{motor torque} = k_m i_a,$$

$$k_m = \text{torque constant} = 0.1 \text{ N} \cdot \text{m/A},$$

$$R_a = \text{armature resistance} = 10\,\Omega,$$

Gear ratio = 1:1,

$$J_L + J_m = \text{total inertia} = 10^{-3} \text{ kg} \cdot \text{m}^2,$$

$$C = 200\,\mu\text{F},$$
$$v_a = K_A(e_i - e_f).$$

a) What is the range of the gain K_A for which the system is stable? Calculate the upper limit graphically using a root-locus plot.

b) Choose a gain K_A that gives roots at $\zeta = 0.7$. Where are all three closed-loop root locations for this value of K_A?

5.44 We wish to design a velocity control for a tape-drive servomechanism. The transfer function from current to velocity (in millimeters per millisecond per ampere) is

$$\frac{V(s)}{I(s)} = \frac{15(s^2 + 0.9s + 0.8)}{(s+1)(s^2 + 1.1s + 1)}.$$

We wish to design a type I feedback system so that the transient satisfies

$$t_r \leqslant 4 \text{ msec}, \qquad t_s \leqslant 15 \text{ msec}, \qquad M_p \leqslant 0.05.$$

a) Use the compensator K/s to achieve type I behavior, and sketch the root-locus with respect to K. Show on the same plot the region of acceptable pole locations corresponding to the specifications.

b) Assume a compensator of the form $K(s + \alpha)/s$, and select the best possible values of K and α you can find. Sketch the root-locus plot of your design, giving values for K and α, and the velocity constant K_v your design achieved. On your plot, indicate the closed-loop poles with $\bullet$, and include the boundary of the region of acceptable root locations.

5.45 Consider the type I system in Fig. 5.73. We would like to design the compensation $D(s)$ to meet the following requirements: (1) The steady-state value of y due to a constant unit disturbance w should be less than $\frac{4}{5}$, and (2) the damping ratio $\zeta = 0.7$. Using root-locus techniques:

a) Show that proportional control alone is not adequate.

b) Show that proportional-derivative control will work.

c) Find values of the gains K and K_d for $D(s) = K + K_d s$ that meet the design specifications.

FIGURE 5.73
Control system for
Problem 5.45

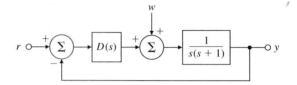

5.46 Consider the 270-ft U.S. Coast Guard cutter *Tampa* (902) shown in Fig. 5.74. Parameter identification based on sea-trials data (Trankle, 1987) was used to estimate the hydrodynamic coefficients in the equations of motion. The result is that

FIGURE 5.74
USCG cutter *Tampa*
(902)

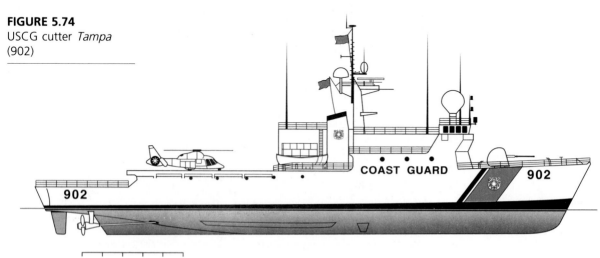

the system may be described by the second-order transfer functions

$$\frac{v(s)}{\delta(s)} = \frac{0.2196(s + 0.0019)}{(s + 0.2647)(s + 0.0063)},$$

$$\frac{r(s)}{\delta(s)} = \frac{-0.0184(s + 0.0068)}{(s + 0.2647)(s + 0.0063)},$$

$$\frac{r(s)}{w(s)} = \frac{0.0000064}{(s + 0.2647)(s + 0.0063)},$$

where

v = lateral velocity, m/sec,

r = yaw rate, rad/sec,

δ = rudder angle, rad,

w = wind speed, m/sec.

a) Determine the open-loop settling time of both v and r for a step change in δ.
b) In order to regulate the heading angle ψ ($\dot{\psi} = r$), design a compensator that uses ψ and the measurement provided by a yaw-rate gyroscope (that is, by r). The settling time is specified to be less than 50 sec, and, for a 5° change in heading, the maximum allowable rudder angle deflection is specified to be less than 10°.
c) Check the response of the closed-loop system you designed in part (b) to a wind gust disturbance of 10 m/sec (model the disturbance as a step input). If the steady-state value of the heading due to the wind gust is more than 0.5°, modify your design so that it meets this specification as well.

5.47 Golden Nugget Airlines has opened a free bar in the tail of their airplanes in an attempt to lure customers. In order to automatically adjust for the sudden weight shift due to passengers rushing to the bar when it first opens, the airline is mechanizing a pitch-attitude auto pilot. Figure 5.75 shows the block diagram of the proposed arrangement. We will model the passenger moment as a step disturbance $M_p(s) = M_0/s$, with a maximum expected value for M_0 of 0.6.

a) What value of K is required to keep the steady-state error in θ to less than 0.02 rad($\cong$ 1°)? (Assume the system is stable.)
b) Draw a root locus with respect to K.

FIGURE 5.75
Golden Nugget Airlines
autopilot

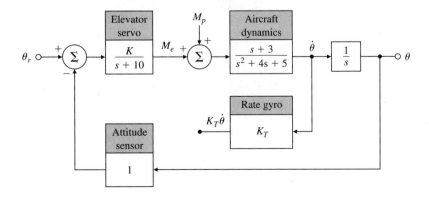

c) Based on your root locus, what is the value of K when the system becomes unstable?

d) Suppose the value of K required for acceptable steady-state behavior is 600. Show that this value yields an unstable system with roots at

$$s = -3, -14, +1 \pm 6.5j.$$

e) You are given a black box with *rate gyro* written on the side and told that when installed, it provides a perfect measure of $\dot{\theta}$, with output $K_T\dot{\theta}$. Assume $K = 600$ as in part (d) and draw a block diagram indicating how you would incorporate the rate gyro into the auto pilot. (Include transfer functions in boxes.)

f) For the rate gyro in part (e), sketch a root locus with respect to K_T.

g) What is the maximum damping factor of the complex roots obtainable with the configuration in part (e)?

h) What is the value of K_T for part (g)?

i) Suppose you are not satisfied with the steady-state errors and damping ratio of the system with a rate gyro in parts (e) through (h). Discuss the advantages and disadvantages of adding an integral term and extra lead networks in the control law. Support your comments using MATLAB or with rough root-locus sketches.

5.48 Consider the instrument servomechanism with the parameters given in Fig. 5.76. For each of the following cases, draw a root locus with respect to the parameter K, and indicate the location of the roots corresponding to your final design.

a) *Lead network:* Let

$$H(s) = 1, \qquad D(s) = K\,\frac{s+z}{s+p}, \qquad \frac{p}{z} = 6.$$

Select z and K so that the roots nearest the origin (the dominant roots) yield

$$\zeta \geqslant 0.4, \qquad -\sigma \leqslant -7, \qquad K_v \geqslant 16\tfrac{2}{3}\ \text{sec}^{-1}.$$

b) *Output-velocity (tachometer) feedback:* Let

$$H(s) = 1 + K_T s \qquad \text{and} \qquad D(s) = K.$$

Select K_T and K so that the dominant roots are in the same location as those of part (a). Compute K_v. If you can, give a physical reason explaining the reduction in K_v when output derivative feedback is used.

FIGURE 5.76
Control system for Problem 5.48

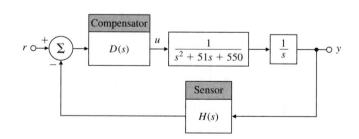

c) *Lag network:* Let

$$H(s) = 1 \quad \text{and} \quad D(s) = K\frac{s+1}{s+p}.$$

Using proportional control, it is possible to obtain a $K_v = 12$ at $\zeta = 0.4$. Select K and p so that the dominant roots correspond to the proportional-control case but with $K_v = 100$ rather than $K_v = 12$.

5.49 Consider the third-order system shown in Fig. 5.77.

a) Sketch the root locus for this system with respect to K, showing your calculations for the asymptote angles, departure angles, and so on.

b) Using graphical techniques, locate carefully the point at which the locus crosses the imaginary axis. What is the value of K at that point?

c) Assume that, due to some unknown mechanism, the amplifier output is given by the following saturation nonlinearity (instead of by a proportional gain K):

$$u = \begin{cases} e, & |e| \leqslant 1; \\ 1, & e > 1; \\ -1, & e < -1. \end{cases}$$

Qualitatively describe how you would expect the system to respond to a unit step input.

FIGURE 5.77
Control system for
Problem 5.49

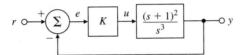

5.50 The block diagram of a positioning servomechanism is shown in Fig. 5.78.

a) Sketch the root locus with respect to K when no tachometer feedback is present ($K_T = 0$).

b) Indicate the root locations corresponding to $K = 4$ on the locus of part (a). For these locations, estimate the transient-response parameters t_r, M_p, and t_s. Compare your estimates to measurements obtained using the step command in MATLAB.

c) For $K = 4$, draw the root locus with respect to K_T.

d) For $K = 4$ and with K_T set so that $M_p = 0.05$ ($\zeta = 0.707$), estimate t_r and t_s. Compare your estimates to the actual values of t_r and t_s obtained using MATLAB.

e) For the values of K and K_T in part (d), what is the velocity constant K_v of this system?

FIGURE 5.78
Control system for
Problem 5.50

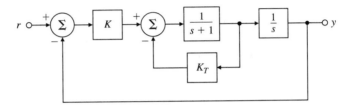

5.51 Consider the mechanical system shown in Fig. 5.79, where g and a_0 are gains. The feedback path containing gs controls the amount of rate feedback. For a fixed value of a_0, adjusting g corresponds to varying the location of a zero in the s-plane.

a) With $g = 0$ and $\tau = 1$, find a value for a_0 such that the poles are complex.

b) Fix a_0 at this value, and construct a root locus that demonstrates the effect of varying g.

FIGURE 5.79

Control system for Problem 5.51

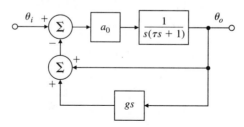

5.52 Sketch the $0°$ (negative) root locus with respect to K for the systems in Problem 5.2.

5.53 Sketch the $0°$ root locus with respect to K for systems in Problem 5.4.

5.54 State and prove an expression for the angles of the asymptotes of the negative ($0°$) root locus.

5.55 Sketch the root locus with respect to K for the system in Fig. 5.80. What is the range of values of K for which the system is unstable?

5.56 Prove that the plant $G(s) = 1/s^3$ *cannot* be made unconditionally stable if pole cancellation is forbidden.

 5.57 For the system

$$G(s) = \frac{1}{s(s + p)[(s + 1)^2 + 4]},$$

use MATLAB to examine the root locus as p varies from $s = -1$ to $s = -10$, making sure to include the point $p = 2$ that was analyzed in Example 5.7.

FIGURE 5.80

Control system for Problem 5.55

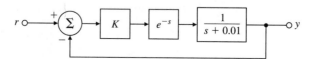

· 6 ·

The Frequency-response Design Method

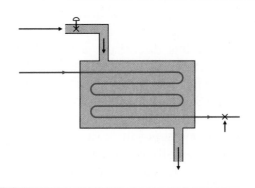

A Perspective on the Frequency-response Design Method

The design of feedback control systems in industry is probably accomplished using frequency-response methods more often than any other. Frequency-response design is popular primarily because it provides good designs in the face of uncertainty in the plant model. For example, for systems with poorly known or changing high-frequency resonances, we can temper their feedback compensation to alleviate the effects of those uncertainties. Currently, this tempering is carried out more easily using frequency-response design than any other method.

Another advantage of using frequency response is the ease with which experimental information can be used for design purposes. Raw measurements of the output amplitude and phase of a plant undergoing a sinusoidal input excitation are sufficient to design a suitable feedback control. No intermediate processing of the data (such as finding poles and zeros or determining system matrices) is required to arrive at the system model. The wide availability of computers has rendered this advantage less important now than it was years ago; however, for relatively simple systems, frequency response is often still the most cost-effective design method. The method is most effective for systems that are stable in open-loop.

Although the underlying theory is somewhat challenging and requires a rather broad knowledge of complex variables, the methodology of frequency-response design is easy, and the insights gained by learning the theory are well worth the struggle.

Chapter Overview

The chapter opens with a discussion of how to obtain the frequency response of a system by analyzing its poles and zeros. An important extension of this discussion is how to use Bode plots to graphically display the frequency response. In Sections 6.2 and 6.3 we discuss stability briefly, and then in more depth the use of the Nyquist stability criterion. In Sections 6.4 through 6.6 we introduce the notion of stability margins, discuss Bode's gain-phase relationship, and study the closed-loop frequency response of dynamic systems. As with our treatment of the root-locus method, we describe how adding dynamic compensation can adjust the frequency response (Section 6.7) and improve system stability and/or error characteristics.

Several alternate methods of displaying frequency-response data have been developed over the years; we present two of them—the Nichols chart and the inverse Nyquist plot—in Section 6.8. In Section 6.9 we discuss issues of sensitivity that relate to the frequency response, including optional material on sensitivity functions and stability robustness. The final two sections—on analyzing time delays in the system and obtaining a pole-zero model from frequency response data—represent additional, somewhat advanced material that may also be considered optional.

6.1   Frequency Response

Frequency response

In this section we will describe how a linear system responds to sinusoidal inputs—called the system's **frequency response**—and how this response can be obtained from knowing its pole and zero locations.

To introduce the ideas, we consider a system described by

$$\frac{Y(s)}{U(s)} = G(s),$$

where the input $u(t)$ is a sine wave with an amplitude U_o:

$$u(t) = U_o \sin \omega t,$$

which has a Laplace transform

$$U(s) = \frac{U_o \omega}{s^2 + \omega^2}.$$

With zero initial conditions, the Laplace transform of the output is

Sinusoidal input causes sinusoidal output at same frequency.

$$Y(s) = G(s) \frac{U_o \omega}{s^2 + \omega^2}. \qquad (6.1)$$

A partial-fraction expansion of Eq. (6.1) [assuming the poles of $G(s)$ are distinct] will result in an equation of the form

$$Y(s) = \frac{\alpha_1}{s - a_1} + \frac{\alpha_2}{s - a_2} + \cdots + \frac{\alpha_n}{s - a_n} + \frac{\alpha_o}{s + j\omega} + \frac{\alpha_o^*}{s - j\omega} \qquad (6.2)$$

where $a_1, a_2, \ldots, a_n$ are the poles of $G(s)$, α_o would be found by performing the partial-fraction expansion as outlined in Section 3.1.5, and α_o^* is the complex conjugate of α_o. The time response that corresponds to $Y(s)$ is

$$y(t) = \alpha_1 e^{a_1 t} + \alpha_2 e^{a_2 t} + \cdots + \alpha_n e^{a_n t} + 2|\alpha_o| \sin(\omega t + \phi) \qquad (6.3)$$

where

$$\phi = \tan^{-1} \frac{\mathrm{Im}(\alpha_o)}{\mathrm{Re}(\alpha_o)}.$$

If all the poles of the system represent stable behavior (the real parts of a_1, $a_2, \ldots, a_n < 0$), the natural unforced response will die out eventually and therefore the steady-state response of the system will be due solely to the sinusoidal term in Eq. (6.3) which is caused by the sinusoidal excitation. For example, Fig. 6.1 shows the response of the system $G(s) = 1/(s + 1)$ to the input

FIGURE 6.1
Response of $G(s) = 1/(s+1)$ to $\sin 10t$

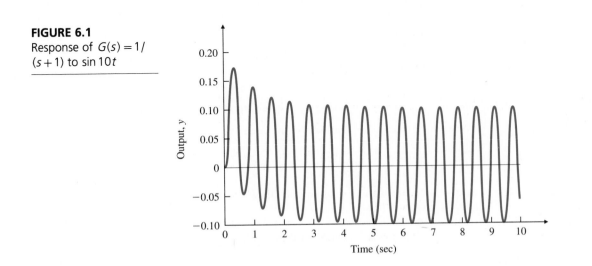

$u = \sin 10t$. It shows that e^{-t}, the natural part of the response associated with $G(s)$, disappears after several time constants and the pure sinusoidal response is essentially all that remains. Example 3.3 showed that the remaining sinusoidal term in Eq. (6.3) can be expressed as

$$y(t) = U_o A \sin(\omega t + \phi), \tag{6.4}$$

where

$$A = |G(j\omega)| = |G(s)|\big|_{s=j\omega} = \sqrt{\{\mathrm{Re}[G(j\omega)]\}^2 + \{\mathrm{Im}[G(j\omega)]\}^2}, \tag{6.5a}$$

$$\phi = \tan^{-1}\frac{\mathrm{Im}[G(j\omega)]}{\mathrm{Re}[G(j\omega)]} = \angle\, G(j\omega). \tag{6.5b}$$

Equation (6.4) shows that a stable system with transfer function $G(s)$ excited by a sinusoid with unit amplitude and frequency ω will, after the response has reached steady state, exhibit a sinusoidal output with a magnitude $A(\omega)$ and a phase $\phi(\omega)$ at the frequency ω. The facts that the output y is a sinusoid with the same frequency as the input u and that the ratio magnitude A and phase ϕ of the output are independent of the amplitude U_o of the input are a consequence of $G(s)$ being a linear constant system. If the system being excited were a nonlinear or time-varying system, the output might contain frequencies other than the input frequency and the output-input ratio might be dependent on the input magnitude.

The **magnitude** A is given by $|G(j\omega)|$, and the **phase** ϕ is given by $\angle[G(j\omega)]$; that is, the magnitude and angle of the complex quantity $G(s)$ are evaluated with s taking on values along the imaginary axis. We are interested in analyzing the frequency response of a system not only because it will help us understand how a system responds to a sinusoidal input, but also because evaluating $G(s)$ with s taking on values along the $j\omega$ axis will prove to be very useful in determining the stability of a closed-loop system. The $j\omega$ axis is the boundary between stability and instability; we will see in Section 6.4 that evaluating $G(j\omega)$ provides information that allows us to determine closed-loop stability from the open-loop $G(s)$.

Frequency response plot

Magnitude and phase

◆ **EXAMPLE 6.1**   *Frequency-response Characteristics of a Capacitor*

Consider the capacitor described by the equation

$$i = C\frac{dv}{dt}$$

where v is the input and i is the output. Determine the sinusoidal steady-state response of the capacitor.

Solution.   The transfer function of this circuit is

$$G(s) = Cs,$$

so

$$G(j\omega) = Cj\omega.$$

Computing the magnitude and phase, we find that

$$A = |Cj\omega| = C\omega \qquad \text{and} \qquad \phi = \angle(Cj\omega) = 90°.$$

For a unit-amplitude sinusoidal input v, the output i will be a sinusoid with magnitude $C\omega$ and in phase will lead the input by 90°. Note for this example that the magnitude is proportional to the input frequency while the phase is independent of frequency.

◆

◆ **EXAMPLE 6.2**   *Frequency-response Characteristics of a Lead Compensator*

Recall from Chapter 5 the transfer function of the lead compensation, Eq. (5.47),

$$D(s) = K\,\frac{Ts + 1}{\alpha Ts + 1}, \qquad \alpha < 1. \tag{6.6}$$

Determine its frequency-response characteristics.

Solution.   Substituting $s = j\omega$ into Eq. (6.6), we get

$$D(j\omega) = K\,\frac{Tj\omega + 1}{\alpha Tj\omega + 1}.$$

From Eqs. (6.5) the amplitude is

$$A = |D| = |K|\,\frac{\sqrt{1 + (\omega T)^2}}{\sqrt{1 + (\alpha\omega T)^2}}$$

and the phase is given by

$$\phi = \angle(1 + j\omega T) - \angle(1 + j\alpha\omega T)$$
$$= \tan^{-1}(\omega T) - \tan^{-1}(\alpha\omega T).$$

Characteristics of a high-pass filter

At very low frequencies, the amplitude is just $|K|$ and at very high frequencies it is $|K/\alpha|$. Therefore, the amplitude increases as a function of frequency. The phase is zero at very low frequencies and goes back to zero at very high frequencies. At intermediate frequencies, evaluation of the $\tan^{-1}$ functions would reveal that ϕ becomes positive. These are the characteristics of a high-pass filter.

◆

In the cases where we do not have a good model of the system and wish to determine the frequency-response magnitude and phase experimentally, we can excite the system with a sinusoid varying in frequency. The magnitude $A(\omega)$ is obtained by measuring the ratio of the output sinusoid to input sinusoid in the

FIGURE 6.2
(a) Magnitude and (b)
phase of Eq. (6.7)

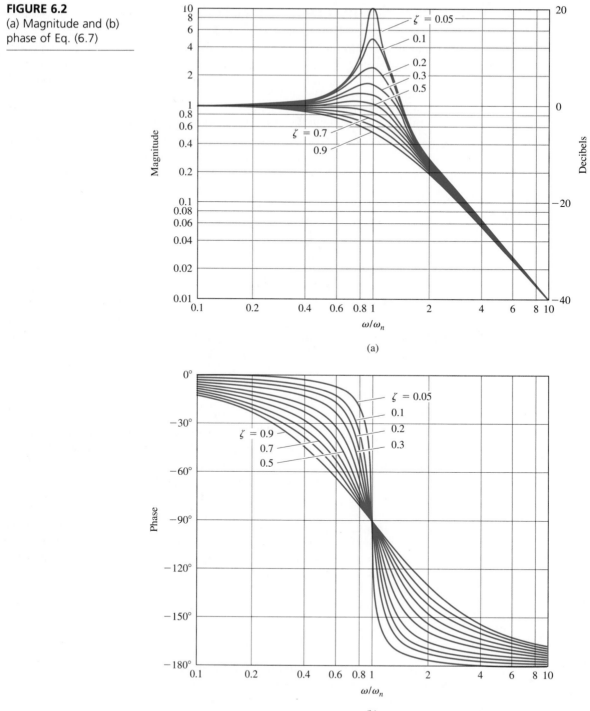

(a)

(b)

steady state at each frequency. The phase $\phi(\omega)$ is the measured difference in phase between input and output signals.[1]

A great deal can be learned about the dynamic response of a system from knowledge of the magnitude $A(\omega)$ and the phase $\phi(\omega)$ of its transfer function. In the obvious case, if the signal is a sinusoid, then A and ϕ completely describe the response. Furthermore, if the input is periodic, then a Fourier series can be constructed to decompose the input into a sum of sinusoids, and again $A(\omega)$ and $\phi(\omega)$ can be used with each component to construct the total response. For transient inputs, our best path to understanding the meaning of A and ϕ is to relate the frequency response $G(j\omega)$ to the transient responses calculated by the Laplace transform. For example, in Fig. 3.12(b) we plotted the step response of a system having the transfer function

$$G(s) = \frac{1}{(s/\omega_n)^2 + 2\zeta(s/\omega_n) + 1} \qquad (6.7)$$

for various values of ζ. These transient curves were normalized with respect to time as $\omega_n t$. In Fig. 6.2 we plot $A(\omega)$ and $\phi(\omega)$ for these same values of ζ to help us see what features of the frequency response correspond to the transient-response characteristics. Figures 3.12(b) and 6.2 indicate the type of system for which we could expect a good closed-loop response. The damping of the system can be determined from the transient response (overshoot) or from the peak in the magnitude of the frequency response (Fig. 6.2a). From the frequency response we see that ω_n is approximately equal to the bandwidth—the frequency where the magnitude starts to fall off from its low-frequency value. (We will define bandwidth more formally in the next paragraph.) Therefore, the rise time can be estimated from the bandwidth. We also see that the peak overshoot in frequency is approximately $1/2\zeta$ for $\zeta < 0.5$, so the peak overshoot in the step response can be estimated from the peak overshoot in the frequency response. Thus we see that essentially the same information is contained in the frequency-response curve as is found in the transient-response curve.

A natural specification for system performance in terms of frequency response is the **bandwidth**, defined to be the maximum frequency at which the output of a system will track an input sinusoid in a satisfactory manner. By convention, for the system shown in Fig. 6.3 with a sinusoidal input r, the bandwidth is the frequency of r at which the output y is attenuated to a factor

Bandwidth

---

1. Hewlett-Packard produces an instrument (model no. 3563A) that automates this experimental procedure and greatly speeds up the process.

FIGURE 6.3
Simplified system definition

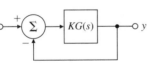

FIGURE 6.4

Definitions of bandwidth and resonant peak

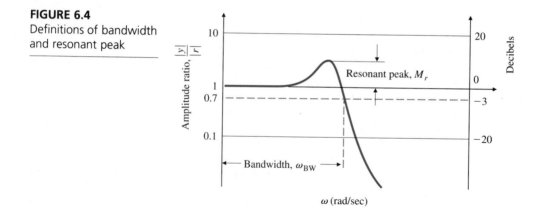

ω (rad/sec)

0.707 times the input. Figure 6.4 depicts the idea graphically for the frequency response of the *closed-loop* transfer function

$$\frac{Y(s)}{R(s)} \triangleq T(s) = \frac{KG(s)}{1 + KG(s)}.$$

The plot is typical of most closed-loop systems in that (1) the output follows the input $[|T| \cong 1]$ at the lower excitation frequencies, and (2) the output ceases to follow the input $[|T| < 1]$ at the higher excitation frequencies. The maximum value of the frequency-response magnitude is referred to as the **resonant peak M_r**.

Bandwidth is a measure of speed of response and is therefore similar to time-domain measures like rise time and peak time or the s-plane measure of dominant-root(s) natural frequency. In fact, if the $KG(s)$ in Fig. 6.3 is such that the closed-loop response is given by Fig. 6.2, we can see that the bandwidth will equal the natural frequency of the closed loop root (that is, $\omega_{Bw} = \omega_n$ or a closed-loop damping ratio of $\zeta = 0.7$). For other damping ratios the bandwidth is approximately equal to the natural frequency of the closed-loop roots, with an error typically less than a factor of 2.

The definition of the bandwidth stated here is meaningful for systems that have a low-pass filter behavior, as is the case for any physical control system. In other applications the bandwidth may be defined differently. Also, if the ideal model of the system does not have a high-frequency roll-off (equal number of poles and zeros), the bandwidth is infinite (for example, in the lead compensation in Example 6.2).

In order to use the frequency response for control systems design, we need to consider an efficient and meaningful form in which to make frequency-response plots as well as find methods to relate the open-loop characteristics of KG to the closed-loop characteristics of T. These are the concerns of Section 6.1.1.

6.1.1   Bode Plot Techniques

Display of frequency response is a problem that has been studied for a long time. The most useful technique for our purposes was developed by H. W. Bode at Bell Laboratories between 1932 and 1942. This technique allows plotting by hand that is quick and yet sufficiently accurate for control systems design. Most control systems designers now have access to computer programs that diminish the need for hand plotting; however, it is still important to develop good intuition so that you can quickly identify erroneous computer results, and for this you need the ability to check results by hand.

The idea in Bode's method is to plot magnitude curves using a logarithmic scale and phase curves using a linear scale. This strategy allows us to plot a high-order $G(j\omega)$ by simply adding the separate terms graphically, as discussed in Appendix B. This addition is possible because a complex expression with zero and pole factors can be written in polar (or phasor) form as

$$G(j\omega) = \frac{\vec{s}_1 \vec{s}_2}{\vec{s}_3 \vec{s}_4 \vec{s}_5} = \frac{r_1 e^{j\theta_1} r_2 e^{j\theta_2}}{r_3 e^{j\theta_3} r_4 e^{j\theta_4} r_5 e^{j\theta_5}} = \left(\frac{r_1 r_2}{r_3 r_4 r_5}\right) e^{j(\theta_1 + \theta_2 - \theta_3 - \theta_4 - \theta_5)}. \quad (6.8)$$

Composite plot from individual terms

(The overhead arrow indicates a phasor.) Note from Eq. (6.8) that the phases of the individual terms are added directly to obtain the phase of the **composite** expression, $G(j\omega)$. Furthermore, since

$$|G(j\omega)| = \frac{r_1 r_2}{r_3 r_4 r_5},$$

then

$$\log_{10}|G(j\omega)| = \log_{10} r_1 + \log_{10} r_2 - \log_{10} r_3 - \log_{10} r_4 - \log_{10} r_5. \quad (6.9)$$

We see that addition of the logarithms of the individual terms provides the logarithm of the magnitude of the composite expression. The frequency response is typically presented as two curves; the logarithm of magnitude versus log ω, and the phase versus log ω. Together these two curves comprise a **Bode plot** of the system. Since

Bode plot

$$\log_{10} A e^{j\phi} = \log_{10} A + j\phi \log_{10} e, \quad (6.10)$$

we see that the Bode plot shows the real and imaginary parts of the logarithm of $G(j\omega)$. In communications it is standard to measure the power gain in decibels (dB):[2]

Decibel

$$|G|_{dB} = 10 \log_{10} \frac{P_2}{P_1}, \quad (6.11)$$

---

2. Researchers at Bell Laboratories first defined the unit of power gain as a **bel** (named for Alexander Graham Bell, the founder of the company). However, this unit proved to be too large, and hence a **decibel** (1/10 of a bel) was selected as a more useful unit.

where P_1 and P_2 are the input and output powers. Since power is proportional to the square of the voltage, the gain is also given by

$$|G|_{dB} = 20 \log_{10} \frac{V_2}{V_1}. \tag{6.12}$$

Hence we can present a Bode plot as the magnitude in decibels versus $\log \omega$ and the phase in degrees versus $\log \omega$.[3] In this book we give Bode plots in the form $\log|G|$ versus $\log \omega$; also we mark an axis in decibels on the right-hand side of the magnitude plot to give you the choice of working with the representation you prefer. If the magnitude data are derived in terms of $\log|G|$, it is conventional to plot them on a log scale but identify the scale in terms of $|G|$ only (without "log"). If the magnitude data are given in decibels, the vertical scale is linear such that each decade of $|G|$ represents 20 dB.

Advantages of Working with Frequency Response in terms of Bode Plots

Advantages of Bode plot representation of frequency response

1. Bode plots of systems in series (or tandem) simply add, which is quite convenient.
2. Bode's important phase-gain relationship is given in terms of logarithms of phase and gain.
3. A much wider range of system behavior—from low- to high-frequency behavior—can be displayed on a single plot.
4. Bode plots can be determined experimentally.
5. Dynamic compensator design can be based entirely on Bode plots.

It is important for the control systems engineer to be able to hand-plot frequency responses for several reasons: This skill not only allows the engineer to deal with simple problems but also to check computer results for more complicated cases. Often approximations can be used to quickly sketch the frequency response and deduce stability as well as determine the form of the needed dynamic compensations. Finally, hand plotting is useful in interpreting frequency-response data that have been generated experimentally.

In Chapter 5 we wrote the open-loop transfer function in the form

$$KG(s) = K \frac{(s - z_1)(s - z_2) \cdots}{(s - p_1)(s - p_2) \cdots} \tag{6.13}$$

---

3. Henceforth we will drop the base of the logarithm; it is understood to be 10.

because it was the most convenient form for determining the degree of stability from the root locus with respect to the gain K. In working with frequency response, it is more convenient to replace s with $j\omega$ and to write the transfer functions in the **Bode form**

Bode form of the transfer function

$$KG(j\omega) = K_o \frac{(j\omega\tau_1 + 1)(j\omega\tau_2 + 1)\cdots}{(j\omega\tau_a + 1)(j\omega\tau_b + 1)\cdots} \qquad (6.14)$$

because the gain K_o in this form is directly related to the transfer-function magnitude at very low frequencies. In fact, for type 0 systems, K_o is the gain at $\omega = 0$ in Eq. (6.14) and is also equal to the DC gain of the system. Although a straightforward calculation will convert a transfer function in the form of Eq. (6.13) to an equivalent transfer function in the form of Eq. (6.14), note that K and K_o will not necessarily have the same value in the two expressions.

Transfer functions can also be rewritten according to Eqs. (6.8) and (6.9). As an example, suppose that

$$KG(j\omega) = K_o \frac{j\omega\tau_1 + 1}{(j\omega)^2(j\omega\tau_a + 1)}. \qquad (6.15)$$

Then

$$\angle KG(j\omega) = \angle K_o + \angle(j\omega\tau_1 + 1) - \angle(j\omega)^2 - \angle(j\omega\tau_a + 1) \qquad (6.16)$$

and

$$\log|KG(j\omega)| = \log|K_o| + \log|j\omega\tau_1 + 1| - \log|(j\omega)^2| - \log|j\omega\tau_a + 1|. \qquad (6.17)$$

In decibels, Eq. (6.17) becomes

$$|KG(j\omega)|_{dB} = 20 \log |K_o| + 20 \log|j\omega\tau_1 + 1| - 20 \log|(j\omega)^2|$$
$$- 20 \log|j\omega\tau_a + 1|. \qquad (6.18)$$

All transfer functions for the kinds of systems we have talked about so far are composed of three classes of terms:

Classes of terms of transfer functions

1. $K_o(j\omega)^n$,
2. $(j\omega\tau + 1)^{\pm 1}$,
3. $\left[\left(\dfrac{j\omega}{\omega_n}\right)^2 + 2\zeta\dfrac{j\omega}{\omega_n} + 1\right]^{\pm 1}$.

First we will discuss the plotting of each individual term and how the terms affect the composite plot including all the terms; then we will discuss how to draw the composite curve

Class 1: singularities at the origin

1. $K_o(j\omega)^n$ Since

$$\log K_o|(j\omega)^n| = \log K_o + n \log|j\omega|,$$

FIGURE 6.5

Magnitude of $(j\omega)^n$

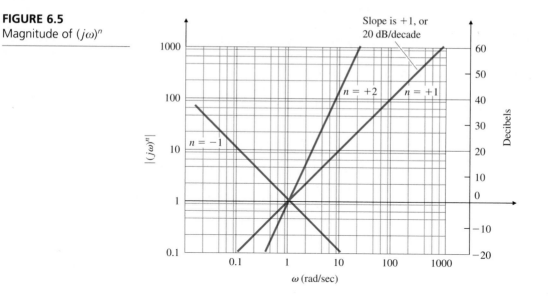

the magnitude plot of this term is a straight line with a slope n (20 dB/decade). Examples for different values of n are shown in Fig. 6.5. $K_o(j\omega)^n$ is the only term that affects the slope at the lowest frequencies because all other terms are constant in that region. The easiest way to draw the curve is to locate $\omega = 1$ and plot log K_o at that frequency. Then draw the line with slope n through that point.[4] The phase of $(j\omega)^n$ is $\phi = n \times 90°$; it is independent of frequency and is thus a horizontal line: $-90°$ for $n = -1$, $-180°$ for $n = -2$, $+90°$ for $n = +1$, and so forth.

Class 2: first-order term

2. **$j\omega\tau + 1$**   The magnitude of this term approaches one asymptote at very low frequencies and another asymptote at very high frequencies:

a) For $\omega\tau \ll 1$, $j\omega\tau + 1 \cong 1$.
b) For $\omega\tau \gg 1$, $j\omega\tau + 1 \cong j\omega\tau$.

Break point

If we call $\omega = 1/\tau$ the **break point**, then we see that below the break point the magnitude curve is approximately constant ($= 1$), while above the break point the magnitude curve behaves approximately like the class 1 term $K_o(j\omega)^n$. The example plotted in Fig. 6.6, $G(s) = 10s + 1$, shows how the two asymptotes cross at the break point and how the actual magnitude curve lies above that point by a factor of 1.4 (or 3 dB). (If the term were in the denominator, it would be below the break point by a factor of 0.707 or -3 dB.) Note that this term will have only a small effect on the composite

---

4. In decibels the slopes are $n \times 20$ dB per decade or $n \times 6$ dB per octave (an octave is a change in frequency by a factor of 2).

FIGURE 6.6
Magnitude plot for
$j\omega\tau + 1j$; $\tau = 0.1$

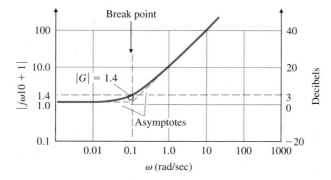

magnitude curve below the break point because its value is equal to 1 ($=0\,\text{dB}$) in this region. The slope at high frequencies is $20\,\text{dB/decade}$.

The phase curve can also be easily drawn by using the following low- and high-frequency asymptotes:

a)   For $\omega\tau \ll 1$, $\angle 1 = 0°$.
b)   For $\omega\tau \gg 1$, $\angle j\omega\tau = 90°$.
c)   For $\omega\tau \cong 1$, $\angle(j\omega\tau + 1) \cong 45°$.

For $\omega\tau \cong 1$, the $\angle(j\omega + 1)$ curve is tangent to an asymptote going from $0°$ at $\omega\tau = 0.2$ to $90°$ at $\omega\tau = 5$ as shown in Fig. 6.7. The figure also illustrates the three asymptotes (dashed lines) used for the phase plot and how the actual curve deviates from the asymptotes by $11°$ at their intersections. Both the composite phase and magnitude curves are unaffected by this class of term at frequencies below the break point by more than a factor of 10 because the term's magnitude is $0\,\text{dB}$ and its phase is $0°$

FIGURE 6.7
Phase plot for
$j\omega\tau + 1j$; $\tau = 0.1$

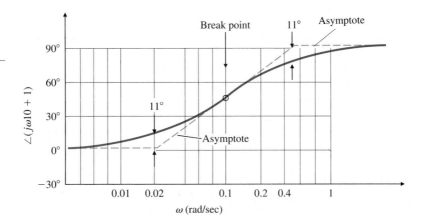

3. $[(j\omega/\omega_n)^2 + 2\zeta(j\omega/\omega_n) + 1]^{\pm 1}$ This term behaves in a manner similar to the class 2 term, with differences in detail: The break point is now $\omega = \omega_n$. The magnitude changes slope by a factor of $+2$ (or $+40\,\mathrm{dB}$ per decade) at the break point, (and -2, or $-40\,\mathrm{dB}$ per decade, when the term is in the denominator). The phase changes by $\pm 180°$, and the transition through the break point region varies with the damping ratio ζ. Figure 6.2 shows the magnitude and phase for several different damping ratios when the term is in the denominator. Note that the magnitude asymptote for frequencies above the break point has a slope of -2, or $-40\,\mathrm{dB}$ per decade, and that the transition through the break-point region has a large dependence on the damping ratio. A rough sketch of this transition can be made by noting that

$$|G(j\omega)| = \frac{1}{2\zeta} \quad \text{at} \quad \omega = \omega_n \tag{6.19}$$

for this class of second-order term in the denominator. If the term was in the numerator, the magnitude would be the reciprocal of the curve plotted in Fig. 6.2(a).

No such handy rule as Eq. (6.19) exists for sketching in the transition for the phase curve; therefore, we would have to resort to Fig. 6.2(b) for an accurate plot of the phase. However, a very rough idea of the transition can be gained by noting that it is a step function for $\zeta = 0$, while it obeys the rule for two first-order (class 2) terms when $\zeta = 1$ with simultaneous break-point frequencies. All intermediate values of ζ fall between these two extremes.

When the system has several poles and several zeros, plotting the frequency response requires that the components be combined into a composite curve. To plot the composite magnitude curve, it is useful to note that the slope of the asymptotes is equal to the sum of the slopes of the individual curves. Therefore, the composite asymptote curve has integer slope changes at each break-point frequency: $+1$ for a first-order term in the numerator, -1 for a first-order term in the denominator, and ± 2 for second-order terms. Furthermore, the lowest-frequency portion of the asymptote has a slope determined by the value of n in the $(j\omega)^n$ term and is located by plotting the point $K_o\omega^n$ at $\omega = 1$. Therefore, the complete procedure consists of plotting the lowest-frequency portion of the asymptote, then sequentially changing the asymptote's slope at each break point in order of ascending frequency, and finally drawing the actual curve by using the transition rules discussed earlier for classes 2 and 3.

The composite phase curve is the sum of the individual curves. Adding of the individual phase curves graphically is made possible by locating the curves so that the composite phase approaches the individual curve as closely as possible. A quick but crude sketch of the composite phase can be found by starting the phase curve below the lowest break point and setting it equal to $n \times 90°$. The phase is then stepped at each break point in order of ascending frequency. The amount of the phase step is $\pm 90°$ for a first-order term and $\pm 180°$ for a second-order term. Break points in the numerator indicate a positive step in phase, while break points in the denominator indicate a negative phase step.[5]

Summary of Bode Plot Rules

1. Manipulate the transfer function into the form given by Eq. (6.14).

2. Determine the value of n for the $K_o(j\omega)^n$ term (class 1). Plot the low-frequency magnitude asymptote through the point K_o at $\omega = 1$ with a slope of n (or $n \times 20$ dB per decade).

3. Complete the composite magnitude asymptotes: Extend the low-frequency asymptote until the first frequency break point. Then step the slope by ± 1 or ± 2, depending on whether the break point is from a first- or second-order term in the numerator or denominator. Continue through all break points in ascending order.

4. Sketch in the approximate magnitude curve: Increase the asymptote value by a factor of 1.4 ($+3$ dB) at first-order numerator break points, and decrease it by a factor of 0.707 (-3 dB) at first-order denominator break points. At second-order break points, sketch in the resonant peak (or valley) according to Fig. 6.2(a) using the relation $|G(j\omega)| = 1/2\zeta$ at the break point.

5. Plot the low-frequency asymptote of the phase curve, $\phi = n \times 90°$.

6. As a guide, sketch in the approximate phase curve by changing the phase by $\pm 90°$ or $\pm 180°$ at each break point in ascending order. For first-order terms in the numerator, the change of phase is $+90°$; for those in the denominator the change is $-90°$. For second-order terms, the change is $\pm 180°$.

7. Locate the asymptotes for each individual phase curve so that their phase change corresponds to the steps in the phase toward or away from the approximate curve indicated by Step 6. Sketch in each individual phase curve as indicated by Fig. 6.7 or Fig. 6.2(b).

8. Graphically add each phase curve. Use grids if an accuracy of about $\pm 5°$ is desired. If less accuracy is acceptable, the composite curve can be done by eye. Keep in mind that the curve will start at the lowest-frequency asymptote and end on the highest-frequency asymptote and will approach the intermediate asymptotes to an extent that is determined by how close the break points are to each other.

---

5. This approximate method was pointed out to us by our Parisian colleagues.

◆ **EXAMPLE 6.3**   *Bode Plot for Real Poles and Zeros*

Plot the Bode magnitude and phase for the system with the transfer function

$$KG(s) = \frac{2000(s + 0.5)}{[s(s + 10)(s + 50)]}.$$

Solution:

STEP 1.   We convert the function to the form of Eq. (6.14):

$$KG(j\omega) = \frac{2[(j\omega/0.5) + 1]}{j\omega\{[(j\omega/10) + 1][(j\omega/50) + 1]\}}.$$

STEP 2.   We note that the term in $j\omega$ is first-order and in the denominator, so $n = -1$. Therefore, the low-frequency asymptote is defined by the first term:

$$KG(j\omega) = \frac{2}{j\omega}.$$

This asymptote is valid for $\omega < 0.1$ because the lowest break point is at $\omega = 0.5$. The magnitude plot of this term has the slope of -1, or $-20\,\mathrm{dB}$ per decade. We locate the magnitude by passing through the value 2 at $\omega = 1$ even though the composite curve will not go through this point because of the break point at $\omega = 0.5$. This is shown in Fig. 6.8(a).

STEP 3.   We obtain the remainder of the asymptotes, also shown in Fig. 6.8(a): First we draw a line with 0 slope that intersects the original -1 slope at $\omega = 0.5$. Then we draw a -1 slope line that intersects the previous one at $\omega = 10$. Finally, we draw a -2 slope line that intersects the previous -1 slope at $\omega = 50$.

STEP 4.   We sketch in the actual curve so that it is approximately tangent to the asymptotes when far away from the break points, a factor of 1.4 ($+3\,\mathrm{dB}$) above the asymptote at the $\omega = 0.5$ break point, and a factor of 0.7 ($-3\,\mathrm{dB}$) below the asymptote at the $\omega = 10$ and $\omega = 50$ break points.

STEP 5.   Since the phase of $2/j\omega$ is $-90°$, the phase curve in Fig. 6.8(b) starts at $-90°$ at the lowest frequencies.

STEP 6.   The result is shown in Fig. 6.8(c).

STEP 7.   The individual phase curves, shown dashed in Fig. 6.8(b), have the correct phase change for each term and are aligned vertically so that their phase change corresponds to the steps in the phase from the approximate curve in Fig. 6.8(c). Note that the composite curve approaches each individual term.

STEP 8.   The graphical addition of each dashed curve results in the solid composite curve in Fig. 6.8(b). As can be seen from the figure, the vertical placement of each individual phase curve makes the required graphical addition particularly easy because the composite curve approaches each individual phase curve in turn.

◆

FIGURE 6.8
Composite plots:
(a) magnitude;
(b) phase;
(c) approximate phase

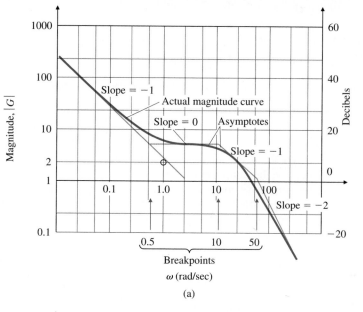

Magnitude, $|G|$

1000

100

Slope = −1

Actual magnitude curve

Slope = 0   Asymptotes

Slope = −1

10

2

⊕

Slope = −2

1

0.1

0.1   1.0   10   100

0.1

0.5   10   50

Breakpoints

ω (rad/sec)

60

40

Decibels

20

0

−20

(a)

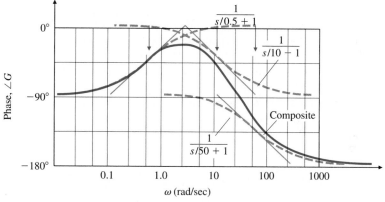

Phase, $\angle G$

$\dfrac{1}{s/0.5 + 1}$

0°

$\dfrac{1}{s/10 - 1}$

−90°

$\dfrac{1}{s/50 + 1}$

Composite

−180°

0.1   1.0   10   100   1000

ω (rad/sec)

(b)

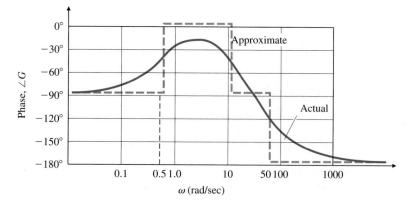

Phase, $\angle G$

0°

Approximate

−30°

−60°

−90°

Actual

−120°

−150°

−180°

0.1   0.5 1.0   10   50 100   1000

ω (rad/sec)

(c)

353

◆ **EXAMPLE 6.4**   *Bode Plot with Complex Poles*

As a second example, draw the frequency response for the system

$$KG(s) = \frac{10}{s(s^2 + 0.4s + 4)}. \tag{6.20}$$

Solution.   A system like this is more difficult to plot than the one in the previous example because the transition between asymptotes is dependent on the damping ratio; however, the same basic ideas illustrated in Example 6.3 apply.

   This system contains a second-order term in the denominator. Proceeding through the steps, we convert Eq. (6.20) to the form of Eq. (6.14):

$$KG(s) = \frac{10}{4} \frac{1}{s(s^2/4 + 0.2s/2 + 1)}.$$

Starting with the low-frequency asymptote, we have $n = -1$ and $|G(j\omega)| \cong 2.5/\omega$. The magnitude plot of this term has a slope of -1 ($-20\,dB$ per decade) and passes through the value of 2.5 at $\omega = 1$ as shown in Fig. 6.9(a). At the break-point frequency of the poles, $\omega = 2$, the slope shifts to -3 ($-60\,dB$ per decade). At the pole break point the magnitude ratio above the asymptote is $1/2\zeta = 1/0.2 = 5$. The phase curve for this case starts at $\phi = -90°$ corresponding to the $1/s$ term, falls to $\phi = -180°$ at $\omega = 2$ due to the pole as shown in Fig. 6.9(b), and then approaches $\phi = -270°$ for higher frequencies. Because the damping is small, the stepwise approximation is a very good one. The true composite phase curve is shown in Fig. 6.9(b).

FIGURE 6.9
Bode plot for a transfer function with complex poles: (a) magnitude; (b) phase

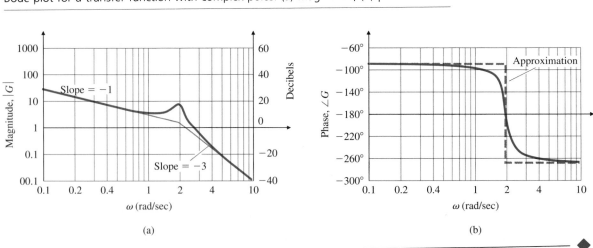

(a)   (b)

◆ **EXAMPLE 6.5**   *Bode Plot for Complex Poles and Zeros*

As a third example, draw the Bode plots for a system with second-order terms. The transfer function represents a mechanical system with two equal masses coupled with a lightly damped spring. The applied force and position measurement are collocated on the same mass. For the transfer function the time scale has been chosen so that the resonant frequency of the complex zeros is equal to 1. The transfer function is

$$KG(s) = \frac{0.01(s^2 + 0.01s + 1)}{s^2[(s^2/4) + 0.02(s/2) + 1]}.$$

Solution.   Proceeding through the steps, we start with the low-frequency asymptote, $0.01/\omega^2$. It has a slope of -2 (-40 dB per decade) and passes through magnitude $= 0.01$ at $\omega = 1$ as shown in Fig. 6.10(a). At the break-point frequency of the zero, $\omega = 1$, the slope shifts to zero until the break point of the pole, which is located at $\omega = 2$, when the slope returns to a slope of -2. To interpolate the true curve, we plot the point at the zero break point, $\omega = 1$, with a magnitude ratio below the asymptote of $2\zeta = 0.01$. At the pole break point the magnitude ratio above the asymptote is $1/2\zeta = 1/0.02 = 50$. The magnitude curve is a "doublet" of a negative pulse followed by a positive pulse. Figure 6.10(b) shows that the phase curve for this system starts at $-180°$ (corresponding to the $1/s^2$ term), jumps $180°$ to $\phi = 0$ at $\omega = 1$ due to the zeros, and then falls $180°$ back to $\phi = -180°$ at $\omega = 2$ due to the pole. With such small damping ratios the stepwise approximation is quite good. (We haven't drawn this on Fig. 6.2(b) because it would not be easily distinguishable from the true phase curve.) Thus the true composite phase curve is a nearly square pulse between $\omega = 1$ and $\omega = 2$.

FIGURE 6.10
Bode plot for a transfer function with complex poles and zeros: (a) magnitude; (b) phase

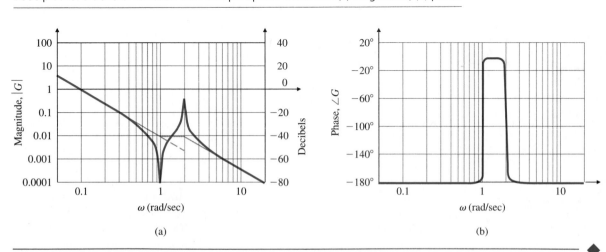

(a)

(b)

In actual designs, most Bode plots are made with the aid of a computer. However, acquiring the ability to quickly sketch Bode plots by hand is a useful skill because it gives the designer insight into how changes in the compensation parameters will affect the frequency response. This allows the designer to iterate to the best designs more quickly.

◆ **EXAMPLE 6.6**   *Computer-aided Bode Plot for Complex Poles and Zeros*

Repeat Example 6.5 using MATLAB.

Solution.   To obtain Bode plots using MATLAB, we call the function bode as follows:

```
numG = 0.01*[1 0.01 1]
denG = [0.25 0.01 1 0 0]
bode(numG, denG)
```

These commands will result in a Bode plot with the magnitude in decibels and the phase in degrees for a range of frequencies.

◆

Nonminimum-phase Systems

Nonminimum phase
systems

A system with a zero in the right half-plane (RHP) undergoes a net change in phase when evaluated for frequency inputs between zero and infinity, which, for a given magnitude plot, is greater than if all poles and zeros were in the left half-plane (LHP). Such a system is called **nonminimum phase**. As can be seen from the construction in Fig. B3, if the zero is in the RHP, then the phase *decreases* at the zero break point instead of exhibiting the usual phase increase that occurs for an LHP zero. Consider the transfer functions

$$G_1(s) = 10 \, \frac{s + 1}{s + 10},$$

$$G_2(s) = 10 \, \frac{s - 1}{s + 10}.$$

Both transfer functions have the same magnitude for all frequencies; that is,

$$|G_1(j\omega)| = |G_2(j\omega)|,$$

as shown in Fig. 6.11(a). But the phases of the two transfer functions are drastically different (Fig. 6.11b). A small amount of change in magnitude produces a small amount of change in the phase of G_1 but a much larger change in the phase of G_2. Thus the deviation in the phase of G_1 is minimal compared with the phase deviation of G_2 for the same change in magnitude. Hence, G_2 is

FIGURE 6.11
Bode plot for minimum and nonminimum phase systems: (a) magnitude; (b) phase

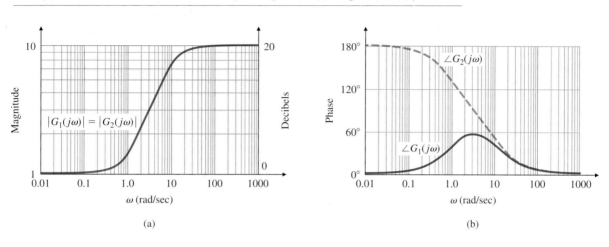

(a)   (b)

nonminimum phase. The discrepancy between G_1 and G_2 in regard to the phase change would be greater if two or more zeros of the plant were in the RHP.

6.1.2   Steady-state Errors

We saw in Section 4.3 that the steady-state error of a feedback system decreases as the gain of the open-loop transfer function increases. In plotting a composite magnitude curve, we saw in Section 6.1.1 that the open-loop low-frequency asymptote is given by

$$KG(j\omega) = K_o(j\omega)^n. \tag{6.21}$$

Therefore, we can conclude that the larger the value of the magnitude on the low-frequency asymptote, the lower the steady-state errors will be for the closed-loop system. This relationship is very useful in the design of compensation: Often we want to evaluate several alternate ways to improve stability and to do so we want to be able to see quickly how changes in the compensation will affect the steady-state errors.

For a system of the form given by Eq. (6.14), that is, where $n = 0$ in Eq. (6.21) (a type 0 system), the low-frequency asymptote is a constant and the gain K_o of the open-loop system is equal to the position-error constant K_p. For a unity feedback system with a unit step input, the steady-state error is

Position error constant

$$e_{ss} = \frac{1}{1 + K_p}.$$

For a unity-feedback system in which $n = -1$ in Eq. (6.21), often referred to as a type I system, the low-frequency asymptote has a slope of -1. The magnitude of the low-frequency asymptote is related to the gain according to Eq. (6.21); therefore, we can again read the gain $K_o\omega$ directly from the Bode magnitude plot. Equation (4.46) tells us that the velocity-error constant K_v is equal to the gain K_o in Eq. (6.21) for this case. For a unity feedback system with a unit ramp input the steady-state error is then

Velocity error coefficient

$$e_{ss} = \frac{1}{K_v}.$$

The easiest way of determining the value of K_v in a type I system is to read the magnitude of the low-frequency asymptote at $\omega = 1$ rad/sec because $K_v = K_o\omega$ at this frequency. In some cases the lowest-frequency break point will be below $\omega = 1$ rad/sec; therefore, the asymptote needs to extend to $\omega = 1$ rad/sec in order to read K_v directly. Alternately, we could read the magnitude at any frequency on the low-frequency asymptote and compute K_v from Eq. (6.21) with $n = -1$.

◆ **EXAMPLE 6.7** *Computation of K_v*

As an example of the determination of steady-state errors, a Bode magnitude plot of plant with compensation and unity feedback is shown in Fig. 6.12. Find the velocity-error constant.

Solution. Because the slope at the low frequencies is -1, we know that the system is type I. The extension of the low-frequency asymptote crosses $\omega = 1$ rad/sec at a magnitude of 10. Therefore, $K_v = 10$ and the steady-state error to a unit ramp for a unity feedback system would be 0.1. Alternatively, at $\omega = 0.01$ we have $|KG(j\omega)| = 1000$; therefore, from Eq. (6.21) we have

$$K_o = K_v \cong \omega|G(j\omega)| = 0.01(1000) = 10.$$

FIGURE 6.12
Determination of K_v from the Bode plot for the system $G(s) = 10/[s(s+1)]$

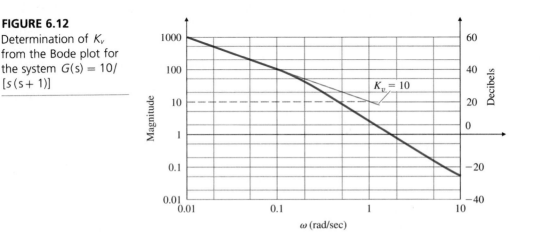

ω (rad/sec)

6.2 Stability

In the early days of electronic communications, most instruments were judged in terms of their frequency response. It is therefore natural that when the feedback amplifier was introduced, techniques to determine stability in the presence of feedback were based on this response.

Suppose the closed-loop transfer function of a system is known. We can determine the stability of a system by simply inspecting the denominator in factored form (since the factors give the system roots) to observe whether the real parts are positive or negative. However, the closed-loop transfer function is usually not known; in fact, the whole purpose behind understanding the root-locus technique is to be able to find the factors of the denominator in the closed-loop transfer function, given only the open-loop transfer function. Another way to determine closed-loop stability is to evaluate the frequency response of the *open-loop* transfer function $KG(j\omega)$ and then perform a test on that response. Note that this method does not require factoring the denominator of the closed-loop transfer function. In this section we will explain the principles of this method.

Suppose we have a system defined by Fig. 6.13(a) and whose root locus behaves as shown in Fig. 6.13(b); that is, instability results if K is larger than 2. The neutrally stable points lie on the imaginary axis, that is, where $K = 2$ and $s = j1.0$. Furthermore, we saw in Section 5.2 that all points on the locus have the property that

$$|KG(s)| = 1 \quad \text{and} \quad \angle G(s) = 180°.$$

At the point of neutral stability we see that these root-locus conditions hold for $s = j\omega$, so

$$|KG(j\omega)| = 1 \quad \text{and} \quad \angle G(j\omega) = 180°. \tag{6.22}$$

Thus a Bode plot of a system that is neutrally stable (that is, with K defined such that a closed-loop root falls on the imaginary axis) will satisfy the conditions

FIGURE 6.13
Stability example:
(a) system definition;
(b) root locus

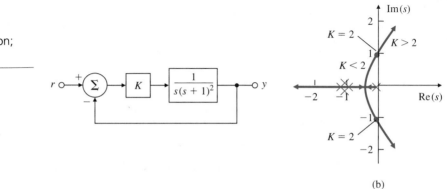

(b)

FIGURE 6.14
Frequency response
magnitude and phase for
the system in Fig. 6.13

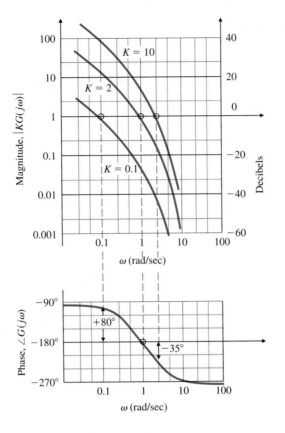

of Eq. (6.22). Figure 6.14 shows the frequency response for the system whose root locus is plotted in Fig. 6.13 for various values of K. The magnitude response corresponding to $K = 2$ passes through 1 at the same frequency ($\omega = 1$ rad/sec) at which the phase passes through 180°, as predicted by Eq. (6.22).

Having determined the point of neutral stability, we turn to a key question: Does increasing the gain increase or decrease the system's stability? We can see from the root locus in Fig. 6.13(b) that any value of K less than the value at the neutrally stable point will result in a stable system. At the frequency ω where the phase $\angle G(j\omega) = -180°$ ($\omega = 1$ rad/sec), the magnitude $|KG(j\omega)| < 1.0$ for stable values of K and > 1 for unstable values of K. Therefore, we have the following trial stability condition based on the character of the open-loop frequency response:

Stability condition

$$|KG(j\omega)| < 1 \quad \text{at} \quad \angle G(j\omega) = -180°. \tag{6.23}$$

This stability criterion holds for all systems where increasing gain leads to instability and $|KG(j\omega)|$ crosses the magnitude $= 1$ once, the most common situation. However, there are systems where an increasing gain can lead from instability to stability; in this case, the stability condition is,

$$|KG(j\omega)| > 1 \qquad \text{at} \quad \angle G(j\omega) = -180°. \tag{6.24}$$

There are also cases when $|KG(j\omega)|$ crosses magnitude $= 1$ more than once. One way to resolve the ambiguity that is usually sufficient is to perform a rough sketch of the root locus. Another, more rigorous, way to resolve the ambiguity is to use the Nyquist stability criterion, the subject of the next section. However, because the Nyquist criterion is fairly complex, it is important while studying it to bear in mind the theme of this section, namely, that for most systems a simple relationship exists between closed-loop stability and the open-loop frequency response.

6.3   The Nyquist Stability Criterion

For most systems, as we saw in the previous section, an increasing gain eventually causes instability. In the very early days of feedback control design, this relationship between gain and stability margins was assumed to be universal. However, designers found occasionally that in the laboratory the relationship reversed itself; that is, the amplifier would become unstable when the gain was decreased. The confusion caused by these conflicting observations motivated Harry Nyquist of the Bell Telephone Laboratories to study the problem in 1932. His study explained the occasional reversals and resulted in a more sophisticated analysis with no loopholes. Not surprisingly, his test has come to be called the **Nyquist stability criterion**. It is based on a result from complex variables theory known as the **argument principle**, as we briefly explain in this section and in more detail in Appendix B, Section B.10.

The Nyquist stability criterion relates the open-loop frequency response to the number of closed-loop poles of the system in the RHP. Study of the Nyquist criterion will allow you to determine stability from the frequency response of a complex system, perhaps with one or more resonances, where the magnitude curve crosses 1 several times and/or the phase crosses 180° several times. It is also very useful in dealing with open-loop, unstable systems, nonminimum-phase systems and systems with pure delays (transportation lags).

Consider the transfer function $H_1(s)$ whose poles and zeros are indicated in the s-plane in Fig. 6.15(a). We wish to evaluate H_1 for values of s on the clockwise contour C_1. (Hence this is called a **contour evaluation**.) We choose the test point s_o for evaluation. The resulting complex quantity has the form $H_1(s_o) = \vec{v} = |\vec{v}|e^{j\alpha}$. The value of the argument of $H_1(s_o)$ is

$$\alpha = \theta_1 + \theta_2 - (\phi_1 + \phi_2).$$

As s traverses C_1 in the clockwise direction starting at s_o, the angle α of $H_1(s)$ in Fig. 6.15(b) will change (decrease or increase), but it will not undergo a net change of 360° as long as there are no poles or zeros within C_1. This is because

FIGURE 6.15
Contour evaluations: (a) s-plane plot of poles and zeros of $H_1(s)$ and the contour C_1;
(b) $H_1(s)$ for s on C_1; (c) s-plane plot of poles and zeros of $H_2(s)$ and the contour C_1;
(d) $H_2(s)$ for s on C_1

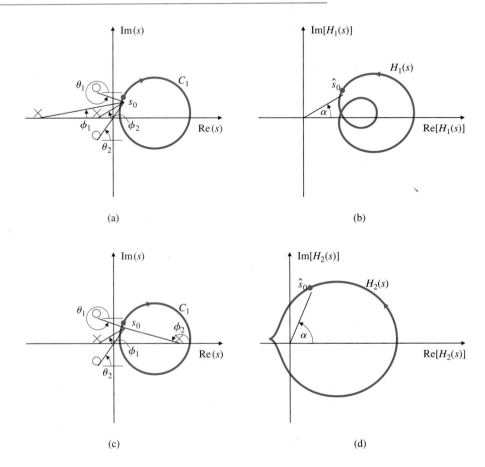

(a)

(b)

(c)

(d)

none of the angles that make up α go through a net revolution. The angles θ_1, θ_2, ϕ_1, and ϕ_2 increase or decrease as s traverses around C_1, but they return to their original values as s returns to s_0 without rotating through $360°$. This means that the plot of $H_1(s)$ (Fig. 6.15b) will not encircle the origin. This conclusion follows from the fact that α is the sum of the angles indicated in Fig. 6.15(a), so the only way that α can be changed by $360°$ after s executes one full traverse of C_1 is for C_1 to contain a pole or zero.

Now consider the function $H_2(s)$, whose pole-zero pattern is shown in Fig. 6.15(c). Note that it has a singularity (pole) within C_1. Again we start at the test point s_0. There is no net change in the angles θ_1, θ_2, or ϕ_1 due to the other singularities. As s traverses in the clockwise direction around C_1, the contributions from these singularities change, but they return to their original values as soon as s returns to s_0. In contrast, ϕ_2, the angle from the pole within C_1, undergoes a net change of 360° after one full traverse of C_1. Therefore, the argument of $H_2(s)$ undergoes the same change, causing H_2 to encircle the origin in the counterclockwise direction, as shown in Fig. 6.15(d). The behavior would be similar if the contour C_1 had enclosed a zero instead of a pole. The mapping of C_1 would again enclose the origin once in the $H_2(s)$-plane except it would do so in the clockwise direction.

Thus we have the essence of the argument principle:

Argument principle

> A contour map of a complex function will only encircle the origin if the contour contains a singularity (pole or zero) of the function.

The principle can be extended by allowing multiple singularities (poles or zeros) within the contour. The number and direction of origin encirclements then change. For example, if the number of poles and zeros within C_1 is the same, the net angles cancel and there will be no net encirclement of the origin.

To apply the principle to control design, we let the C_1 contour in the s-plane encircle the entire RHP (Fig. 6.16). The resulting evaluation of $H(s)$ will only encircle the origin if $H(s)$ has a RHP pole or zero.

As stated earlier, what makes all this contour behavior useful is that a contour evaluation of an *open-loop* $KG(s)$ can be used to determine stability of

FIGURE 6.16
An $s =$ plane plot of a contour C_1 that encircles the entire RHP

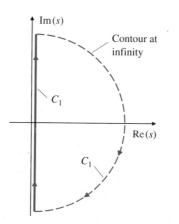

FIGURE 6.17
Block diagram for $Y(s)/$
$R(s) = KG(s)/$
$[1 + KG(s)]$

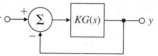

the *closed-loop* system. Specifically, for the system in Fig. 6.17, the closed-loop transfer function is

$$\frac{Y(s)}{R(s)} = T(s) = \frac{KG(s)}{1 + KG(s)}.$$

Therefore, the closed-loop roots are the solutions of

$$1 + KG(s) = 0.$$

If the evaluation contour of s enclosing the entire RHP contains a zero or pole of $1 + KG(s)$, then the evaluated contour of $1 + KG(s)$ will encircle the origin. Notice that $1 + KG(s)$ is simply $KG(s)$ shifted to the right 1 unit (Fig. 6.18), and we can plot $KG(s)$. Therefore, if the plot of $1 + KG(s)$ encircles the origin, the plot of $KG(s)$ will encircle -1 on the real axis. Therefore, -1 is the critical point. Presentation of the evaluation of $KG(s)$ in this manner is often referred to as a **Nyquist**, or **polar plot** since we plot the magnitude of $KG(j\omega)$ and the angle of $KG(j\omega)$.

Nyquist plot; polar plot

To determine whether an encirclement is due to a pole or zero, we write $1 + KG(s)$ in terms of poles and zeros of $KG(s)$:

$$1 + KG(s) = 1 + K\frac{b(s)}{a(s)} = \frac{a(s) + Kb(s)}{a(s)}. \qquad (6.25)$$

FIGURE 6.18
Evaluations of $KG(s)$ and $1 + KG(s)$: Nyquist plots

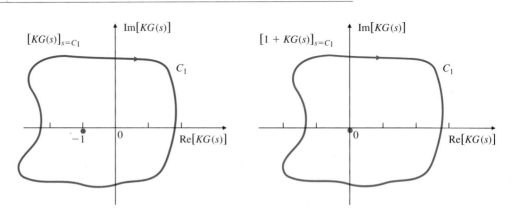

Equation (6.25) shows that the poles of $1+KG(s)$ are also the poles of $G(s)$. Since it is safe to assume that the poles of $G(s)$ [or factors of $a(s)$] are known, the (rare) existence of any of these poles in the RHP can be accounted for. Assuming for now that there are no poles of $G(s)$ in the RHP, an encirclement of -1 by $KG(s)$ indicates a zero of $1 + KG(s)$ in the RHP and thus an unstable root of the closed-loop system.

We can generalize this basic idea by noting that a clockwise contour C_1 enclosing a zero of $1 + KG(s)$–that is, a closed-loop system root—will result in $KG(s)$ encircling the -1 point in a clockwise direction. Likewise, if C_1 encloses a pole of $1 + KG(s)$—that is, if there is an unstable open-loop pole—there will be a counterclockwise $KG(s)$ encirclement of -1. Furthermore, if two poles or two zeros are in the RHP, $KG(s)$ will encircle -1 twice, and so on. The net number of encirclements, N, equals the number of zeros (closed-loop system roots) in the RHP, Z, minus the number of open-loop poles in the RHP, P:

$$N = Z - P.$$

A simplification in the plotting of $KG(s)$ results from the fact tht any $KG(s)$ that represents a physical system will have zero response at infinite frequency (i.e., has more poles than zeros). This means that the big arc of C_1 corresponding to s at infinity (Fig. 6.16) results in $KG(s)$ being a point of infinitesimally small value near the origin for that portion of C_1. Therefore, we accomplish a complete evaluation of a physical system $KG(s)$ by letting s traverse the imaginary axis from $-j\infty$ to $+j\infty$. The evaluation of $KG(s)$ from $s = 0$ to $s = +j\infty$ has already been discussed in Section 6.1 under the context of finding the frequency response of $KG(s)$. Because $G(-j\omega)$ is the complex conjugate of $G(j\omega)$, we can easily obtain the entire plot of $KG(s)$ by reflecting the $0 \leqslant s \leqslant +j\infty$ portion about the real axis, to get the $(-j\infty \leqslant s < 0)$ portion. Hence we see that closed loop stability can be determined in all cases by examination of the frequency response of the open-loop transfer function on a polar plot. In some applications, models of physical systems are simplified so as to eliminate some high-frequency dynamics. The resulting reduced-order transfer function has an equal number of poles and zeros.[6] In that case the big arc of C_1 at infinity needs to be considered.

In practice, many systems behave like those discussed in Section 6.2, so you need not carry out a complete evaluation of $KG(s)$ with subsequent inspection of the -1 encirclements; a simple look at the frequency response may suffice to determine stability based on the gain and phase margins. However, in the case of a complex system for which the simplistic rules given in Section 6.3 become

---

6. An example is a model of a turbofan jet engine.

ambiguous, you will want to perform the complete analysis, summarized as follows:

Procedure for Plotting the Nyquist Plot

1. Plot $KG(s)$ for $-j\infty \leqslant s \leqslant +j\infty$. Do this by first evaluating $KG(j\omega)$ for $\omega = 0$ to ω_h, where ω_h is so large that the magnitude of $KG(j\omega)$ for $\omega > \omega_h$ is always less than a small number such as 0.1, then reflecting the image about the real axis and adding it to the preceding image. If $KG(j\omega)$ does not get small for very large ω, then you must plot the semicircle of infinite radius with the center at the origin covering the RHP as shown in Figure 6.16. The Nyquist plot will always be symmetric with respect to the real axis.

2. Evaluate the number of clockwise encirclements of -1, and call that number N. Do this by drawing a straight line in *any* direction from -1 to ∞. Then count the net number of left-to-right crossings of the straight line by $KG(s)$. If encirclements are in the counterclockwise direction, N is negative.

3. Determine the number of unstable (RHP) poles of $G(s)$, and call that number P.

4. Calculate the number of unstable closed-loop roots, Z:

$$Z = N + P. \qquad (6.26)$$

For stability we wish to have $Z = 0$; that is, no closed-loop poles in the RHP.

Let us now examine a rigorous application of the procedure for drawing Nyquist plots for some examples.

◆ **EXAMPLE 6.8**   *Nyquist Plot for a Second-order System*

Determine the stability properties of the system defined in Fig. 6.19.

FIGURE 6.19
Control system for
Example 6.8

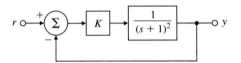

FIGURE 6.20
Root locus of $G(s) = 1/(s + 1)^2$ with respect to K

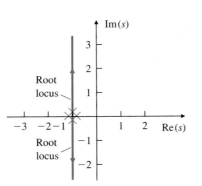

Solution. The root locus of the system in Fig. 6.19 is shown in Fig. 6.20. It is stable for all values of K. The magnitude of the frequency response of $KG(s)$ is plotted in Fig. 6.21(a) for $K = 1$, and the phase is plotted in Fig. 6.21(b); this is the typical Bode method of presenting frequency response and represents the evaluation of $G(s)$ over the interesting range of frequencies. The same information is replotted in Fig. 6.22 in the Nyquist (polar) plot form. Note how the points A, B, C, D, and E are mapped from the Bode plot to the Nyquist plot in Fig. 6.22. The arc from $G(s) = +1$ ($\omega = 0$) to $G(s) = 0$ ($\omega = \infty$) that lies below the real axis is derived from Fig. 6.21. The portion of the C_1 arc at infinity from Fig. 6.16 transforms into $G(s) = 0$ in Fig. 6.22; therefore, a continuous evaluation of $G(s)$ with s traversing C_1 is completed by simply reflecting the lower arc about the real axis. This creates the portion of the contour above the real axis and completes the Nyquist (polar) plot. Since the plot does not encircle -1, $N = 0$. Also, there are no poles of $G(s)$ in the RHP, so $P = 0$. From Eq. (6.26), we conclude that $Z = 0$, which indicates there are no unstable roots of the closed-loop system for $K = 1$.

FIGURE 6.21
Open loop Bode plot for $G(s) = 1/(s + 1)^2$

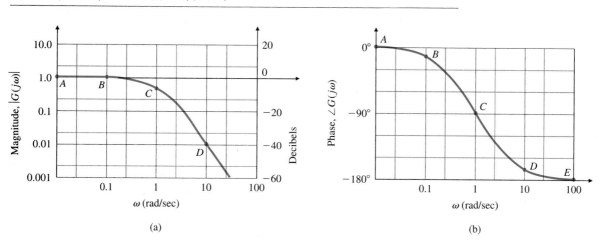

(a)

(b)

FIGURE 6.22

Nyquist plot of the evaluation of $KG(s)$ for $s = C_1$ and $K = 1$

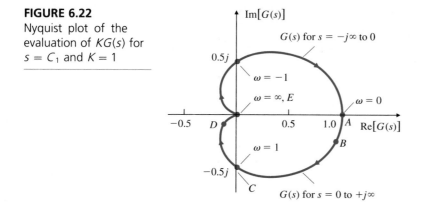

The MATLAB statements that will produce this Nyquist plot are

numG = 1

denG = [1 2 1]

nyquist(numG, denG)

Stability for a range of gains

Often the control systems engineer is more interested in determining a range of gains K for which the system is stable than in testing for stability at a specific value of K. To accommodate this requirement but to avoid drawing multiple Nyquist plots, each for a specific value of the gain, the test can be slightly modified. To do so we scale $KG(s)$ by K and examine $G(s)$ to determine stability for a range of gains K. This is possible because an encirclement of -1 by $KG(s)$ is equivalent to an encirclement of $-1/K$ by $G(s)$. Therefore, instead of having to deal with $KG(s)$, we need only consider $G(s)$, and count the number of the encirclements of the point $-1/K$.

We illustrate this idea by applying it to Example 6.8. There are no poles of $G(s)$ in the RHP, so $P = 0$. For stability we wish to have $Z = 0$. Therefore, $N = 0$, which means that the Nyquist plot cannot encircle the $-1/K$ point. We note from the Nyquist plot that this is true so long as

$$-\frac{1}{K} < 0 \qquad \text{or} \qquad -\frac{1}{K} > 1.$$

This yields $K > -1$ for stability.

For this example, $G(s)$ never crosses the negative real axis except at $G(s) = 0$; therefore, it will never encircle $-1/K$ as long as $K > 0$. This result is identical to that shown by the root locus, as it should be. However, the gain K can actually be negative, as seen from the above relation, a result which may be verified by drawing the $0°$ locus.

◆ **EXAMPLE 6.9**   *Nyquist Plot for a Third-order System*

As a second example, consider the system $G(s) = 1/s(s + 1)^2$ for which the closed-loop system is defined in Fig. 6.23. Determine its stability properties using the Nyquist criterion.

Solution.   This is the same system discussed in Section 6.2. The root locus in Fig. 6.13(b) shows that this system is stable for small values of K but unstable for large values of K. The magnitude and phase of $G(s)$ in Fig. 6.24 are transformed into the Nyquist plot shown in Fig. 6.25. Note how the points A, B, C, D and E on the Bode plot of Fig. 6.24 map into those on the Nyquist plot of Fig. 6.25. Note the large arc at infinity that arises from the open-loop pole at $s = 0$. This pole creates an infinite magnitude of $G(s)$ at $\omega = 0$; in fact, any pole or zero on the imaginary axis will create an arc at infinity. To correctly determine the number of $-1/K$ point encirclements, we must draw this arc in the proper half-plane: Should it cross the *positive* real axis, as shown in Fig. 6.25, or the negative one? It is also necessary to assess whether the arc should sweep out 180° (as in Fig. 6.25), 360°, or 540°.

A simple artifice suffices to answer these questions. We modify the C_1 contour to take a small detour around the pole either to the right (Fig. 6.26), or to the left. It makes no difference to the final stability question which way, but it is more convenient to go to the right because then no poles are introduced within the C_1 contour, keeping the value of P equal to 0. Since the phase of $G(s)$ is the negative of the sum of the angles from all of the poles, we see that the evaluation results in a Nyquist plot moving from

FIGURE 6.23
Control system for
Example 6.9

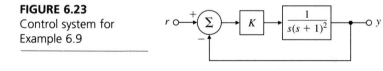

FIGURE 6.24
Bode plot for $G(s) = 1/[s(s + 1)^2]$

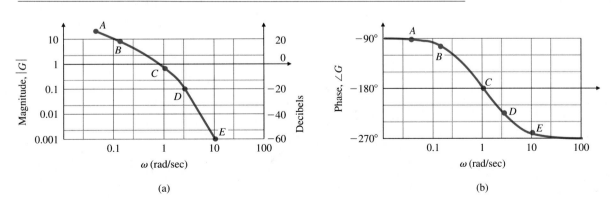

(a)   (b)

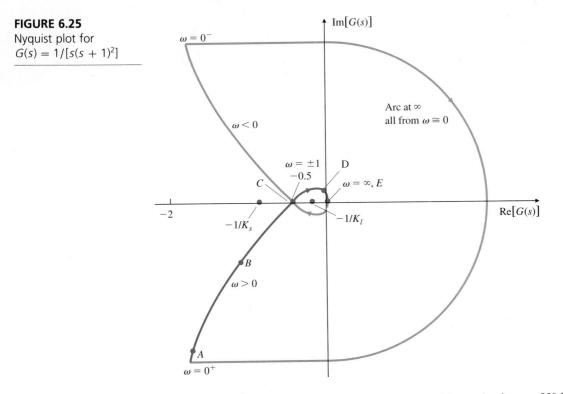

FIGURE 6.25
Nyquist plot for
$G(s) = 1/[s(s + 1)^2]$

$+90°$ for s just below the pole at $s = 0$, across the positive real axis to $-90°$ for s just above the pole. Had there been two poles at $s = 0$, the Nyquist plot at infinity would have executed a full $360°$ arc, and so on for three or more poles. Furthermore, for a pole elsewhere on the imaginary axis, a $180°$ clockwise arc would also result but would be oriented differently than the example in Fig. 6.25.

The Nyquist plot crosses the real axis at $\omega = 1$ with $|G| = 0.5$, as indicated by the Bode plot. The frequency response of the system is

$$G(j\omega) = \frac{1}{j\omega(j\omega + 1)^2} = \frac{1}{-2\omega^2 + j\omega(1 - \omega^2)}$$

$$= \frac{-2\omega^2 - j\omega(1 - \omega^2)}{4\omega^4 + \omega^2(1 - \omega^2)^2}.$$

The real part of $G(j\omega)$ is

$$\mathrm{Re}[G(j\omega)] = -\frac{2}{1 + \omega^4 + 2\omega^2},$$

which for small values of ω equals approximately -2, as shown in Fig. 6.25.

There are three possibilities for the location of $-1/K$: inside the two loops of the Nyquist plot, inside the large loop, or outside the Nyquist contour completely. For larger values of K (K_l in Fig. 6.25), $-0.5 < -1/K_l < 0$ will lie inside the two loops; hence $N = 2$, and therefore, $Z = 2$, indicating there are two unstable roots. This happens for

FIGURE 6.26
C_1 contour enclosing the
RHP for system in
Example 6.9

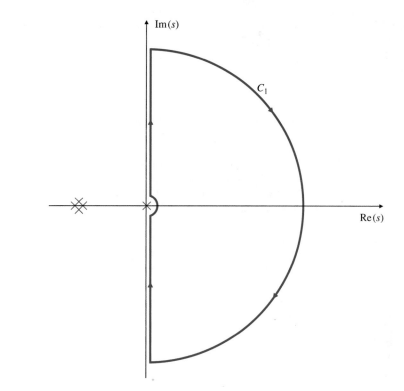

$K_l > 2$. If $-1/K$ lies on the positive real axis (so $-1/K > 0$, then $N = 1$, which means $Z = 1$ and the system has one unstable root. This is the case for $K < 0$. For small values of K (K_s in Fig. 6.25), $-1/K$ lies outside the loops; thus $N = 0$, and all roots are stable. All this information is in agreement with the root locus in Figure 6.13(b). (Of course, to verify the $K < 0$ result, we should draw the $0°$ root locus.)

For this and many similar systems, we can see that the encirclement criterion reduces to a very simple test for stability based on the open-loop frequency response: The system is stable if $|KG(j\omega)| < 1$ when the phase of $G(j\omega)$ is $180°$. Note that this relation is identical to the stability criterion given in Eq. (6.23); however, by using the Nyquist criterion, we don't require the root locus to determine whether $|KG(j\omega)| < 1$ or $|KG(j\omega)| > 1$.

Nyquist plot via MATLAB   To draw the Nyquist plot using MATLAB, we use the following commands:

```
numG = 1

denG = [1 2 1 0]

axis([−5 5 −5 5])

nyquist(numG, denG)
```

The axis command scales the plot so that only points between $+5$ and -5 on the real and imaginary axes are included. Without manual scaling, points between $\pm\infty$ are plotted and the essential features are lost.

◆

For systems that are open-loop unstable, care must be taken because now $P \neq 0$ in Eq. (6.26).

◆ **EXAMPLE 6.10**   *Nyquist Plot for an Open-loop Unstable System*

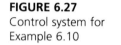

The third example is defined in Fig. 6.27. Determine its stability properties using the Nyquist criterion.

Solution.   The root locus for this system is sketched in Fig. 6.28. The open-loop system is unstable since it has a pole in the RHP. The open-loop Bode plot is shown in Fig. 6.29. By examining $|KG(j\omega)|$ and $\angle G(j\omega)$ graphically, we can see that the pole in the RHP causes the magnitude to behave as if the pole were in the LHP while causing the phase to increase by 90° instead of the usual decrease at a pole. Any system with a pole in the RHP is unstable; hence it would be impossible to determine its frequency response experimentally because the system would never reach a steady-state sinusoidal response for a sinusoidal input. It is, however, possible to compute the magnitude and phase of the

FIGURE 6.27
Control system for
Example 6.10

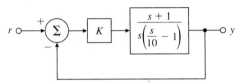

FIGURE 6.28
Root locus for
$G(s) = (s + 1)/[s(s/10 - 1)]$

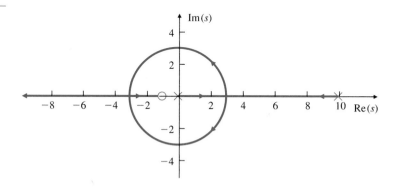

FIGURE 6.29
Bode plot for $G(s) = (s + 1)/[s(s/10 - 1)]$

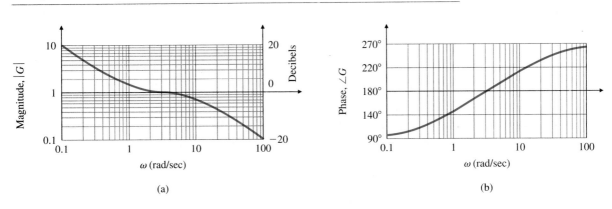

(a)

(b)

transfer function according to the rules in Section 6.1. The pole in the RHP affects the Nyquist encirclement criterion because the value of P in Eq. (6.26) is -1.

We convert the frequency-response information of Fig. 6.29 into the Nyquist plot (Fig. 6.30) as in the previous examples. As before, the C_1 detour around the pole at $s = 0$ in Fig. 6.31 creates a large arc at infinity in Fig. 6.30. This arc crosses the *negative* real axis because of the $180°$ phase contribution of the pole in the RHP.

FIGURE 6.30
Nyquist plot for $G(s) = (s + 1)/[s(s/10 - 1)]$

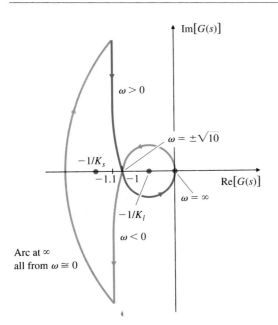

FIGURE 6.31
C_1 contour for Example 6.10

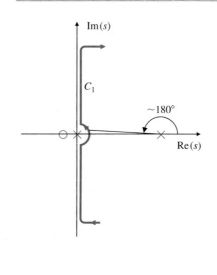

The real-axis crossing occurs at $|G(s)| = 1$ because in the Bode plot $|G(s)| = 1$ when $\angle G(s) = 180°$, which happens to be at $\omega \cong 3\,\text{rad/sec}$. The frequency response of the system is

$$G(s) = \frac{10(j\omega + 1)}{j\omega(j\omega - 10)}$$

$$= \frac{-110\omega - 10j(\omega^2 - 10)}{\omega^3 + 100\omega}.$$

The real part of $G(j\omega)$,

$$\text{Re}[G(j\omega)] = \frac{-110}{\omega^2 + 100},$$

is approximately -1.1 for small values of ω, which explains the corresponding behavior in Fig. 6.30.

The contour shows three different behaviors depending on the values of K. For large values of K (K_l in Fig. 6.30), where $-1 < -1/K < 0$, there is one counterclockwise encirclement; hence $N = -1$. However, since $P = 1$ from the RHP pole, $Z = N + P = 0$, so there are no unstable system roots. Thus the system is stable for $K_l > 1$. For small values of K (K_s in Fig. 6.30), $N = +1$ because of the clockwise encirclement and $Z = 2$, indicating two unstable roots. This happens if $-1/K_s < -1$ (or $K_s < 1$). Finally, if $-1/K > 0$ (or $K < 0$), then $N = 0$ and $Z = 1$, so the system will have one unstable closed-loop pole. These results agree with the root locus, as they should, but we need to draw the $0°$ root locus to verify the case where the gain is negative.

As with all systems, the stability boundary occurs at $|KG(j\omega)| = 1$ for the phase of $\angle G(j\omega) = 180°$. However, in this case, $|KG(j\omega)|$ must be greater than 1 to yield the correct number of -1 point encirclements to achieve stability.

Nyquist plot via MATLAB   To draw the Nyquist plot using MATLAB, we use the following commands:

```
numG =[1  1]

denG =[0.1  −1  0]

axis([−5  5 −5  5])

nyquist(numG, denG)
```

◆

The existence of the RHP pole in Example 6.10 affected the Bode plotting rules of the phase curve and affected the relationship between encirclements and unstable closed-loop roots because $P = 1$ in Eq. (6.26). But we apply the Nyquist stability criterion without any modifications. The same is true for systems with a RHP zero; that is, a nonminimum-phase zero has no effect on the Nyquist stability criterion, but the Bode plotting rules are affected.

6.4 Stability Margins

A large fraction of control system designs behave in a pattern roughly similar to that of the system in Section 6.2 and Example 6.9 in Section 6.3; that is, the system becomes unstable if the gain increases past a certain critical point. Two quantities that measure the stability margin of a system are directly related to the stability criterion of Eq. (6.23): gain margin and phase margin. In this section we will define and use these two concepts to study system design.

Gain margin

The **gain margin** (GM) is the factor by which the gain is less than the neutral stability value. For the typical case it can be read directly from the Bode plot in Fig. 6.14 by measuring the vertical distance between the $|KG(j\omega)|$ curve and the $|KG(j\omega)| = 1$ line at the frequency where $\angle G(j\omega) = 180°$. We see from the figure that when $K = 0.1$, the system is stable and GM = 20 (or 26 dB). When $K = 2$, the system is neutrally stable with GM = 1 (0 dB), while $K = 10$ results in an unstable system with GM = 0.125 (-18 dB). Note that GM is the factor by which the gain K can be raised before instability results. $|GM| < 1$ (or $|GM| < 0$ dB) indicates an unstable system. The GM can also be determined from a root locus with respect to K by noting two values of K: at the point where the locus crosses the $j\omega$-axis, and at the nominal closed-loop poles. The GM is the ratio of these two values.

Phase margin

Another measure that is used to indicate the stability margin in a system is the **phase margin** (PM). It is the amount by which the phase of $G(j\omega)$ exceeds $-180°$ when $|KG(j\omega)| = 1$, which is an alternate way of measuring the degree to which the stability conditions of Eq. (6.23) are met. From Fig. 6.14 we see that PM $\cong 80°$ for $K = 0.1$ and $-35°$ for $K = 10$. A positive PM is required for stability.

Note that the two stability measures, PM and GM, together determine how far the complex quantity $G(j\omega)$ passes from the -1 point, which is another way of stating the neutral-stability point specified by Eq. (6.22).

The stability margins may also be defined in terms of the Nyquist plot. Figure 6.32 shows that GM and PM are measures of how close the Nyquist plot

FIGURE 6.32
Nyquist plot for defining GM and PM

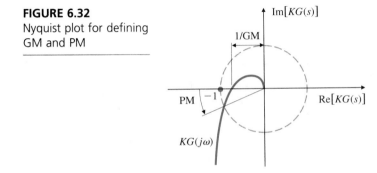

comes to encircling the point -1. Again we can see that the GM indicates how much the gain can be raised before instability results in a system like the one in Example 6.9. The PM is the difference between the phase of $G(j\omega)$ and $180°$ when $KG(j\omega)$ crosses the circle $|KG(s)| = 1$; the positive value of PM is assigned to the stable case (that is, with no Nyquist encirclements).

Crossover frequency

It is easier to determine these margins directly from the Bode plot than from the Nyquist plot. The term **crossover frequency** is often used to refer to the frequency at which the gain is unity, or 0 dB. Figure 6.33 shows the same data plotted in Fig. 6.24, but for the case with $K = 1$. The same values of PM ($22°$) and GM(2) may be obtained from the Nyquist plot shown in Fig. 6.25: The real-axis crossing at -0.5 corresponds to a GM of $1/0.5$ or 2. The PM can be computed graphically, or it can be computed as $\tan^{-1}(0.36/0.93) = 22°$.

One of the useful aspects of frequency-response design is the ease with which we can evaluate the effects of gain changes. In fact, we can determine the PM

FIGURE 6.33
GM and PM from the magnitude and phase plots

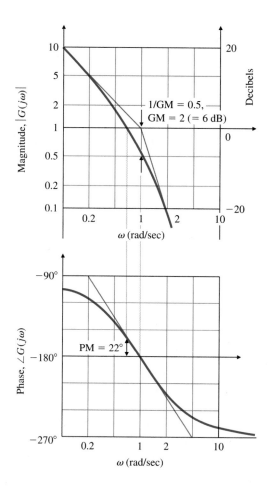

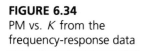

FIGURE 6.34

PM vs. K from the frequency-response data

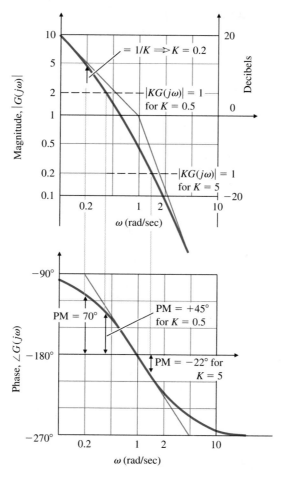

from Fig. 6.33 for any value of K without redrawing the magnitude or phase information. We need only indicate on the figure where $|KG(j\omega)| = 1$ for selected trial values of K, as has been done with dashed lines in Fig. 6.34. Now we can see that $K = 5$ yields an unstable PM of $-22°$, while a gain of $K = 0.5$ yields a PM of $+45°$. Furthermore, if we wish a certain PM (say $60°$), we simply read the value of $|G(j\omega)|$ corresponding to the frequency that would create the desired PM (here $\omega = 0.2\,\text{rad/sec}$ yields $70°$, where $|G(j\omega)| = 5$), and note that the magnitude at this frequency is $1/K$. Therefore, a PM of $70°$ will be achieved with $K = 0.2$.

Because the PM is related to damping, it is sometimes used directly to specify control system performance. It is therefore of interest to relate the margins to other measures of performance. For simplicity, let us take the open-loop second-order system

$$G(s) = \frac{\omega^2}{s(s + 2\zeta\omega)}.$$

With unity feedback the closed-loop system

$$T(s) = \frac{\omega^2}{s^2 + 2\zeta\omega s + \omega^2} \qquad (6.27)$$

results. The relationship between the PM and ζ in this system (as shown in Problem 6.25) is

$$PM = \tan^{-1} \frac{2\zeta}{\sqrt{\sqrt{1 + 4\zeta^4} - 2\zeta^2}} \qquad (6.28)$$

and is plotted in Fig. 6.35. Note that the function is approximately a straight line up to about PM = 60°. The dashed line shows a straight-line approximation to the function where

$$\zeta \cong \frac{PM}{100}. \qquad (6.29)$$

It is clear that the approximation only holds for phase margins below about 70°. Furthermore, Eq. (6.28) is only accurate for the second-order system of Eq. (6.27). In spite of these limitations Eq. (6.29) is often used as a rule of thumb for relating the damping ratio to PM. It is useful as a starting point; however, it is important always to check the actual damping of a design as well as other aspects of the performance before calling the design complete.

The gain margin for this second-order system is infinite (GM = ∞) since the phase curve does not cross −180° as the frequency increases. The Nyquist plot does not cross the real axis but converges to the origin for high frequencies, resulting in the infinite gain margin.

Additional data to aid in evaluating a control system based on its PM can be derived from the relationship between the resonant peak M_r and ζ seen in Fig.

FIGURE 6.35

Damping ratio vs. phase margin (PM)

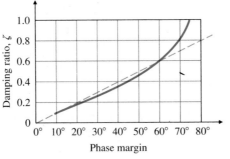

FIGURE 6.36
Transient-response overshoot and frequency response: resonant peak versus phase margin (PM) for $T(s) = \omega^2/(s^2 + 2\zeta\omega s + \omega^2)$

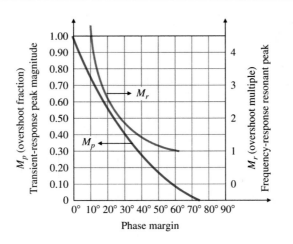

6.2. Note that this figure was derived for the same system (Eq. 6.7) as Eq. (6.27). We can convert the information in Fig. 6.35 into a form relating M_r to the PM. This is depicted in Fig. 6.36, along with the step-response overshoot M_p.

Many engineers think directly in terms of the PM when judging whether a control system is adequately stabilized, with PM = 30° often judged to be the lowest acceptable value. In addition to testing the stability of a system design using the PM, a designer would typically also be concerned with meeting a speed-of-response specification like bandwidth, as discussed in Section 6.1. In terms of the frequency-response parameters discussed so far, the crossover frequency would best describe a system's speed of response. This idea will be discussed further in Sections 6.6 and 6.7.

In some cases the PM and GM are not helpful indicators of stability. For first- and second-order systems, the phase never crosses the 180° line; hence the GM is always ∞ and not a useful design parameter. For higher-order systems it is possible to have more than one frequency where $|KG(j\omega)| = 1$ or where $\angle KG(j\omega) = 180°$, and the margins as previously defined need clarification. An example of this can be seen in Fig. 9.12, where the magnitude crosses 1 three times. A decision was made to define PM by the first crossing, because the PM at this crossing was the smallest of the three values and thus the most conservative assessment of stability. A Nyquist plot based on the data in Fig. 9.12 would show that the portion of the Nyquist curve closest to the -1 point was the critical indicator of stability, and therefore that use of the crossover frequency yielding the minimum value of PM was the logical choice.

FIGURE 6.37
Root locus for a conditionally stable system

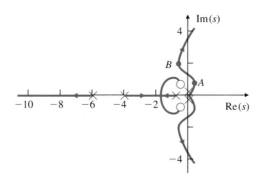

There are certain practical examples where an increase in the gain can make the system stable. As we saw in Chapter 5, these systems are called **conditionally stable**. A representative root-locus plot for such systems is shown in Fig. 6.37. For a point on the root locus such as A, an increase in the gain would make the system stable by bringing the unstable roots into the LHP. For point B, either a gain increase or decrease could make the system become unstable. Thus stability is here strictly dependent on the particular value of the gain. Therefore, several gain margins exist that correspond to either gain reduction or gain increase.

◆ **EXAMPLE 6.11**   *Stability Properties for a Conditionally Stable System*

Determine the stability properties as a function of the gain K for the system with the open-loop transfer function

$$G(s) = \frac{K(s + 10)^2}{s^3}.$$

Solution.   This is a system for which increasing gain causes a transition from instability to stability. The root locus in Fig. 6.38(a) shows that the system is unstable for $K < 5$ and stable for $K > 5$. The Nyquist plot in Fig. 6.38(b) was drawn for the stable value $K = 7$. Determination of the margins according to Fig. 6.32 yields PM $= +10°$ and GM $= 0.7$. According to the rules for stability discussed earlier, these two margins yield conflicting signals on the system's stability.

We resolve the conflict by counting the Nyquist encirclements in Fig. 6.38(b). There is one clockwise encirclement and one counterclockwise encirclement of the -1 point. Hence there are no net encirclements, which confirms that the system is stable for $K = 7$. For systems like this it is best to resort to the root locus and Nyquist plot (rather than the Bode plot) to determine stability (and the GM and PM).

FIGURE 6.38
System in which increasing gain leads from instability to stability: (a) root locus;
(b) Nyquist plot

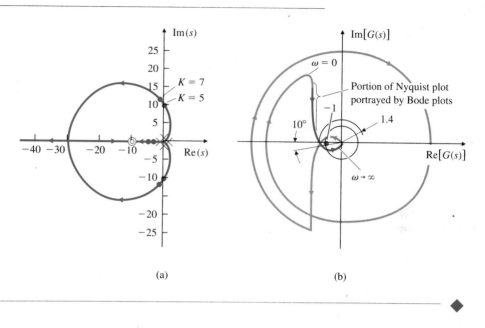

(a)   (b)

◆ **EXAMPLE 6.12**   *Nyquist Plot for a System with Multiple Crossover Frequencies*

Draw the Nyquist plot for the system

$$G(s) = \frac{85(s + 1)(s^2 + 2s + 43.25)}{s^2(s^2 + 2s + 82)(s^2 + 2s + 101)}$$

$$= \frac{85(s + 1)(s + 1 \pm 6.5j)}{s^2(s + 1 \pm 9j)(s + 1 \pm 10j)},$$

and determine the stability margins.

Solution.   The Nyquist plot (Fig. 6.39) shows there are three crossover frequencies with three candidate PM values 36.7°, 71.2°, and 42.8° corresponding to $\omega = 0.7$, 9.5, and 9.8 rad/sec, respectively. However, the key indicator of stability is the proximity of the Nyquist plot as it approaches the -1 point while crossing the real axis. In this case only the GM indicates the marginal stability of this system. Of the three possible phase margins, we choose PM = 36°, since this value requires the least phase perturbation to drive the system to instability.

   For comparison it is interesting to consider the Bode plot for this system (Fig. 6.40). It shows three crossings of magnitude = 1: at 0.74, 9.5, and 9.8 rad/sec. The candidate

FIGURE 6.39
Nyquist plot of the
complex system in
Example 6.12

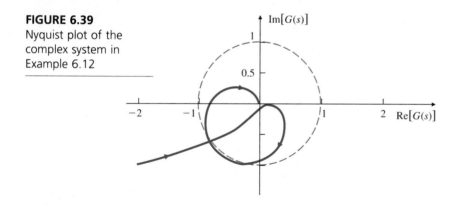

FIGURE 6.40
Bode plot of the system in Example 6.12

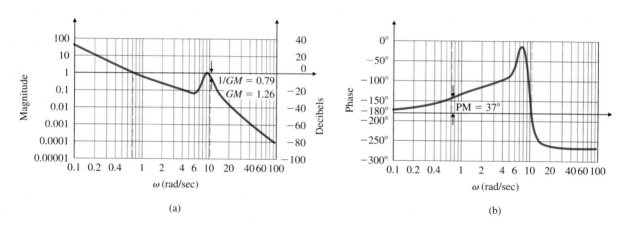

(a)   (b)

PM values are 37°, 43°, and 71°, respectively. Again we must choose the minimum additional phase lag that will drive the system to instability, so PM = 37°. The gain margin value 1.26 from the Bode plot corresponding to $\omega = 10.4$ rad/sec also agrees with the GM of the Nyquist plot. This example illustrates the need for the Nyquist plot in determining the PM.

◆

Unstable open-loop systems (such as Example 6.10) exhibit stability criteria different from those defined by Fig. 6.32; their Nyquist plots need to show one encirclement of −1 for each unstable pole in order to be stable. Therefore, we need to verify the GM and PM as previously defined, and perhaps to modify them by reverting back to the Nyquist stability criterion.

FIGURE 6.41
Definition of the vector margin on the Nyquist plot (M circle is explained in Section 6.8)

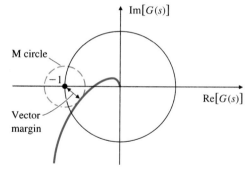

It should be clear that a designer needs to be judicious when applying the margin definitions described in Fig. 6.32. The stability margin of a system can only be rigorously assessed by examining the Nyquist plot to determine its closest approach to the -1 point. To aid in this analysis, O. J. M. Smith (1958) introduced the **vector margin**, which he defined to be the distance to the -1 point from the closest approach of the Nyquist plot.

Vector margin

Figure 6.41 illustrates the idea graphically. Because the vector margin is a single margin parameter, it removes all the ambiguities in assessing stability that come with using GM and PM in combination. In the past it has not been used extensively due to difficulties in computing it. However, with the widespread availability of computer aids, the idea of using the vector margin to describe the degree of stability is much more feasible.

6.5   Bode's Gain-phase Relationship

One of Bode's important contributions is the theorem that states:

For any stable minimum-phase system (that is, one with no RHP zeros *or poles*), the phase of $G(j\omega)$ is uniquely related to the magnitude of $G(j\omega)$.

When the slope of $|G(j\omega)|$ versus ω on a log-log scale persists at a constant value for approximately a decade of frequency, the relationship is particularly simple:

Bode's gain-phase relationship

$$\angle G(j\omega) \cong n \times 90°, \tag{6.30}$$

where n is the slope of $|G(j\omega)|$ in units of decade of amplitude per decade of frequency. For example, in considering the magnitude curve alone in Fig. 6.42, we see that Eq. (6.30) can be applied to the two frequencies $\omega_1 = 0.1$, and $\omega_2 = 10$ which are a decade removed from the change in slope, to yield the approximate values of phase, $-180°$ and $-90°$. The exact phase curve shown in

FIGURE 6.42

An approximate gain-phase relationship demonstration

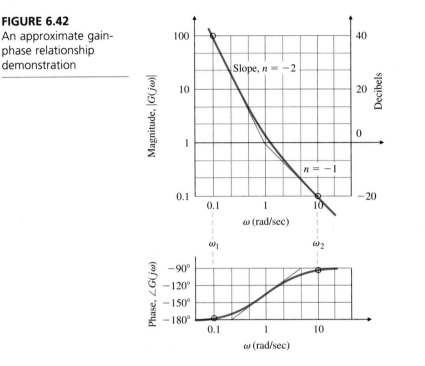

the figure verifies that indeed the approximation is quite good. It also shows that the approximation will degrade if the evaluation is performed at frequencies closer to the change in slope.

An exact statement of the Bode gain-phase theorem is

$$\angle\, G(j\omega_0) = \frac{1}{\pi} \int_{-\infty}^{+\infty} \left(\frac{dM}{du} \right) W(u)\, du \qquad \text{(in radians)}, \qquad (6.31)$$

where

$$M = \text{log magnitude} = \ln|G(j\omega)|,$$

$$u = \text{normalized frequency} = \ln(\omega/\omega_0),$$

$$dM/du \cong \text{slope } n, \text{ as defined in Eq. (6.30)},$$

$$W(u) = \text{weighting function} = \ln(\coth|u|/2).$$

Figure 6.43 is a plot of the weighting function $W(u)$ and shows how the phase is most dependent on the slope at ω_0; it is also dependent, though to a lesser degree, on slopes at neighboring frequencies. The figure also suggests that the weighting could be approximated by an impulse function centered at ω_0. We may approximate the weighting function as

$$W(u) \cong \frac{\pi^2}{2}\, \delta(u),$$

FIGURE 6.43
Weighting function in Bode's gain-phase theorem (*from Clark, 1962*)

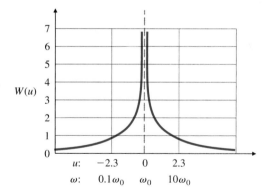

which is precisely the approximation made to arrive at Eq. (6.30) using the "sifting" property of the impulse function (and conversion from radians to degrees).

In practice, Eq. (6.31) is never used, but Eq. (6.30) *is* used—as a guide to infer stability from $|G(j\omega)|$ alone. When $|KG(j\omega)| = 1$,

$$\angle G(j\omega) \cong -90° \qquad \text{if} \qquad n = -1,$$

$$\angle G(j\omega) \cong -180° \qquad \text{if} \qquad n = -2.$$

For stability we want $\angle G(j\omega) > -180°$ for PM > 0. Therefore, we adjust the $|KG(j\omega)|$ curve so that it has a slope of -1 at the crossover frequency yielding $|KG(j\omega)| = 1$. If the slope is -1 for a decade above and below the crossover frequency, then PM $\cong 90°$; however, to ensure a reasonable PM, it is usually only necessary to insist that a -1 slope ($-20\,\text{dB}$ per decade) persist for a decade in frequency that is centered at the crossover frequency.

◆ **EXAMPLE 6.13**   *Bode's Gain-phase Relationship in Spacecraft Attitude Control*

❖   For the spacecraft-attitude-control problem defined in Fig. 6.44, find a suitable expression for $KD(s)$ that will provide good damping and a bandwidth of approximately 0.2 rad/sec.

FIGURE 6.44
Spacecraft-attitude control system

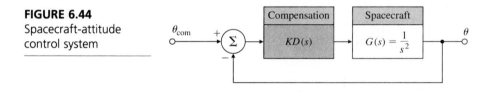

FIGURE 6.45

Magnitude of the
spacecraft's frequency
response

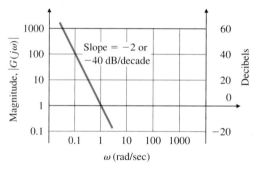

Solution. The magnitude of the frequency response of the spacecraft (Fig. 6.45) clearly requires some reshaping since it has a slope of -2 or -40 dB per decade everywhere. The simplest compensation to do the job consists of using proportional and derivative terms (a PD compensator), which produces the relation

$$KD(s) = K(Ts + 1). \tag{6.32}$$

We will adjust the gain K to produce the desired bandwidth, and adjust break point $\omega_1 = 1/T$ to provide the -1 or -20 dB per decade slope at the crossover frequency. The actual design process to achieve the desired specifications is now very simple: We pick a value of K to provide a crossover at 0.2 rad/sec and choose a value of ω_1 that is about 4 times lower than the crossover frequency so the slope will be -1 in the vicinity of the crossover. Figure 6.46 shows the steps we take to arrive at the final compensation:

FIGURE 6.46

Compensated open-loop
transfer function

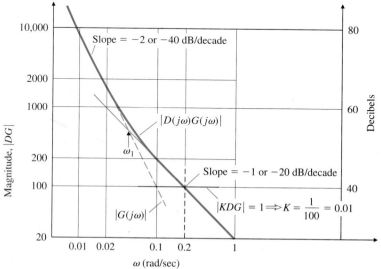

FIGURE 6.47
Closed-loop frequency response

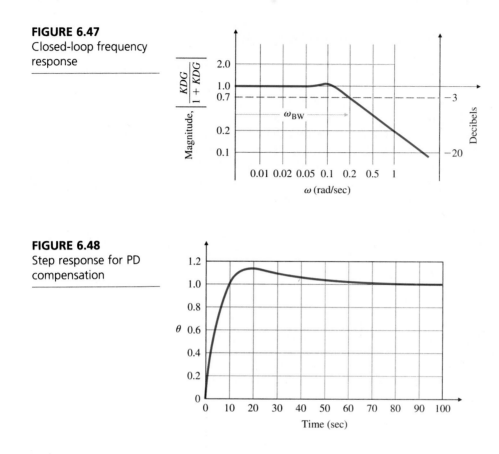

FIGURE 6.48
Step response for PD compensation

STEP 1.   Plot $|G(j\omega)|$.

STEP 2.   Modify the plot to include $|D(j\omega)|$, with $\omega_1 = 0.05\,\text{rad/sec}$ ($T = 20$).

STEP 3.   Determine that $|DG| = 100$ where the $|DG|$ curve crosses the line $\omega = 0.2\,\text{rad/sec}$.

STEP 4.   Compute

$$K = \frac{1}{[|DG|]_{\omega = 0.2}} = \frac{1}{100} = 0.01.$$

Therefore, $KD(s) = 0.01(20s + 1)$ will meet the specifications, so the design is complete.

If we were to draw the phase curve of KDG above, we would find that PM $= 75°$, which is certainly quite adequate. A plot of the closed-loop frequency-response magnitude (Fig. 6.47) shows that, indeed, the crossover frequency and the bandwidth are almost identical in this case. The step response of the closed-loop system is shown in Fig. 6.48 and its 14% overshoot confirms the adequate damping.

6.6   Closed-loop Frequency Response

The closed-loop bandwidth was defined in Section 6.1 and in Fig. 6.4. Figure 6.2 showed that the natural frequency is always within a factor of two of the bandwidth for a second-order system. In Example 6.13, we designed the compensation so that the crossover frequency was at the desired bandwidth and verified by computation that the bandwidth was identical to the crossover frequency. Generally, the match between the crossover frequency and the bandwidth is not as good as in Example 6.13. We can help establish a more exact correspondence by making a few observations. Consider the simplified system shown in Fig. 6.40 in which $|KG(j\omega)|$ shows the typical behavior

$$|KG(j\omega)| \gg 1 \qquad \text{for} \quad \omega \ll \omega_c,$$

$$|KG(j\omega)| \ll 1 \qquad \text{for} \quad \omega \gg \omega_c,$$

where ω_c is the crossover frequency. The closed-loop frequency-response magnitude is approximated by

$$|T| = \left| \frac{KG(j\omega)}{1 + KG(j\omega)} \right| \cong \begin{cases} 1, & \omega \ll \omega_c, \\ |KG|, & \omega \gg \omega_c. \end{cases} \tag{6.33}$$

In the vicinity of crossover where $|KG(j\omega)| = 1$, $|T|$ depends heavily on the PM. A PM of $90°$ means that $\angle G(j\omega_c) = -90°$, and therefore $|T(j\omega_c)| = 0.707$. On the other hand, PM $= 45°$ yields $|T(j\omega_c)| = 1.31$.

The approximations in Eq. (6.33) were used to generate the curves of $|T(j\omega)|$ in Fig. 6.49. It shows that the bandwidth for smaller values of PM is typically somewhat greater than ω_c, though usually it is less than $2\omega_c$; thus

$$\omega_c \leqslant \omega_{BW} \leqslant 2\omega_c.$$

FIGURE 6.49
Closed-loop bandwidth with respect to PM

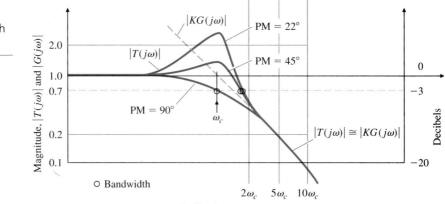

Another specification related to the frequency response is the resonant-peak magnitude M_r, defined in Fig. 6.4. Figure 6.2 shows that M_r is generally related to the damping of a system. In practice, M_r is rarely used; most designers prefer to use the gain and/or phase margins to specify the damping of a system.

6.7   Compensation

As we discussed in Chapters 4 and 5, dynamic elements (or compensation) are typically added to feedback control systems to improve their stability and error characteristics.

Section 4.2 discussed the basic types of feedback: proportional, derivative, and integral. Section 5.6 discussed the two kinds of dynamic compensation: the lead network, which approximates proportional-derivative (PD) feedback, and the lag network, which approximates proportional-integral (PI) control. In this section we discuss these and other kinds of compensation in terms of their frequency-response characteristics. In many cases the compensation will be implemented in a microprocessor. Techniques for converting the continuous compensation $D(s)$ into a form that can be coded in the computer will be discussed in Chapter 8.

The frequency response stability analysis to this point has considered the closed-loop system to have the characteristic equation $1 + KG(s) = 0$. With the introduction of compensation, the closed-loop characteristic equation becomes $1 + KD(s)G(s) = 0$ and all the previous discussion pertaining to the frequency response of $KG(s)$ applies directly to the compensated case if we analyze the frequency response of $KD(s)G(s)$.

6.7.1   PD Compensation

We will start the discussion of compensation design by using the frequency response with PD control. The compensator transfer function, given by

PD compensation

$$D(s) = K(T_D s + 1), \tag{6.34}$$

was shown in Fig. 5.26 to have a stabilizing effect on the root locus of a second-order system. The frequency-response characteristics of Eq. (6.34) are shown in Fig. 6.50. A stabilizing influence is apparent in the increase in phase at frequencies above the break point $1/T_D$. We use this compensation by locating $1/T_D$ so that the increased phase occurs in the vicinity of crossover (that is, where $|D(s)G(s)| = 1$), thus increasing the phase margin.

Note that the magnitude of the compensation continues to grow with increasing frequency. This feature is undesirable since it amplifies the high-frequency noise that is typically present in any real system and, as a continuous transfer function, cannot be exactly realized with physical elements. It is also the reason we stated in Section 5.6 that pure derivative compensation gives trouble.

FIGURE 6.50
Frequency response of PD control

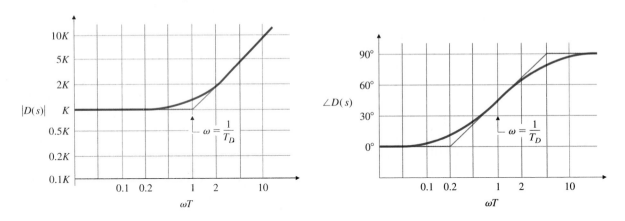

6.7.2 Lead Compensation

In order to alleviate the high-frequency amplification of the PD compensation, a first-order pole is added in the denominator at frequencies higher than the breakpoint of the PD compensator. Thus the phase increase (or lead) still occurs, but the amplification at high frequencies is limited. The resulting **lead compensation** has a transfer function of

Lead compensation

$$D(s) = K \frac{Ts + 1}{\alpha Ts + 1},$$

(6.35)

where $\alpha < 1$. Figure 6.51 shows the frequency response of this lead compensation. Note that a significant amount of phase lead is still provided, but with much less amplification at high frequencies. A lead compensator is generally used whenever a substantial improvement in damping of the system is required.

The phase contributed by the lead compensation in Eq. (6.35) is given by

$$\phi = \tan^{-1}(T\omega) - \tan^{-1}(\alpha T\omega).$$

It can be shown (see Problem 6.42) that the frequency where the phase is maximum is given by

$$\omega_{max} = \frac{1}{T\sqrt{\alpha}}.$$

(6.36)

The maximum phase contribution, that is, the peak of the $\angle D(s)$ curve in Fig.

6.51, corresponds to

$$\sin \phi_{max} = \frac{1 - \alpha}{1 + \alpha} \tag{6.37}$$

or

$$\alpha = \frac{1 - \sin \phi_{max}}{1 + \sin \phi_{max}}.$$

Another way to look at this is the following: The maximum frequency occurs midway between the two break-point frequencies (sometimes called corner frequencies) on a logarithmic scale.

$$\log \omega_{max} = \log \frac{1/\sqrt{T}}{\sqrt{\alpha T}}$$

$$= \log \frac{1}{\sqrt{T}} + \log \frac{1}{\sqrt{\alpha T}}$$

$$= \frac{1}{2} \left[\log \left(\frac{1}{T} \right) + \log \left(\frac{1}{\alpha T} \right) \right], \tag{6.38}$$

as shown in Fig. 6.51. Alternatively, we may state these results in terms of the pole-zero locations. Rewriting $D(s)$ as

$$D(s) = K \frac{(s + z)}{(s + p)}, \tag{6.39}$$

FIGURE 6.51
Lead-compensation frequency response with $1/\alpha = 10$

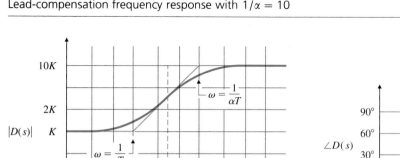
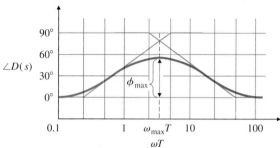

we get

$$\omega_{max} = \sqrt{|z||p|} \qquad (6.40)$$

and

$$\log \omega_{max} = \tfrac{1}{2}(\log|z| + \log|p|). \qquad (6.41)$$

These results agree with the previous ones if we let $z = -1/T$ and $p = -1/\alpha T$ in Eqs. (6.36) and (6.38).

For example, a lead compensator with a zero at $s = -2$ ($T = 0.5$) and a pole at $s = -10$ ($\alpha T = 0.1$), thus $\alpha = \tfrac{1}{5}$ would yield the maximum phase lead at

$$\omega_{max} = \sqrt{2 \cdot 10} = 4.47 \, \text{rad/sec}.$$

The amount of phase lead at the midpoint depends only on α in Eq. (6.37) and is plotted in Fig. 6.52. For $\alpha = \tfrac{1}{5}$, Fig. 6.52 shows that $\phi_{max} = 40°$. Note from the figure that we could increase the phase lead up to $90°$ using higher values of the **lead ratio** $1/\alpha$; however, Fig. 6.51 shows that increasing values of $1/\alpha$ also produces higher amplifications at higher frequencies. Thus our task is to select a value of $1/\alpha$ that is a good compromise between an acceptable phase margin and an acceptable noise sensitivity at high frequencies. Usually a single lead compensation can contribute a maximum of $60°$ to the phase. If a greater phase lead is needed, then a double lead compensation would be required, where

Lead ratio $= \dfrac{1}{\alpha}$

$$D(s) = K \left(\frac{Ts + 1}{\alpha Ts + 1} \right)^2.$$

Even if a system had negligible amounts of noise present and the pure derivative compensation of Eq. (6.34) were acceptable, a continuous compensation would look more like Eq. (6.35) than Eq. (6.34) because of the impossibility of building a pure differentiator. No physical system—mechanical

FIGURE 6.52
Maximum phase increase for lead compensation

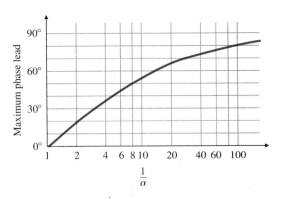

or electrical—responds with infinite amplitude at infinite frequencies, so there will be a limit in the frequency range (or bandwidth) for which derivative information (or phase lead) can be provided.

◆ **EXAMPLE 6.14**   *Lead Compensation via Frequency Response*

❖   As an example of designing a lead compensator, let us repeat the design of a compensation for

$$G(s) = \frac{1}{s(s + 1)}$$

that was carried out in Section 5.6. This represents the model of a DC motor (as discussed in Chapters 2 and 4) as well as a satellite antenna (see Fig. 3.50). This time we wish to obtain a steady-state error of less than 0.1 for a unit ramp input. Furthermore, we desire an overshoot $M_p < 25\%$.

Solution.   The steady-state error is given by

$$e_{ss} = \lim_{s \to 0} s \left[\frac{1}{1 + D(s)G(s)} \right] R(s), \tag{6.42}$$

where $R(s) = 1/s^2$ for a unit ramp, so Eq. (6.42) reduces to

$$e_{ss} = \lim_{s \to 0} \left\{ \frac{1}{s + D(s)[1/(s + 1)]} \right\} = \frac{1}{D(0)}.$$

Therefore, we find that $D(0)$, the steady-state gain of the compensation, cannot be less than 10 if it is to meet the error criterion. We pick $K = 10$ in Eq. (6.35). To relate the overshoot requirement to phase margin, Fig. 6.36 shows that a PM of 45° should suffice. The frequency response of $KG(s)$ in Fig. 6.53 shows that PM = 20° results if no phase lead is added by compensation. If it were possible to simply add phase without affecting the magnitude, we would need an additional phase of only 25° at the $KG(s)$ crossover frequency of $\omega = 3$ rad/sec. However, maintaining the same low-frequency gain and adding a compensator zero would increase the crossover frequency; hence more than a 25° phase contribution will be required from the lead network. To be safe we will design the lead compensator so that it supplies a maximum phase lead of 40°. Fig. 6.52 shows that $1/\alpha = 5$ will accomplish that goal. We will derive the greatest benefit from the compensation if the maximum phase lead from the compensator occurs at the crossover frequency corresponding to a magnitude of 1. With some trial and error, we determine that placing the zero at $\omega = 2$ rad/sec and the pole at $\omega = 10$ rad/sec, causes the maximum phase lead to be at the crossover frequency. The compensation, therefore, is

$$D(s) = 10 \, \frac{s/2 + 1}{s/10 + 1}.$$

The frequency-response characteristics of $D(s)G(s)$ in Fig. 6.53 can be seen to yield a phase margin of 45°, which satisfies the design goals.

FIGURE 6.53
Frequency response for
lead-compensation
design

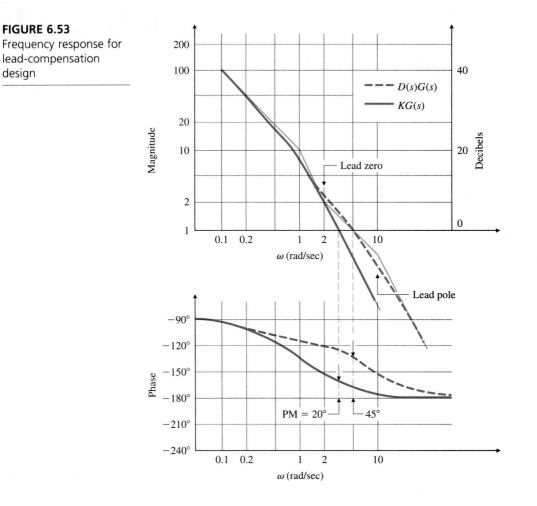

FIGURE 6.54
Root locus for lead
compensation design

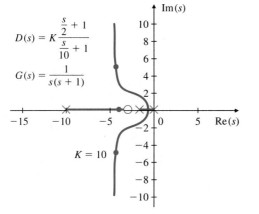

$$D(s) = K \frac{\frac{s}{2} + 1}{\frac{s}{10} + 1}$$

$$G(s) = \frac{1}{s(s+1)}$$

FIGURE 6.55
Transient response for a lead-compensation design: (a) step response; (b) ramp response

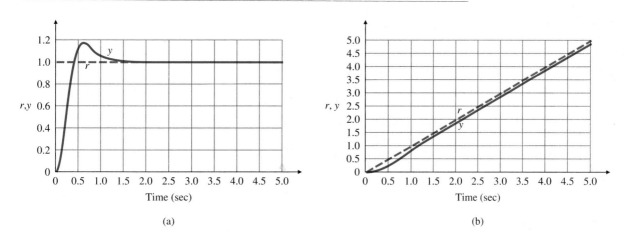

(a)

(b)

The root locus for this design, originally given as Fig. 5.28, is repeated here as Fig. 6.54 with the root locations marked for $K = 10$. The locus verifies that this choice of compensation parameters yields the desired damping ratio. The closed-loop poles of the system are located at $-4.2 \pm 4.5j$ and -2.6. The step and ramp responses of the closed-loop system are shown in Fig. 6.55. We have met the specifications, since the overshoot is 18.4% and the steady-state error to a ramp input is less than 0.1.

The design procedure used in Example 6.14 can be summarized as follows:

1. Determine the low-frequency gain so that the steady-state errors are within specification.
2. Select the combination of lead ratio $1/\alpha$ and zero values that achieves an acceptable phase margin at crossover.

This design procedure will apply to many cases; however, keep in mind that the specific procedure followed in any particular design may need to be tailored to its particular set of specifications.

In Example 6.14 there were two specifications: peak overshoot and steady-state error. We transformed the overshoot specification into a PM, but the steady-state error specification we used directly. No speed-of-response type of specification was given; however, it would have impacted the design in the same way that the steady-state error specification did. The speed of response or bandwidth of a system is directly related to the crossover frequency, as we pointed out earlier in Section 6.6. Fig. 6.53 shows that the crossover frequency was 4 rad/sec. We could have increased it by raising the gain K and increasing the frequency of the lead compensator pole and zero in order to keep the slope of -1 at the crossover frequency. Raising the gain would also have decreased the steady-state error to less than the specified limit. The gain margin was never introduced into the problem because the stability was adequately specified by the phase margin alone. Furthermore, the gain margin would have not been useful for this system because the phase never crossed the $180°$ line and the GM was always infinite.

In lead-network designs there are three primary design parameters:

Design parameters for lead networks

1. The crossover frequency ω_c, which determines bandwidth ω_{BW}, rise time t_r, and settling time t_s;
2. The phase margin (PM), which determines the damping coefficient ζ and the overshoot M_p;
3. The low-frequency gain, which determines the steady-state error characteristics.

The design problem reduces to assuming a fixed value of one of these three design parameters and then adjusting the other two repeatedly to meet specifications. One approach is to set the low-frequency gain to meet the specifications and add a lead compensator to increase the bandwidth by increasing the crossover frequency (shifting it to the right) to increase the PM. The alternative is to allow the gain to decrease and lower the crossover frequency until the PM specification is met.

In lead-compensator design it is possible to summarize a step-by-step procedure that applies to a sizable class of problems for which a single lead is sufficient. The following procedure assumes that it is desirable to realize the phase margin by moving the crossover frequency to the right. As with all such itemized design procedures, it provides only a starting point; the designer will typically find it necessary to go through several design iterations in order to meet all the specifications.

Design Procedure for Lead Compensation

CASE 1: Design for low-frequency gain (e_{ss} error specification).

1. Determine open-loop gain K to satisfy requirements on error constants (K_p, K_v, and so forth).

CASE 2: Design for closed-loop bandwidth.

1. Determine open-loop crossover frequency to be a factor of two below the desired closed-loop bandwidth.

2. Evaluate the phase margin (PM) of the uncompensated system using the value of K obtained from Step 1.
3. Allow for a small amount of extra margin (5° to 12°), and determine the needed phase lead ϕ_{max}.
4. Determine α from Eq. (6.37) or Fig. 6.52.
5. Set the new gain crossover frequency at ω_{max}, and by trial and error determine the corner frequencies of the lead compensator: $1/T = \omega_{max}\sqrt{\alpha}$ and $1/\alpha T = \omega_{max}/\sqrt{\alpha}$.
6. Draw the compensated frequency response, check the PM, and iterate on the compensator design if necessary.
7. Iterate on the design. Adjust compensator parameters (poles, zeros, and gain) until all specifications are met. Add an additional lead compensator if necessary.

While these guidelines will not apply to all the systems you will encounter in practice, they do indicate that the trial-and-error process to search for a satisfactory compensator can be pursued in a systematic fashion.

◆ **EXAMPLE 6.15** *Lead Compensator for a Type 0 System*

Consider the third-order system

$$KG(s) = \frac{K}{(s/0.5 + 1)(s + 1)(s/2 + 1)}.$$

Design a lead compensator such that $K_p = 9$ and the phase margin is at least 25°.

Solution. Let us follow the design procedure earlier:

STEP 1. Given the specification for K_p, we solve for K:

$$K_p = \lim_{s \to 0} KG(s) = K = 9.$$

STEP 2. The Bode plot of the uncompensated system with $K = 9$ is shown in Fig. 6.56. The phase margin of the uncompensated system is approximately 7°.

STEP 3. Allowing for 5° of extra margin, we want the lead compensator to contribute $25° + 5° - 7° = 23°$.

STEP 4. From Eq. (6.37) or Fig. 6.52 we have

$$\alpha = \frac{1 - \sin 23°}{1 + \sin 23°} = 0.43.$$

The gain is $\sqrt{\alpha} = 0.66$ or $-3.6\,\text{dB}$ (from Fig. 6.52), so that the new gain crossover frequency is at $\omega_c = 2.1$ rad/sec.

FIGURE 6.56
Bode plot for the lead-compensation design in Example 6.15

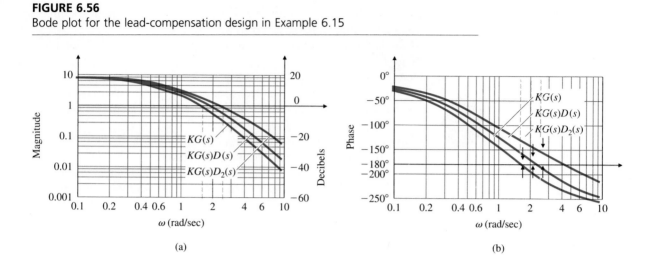

(a)   (b)

STEP 5.   The pole and zero of the lead network are at $1/T = \omega_c\sqrt{\alpha} = 1.3$ rad/sec and $1/\alpha T = 3$ rad/sec. The lead compensator is

$$D(s) = 9\,\frac{1}{0.44}\left(\frac{s+1.3}{s+3}\right) = 9\,\frac{s/1.3+1}{s/3+1}.$$

STEP 6.   The Bode plot of the compensated system, (Fig. 6.56, middle curve) has a PM of 18°. This is not surprising because of the high-frequency slope of -3. The closed-loop poles are at -4.7, $-0.25 \pm 2.2j$, and -1.31. There is a closed-loop zero at -1.3. The step response of the system (Fig. 6.57) shows an oscillatory response, as we might expect from the low damping value and the low PM of 18°.

STEP 7.   We should repeat the design starting with a higher extra phase margin in order to meet the specification. An increased PM can be accomplished if α is reduced. We

FIGURE 6.57
Step response for lead-compensation design

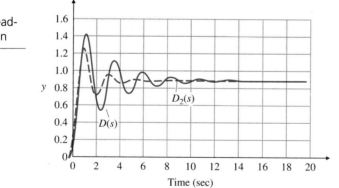

choose $\alpha = \frac{1}{10}$ with the zero at $s = -1$, so

$$D_2(s) = \frac{1}{0.1}\left(\frac{s+1}{s+10}\right) = \frac{s+1}{s/10+1},$$

and PM = 34°. Figure 6.56 (upper curve) shows the frequency response of the revised design. Figure 6.57 shows a substantial reduction in the oscillations in the revised design, which is reasonable because of the higher damping and PM values.

◆ **EXAMPLE 6.16**   *Lead-compensator Design for a Type I System*

Consider the third-order system

$$G(s) = \frac{10}{s(s/2.5 + 1)(s/6 + 1)}$$

Design a lead compensator so that the PM = 45° and $K_v = 10$.

Solution.   Again we follow the design procedure given earlier:

STEP 1. K as given will yield $K_v = 10$ if $D(0) = 1$.

STEP 2.   The Bode plot of the system is shown in Fig. 6.58. The phase margin of the uncompensated system (lower curve) is approximately −4°.

STEP 3.   Allowing for 5° of extra phase margin, we need PM = 45° + 5° − (−4°) = 54° to be contributed by the lead compensator.

FIGURE 6.58
Bode plot for the lead-compensation design in Example 6.16

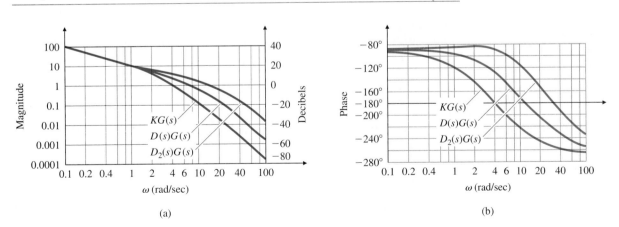

(a)   (b)

STEP 4.   From Eq. (6.37) we get

$$\alpha = \frac{1 - \sin 54°}{1 + \sin 54°} = 0.1,$$

and so gain is $\sqrt{\alpha} = 9.76\,\text{dB}$. The new gain crossover frequency is at $\omega_c = 6.9\,\text{rad/sec}$.
STEP 5.   The pole and zero of the lead network are at $1/\alpha T = \omega_c/\sqrt{\alpha} = 20$ and $1/T = 2\,\text{rad/sec}$, respectively. The compensator is

$$D(s) = \frac{1}{0.1}\frac{s + 2}{s + 20} = \frac{s/2 + 1}{s/20 + 1}.$$

STEP 6.   The Bode plot of the compensated system (Fig. 6.58, dashed curve) shows a PM of 22°. So a single lead compensator cannot meet the specification because of the high-frequency slope of -3.
STEP 7.   We need a double lead compensator in this system. If we try a compensator of the form

$$D_2(s) = \frac{1}{(0.1)^2}\frac{(s + 2)(s + 4)}{(s + 20)(s + 40)} = \frac{(s/2 + 1)(s/4 + 1)}{(s/20 + 1)(s/40 + 1)},$$

we obtain PM = 45°. The Bode plot for this case is shown as the upper curve in Fig. 6.58.

◆

Both Examples 6.15 and 6.16 are third order. Example 6.16 was more difficult to compensate because the error requirement, K_v, forced ω_c to be so high that a single lead could not provide enough PM.

6.7.3   PI Compensation

In many problems it is important to keep the bandwidth low and also to reduce the steady-state error. For this purpose a proportional-integral (PI) or lag compensator is useful. By letting $T_D = 0$ in Eq. (4.31), we see that PI control has the transfer function

PI compensation

$$D(s) = \frac{K}{s}\left(s + \frac{1}{T_I}\right), \tag{6.43}$$

which results in the frequency-response characteristics shown in Fig. 6.59. The desirable aspect of this compensation is the infinite gain at zero frequency, which reduces the steady-state error to a polynomial time input. This is accomplished, however, at the cost of a phase decrease below the break point at $\omega = 1/T_I$. Therefore, $1/T_I$ is usually located at a frequency substantially less than the crossover frequency so that the system's phase margin is not affected very much.

FIGURE 6.59
Frequency response of PI control

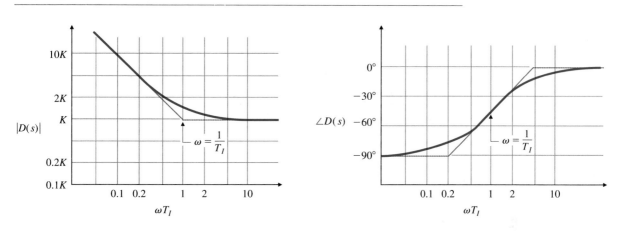

6.7.4 Lag Compensation

As we discussed in Section 5.6, **lag compensation** approximates PI control. Its transfer function was given by Eq. (5.45) for root-locus design, but for frequency-response design, it is more convenient to write the transfer function of the lag compensation *alone* in the form

Lag compensation

$$D(s) = \alpha \frac{Ts + 1}{\alpha Ts + 1}, \tag{6.44}$$

where $\alpha > 1$. The complete compensation will almost always include an overall gain K and perhaps other dynamics in addition to the lag compensation. Although Eq. (6.44) looks identical to the lead compensation in Eq. (6.35), the fact that $\alpha > 1$ causes the pole to have a lower break-point frequency than the zero. This relationship produces the low-frequency increase in amplitude and phase decrease (lag) apparent in the frequency-response plot in Fig. 6.60 and gives the compensation the essential feature of integral control: a high low-frequency gain. The primary objective of lag-network design is to provide additional gain of $+20 \log \alpha$ in the low-frequency range and to leave the system sufficient phase margin (PM). Of course, phase lag is not the useful effect, and the pole and zero of the lag compensator are selected to be much smaller than the uncompensated system crossover frequency in order to keep the effect on the PM to a minimum. Thus, the lag compensator increases the open-loop DC gain thereby improving the steady-state response characteristics, without changing the transient response characteristics significantly. If the pole and zero are relatively close together and near the origin (that is, if

FIGURE 6.60
Frequency response of lag compensation with $\alpha = 10$

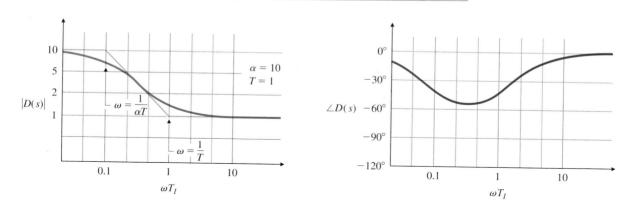

the value of T is large), we can increase K_v by a factor α without moving the closed-loop poles appreciably. Hence the transient response remains the same while the steady-state response is improved. The lag compensator provides a gain of α (or $+20 \log \alpha$ dB) in the low-frequency range, as seen in Fig. 6.60.

◆ **EXAMPLE 6.17**   *Lag Compensation via Frequency Response*

Repeat Example 6.14, this time using lag compensation.

Solution.   The frequency response of the system $KG(s)$, with the required steady-state gain of $K = 10$, is shown in Fig. 6.61. The designer's task is to select the lag compensation break points so that the crossover frequency is lowered and more favorable PM results. To prevent detrimental effects from the compensation phase lag, the pole and zero position values of the compensation need to be substantially lower than the new crossover frequency. One possible choice is shown in Fig. 6.61: The lag zero is at 0.1 rad/sec, and the lag pole is at 0.01 rad/sec. This selection of parameters, produces a PM of 50°, thus satisfying the specifications. Here the stabilization is achieved by keeping the crossover frequency to a region where $G(s)$ has favorable phase characteristics. The criterion for selecting the pole and zero locations $1/T$ is to make them low enough to minimize the effects of the phase lag on the compensation at the crossover frequency. Generally, however, the pole and zero are located no lower than necessary because the additional system root (compare with the root locus of a similar system design in Fig. 5.32b) introduced by the lag will be in the same frequency range as the compensation zero and will have some effect on the output response, especially the response to disturbance inputs.

The response of the system to a step reference input is shown in Fig. 6.62. The

FIGURE 6.61
Frequency response of
lag-compensation design
in Example 6.17

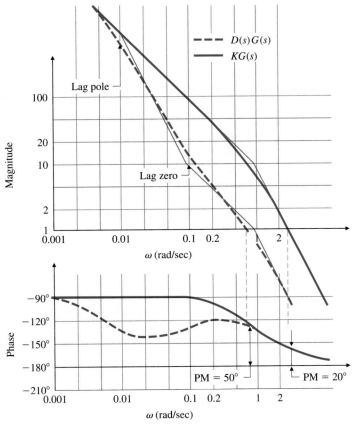

FIGURE 6.62
Step response of lag-
compensation design in
Example 6.17

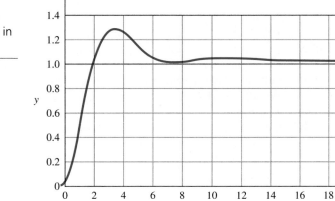

closed-loop poles are at $-0.45 \pm 0.84j$ and -0.11, and there is a closed-loop zero at -0.1. Note the near cancellation of a pole-zero pair near $s = -0.1$ in the closed-loop transfer function and the lack of any noticeable time response with a time constant of 10 sec.

◆

We can interpret the beneficial effects of lag compensation in two ways: It reduces high-frequency gain for a better phase margin, or it increases low-frequency gain for better error characteristics. The design shown in Fig. 6.61 illustrates the first interpretation. However, if we were given the same system with a proportional feedback gain $K = 1$, and told to meet the error specifications while preserving stability and bandwidth, we could accomplish the task by introducing the lag compensation shown in Fig. 6.61. This would simply increase the low-frequency gain by a factor of 10 with essentially no change in gain or phase at crossover. This situation illustrates the second interpretation.

Examples 6.14 and 6.17 meet an identical set of specifications for the same plant in very different ways. In the first case the specifications are met with a lead network, and a crossover frequency $\omega_c = 5$ rad/sec ($\omega_{BW} \cong 6$ rad/sec) results. In the second case the same specifications are met with a lag network, and $\omega_c \cong 0.8$ rad/sec ($\omega_{BW} \cong 1$ rad/sec) results. In general, had there been specifications for rise time or bandwidth, they would have influenced the choice of compensation (lead or lag); however, in this particularly simple example, any bandwidth or steady-state error specification could be met using either compensation simply by raising the low-frequency gain sufficiently high.

In more realistic systems dynamic elements usually represent the actuator and sensor as well as the system itself, so it is typically impossible to raise the crossover frequency much beyond the value representing the speed of response of the components being used. Although linear analysis seems to suggest that almost any system can be compensated, in fact, if we attempt to drive a set of components much faster than their natural frequencies, the system will saturate, and the linearity assumptions will be no longer valid, and the linear design will represent little more than wishful thinking. With this behavior in mind, we see that simply increasing the gain of a system with a lead compensator may not always be possible. In this case, use of the lag compensator to increase the low-frequency gain while holding the crossover frequency constant may be the only viable option.

We now summarize a step-by-step procedure for lag-compensator design.

Design Procedure for Lag Compensation

1. Determine the open-loop gain K that will meet the phase-margin requirement without compensations.
2. Draw the Bode plot of the uncompensated system with crossover frequency from 1, and evaluate the low-frequency gain.
3. Determine α to meet the low-frequency gain.
4. Choose the corner frequency $\omega = 1/T$ (the zero of the lag compensator) to be one octave to one decade below the new crossover frequency ω_c.
5. The other corner frequency (the pole location of the lag compensator) is then $\omega = 1/\alpha T$.
6. Iterate on the design. Adjust compensator parameters (poles, zeros, and gain) to meet all the specifications.

◆ **EXAMPLE 6.18** *Lag Compensator Design for a Type 0 System*

Again consider the third-order system of Example 6.15:

$$G(s) = \frac{K}{(\frac{1}{0.5}s + 1)(s + 1)(\frac{1}{2}s + 1)}.$$

Design a lag compensator so the phase margin is at least $25°$ and $K_p = 9$.

Solution. We follow the design procedure enumerated above.

STEP 1. If we set the gain K to be 9/2, the PM requirement would be satisfied. For PM $= 25°$, the new crossover frequency should be around $\omega_c = 1$ rad/sec. To be on the safe side, we choose $\omega_c = 1.2$ rad/sec, which corresponds to PM $= 30°$, as shown in Fig. 6.63.

STEP 2. The low-frequency gain is now 4.5.

STEP 3. The low-frequency gain should be raised by a factor of 2, which means $\alpha = 2$.

STEP 4. We choose the corner frequency for the zero to be at 0.2 rad/sec; that is, $1/T = 0.2$, or $T = 5$.

STEP 5. We then have the value for the other corner frequency: $\omega = 1/\alpha T = 1/(2 \cdot 5) = \frac{1}{10}$. The compensator is thus

$$D(s) = 2 \frac{5s + 1}{10s + 1}.$$

The compensated frequency response is also shown in Fig. 6.63. The system has closed-loop poles at -2.9, -0.19, and $-0.24 \pm 1.3j$. The step response of the system, shown in

FIGURE 6.63
Frequency response of lag-compensation design in Example 6.18

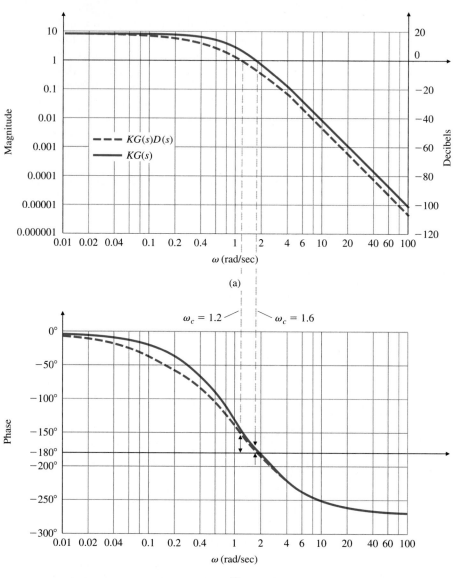

(a)

(b)

FIGURE 6.64
Step response of lag-compensation design in Example 6.18

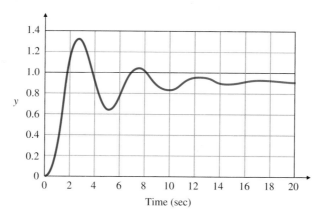

Fig. 6.64, illustrates the light damping that we would expect from PM = 28° and roots with a damping of 0.18.

STEP 6.   We could iterate on the design to increase the damping by adjusting the compensator parameters.

◆

6.7.5   PID Compensation

For problems that need phase-margin improvement at ω_c and low-frequency gain improvement, it is effective to use both lead and lag compensation. By combining the derivative and integral feedback, Eqs. (6.34) and (6.37), we obtain PID control. Its transfer function is

PID compensation

$$D(s) = \frac{K}{s}\left[(T_D s + 1)\left(s + \frac{1}{T_I}\right)\right], \tag{6.45}$$

and its frequency-response characteristics are shown in Fig. 6.65. This form is slightly different from that given by Eq. (4.31); however, the effect of the difference is inconsequential. This compensation is roughly equivalent to combining lead and lag compensators in the same design, and so is sometimes referred to as a **lead-lag compensator**. Hence, it can provide simultaneous improvement in transient and steady-state responses.

FIGURE 6.65
Frequency response of PID compensation with $T_I/T_D = 20$

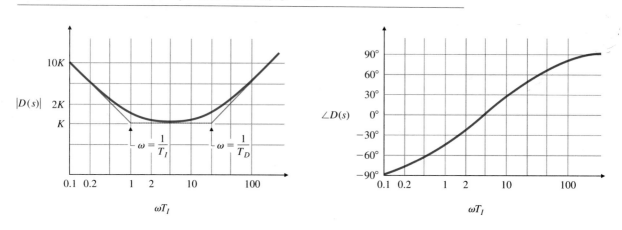

◆ **EXAMPLE 6.19**   *PID Compensation Design for Spacecraft Attitude Control*

Spacecraft
attitude control

❖ As an example of PID compensation design using frequency-response methods, consider the spacecraft-attitude-control problem. A simplified design was presented in Section 6.5; however, here we have a more realistic situation that includes a sensor lag and a disturbing torque. Figure 6.66 defines the system. Design a PID controller to have zero steady-state error to a constant-disturbance torque, a phase margin of 65°, and as high a bandwidth as is reasonably possible.

Solution.   First, let us take care of the steady-state error. For the spacecraft to be at a steady final value, the total input torque, $T_d + T_c$, must equal zero. Therefore, if $T_d \neq 0$, then $T_c = -T_d$. The only way this can be true with no error ($e = 0$) is for $D(s)$ to contain an integral term, hence the necessity for integral control in the compensation. This could also be verified mathematically by use of the Final Value Theorem (see Problem 6.45).

FIGURE 6.66
Block diagram for PID
design example

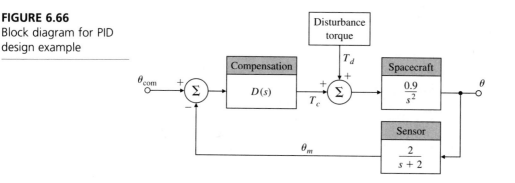

The frequency response of the spacecraft and sensor,

$$G(s) = \frac{0.9}{s^2}\left(\frac{2}{s+2}\right),\qquad(6.46)$$

is shown in Fig. 6.67. The slopes of -2 (that is, $-40\,$dB per decade) and $-3\,(-60\,$dB per decade) show that the system would be unstable for any value of K if no derivative feedback were used. Therefore, derivative control is also required to bring the slope to

FIGURE 6.67
Compensation for PID
design in Example 6.19

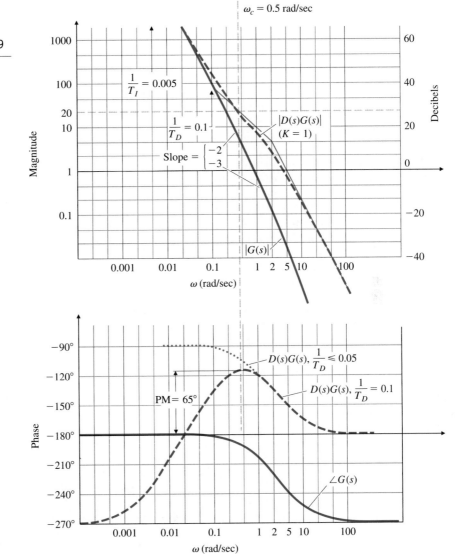

-1 ($-20\,$dB per decade) at the crossover frequency. The problem now is to pick values for the three parameters in Eq. (6.45)—K, T_D, and T_I—that will satisfy the specifications.

The easiest approach is to work first on the phase so that PM = 65° is achieved at a reasonably high frequency. This can be accomplished primarily by adjusting T_D, noting that T_I has a minor effect if sufficiently larger than T_D. Once the phase is adjusted, we establish the crossover frequency and then can easily determine the gain K.

We examine the phase of the PID controller in Fig. 6.65, to determine what would happen to the compensated spacecraft system, $D(s)G(s)$, as T_D is varied. If $1/T_D \geqslant 2\,$rad/sec, the phase lead from the PID control would simply cancel the sensor phase lag, and the composite phase would never exceed $-180°$, an unacceptable situation. If $1/T_D \leqslant 0.01$, the composite phase would approach $-90°$ for some range of frequencies and would exceed $-115°$ for an even wider range of frequencies; the latter threshold would provide a PM of 65°. In the compensated phase curve shown in Fig. 6.67, $1/T_D = 0.1$, which is the largest value of $1/T_D$ that could provide the required PM of 65°. The phase would never cross the $-115°$ (65° PM) line for any $1/T_D > 0.1$. For $1/T_D = 0.1$, the crossover frequency ω_c that produces the 65° PM is 0.5 rad/sec. For a value of $1/T_D \ll 0.05$, the phase essentially follows the dotted curve in Fig. 6.67, which indicates that the maximum possible ω_c is approximately 1 rad/sec and is provided by $1/T_D = 0.05$. Therefore, $0.05 < 1/T_D < 0.1$ is the only sensible range for $1/T_D$; anything less than 0.05 would provide no significant increase in bandwidth, while anything more than 0.1 could not meet the PM specification. Although the final choice is somewhat arbitrary, we have chosen $1/T_D = 0.1$ for our final design.

Our choice for $1/T_I$ is a factor of 20 lower than $1/T_D$; that is, $1/T_I = 0.005$. A factor less than 20 would negatively impact the phase at crossover, thus lowering the PM. Furthermore, it is generally desirable to keep the compensated magnitude as large as possible at frequencies below ω_c in order to have a faster transient response and smaller errors; maintaining $1/T_D$ and $1/T_I$ at the highest-possible frequencies will bring this about.

The only remaining task is to determine the proportional part of the PID controller, or K. Unlike the system in Example 6.18, where we selected K in order to meet a steady-state error specification, here we select a value of K that will yield a crossover frequency at the point corresponding to the required PM of 65°. The basic procedure for finding K, discussed in Section 6.6, consists of plotting the compensated system amplitude with $K = 1$, finding its value at crossover, then setting $1/K$ equal to that value. Figure 6.67 shows that when $K = 1$, $|D(s)G(s)| = 20$ at the desired crossover frequency $\omega_c = 0.5\,$rad/sec. Therefore,

$$\frac{1}{K} = 20, \quad \text{so} \quad K = \frac{1}{20} = 0.05.$$

The compensation equation that satisfies all of the specifications is now complete:

$$D(s) = \frac{0.05}{s}\,[(10s + 1)(s + 0.005)].$$

It is interesting to note that this system would become unstable if the gain were lowered so that $\omega_c \leqslant 0.02\,$rad/sec, the region in Fig. 6.67 where the phase of the compensated system is less than $-180°$. As mentioned in Section 6.4 this situation is

FIGURE 6.68
Transient response for PID example: (a) step response; (b) step-disturbance response

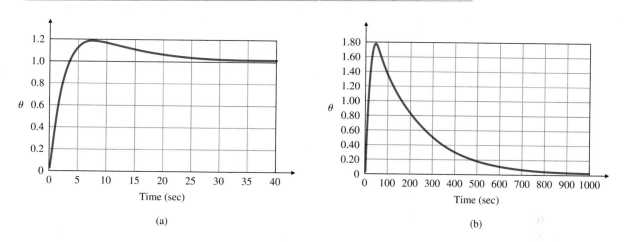

(a)

(b)

referred to as a conditionally stable system. A root locus with respect to K for this and any conditionally stable system would show the portion of the locus corresponding to very low gains in the RHP. The response of the system for a unit step θ_{com} is shown in Fig. 6.68(a). The closed-loop poles of the system are at -1.41, -0.442, -0.144, and -0.005, and the closed-loop system zeros are at -2, -0.1, and -0.005. As a result there is a pole-zero cancellation in the closed-loop transfer function at -0.005.

The response of the system for a step disturbance torque $T_d = 0.1$ N is shown in Fig. 6.68(b). The closed-loop poles of the system are the same as in Fig. 6.68(a), and the closed-loop system zeros from T_d to θ are at -2 and 0. As a result there is a blocking zero at the origin that provides rejection of the disturbance. Note that the slow disturbance-rejection response is due to the presence of the closed-loop pole at -0.005, which is *not* canceled in the transfer function from T_d to θ. If the slow disturbance response is not acceptable, this pole must be moved.

◆

Summary of Compensation Characteristics

1. *PD Control* adds phase lead at all frequencies above the break point. If there is no change in gain on the low-frequency asymptote, PD compensation will increase the crossover frequency and the speed of response. The increase in magnitude of the frequency response at the higher frequencies will increase the system's sensitivity to noise.

2. *Lead Compensation* adds phase lead at a frequency band between the two break points, which are usually selected to bracket the crossover frequency. If there is no change in gain on the low-frequency asymptote, lead compensation will increase both the crossover frequency and the speed of response over the uncompensated system. if K is selected so that the low-frequency magnitude is unchanged, then the steady-state errors of the system will increase.

3. *PI Control* increases the frequency-response magnitude at frequencies below the break point thereby decreasing steady-state errors. It also contributes phase lag below the break point, which must be kept at a low enough frequency to avoid degrading the stability excessively.

4. *Lag Compensation* increases the frequency-response magnitude at frequencies below the two break points, thereby decreasing steady-state errors. Alternatively, with suitable adjustments in K, lag compensation can be used to decrease the frequency-response magnitude at frequencies above the two break points so that ω_c yields an acceptable phase margin. Lag compensation also contributes phase lag between the two break points, which must be kept at frequencies low enough to keep the phase decrease from degrading the PM excessively.

6.8   Alternate Presentations of Data

Before computers were widely available, other ways to present frequency-response data were developed over the years to aid both in understanding design and in easing the designer's work load. The widespread availability of computers has reduced the need for some of these methods. Two techniques still in use are the Nichols chart and the inverse Nyquist plot, both of which we examine in this section.

6.8.1   Nichols Chart

A rectangular plot of $log\ |G(j\omega)|$ versus $\angle G(j\omega)$ can be drawn by simply transferring the information directly from the separate magnitude and phase portions in a Bode plot; one point on the new curve thus results from a given value of the frequency ω. This means that the new curve is parameterized as a function of frequency. As with the Bode plots, the magnitude information is plotted on a logarithmic scale, while the phase information is plotted on a linear scale. This template was suggested by N. Nichols and is usually referred to as a **Nichols chart**. The idea of plotting the magnitude of $G(j\omega)$ versus its phase is similar to the concept of plotting the real and imaginary parts of $G(j\omega)$, which formed the basis for the Nyquist plots shown in Sections 6.3 and 6.4. However, it is difficult to capture all the pertinent characteristics of $G(j\omega)$ on the linear scale of the Nyquist plot. The log scale for magnitude in the Nichols chart alleviates this difficulty, allowing this kind of presentation to be useful for design.

For any value of the complex transfer function $G(j\omega)$, Section 6.6 showed that there is a unique mapping to the unity-feedback closed-loop transfer function

$$T(j\omega) = \frac{G(j\omega)}{1 + G(j\omega)}, \tag{6.47}$$

or

$$T(j\omega) = M(\omega)e^{j\alpha(\omega)}, \tag{6.48}$$

where $M(\omega)$ is the magnitude of the closed-loop transfer function and $\alpha(\omega)$ is the phase of the closed-loop transfer function. Specifically,

$$M = \left| \frac{G}{1 + G} \right|, \tag{6.49}$$

$$\alpha = \tan^{-1}N = \left/ \frac{G}{1 + G} \right. . \tag{6.50}$$

It can be proven (see Problem 6.60) that the contours of constant closed-loop magnitude and phase are circles when $G(j\omega)$ is presented in the Nyquist plot. These circles are referred to as the **M and N circles**, respectively. An M circle was shown in Fig. 6.41.

M and N circles

The Nichols chart also contains contours of constant *closed-loop* magnitude and phase based on these relationships, as shown in Fig. 6.69. A designer can therefore graphically determine the bandwidth of a closed-loop system from the plot of the open-loop data on a Nichols chart by noting where the open-loop curve crosses the 0.70 contour of the closed-loop magnitude and determining the frequency of the corresponding data point. Likewise, a designer can determine the resonant peak amplitude M_r by noting the value of the magnitude of the highest closed-loop contour tangent to the curve. The frequency associated with the magnitude and phase at the point of tangency is sometimes referred to as the **resonant frequency** ω_r. Similarly, a designer can determine the gain margin (GM) by observing the value of the gain where the Nichols plot crosses the $-180°$ line, and the phase margin (PM) by observing the phase where the plot crosses the amplitude-1 line.

Resonant frequency

7. James, H. M., N. B. Nichols, and R. S. Phillips (1947).

FIGURE 6.69
Nichols chart

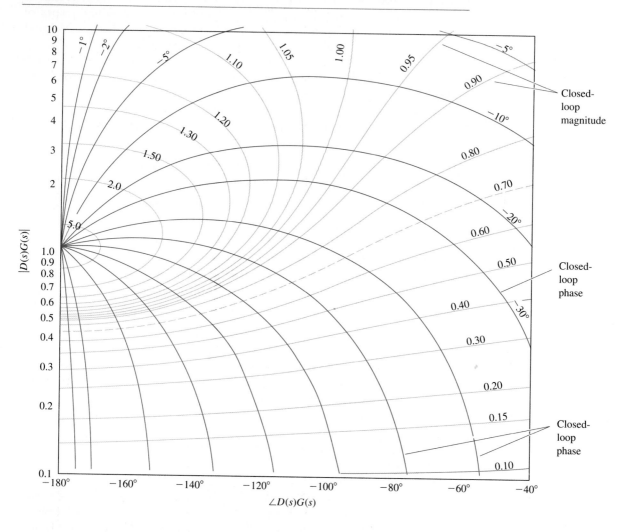

EXAMPLE 6.20 *Nichols Chart for PID Example*

Determine the bandwith and resonant peak magnitude of the compensated system whose frequency response is shown in Fig. 6.70.

Solution. The compensated design example seen in Fig. 6.67 is shown on a Nichols chart in Fig. 6.70. When comparing the two figures, it is important to divide the magnitudes in Fig. 6.67 by a factor of 20 in order to obtain $|D(s)G(s)|$ rather than the normalized values used in Fig. 6.67. Since the curve crosses the closed-loop magnitude 0.70 contour at $\omega = 0.8$ rad/sec, we see that the bandwidth of this system is 0.8 rad/sec. Since the largest-magnitude contour touched by the curve is 1.20, we also see that $M_r = 1.2$.

FIGURE 6.70
Example plot on the Nichols chart for determining bandwidth and M_r

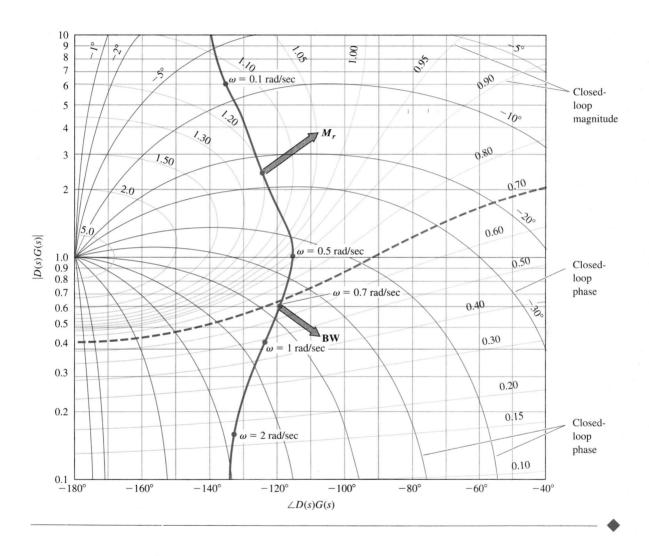

This presentation of data was particularly valuable when a designer had to generate plots and perform calculations by hand. A change in gain, for example, could be evaluated by sliding the curve vertically on transparent paper over a standard Nichols chart. The GM, PM, and bandwidth were then easy to read off the chart, thus allowing evaluations of several values of gain with a minimal amount of effort. With access to computer-aided methods, however, we can now calculate the bandwidth and perform many repetitive evaluations of the gain or any other parameter with a few key strokes. Today the Nichols chart is used primarily as an alternate way to present the information in a Nyquist plot. For complex systems where the −1 encirclements need to be evaluated, the

FIGURE 6.71
Nichols chart of the complex system in Examples 6.12 and 6.21

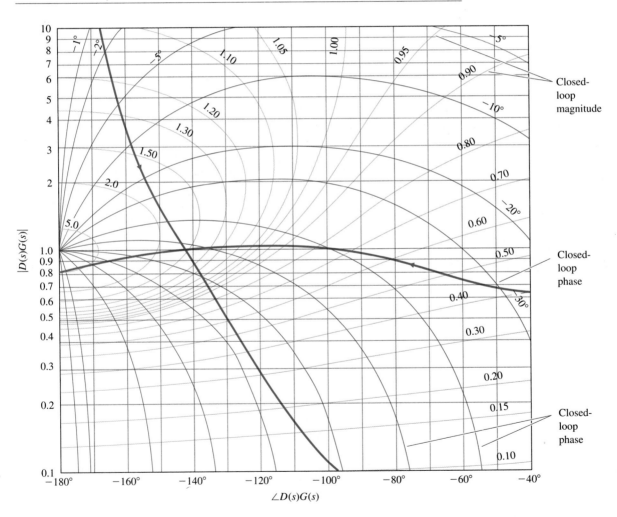

magnitude log scale of the Nichols chart enables us to examine a wider range of frequencies than a Nyquist plot does, as well as allowing us to read the gain and phase margins directly.

◆ **EXAMPLE 6.21**   *Stability Margins from Nichols Chart*

Determine the PM and GM for the system with the Nyquist plot shown in Fig. 6.39.

Solution.   Figure 6.71 shows a Nichols chart with the data from the same system shown in Fig. 6.39. Note that the PM for the most critical magnitude-1 crossover frequency is 36° and the GM is 1.25 (= 1/0.8). It is clear from this presentation of the data that the most critical portion of the curve is where it crosses the −180° line; hence the GM is the most relevant stability margin in this example.

◆

6.8.2   Inverse Nyquist

The **inverse Nyquist plot** is simply the reciprocal of the Nyquist plot described in Section 6.3 and used in Section 6.4 for the definition and discussion of stability margins. It is obtained most easily by computing the inverse of the magnitude from the Bode plot and plotting that quantity at an angle in the complex plane as indicated by the phase from the Bode plot. It can be used to find the PM and GM in the same way that the Nyquist plot was used. When $|G(j\omega)| = 1$, $|G^{-1}(j\omega)| = 1$ also, so the definition of PM is identical on the two plots. However, when the phase is −180° or +180°, the value of $|G^{-1}(j\omega)|$ is the GM directly; no calculation of an inverse is required, as was the case for the Nyquist plot.

The inverse Nyquist plot for the system in Fig. 6.23 (Example 6.9) is shown in Fig. 6.72 for the case where $K = 1$ and the system is stable. Note that

FIGURE 6.72
Inverse Nyquist plot for Example 6.9

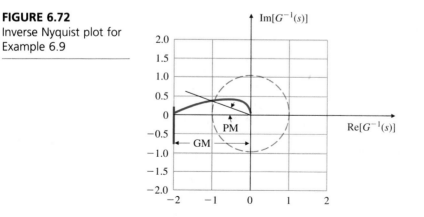

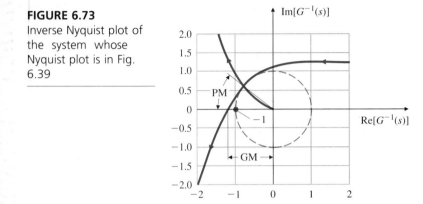

GM = 2 and PM $\cong$ 20°. As an example of a more complex case, Fig. 6.73 shows an inverse Nyquist plot for the sixth-order case whose Nyquist plot was shown in Fig. 6.39 and whose Nichols chart was shown in Fig. 6.71. Note here that GM = 1.2 and PM $\cong$ 35°. Had the two crossings of the unit circle not occurred at the same point, the crossing with the smallest PM would have been the appropriate one to use.

6.9 Sensitivity

In this section we will develop conditions on the Bode plot of the open-loop transfer function DG that will ensure good performance with respect to sensitivity, steady-state errors, and sensor noise.

One justification for feedback, developed in Chapter 4, is to reduce the effect of disturbances and parameter changes on the performance of a control system. One aspect of the sensitivity issue has been the consideration of steady-state errors due to command inputs and disturbances. This has been an important design component in different design methods. Design for acceptable steady-state errors can be thought of as placing a lower bound on the very-low-frequency gain of the system.

Another aspect of the sensitivity issue concerns the high-frequency portion of the system. So far, Chapter 4 and Sections 5.6 and 6.7 have briefly discussed the idea that, to alleviate the effects of sensor noise, the gain of the system at high frequencies must be kept low. In fact, in the development of lead compensation, we placed a pole in the system specifically to reduce the effects of sensor noise at the higher frequencies. It is not unusual for designers to place extra poles in the compensation, that is, to use the relation

$$D(s) = K \frac{Ts + 1}{(\alpha Ts + 1)^2},$$

in order to introduce even more attenuation for noise reduction.

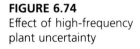

FIGURE 6.74
Effect of high-frequency
plant uncertainty

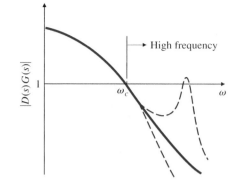

Amplitude and phase
stabilization

A second consideration affecting high-frequency gains is that many systems have high-frequency dynamic phenomena such as mechanical resonances that could have an impact on the stability of a system. In very-high-performance designs, these high-frequency dynamics are included in the plant model, and a compensator is designed with a specific knowledge of those dynamics. A standard approach to designing for unknown high-frequency dynamics is to keep the high-frequency gain low, just as we did for sensor-noise reduction. The reason for this can be seen from the gain–frequency relationship of a typical system, shown in Fig. 6.74. The only way instability can result from high-frequency dynamics is if an unknown high-frequency resonance causes the magnitude to rise above 1. Conversely, if all unknown high-frequency phenomena are guaranteed to remain below a magnitude of 1, stability can be guaranteed. The likelihood of an unknown resonance in the plant G rising above 1 can be reduced if the nominal high-frequency gain of DG is lowered by the addition of extra poles in D. When the stability of a system with resonances is assured by tailoring the high-frequency magnitude never to exceed 1, we refer to this process as **amplitude stabilization**. Of course, if the resonance characteristics are known exactly, a specially tailored compensation, such as one with a notch at the resonant frequency, can be used to reduce the gain at a specific frequency. This method of stabilization is referred to as **phase stabilization**. A drawback to phase stabilization is that the resonance information is often not available with adequate precision; therefore, the method is more susceptible to errors in the plant model used in the design. Thus we see that sensitivity to plant uncertainty and sensor noise are both reduced by sufficiently low gain at high frequency.

These two aspects of sensitivity—high- and low-frequency behavior— can be depicted graphically, as shown in Fig. 6.75. There is a minimum low-frequency gain allowable for acceptable steady-state error performance and a maximum high-frequency gain allowable for acceptable noise performance and for low probability of instabilities caused by plant-modeling errors. Between these two bounds the control engineer must achieve a gain crossover near the required bandwidth; as we have seen, the crossover must occur at essentially a slope of -1.

FIGURE 6.75
Design criteria for low
sensitivity

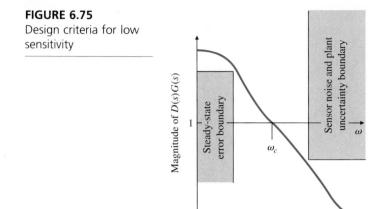

■ 6.9.1   Sensitivity Functions

In order to aid the designer in making a control system as insensitive as possible, special functions have been defined to show the sensitivity of a closed-loop control system.

Consider the unity feedback system in Fig. 6.76 with the plant G and the compensator D. One of the main objectives of the control system is to keep the tracking error ($e = r - y$) small for a reference input excitation and to keep the output y small for a disturbance input w. The measurement noise is represented by n. From the block diagram we may write

$$Y = W + GD(R - Y - N), \tag{6.51}$$

$$(1 + GD)Y = W + GD(R - N), \tag{6.52}$$

or

$$Y = (1 + GD)^{-1}W + (1 + GD)^{-1}GD(R - N). \tag{6.53}$$

For the tracking error we have

$$E \triangleq R - Y, \tag{6.54}$$

$$E = R - (1 + GD)^{-1}W + (1 + GD)^{-1}GD(R - N)$$

$$= (1 + GD)^{-1}(R - W) - (1 + GD)^{-1}GDN. \tag{6.55}$$

Sensitivity function

The **sensitivity function** is defined as

$$S(s) = (1 + GD)^{-1}, \tag{6.56}$$

Complementary
sensitivity function

which Eq. (6.55) shows to be the transfer function from r to e. We also see from Eq. (6.55) that S is the transfer function from w to $-e$. The **complementary**

FIGURE 6.76
Unity feedback control
system

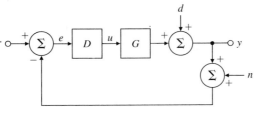

sensitivity function is defined as[9]

$$T(s) = (1 + GD)^{-1}GD \qquad (6.57)$$

and is the transfer function between the reference input r and the output y; that is, it is the *closed-loop*-system transfer function and the transfer function from $-n$ to e. With some manipulation, we can also express the complementary sensitivity function as

$$T(s) = [1 + (GD)^{-1}]^{-1} \qquad (6.58)$$

and can show that the sum of the two sensitivity functions obeys the relation

$$S(s) + T(s) = 1. \qquad (6.59)$$

Therefore, with $n = 0$, we may then rewrite Eq. (6.53) as

$$Y = SW + TR \qquad (6.60)$$

and rewrite Eq. (6.55) as

$$E = S(R - W). \qquad (6.61)$$

Equation (6.59) states that, regardless of the compensator used, the sum of the sensitivity and the complementary sensitivity will always be unity. Therefore, Eq. (6.59) establishes the inherent restriction imposed by nature and the fundamental tradeoff available to the control system designer. Finally, we may refer to the transfer function from r to u as the **control sensitivity function**:

$$\frac{U(s)}{R(s)} = \frac{D(s)}{1 + D(s)G(s)}. \qquad (6.62)$$

The sensitivity function discussed here is related to the one in Chapter 4 (Eq. 4.19). If the closed-loop transfer function is $T(s)$, then its sensitivity with respect

---

8. It is traditional to use the symbol T for both the complementary sensitivity function and the break point in the lead and lag compensators discussed in Section 6.7. Because these two items are unrelated, the meaning of the symbol should be clear from the context.

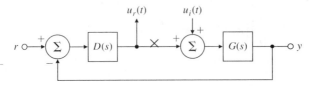

to the plant perturbations is

$$S_G^T = \frac{G}{T}\left(\frac{\partial T}{\partial G}\right) = \frac{G(s)[1 + D(s)G(s)]}{G(s)}\frac{1 + D(s)G(s) - D(s)G(s)}{[1 + D(s)G(s)]^2}$$

$$= \frac{1}{1 + D(s)G(s)}, \qquad (6.63)$$

Return difference
Loop breaking

which is equivalent to Eq. (6.56). The term $1 + D(s)G(s)$ is called the **return difference**. The name comes from the physical situation depicted in Fig. 6.77. Here we have broken the loop at x and inserted a fictitious input signal $u_i(t)$. The signal coming back—the "returned" signal—is $u_r(t)$, and the difference between the injected signal and the returned signal is

$$U_i(s) - U_r(s) = [1 + D(s)G(s)]U_i(s). \qquad (6.64)$$

This idea of "breaking the loop" is useful in determining the robustness of systems; that is, the response characteristics due to various perturbations in the system.

The sensitivity function S is then the primary measure of performance as far as it relates to tracking performance (making the value of e small for a given value of r) and disturbance rejection (making y small for a given d). Thus it is important to make the value of $S(s)$ small according to some measure. For physically realizable systems, the value of the **loop gain**[9]

$$L = |GD| \qquad (6.65)$$

must become small for high frequencies. This means that

$$\lim_{s \to \infty} S(s) = 1. \qquad (6.66)$$

Therefore, it is only possible to make the sensitivity function small over low and midrange frequencies but not at very high frequencies. Figure 6.78(a) shows a typical plot of the sensitivity function S. The function is small at low frequencies, which is desirable for good tracking performance and disturbance rejection. At the midrange frequencies, S approaches unity and crosses it, before approaching unity again for high frequencies.

At the same time, we ideally wish to make the complementary sensitivity function equal to 1 at all frequencies. However, this is not physically possible

9. This is the same quantity defined in Chapter 3 as the product of gains in a loop.

FIGURE 6.78

Typical plots of (a) the sensitivity function S and (b) the complementary sensitivity function T

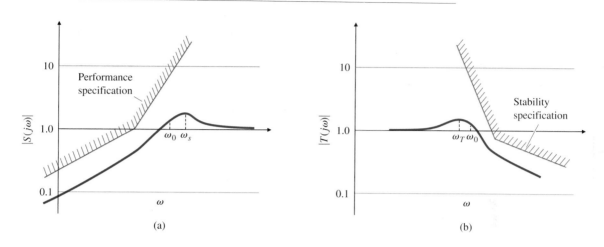

(a)

(b)

since

$$\lim_{s \to \infty} T(s) = 0. \tag{6.67}$$

Therefore, we can only make $T(s)$ close to unity at low and midrange frequencies, but the complementary sensitivity function must roll off (decrease) at high frequencies, as suggested by Eq. (6.59). Figure 6.78(b) shows a typical plot of the complementary sensitivity function. It is unity at low and midrange frequencies but rolls off where $S(s)$ approaches 1.

From Eq. (6.53), with $d = r = 0$, we see that the transfer function from the measurement noise n to the output y is

$$Y = -(1 + GD)^{-1}GDN. \tag{6.68}$$

Therefore, we also see that the complementary sensitivity function is

$$T(s) = -\frac{Y}{N} = (1 + GD)^{-1}GD. \tag{6.69}$$

We would like to make this transfer function as small as possible. This brings out the classical tradeoff in feedback control: Good tracking and disturbance rejection (S small and T large) must be balanced by minimizing the effect of sensor noise (S large and T small).

The complementary sensitivity function $T(s)$ also plays a key role in determining the stability properties of the system. Consider the Nyquist plot in Fig. 6.79, which shows the phase and gain margins defined in Section 6.4. The sensitivity function is related to these margins as follows: According to Eq.

FIGURE 6.79

Nyquist plot showing
stability margins

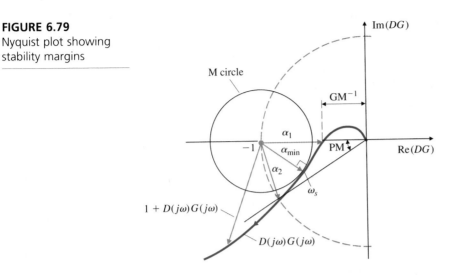

(6.56), S represents the inverse of the distance from $G(j\omega)D(j\omega)$ to the -1 point. For convenience we define α to be the distance from $G(j\omega)D(j\omega)$ to the -1 point. Therefore,

$$\alpha_{min} = \frac{1}{\max\limits_{\omega}|S(j\omega)|} = \min_{\omega}\frac{1}{|S(j\omega)|}. \tag{6.70}$$

Note that α_{min} is the radius of the circle whose center is at -1 and is tangent to the Nyquist plot, as shown in Fig. 6.79. As noted in Section 6.8, this is sometimes referred to as an M circle. Note also that α_{min} is the same quantity that Smith (1958) referred to as the vector margin (see Section 6.4).

From Fig. 6.79 we see that

$$GM = \frac{1}{1 - \alpha_1} \tag{6.71}$$

and

$$PM = 2\arcsin\frac{\alpha_2}{2}. \tag{6.72}$$

Because α_1 and α_2 must both be greater than α_{min}, it is also clear that the maximum value of the sensitivity function yields information about the phase and gain margins. We could use α_{min} in Eqs. (6.71) and (6.72) to yield conservative estimates of the PM and GM.

The complementary sensitivity function may also be related to the phase and gain margins. Consider the inverse Nyquist plot shown in Fig. 6.80. From Eq. (6.58) it is apparent that $T(j\omega)$ is the reciprocal of the distance between the -1 point and $[DG]^{-1}(j\omega)$. We can therefore define the quantity β_{min}, sometimes referred to as the **complex margin**, as

β_{min}: complex margin

$$\beta_{min} = \frac{1}{\max\limits_{\omega} |T(j\omega)|} = \min\limits_{\omega} \frac{1}{|T(j\omega)|} \tag{6.73}$$

The gain margin is

$$GM \geqslant 1 + \beta_{min}, \tag{6.74}$$

and the phase margin is

$$PM \geqslant 2 \arcsin \frac{\beta_{min}}{2}. \tag{6.75}$$

Note that β_{min} is the radius of the circle with center at -1 and tangent to the inverse Nyquist plot; that is, β_{min} defines the minimum distance from the -1 point.

Use of β_{min} to compute the GM and PM will obviously yield conservative estimates. The GM estimate will be exact only if the vector defining the minimum distance lies along the negative real axis. Likewise, the PM estimate will be exact only if the vector defining the minimum distance coincides with the point where GD crosses the unit circle.

Even though the PM and GM estimates discussed in this section may be conservative, they are very attractive from a computational point of view: It is much easier to search for the maximum or minimum of a function than to locate specific, possibly multiple values of crossover frequencies. Therefore, we could compute PM and GM estimates based on both the sensitivity and complementary sensitivity functions and then select the least conservative answer.

FIGURE 6.80

Inverse Nyquist plot showing stability margins

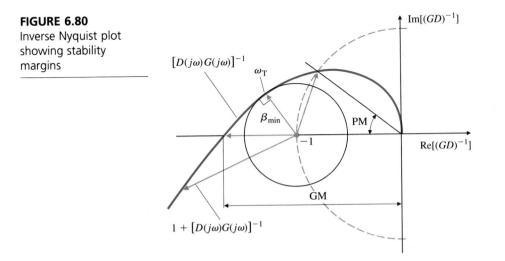

◆ **EXAMPLE 6.22** *Sensitivity and Loop Gain for Satellite Antenna*

Consider the system in Example 6.14. Compute the sensitivity functions and the loop gain for the compensated system. Determine estimates of the PM and GM from the sensitivity functions.

Solution. For this example $G(s) = 1/s(s + 1)$ and $D(s) = 10(0.5s + 1)/(0.1s + 1)$. The sensitivity function is

$$S(s) = \frac{s(s + 1)(s + 10)}{s^3 + 11s^2 + 60s + 100},$$

and the complementary sensitivity function is

$$T(s) = 1 - S(s) = \frac{50(s + 2)}{s^3 + 11s^2 + 60s + 100}.$$

The control sensitivity function is

$$\frac{U(s)}{\Theta_{com}(s)} = \frac{50s(s + 1)(s + 2)}{s^3 + 11s^2 + 60s + 100},$$

and the loop gain is

$$L(s) = D(s)G(s) = -\frac{50(s + 2)}{s(s + 1)(s + 10)}.$$

The plots of the sensitivity functions are shown in Fig. 6.81. From Fig. 6.81(a) we see that the maximum of the sensitivity function is 1.4 at $\omega_s = 7.56$ rad/sec. Therefore,

$$\alpha_{min} = 0.72,$$

FIGURE 6.81
(a) Sensitivity function and (b) complementary sensitivity function for Example 6.22

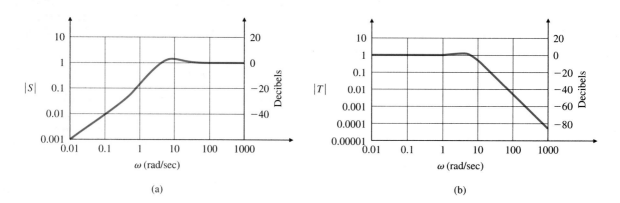

(a)

(b)

which results in the following estimates of the gain and phase margins:

$$\text{GM} \cong \frac{1}{1 - 0.72} = 3.6,$$

$$\text{PM} \cong 2 \arcsin \frac{\alpha_{min}}{2} = 43°.$$

In actuality, $\text{GM} = \infty$ and $\text{PM} = 53.5°$, so the estimates are obviously conservative. The maximum of the complementary sensitivity function is 1.18 and occurs at $\omega_T = 3.35$ rad/sec, as shown in Fig. 6.81(b). Hence,

$$\beta_{min} = 0.85,$$

and the margin estimates are

$$\text{GM} \geqslant 1 + \beta_{min} = 1.85,$$

$$\text{PM} \geqslant 2 \arcsin \frac{\beta_{min}}{2} = 50°.$$

Again the estimates are conservative. However, the PM estimate is close to the true value.

$\blacklozenge$

■ 6.9.2 Stability Robustness

Model uncertainty

All our control systems designs are based on a model of the plant; however, it is inevitable that the model we use is only an approximation of the true dynamics of the system. The difference between the model on which the design is based and the actual plant is referred to generally as **model uncertainty**. If the design performs well for substantial variations in the dynamics of the plant from the design values, we say the design is **robust**. We turn now to consider the effect of model uncertainty on the stability of control systems. Typical sources of uncertainty include unmodeled (high-frequency) dynamics, neglected non-linearities, effects of deliberate reduced-order modeling, and plant-parameter perturbations due to environmental factors such as temperature, air speed, and age. Uncertainty can be represented in the form of an **additive** or **multiplicative perturbation**, as shown in Fig. 6.82. For the multiplicative case of Fig. 6.82(b), the true transfer function is

$$G(s) = G_0(s, \theta)[1 + l(s)],$$

where $G_0(s, \theta)$ is a parameterized model of the plant with the **structurecd uncertainty** θ, which represents the plant-parameter variations. $G_0(s, \theta)$ is a known function (known structure), but the values of the parameter θ are uncertain. The function $l(s)$ is an **unstructured uncertainty** and is entirely unknown, except that it is limited in magnitude to

$$|l(j\omega)| \leqslant l_0(\omega), \tag{6.76}$$

FIGURE 6.82
Representation of uncertainty: (a) additive perturbation; (b) multiplicative perturbation

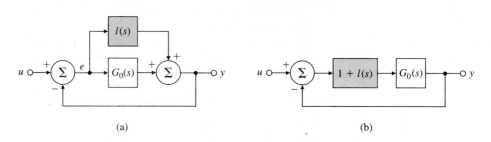

(a)   (b)

where $l_0(\omega)$ is a known real scalar function. The bound can be viewed as a frequency-dependent "radius" of uncertainty of the true plant transfer functions $G(s)$ about some model $G_0(s, \theta)$ for a given θ. Fig. 6.83 shows how such a bound can be calculated experimentally by simply comparing the actual plant with the model. In general, a good model will be well known at low frequencies, which result in small values of $l(\omega)$ for $\omega \ll \omega_1$, and less well known at high frequencies, where the value of $l(\omega)$ is large. The curve of Fig. 6.83 is characteristic of unstructured uncertainty.

Let us now assume that all structured and unstructured uncertainties can be lumped into a **stable multiplicative perturbation**, that is,

$$G(s) = G_0(s)[1 + l(s)], \qquad (6.77)$$

FIGURE 6.83
Frequency-dependent uncertainty bound

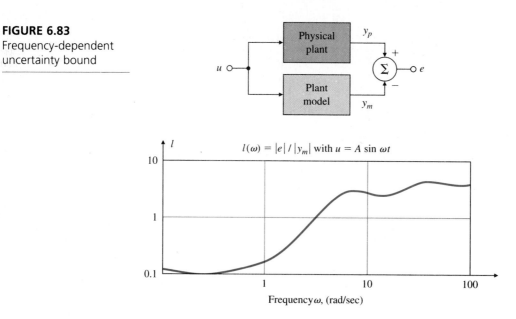

FIGURE 6.84
Compensated feedback
system

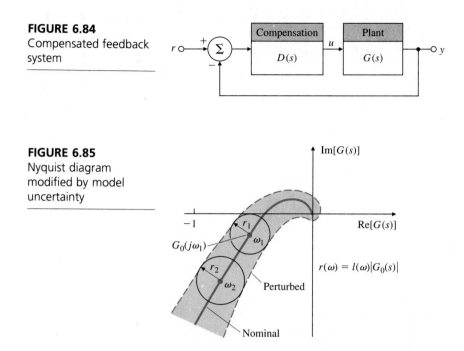

FIGURE 6.85
Nyquist diagram
modified by model
uncertainty

where $G_0(s)$ represents the nominal plant transfer function and $l(s)$ is the model
error. The system with compensation $D(s)$ is shown in Fig. 6.84, and the nominal
closed-loop system $[l(s) \equiv 0]$ is designed to be stable.[10] A typical Nyquist plot of
the compensated system is shown in Fig. 6.85. As $G_0(s)$ is perturbed toward $G(s)$,
the Nyquist plot moves around within an envelope, as shown in the figure. The
system remains stable as long as the number of encirclements of -1 remains
unchanged, which will be true if, for all l,

$$|1 + D(j\omega)G(j\omega)| \neq 0; \tag{6.78}$$

that is, as long as the perturbed Nyquist diagram does not pass through the -1
point. The stability condition of Eq. (6.78) can be rewritten as

$$|1 + D(j\omega)G_0(j\omega)(1 + \varepsilon l)| \neq 0 \quad \text{for} \quad \begin{cases} 0 \leqslant \varepsilon \leqslant 1, \\ 0 \leqslant \omega \leqslant \infty, \end{cases} \tag{6.79}$$

which is true if and only if

$$|1 + D(j\omega)G_0(j\omega)(1 + \varepsilon l)| > 0. \tag{6.80}$$

If we take out a factor of DG_0, this can be expressed as

$$|DG_0[(DG_0)^{-1} + 1 + \varepsilon l]| > 0, \tag{6.81}$$

---

10. $[D(s)G(s)]/[1 + D(s)G(s)]$ is stable.

which is true if $|DG_0| \neq 0$ and

$$|(DG_0)^{-1} + 1 + \varepsilon l| > 0. \tag{6.82}$$

We would like to be able to express the requirement in terms of the open-loop gain of the nominal system DG_0. We can do this if we notice that, for $0 \leqslant \varepsilon \leqslant 1$,

$$|(DG_0)^{-1} + 1 + \varepsilon l| > |1 + (DG_0)^{-1}| - |l| > 0, \tag{6.83}$$

or

$$|1 + (DG_0)^{-1}| > |l|. \tag{6.84}$$

The system is then guaranteed to be stable as long as Eq. (6.84) is satisfied. We define the **stability robustness measure** to be

$$\delta_{SR} = |1 + (DG_0)^{-1}|. \tag{6.85}$$

Then as long as the model uncertainty remains below δ_{SR} for all frequencies, the system is guaranteed to remain stable in spite of the perturbations $l(s)$.[11] An alternative derivation of this result is possible using the small-gain theorem (see Problem 6.71). Note that δ_{SR} is simply the inverse of the closed-loop magnitude frequency response; that is, the inverse of the complementary sensitivity function $T(s)$. In terms of the complementary sensitivity function,

$$|T(s)^{-1}| > |l| \tag{6.86}$$

or

$$|T(s)| < \frac{1}{|l|}. \tag{6.87}$$

This means that model uncertainty dictates an upper bound on the magnitude of $T(s)$. The quantity l defines the minimum distance from the -1 point to the inverse Nyquist plot and, for the case where Eq. (6.86) is an equality, l would be the quantity β_{min} as in Fig. 6.80. For high frequencies that is, $|GD| \ll 1$, $|T| \sim |GD|$, and Eq. (6.86) suggests that

$$|GD| < \frac{1}{|l|}. \tag{6.88}$$

In other words, the (multiplicative) modeling error defines an upper bound on the loop gain.

Our discussion in this section so far suggests the introduction of the following bounds on the sensitivity and complementary sensitivity functions

---

11. For the extension of this result to the multi-input, multi-output case, see Doyle and Stein (1981).

Stability robustness (margin note)

FIGURE 6.86
Constraint boundaries on the loop gain ($W_1^{-1} \cong S$, $W_2^{-1} = T$)

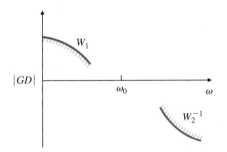

depending on the tracking accuracy, bandwidth, and rolloff required:[12]

$$|S| < |W_1^{-1}| \tag{6.89}$$

$$|T| < |W_2^{-1}|. \tag{6.90}$$

Figure 6.86 shows these bounds in terms of the loop gain. Note that it illustrates essentially the same idea as that shown in Fig. 6.75 and used in the discussion at the beginning of Section 6.9 to relate the ideas from compensation design to the magnitude portion of a Bode plot.

◆ **EXAMPLE 6.23**   *Stability Robustness of Spacecraft Attitude Control*

Consider the closed-loop system in Example 6.14. It is known from experiments that a bound on the multiplicative model error to the plant model is,

$$l_0(s) = \frac{2}{s^2 + 12s + 144}.$$

Determine whether or not the closed-loop system remains stable with this error bound.

Solution.   The complementary sensitivity function (closed-loop transfer function) is

$$T(s) = \frac{50(s + 2)}{s^3 + 11s^2 + 60s + 100}$$

12. These bounds are used in H_∞ control theory as weightings on the sensitivity and complementary sensitivity functions to design compensators (see Doyle, J. C. *et al.*, 1992.)

FIGURE 6.87
Stability robustness
measure

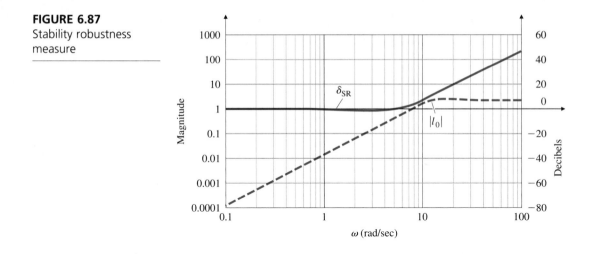

Therefore,

$$\delta_{SR}(\omega) = \left| \frac{1}{T(j\omega)} \right|.$$

The stability robustness measure is overplotted with $|l_0|$ in Fig. 6.87. Since the latter curve always remains below the inverse magnitude of the closed-loop transfer function, the closed-loop system remains stable for the class of perturbations $|l(s)| \leqslant l_0(\omega)$.

◆

■ 6.10   Time Delay

The Laplace transform of a pure time delay is $G_D(s) = e^{-sT}$ and was approximated by a rational function (Padé approximate) in our earlier discussion of root-locus analysis. Although this same approximation could be used with frequency-response methods, an exact analysis of the delay is possible with the Nyquist criterion.

The frequency response of the delay is given by the magnitude and phase of $e^{-sT}|_{s=j\omega}$. The magnitude is

Time-delay magnitude

$$|G_D(j\omega)| = |e^{-j\omega T}| = |\cos \omega T - j \sin \omega T| = 1 \qquad \text{for all } \omega. \qquad (6.91)$$

This result is expected, since a time delay merely shifts the signal in time and has no effect on its magnitude. The phase is

Time-delay phase

$$\angle G_D(j\omega) = -\omega T \qquad (6.92)$$

in radians, and it grows increasingly negative in proportion to the frequency. This, too, is expected since a fixed time delay T becomes a larger fraction or multiple of a sine-wave period as the period drops due to increasing frequency. A

FIGURE 6.88

Phase lag due to pure time delay

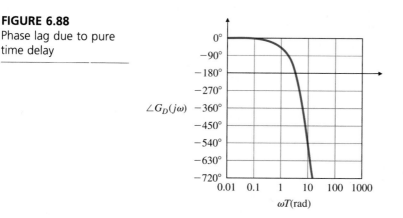

plot of $\angle G_D(j\omega)$ is drawn in Fig. 6.88. Note that the phase lag is greater than 270° for values of ωT greater than about 5 rad. This trend implies that it would be virtually impossible to stabilize a system (or to achieve a positive PM) with a crossover frequency greater than $\omega = 5/T$, and it would be difficult for frequencies greater than $\omega \cong 3/T$. These characteristics essentially place a constraint on the achievable bandwidth of any system with a time delay. (See Problem 6.73 for an illustration of this constraint.)

The frequency domain concepts such as the Nyquist criterion apply directly to systems with pure time delay. This means that no approximations (Padé type or otherwise) are needed, as shown in the following example.

◆ **EXAMPLE 6.24**   *Nyquist Plot for System with Time Delay*

Consider the system with

$$KG(s) = \frac{Ke^{-s}}{s}.$$

Determine the range of K for which the system is stable.

Solution.   Since the Bode plotting rules do not apply for a time delay term, we will use an analytical approach to determine the key features of the plot. As just discussed, the magnitude of the frequency response of the delay term is unity and its phase is $-\omega$ radians. The magnitude of the frequency response of the pure integrator is $1/\omega$ with a constant phase of $-\pi/2$. Therefore,

$$G(j\omega) = \frac{1}{\omega} e^{-j(\omega + \pi/2)}.$$

$$= \frac{1}{\omega} (-\sin \omega - j \cos \omega). \tag{6.93}$$

FIGURE 6.89
Nyquist plot for Example 6.24

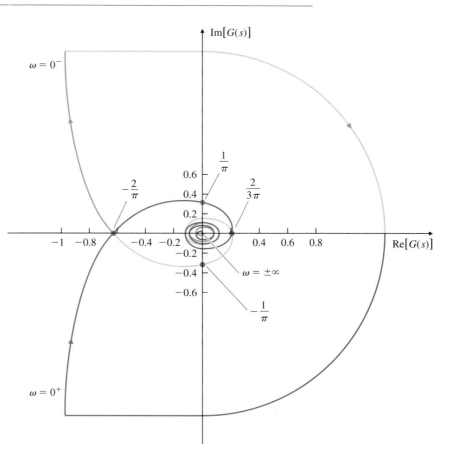

Using Eq. (6.93) and substituting in different values of ω, we can make the Nyquist plot, which is the spiral shown in Fig. 6.89.

Let us examine the shape of the spiral in more detail. We pick a Nyquist path with a small detour to the right of the origin. The effect of the pole at the origin is the large arc at infinity with a 180° sweep, as shown in Fig. 6.89. From Eq. (6.93), for small values of $\omega > 0$, the real part of the frequency response is close to -1 because $\sin\omega \cong \omega$ and $\mathrm{Re}[G(j\omega)] \cong -1$. Similarly, for small values of $\omega > 0$, $\cos\omega \cong 1$ and $\mathrm{Im}[G(j\omega)] \cong -1/\omega$, that is, very large, negative values as shown in Fig. 6.89. To obtain the crossover points on the real axis, we set the imaginary part equal to zero:

$$\frac{\cos \omega}{\omega} = 0. \tag{6.94}$$

The solution is then

$$\omega_o = \frac{(2n + 1)\pi}{2}, \qquad n = 0, 1, 2, \ldots . \tag{6.95}$$

After substituting Eq. (6.95) back into Eq. (6.94), we find

$$G(j\omega_o) = \frac{(-1)^n}{(2n + 1)} \left(\frac{2}{\pi}\right), \qquad n = 0, 1, 2, \ldots . $$

So the first crossover of the negative real axis is at $-2/\pi$ corresponding to $n = 0$. The first crossover of the positive real axis occurs for $n = 1$ and is located at $2/3\pi$. As we can infer from Fig. 6.89, there are an infinite number of other crossings of the real axis. Finally, for $\omega = \infty$, the Nyquist plot converges to the origin. Note that the Nyquist plot for $\omega < 0$ is the mirror image of the one for $\omega > 0$.

The number of poles in the RHP is zero ($P = 0$), so for closed-loop stability we need $Z = N = 0$. Therefore, the Nyquist plot cannot be allowed to encircle the $-1/K$ point. It will not do so as long as

$$-\frac{1}{K} < -\frac{2}{\pi}, \tag{6.96}$$

which means for stability we must have $0 < K < \pi/2$.

◆

■ 6.11   Obtaining a Pole-zero Model from Frequency-response Data

As we pointed out earlier, it is relatively easy to obtain the frequency response of a system experimentally. Sometimes it is desirable to obtain an approximate model, in terms of a transfer function, directly from the frequency response. The derivation of such a model can be done to various degrees of accuracy. The method described in this section is usually adequate and is widely used in practice.

There are two ways to obtain a model from frequency-response data. In the first case we can introduce a sinusoidal input, measure the gain (logarithm of the amplitude ratio of output to input) and the phase difference between output and input, and accept the curves plotted from this data as the model. Using the methods given in previous sections, we can derive the design directly from this information. In the second case, we wish to use the frequency data to verify a mathematical model obtained by other means. To do so we need to extract an approximate transfer function from the plots, again by fitting straight lines to the data, estimating break points (that is, finding the poles and zeros), and using Fig. 6.2 to estimate the damping ratios of complex factors from the frequency overshoot. The next example illustrates the second case.

◆ **EXAMPLE 6.25** *Transfer Function from Measured Frequency Response*

Determine a transfer function from the frequency response plotted in Fig. 6.90, where frequency f is plotted in hertz.

Solution. Drawing an asymptote to the final slope of -2 (or $-40\,$dB per decade), we assume a break point at the frequency where the phase is $-90°$. This occurs at $f_1 \cong 1.66\,$Hz ($\omega_1 = 2\pi f_1 = 10.4\,$rad/sec). We need to know the damping ratio in order to subtract out this second-order pole. For this the phase curve may be of more help. Since the phase around the break-point frequency is symmetric, we draw a line at the slope of

FIGURE 6.90

An experimental frequency response

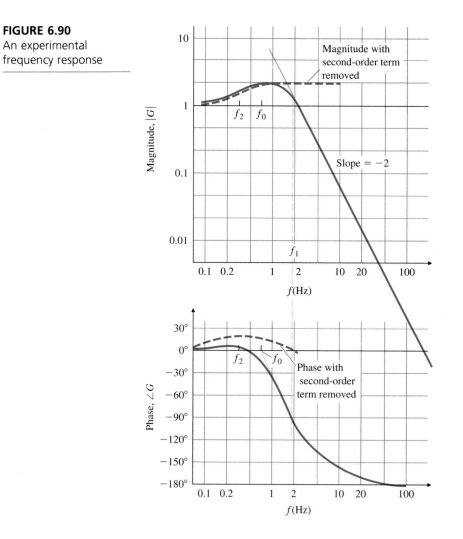

the phase curve at f_1 to find that the phase asymptote intersects the $0°$ line at $f_0 \cong 0.71$ Hz (or 4.46 rad/sec). This corresponds to $f_1/f_0 \cong 2.34$, which in time corresponds to $\zeta \cong 0.5$, as seen on the normalized response curves in Fig. 6.2(b). The magnitude curve with the second-order factor taken out shows an asymptotic amplitude gain of about 6.0 dB, or a factor of $10^{6.0/20} = 2.0$. Since this is a gain rise, it occurs because of a lead network of the form

$$\frac{s/z + 1}{s/p + 1},$$

where $p/z = 2.0$. If we remove the second-order terms in the phase curve, we obtain a phase curve with a maximum phase of about $20°$, which also corresponds to a frequency separation of about 2. To locate the center of the lead network, we must estimate the point of maximum phase based on the lead term alone, which occurs at the geometric mean of the two break-point frequencies. The lead center seems to occur at $f_2 \cong 0.3$ Hz (or $\omega_2 = 1.88$ rad/sec).

Thus we have the relations

$$zp = (1.88)^2 = 3.55,$$

$$\frac{p}{z} = 2,$$

from which we can solve

$$2z^2 = 3.55,$$

$$z = 1.33,$$

$$p = 2.66.$$

Our final model is given by

Model from measured response

$$\hat{G}(s) = \frac{(s/1.33) + 1}{[(s/2.66) + 1][(s/10.4)^2 + (s/10.4) + 1]}. \tag{6.97}$$

The actual data were plotted from

$$G(s) = \frac{(s/2) + 1}{[(s/4) + 1][(s/10)^2 + (s/10) + 1]}.$$

As can be seen, we found the second-order term quite easily, but the location of the lead network is off in center frequency by a factor of $4/2.66 \cong 1.5$. However, the subtraction of the second-order term from the composite curve was not done with great accuracy, rather, by reading the curves. Again, as with the transient response, we conclude that by a bit of approximate plotting we can obtain a crude model (usually within ± 3 dB in amplitude and $\pm 10°$ in phase) that can be used for control design.

◆

Refinements on these techniques with computer aids are rather obvious, and an interactive program for removing standard first- and second-order terms and accurately plotting the residual function would greatly improve the speed and accuracy of the process. It is also common to have computer tools that can find the parameters of an assumed model structure by minimizing the sum of squares of the difference between the model's frequency response and the experimental frequency response.

Summary

- The frequency-response **Bode plot** is a graph of the transfer function magnitude in logarithmic scale and the phase in linear scale versus frequency in logarithmic scale. For a transfer function $G(s)$,

$$A = |G(j\omega)| = |G(s)|\big|_{s=j\omega}$$
$$= \sqrt{\{\text{Re}[G(j\omega)]\}^2 + \{\text{Im}[G(j\omega)]\}^2}, \qquad (6.5a)$$

$$\phi = \tan^{-1}\frac{\text{Im}[G(j\omega)]}{\text{Re}[G(j\omega)]} = \angle G(j\omega). \qquad (6.5b)$$

- For a transfer function in Bode form,

$$KG(\omega) = K_o \frac{(j\omega\tau_1 + 1)(j\omega\tau_2 + 1)\cdots}{(j\omega\tau_a + 1)(j\omega\tau_b + 1)\cdots}, \qquad (6.14)$$

the Bode frequency response can be easily plotted by hand using the rules described in Section 6.1.1.

- Bode plots can be obtained using computer algorithms (bode in MATLAB), but hand-plotting skills are still extremely helpful.

- For a second-order system the peak magnitude of the Bode plot is related to the damping by

$$|G(j\omega)| = \frac{1}{2\zeta} \qquad \text{at } \omega = \omega_n. \qquad (6.19)$$

- A method of determining the stability of a closed-loop system based on the frequency response of the system's open-loop transfer function is the **Nyquist stability criterion.** Rules for plotting the **Nyquist plot** are described in Section 6.3. The number of RHP closed-loop roots is given by

$$Z = N + P, \qquad (6.26)$$

where

N = number of clockwise encirclements of the -1 point,

P = number of open-loop poles in the RHP.

- The Nyquist plot may be obtained using computer algorithms (nyquist in MATLAB).

- The **gain margin** (GM) and **phase margin** (PM), can be determined directly by inspecting the open-loop Bode plot or the Nyquist plot.

- For a standard second-order system,

$$\zeta \cong \frac{PM}{100}. \tag{6.29}$$

- The **vector margin** is a single-parameter stability margin and may be determined directly from the Nyquist plot.

- For a stable minimum-phase system, Bode's gain-phase relationship uniquely relates the Bode gain and phase of the system,

$$\angle G(j\omega) \cong n \times 90°, \tag{6.30}$$

where n is the slope of $|G(j\omega)|$ in units of decade of amplitude per decade of frequency.

- Experimental frequency response data can be used directly for analysis and design.

- For the system

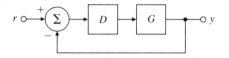

the open-loop Bode plot is the frequency response of GD, and the closed-loop frequency response is obtained from $T = GD/(1 + GD)$.

- The **bandwidth** of the system is a measure of speed of response. For control systems it is defined as the frequency corresponding to 0.707 (-3 dB) in the closed-loop magnitude Bode plot.

- The frequency response characteristics of several types of compensation have been described, and examples of design using these characteristics have been discussed. Design procedures were given for lead and lag compensators in Section 6.7. The examples in that section show the ease of selecting specific values of design variables, a result of using frequency-response methods. A summary was provided in Section 6.7.6.

- **Lead compensation**, given by

$$D(s) = K \frac{Ts + 1}{\alpha Ts + 1}, \qquad \alpha < 1, \tag{6.35}$$

is a high-pass filter and approximates PD control. It is used whenever substantial improvement in damping of the system is required.

- **Lag compensation**, given by

$$D(s) = \alpha \, \frac{Ts + 1}{\alpha Ts + 1}, \qquad \alpha > 1, \tag{6.44}$$

  is a low-pass filter and approximates PI control. It is used to increase the low-frequency gain of the system so as to improve steady-state response. The overall gain K is also designed and is part of the complete compensation.

- The **Nichols plot** is an alternate representation of the frequency response and is parameterized as a function of frequency.

- Tracking-error reduction and disturbance rejection can be specified in terms of the low-frequency gain of the Bode plot. Sensor-noise rejection can be specified in terms of high-frequency attenuation of the Bode plot (see Fig. 6.67).

- Time delay can be analyzed directly by sketching the Nyquist plot.

- Transfer-function poles and zeros can be estimated from the measured frequency response.

Problems

6.1   **a)** Show that α_o in Eq. (6.2) is given by

$$\alpha_o = \left[G(s) \, \frac{U_o \omega}{s - j\omega} \right] \Bigg|_{s = -j\omega} = -U_o G(-j\omega) \, \frac{1}{2j}$$

and

$$\alpha_o^* = \left[G(s) \, \frac{U_o \omega}{s + j\omega} \right] \Bigg|_{s = +j\omega} = U_o G(j\omega) \, \frac{1}{2j}.$$

(b) By assuming the output can be written as

$$y(t) = \alpha_o e^{-j\omega t} + \alpha_o^* e^{j\omega t},$$

derive Eqs. (6.4) and (6.5).

6.2   **a)** Calculate the magnitude and phase of

$$G(s) = \frac{1}{s + 1}$$

for $\omega = 0.1, 0.2, 0.5, 1, 2, 5,$ and $10 \, \text{rad/sec}$.

b) Sketch the asymptotes for $G(s)$, and compare these with your computed results from part (a).

6.3 Draw a Bode plot for each of the following systems. Compare your sketches with the plots obtained using the bode command in MATLAB.

a) $G(s) = \dfrac{1}{(s+1)^2(s^2+s+4)}$

f) $G(s) = \dfrac{(s+5)(s+3)}{s(s+1)(s^2+s+4)}$

b) $G(s) = \dfrac{s}{(s+1)(s+10)(s^2+5s+2500)}$

g) $G(s) = \dfrac{4000}{s(s+40)}$

c) $G(s) = \dfrac{4s(s+10)}{(s+50)(4s^2+5s+4)}$

h) $G(s) = \dfrac{100}{s(1+0.1s)(1+0.5s)}$

d) $G(s) = \dfrac{10(s+4)}{s(s+1)(s^2+2s+5)}$

i) $G(s) = \dfrac{1}{s(1+s)(1+0.02s)}$

e) $G(s) = \dfrac{1000(s+1)}{s(s+2)(s^2+8s+64)}$

6.4 A certain system is represented by the asymptotic Bode diagram shown in Fig. 6.91. Find and sketch the response of this system to a unit step input (assuming zero initial conditions).

FIGURE 6.91
Magnitude portion
of Bode plot for
Problem 6.4

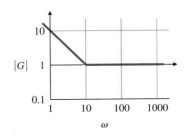

6.5 Prove that a magnitude slope of -1 in a Bode plot corresponds to $-20\,\mathrm{dB}$ per decade.

6.6 In this problem you will study the Bode plots of normalized second-order systems with a damping ratio $\zeta = 0.5$ and an additional zero. (a) Draw Bode plots of

$$G(s) = \dfrac{s/z + 1}{s^2 + s + 1}$$

with $z = \alpha/2$ for $\alpha = 0.1$, 1, and 10. Do the plots lead you to expect extra transient overshoot? How? (b) Use MATLAB to draw the Bode plots for several (at least seven) other values of α.

6.7 In this problem you will study the Bode plots of normalized second-order systems with $\zeta = 0.5$ and an additional pole. (a) Draw Bode plots of

$$G(s) = \dfrac{1}{[(s/p) + 1](s^2 + s + 1)}$$

with $p = \alpha/2$ for $\alpha = 0.1$, 1, and 10. Do the plots lead you to expect additional rise times? How? (b) Use MATLAB to draw the Bode plots for several (at least seven) other values of α.

6.8   For the closed-loop transfer function

$$T(s) = \frac{\omega_n^2}{s^2 + 2\zeta\omega_n s + \omega_n^2},$$

derive the following expression for the bandwidth ω_{BW} of $T(s)$ in terms of ω_n and ζ:

$$\omega_{BW} = \omega_n \sqrt{1 - 2\zeta^2 + \sqrt{2 + 4\zeta^4 - 4\zeta^2}}.$$

Assuming $\omega_n = 1$, plot ω_{BW} for $0 \leqslant \zeta \leqslant 1$.

6.9   Consider the system whose transfer function is

$$G(s) = \frac{A_o \omega_o s}{Q s^2 + \omega_o s + \omega_o^2 Q}.$$

This is a model of a tuned circuit with *quality factor Q*. Compute the magnitude and phase of the transfer function analytically, and plot them for $Q = 0.5, 1, 2,$ and 5 as a function of the normalized frequency ω/ω_o. Define the bandwidth as the distance between the frequencies on either side of ω_o where the magnitude drops to 3 dB below its value at ω_o. Show that the bandwidth is given by

$$BW = \frac{1}{2\pi}\left(\frac{\omega_o}{Q}\right).$$

6.10   A DC voltmeter schematic is shown in Fig. 6.92. The pointer is damped so that its maximum overshoot to a step input is 10%.

a) What is the undamped natural frequency of the system?
b) What is the damped natural frequency of the system?
c) What input frequency will produce the largest magnitude output?
d) Suppose this meter is now used to measure a 1-V AC input with a frequency of 2 rad/sec. What amplitude will the meter indicate after initial transients have died out? What is the phase lag of the output with respect to the input? Use a Bode plot analysis to answer these questions. Use the lsim command in MATLAB to verify your answer in part (d).

FIGURE 6.92
Voltmeter schematic

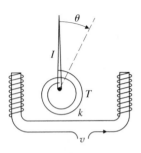

$I = 40 \times 10^{-6}$ kg $\cdot$ m^2
$k = 4 \times 10^{-6}$ kg $\cdot$ m^2/sec^2
$T = $ input torque $= K_m v$
$v = $ input voltage
$K_m = 1$ N $\cdot$ m/V

6.11   Determine the range of K for which each of the following systems is stable by making a Bode plot for $K = 1$ and imagining the magnitude plot sliding up or down until instability results. Verify your answers using a root-locus plot (generated by hand or computer).

a) $KG(s) = \dfrac{K(s + 2)}{s + 10}$

b) $KG(s) = \dfrac{K}{(s + 10)(s + 2)^2}$

c) $KG(s) = \dfrac{K(s + 10)(s + 1)}{(s + 100)(s + 2)^3}.$

6.12   Determine the range of K for which each of the following systems is stable by making a Bode plot for $K = 1$ and imagining the magnitude plot sliding up or down until instability results. Verify your answers using a root-locus plot (generated by hand or computer).

a) $KG(s) = \dfrac{K(s + 1)}{s^2(s + 10)}$

b) $KG(s) = \dfrac{K(s + 1)}{s(s + 2)}$

c) $KG(s) = \dfrac{K}{(s + 2)(s^2 + 9)}$

d) $KG(s) = \dfrac{K(s + 1)^2}{s^3(s + 10)}.$

6.13 a) Sketch the polar plot for an open-loop system with transfer function $1/s^2$; that is, sketch

$$\left. \frac{1}{s^2} \right|_{s = C_1},$$

where C_1 is a contour enclosing the entire RHP, as shown in Fig. 6.16. (*Hint*: Assume C_1 takes a small detour around the poles at $s = 0$, as shown in Fig. 6.26.)

b) Repeat part (a) for an open-loop system whose transfer function is $G(s) = 1/(s^2 + \omega_0^2)$.

6.14 Draw a Nyquist diagram for each of the following systems, and compare your result with that obtained using the MATLAB command nyquist:

a) $KG(s) = \dfrac{K(s + 2)}{s + 10}$

b) $KG(s) = \dfrac{K}{(s + 10)(s + 2)^2}$

c) $KG(s) = \dfrac{K(s + 10)(s + 1)}{(s + 100)(s + 2)^3}$

Using your plots, estimate the range of K for which each system is stable, and qualitatively verify your result using a root-locus plot (generated by hand or using MATLAB).

6.15 Draw a Nyquist diagram for each of the following systems in parts (a) through (c), choosing the contour to be to the right of any singularities on the $j\omega$-axis.

a) $KG(s) = \dfrac{K(s + 1)}{s^2(s + 10)}.$

b) $KG(s) = \dfrac{K(s + 1)}{s(s + 2)}.$

c) $KG(s) = \dfrac{K}{(s + 2)(s^2 + 9)}.$

d) Redo the Nyquist plots in parts (b) and (c), this time choosing the contour to be to the left of all singularities on the imaginary axis.

e) Use the Nyquist stability criterion to determine the range of K for which each system is stable, and verify your answer using a root-locus plot (generated by hand or computer).

FIGURE 6.93
Control system for
Problem 6.16

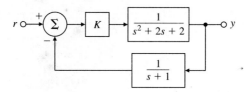

6.16 Draw the Nyquist diagram for the system in Fig. 6.93. Using the Nyquist stability criterion, determine the range of K for which the system is stable.

6.17 a) For $\omega = 0.1$ to $100 \, \text{rad/sec}$, sketch the phase of the minimum-phase system

$$G(s) = \left. \frac{s + 1}{s + 10} \right|_{s = j\omega}$$

and the nonminimum-phase system

$$G(s) = \left. -\frac{s - 1}{s + 10} \right|_{s = j\omega},$$

noting that $\angle(j\omega - 1)$ decreases with ω rather than increasing. Does a RHP zero affect the relationship between the -1 encirclements on a polar plot and the number of unstable closed-loop roots in Eq. (6.26)?

b) Sketch the phase of the following unstable system for $\omega = 0.1$ to $100 \, \text{rad/sec}$:

$$G(s) = \left. \frac{s + 1}{s - 10} \right|_{s = j\omega}.$$

c) Check the stability of the systems in (a) and (b) using the Nyquist criterion on $KG(s)$. Determine the range of K for which the closed-loop system is stable, and check your results qualitatively using a root-locus sketch (generated by hand or computer).

6.18 The Nyquist plot for some actual control systems resembles the one shown in Fig. 6.94. You may wish to compare it with Fig. 6.32. Determine the gain and phase

FIGURE 6.94
Nyquist plot for
Problem 6.18

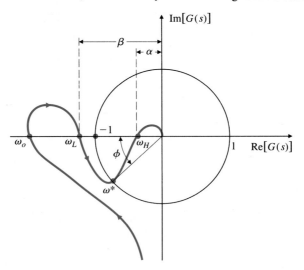

margin(s) for Fig. 6.94. What does the diagram imply about the stability of the system? Sketch what a typical root locus would look like for such a system, and discuss what it tells about the system's stability. Also sketch what a typical Bode frequency response would look like for the system, indicating on the plot its important features.

6.19 The Bode plot for

$$G(s) = \frac{100[(s/10) + 1]}{s[(s/1) - 1][(s/100) + 1]}$$

is shown in Fig. 6.95.

a) Sketch the Nyquist plot for $G(s)$.

b) Is the closed-loop system shown in Fig. 6.95 stable?

FIGURE 6.95
Bode plot for
Problem 6.19

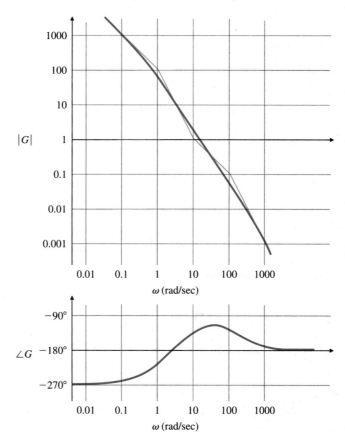

6.20 Suppose that in Fig. 6.96,

$$G(s) = \frac{25(s + 1)}{s(s + 2)(s^2 + 2s + 16)}.$$

Carefully draw a Bode diagram of the frequency response $G(j\omega)$, and use this data to

FIGURE 6.96
Control system for Problem 6.20

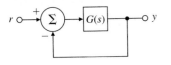

FIGURE 6.97
Control system for Problem 6.21

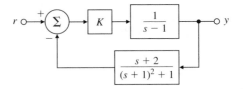

plot the corresponding Nyquist diagram. Conclude whether or not the loop is stable, and verify your conclusion using a root-locus plot (drawn by hand or computer).

6.21 Consider the system given in Fig. 6.97.
  a) Using Routh's criterion, determine the closed-loop stability of this system for all values of K.
  b) Sketch the locus of the closed-loop poles as K varies from zero to infinity. (Include angles of departure or arrival as well as values for K and s at all break points and imaginary-axis crossings.)
  c) Sketch the Nyquist plot of the system, and use it to verify the results of parts (a) and (b).

6.22 For each of the open-loop frequency-response plots shown in Fig. 6.98, determine the following:
  • the minimum number of open-loop poles n,
  • the minimum number of zeros m,
  • the pole-zero excess $n - m$,
  • a pole-zero pattern in the s-plane that could correspond to the frequency response.

6.23 Associate the Nyquist plots, step responses, and frequency responses shown in Fig. 6.99. Justify your conclusions.

FIGURE 6.98
Nyquist plots for Problem 6.22

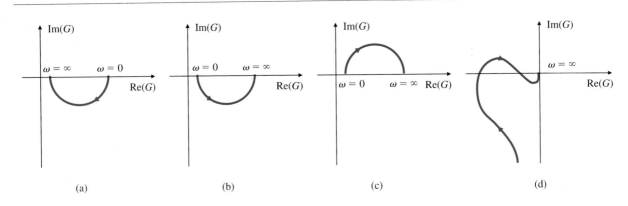

(a)   (b)   (c)   (d)

FIGURE 6.99
Plots for Problem 6.23

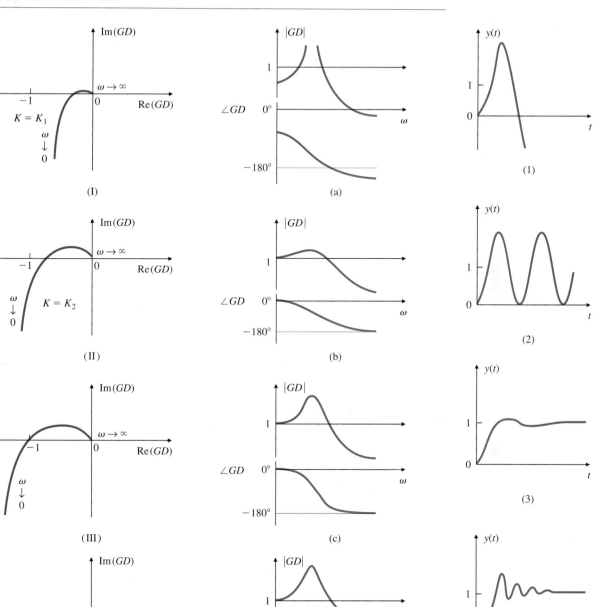

(I)   (a)   (1)

(II)   (b)   (2)

(III)   (c)   (3)

(IV)   (d)   (4)

447

6.24 For a given system, show that the ultimate period P_u and the corresponding ultimate gain K_u for the Zeigler–Nichols method can be found using the following:
a) Nyquist diagram
b) Bode plot
c) root locus.

6.25 a) If a system has the open-loop transfer function

$$G(s) = \frac{\omega_n^2}{s(s + 2\zeta\omega_n)}$$

with unity feedback, then the closed-loop transfer function is given by

$$T(s) = \frac{\omega_n^2}{s^2 + 2\zeta\omega_n s + \omega_n^2}.$$

Verify the values of the PM shown in Fig. 6.36 for $\zeta = 0.1$, 0.4, and 0.7. If G is plotted versus s/ω_n, the value of ω_n does not enter into the problem. If the loop gain is some value other than 1, does the relationship between the PM and the closed-loop system damping change?

b) Show that

$$PM = \tan^{-1} \frac{2\zeta}{\sqrt{\sqrt{1 + 4\zeta^4} - 2\zeta^2}}.$$

6.26 In this problem you will see that it is possible to determine the GM and PM from the root locus. Consider a unity feedback system with the nominal forward transfer function $K_o G(s)$, so that the characteristic polynomial is $1 + K_o G(s)$. Explain how this equation may be used to compute the GM and PM, and specify a procedure to compute each margin.

6.27 Consider the unity feedback system with the open-loop transfer function

$$G(s) = \frac{K}{s(s + 1)[(s^2/25) + 0.4(s/5) + 1]}.$$

a) Draw the Bode plot (gain and phase) for $G(j\omega)$ assuming $K = 1$.
b) What gain K is required for a PM of $45°$? What is the GM for this value of K?
c) What is K_v when the gain K is set for PM $= 45°$?
d) Sketch the root locus with respect to K, and indicate the roots for a PM of $45°$.

6.28 Consider the Nyquist plot of Fig. 6.42. Show that

$$PM \geqslant 2\sin^{-1}\left(\frac{M}{2}\right)$$

$$GM = 20\log(1 + M) \text{ decibels.}$$

(*Hint:* You may wish to use the law of cosines.)

6.29 For the system depicted in Fig. 6.100(a), the transfer-function blocks are defined by

$$G(s) = \frac{1}{(s + 2)^2(s + 4)} \quad \text{and} \quad H(s) = \frac{1}{s + 1},$$

and the root locus is shown in Fig. 6.100(b).

a) Using root-locus methods, determine the value of K at the stability boundary.

FIGURE 6.100
Control system for Problem 6.29: (a) block diagram; (b) root locus

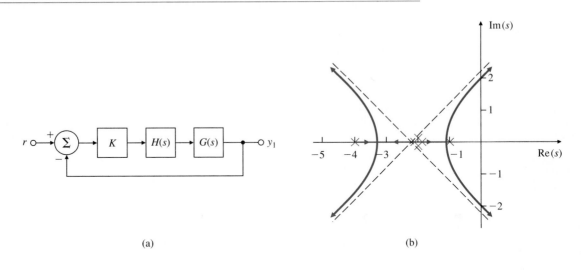

(a)   (b)

b) Using root-locus methods, determine the value of K that will produce roots with damping corresponding to $\zeta = 0.707$.

c) What is the gain margin of the system if the gain is set to the value determined in part (b)? Answer this question *without* drawing Bode or Nyquist plots.

d) Draw the Bode plots for the system, and determine the gain margin that results for PM = 65°. What damping ratio would you expect for this PM?

e) Draw a root locus for the system shown in Fig. 6.101. How does it differ from the one in Fig. 6.100(b)?

f) For the systems in Figs. 6.100 and 6.101, how does the transfer function $Y_2(s)/R(s)$ differ from $Y_1(s)/R(s)$?

FIGURE 6.101
Control system for Problem 6.29(e, f)

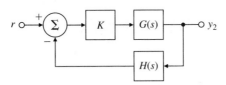

6.30 For the system shown in Fig. 6.102, use Bode and root-locus plots to determine the gain and frequency at which instability occurs. What gain (or gains) gives a PM of 20°? What is the gain margin when PM = 20°?

FIGURE 6.102
Control system for Problem 6.30

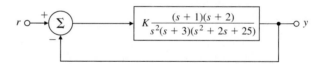

6.31 A magnetic tape-drive speed-control system is shown in Fig. 6.103. The speed sensor is slow enough that its dynamics must be included. The speed-measurement time constant is $\tau_m = 0.5$ sec; the reel time constant is $\tau_r = J/b = 4$ sec, where $b = $ the output shaft damping constant $= 1\,\text{N}\cdot\text{m}\cdot\text{sec}$; and the motor time constant is $\tau_1 = 1$ sec.

a) Determine the gain K required to keep the steady-state speed error to less than 7% of the reference-speed setting.

b) With the gain determined in part (a), use the Nyquist criterion to investigate the stability of the system.

c) Estimate the gain and phase margins of the system. Is this a good system design? Why or why not?

FIGURE 6.103
Magnetic tape-drive
speed control

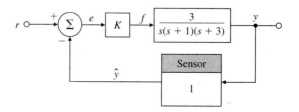

6.32 For the system in Fig. 6.104, apply the formal Nyquist criterion
- to determine the range of values of K (positive and negative) for which the system will be stable, and
- to determine the number of roots in the RHP for those values of K for which the system is unstable.

Check your answer using a root-locus plot (drawn by hand or computer).

FIGURE 6.104
Control system for
Problems 6.32, 6.61,
and 6.62

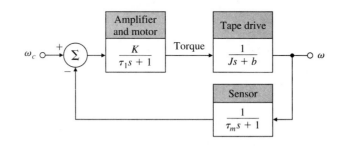

6.33 Repeat Problem 6.32 for the system shown in Fig. 6.105.

FIGURE 6.105
Control system for
Problem 6.33

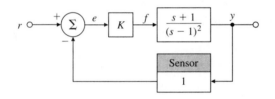

6.34 Repeat Problem 6.32 for the system shown in Fig. 6.106.

FIGURE 6.106
Control system for Problem 6.34

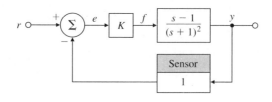

6.35 The Nyquist diagrams for two stable, open-loop systems are sketched in Fig. 6.107. The proposed operating gain is indicated as K_0, and arrows indicate increasing frequency. In each case give your best estimates of the following quantities for the closed-loop (unity feedback) system:
a) phase margin
b) damping ratio
c) closed-loop bandwidth
d) range of gain for stability (if any)
e) system type (0, I, or II).

FIGURE 6.107
Nyquist plots for Problem 6.35

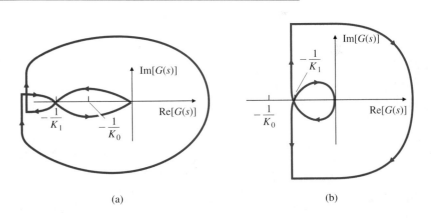

(a)   (b)

6.36 The theoretical Bode plot for a system is given in Fig. 6.108.
a) Sketch the (minimum) phase curve you would expect to accompany the given magnitude curve. Compare this with the given phase plot.
b) Draw the Nyquist plot, and consider the closed-loop stability. The system is known to have one open-loop pole in the RHP.

FIGURE 6.108
Bode plot for Problem 6.36

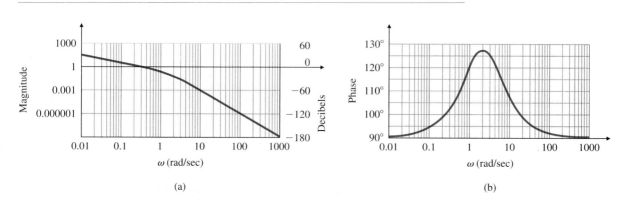

(a)

(b)

6.37 The steering dynamics of a ship are represented by the transfer function

$$\frac{V(s)}{\delta_r(s)} = G(s) = \frac{K[-(s/0.142) + 1]}{s(s/0.325 + 1)(s/0.0362 + 1)},$$

where v is the ship's lateral velocity in meters per second, and δ_r is the rudder angle in radians.

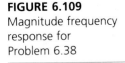

a) Use the MATLAB command bode to plot the log magnitude and phase of $G(j\omega)$ for $K = 2$.

b) On your plot, indicate the following:
   - the crossover frequency,
   - the phase margin at crossover,
   - the phase-crossover frequency where $\phi = 180°$.

FIGURE 6.109
Magnitude frequency
response for
Problem 6.38

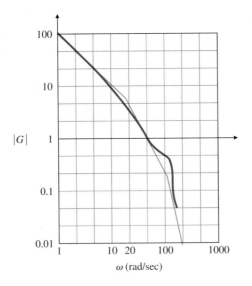

6.38 The frequency response of a plant in a unity feedback configuration is sketched in Fig. 6.109. Assume the plant is open-loop stable and minimum phase.

a) What is the velocity constant K_v for the system as drawn?

b) What is the damping ratio of the complex poles at $\omega = 100$?

c) Approximately what is the system error in tracking (following) a sinusoidal input of $\omega = 3$ rad/sec?

d) What is the phase margin of the system as drawn? (Estimate to within $\pm 10°$.)

6.39 Determine the relationship between the gain and phase for the transfer function $G(s) = 1/s$ using Bode's gain-phase relationship. *Hint:* Use the following result obtained from a table of integrals:

$$\int_0^1 \left(\ln \frac{1+z}{1-z} \right) \frac{1}{z} \, dz = \frac{\pi^2}{4}.$$

6.40 For the system

$$G(s) = \frac{100(s/a + 1)}{s(s+1)(s/b + 1)},$$

where $b = 10a$, find a value of a that will yield good stability by plotting only the magnitude of the frequency response.

6.41 a) Show that

$$\left| \frac{1}{(j\omega/\omega_n)^2 + 2\zeta(j\omega/\omega_n) + 1} \right| = \frac{1}{2\zeta} \quad \text{at} \quad \omega = \omega_n.$$

b) For a unity feedback system, show that

$$|T(j\omega_c)| = \frac{1}{2\sin(\text{PM}/2)},$$

where T is the closed-loop transfer function, and ω_c is the crossover frequency of the open-loop system.

c) Using the results of parts (a) and (b), show that

$$\zeta \cong \frac{\pi \text{PM}}{360} \cong \frac{\text{PM}}{100}$$

for small values of ζ and PM, where PM is given in degrees. (*Hint:* Approximate the closed-loop system with a second-order transfer function.)

6.42 Consider the lead compensator

$$D(s) = K \frac{Ts + 1}{\alpha Ts + 1},$$

where $\alpha < 1$.

a) Show that the phase of the lead compensator is given by

$$\phi = \tan^{-1}(T\omega) - \tan^{-1}(\alpha T\omega).$$

b) Show that the frequency where the phase is maximum is given by

$$\omega_{\max} = \frac{1}{T\sqrt{\alpha}},$$

and that the maximum phase corresponds to

$$\sin \phi_{max} = \frac{1 - \alpha}{1 + \alpha}.$$

c) Rewrite your expression for ω_{max} to show that the maximum-phase frequency occurs at the geometric mean of the two corner frequencies on a logarithmic scale:

$$\log \omega_{max} = \frac{1}{2} \left(\log \frac{1}{T} + \log \frac{1}{\alpha T} \right).$$

d) To derive the same results in terms of the pole-zero locations, rewrite $D(s)$ as

$$D(s) = K' \frac{s + z}{s + p},$$

and then show that the phase is given by

$$\phi = \tan^{-1} \left(\frac{\omega}{|z|} \right) - \tan^{-1} \left(\frac{\omega}{|p|} \right),$$

such that

$$\omega_{max} = \sqrt{|z||p|}.$$

Hence the frequency at which the phase is maximum is the square root of the product of the pole and zero locations.

6.43 Consider the third-order servo system

$$G(s) = \frac{34{,}381}{s(s + 15.5)(s + 35.5)}.$$

Design a lead compensator so that PM = 52° and GM = 4.4. Verify that your design works using MATLAB.

6.44 For the system shown in Fig. 6.110, suppose that

$$G(s) = \frac{5}{s(s/1.4 + 1)(s/3 + 1)}.$$

Design a lead compensation $D(s)$ with unity DC gain so that PM $\geqslant$ 40°. What is the approximate bandwidth of the system? Use MATLAB to verify that your design works.

FIGURE 6.110
Control system for
Problem 6.44

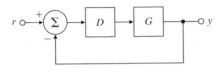

$r \circ \xrightarrow{+} \Sigma \longrightarrow \boxed{D} \longrightarrow \boxed{G} \longrightarrow \circ y$

6.45 Derive the transfer function from T_d to θ for the system in Fig. 6.66. Then apply the Final Value Theorem (assuming T_d = constant) to determine whether $\theta(\infty)$ is nonzero for the following two cases:

a) When $D(s)$ has no integral term: $\lim_{s \to 0} D(s) = $ constant;

b) When $D(s)$ has an integral term:

$$D(s) = \frac{D'(s)}{s},$$

where $\lim_{s \to 0} D'(s) = $ constant.

6.46 Consider the following transfer function, which is representative of an inverted pendulum balancer:

$$G(s) = \frac{1}{s^2(s^2 - 1)}.$$

Design a dynamic compensator to achieve a PM of 30°. Draw the root locus, and correlate it with the Bode plot of the system.

6.47 The open-loop transfer function of a unity feedback system is

$$G(s) = \frac{K}{s(1 + s/5)(1 + s/60)}.$$

Design a lag compensator for $G(s)$ so that the closed-loop system satisfies the following specifications:

- The steady-state error to a unit ramp reference input is less than 0.01.
- PM $= 40° \pm 3°$.

Verify your design using MATLAB.

6.48 The open-loop transfer function of a unity feedback system is

$$G(s) = \frac{K}{s(1 + s/5)(1 + s/250)}.$$

Design a lead compensator for $G(s)$ so that the closed-loop system satisfies the following specifications:

- The steady-state error to a unit ramp input is less than 0.01.
- For the dominant closed-loop poles the damping ratio is $\zeta = 0.4$.

Verify your design using MATLAB.

6.49 A DC motor with negligible armature inductance is to be used in a position control system. Its open-loop transfer function is given by

$$G(s) = \frac{50}{s(1 + s/2)}.$$

Design a compensator for the motor so that the closed-loop system satisfies the following specifications:

- The steady-state error to a unit ramp input is less than $1/150$.
- The unit step response has an overshoot of less than 25%.
- The bandwidth of the compensated system is no less than that of the uncompensated system.

Verify your design using MATLAB.

6.50 The open-loop transfer function of a unity feedback system is

$$G(s) = \frac{K}{s(1 + s/5)(1 + s/20)}.$$

Design a compensator for $G(s)$ so that the closed-loop system satisfies the following specifications:

- The steady-state error to a unit ramp input is less than 0.01.
- PM = $45° \pm 3°$.
- The steady-state error for sinusoidal inputs with $\omega < 0.2$ rad/sec is less than $1/250$.
- Noise components introduced with the input signal at frequencies greater than 100 rad/sec are to be attenuated at the output by at least a factor of 100, and no special noise filter is to be added.

Verify your design using MATLAB.

6.51 a) Sketch the locus of the closed-loop roots for the system whose open-loop transfer function is

$$G(s) = \frac{K}{s(s + 1.33)^3}.$$

b) Based on your root-locus sketch from part (a), find the value of the gain K at the stability boundary.

c) Is there a compensator that would stabilize this system and yield a closed-loop bandwidth of 10 rad/sec or greater? If so, find such a compensator.

6.52 Consider a type I unity feedback system with

$$G(s) = \frac{K}{s(s + 1)}.$$

Design a lead compensator so that $K_v = 12 \text{ sec}^{-1}$ and PM > 40°. Use MATLAB to verify that your design meets the specifications.

6.53 Consider a satellite-attitude control system with the transfer function

$$G(s) = \frac{0.036(s + 25)}{s^2(s + 0.02 \pm 1j)}.$$

Amplitude-stabilize the system using a double-lead network so that GM = 2(6 dB), and choose the compensator parameters so that PM = 45°.

6.54 In one mode of operation the autopilot of a jet transport is used to control altitude. For the purpose of designing the autopilot loop, only the long-period airplane dynamics are important. The linearized relationship between altitude and elevator angle is therefore third-order. Specifically,

$$G(s) = \frac{h(s)}{\delta(s)} = \frac{20(s + 0.01)}{s(s^2 + 0.01s + 0.0025)} \frac{\text{ft/sec}}{\text{deg}}.$$

The autopilot receives from the altimeter an electrical signal proportional to altitude. This signal is compared with a command signal (proportional to the altitude selected by the pilot), and the difference provides an error signal. The error signal is processed through ,a compensation network, and the result is used to

FIGURE 6.111
Control system for
Problem 6.54

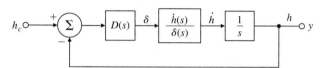

command the elevator actuators. A block diagram of this system is shown in Fig. 6.111. You have been given the task of designing the compensation. Begin by considering a proportional control law $D(s) = K$.

a) Use MATLAB to draw a Bode plot of the open-loop system for $D(s) = K = 1$.

b) What value of K would provide a crossover frequency of 0.16 rad/sec?

c) For this value of K, would the system be stable if the loop were closed?

d) What is the phase margin for this value of K?

e) Sketch the Nyquist plot of the system, and locate carefully any points where the phase angle is $180°$ or the magnitude is unity.

f) Use MATLAB to plot the root locus with respect to K, and locate the roots for your value of K from part (b).

g) What steady-state error would result if the command were a ramp input of 1 ft/sec?

For parts (h) through (k), assume a compensator of the form

$$D(s) = K \frac{Ts + 1}{\alpha Ts + 1}.$$

h) Choose the parameters K, T, and α so that the crossover frequency is 0.16 rad/sec and the phase margin is greater that $50°$. Verify your design by superimposing a Bode plot of $D(s)G(s)/K$ on top of the Bode plot you obtained for part (a), and measure the phase margin directly.

i) Use MATLAB to plot the root locus with respect to K for the system including the compensator you designed in part (h). Locate the roots for your value of K from part (h).

j) What steady-state error would result for your design if the command were a ramp input of 1 ft/sec?

k) If the error in part (j) is too large, what type of compensation would you add? At what frequencies would you put its break points?

6.55 For a system with open-loop transfer function

$$G(s) = \frac{10}{s[(s/1.4) + 1][(s/3) + 1]},$$

design a lag compensator with unity DC gain so that $PM \cong 40°$. What is the approximate bandwidth of this system? Use MATLAB to verify that your design meets the PM specification.

6.56 a) Design a compensator for the ship-steering system in Problem 6.37 that meets the following specifications:

- velocity constant $K_v = 2$,
- $PM \geqslant 50°$,
- unconditional stability ($PM > 0$ for all $\omega \leqslant \omega_c$, the crossover frequency).

b) For your final design, draw a root locus with respect to K, and indicate the location of the closed-loop poles.

6.57 Consider the system frequency-response characteristics shown for Problem 6.38.

a) A lead compensator is introduced with $\alpha = 1/5$ and a zero at $1/T = 20$. How must the gain be changed to obtain crossover at $\omega_c = 31.6$ rad/sec, and what is the resulting value of K_v?

b) With the lead compensator in place, what is the required value of K for a lag compensator that will readjust the gain to the original K_v value of 100?

c) Place the pole of the lag compensator at 3.16 rad/sec, and determine the zero location that will maintain the crossover frequency at $\omega_c = 31.6$ rad/sec. Plot the compensated frequency response on the same graph.

d) Estimate the phase margin of the compensated design, and then calculate it explicitly using MATLAB.

6.58 A feedback control system is shown in Fig. 6.112. The closed-loop system is specified to have an overshoot of less than 30% to a step input.

a) Determine the corresponding specification in the frequency domain for the closed-loop system in terms of the resonant peak value M_r.

b) Determine the resonant frequency ω_r for the maximum value of K that satisfies the specifications.

c) Determine the bandwidth of the closed-loop system resulting from part (b). (*Hint:* Plot the open-loop magnitude versus phase, as shown in Fig. 6.67, on transparent paper; then slide the paper vertically on the Nichols chart in Fig. 6.69 to obtain your answers.)

FIGURE 6.112
Control system for
Problem 6.58

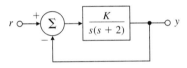

6.59 The Nichols plot of an uncompensated and a compensated system are shown in Fig. 6.113.

a) What are the resonance peaks of each system?

b) What are the phase and gain margins of each system?

c) What are the bandwidths of each system?

d) What type of compensation is used?

6.60 *M and N Circles.* Consider the closed-loop magnitude of a system,

$$\frac{|G|}{|1 + G|} = M, \qquad (6.98)$$

where M is a constant.

a) Show that this is the equation of a family of circles centered on a straight line that bisects the line between -1 and 0, that is, which goes through $(0, \frac{1}{2})$. To see this, let $G = x + jy$ in Eq. (6.98), and show that for $M \neq 1$,

$$(x - x_o)^2 + y^2 = R^2, \qquad (6.99)$$

where

$$x_o = -\frac{M^2}{M^2 - 1} \qquad (6.100)$$

FIGURE 6.113
Nichols plot for Problem 6.59

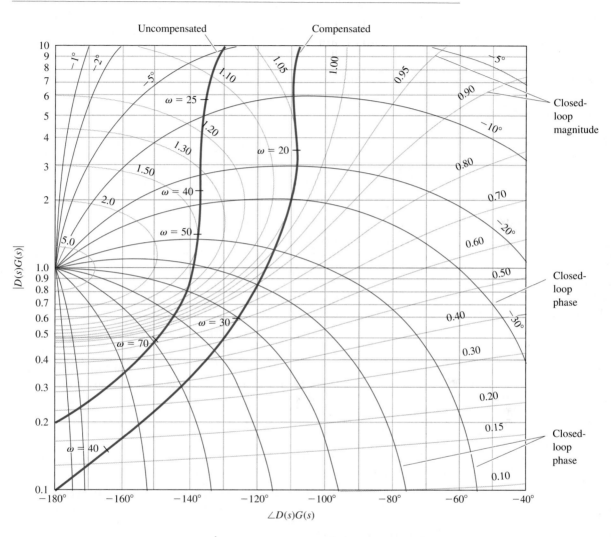

and

$$R = \left| \frac{M}{M^2 - 1} \right| \qquad \text{for } M \neq 1. \qquad (6.101)$$

Also show that for $M = 1$ the result is the straight line

$$x = -\frac{1}{2} \qquad \text{for } M = 1. \qquad (6.102)$$

In the xy-plane, Eqs. (6.99) to (6.102) describe a family of circles with centers at

$(x_o, 0)$ and radii R. Referred to as the M circles, they are loci of constant closed-loop magnitude. If $M < 1$, then $x_o > -\frac{1}{2}$; if $M > 1$, then $x_o < -\frac{1}{2}$.

b) A similar approach leads to loci of constant closed-loop phase loci,

$$\alpha = \tan^{-1} N = \left| \frac{x + jy}{1 + x + jy} \right| \tag{6.103}$$

where N is a constant. Show that

$$(x - x_o)^2 + (y - y_o)^2 = R^2, \tag{6.104}$$

where

$$x_o = -\frac{1}{2}, \qquad y_o = \frac{1}{2N}, \qquad R = \left| \frac{\sqrt{N^2 + 1}}{2N} \right|. \tag{6.105}$$

Equations (6.103) to (6.105) describe a family of circles called N circles. They are the loci of constant closed-loop phase with centers at (x_o, y_o) and radii R. All N circles go through the points $(x, y) = (0, 0)$ and $(-1, 0)$.

c) A useful alternate representation of the M circles is possible. Starting from the closed-loop transfer function

$$T = \frac{1}{1 + |G|^{-1} e^{-j\phi}}, \tag{6.106}$$

show that

$$\left(1 - \frac{1}{M^2} \right) |G|^2 + 2|G| \cos \phi + 1 = 0, \tag{6.107}$$

or

$$|G| = \frac{-\cos \phi \pm \sqrt{\cos^2 \phi - (1 - 1/M^2)}}{1 - 1/M^2} \qquad \text{for } M \neq 1. \tag{6.108}$$

Also show that for $M = 1$,

$$|G| = -\frac{1}{2 \cos \phi} \qquad \text{for } M = 1. \tag{6.109}$$

Equations (6.106) to (6.109) describe the family of curves for constant closed-loop magnitude shown in Fig. 6.69. (In log magnitude–phase coordinates these curves appear to be distorted circles.) Show that the curves are symmetric with respect to the $-180°$ line.

d) Similarly, show that

$$|G| = -\cos \phi + \frac{1}{N} \sin \phi. \tag{6.110}$$

Again Eq. (6.110) describes the family of N circles shown as the curves of constant closed-loop phase loci in Fig. 6.69. (As before, in the log magnitude–phase coordinates, the curves do not look like true circles.) Show that the curves are symmetric with respect to $-180°$.

6.61 Consider the system shown in Fig. 6.104.

a) Construct an inverse Nyquist plot of $[Y(j\omega)/E(j\omega)]^{-1}$.

b) Show how the value of K for neutral stability can be read directly from the inverse Nyquist plot.

c) For $K = 4$, 2, and 1, determine the gain and phase margins.

d) Construct a root-locus plot for the system, and identify corresponding points in the two plots. To what damping ratios ζ do the GM and PM of part (c) correspond?

6.62 An unstable plant has the transfer function

$$\frac{Y(s)}{F(s)} = \frac{s+1}{(s-1)^2}.$$

A simple control loop is to be closed around it, as shown in Fig. 6.104.

a) Construct an inverse Nyquist plot of Y/F.

b) Choose a value of K to provide a PM of 45°. What is the corresponding GM?

c) What can you infer from your plot about the stability of the system when $K < 0$?

d) Construct a root-locus plot for the system, and identity corresponding points in the two plots. In this case, to what value of ζ does PM = 45° correspond?

6.63 Consider the system shown in Fig. 6.114(a).

a) Construct a Bode plot for the system.

b) Use your Bode plot to sketch an inverse Nyquist plot.

c) Consider closing a control loop around $G(s)$, as shown in Fig. 6.114(b). Using the inverse Nyquist plot as a guide, read from your Bode plot the values of GM and PM when $K = 0.7$, 1.0, 1.4, and 2. What value of K yields PM = 30°?

d) Construct a root-locus plot, and label the same values of K on the locus. To what value of ζ does each pair of PM/GM values correspond?

FIGURE 6.114
Control system for
Problem 6.63

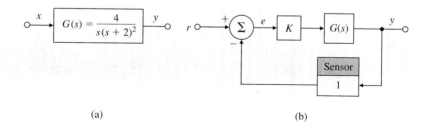

(a)   (b)

6.64 Repeat Problem 6.63 for the system shown in Fig. 6.115, using Fig. 6.114(b) again for part (c). For this case, use values of $K = 0.5$, 1, 2, and 4.

FIGURE 6.115
Control system for Problem 6.64

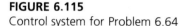

6.65 Draw the inverse Nyquist plot for the open-loop system

$$KG(s) = \frac{K(s + 1)}{s^2(s + 10)^2}.$$

Determine the values for K at the stability boundary and at the point where PM = 30°.

6.66 Augment the system of Problem 6.63 to include a lead compensator, as shown in Fig. 6.116.

a) Superimpose the new root locus, Bode, and inverse Nyquist plots on those for Problem 6.63.

b) What value of K now produces PM = 30°? What are the values of ζ and GM for this K?

c) Compare the closed-loop system's lowest natural frequency, with and without the lead compensator, when K is set to produce PM = 30° in each case.

FIGURE 6.116
Control system for
Problem 6.66

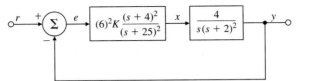

6.67 Consider a system with the loop gain

$$G(s) = \frac{1}{s(s + 1)}.$$

a) Draw Bode and Nyquist plots for the system, and find the gain and phase margins.

b) Compute the sensitivity and complementary sensitivity functions, and plot their magnitude frequency responses.

c) Find estimates of the gain and phase margins from part (b). Compare the results to those of part (a), and comment on their conservatism.

d) Draw the inverse Nyquist plot for the system, and identify the complex margin. Compute approximate values for the PM and GM based on the complex margin, and compare your answer with part (a).

6.68 Consider the unity feedback system with

$$G(s) = \frac{0.036(s + 25)}{s^2(s + 0.02 \pm 1j)},$$

$$D(s) = 0.5(1.4s + 1)\frac{(s/0.9)^2 + 1}{[(s/25) + 1]^2}.$$

Determine the sensitivity, complementary sensitivity, and control sensitivity functions for the system. Plot the magnitude of the sensitivity functions on the same graph, and determine estimates of the gain and phase margins.

6.69 Prove that the sensitivity function $S(s)$ has magnitude greater than 1 inside a circle with a radius of 1 centered at the -1 point. What does this imply about the shape of the Nyquist plot if closed-loop control is to outperform open-loop control at all frequencies?

6.70 a) Prove that, for a stable open-loop system with at least two more poles than zeros,

$$\int_0^\infty \ln |S(j\omega)|\, d\omega = 0, \qquad (6.111)$$

where S is the sensitivity function. Equation (6.111) indicates that if the sensitivity is reduced over some frequency range, it must increase at other frequencies.

b) Prove that if an open-loop system has RHP poles $\{p_i, i = 1, \ldots, P\}$ (including multiplicities), and if the magnitude of its frequency response rolls off at high frequencies, then

$$\int_0^\infty \ln |S(j\omega)|\, d\omega = \pi \sum_{i=1}^{P} \mathrm{Re}\{p_i\}.$$

Hint: see Doyle, *et al.*, 1992.

FIGURE 6.117
Nonlinear feedback with inputs r_1 and r_2 and outputs u_1, y_1, u_2, and y_2

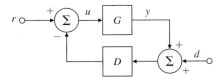

6.71 *Small Gain Theorem*: An alternative derivation of the stability robustness result is possible using the Small Gain Theorem (Zames, 1966; Mees, 1981). Consider the system shown in Fig. 6.117, where $\mathbf{H}_1$ and $\mathbf{H}_2$ are not necessarily linear. The Small Gain Theorem states that if the inputs (r_1 and r_2) are bounded, and the loop gain is less than 1 (that is, $|\mathbf{H}_2||\mathbf{H}_1| < 1$), then the closed-loop system is bounded input–bounded output (BIBO) stable; that is, u_1, u_2, y_1, and y_2 are all bounded. Now consider the typical feedback system shown in Fig. 6.118, where it is known that

$$\left| \frac{G - G_0}{G_0} \right| < l(\omega).$$

Prove that the system shown in Fig. 6.118 is equivalent to that of Fig. 6.119. Apply the Small Gain Theorem to the system shown in Fig. 6.119 to obtain the stability-

FIGURE 6.118
Typical feedback system with plant dynamics G, feedback dynamics D, reference input r, and disturbance input d

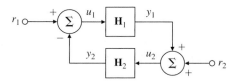

FIGURE 6.119
Equivalent feedback representation of Fig. 6.118

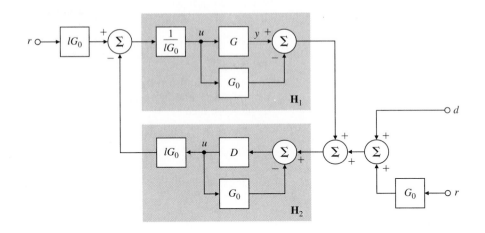

robustness result derived in this chapter. Explain why is it not meaningful to apply the theorem to the system in Fig. 6.118 directly.

6.72 Assume that the system

$$G(s) = \frac{e^{-0.2s}}{s + 10},$$

has a 0.2-sec time delay ($T = 0.2$ sec). While maintaining a phase margin of about 40°, find the maximum possible bandwidth using the following:

a) One lead-compensator section

$$D(s) = K \frac{s + a}{s + b},$$

where $b/a = 100$;

b) Two lead-compensator sections

$$D(s) = K \left(\frac{s + a}{s + b} \right)^2,$$

where $b/a = 10$.

6.73 Determine the range of K for which the following systems are stable:

a) $G(s) = K \dfrac{e^{-4s}}{s}$

b) $G(s) = K \dfrac{e^{-s}}{s(s + 2)}$

6.74 The pure time delay $x_{out}(t) = x_{in}(t - \tau)$ is represented by the transfer function $H(s) = e^{-s\tau}$.

   a) Graph $|H(\omega)|$ and $\angle H(\omega)$ versus ω.

   b) $H(s)$ cannot be represented exactly as a ratio of polynomials in s. Therefore, $H(s)$ cannot be described by a collection of poles and zeros. However, $H(s)$ may be approximated by a rational transfer function. Consider the first order Padé

$$H_1(s) = \frac{1 - \tau s/2}{1 + \tau s/2}.$$

Graph $|H_1(\omega)|$ and $\angle H_1(\omega)$, and compare your results with those from part (a).

6.75 Consider the heat exchanger of Example 2.16 with the open-loop transfer function

$$G(s) = \frac{e^{-5s}}{(10s + 1)(60s + 1)}.$$

   a) Design a lead compensator that yields PM = 45° and the maximum possible closed-loop bandwidth.

   b) Design a PI compensator that yields PM = 45° and the maximum possible closed-loop bandwidth.

6.76 The Bode plot in Fig. 6.120 is for a transfer function of the form

$$G(s) = \frac{Ks}{(1 + T_1 s)(1 + T_2 s)^2},$$

where K, T_1, and T_2 are positive constants. Determine values for these three constants from the Bode plot.

FIGURE 6.120
Bode plot for Problem 6.76

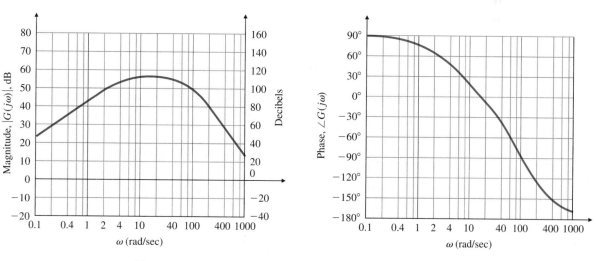

(a)   (b)

FIGURE 6.121
Bode plot for Problem 6.77

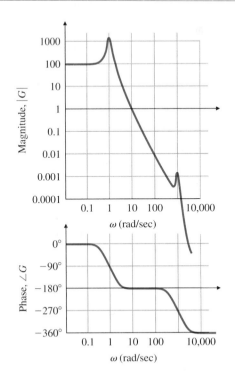

6.77 You are given the experimentally determined Bode plot shown in Fig. 6.121. Design a compensation that will yield a crossover frequency of $\omega_c = 10$ rad/sec with PM $> 75°$.

6.78 Consider the frequency-response plot in Fig. 6.122, which was obtained from a finite-element analysis. Determine a transfer function that approximately matches the measured frequency response up to 60 Hz.

6.79 The frequency response data shown in Table 6.1 were taken from a DC motor that is to be used in a position control system. Assume that the motor is linear and minimum-phase.

a) Estimate $G(s)$, the transfer function of the system.

b) Design a series compensator for the motor so that the closed-loop system meets the following specifications:

- The steady-state error to a unit ramp input is less than 0.01.
- PM $= 45° \pm 5°$.

FIGURE 6.122
Frequency-response plot for Problem 6.78

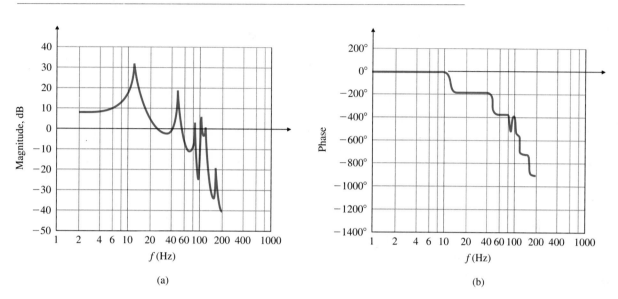

(a)

(b)

TABLE 6.1   Frequency-response data for Problem 6.79

ω (rad/sec)	|G(s)| (dB)	ω (rad/sec)	|G(s)| (dB)	ω (rad/sec)	|G(s)| (dB)
0.1	60.0	3.0	30.5	60.0	−20.0
0.2	54.0	4.0	27.0	65.0	−21.0
0.3	50.0	5.0	23.0	80.0	−24.0
0.5	46.0	7.0	19.5	100.0	−30.0
0.8	42.0	10.0	14.0	200.0	−48.0
1.0	40.0	20.0	2.0	300.0	−59.0
2.0	34.0	40.0	−10.0	500.0	−72.0

· 7 ·

State-space Design

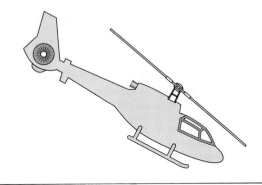

A Perspective on State-space Design

In addition to the transform techniques of root locus and frequency response, there is a third major method of designing feedback control systems: the state-space method. In Chapter 2 we introduced the state-variable method of describing differential equations. In state-space design, the control engineer designs a dynamic compensation by working directly with the state-variable description of the system. Like the transform techniques, the aim of the state-space method is to find a compensation $D(s)$, such as that shown in Fig. 7.1, that satisfies the design specifications. Because the state-space method of describing the plant and computing the compensation is so different from the transform techniques, it may seem at first to be solving an entirely different problem. We selected the examples and analysis given toward the end of this chapter to help convince you that, indeed, state-space design

FIGURE 7.1
A control system design definition

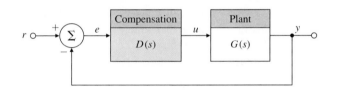

469

results in a compensator with a transfer function $D(s)$ that is equivalent to those $D(s)$ compensators obtained with the other two methods.

Because it is particularly well suited to the use of computer techniques, state-space design is increasingly studied and used today by control engineers.

Chapter Overview

This chapter begins by considering the purposes and advantages of using state-space design. In Section 7.2 we review the development of state-variable equations from block diagrams. We then solve for the dynamic response, using state equations for both hand and computer analysis. Having covered these fundamentals, we next proceed to the major steps of the state-space design method:

1. Select pole locations and develop the control law for the closed-loop system that corresponds to satisfactory dynamic response (Sections 7.3 and 7.4);
2. Design an estimator (Section 7.5);
3. Combine the control law and the estimator (Section 7.6);
4. Introduce the reference input (Section 7.7).

After working through the central design steps, we briefly explore the use of integral control in state space (Section 7.8). The final three sections of this chapter consider briefly some additional concepts pertaining to the state-space method; because they are relatively advanced, they may be considered optional to some courses or readers.

7.1   Advantages of State Space

The idea of **state space** comes from the state-variable method of describing differential equations, which we introduced in Chapter 2. In this method the differential equations describing a dynamic system are organized as a set of first-order differential equations in the vector-valued state of the system, and the solution is visualized as a trajectory of this state vector in space. **State-space control design** is the technique in which the control engineer designs a dynamic compensation by working directly with the state-variable description of the system. Thus far we have seen that the ordinary differential equations (ODEs) of physical dynamic systems can be manipulated into state-variable form. In the field of mathematics, where ODEs are studied, the state-variable form is called the **normal form** for the equations. There are several good reasons for studying equations in this form, three of which are listed here:

- *To study more general models:* The ODEs do not have to be linear or stationary. Thus by studying the equations themselves, we can develop methods that are very general. Having them in state-variable form gives us a compact, standard form for study. Of course, in this text we study mainly linear constant models (for the reasons given earlier).

- *To introduce the ideas of geometry into differential equations:* In physics the plane of position versus velocity of a particle or rigid body is called the **phase plane**, and the trajectory of the motion can be plotted as a curve in this plane. The state is a generalization of that idea to include more than two dimensions. While we cannot plot more than three dimensions, the concepts of distance, of orthogonal and parallel lines, and other concepts from geometry can be useful in visualizing the solution of an ODE as a path in state space.

- *To connect internal and external descriptions:* The state of a dynamic system often directly describes the distribution of internal energy in the system. For example, it is common to select the following as state variables: position (potential energy), velocity (kinetic energy), capacitor voltage (electric energy), and inductor current (magnetic energy). The internal energy can always be computed from the state variables. By a system of analysis to be described shortly, we can relate the state to the system inputs and outputs and thus connect the internal variables to the external inputs and to the sensed outputs. In contrast, the transfer function relates only the input to the output and does not show the internal behavior. The state form keeps the latter information, which is sometimes important.

Use of the state-space approach has often been referred to as **modern control design**, and use of transfer-function-based methods such as root locus and frequency response referred to as **classical control design**. However, since the state-space method of description for ODEs has been in use for over 100 years and was introduced to control design in the late 1950s, it seems somewhat misleading to refer to it as modern. We prefer to refer to the two approaches to design as state-space methods and transform methods.

Advantages of state-space design are especially apparent when the system to be controlled has more than one control input or more than one sensed output. However, in this book we will examine the ideas of state-space design using the simpler single input–single output systems. The design approach used for systems described in state form is "divide and conquer." First we design the control as if all of the state were measured and available for use in the control law. Having a satisfactory control law based on full state feedback, we introduce the concept of an observer and construct estimates of the state based on the sensed output. We then show that these estimates can be used in place of the actual state variables. Finally, we introduce the external reference-command inputs, and the structure is complete. Only at this point can we recognize that the resulting compensation has the same essential structure as that developed with transform methods.

Before we can begin the design using state descriptions, it is necessary to develop some analytical results and tools from matrix linear algebra for use throughout the chapter. We assume that you are familiar with such elementary matrix concepts as the identity matrix, triangular and diagonal matrices, and the transpose of a matrix. We also assume you have some familiarity with the mechanics of matrix algebra, including adding, multiplying, and inverting matrices. More advanced results will be developed in Section 7.2 in the context of the dynamic response of a linear system. All of the linear algebra results used in this chapter are repeated in Appendix C for your reference and review.

7.2   Analysis of the State Equations

In Chapter 2 we introduced and illustrated the process of selecting a state and organizing the equations in state form. In this section we review that process and describe how to analyze the dynamic response using the state description. In Section 7.2.1 we begin by relating the state description to block diagrams and the Laplace transform description and to consider the fact that for a given system the choice of state is not unique. We show how to use this nonuniqueness to select among several canonical forms for the one that will help solve the particular problem at hand; a control canonical form makes feedback gains of the state easy to design. After studying the structure of state equations in Section 7.2.2, we consider the dynamic response and show how transfer-function poles and zeros are related to the matrices of the state descriptions. To illustrate the results with hand calculations, we offer a simple nonphysical example. For more realistic examples, a computer-aided control-systems-design (CACSD) software package such as MATLAB is especially helpful; relevant MATLAB commands will be described from time to time.

7.2.1   Block Diagrams and Canonical Forms

We begin with a system that has a simple transfer function

$$G = \frac{b(s)}{a(s)} = \frac{s + 2}{s^2 + 7s + 12} = \frac{2}{s + 4} + \frac{-1}{s + 3} \tag{7.1}$$

The roots of the numerator polynomial $b(s)$ are the zeros of the transfer function, and the roots of the denominator polynomial $a(s)$ are the poles. Notice that we have represented the transfer function in two forms, as a ratio of polynomials and as the result of a partial-fraction expansion. In order to develop a state description of this system (and this is a generally useful

FIGURE 7.2
A block diagram
representing Eq. (7.1) in
control form

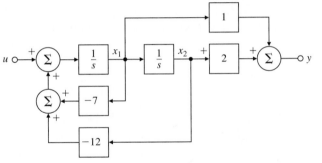

technique), we construct a block diagram that corresponds to the transfer function (and the differential equations) *using only isolated integrators as the dynamic elements.* One such block diagram, structured in **control canonical form**, is drawn in Fig. 7.2. The central feature of this structure is that each state variable is connected by the feedback to the control input.

Once we have drawn the block diagram in this form, we can identify the state description matrices simply by inspection; this is possible because when the output of an integrator is a state variable, the input of that integrator is the derivative of that variable. For example, in Fig. 7.2, the equation for the first state variable is

$$\dot{x}_1 = -7x_1 - 12x_2 + u.$$

Continuing in this fashion, we get

$$\dot{x}_2 = x_1,$$

$$y = x_1 + 2x_2.$$

These three equations can then be rewritten in the matrix form

$$\dot{\mathbf{x}} = \mathbf{A}_c\mathbf{x} + \mathbf{B}_c u, \tag{7.2}$$

$$y = \mathbf{C}_c\mathbf{x}, \tag{7.3}$$

where

$$\mathbf{A}_c = \begin{bmatrix} -7 & -12 \\ 1 & 0 \end{bmatrix}, \quad \mathbf{B}_c = \begin{bmatrix} 1 \\ 0 \end{bmatrix},$$

$$\mathbf{C}_c = [1 \quad 2], \quad\quad D_c = 0, \tag{7.4}$$

and where the subscript c refers to control canonical form.

Two significant facts about this form are that the coefficients 1 and 2 of the numerator polynomial $b(s)$ appear in the $\mathbf{C}_c$ matrix, and (except for the leading term) the coefficients 7 and 12 of the denominator polynomial $a(s)$ appear (with opposite signs) as the first row of the $\mathbf{A}_c$ matrix. Armed with this knowledge, we can thus write down *by inspection* the state matrices in control canonical form for any system whose transfer function is known as a ratio of numerator and denominator polynomials. If $b(s) = b_1 s^{n-1} + b_2 s^{n-2} + \cdots + b_n$ and $a(s) = s^n + a_1 s^{n-1} + a_2 s^{n-2} + \cdots + a_n$, then the MATLAB steps are:

$$b = [b_1 \quad b_2 \quad \cdots \quad b_n]$$

$$a = [1 \quad a_1 \quad a_2 \quad \cdots \quad a_n]$$

MATLAB tf2ss

$$[\mathbf{A}_c, \quad \mathbf{B}_c, \quad \mathbf{C}_c, D_c] = \text{tf2ss}(b, a).$$

We read tf2ss as "transfer function to state space." The result will be

Control canonical form

$$\mathbf{A}_c = \begin{bmatrix} -a_1 & -a_2 & \cdots & -a_n \\ 1 & 0 & \cdots & 0 \\ \vdots & & & \\ 0 & & & \vdots \\ 0 & \cdots & 0 & 1 & 0 \end{bmatrix}, \quad \mathbf{B}_c = \begin{bmatrix} 1 \\ 0 \\ \vdots \\ 0 \end{bmatrix}, \tag{7.5}$$

$$\mathbf{C}_c = [b_1 \quad b_2 \quad \cdots \quad b_n], \qquad D_c = 0.$$

The block diagram of Fig. 7.2 and the corresponding matrices of Eqs. (7.4) are not the only way to represent the transfer function $G(s)$. A block diagram corresponding to the partial-fraction expansion of $G(s)$ is given in Fig. 7.3. Using the same technique as before, with the state variables marked as shown on the figure, we can determine the matrices directly from the block diagram as being

$$\dot{\mathbf{z}} = \mathbf{A}_m \mathbf{z} + \mathbf{B}_m u,$$

$$y = \mathbf{C}_m \mathbf{z} + D_m u,$$

where

$$\mathbf{A}_m = \begin{bmatrix} -4 & 0 \\ 0 & -3 \end{bmatrix}, \quad \mathbf{B}_m = \begin{bmatrix} 1 \\ 1 \end{bmatrix}, \tag{7.6}$$

$$\mathbf{C}_m = [2 \quad -1], \qquad D_m = 0,$$

FIGURE 7.3
Block diagram for Eq. (7.1) in modal canonical form

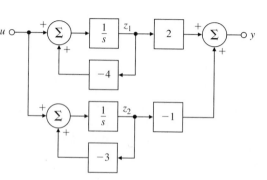

modal form

and the subscript m refers to **modal canonical form**. The name for this form derives from the fact that the poles of the system transfer function are sometimes called the **normal modes** of the system. The important fact about the matrices in this form is that the system poles (here -4 and -3) appear as the elements along the diagonal of the $\mathbf{A}_m$ matrix, and the residues, the numerator terms in the partial-fraction expansion, (here 2 and -1) appear in the $\mathbf{C}_m$ matrix.

Expressing a system in modal canonical form can be complicated by two factors: (1) the elements of the matrices will be complex when the poles of the system are complex, and (2) the system matrix *cannot* be diagonal when the partial-fraction expansion has repeated poles. To solve the first problem we express the complex poles of the partial-fraction expansion as conjugate pairs in second-order terms so that all the elements remain real. The corresponding $\mathbf{A}_m$ matrix will then have 2×2 blocks along the main diagonal representing the local coupling between the variables of the complex-pole set. To handle the second difficulty we also couple the corresponding state variables, so that the poles appear along the diagonal with off-diagonal terms indicating the coupling. A simple example of this latter case is the system from Example 2.7, whose transfer function is $G(s) = 1/s^2$. The system matrices for this transfer function in a modal form are

$$\mathbf{F} = \begin{bmatrix} 0 & 1 \\ 0 & 0 \end{bmatrix}, \qquad \mathbf{G} = \begin{bmatrix} 0 \\ 1 \end{bmatrix},$$

$$\mathbf{H} = [1 \quad 0], \qquad J = 0. \tag{7.7}$$

◆ **EXAMPLE 7.1**   *State Equations in Modal Canonical Form*

A simple model for a disk drive servo with one resonant mode has a transfer function given by

$$G(s) = \frac{2s + 4}{s^2(s^2 + 2s + 4)} = \frac{1}{s^2} - \frac{1}{s^2 + 2s + 4}. \tag{7.8}$$

Find state matrices in modal form describing this system.

Solution.   The transfer function has been given in real partial-fraction form. To get state-description matrices, we draw a corresponding block diagram with integrators only, assign the state, and write down the corresponding matrices. This process is not unique, so there are several acceptable solutions to the problem as stated, but they will differ in only trivial ways. A block diagram with a satisfactory assignment of variables is given in Fig. 7.4.

Notice that the second-order term to represent the complex poles has been realized in control canonical form. There are a number of other possibilities that can be used as alternatives for this part. This particular form allows us to write down the system

FIGURE 7.4
Block diagram for a fourth-order system in modal canonical form with shading
indicating portion in control canonical form

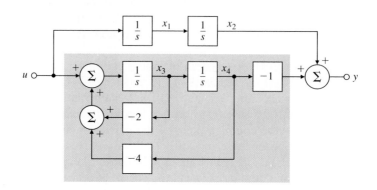

matrices by inspection:

$$F = \begin{bmatrix} 0 & 0 & 0 & 0 \\ 1 & 0 & 0 & 0 \\ 0 & 0 & -2 & -4 \\ 0 & 0 & 1 & 0 \end{bmatrix}, \qquad G = \begin{bmatrix} 1 \\ 0 \\ 1 \\ 0 \end{bmatrix},$$

$$H = [0 \quad 1 \quad 0 \quad -1], \qquad J = 0.$$

(7.9)

Thus far we have seen that we can obtain the state description from a transfer function in either control or modal form. Since these matrices represent the same dynamic system, we might ask: What is the relationship between the matrices in the two forms (and their corresponding state variables)? More generally, suppose we have a set of state equations that describe some physical system in no particular form and we are given a problem for which the control canonical form would be helpful. (We will see such a problem in Section 7.3.) Is it possible to calculate the desired canonical form without obtaining the transfer function first? To answer these questions requires a look at the topic of state transformations.

Consider a system described by the state equations

State description and
output equation

$$\dot{x} = Fx + Gu, \tag{7.10a}$$

$$y = Hx + Ju. \tag{7.10b}$$

As we have seen, this is not a unique description of the dynamic system. We consider a change of state from x to a new state z that is a linear transformation of $\mathbf{x}$. For a nonsingular matrix $\mathbf{T}$ we let

$$\mathbf{x} = \mathbf{Tz}. \tag{7.11}$$

By substituting Eq. (7.11) into Eq. (7.10a), we have the equations of motion in terms of the new state $\mathbf{z}$:

$$\dot{\mathbf{x}} = \mathbf{T}\dot{\mathbf{z}} = \mathbf{FTz} + \mathbf{G}u,$$

$$\dot{\mathbf{z}} = \mathbf{T}^{-1}\mathbf{FTz} + \mathbf{T}^{-1}\mathbf{G}u, \tag{7.12}$$

$$\dot{\mathbf{z}} = \mathbf{A}z + \mathbf{B}u,$$

where

Transformation of state

$$\mathbf{A} = \mathbf{T}^{-1}\mathbf{FT}, \tag{7.13a}$$

$$\mathbf{B} = \mathbf{T}^{-1}\mathbf{G}. \tag{7.13b}$$

Then we substitute Eq. (7.11) into Eq. (7.10b) to get the output in terms of the new state $\mathbf{z}$:

$$y = \mathbf{HTz} + Ju$$

$$= \mathbf{C}z + Du,$$

where

$$\mathbf{C} = \mathbf{HT}, \qquad D = J. \tag{7.14}$$

Given the general matrices $\mathbf{F}$, $\mathbf{G}$, and $\mathbf{H}$ and scalar J, we would like to find the transformation matrix $\mathbf{T}$ such that $\mathbf{A}$, $\mathbf{B}$, $\mathbf{C}$, and D are in a particular form, for example, control canonical form. To find such a $\mathbf{T}$ we assume that $\mathbf{A}$, $\mathbf{B}$, $\mathbf{C}$, and D are already in the required form, further assume that the transformation $\mathbf{T}$ has a general form, and match terms. Here we will work out the third-order case; how to extend the analysis to the general case should be clear from the development. It goes like this.

First we rewrite Eq. (7.13a) as

$$\mathbf{AT}^{-1} = \mathbf{T}^{-1}\mathbf{F}.$$

If $\mathbf{A}$ is in control canonical form, and we describe $\mathbf{T}^{-1}$ as a matrix with rows $\mathbf{t}_1$, $\mathbf{t}_2$, and $\mathbf{t}_3$, then

$$\begin{bmatrix} -a_1 & -a_2 & -a_3 \\ 1 & 0 & 0 \\ 0 & 1 & 0 \end{bmatrix} \begin{bmatrix} \mathbf{t}_1 \\ \mathbf{t}_2 \\ \mathbf{t}_3 \end{bmatrix} = \begin{bmatrix} \mathbf{t}_1\mathbf{F} \\ \mathbf{t}_2\mathbf{F} \\ \mathbf{t}_3\mathbf{F} \end{bmatrix}. \tag{7.15}$$

Working out the third and second rows gives the matrix equations

$$\mathbf{t}_2 = \mathbf{t}_3\mathbf{F}, \tag{7.16a}$$

$$\mathbf{t}_1 = \mathbf{t}_2\mathbf{F} = \mathbf{t}_3\mathbf{F}^2. \tag{7.16b}$$

From Eq. (7.13b), assuming that **B** is also in control canonical form, we have the relation

$$\mathbf{T}^{-1}\mathbf{G} = \mathbf{B},$$

or

$$\begin{bmatrix} \mathbf{t}_1\mathbf{G} \\ \mathbf{t}_2\mathbf{G} \\ \mathbf{t}_3\mathbf{G} \end{bmatrix} = \begin{bmatrix} 1 \\ 0 \\ 0 \end{bmatrix}. \tag{7.17}$$

Combining Eqs. (7.16) and (7.17), we get

$$\mathbf{t}_3\mathbf{G} = 0,$$

$$\mathbf{t}_2\mathbf{G} = \mathbf{t}_3\mathbf{F}\mathbf{G} = 0,$$

$$\mathbf{t}_1\mathbf{G} = \mathbf{t}_3\mathbf{F}^2\mathbf{G} = 1.$$

These equations can in turn be written in matrix form as

$$\mathbf{t}_3[\mathbf{G} \quad \mathbf{F}\mathbf{G} \quad \mathbf{F}^2\mathbf{G}] = [0 \quad 0 \quad 1]$$

or

$$\mathbf{t}_3 = [0 \quad 0 \quad 1]\mathscr{C}^{-1}, \tag{7.18}$$

Controllability matrix transformation to control canonical form

where the **controllability matrix** $\mathscr{C} = [\mathbf{G} \quad \mathbf{F}\mathbf{G} \quad \mathbf{F}^2\mathbf{G}]$. Having $\mathbf{t}_3$, we can now go back to Eq. (7.16) and construct all the rows of $\mathbf{T}^{-1}$.

To sum up, the recipe for converting a general state description of dimension n to control canonical form is as follows:

- From **F** and **G**, form the controllability matrix $\mathscr{C}$, as

$$\mathscr{C} = [\mathbf{G} \quad \mathbf{F}\mathbf{G} \quad \cdots \quad \mathbf{F}^{n-1}\mathbf{G}]. \tag{7.19}$$

- Compute the last row of the inverse of the transformation matrix as

$$\mathbf{t}_n = [0 \quad 0 \quad \cdots \quad 1]\mathscr{C}^{-1}. \tag{7.20}$$

- Construct the entire transformation matrix as

$$\mathbf{T}^{-1} = \begin{bmatrix} \mathbf{t}_n\mathbf{F}^{n-1} \\ \mathbf{t}_n\mathbf{F}^{n-2} \\ \vdots \\ \mathbf{t}_n \end{bmatrix}. \tag{7.21}$$

- Compute the new matrices from $\mathbf{T}^{-1}$ using Eqs. (7.13) and Eq. (7.14).

Controllable systems

When the controllability matrix $\mathscr{C}$ is nonsingular, the corresponding **F** and **G** matrices are said to be **controllable**. This is a technical property that usually holds for physical systems and will be important when we consider feedback of the state in Section 7.3. We will also consider a few physical illustrations of loss of controllability at that time.

Because computing the transformation given by Eq. (7.21) is numerically difficult to do accurately, it is almost never done. The reason for developing this transformation in some detail is to show how such changes of state could be done in theory and to make the following important observation:

One can *always* transform a given state description to control canonical form if (and only if) the controllability matrix $\mathscr{C}$ is nonsingular.

If we need to test for controllability in a real case with numbers, we use a numerically stable method that depends on converting the system matrices to "staircase" form rather than on trying to compute the controllability matrix. Problem 7.30 at the end of the chapter calls for consideration of this method.

An important question regarding controllability follows directly from our discussion so far: What is the effect of a state transformation on controllability? We can show the result by using Eqs. (7.19) and (7.13). The controllability matrix of the system $(\mathbf{F}, \mathbf{G})$ is

$$\mathscr{C}_\mathbf{x} = [\mathbf{G} \quad \mathbf{FG} \quad \cdots \quad \mathbf{F}^{n-1}\mathbf{G}]. \tag{7.22}$$

After the state transformation, the new description matrices are given by Eqs. (7.13), and the controllability matrix changes to

$$
\begin{aligned}
\mathscr{C}_\mathbf{z} &= [\mathbf{B} \quad \mathbf{AB} \quad \cdots \quad \mathbf{A}^{n-1}\mathbf{B}] \\
&= [\mathbf{T}^{-1}\mathbf{G} \quad \mathbf{T}^{-1}\mathbf{F}\mathbf{T}\mathbf{T}^{-1}\mathbf{G} \quad \cdots \quad \mathbf{T}^{-1}\mathbf{F}^{n-1}\mathbf{T}\mathbf{T}^{-1}\mathbf{G}] \\
&= \mathbf{T}^{-1}\mathscr{C}_\mathbf{x}.
\end{aligned} \tag{7.23}
$$

Thus we see that $\mathscr{C}_\mathbf{z}$ is nonsingular if and only if $\mathscr{C}_\mathbf{x}$ is nonsingular, yielding the following observation:

A change of state by a nonsingular linear transformation does *not* change controllability.

Observer canonical form

We return once again to the transfer function of Eq. (7.1), this time to represent it with the block diagram having the structure known as **observer canonical form** (Fig. 7.5). The corresponding matrices for this form are

$$
\mathbf{A}_o = \begin{bmatrix} -7 & 1 \\ -12 & 0 \end{bmatrix}, \qquad \mathbf{B}_o = \begin{bmatrix} 1 \\ 2 \end{bmatrix},
$$
$$
\mathbf{C}_o = [1 \quad 0], \qquad D_o = 0. \tag{7.24}
$$

The significant fact about this canonical form is that all the feedback is from the output to the state variables.

FIGURE 7.5
Observer canonical form

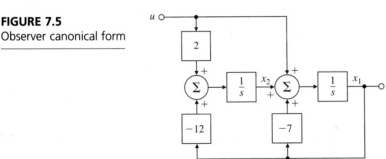

Let us now consider what happens to the controllability of this system as the zero at -2 is varied. For this purpose we replace the second element 2 of $\mathbf{B}_o$ with the variable zero location $-z_o$ and form the controllability matrix:

$$\mathscr{C}_x = [\mathbf{B}_o \quad \mathbf{A}_o \mathbf{B}_o]$$

$$= \begin{bmatrix} 1 & -7-z_o \\ -z_o & -12 \end{bmatrix}. \qquad (7.25)$$

The determinant of this matrix is a function of z_o:

$$\det(\mathscr{C}_x) = -12 + (z_o)(-7 - z_o)$$

$$= -(z_o^2 + 7z_o + 12).$$

This polynomial is zero for $z_o = -3$ or -4, implying that controllability is lost for these values. What does this mean? In terms of the parameter z_o, the transfer function is

$$G(s) = \frac{s - z_o}{(s + 3)(s + 4)}.$$

If $z_0 = -3$ or -4, there is a pole-zero cancellation and the transfer function reduces from a second-order system to a first-order one. When $z_o = -3$, for example, the mode at -3 is decoupled from the input and control of this mode is lost.

Notice that we have taken the transfer function given by Eq. (7.1) and given it two realizations, one in control canonical form and one in observer canonical form. The control form is always controllable for any value of the zero, while the observer form loses controllability if the zero cancels either of the poles. Thus, these two forms may represent the same transfer function, but it may not be possible to transform the state of one to the state of the other (in this case, from observer to control canonical form). While a transformation of state cannot affect controllability, the particular state selected from a transfer function can:

Controllability is a function of the *state* of the system and cannot be decided from a transfer function.

To discuss controllability more at this point would take us too far afield. The closely related property of observability and the observer canonical form will be taken up in Section 7.5.1. A more detailed discussion of these properties of dynamic systems is given in the Appendix D, for those who would like to learn more.

We return now to the modal form for the equations, given by Eqs. (7.6) for the example transfer function. As mentioned before, it is not always possible to find a modal form for transfer functions that have repeated poles, so we assume our system has only distinct poles. Furthermore, we assume that the general state equations given by Eqs. (7.10) apply. We want to find a transformation matrix $\mathbf{T}$ defined by Eq. (7.11) such that the transformed Eqs. (7.13) and (7.14) will be in modal form. In this case, we assume that the $\mathbf{A}$ matrix is diagonal and that $\mathbf{T}$ is composed of the *columns* $\mathbf{t}_1$, $\mathbf{t}_2$, and $\mathbf{t}_3$. With this assumption, the state transformation Eq. (7.13a) becomes

$$\mathbf{TA} = \mathbf{FT}$$

$$[\mathbf{t}_1 \quad \mathbf{t}_2 \quad \mathbf{t}_3] \begin{bmatrix} p_1 & 0 & 0 \\ 0 & p_2 & 0 \\ 0 & 0 & p_3 \end{bmatrix} = \mathbf{F}[\mathbf{t}_1 \quad \mathbf{t}_2 \quad \mathbf{t}_3]. \qquad (7.26)$$

Transformation to modal form

Equation (7.26) is equivalent to the three vector-matrix equations

$$p_i \mathbf{t}_i = \mathbf{F} \mathbf{t}_i, \qquad i = 1, 2, 3. \qquad (7.27)$$

In matrix algebra Eq. (7.27) is a famous equation, whose solution is known as the **eigenvector/eigenvalue problem**. Recall that $\mathbf{t}_i$ is a vector, $\mathbf{F}$ is a matrix, and p_i is a scalar. The vector $\mathbf{t}_i$ is called an **eigenvector** of $\mathbf{F}$, and p_i is called the corresponding **eigenvalue**. Since we saw earlier that the modal form is equivalent to a partial-fraction-expansion representation with the system poles along the diagonal of the state matrix, it should be clear that these "eigenvalues" are precisely the poles of our system. The transformation matrix that will convert the state description matrices to modal form has as its columns the eigenvectors of $\mathbf{F}$ as shown in Eq. (7.26) for the third-order case. As it happens, there are robust, reliable computer algorithms to compute eigenvalues and the eigenvectors of quite large systems using the QR algorithm.[1] In MATLAB the

Eigenvectors
Eigenvalues

---

1. This algorithm is part of MATLAB and all other well-known CACSD packages. It is a carefully documented in the software package EISPAK (Garbow *et al.*, 1977). See also Strang (1988).

command $p = \text{eig}(\mathbf{F})$ is the way to compute the poles if the system equations are in state form.

Notice also that Eq. (7.27) is homogeneous in that, if $\mathbf{t}_i$ is an eigenvector, so is $\alpha \mathbf{t}_i$ for any scalar α. In most cases the scale factor is selected so that the length (square root of the sum of squares of the magnitudes of the elements) is unity. MATLAB will perform this operation. Another option is to select the scale factors so that the input matrix $\mathbf{B}$ is composed of all 1's. The latter choice is suggested by a partial-fraction expansion with each part realized in control canonical form. If the system is real, then each element of $\mathbf{F}$ is real, and if $p = \sigma + j\omega$ is a pole, so is the conjugate, $p^* = \sigma - j\omega$. For these eigenvalues the eigenvectors are also complex and conjugate. It is possible to compose the transformation matrix using the real and complex parts of the eigenvectors separately so the modal form is real but has 2×2 blocks for each pair of complex poles. Later we will see the result of the MATLAB function that does this, but first let us look at the simple real-poles case.

◆ **EXAMPLE 7.2**   *Transformation from Control to Modal Form*

Find the matrix to transform the control form matrices in Eq. (7.4) into the modal form of Eq. (7.6).

Solution.   According to Eqs. (7.26) and (7.27), we need first to find the eigenvectors and eigenvalues of the $\mathbf{A}_c$ matrix. We take the eigenvectors to be

$$\begin{bmatrix} t_{11} \\ t_{21} \end{bmatrix} \quad \text{and} \quad \begin{bmatrix} t_{12} \\ t_{22} \end{bmatrix}.$$

The equations using the eigenvector on the left are:

$$\begin{bmatrix} -7 & -12 \\ 1 & 0 \end{bmatrix} \begin{bmatrix} t_{11} \\ t_{21} \end{bmatrix} = p \begin{bmatrix} t_{11} \\ t_{21} \end{bmatrix} \tag{7.28a}$$

$$-7t_{11} - 12t_{21} = pt_{11} \tag{7.28b}$$

$$t_{11} = pt_{21} \tag{7.28c}$$

Substituting Eq. (7.28c) into (7.28b) results in

$$-7pt_{21} - 12t_{21} = p^2 t_{21}$$

$$p^2 t_{21} + 7pt_{21}\; 12t_{21} = 0 \tag{7.29a}$$

$$p^2 + 7p + 12 = 0 \tag{7.29b}$$

$$p = -3, -4 \tag{7.29c}$$

We have found (again!) that the eigenvalues (poles) are -3 and -4; furthermore, Eq. (7.28c) tells us that the two eigenvectors are

$$\begin{bmatrix} -4t_{21} \\ t_{21} \end{bmatrix} \quad \text{and} \quad \begin{bmatrix} -3t_{22} \\ t_{22} \end{bmatrix},$$

where t_{21} and t_{22} are arbitrary nonzero scale factors. We want to select the two scale factors such that both elements of $\mathbf{B}_m$ in Eq. (7.6) are unity. The equation for $\mathbf{B}_m$ in terms of $\mathbf{B}_c$ is $\mathbf{TB}_m = \mathbf{B}_c$, and its solution is $t_{21} = -1$ and $t_{22} = 1$. Therefore, the transformation matrix and its inverse[2] are

$$\mathbf{T} = \begin{bmatrix} 4 & -3 \\ -1 & 1 \end{bmatrix}, \qquad \mathbf{T}^{-1} = \begin{bmatrix} 1 & 3 \\ 1 & 4 \end{bmatrix}. \tag{7.30}$$

Elementary matrix multiplication shows that, using $\mathbf{T}$ as defined by Eq. (7.30), the matrices of Eqs. (7.4) and (7.6) are related as follows:

$$\mathbf{A}_m = \mathbf{T}^{-1}\mathbf{A}_c\mathbf{T}, \qquad \mathbf{B}_m = \mathbf{T}^{-1}\mathbf{B}_c,$$
$$\mathbf{C}_m = \mathbf{C}_c\mathbf{T}, \qquad D_m = D_c.$$

2. To find the inverse of a 2×2 matrix, you need only interchange the elements subscripted "11" and "22", change the signs of the "12" and the "21" elements and divide by the determinant ($=1$ in Eq. 7.30).

In Chapter 9 we will present several case studies of control problems that include study of a servomechanism for the control of position and tension of a magnetic tape used as a data storage device. This system has five state variables and in state-variable form is too complicated for hand calculations. However, it is a good example for illustrating the use of computer software designed for the purpose. The model we will use is based on a physical state after amplitude and time scaling have been done.

◆ **EXAMPLE 7.3**   *Using MATLAB to Find Poles and Zeros*

Find the eigenvalues of the system matrix for the tape-drive control as given in Section 9.5. Also compute the transformation of the equations of the tape drive in their given form to modal canonical form. The system matrices are

$$\mathbf{F} = \begin{bmatrix} 0 & 2 & 0 & 0 & 0 \\ -0.1 & -0.35 & 0.1 & 0.1 & 0.75 \\ 0 & 0 & 0 & 2 & 0 \\ 0.4 & 0.4 & -0.4 & -1.4 & 0 \\ 0 & -0.03 & 0 & 0 & -1 \end{bmatrix}, \qquad \mathbf{G} = \begin{bmatrix} 0 \\ 0 \\ 0 \\ 0 \\ 1 \end{bmatrix},$$

$$\mathbf{H}_2 = [0.0 \quad 0.0 \quad 1.0 \quad 0.0 \quad 0.0],$$
$$\mathbf{H}_3 = [0.5 \quad 0 \quad 0.5 \quad 0 \quad 0],$$
$$\mathbf{H}_T = [-0.2 \quad -0.2 \quad 0.2 \quad 0.2 \quad 0], \qquad J = 0.$$

The matrix $\mathbf{H}_3$ corresponds to making x_3 (the position of the tape over the read/write head) the output, and the matrix $\mathbf{H}_T$ corresponds to making tension the output.

Solution. To compute the eigenvalues using MATLAB, we write

$$P = \text{eig}(F)$$

which results in

$$P = \begin{bmatrix} -0.6371 + 0.6669i \\ -0.6371 - 0.6669i \\ 0.0000 \\ -0.5075 \\ -0.9683 \end{bmatrix}.$$

Notice that the system has all poles in the left half-plane (LHP) except for one pole at the origin. This means that a step input will result in a ramp output, so we conclude the system has type I behavior.

MATLAB canon

To transform to modal form, we use the MATLAB function canon:

$$[Am,\ Bm,\ Cm,\ Dm,\ Ti] = \text{canon}(F,\ G,\ H3,\ J,\ 'modal')$$

The result of this calculation is

$$Am = \mathbf{A}_m = \begin{bmatrix} -0.6371 & 0.6669 & -0.0000 & -0.0000 & 0.0000 \\ -0.6669 & -0.6371 & 0.0000 & -0.0000 & 0.0000 \\ -0.0000 & -0.0000 & 0.0000 & 0.0000 & 0.0000 \\ 0.0000 & -0.0000 & 0.0000 & -0.5075 & -0.0000 \\ -0.0000 & -0.0000 & 0.0000 & 0.0000 & -0.9683 \end{bmatrix}.$$

Notice that the complex poles appear in the 2×2 block in the upper left corner of $\mathbf{A}_m$, and the real poles fall on the main diagonal of this matrix. The rest of the calculations from canon are:

$$Bm = \mathbf{B}_m = \begin{bmatrix} 3.0868 \\ 1.2388 \\ 4.0599 \\ 9.9016 \\ 3.9341 \end{bmatrix}, \qquad Cm = \mathbf{C}_m = [0.3268 \quad 0.2190 \quad 0.7071 \quad -0.65625 \quad 0.6124],$$

$$Dm = \mathbf{D}_m = 0, \qquad Ti = \mathbf{T}^{-1} = \begin{bmatrix} -1.1369 & 2.5951 & 1.1369 & 0.4764 & 3.0868 \\ -1.1895 & -2.1453 & 1.1895 & 3.2536 & 1.2388 \\ 0.73376 & 5.54133 & 0.6767 & 1.3533 & 4.0599 \\ -0.9227 & 6.5022 & 0.927 & 2.7962 & 9.9016 \\ -0.0014 & 0.1663 & 0.0014 & 0.0449 & 3.9341 \end{bmatrix}.$$

It happens that canon was written to compute the *inverse* of the transformation we are working with (as you can see from Ti above), so we need to invert our MATLAB results. The inverse is calculated from

MATLAB inv

$$T = \text{inv}(Ti)$$

and results in

$$
T = \mathbf{T} =
\begin{bmatrix}
-0.1748 & 0.1420 & 0.7071 & -0.4871 & 0.5887 \\
0.0083 & -0.1035 & -0.0000 & 0.1236 & -0.2850 \\
0.8284 & 0.2960 & 0.7071 & -0.837M9 & 0.6360 \\
-0.3626 & 0.1820 & 0.0000 & 0.2126 & -0.3079 \\
0.0034 & 0.0022 & -0.0000 & -0.0075 & 0.2697
\end{bmatrix}.
$$

The eigenvectors computed with eig are

$$
V = \mathbf{V} =
\begin{bmatrix}
-0.1748+0.1420i & -0.1748-0.1420i & 0.7071 & -0.4871 & 0.5887 \\
0.0083-0.1035i & 0.0083+0.1035i & 0.0000 & 0.1236 & -0.2850 \\
0.8284+0.2960i & 0.8284-0.2960i & 0.7071 & -0.8379 & 0.6360 \\
-0.3626+0.1820i & -0.3626-0.1820i & 0.0000 & 0.2126 & -0.3079 \\
0.0034+0.0022i & 0.0034-0.0022i & -0.0000 & -0.0075 & 0.2697
\end{bmatrix}.
$$

Notice that the first two columns of the real transformation $\mathbf{T}$ are composed of the real and the imaginary parts of the first eigenvector in the first column of $\mathbf{V}$. It is this step that causes the complex roots to appear in the 2×2 block in the upper left of the $\mathbf{A}_m$ matrix. The vectors in $\mathbf{V}$ are normalized to unit length which results in nonnormalized values in $\mathbf{B}_m$ and $\mathbf{C}_m$. If we found it desirable to do so, we could readily find further transformations to make each element of $\mathbf{B}_m$ equal 1 or to interchange the order in which the poles appear.

◆

7.2.2   Dynamic Response from the State Equations

Having considered the structure of the state-variable equations, we now turn to finding the dynamic response from the state description and to the relationships between the state description and our earlier discussion in Chapter 6 of the frequency response and poles and zeros. Let us begin with the general equations of state given by Eqs. (7.10), and consider the problem in the frequency domain. Taking the Laplace transform of

$$\dot{\mathbf{x}} = \mathbf{F}\mathbf{x} + \mathbf{G}u, \tag{7.32}$$

we obtain

$$s\mathbf{X}(s) - \mathbf{x}(0) = \mathbf{F}\mathbf{X}(s) + \mathbf{G}U(s), \tag{7.33}$$

which is now an algebraic equation. If we collect the terms involving $\mathbf{X}(s)$ on the left side of Eq. (7.33), keeping in mind that in matrix multiplication order is very important, we find that[3]

$$(s\mathbf{I} - \mathbf{F})\mathbf{X}(s) = \mathbf{G}U(s) + \mathbf{x}(0).$$

---

3. The identity matrix $\mathbf{I}$ is a matrix of 1s on the main diagonal and zeros everywhere else; therefore, $\mathbf{I}\mathbf{x} = \mathbf{x}$.

If we premultiply both sides by the inverse of $s\mathbf{I} - \mathbf{F}$, then

$$\mathbf{X}(s) = (s\mathbf{I} - \mathbf{F})^{-1}\mathbf{G}U(s) + (s\mathbf{I} - \mathbf{F})^{-1}\mathbf{x}(0). \tag{7.34}$$

The output of the system is

$$Y(s) = \mathbf{H}\mathbf{X}(s) + JU(s),$$

$$= \mathbf{H}(s\mathbf{I} - \mathbf{F})^{-1}\mathbf{G}U(s) + \mathbf{H}(s\mathbf{I} - \mathbf{F})^{-1}\mathbf{x}(0) + JU(s). \tag{7.35}$$

This equation expresses the output response to both an initial condition and an external forcing input. The coefficient of the external input is the transfer function of the system, which in this case is given by

Transfer function from
state equations

$$G(s) = \frac{Y(s)}{U(s)} = \mathbf{H}(s\mathbf{I} - \mathbf{F})^{-1}\mathbf{G} + J. \tag{7.36}$$

◆ **EXAMPLE 7.4**   *Transfer Function from the State Description*

Use Eq. (7.36) to find the transfer function of the system described by Eq. (7.4).

Solution.   The state-variable description matrices of the system are

$$\mathbf{F} = \begin{bmatrix} -7 & -12 \\ 1 & 0 \end{bmatrix}, \qquad \mathbf{G} = \begin{bmatrix} 1 \\ 0 \end{bmatrix},$$

$$\mathbf{H} = \begin{bmatrix} 1 & 2 \end{bmatrix} \qquad\qquad J = 0.$$

To compute the transfer function according to Eq. (7.36), we form

$$s\mathbf{I} - \mathbf{F} = \begin{bmatrix} s+7 & 12 \\ -1 & s \end{bmatrix},$$

and compute

$$(s\mathbf{I} - \mathbf{F})^{-1} = \frac{\begin{bmatrix} s & -12 \\ 1 & s+7 \end{bmatrix}}{s(s+7) + 12}. \tag{7.37}$$

We then substitute Eq. (7.37) into Eq. (7.36) to get

$$G(s) = \frac{\begin{bmatrix} 1 & 2 \end{bmatrix}\begin{bmatrix} s & -12 \\ 1 & s+7 \end{bmatrix}\begin{bmatrix} 1 \\ 0 \end{bmatrix}}{s(s+7) + 12}$$

$$= \frac{\begin{bmatrix} 1 & 2 \end{bmatrix}\begin{bmatrix} s \\ 1 \end{bmatrix}}{s(s+7) + 12}$$

$$= \frac{(s+2)}{(s+3)(s+4)}. \tag{7.38}$$

We saw earlier that a state description is not unique, so it should follow that the transfer function is not dependent on the transformation of a change of state. To demonstrate that this is true, we can compute the transfer function in the transformed domain using the new matrices as given by Eqs. (7.13) and (7.14). In terms of $\mathbf{A}$, $\mathbf{B}$, $\mathbf{C}$, and D, the transfer function is

$$G(s) = \mathbf{C}(s\mathbf{I} - \mathbf{A})^{-1}\mathbf{B} + D. \qquad (7.39)$$

If we substitute Eqs. (7.13) and (7.14) into this expression, we have

$$G(s) = \mathbf{HT}(s\mathbf{I} - \mathbf{T}^{-1}\mathbf{FT})^{-1}\mathbf{T}^{-1}\mathbf{G} + J. \qquad (7.40)$$

If we use the identity $\mathbf{I} = \mathbf{T}^{-1}\mathbf{T}$, we can write Eq. (7.40) in the form

$$G(s) = \mathbf{HT}[\mathbf{T}^{-1}(s\mathbf{I} - \mathbf{F})\mathbf{T}]^{-1}\mathbf{T}^{-1}\mathbf{G} + J.$$

Finally, it is necessary to express the inverse of the three matrices in brackets. For this we need the result that the inverse of a product of matrices is the product of their respective inverses in *reverse* order:

$$G(s) = \mathbf{HTT}^{-1}(s\mathbf{I} - \mathbf{F})^{-1}\mathbf{TT}^{-1}\mathbf{G} + J.$$

Now the transformation matrix cancels out, leaving

$$G(s) = \mathbf{H}(s\mathbf{I} - \mathbf{F})^{-1}\mathbf{G} + J.$$

Note that this is the same transfer function we would have derived had we started with $\mathbf{F}$, $\mathbf{G}$, $\mathbf{H}$, and J in the first place.

The transfer function is not changed by a linear transformation of state.

Since Eq. (7.36) expresses the transfer function in terms of the general state-space descriptor matrices $\mathbf{F}$, $\mathbf{G}$, $\mathbf{H}$, and J, we are able to express poles and zeros in terms of these matrices. We saw earlier that by transforming the state matrices to diagonal form, the poles appear as the eigenvalues on the main diagonal of the $\mathbf{F}$ matrrix. We now take a systems theory point of view to look at the poles and zeros as they are involved in the transient response of a system.

As we saw in Chapter 3, a pole of the transfer function $G(s)$ is a value of generalized frequency s such that, if $s = p_i$, then the system can respond to an initial condition as $K_i e^{p_i t}$, with *no forcing function*. In this context, p_i is called a **natural frequency** or **natural mode** of the system. If we take the state-space equations (7.10) and set the forcing function u to zero, we have

$$\dot{\mathbf{x}} = \mathbf{F}\mathbf{x}. \qquad (7.41)$$

If we assume some (as yet unknown) initial condition

$$\mathbf{x}(0) = \mathbf{x}_0 \qquad (7.42)$$

and that the entire state motion behaves according to the same natural frequency, then the state can be written as $\mathbf{x}(t) = e^{p_i t}\mathbf{x}_0$. It follows from Eq. (7.41) that

$$\dot{\mathbf{x}}(t) = p_i e^{p_i t}\mathbf{x}_0 = \mathbf{Fx} = \mathbf{F}e^{p_i t}\mathbf{x}_0, \tag{7.43}$$

or

$$\mathbf{Fx}_0 = p_i\mathbf{x}_0. \tag{7.44}$$

We can rewrite Eq. (7.44) as

$$(p_i\mathbf{I} - \mathbf{F})\mathbf{x}_0 = 0. \tag{7.45}$$

Equations (7.44) and (7.45) comprise the eigenvector/eigenvalue problem we saw in Eq. (7.27) with eigenvalues p_i and in this case eigenvectors $\mathbf{x}_0$ of the matrix $\mathbf{F}$. If we are just interested in the eigenvalues, we can use the fact that for a nonzero $\mathbf{x}_0$ Eq. (7.45) has a solution if and only if

Transfer function poles from state equations

$$\det(p_i\mathbf{I} - \mathbf{F}) = 0. \tag{7.46}$$

These equations show again that the *poles* of the transfer function are the eigenvalues of the system matrix $\mathbf{F}$. The determinant equation (7.46) is a polynomial in the eigenvalues p_i known as the **characteristic equation**. In Example 7.2 we computed the eigenvalues and eigenvectors of a particular matrix in control canonical form. As an alternative computation for the poles of that system, we could solve the characteristic equation (7.46). For the system described by Eq. (7.4), we can find the poles from Eq. (7.46) by solving

$$\det(s\mathbf{I} - \mathbf{F}) = 0, \tag{7.47a}$$

$$\det\begin{bmatrix} s + 7 & 12 \\ -1 & s \end{bmatrix} = 0, \tag{7.47b}$$

$$s(s + 7) + 12 = (s + 3)(s + 4) = 0. \tag{7.47c}$$

This confirms again that the poles of the system are the eigenvalues of $\mathbf{F}$.

We can also determine the zeros of a system from the state-variable description matrices $\mathbf{F}$, $\mathbf{G}$, $\mathbf{H}$, and J using a systems theory point of view. From this perspective a zero is a value of generalized frequency s such that the system can have a nonzero input and state and yet have an output of zero. If the input is exponential at the zero frequency z_i, given by

$$u(t) = u_0 e^{z_i t}, \tag{7.48}$$

then the output is identically zero:

$$y(t) \equiv 0. \tag{7.49}$$

The state-space description of Eqs. (7.48) and (7.49) would be

$$u = u_0 e^{z_i t}, \qquad \mathbf{x}(t) = \mathbf{x}_0 e^{z_i t}, \qquad y(t) \equiv 0. \tag{7.50}$$

Thus

$$\dot{\mathbf{x}} = z_i e^{z_i t} \mathbf{x}_0 = \mathbf{F} e^{z_i t} \mathbf{x}_0 + \mathbf{G} u_0 e^{z_i t}, \tag{7.51}$$

or

$$[z_i \mathbf{I} - \mathbf{F} \quad -\mathbf{G}] \begin{bmatrix} \mathbf{x}_0 \\ u_0 \end{bmatrix} = \mathbf{0} \tag{7.52}$$

and

$$y = \mathbf{H}x + Ju = \mathbf{H} e^{z_i t} \mathbf{x}_0 + J u_0 e^{st} \equiv 0 \tag{7.53}$$

Combining Eqs. (7.52) and (7.53), we get

$$\begin{bmatrix} z_i \mathbf{I} - \mathbf{F} & -\mathbf{G} \\ \mathbf{H} & J \end{bmatrix} \begin{bmatrix} \mathbf{x}_0 \\ u_0 \end{bmatrix} = \begin{bmatrix} \mathbf{0} \\ 0 \end{bmatrix}. \tag{7.54}$$

Transfer function zeros from state equations

From Eq. (7.54) we can conclude that a zero of the state-space system is a value of z_i where Eq. (7.54) has a nontrivial solution. With one input and one output, the matrix is square, and a solution to Eq. (7.54) is equivalent to a solution to

$$\det \begin{bmatrix} z_i \mathbf{I} - \mathbf{F} & -\mathbf{G} \\ \mathbf{H} & J \end{bmatrix} = 0. \tag{7.55}$$

◆ **EXAMPLE 7.5**   *Zeros from a State Description*

Compute the zero(s) of the system described by Eq. (7.4).

Solution.   We use Eq. (7.55) to compute the zeros:

$$\det \begin{bmatrix} s+7 & 12 & -1 \\ 1 & s & 0 \\ 1 & 2 & 0 \end{bmatrix} = 0,$$

$$-2 - s = 0,$$

$$s = -2.$$

Note that this result agrees with the zero of the transfer function given by Eq. (7.1).

Equation (7.46) for the characteristic equation and Eq. (7.55) for the zeros polynomial can be combined to express the transfer function in a compact form from state-description matrices as

$$G(s) = \frac{\det \begin{bmatrix} s\mathbf{I} - \mathbf{F} & -\mathbf{G} \\ \mathbf{H} & J \end{bmatrix}}{\det(s\mathbf{I} - \mathbf{F})}. \tag{7.56}$$

(See Appendix C for more details.) While Eq. (7.56) is a compact formula for theoretical studies, it is very sensitive to numerical errors. A numerically stable algorithm for computing the transfer function is described in Emami-Naeini and Van Dooren (1982). Given the transfer function, we can compute the frequency response as $G(j\omega)$, and as discussed earlier, we can use Eqs. (7.45) and (7.54) to find the poles and zeros, upon which the transient response depends, as we saw in Chapter 3.

◆ **EXAMPLE 7.6**   *Analysis of the State Equations of a Tape Drive*

❖   Compute the poles, zeros, and transfer function for the equations of the tape-drive servomechanism given in Example 7.3.

Solution.   The state equations are described by the system matrices:

$$\mathbf{F} = \begin{bmatrix} 0 & 2 & 0 & 0 & 0 \\ -0.1 & -0.35 & 0.1 & 0.1 & 0.75 \\ 0 & 0 & 0 & 2 & 0 \\ 0.4 & 0.4 & -0.4 & -1.4 & 0 \\ 0 & -0.03 & 0 & 0 & -1 \end{bmatrix}, \qquad \mathbf{G} = \begin{bmatrix} 0 \\ 0 \\ 0 \\ 0 \\ 1 \end{bmatrix},$$

$\mathbf{H}_2 = [0.0 \quad 0.0 \quad 1.0 \quad 0.0 \quad 0.0]$ servo motor position output,
$\mathbf{H}_3 = [0.5 \quad 0.0 \quad 0.5 \quad 0.0 \quad 0.0]$ position at read/write head as output,
$\mathbf{H}_T = [-0.2 \quad -0.2 \quad 0.2 \quad 0.2 \quad 0.0]$ tension output,
$J = 0.0.$

MATLAB ss2tf

There are two different ways to compute the answer to this problem. The most direct is to use the MATLAB function ss2tf (state space to transfer function), which will give the numerator and denominator polynomials directly. This function permits multiple inputs and outputs; the fifth argument of the function tells which input is to be used. We have only one input here but must still provide the argument. The computation of the transfer function from motor-current input to the servomotor position output is

[N2, D2] = ss2tf(F, G, H2, J, 1)

which results in

N2 = [0   0.0000   0   0   0.6000   1.2000]

D2 = [1.0000   2.7500   3.2225   1.8815   0.4180   −0.000].

Similarly, for the position at the read/write head, the transfer function polynomials are computed by

[N3, D3] = ss2tf(F, G, H3, J, 1),

which results in

$$N3 = [0 \quad -0.0000 \quad -0.0000 \quad 0.7500 \quad 1.3500 \quad 1.2000],$$

$$D3 = [1.0000 \quad 2.7500 \quad 3.2225 \quad 1.8815 \quad 0.4180 \quad -0.0000].$$

Finally, the transfer function to tension is

$$[NT, DT] = ss2tf(F, G, Ht, J, 1)$$

producing

$$NT = [0 \quad -0.0000 \quad -0.1500 \quad -0.4500 \quad -0.3000 \quad 0.0000],$$

$$DT = [1.0000 \quad 2.7500 \quad 3.2225 \quad 1.8815 \quad 0.4180 \quad -0.0000].$$

MATLAB roots

It is interesting to check to see whether the poles and zeros determined this way agree with those found by other means. For a polynomial we use the function roots:

$$roots(D3) = \begin{bmatrix} -0.6371 + 0.6669i \\ -0.6371 - 0.6669i \\ -0.9683 \\ -0.5075 \\ 0.0000 \end{bmatrix}.$$

Checking with Example 7.3, we confirm that they agree.

How about the zeros? We can find these by finding the roots of the numerator polynomial. We compute the roots of the polynomial N3:

$$roots(N3) = \begin{bmatrix} 2.0548 \times 10^7 \\ -2.0548 \times 10^7 \\ -0.0000 + 0.0000i \\ -0.0000 - 0.0000i \end{bmatrix}.$$

Here we notice that roots are given with a magnitude of 10^7, which seems inconsistent with the values given for the polynomial. The problem is that MATLAB has used the very small leading terms in the polynomial as real values and thereby introduced extraneous roots that are for all practical purposes at infinity. The true zeros are found by truncating the polynomial to the significant values using the statement

$$N3R = N3(4:6)$$

to get

$$N3R = [0.7500 \quad 1.3500 \quad 1.200],$$

$$roots(N3R) = \begin{bmatrix} -0.9000 + 0.8888i \\ -0.900 \quad -0.8888i \end{bmatrix}.$$

The other approach is to compute the poles and zeros separately and, if desired, combine these into a transfer function. The poles were computed with eig in Example

7.3 and are

$$
P = \begin{bmatrix} -0.6371+0.6669\,i \\ -0.6371-0.6669\,i \\ 0.0000 \\ -0.5075 \\ -0.9683 \end{bmatrix}.
$$

MATLAB tzero

The zeros can be computed by the equivalent of Eq. (7.54) with the function tzero (transmission zeros). The zeros depend on which output is being used, of course, and are respectively given below. For the position of the tape at the servomotor as the output, the statement:

$$
ZER2 = tzero(F, G, H2, J)
$$

yields

$$
ZER2 = -2.0000.
$$

For the position of the tape over the read/write head as the output, the statement

$$
ZER3 = tzero(F, G, H3, J)
$$

$$
ZER3 = \begin{bmatrix} -0.900 & +0.8888\,i \\ -0.9000 & -0.8888\,i \end{bmatrix}.
$$

We note that these results agree with the values computed from the numerator polynomial N3 above. And, finally, for the tension as output we use

$$
ZERT = tzero(F, G, HT, J)
$$

to get

$$
ZERT = \begin{bmatrix} -1.0000 \\ -0.0000 \\ -2.0000 \end{bmatrix}.
$$

From these results we can write down, for example, the transfer function to position x_3 as

$$
G(s) = \frac{X_3(s)}{E_1(s)}
$$

$$
= \frac{0.75s^2 + 1.35s + 1.2}{s^5 + 2.75s^4 + 3.22s^3 + 1.88s^2 + 0.418s}
$$

$$
= \frac{0.75(s + 0.9 \pm 0.8888\,j)}{s(s + 0.507)(s + 0.968)(s + 0.637 \pm 0.667\,j)}. \tag{7.57}
$$

7.3   Control-law Design for Full State Feedback

One of the attractive features of the state-space design method is that it consists of a sequence of independent steps, as mentioned in the chapter overview. The first step, discussed in Section 7.3.1, is to determine the control. The purpose of the control law is to allow us to assign a set of pole locations for the closed-loop system that will correspond to satisfactory dynamic response in terms of rise time and other measures of transient response. In Section 7.3.2 we will show how to introduce the reference input with full state feedback, and in Section 7.4 we will describe the process of finding the poles for good design.

Estimator/observer

The second step — necessary if the full state is not available — is to design an **estimator** (sometimes called an **observer**), which computes an estimate of the entire state vector when provided with the measurements of the system indicated by Eq. (7.10b). We will examine estimator design in Section 7.5.

The third step consists of combining the control law and the estimator. Figure 7.6 shows how the control law and the estimator fit together and how the combination takes the place of what we have been previously referring to as **compensation**. At this stage the control-law calculations are based on the estimated state rather than the actual state. In Section 7.6 we will show that this substitution is reasonable, and also that using the combined control law and estimator results in closed-loop pole locations that are the same as those determined when designing the control and estimator separately.

The control law and the estimator together form the compensation.

The fourth and final step of state-space design is to introduce the reference input in such a way that the plant output will track external commands with acceptable rise-time, overshoot, and settling-time values. At this point in the design, all the closed-loop poles have been selected, and the designer is concerned with the zeros of the overall transfer function. Figure 7.6 shows the command input r introduced in the same relative position as was done with the transform design methods; however, in Section 7.7 we will show how to introduce the reference at another location, resulting in different zeros and (usually) superior control.

FIGURE 7.6
Schematic diagram of state-space design elements

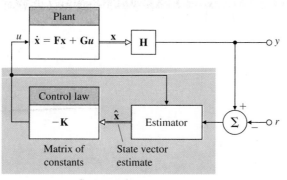

Compensation

7.3.1   Finding the Control Law

The first step in the state-space design method, as mentioned earlier, is to find the control law as feedback of a linear combination of the state variables — that is,

Control law

$$u = -\mathbf{K}\mathbf{x} = -[K_1 \quad K_2 \quad \cdots \quad K_n] \begin{bmatrix} x_1 \\ x_2 \\ \vdots \\ x_n \end{bmatrix}. \tag{7.58}$$

Control law

We assume for feedback purposes that all the elements of the state vector are at our disposal. In practice, of course, this would usually be a ridiculous assumption; moreover, a well-trained control designer knows that other design methods do not require so many sensors. The assumption that all state variables are available merely allows us to proceed with this first step.

Equation (7.58) tells us that the system has a constant matrix in the state-vector feedback path, as shown in Fig. 7.7. For an nth-order system there will be n feedback gains, $K_1, \ldots, K_n$, and since there are n roots of the system, it is possible that there are enough degrees of freedom to select *arbitrarily* any desired root location by choosing the proper values of K_i. This freedom contrasts sharply with root-locus design, in which we have only one parameter and the closed-loop poles are restricted to the locus.

Substituting the feedback law given by Eq. (7.58) into the system described by Eq. (7.10) yields

$$\dot{\mathbf{x}} = \mathbf{F}\mathbf{x} - \mathbf{G}\mathbf{K}\mathbf{x}. \tag{7.59}$$

The characteristic equation of this closed-loop system is

Control characteristic equation

$$\det[s\mathbf{I} - (\mathbf{F} - \mathbf{G}\mathbf{K})] = 0. \tag{7.60}$$

When evaluated, this yields an nth-order polynomial in s containing the gains $K_1, \ldots, K_n$. The control-law design then consists of picking the gains $\mathbf{K}$ so that the roots of Eq. (7.60) are in desirable locations. Selecting desirable root locations is an inexact science that may require some iteration by the designer. Issues in their selection are considered in Examples 7.7 to 7.9 as well as in Section 7.4. For now we assume that the desired locations are known, say,

$$s = s_1, s_2, \ldots, s_n.$$

FIGURE 7.7
Assumed system for control-law design

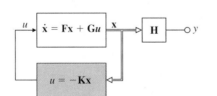

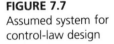

Then the corresponding desired (control) characteristic equation is

$$\alpha_c(s) = (s - s_1)(s - s_2) \cdots (s - s_n) = 0. \tag{7.61}$$

Hence the required elements of $\mathbf{K}$ are obtained by matching coefficients in Eqs. (7.60) and (7.61). This forces the system's characteristic equation to be identical to the desired characteristic equation and the closed-loop poles to be placed at the desired locations.

◆ **EXAMPLE 7.7** *Control Law for a Second-order System*

Suppose you have an undamped oscillator with frequency ω_0 and a state-space description given by

$$\begin{bmatrix} \dot{x}_1 \\ \dot{x}_2 \end{bmatrix} = \begin{bmatrix} 0 & 1 \\ -\omega_0^2 & 0 \end{bmatrix} \begin{bmatrix} x_1 \\ x_2 \end{bmatrix} + \begin{bmatrix} 0 \\ 1 \end{bmatrix} u. \tag{7.62}$$

Find the control law that places the closed-loop poles of the system so that they are both at $-2\omega_0$. In other words, you wish to double the natural frequency and increase the damping ratio ζ from 0 to 1.

Solution. From Eq. (7.61) we find that

$$\alpha_c(s) = (s + 2\omega_0)^2$$
$$= s^2 + 4\omega_0 s + 4\omega_0^2. \tag{7.63}$$

Equation (7.60) tells us that

$$\det[s\mathbf{I} - (\mathbf{F} - \mathbf{GK})] = \det \left\{ \begin{bmatrix} s & 0 \\ 0 & s \end{bmatrix} - \left(\begin{bmatrix} 0 & 1 \\ -\omega_0^2 & 0 \end{bmatrix} - \begin{bmatrix} 0 \\ 1 \end{bmatrix} [K_1 \quad K_2] \right) \right\},$$

or

$$s^2 + K_2 s + \omega_0^2 + K_1 = 0. \tag{7.64}$$

Equating the coefficients with like powers of s in Eqs. (7.63) and (7.64) yields the system of equations

$$K_2 = 4\omega_0,$$
$$\omega_0^2 + K_1 = 4\omega_0^2,$$

and therefore,

$$K_1 = 3\omega_0^2,$$
$$K_2 = 4\omega_0.$$

More concisely, the control law is

$$\mathbf{K} = [K_1 \quad K_2] = [3\omega_0^2 \quad 4\omega_0].$$

FIGURE 7.8
Impulse response of the undamped oscillator with full-state feedback ($\omega_0 = 1$)

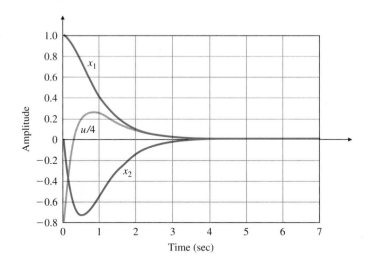

Figure 7.8 shows the response of the closed-loop system to the initial conditions $x_1 = 1.0$, $x_2 = 0.0$, and $\omega_0 = 1$. It shows a very well damped response, as would be expected from having two roots at $s = -2$.

◆

Calculating the gains using the technique illustrated in Example 7.7 becomes rather tedious when the order of the system is higher than 3. Furthermore, it is not clear whether a solution always exists to the equations that result from matching coefficients of the system's characteristic equation to a desired equation. In contrast, when the system has been transformed into control canonical form, the equations have a solution that is trivial to find. From Eq. (7.5) the matrices describing control canonical form are

$$\mathbf{F}_c = \begin{bmatrix} -a_1 & -a_2 & \cdots & -a_n \\ 1 & 0 & \cdots & 0 \\ 0 & 1 & \cdots & 0 \end{bmatrix}, \qquad \mathbf{G}_c = \begin{bmatrix} 1 \\ 0 \\ 0 \end{bmatrix},$$

$$\mathbf{H}_c = [b_1 \quad b_2 \quad \cdots \quad b_n], \qquad\qquad J_c = 0. \tag{7.65}$$

Companion form matrix   The special structure of this system matrix is referred to as the **upper companion form** because the characteristic equation is $a(s) = s^n + a_1 s^{n-1} + a_2 s^{n-2} \cdots + a_n$ and the coefficients of this monic "companion" polynomial are the elements in the first row of $\mathbf{F}_c$. If we now form the closed-loop system matrix $\mathbf{F}_c - \mathbf{G}_c \mathbf{K}_c$, we find that

$$\mathbf{F}_c - \mathbf{G}_c \mathbf{K}_c = \begin{bmatrix} -a_1 - K_1 & -a_2 - K_2 & \cdots & -a_n - K_n \\ 1 & 0 & \cdots & 0 \\ 0 & 1 & \cdots & 0 \end{bmatrix}. \tag{7.66}$$

By visually comparing Eqs. (7.65) and (7.66) we see that the closed-loop characteristic equation is

$$s^n + (a_1 + K_1)s^{n-1} + (a_2 + K_2)s^{n-2} + \cdots + (a_n + K_n) = 0. \tag{7.67}$$

Therefore, if the desired pole locations result in the characteristic equation given by

$$\alpha_c(s) = s^n + \alpha_1 s^{n-1} + \alpha_2 s^{n-2} + \cdots + \alpha_n = 0, \tag{7.68}$$

then the necessary feedback gains can be found by equating the coefficients in Eqs. (7.67) and (7.68):

$$K_1 = -a_1 + \alpha_1, \quad K_2 = -a_2 + \alpha_2, \quad \cdots, \quad K_n = -a_n + \alpha_n. \tag{7.69}$$

We now have the basis for a design procedure. Given a system of order n described by an arbitrary $(\mathbf{F}, \mathbf{G})$ and given a desired nth-order monic characteristic polynomial $\alpha_c(s)$, we (1) transform $(\mathbf{F}, \mathbf{G})$ to control canonical form $(\mathbf{F}_c, \mathbf{G}_c)$ by changing the state $\mathbf{x} = \mathbf{Tz}$, and we (2) solve for the control gains by inspection using Eq. (7.69) to give the control law $u = -\mathbf{K}_c \mathbf{z}$. Since this gain is for the state in the *control* form, we must (3) transform the gain back to the original state to get $\mathbf{K} = \mathbf{K}_c \mathbf{T}^{-1}$.

Ackermann's formula for pole placement

An alternative to this transformation method is given by **Ackermann's formula** (1972), which organizes the three-step process of converting to $(\mathbf{F}_c, \mathbf{G}_c)$, solving for the gains, and converting back again into the following very compact form:

$$\mathbf{K} = [0 \quad \cdots \quad 0 \quad 1]\mathscr{C}^{-1}\alpha_c(\mathbf{F}), \tag{7.70}$$

such that

$$\mathscr{C} = [\mathbf{G} \quad \mathbf{FG} \quad \mathbf{F}^2\mathbf{G} \quad \cdots \quad \mathbf{F}^{n-1}\mathbf{G}] \tag{7.71}$$

where $\mathscr{C}$ is the controllability matrix we saw in Section 7.2, n gives the order of the system and the number of state variables, and $\alpha_c(\mathbf{F})$ is a matrix defined as

$$\alpha_c(\mathbf{F}) = \mathbf{F}^n + \alpha_1 \mathbf{F}^{n-1} + \alpha_2 \mathbf{F}^{n-2} + \cdots + \alpha_n \mathbf{I}, \tag{7.72}$$

where the α_i are the coefficients of the desired characteristic polynomial Eq. (7.68). Note that Eq. (7.72) is a matrix equation.

◆ **EXAMPLE 7.8**   *Ackermann's Formula*

Use Ackermann's formula to solve for the gains for the undamped oscillator of Example 7.7.

Solution.   Substituting the desired characteristic polynomial coefficients,

$$\alpha_1 = 4\omega_0, \qquad \alpha_2 = 4\omega_0^2,$$

into Eq. (7.72) results in

$$
\alpha_c(\mathbf{F}) = \begin{bmatrix} -\omega_0^2 & 0 \\ 0 & -\omega_0^2 \end{bmatrix} + 4\omega_0 \begin{bmatrix} 0 & 1 \\ -\omega_0^2 & 0 \end{bmatrix} + 4\omega_0^2 \begin{bmatrix} 1 & 0 \\ 0 & 1 \end{bmatrix}
$$

$$
= \begin{bmatrix} 3\omega_0^2 & 4\omega_0 \\ -4\omega_0^3 & 3\omega_0^2 \end{bmatrix}. \tag{7.73}
$$

The controllability matrix is

$$
\mathscr{C} = [\mathbf{G} \quad \mathbf{FG}] = \begin{bmatrix} 0 & 1 \\ 1 & 0 \end{bmatrix},
$$

which yields

$$
\mathscr{C}^{-1} = \begin{bmatrix} 0 & 1 \\ 1 & 0 \end{bmatrix}. \tag{7.74}
$$

Finally, we substitute Eqs. (7.74) and (7.73) into Eq. (7.70) to get

$$
\mathbf{K} = [K_1 \quad K_2]
$$

$$
= [0 \quad 1] \begin{bmatrix} 0 & 1 \\ 1 & 0 \end{bmatrix} \begin{bmatrix} 3\omega_0^2 & 4\omega_0 \\ -4\omega_0^3 & 3\omega_0^2 \end{bmatrix}
$$

Therefore

$$
\mathbf{K} = [3\omega_0^2 \quad 4\omega_0],
$$

which is the same result we obtained previously.

◆

MATLAB acker, place

As was mentioned earlier, computation of the controllability matrix has very poor numerical accuracy, and this carries over to Ackermann's formula. Equation (7.70), implemented in MATLAB with the function acker, can be used for the design of single input–single output systems with a small (≤ 10) number of state variables. For more complex cases a more reliable formula is available, implemented in MATLAB with place. A modest limitation on place is that, because it is based on assigning closed-loop eigenvectors, none of the desired closed-loop poles may be repeated; that is, the poles must be distinct, a requirement that does not apply to acker.

The fact that we can shift the poles of a system by state feedback to any desired location is a rather remarkable result. The development in this section reveals that this shift is possible if we can transform $(\mathbf{F}, \mathbf{G})$ to the control form $(\mathbf{F}_c, \mathbf{G}_c)$, which in turn is possible if the system is controllable. In rare instances the system may be uncontrollable, in which case no possible control will yield arbitrary pole locations. **Uncontrollable systems** have certain modes, or subsystems, that are unaffected by the control. This usually means that parts of the system are physically disconnected from the input. For example, in modal canonical form for a system with distinct poles, one of the modal state variables is not connected to the input if there is a zero entry in the $\mathbf{B}_m$ matrix. A good physical understanding of the system being controlled would prevent any attempt to design a controller for an uncontrollable system. As we saw earlier, there are algebraic tests for controllability; however, no mathematical test can replace the control engineer's understanding of the physical system. Often the physical situation is such that every mode is controllable to some degree, and, while the mathematical tests indicate the system is controllable, certain modes are so weakly controllable that designs to control them are virtually useless.

An example of weak controllability

Airplane control is a good example of weak controllability of certain modes. Pitch plane motion $\mathbf{x}_p$ is primarily affected by the elevator δ_e and weakly affected by rolling motion $\mathbf{x}_r$. Rolling motion is essentially only affected by the ailerons δ_a. The state-space description of these relationships is

$$\begin{bmatrix} \dot{\mathbf{x}}_p \\ \dot{\mathbf{x}}_r \end{bmatrix} = \begin{bmatrix} \mathbf{F}_p & \boldsymbol{\varepsilon} \\ 0 & \mathbf{F}_r \end{bmatrix} \begin{bmatrix} \mathbf{x}_p \\ \mathbf{x}_r \end{bmatrix} + \begin{bmatrix} \mathbf{G}_p & 0 \\ 0 & \mathbf{G}_r \end{bmatrix} \begin{bmatrix} \delta_e \\ \delta_a \end{bmatrix}, \tag{7.75}$$

where the matrix of small numbers ε represents the weak coupling from rolling motion to pitching motion. A mathematical test of controllability for this system would conclude that pitch plane motion (and therefore altitude) is controllable by the ailerons as well as by the elevator! However, it is impractical to attempt to control an airplane's altitude by rolling the aircraft with the ailerons.

Another example will illustrate some of the properties of pole placement by state feedback and the effects of loss of controllability on the process.

◆ **EXAMPLE 7.9** *How Zero Location Can Affect the Control Law*

A specific system is described by Eq. (7.24) in observer canonical form with a zero at $s = z_o$. Find the state feedback gains necessary for placing the poles of this system at the roots of $s^2 + 2\zeta\omega_n s + \omega_n^2$.

Solution. The state description matrices are

$$\mathbf{A}_o = \begin{bmatrix} -7 & 1 \\ -12 & 0 \end{bmatrix}, \qquad \mathbf{B}_o = \begin{bmatrix} 1 \\ -z_o \end{bmatrix},$$

$$\mathbf{C}_o = [1 \quad 0], \qquad D_o = 0.$$

First we substitute these matrices into Eq. (7.60) to get the closed-loop characteristic equation in terms of the unknown gains and the zero position:

$$s^2 + (7 + K_1 - z_o K_2)s + 12 - K_2(7z_o + 12) - K_1 z_o = 0.$$

Next we equate this equation to the desired characteristic equation to get the equations

$$K_1 - z_o K_2 = 2\zeta\omega_n - 7,$$

$$-z_o K_1 - (7z_o + 12)K_2 = \omega_n^2 - 12.$$

The solutions to these equations are

$$K_1 = \frac{z_o(14\zeta\omega_n - 37 - \omega_n^2) + 12(2\zeta\omega_n - 7)}{(z_o + 3)(z_o + 4)},$$

$$K_2 = \frac{z_o(7 - 2\zeta\omega_n) + 12 - \omega_n^2}{(z_o + 3)(z_o + 4)}.$$

◆

Two important observations should be made from this example. The first is that the gains grow as the zero z_o approaches either -3 or -4, the values where this system loses controllability. In other words, as controllability is almost lost, the control gains become very large.

The system has to work harder and harder to achieve control as controllability slips away.

The second important observation illustrated by the example is that both K_1 and K_2 grow as the desired closed-loop bandwidth given by ω_n is increased. From this, we conclude that

---

To move the poles a long way requires large gains.

These observations lead us to a discussion of how we might go about selecting desired pole locations in general. Before we begin that topic we will complete the design with full state feedback by showing how the reference input might be applied to such a system and what the resulting response characteristics are.

7.3.2 Introducing the Reference Input with Full State Feedback

Thus far, the control has been given by Eq. (7.58), or $u = -\mathbf{Kx}$. In order to study the transient response of the pole-placement designs to input commands, it is necessary to introduce the reference input into the system. An obvious way to do this is to change the control to $u = -\mathbf{Kx} + r$. However, the system will now almost surely have a nonzero steady-state error to a step input. The way to correct this problem is to compute the steady-state values of the state and the control input that will result in zero output error and then force them to take these values. If the desired final values of the state and the control input are $\mathbf{x}_{ss}$ and u_{ss}, then the new control formula should be

$$u = u_{ss} - \mathbf{K}(\mathbf{x} - \mathbf{x}_{ss}), \tag{7.76}$$

so that when $\mathbf{x} = \mathbf{x}_{ss}$ (no error), $u = u_{ss}$. To pick the correct final values, we must solve the equations so that the system will have zero steady-state error to *any* constant input. The system differential equations are the standard ones:

$$\dot{\mathbf{x}} = \mathbf{F}\mathbf{x} + \mathbf{G}u, \tag{7.77a}$$

$$y = \mathbf{H}\mathbf{x} + Ju. \tag{7.77b}$$

In the steady state, Eqs. (7.77a) and (7.77b) reduce to the pair

$$\mathbf{0} = \mathbf{F}\mathbf{x}_{ss} + \mathbf{G}u_{ss}, \tag{7.78a}$$

$$y_{ss} = \mathbf{H}\mathbf{x}_{ss} + Ju_{ss}. \tag{7.78b}$$

We want to solve for the values for which $y_{ss} = r_{ss}$ for any value of r_{ss}. To do this we make $\mathbf{x}_{ss} = \mathbf{N}_x r_{ss}$ and $u_{ss} = N_u r_{ss}$. With these substitutions we can write Eqs. (7.78) as a matrix equation; the common factor of r_{ss} cancels out to give the equation for the gains:

Gain calculation for reference input

$$\begin{bmatrix} \mathbf{F} & \mathbf{G} \\ \mathbf{H} & J \end{bmatrix} \begin{bmatrix} \mathbf{N}_x \\ N_u \end{bmatrix} = \begin{bmatrix} \mathbf{0} \\ 1 \end{bmatrix}. \tag{7.79}$$

This equation can be solved for $\mathbf{N}_x$ and N_u to get

$$\begin{bmatrix} \mathbf{N}_x \\ N_u \end{bmatrix} = \begin{bmatrix} \mathbf{F} & \mathbf{G} \\ \mathbf{H} & J \end{bmatrix}^{-1} \begin{bmatrix} \mathbf{0} \\ 1 \end{bmatrix}.$$

With these values we finally have the basis for introducing the reference input so as to get zero steady-state error to a step input:

Control equation with reference input

$$u = N_u r - \mathbf{K}(\mathbf{x} - \mathbf{N}_x r)$$
$$= -\mathbf{K}\mathbf{x} + (N_u + \mathbf{K}\mathbf{N}_x)r. \tag{7.80}$$

The coefficient of r in parentheses is a constant that can be computed beforehand. We give it the symbol $\bar{N}$, so

$$u = -\mathbf{K}\mathbf{x} + \bar{N}r. \tag{7.81}$$

The block diagram of the system is shown in Fig. 7.9.

FIGURE 7.9

Block diagram for introducing the reference input with full-state feedback: (a) with state and control gains; (b) with a single composite gain

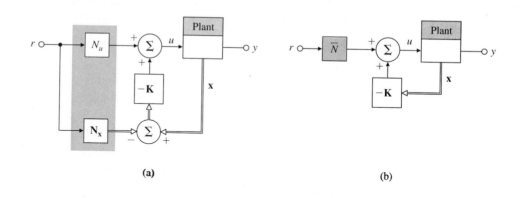

(a)   (b)

◆ **EXAMPLE 7.10**   *Introducing the Reference Input*

Compute the necessary gains for zero steady-state error to a step command at x_1, and plot the resulting unit step response for the oscillator in Example 7.7 with $\omega_0 = 1$.

Solution.   We substitute the matrices of Eq. (7.62) (with $\omega_0 = 1$ and $\mathbf{H} = [1 \quad 0]$ because $y = x_1$) into Eq. (7.79) to get

$$\begin{bmatrix} 0 & 1 & 0 \\ -1 & 0 & 1 \\ 1 & 0 & 0 \end{bmatrix} \begin{bmatrix} \mathbf{N_x} \\ N_u \end{bmatrix} = \begin{bmatrix} 0 \\ 0 \\ 1 \end{bmatrix}. \tag{7.82}$$

FIGURE 7.10

Step response of oscillator to a reference input

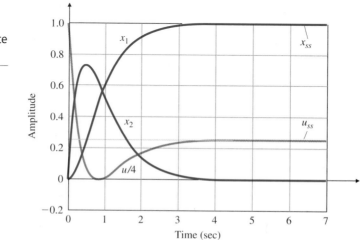

The solution is

$$\mathbf{N}_x = \begin{bmatrix} 1 \\ 0 \end{bmatrix},$$

$$N_u = 1,$$

and, for the given control law, **K**,

$$\bar{N} = 4. \tag{7.83}$$

The corresponding step response is plotted in Fig. 7.10 for $\mathbf{K} = \begin{bmatrix} 3 & 4 \end{bmatrix}$.

◆

Note that there are two equations for the control—Eqs. (7.80) and (7.81). While these expressions are equivalent in theory, they differ in practical implementation in that Eq. (7.80) is usually more robust to parameter errors than Eq. (7.81), particularly when the plant includes a pole at the origin and type I behavior is possible. The difference is most clearly illustrated by the following example.

◆ **EXAMPLE 7.11**   *Reference Input to a Type I System*

Compute the input gains necessary to introduce a reference input with zero steady-state error to a step for the DC motor of Example 5.1, which in state-variable form is described by the matrices:

$$\mathbf{F} = \begin{bmatrix} 0 & 1 \\ 0 & -1 \end{bmatrix}, \qquad \mathbf{G} = \begin{bmatrix} 0 \\ 1 \end{bmatrix},$$

$$\mathbf{H} = \begin{bmatrix} 1 & 0 \end{bmatrix}, \qquad J = 0.$$

Assume that the state feedback gain is $\begin{bmatrix} K_1 & K_2 \end{bmatrix}$.

Solution.   If we substitute the system matrices of this example into the equation for the input gains, Eq. (7.79), we find that the solution is

$$\mathbf{N}_x = \begin{bmatrix} 1 \\ 0 \end{bmatrix},$$

$$N_u = 0,$$

$$\bar{N} = K_1.$$

With these values the expression for the control using $\mathbf{N}_x$ and N_u (Eq. 7.80) reduces to

$$u = -K_1(x_1 - r) - K_2 x_2,$$

while the one using $\bar{N}$ (Eq. 7.81) becomes

$$u = -K_1 x_1 - K_2 x_2 + K_1 r.$$

FIGURE 7.11
Alternative structures for
introducing the
reference input: (a) Eq.
(7.80); (b) Eq. (7.81)

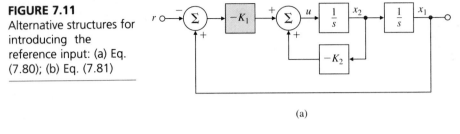

(a)

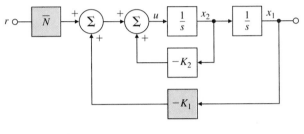

(b)

The block diagrams for the systems using each of the control equations are given in Fig. 7.11. When using Eq. (7.81), as shown in Fig. 7.11b, it is necessary to multiply the input by a gain $K_1 \, (=\bar{N})$ exactly equal to that used in the feedback. If these two gains do not match exactly, there will be a steady-state error. On the other hand, if we use Eq. (7.80) as shown in Fig. 7.11a, there is only one gain to be used on the difference between the reference input and the first state, and zero steady-state error will result even if this gain is slightly in error. The system of Fig. 7.11(a) is more robust than the system of Fig. 7.11(b).

♦

With the reference input in place, the closed-loop system has input r and output y. From the state description we know that the system poles are at the eigenvalues of the closed-loop system matrix, $\mathbf{F} - \mathbf{GK}$. In order to compute the closed-loop transient response, it is necessary to know where the closed-loop zeros of the transfer function from r to y are. They are to be found by applying Eq. (7.55) to the closed-loop description, which we assume has no direct path from input u to output y so that $J = 0$. The zeros are values of s such that

$$\det \begin{bmatrix} s\mathbf{I} - (\mathbf{F} - \mathbf{GK}) & -\bar{N}\mathbf{G} \\ \mathbf{H} & 0 \end{bmatrix} = 0. \tag{7.84}$$

We can use two elementary facts about determinants to simplify Eq. (7.84). In the first place, if we divide the last column by $\bar{N}$, which is a scalar, then the point where the determinant is zero remains unchanged. The determinant is also not changed if we multiply the last column by $\mathbf{K}$ and add it to the first (block) column, with the result that the **GK**-term is cancelled out. Thus the matrix equation for the zeros reduces to

$$\det \begin{bmatrix} s\mathbf{I}-\mathbf{F} & -\mathbf{G} \\ \mathbf{H} & 0 \end{bmatrix} = 0. \tag{7.85}$$

Equation (7.85) is the same as Eq. (7.55) for the zeros of the plant *before* the feedback was applied. The important conclusion is that

---

When full state feedback is used as in Eq. (7.80) or (7.81), the zeros remain unchanged by the feedback.

7.4   Selection of Pole Locations for Good Design

The first step in the pole-placement design approach is to decide on the closed-loop pole locations. When selecting pole locations, it is always useful to keep in mind that the control effort required is related to how far the open-loop poles are moved by the feedback. Furthermore, when a zero is near a pole, the system may be nearly uncontrollable and as we saw in Section 7.3, moving such poles requires large control gains and thus large control effort. Therefore, a pole-placement philosophy that aims to fix only the undesirable aspects of the open-loop response and avoids either large increases in bandwidth or efforts to move poles that are near zeros will typically allow smaller gains and thus smaller control actuators than a philosophy that arbitrarily picks all the poles without regard to the original open-loop pole and zero locations.

Three methods of pole selection

In this section we discuss three techniques to aid in the pole-selection process. The first two approaches—dominant second-order poles and prototype design—deal with pole selection without explicit regard for their effect on control effort. The third method (called optimal control, or symmetric root locus) does specifically address the issue of achieving a balance between good system response and control effort.

7.4.1   Dominant Second-order Poles

The step response corresponding to the second-order transfer function with complex poles at radius ω_n and damping ratio ζ was discussed in Chapter 3. The rise time, overshoot, and settling time can be deduced directly from the pole locations. We can choose the closed-loop poles for a higher-order system as a desired pair of dominant second-order poles, and select the rest of the poles to have real parts corresponding to sufficiently damped modes, so that the system will mimic a second-order response with reasonable control effort. We also must make sure that the zeros are far enough into the left half-plane

(LHP) to avoid having any appreciable effect on the second-order behavior. A system with several lightly damped high-frequency vibration modes plus two rigid-body low-frequency modes lends itself to this philosophy. Here we can pick the low-frequency modes to achieve desired values of ω_n and ζ and select the rest of the poles to increase the damping of the high-frequency modes, while holding their frequency constant in order to minimize control effort. To illustrate this design method we obviously need a system of higher than second order; we will use the tape drive servomotor described in Example 7.3.

◆ **EXAMPLE 7.12**   *Pole Placement as a Dominant Second-order System*

Design the tape servomotor by the dominant second-order poles method to have no more than 5% overshoot and a rise time of no more than 4 sec. Keep the peak tension as low as possible.

Solution.   From the plots of the second-order transients in Fig. 3.12, a damping ratio $\zeta = 0.7$ will meet the overshoot requirement and, for this damping ratio, a rise time of 4 sec suggests a natural frequency of about 1/1.5. There are five poles in all, so the other three need to be placed far to the left of the dominant pair. For our purposes, "far" means the transients due to the fast poles should be over well before the transients due to the dominant poles, and we assume a factor of 4 in the respective undamped natural frequencies to be adequate. From these considerations, the desired poles are given by

$$p = \frac{-0.707 + 0.707j, \ -0.707 - 0.707j, \ -4, \ -4, \ -4}{1.5} \tag{7.86}$$

FIGURE 7.12
Step responses of the tape servomotor designs

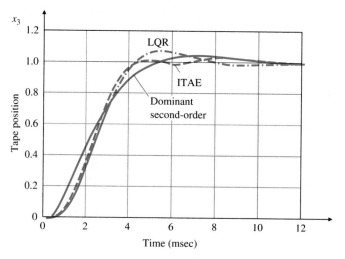

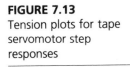

FIGURE 7.13
Tension plots for tape
servomotor step
responses

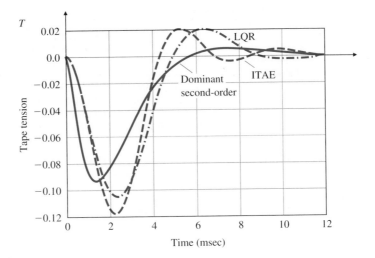

MATLAB acker

With these desired poles, we can use the function acker with **F** and **G** from Eq. (7.31) to find the control gains

$$\mathbf{K_2} = [8.5123 \quad 20.3457 \quad -1.4911 \quad -7.8821 \quad 6.1927]. \tag{7.87}$$

The step response and the corresponding tension plots for this and two other designs (to be discussed in Sections 7.4.2 and 7.4.3) are given in Figs. 7.12 and Fig. 7.13. Notice that the rise time is approximately 4 sec and the overshoot is about 5%, as specified.

Because the design process is iterative, the poles we selected should be seen as only a first step, to be followed by further modifications to meet the specifications as accurately as necessary.

For this example we happened to select adequate pole locations on the first try.

◆

7.4.2 Prototype Design

An alternative for higher-order systems is to select prototype responses with desirable dynamics that the system can be given. There are several sets of these, of which the ITAE and Bessel responses are most relevant to control.

ITAE prototype design

Graham and Lathrop (1953) worked out a prototype set of transient responses to minimize the integral of the time multiplied by the absolute value of the error (**ITAE**)—that is, to minimize

$$\mathscr{I} = \int_0^\infty t|e|\, dt. \tag{7.88}$$

Their evaluation was done numerically, and the resulting pole locations and step responses are shown in Table 7.1 and Fig. 7.14, respectively, for a nominal cutoff frequency $\omega_0 = 1 \text{ rad/sec}$.

TABLE 7.1 Prototype Response Poles

	k	Pole Locations for $\omega_0 = 1$ rad/sec[a]
(a) ITAE transfer function poles	1	$s+1$
	2	$s+0.7071\pm0.7071j$[b]
	3	$(s+0.7081)(s+0.5210\pm1.068j)$
	4	$(s+0.4240\pm1.2630j)(s+0.6260\pm0.4141j)$
	5	$(s+0.8955)(s+0.3764\pm1.2920j)(s+0.5758\pm0.5339j)$
	6	$(s+0.3099\pm1.2634j)(s+0.5805\pm0.7828j)(s+0.7346\pm0.2873j)$
(b) Bessel transfer function poles	1	$s+1$
	2	$s+0.8660\pm0.5000j$[b]
	3	$(s+0.9420)(s+0.7455\pm0.7112j)$
	4	$(s+0.6573\pm0.8302j)(s+0.9047\pm0.2711j)$
	5	$(s+0.9264)(s+0.5906\pm0.9072j)(s+0.8516\pm0.4427j)$
	6	$(s+0.5385\pm0.9617j)(s+0.7998\pm0.5622j)(s+0.9093\pm0.1856j)$

[a]Pole locations for other values of ω_0 can be obtained by substituting s/ω_0 for s everywhere.
[b]The factors $(s+a+bj)(s+a-bj)$ are written as $(s+a\pm bj)$ to conserve space.

FIGURE 7.14
Step response of ITAE prototypes for $\omega_0 = 1$ rad/sec

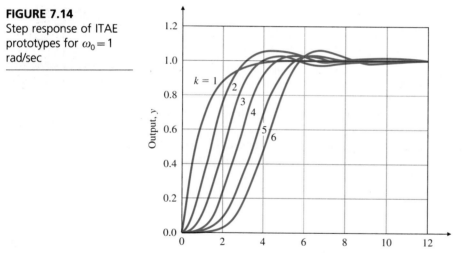

For those applications where overshoot must be avoided altogether, we can consider the **Bessel filter prototype**. These transfer functions are given by $1/B_n(s)$, where $B_n(s)$ is the nth-degree Bessel polynomial. The pole locations are given in Table 7.1, and the step responses are shown in Fig. 7.15.

Comparing the transient responses shown in Fig. 7.14 with those shown in Fig. 7.15, we see that the ITAE prototype has some overshoot, whereas the Bessel prototype has almost none. However, the ITAE responses have faster rise times compared with Bessel prototypes. Further insight into the two

Bessel filter prototype design

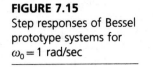

FIGURE 7.15
Step responses of Bessel prototype systems for $\omega_0 = 1$ rad/sec

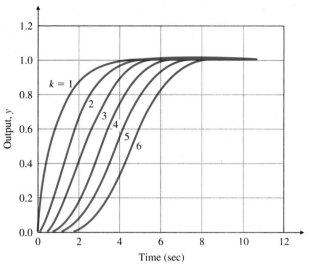

prototype designs can be gained from studying their frequency responses, shown in Fig. 7.16 for $k = 4$. It shows that, given the same value of ω_0, the ITAE prototype has a higher bandwidth for the same attenuation at higher frequencies. Therefore, by selecting ω_0 to give the same bandwidths, we can achieve reduced sensitivity to sensor noise at high frequencies with the ITAE prototype.

Note that all the poles in a particular prototype response have similar natural frequencies. Therefore, direct application of a prototype design to a system with a large spread in the frequencies of its open-loop modes will

FIGURE 7.16
Frequency-response magnitude of ITAE and Bessel prototype systems for $k = 4$

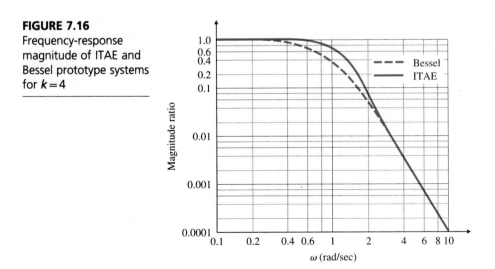

require that some modes be moved a long distance, likely resulting in excessive control effort. An alternative approach would be to use a prototype pole constellation for the lower-frequency modes of the system and simply to add damping as required to the higher-frequency modes. Examples of these approaches are given in Section 7.6. Here we give the results of a sample design using the tape servomotor.

◆ **EXAMPLE 7.13**   *Pole Placement Using an ITAE Prototype*

Design the tape drive servomotor with ITAE-prototype poles selected to have a rise time of no more than 4 sec. Keep the peak tension as low as possible.

Solution.   From the step responses, it would appear that the fifth-order ITAE reponse has a normalized rise time of about 5. However, recall that the tape transfer function has two zeros at $\omega_n \cong 1$, and these will tend to reduce the rise time. (It is not unusual for the designer to notice this kind of behavior *after* a first-pass design). A first choice is to take the desired characteristic equation to have the roots of the fifth-order entry in Table 7.1 with $\omega_0 = 1$. With this choice, the control gain from acker is

$$\mathbf{K}_{\text{ITAE}} = [0.8333 \quad 2.2534 \quad 0.0001 \quad -0.2336 \quad 0.0499]. \tag{7.89}$$

The step response with this gain is also given in Fig. 7.12, and the tension response in Fig. 7.13.

◆

7.4.3 Symmetric Root Locus

LQR design

A most effective and widely used technique of linear control systems design is the optimal **linear quadratic regulator (LQR)**. The simplified version of the LQR problem is to find the control such that the performance index

$$\mathscr{I} = \int_0^\infty \left[\rho z^2(t) + u^2(t) \right] dt \tag{7.90}$$

is minimized for the system

$$\dot{\mathbf{x}} = \mathbf{F}\mathbf{x} + \mathbf{G}u, \tag{7.91a}$$

$$z = \mathbf{H}_1\mathbf{x}, \tag{7.91b}$$

where ρ in Eq. (7.90) is a weighting factor of the designer's choice. The control law that minimizes $\mathscr{I}$ is given by linear-state feedback

$$u = -\mathbf{K}\mathbf{x}. \tag{7.92}$$

Here the optimal value of **K** is that which places the closed-loop poles at the stable roots (those in the LHP) of the root-locus equation

$$1 + \rho G_0(-s)G_0(s) = 0, \tag{7.93}$$

where G_0 is the open-loop transfer function from u to z:

$$G_0(s) = \frac{z(s)}{u(s)} = \mathbf{H}_1(s\mathbf{I} - \mathbf{F})^{-1}\mathbf{G} = \frac{N(s)}{D(s)}. \tag{7.94}$$

Note that this is a root-locus problem as discussed in Chapter 5 with respect to the parameter ρ, which weights the cost (tracking error) z^2 with respect to the control effort u^2 in the performance index Eq. (7.90). Note also that s and $-s$ affect Eq. (7.93) in an identical manner; therefore, for any root s_0 of Eq. (7.93), there will also be a root at $-s_0$. We call the resulting root locus a **symmetric root locus** (SRL), since the locus in the LHP will have a mirror image in the right-half-plane (RHP); that is, there is symmetry with respect to the imaginary axis. We may thus choose the optimal closed-loop poles by first selecting the matrix $\mathbf{H}_1$, which defines the tracking error and which the designer wishes to keep small, and then choosing ρ, which balances the importance of this tracking error against the control effort. Notice that the output we select as tracking error does *not* need to be the plant sensor output. That is why we call the output in Eq. (7.91) z rather than y.

Selecting a set of stable poles from the solution of Eq. (7.93) results in desired closed-loop poles, which we can then use in a pole-placement calculation such as Ackermann's formula (Eq. 7.70) to obtain $\mathbf{K}$. As with all root loci for real transfer functions G_0, the locus is also symmetric with respect to the real axis; thus there is symmetry with respect to both the real and imaginary axes. We can write the SRL equation in the standard root-locus form

$$1 + \rho \frac{N(-s)N(s)}{D(-s)D(s)} = 0, \tag{7.95}$$

obtain the locus poles and zeros by reflecting the open-loop poles and zeros of the transfer function from u to z across the imaginary axis (which doubles the number of poles and zeros), and then sketch the locus. Note that the locus could be either a $0°$ or $180°$ locus, depending on the sign of $G_0(-s)G_0(s)$ in Eq. (7.93). A quick way to determine which type of locus to use ($0°$ or $180°$) is to pick the one that has no part on the imaginary axis. The real-axis rule of root locus plotting will reveal this right away. For the controllability assumptions we have made here, plus the assumption that all the system modes are present in the chosen output z, the optimal closed-loop system is guaranteed to be stable; thus no part of the locus can be on the imaginary axis. Examination of the SRL can provide us with pole locations that achieve varying balances (by choosing different values of ρ) between a fast response (small values of z) and a low control effort (small values of u). The following examples illustrate the use of the SRL.

◆ **EXAMPLE 7.14** *A First-order Symmetric Root Locus*

Plot the SRL for the following first-order system with $z = y$:

$$\dot{y} = -ay + u,$$

$$G_0(s) = \frac{1}{s+a}. \tag{7.96}$$

Solution. The SRL equation (Eq. 7.93) for this example is

$$1 + \rho \frac{1}{(-s+a)(s+a)} = 0. \tag{7.97}$$

The SRL, shown in Fig. 7.17, is a $0°$ locus. The optimal (stable) pole can be determined explicitly in this case as

$$s = -\sqrt{a^2 + \rho}. \tag{7.98}$$

Thus the closed-loop root location that minimizes the performance index of Eq. (7.90) lies on the real axis at the distance given by Eq. (7.98) and is always to the left of the open-loop root.

FIGURE 7.17
Symmetric root locus for
a first-order system

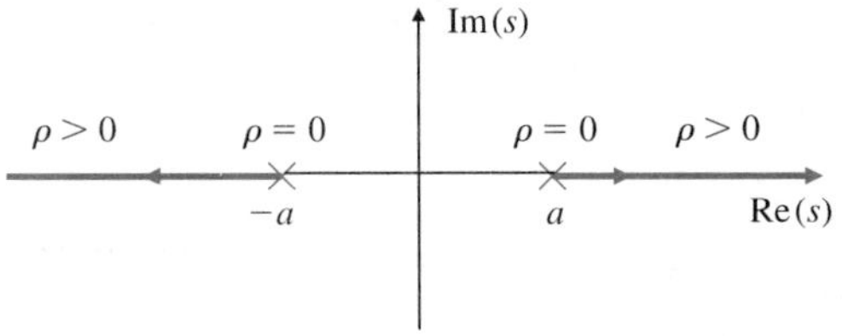

It is also possible to locate optimal pole locations for the design of an unstable system using the SRL and LQR method.

◆ **EXAMPLE 7.15** *SRL Design for an Unstable System*

Draw the SRL for the linearized equations of the simple inverted pendulum with $\omega_0 = 1$. Take the output to be the sum of twice the position plus the velocity.

Solution. The equations of motion are

$$\dot{\mathbf{x}} = \begin{bmatrix} 0 & 1 \\ \omega_0^2 & 0 \end{bmatrix} \mathbf{x} + \begin{bmatrix} 0 \\ -1 \end{bmatrix} u. \tag{7.99}$$

FIGURE 7.18
Symmetric root locus for
the inverted pendulum

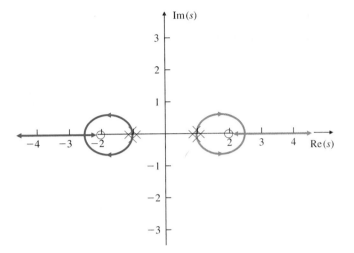

For the specified output of $2 \times$ position plus velocity we let the tracking error be

$$z = [2 \quad 1]\mathbf{x}. \tag{7.100}$$

We then calculate from Eqs. (7.99) and (7.100) that

$$G_0(s) = -\frac{s+2}{s^2 - \omega_0^2}. \tag{7.101}$$

The symmetric $0°$ loci are shown in Fig. 7.18. We would choose two stable roots for a given value of ρ and use them for pole-placement and control-law design.

◆

As a final example in this section, we consider again the tape servomotor and introduce LQR design using the computer directly to solve for the optimal control law. From Eqs. (7.90) and (7.92) we know the information needed to find the optimal control is given by the system matrices $\mathbf{F}$ and $\mathbf{G}$ and the output matrix $\mathbf{H}_1$. Most CACSD packages, including MATLAB, use a more general form of Eq. (7.90):

$$\mathscr{J} = \int_0^\infty (\mathbf{x}^T \mathbf{Q} \mathbf{x} + \mathbf{u}^T \mathbf{R} \mathbf{u}) \, dt. \tag{7.102}$$

Equation (7.102) reduces to the simpler form of Eq. (7.90) if we take $\mathbf{Q} = \rho \mathbf{H}_1^T \mathbf{H}_1$ and $\mathbf{R} = 1$. The direct solution for the optimal control gain is the MATLAB statement

MATLAB lqr

$$K = \mathrm{lqr}(F, G, Q, R). \tag{7.103}$$

EXAMPLE 7.16   *LQR Design for a Tape Drive*

Find the optimal control for the tape drive of example 7.6, using the position x_3 as the output for the performance index. Let $\rho = 1$.

Solution.   All we need to do here is to substitute the matrices into Eq. (7.103), form the feedback system, and plot the response. The performance index matrix is the scalar $R = 1$; the most difficult part of the problem is finding the state-cost matrix **Q**. With the output-cost variable $z = x_3$, the output matrix from Example 7.3 is

$$\mathbf{H}_3 = [0.5 \quad 0 \quad 0.5 \quad 0 \quad 0],$$

and with $\rho = 1$, then the required matrix is

$$\mathbf{Q} = \mathbf{H}_3^T \mathbf{H}_3$$

$$= \begin{bmatrix} 0.25 & 0 & 0.25 & 0 & 0 \\ 0 & 0 & 0 & 0 & 0 \\ 0.25 & 0 & 0.25 & 0 & 0 \\ 0 & 0 & 0 & 0 & 0 \end{bmatrix}.$$

The gain is given by MATLAB as

$$\mathbf{K} = [0.6526 \quad 2.1667 \quad 0.3474 \quad 0.5976 \quad 1.0616]. \tag{7.104}$$

The results of a position step and the corresponding tension are plotted in Figs. 7.12 and 7.13 with the earlier responses for comparison. Obviously, there is a vast range of choice for the elements of **Q** and **R**, so substantial experience is needed in order to use the LQR method effectively.

7.4.4   Comments on the Methods

The three methods of pole selection described in Sections 7.4.1 to 7.4.3 are alternatives the designer can use for an initial design by pole placement. Note that the first two methods (dominant second-order and prototype design) suggest selecting closed-loop poles without regard to the effect on the control effort required to achieve that response. In some cases, therefore, the resulting control effort may be ridiculously high. The third method (SRL), on the other hand, selects poles that result in some balance between system errors and control effort. The designer can easily examine the relationship between shifts in that balance (by changing ρ) and system root locations, time response, and feedback gains. However, the SRL method does not indicate directly the shape of the transient response in the way the other methods do since the poles with the SRL method are restricted to the locus. Whatever initial pole-selection method we use, some modification is almost always necessary to achieve the desired balance of bandwidth, overshoot, sensitivity, control effort, and other practical design requirements. Further insight into pole selection will be gained from the examples that illustrate compensation in Section 7.6 and from the case studies in Chapter 9.

7.5   Estimator Design

The control law designed in Section 7.3 assumed that all the state variables are available for feedback. In most cases, not all the state variables are measured. The cost of the required sensors may be prohibitive, or it may be physically impossible to measure all the state variables, as in, for example, a nuclear power plant. In this section we demonstrate how to reconstruct all the state variables of a system from a few measurements. If the estimate of the state is denoted by $\hat{\mathbf{x}}$, it would be convenient if we could replace the true state in the control law given by Eq. (7.81) with the estimates so that the control becomes $u = -\mathbf{K}\hat{\mathbf{x}} + \bar{N}r$. This is indeed possible, as we shall see in Section 7.6, so construction of a state estimate is a key part of state-space control design.

7.5.1   Full-order Estimators

One method of estimating the state is to construct a full-order model of the plant dynamics,

$$\dot{\hat{\mathbf{x}}} = \mathbf{F}\hat{\mathbf{x}} + \mathbf{G}u, \tag{7.105}$$

where $\hat{\mathbf{x}}$ is the estimate of the actual state $\mathbf{x}$. We know $\mathbf{F}$, $\mathbf{G}$, and $u(t)$. Hence this estimator will be satisfactory if we can obtain the correct initial condition $\mathbf{x}(0)$ and set $\hat{\mathbf{x}}(0)$ equal to it. Figure 7.19 depicts this open-loop estimator. However, it is precisely the lack of information about $\mathbf{x}(0)$ that requires the construction of an estimator. Otherwise, the estimated state would track the true state exactly. Thus, if we made a poor estimate for the initial condition, the estimated state would have a continually growing error or an error that goes to zero too slowly to be of use. Furthermore, small errors in our knowledge of the system $(\mathbf{F}, \mathbf{G})$ would also cause the estimate to diverge from the true state.

To study the dynamics of this estimator we define the error in the estimate to be

$$\tilde{\mathbf{x}} \triangleq \mathbf{x} - \hat{\mathbf{x}}. \tag{7.106}$$

FIGURE 7.19
Open-loop estimator

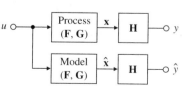

Then the dynamics of this error system are given by

$$\dot{\tilde{\mathbf{x}}} = \mathbf{F}\tilde{\mathbf{x}}, \qquad \tilde{\mathbf{x}}(0) = \mathbf{x}(0) - \hat{\mathbf{x}}(0). \tag{7.107}$$

The error converges to zero for a stable system (**F** stable), but we have no ability to influence the rate at which the state estimate converges to the true state. Furthermore, the error is converging to zero at the same rate as the natural dynamics of **F**. If this convergence rate were satisfactory, no control or estimation would be required.

We now invoke the golden rule: When in trouble, use feedback. Consider feeding back the difference between the measured and estimated outputs and correcting the model continuously with this error signal. The equation for this scheme, shown in Fig. 7.20, is

$$\dot{\hat{\mathbf{x}}} = \mathbf{F}\hat{\mathbf{x}} + \mathbf{G}u + \mathbf{L}(y - \mathbf{H}\hat{\mathbf{x}}). \tag{7.108}$$

Here **L** is a proportional gain defined as

$$\mathbf{L} = [l_1, l_2, \ldots, l_n]^T \tag{7.109}$$

and is chosen to achieve satisfactory error characteristics. The dynamics of the error can be obtained by subtracting the estimate (Eq. 7.108) from the state (Eq. 7.32), to get the error equation

$$\dot{\tilde{\mathbf{x}}} = (\mathbf{F} - \mathbf{LH})\tilde{\mathbf{x}}. \tag{7.110}$$

The characteristic equation of the error is now given by

$$\det[s\mathbf{I} - (\mathbf{F} - \mathbf{LH})] = 0. \tag{7.111}$$

If we can choose **L** so that $\mathbf{F} - \mathbf{LH}$ has stable and reasonably fast eigenvalues, then $\tilde{\mathbf{x}}$ will decay to zero and remain there — independent of the known forcing function $u(t)$ and its effect on the state $\mathbf{x}(t)$ and irrespective of the initial condition $\tilde{\mathbf{x}}(0)$. This means that $\hat{\mathbf{x}}(t)$ will converge to $\mathbf{x}(t)$, regardless of the value of $\hat{\mathbf{x}}(0)$; furthermore, we can choose the dynamics of the error to be stable as well as much faster than the open-loop dynamics determined by **F**.

Note that in obtaining Eq. (7.110) we have assumed that **F**, **G**, and **H** are identical in the physical plant and in the computer implementation of the estimator. If we do not have an accurate model of the plant (**F**, **G**, **H**), the

Feed back the output error to correct the state estimate equation.

Estimate-error characteristic equation

FIGURE 7.20
Closed-loop estimator

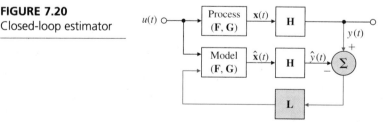

dynamics of the error are no longer governed by Eq. (7.110). However, we can typically choose **L** so that the error system is still at least stable and the error remains acceptably small, even with (small) modeling errors and disturbing inputs. It is important to emphasize that the nature of the plant and the estimator are quite different. The plant is a physical system such as a chemical process or servomechanism, whereas the estimator is usually an electronic unit computing the estimated state according to Eq. (7.108).

The selection of **L** can be approached in exactly the same fashion as **K** is selected in the control-law design. If we specify the desired location of the estimator error poles as

$$s_i = \beta_1, \beta_2, \ldots, \beta_n,$$

then the desired estimator characteristic equation is

$$\alpha_e(s) \triangleq (s - \beta_1)(s - \beta_2) \cdots (s - \beta_n). \tag{7.112}$$

We can then solve for **L** by comparing coefficients in Eqs. (7.111) and (7.112).

◆ **EXAMPLE 7.17**   *An Estimator Design for a Second-order System*

Design an estimator for the simple pendulum. Compute the estimator gain matrix that will place both the estimator error poles at $-10\omega_0$ (five times as fast as the controller poles selected in Example 7.7).

Solution.   The equations of motion are

$$\dot{\mathbf{x}} = \begin{bmatrix} 0 & 1 \\ -\omega_0^2 & 0 \end{bmatrix} \mathbf{x} + \begin{bmatrix} 0 \\ 1 \end{bmatrix} u, \tag{7.113a}$$

$$y = [1 \quad 0]\mathbf{x}. \tag{7.113b}$$

We are asked to place the two estimator error poles at $-10\omega_0$. The corresponding characteristic equation is

$$\alpha_e(s) = (s + 10\omega_0)^2 = s^2 + 20\omega_0 s + 100\omega_0^2. \tag{7.114}$$

From Eq. (7.111) we get

$$\det[s\mathbf{I} - (\mathbf{F} - \mathbf{LH})] = s^2 + l_1 s + l_2 + \omega_0^2. \tag{7.115}$$

Comparing coefficients in Eqs. (7.114) and (7.115), we find that

$$\mathbf{L} = \begin{bmatrix} l_1 \\ l_2 \end{bmatrix} = \begin{bmatrix} 20\omega_0 \\ 99\omega_0^2 \end{bmatrix}. \tag{7.116}$$

This completes the design of the estimator for this example.

Performance of the estimator can be tested by adding state feedback to the plant and plotting the estimate errors. Combining Eq. (7.59) of the plant with state feedback with Eq. (7.108) of the estimator with output feedback results in the following overall

FIGURE 7.21
Estimator connected to the plant

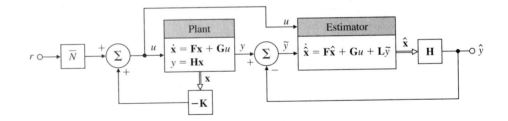

system equations:

$$\begin{bmatrix} \dot{\mathbf{x}} \\ \dot{\hat{\mathbf{x}}} \end{bmatrix} = \begin{bmatrix} \mathbf{F} - \mathbf{GK} & 0 \\ \mathbf{LH} - \mathbf{GK} & \mathbf{F} - \mathbf{LH} \end{bmatrix} \begin{bmatrix} \mathbf{x} \\ \hat{\mathbf{x}} \end{bmatrix},$$ (7.117)

$$y = \begin{bmatrix} \mathbf{H} & 0 \end{bmatrix} \begin{bmatrix} \mathbf{x} \\ \hat{\mathbf{x}} \end{bmatrix},$$ (7.118)

$$\tilde{y} = \begin{bmatrix} \mathbf{H} & -\mathbf{H} \end{bmatrix} \begin{bmatrix} \mathbf{x} \\ \hat{\mathbf{x}} \end{bmatrix}.$$ (7.119)

A block diagram of the setup is drawn in Fig. 7.21.

The response of this closed-loop system with $\omega_0 = 1$ to an initial condition $\mathbf{x}_0 = [1.0, 0.0]^T$ and $\hat{\mathbf{x}}_0 = [0, 0]^T$ is shown in Fig. 7.22, where **K** is obtained from Example 7.7 and **L** comes from Eq. (7.116). Note that the state estimates converge to the actual state variables after an initial transient even though the initial value of $\hat{\mathbf{x}}$ had a large error. Also note that the estimate error decays approximately five times faster than the decay of the state itself, as we designed it to do.

FIGURE 7.22
Initial-condition response of oscillator showing x and $\hat{x}$

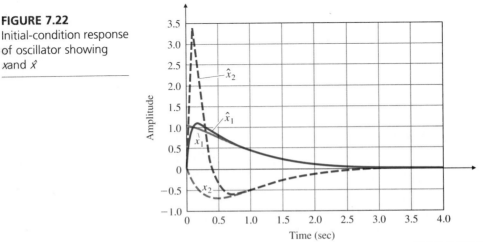

Observer Canonical Form

As was the case for control-law design, there is a canonical form for which the estimator-gain design equations are particularly simple and the existence of a solution is obvious. We introduced this form in Section 7.2.1: The equations are in the observer canonical form and have the following structure:

$$\dot{\mathbf{x}}_o = \mathbf{F}_o\mathbf{x}_o + \mathbf{G}_o u, \tag{7.120a}$$

$$y = \mathbf{H}_o\mathbf{x}_o, \tag{7.120b}$$

where

Observer canonical form

$$\mathbf{F}_o = \begin{bmatrix} -a_1 & 1 & 0 & 0 & \cdots & 0 \\ -a_2 & 0 & 1 & 0 & \cdots & \vdots \\ \vdots & \vdots & & \ddots & & 1 \\ -a_n & 0 & & 0 & & 0 \end{bmatrix}, \qquad \mathbf{G}_o = \begin{bmatrix} b_1 \\ b_2 \\ \vdots \\ b_n \end{bmatrix},$$

$$\mathbf{H}_o = \begin{bmatrix} 1 & 0 & 0 & \cdots & 0 \end{bmatrix}..$$

A block diagram for the third-order case is shown in Fig. 7.23. In observer canonical form, all the feedback loops come from the output, or observed signal. Like control canonical form, observer canonical form is a "direct" form because the values of the significant elements in the matrices are obtained directly from the coefficients of the numerator and denominator polynomials of the corresponding transfer function $G(s)$. The matrix $\mathbf{F}_o$ is called a **left companion matrix** to the characteristic equation because the coefficients of the equation appear on the left side of the matrix.

One of the advantages of the observer canonical form is that the estimator gains can be obtained from it by inspection. The estimator error closed-loop matrix for the third-order case is

$$\mathbf{F}_o - \mathbf{L}\mathbf{H}_o = \begin{bmatrix} -a_1-l_1 & 1 & 0 \\ -a_2-l_2 & 0 & 1 \\ -a_3-l_3 & 0 & 0 \end{bmatrix}, \tag{7.121}$$

FIGURE 7.23
Block diagram for observer canonical form of a third-order system

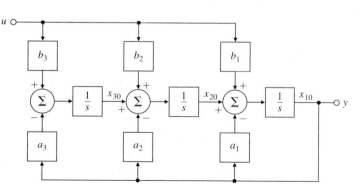

which has the characteristic equation

$$s^3 + (a_1 + l_1)s^2 + (a_2 + l_2)s + (a_3 + l_3) = 0, \tag{7.122}$$

and the estimator gain can be found by comparing the coefficients of Eq. (7.122) with $\alpha_e(s)$ from Eq. (7.112).

In a development exactly parallel with the control-law case, we can find a transformation to take a given system to observer canonical form if and only if the system has a structural property that in this case we call **observability**. Roughly speaking, observability refers to our ability to deduce information about all the modes of the system by monitoring only the sensed outputs. Unobservability results when some mode or subsystem is disconnected physically from the output and therefore no longer appears in the measurements. For example, if only derivatives of certain state variables are measured, and these state variables do not affect the dynamics, a constant of integration is obscured. This situation occurs with a plant having the transfer function $1/s^2$ if only velocity is measured, for then it is impossible to deduce the initial value of the position. On the other hand, for an oscillator a velocity measurement is sufficient to estimate position because the acceleration, and consequently the velocity observed, are affected by position. The mathematical test for determining observability is that the **observability matrix**

Observability *(margin)*

Observability matrix *(margin)*

$$\mathcal{O} = \begin{bmatrix} \mathbf{H} \\ \mathbf{HF} \\ \vdots \\ \mathbf{HF}^{n-1} \end{bmatrix} \tag{7.123}$$

must have independent columns. In the one output case we will study, $\mathcal{O}$ is square, so the requirement is that $\mathcal{O}$ be nonsingular or have a nonzero determinant. In general, we can find a transformation to observer canonical form if and only if the observability matrix is nonsingular. Note that this is analogous to our earlier conclusions for transforming system matrices to control canonical form.

As with control-law design, we could find the transformation to observer form, compute the gains from the equivalent of Eq. (7.122), and transform back. An alternative method of computing $\mathbf{L}$ is to use Ackermann's formula in estimator form, which is

Ackermann's estimator formula *(margin)*

$$\mathbf{L} = \alpha_e(\mathbf{F})\mathcal{O}^{-1} \begin{bmatrix} 0 \\ 0 \\ \vdots \\ 1 \end{bmatrix}, \tag{7.124}$$

where $\mathcal{O}$ is the observability matrix given in Eq. (7.123).

TABLE 7.2 Duality

Control	Estimation
F	F^T
G	H^T
H	G^T

Duality

Duality of estimation and control

You may already have noticed from this discussion the considerable resemblance between estimation and control problems. In fact, the two problems are mathematically equivalent. This property is called **duality**. Table 7.2 shows the duality relationships between the estimation and control problems. For example, Ackermann's control formula (Eq. 7.70) becomes the estimator formula Eq. (7.124) if we make the substitutions given in Table 7.2. We can demonstrate this directly using matrix algebra. The control problem is to select the row matrix $\mathbf{K}$ for satisfactory placement of the poles of the system matrix $\mathbf{F} - \mathbf{GK}$; the estimator problem is to select the column matrix $\mathbf{L}$ for satisfactory placement of the poles of $\mathbf{F} - \mathbf{LH}$. However, the poles of $\mathbf{F} - \mathbf{LH}$ equal those of $(\mathbf{F} - \mathbf{LH})^T = \mathbf{F}^T - \mathbf{H}^T \mathbf{L}^T$, and in this form, the algebra of the design for $\mathbf{L}^T$ is identical to that for $\mathbf{K}$. Therefore, we can use Ackermann's formula or the

MATLAB acker, place

place algorithm in the forms

$$\mathbf{L} = \mathsf{acker}(F^T, H^T, p_e)^T,$$

$$\mathbf{L} = \mathsf{place}(F^T, H^T, p_e)^T.$$

where p_e is a vector containing the desired estimator error poles.

Thus duality allows us to use the same design tools for estimator problems as for control problems with proper substitutions. The two canonical forms are also dual, as we can see from comparing the triples $(\mathbf{F}_c, \mathbf{G}_c, \mathbf{H}_c)$ and $(\mathbf{F}_o, \mathbf{G}_o, \mathbf{H}_o)$.

7.5.2 Reduced-order Estimators

The estimator design method described in Section 7.5.1 reconstructs the entire state vector using measurements of some of the state variables. If the sensors have no noise, then a full-order estimator contains redundancies, and it seems reasonable to question the necessity for estimating state variables that are measured directly. Can we reduce the complexity of the estimator by using the state variables that are measured directly and exactly? Yes. However, it is better to implement a full-order estimator if there is significant noise on the measurements because, in addition to estimating unmeasured state variables, the estimator filters the measurements.

The **reduced-order estimator** reduces the order of the estimator by the number (1 in this text) of sensed outputs. To derive this estimator we start with the assumption that the output equals the first state as, for example, $y = x_a$. If this is not so, a preliminary step is required. Transforming to observer form is possible but overkill; any nonsingular transformation with $\mathbf{H}$ as the first row will do. We now partition the state vector into two parts: x_a, which is directly measured, and $\mathbf{x}_b$, which represents the remaining state variables that need to be estimated. If we partition the system matrices accordingly, the complete description of the system is given by

$$\begin{bmatrix} \dot{x}_a \\ \dot{\mathbf{x}}_b \end{bmatrix} = \begin{bmatrix} F_{aa} & \mathbf{F}_{ab} \\ \mathbf{F}_{ba} & \mathbf{F}_{bb} \end{bmatrix} \begin{bmatrix} x_a \\ \mathbf{x}_b \end{bmatrix} + \begin{bmatrix} G_a \\ \mathbf{G}_b \end{bmatrix} u$$

$$y = \begin{bmatrix} 1 & 0 \end{bmatrix} \begin{bmatrix} x_a \\ \mathbf{x}_b \end{bmatrix}. \tag{7.125}$$

The dynamics of the unmeasured state variables are given by

$$\dot{\mathbf{x}}_b = \mathbf{F}_{bb}\mathbf{x}_b + \underbrace{\mathbf{F}_{ba}x_a + \mathbf{G}_b u}_{\text{known input}}, \tag{7.126}$$

where the rightmost two terms are known and can be considered as an input into the $\mathbf{x}_b$ dynamics. Since $x_a = y$, the measured dynamics are given by the scalar equation

$$\dot{x}_a = \dot{y} = F_{aa}y + \mathbf{F}_{ab}\mathbf{x}_b + G_a u. \tag{7.127}$$

If we collect the known terms of Eq. (7.127) on one side,

$$\underbrace{\dot{y} - F_{aa}y - G_a u}_{\text{known measurement}} = \mathbf{F}_{ab}\mathbf{x}_b, \tag{7.128}$$

we obtain a relationship between known quantities on the left side, which we consider measurements, and unknown state variables on the right. Therefore, Eqs. (7.127) and (7.128) have the same relationship to the state $\mathbf{x}_b$ that the original equation (Eq. 7.125) had to the entire state $\mathbf{x}$. Following this line of reasoning, we can establish the following substitutions in the original estimator equations to obtain a (reduced-order) estimator of $\mathbf{x}_b$:

$$\mathbf{x} \leftarrow \mathbf{x}_b,$$

$$\mathbf{F} \leftarrow \mathbf{F}_{bb},$$

$$Gu \leftarrow \mathbf{F}_{ba}y + \mathbf{G}_b u, \tag{7.129}$$

$$y \leftarrow \dot{y} - F_{aa}y - G_a u,$$

$$\mathbf{H} \leftarrow \mathbf{F}_{ab}.$$

Therefore, the reduced-order estimator equations are obtained by substituting Eqs. (7.129) into the full-order estimator (Eq. 7.108):

$$\dot{\hat{x}}_b = F_{bb}\hat{x}_b + \underbrace{F_{ba}y + G_bu}_{\text{input}} + L(\underbrace{\dot{y} - F_{aa}y - G_au - F_{ab}\hat{x}_b}_{\text{measurement}}). \tag{7.130}$$

If we define the estimator error to be

$$\tilde{x}_b \overset{\Delta}{=} x_b - \hat{x}_b, \tag{7.131}$$

then the dynamics of the error are given by subtracting Eq. (7.130) from Eq. (7.126) to get

$$\dot{\tilde{x}}_b = (F_{bb} - LF_{ab})\tilde{x}_b, \tag{7.132}$$

and its characteristic equation is given by

$$\det[sI - (F_{bb} - LF_{ab})] = 0. \tag{7.133}$$

We design the dynamics of this estimator by selecting L so that Eq. (7.133) matches a reduced order $\alpha_e(s)$. Now Eq. (7.130) can be rewritten as

$$\dot{\hat{x}}_b = (F_{bb} - LF_{ab})\hat{x}_b + (F_{ba} - LF_{aa})y + (G_b - LG_a)u + L\dot{y}. \tag{7.134}$$

The fact that we must form the derivative of the measurements in Eq. (7.134) appears to present a practical difficulty. It is known that differentiation amplifies noise, so if y is noisy, the use of $\dot{y}$ is unacceptable. To get around this difficulty, we define the new controller state to be

$$x_c \overset{\Delta}{=} \hat{x}_b - Ly. \tag{7.135}$$

In terms of this new state the implementation of the reduced-order estimator is given by

Reduced-order estimator

$$\dot{x}_c = (F_{bb} - LF_{ab})\hat{x}_b + (F_{ba} - LF_{aa})y + (G_b - LG_a)u, \tag{7.136}$$

and $\dot{y}$ no longer appears directly. A block-diagram representation of the reduced-order estimator is shown in Fig. 7.24.

FIGURE 7.24
Reduced-order estimator structure

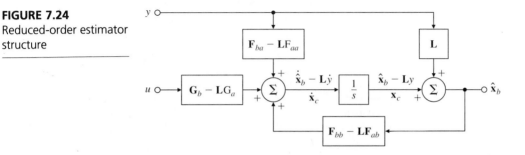

◆ **EXAMPLE 7.18**   *A Reduced-order Estimator Design*

Design a reduced-order estimator for the pendulum that has the error pole at $-10\omega_0$.

Solution.   We are given the system equations

$$
\begin{bmatrix} \dot{x}_1 \\ \dot{x}_2 \end{bmatrix} = \begin{bmatrix} 0 & 1 \\ -\omega_0^2 & 0 \end{bmatrix} \begin{bmatrix} x_1 \\ x_2 \end{bmatrix} + \begin{bmatrix} 0 \\ 1 \end{bmatrix} u,
$$

$$
y = \begin{bmatrix} 1 & 0 \end{bmatrix} \begin{bmatrix} x_1 \\ x_2 \end{bmatrix}.
$$

The partitioned matrices are

$$
\begin{bmatrix} F_{aa} & F_{ab} \\ F_{ba} & F_{bb} \end{bmatrix} = \begin{bmatrix} 0 & 1 \\ -\omega_0^2 & 0 \end{bmatrix},
$$

$$
\begin{bmatrix} G_a \\ G_b \end{bmatrix} = \begin{bmatrix} 0 \\ 1 \end{bmatrix}.
$$

From Eq. (7.133) we find the characteristic equation in terms of L:

$$
s - (0 - L) = 0.
$$

We compare it with the desired equation,

$$
\alpha_e(s) = s + 10\omega_0 = 0,
$$

which yields

$$
L = 10\omega_0.
$$

FIGURE 7.25
Initial-condition response
of the reduced-order
estimator

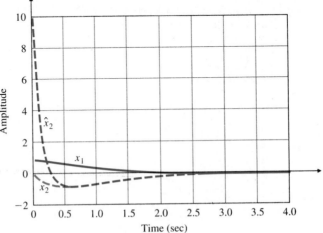

The estimator equation, from Eq. (7.136), is

$$\dot{x}_c = -10\omega_0 \hat{x}_2 - \omega_0^2 y + u,$$

and the state estimate, from Eq. (7.135), is

$$\hat{x}_2 = x_c + 10\omega_0 y.$$

We use the control law given in the earlier examples. The response of the estimator to a plant initial condition $\mathbf{x}_0 = [1.0, 0.0]^T$ and an estimator initial condition $x_{c0} = 0$ is shown in Fig. 7.25 for $\omega_0 = 1$. Note the similarity of the initial-condition response to that of the full-order estimator plotted in Fig. 7.22.

◆

The reduced-order estimator gain can also be found from Ackermann's formula (Eq. 7.124), with appropriate substitutions from Eq. (7.129), as follows:

$$\mathbf{L} = \alpha_e(\mathbf{F}_{bb})\mathcal{O}_b^{-1} \begin{bmatrix} 0 \\ 0 \\ \vdots \\ 1 \end{bmatrix}, \tag{7.137}$$

where

$$\mathcal{O}_b = \begin{bmatrix} \mathbf{F}_{ab} \\ \mathbf{F}_{ab}\mathbf{F}_{bb} \\ \mathbf{F}_{ab}\mathbf{F}_{bb}^2 \\ \vdots \\ \mathbf{F}_{ab}\mathbf{F}_{bb}^{n-2} \end{bmatrix}.$$

The conditions for the existence of the reduced-order estimator are the same as for the full-order estimator — namely, observability of $(\mathbf{F}, \mathbf{H})$.

7.5.3   Estimator Pole Selection

We can base our selection of estimator-pole locations on the techniques discussed in Section 7.4 for the case of controller poles. As a rule of thumb, the estimator poles can be chosen to be faster than the controller poles by a factor of 2 to 6. This ensures a faster decay of the estimator errors compared with the desired dynamics, thus causing the controller poles to dominate the total

Design rules of thumb for selecting estimator poles

response. If sensor noise is large enough to be a major concern, we may choose the estimator poles to be slower than two times the controller poles, which would yield a system with lower bandwidth and more noise smoothing. However, we would expect the total system response in this case to be strongly influenced by the location of the estimator poles. If the estimator poles are slower than the controller poles, we would expect the system response to disturbances to be dominated by the dynamic characteristics of the estimator rather than by those selected by the control law.

In comparison with the selection of controller poles, estimator pole selection requires us to be concerned with a much different relationship than with control effort. As in the controller, there is a feedback term in the estimator that grows in magnitude as the requested speed of response increases. However, this feedback is in the form of an electronic signal or a digital word in a computer, so its growth causes no special difficulty. In the controller, increasing the speed of response increases the control effort; this implies the use of a larger actuator, which in turn increases size, weight, and cost. The important consequence of increasing the speed of response of an estimator is that the bandwidth of the estimator becomes higher, thus causing more sensor noise to pass on to the control actuator. There is a limit, of course, if $(\mathbf{F}, \mathbf{H})$ are not observable, for then no amount of estimator gain can produce a reasonable state estimate. Thus, as with controller design, the best estimator design is a balance between good transient response and low-enough bandwidth that sensor noise does not signficantly impair actuator activity. Both dominant second-order and prototype characteristic equation ideas can be used to meet the requirements.

There is also a result for estimator gain design based on the SRL. In optimal estimation theory, the best choice for estimator gain is dependent on the ratio of sensor noise intensity v to process (disturbance) noise intensity (w in Eq. (7.139) below). This is best understood by reexamining the estimator equation

$$\dot{\hat{\mathbf{x}}} = \mathbf{F}\hat{\mathbf{x}} + \mathbf{G}u + \mathbf{L}(y - \mathbf{H}\hat{\mathbf{x}}) \tag{7.138}$$

Process noise   to see how it interacts with the system when process noise w is present. The plant with process noise is described by

$$\dot{\mathbf{x}} = \mathbf{F}\mathbf{x} + \mathbf{G}u + \mathbf{G}_1 w, \tag{7.139}$$

Sensor noise   and the measurement equation with sensor noise v is described by

$$y = \mathbf{H}\mathbf{x} + v. \tag{7.140}$$

The estimator error equation with these additional inputs is found directly by subtracting Eq. (7.138) from Eq. (7.139) and substituting Eq. (7.140) for y:

$$\dot{\tilde{\mathbf{x}}} = (\mathbf{F} - \mathbf{L}\mathbf{H})\tilde{\mathbf{x}} + \mathbf{G}_1 w - \mathbf{L}v. \tag{7.141}$$

In Eq. (7.141) the sensor noise is multiplied by $\mathbf{L}$ and the process noise is not. If $\mathbf{L}$ is very small, then the effect of sensor noise is removed but the estimator's dynamic response will be "slow," so the error will not reject effects of w very well. The state of a low-gain estimator will not track uncertain plant inputs very well. These results can, with some success, also be applied to model errors in, for example, $\mathbf{F}$ or $\mathbf{G}$. Such model errors will add terms to Eq. (7.141) and act like additional process noise. On the other hand, if $\mathbf{L}$ is large, then the estimator response will be fast and the disturbance or process noise will be rejected, but the sensor noise, multiplied by $\mathbf{L}$, results in large errors. Clearly, a balance between these two effects is required. It turns out that the optimal solution to this balance can be found under very reasonable assumptions by solving an SRL equation for the estimator that is very similar to the one established for the optimal control formulation (Eq. 7.93). The estimator SRL equation is

$$1 + qG_e(-s)G_e(s) = 0, \tag{7.142}$$

where q is the ratio of input disturbance noise intensity to sensor noise intensity and G_e is the transfer function from the process noise to the sensor output and is given by

$$G_e(s) = \mathbf{H}(s\mathbf{I} - \mathbf{F})^{-1}\mathbf{G}_1. \tag{7.143}$$

Note from Eqs. (7.93) and (7.142) that $G_e(s)$ is similar to $G_0(s)$. However, a comparison of Eqs. (7.94) and (7.143) shows that $G_e(s)$ has the input matrix $\mathbf{G}_1$ instead of $\mathbf{G}$, and that G_0 is the transfer function from the control input u to *cost* output z and has output matrix $\mathbf{H}_1$ instead of $\mathbf{H}$.

The use of the estimator SRL (Eq. 7.142) is identical to the use of the controller SRL. A root locus with respect to q is generated, thus yielding sets of optimal estimator poles corresponding to more or less for the ratio of process noise intensity to sensor noise intensity. The designer then picks the set of (stable) poles that seems best considering all aspects of the problem. An important advantage of using the SRL technique is that after the process noise input matrix $\mathbf{G}_1$ has been selected, the arbitrariness is reduced to one degree of freedom, the selection q, instead of the many degrees of freedom required to select the poles directly in a higher-order system.

A final comment concerns the reduced-order estimator. Because of the presence of a direct transmission term from y through $\mathbf{L}$ to $\mathbf{x}_b$ (see Fig. 7.24), the reduced-order estimator has a much higher bandwidth from sensor to control when compared with the full-order estimator. Therefore, if sensor noise is a significant factor, the reduced-order estimator is less attractive since the potential savings in complexity is more than offset by the increased sensitivity to noise.

7.6   Compensator Design: Combined Control Law and Estimator

regulator

If we take the control-law design described in Section 7.3, combine it with the estimator design described in Section 7.5, and implement the control law using the estimated state variables, the design is complete for a **regulator** that is able to reject disturbances but has no reference input to be tracked. However, since the control law was designed for feedback of the actual (not the estimated) state, you may wonder what effect using $\hat{x}$ in place of x has on the system dynamics. In this section we compute this effect. In doing so we will compute the closed-loop characteristic equation and the open-loop compensator transfer function. We will use these results to compare the state-space designs with root-locus and frequency-response designs.

The plant equation with feedback is now

$$\dot{x} = Fx - GK\hat{x}, \tag{7.144}$$

which can be rewritten in terms of the state error $\tilde{x}$ as

$$\dot{x} = Fx - GK(x - \tilde{x}). \tag{7.145}$$

The overall system dynamics in state form are obtained by combining Eq. (7.145) with the estimator error (Eq. 7.110) to get

$$\begin{bmatrix} \dot{x} \\ \dot{\tilde{x}} \end{bmatrix} = \begin{bmatrix} F - GK & GK \\ 0 & F - LH \end{bmatrix} \begin{bmatrix} x \\ \tilde{x} \end{bmatrix} \tag{7.146}$$

The characteristic equation of this closed-loop system is

$$\det \begin{bmatrix} sI - F + GK & -GK \\ 0 & sI - F + LH \end{bmatrix} = 0. \tag{7.147}$$

Because the matrix is block triangular (see Appendix C), we can rewrite Eq. (7.147) as

$$\det(sI - F + GK) \cdot \det(sI - F + LH) = \alpha_c(s)\alpha_e(s) = 0. \tag{7.148}$$

Poles of the combined control law and estimator

In other words, the set of poles of the combined system consists of the union of the control poles and the estimator poles. This means that the designs of the control law and the estimator can be carried out independently, yet when they are used together in this way, the poles remain unchanged.[4]

To compare the state-variable method of design with the transform methods discussed in Chapters 5 and 6, we note from Fig. 7.26 that the shaded portion corresponds to a compensator. The state equation for this compensator is obtained by including the feedback law $u = -K\hat{x}$ (since it is part of the compensator) in the estimator Eq. (7.108) to get

$$\dot{\hat{x}} = (F - GK - LH)\hat{x} + Ly, \tag{7.149a}$$

$$u = -K\hat{x}, \tag{7.149b}$$

Note that Eq. (7.149a) has the same structure as Eq. (7.10a), which we repeat here:

$$\dot{x} = Fx + Gu. \tag{7.150}$$

---

4. This is a special case of the **separation principle**, which holds in much more general contexts and allows us to obtain an overall optimal design by combining the separate designs of control law and estimator in certain stochastic cases.

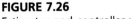

FIGURE 7.26
Estimator and controller mechanization

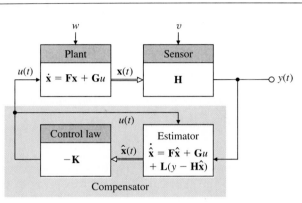

Since the characteristic equation of Eq. (7.150) is

$$\det(s\mathbf{I} - \mathbf{F}) = 0, \tag{7.151}$$

the characteristic equation of the compensator is found by comparing Eqs. (7.149a) and (7.150) and substituting the equivalent matrices into Eq. (7.151) to get

$$\det(s\mathbf{I} - \mathbf{F} + \mathbf{GK} + \mathbf{LH}) = 0. \tag{7.152}$$

Note that we never specified the roots of Eq. (7.152) nor used them in our discussion of the state-space design technique. (Note also that the compensator is not guaranteed to be stable; the roots of Eq. (7.152) can be in the RHP.) The transfer function from y to u representing the dynamic compensator is obtained by inspecting Eq. (7.36) and substituting in the corresponding matrices from Eq. (7.149):

Compensator transfer
function

$$D_c(s) = \frac{U(s)}{Y(s)} = -\mathbf{K}(s\mathbf{I} - \mathbf{F} + \mathbf{GK} + \mathbf{LH})^{-1}\mathbf{L}. \tag{7.153}$$

The same development can be carried out for the reduced-order estimator. Here the control law is

$$u = -\begin{bmatrix} K_a & K_b \end{bmatrix} \begin{bmatrix} x_a \\ \hat{\mathbf{x}}_b \end{bmatrix} = -K_a y - K_b \hat{\mathbf{x}}_b. \tag{7.154}$$

Substituting Eq. (7.154) into Eq. (7.150) and using Eq. (7.136) and some algebra, we obtain

$$\dot{\mathbf{x}}_c = \mathbf{A}_r \mathbf{x}_c + \mathbf{B}_r y,$$
$$u = \mathbf{C}_r \mathbf{x}_c + D_r y, \tag{7.155}$$

where

$$
\begin{aligned}
\mathbf{A}_r &= \mathbf{F}_{bb} - \mathbf{L}\mathbf{F}_{ab} - (\mathbf{G}_b - \mathbf{L}\mathbf{G}_a)\mathbf{K}_b, \\
\mathbf{B}_r &= \mathbf{A}_r\mathbf{L} + \mathbf{F}_{ba} - \mathbf{L}\mathbf{F}_{aa} - (\mathbf{G}_b - \mathbf{L}\mathbf{G}_a)\mathbf{K}_a, \\
\mathbf{C}_r &= -\mathbf{K}_b, \\
\mathbf{D}_r &= -K_a - \mathbf{K}_b\mathbf{L}.
\end{aligned}
\tag{7.156}
$$

Reduced-order compensator transfer function

The dynamic compensator now has the transfer function

$$
D_{cr}(s) = \frac{U(s)}{Y(s)} = \mathbf{C}_r(s\mathbf{I} - \mathbf{A}_r)^{-1}\mathbf{B}_r + D_r.
\tag{7.157}
$$

When we compute $D_c(s)$ or $D_{cr}(s)$ for a specific case, we will find that they are very similar to the classical compensators given in Chapters 5 and 6, in spite of the fact that they are arrived at by entirely different means.

◆ **EXAMPLE 7.19**   *Full-order Compensator Design for a Second-order System*

Design a compensator using pole placement for the plant with transfer function $1/s^2$. Place the control poles at $s = -0.707 \pm 0.707j$ ($\omega_n = 1$ rad/sec, $\zeta = 0.707$) and place the estimator poles at $\omega_n = 5$ rad/sec, $\zeta = 0.5$.

Solution.   A state-variable description for the given transfer function $G(s) = 1/s^2$ is

$$
\dot{\mathbf{x}} = \begin{bmatrix} 0 & 1 \\ 0 & 0 \end{bmatrix}\mathbf{x} + \begin{bmatrix} 0 \\ 1 \end{bmatrix}u,
$$

$$
y = \begin{bmatrix} 1 & 0 \end{bmatrix}\mathbf{x}.
$$

If we place the control roots at $s = -0.707 \pm 0.707j$ ($\omega_n = 1$ rad/sec, $\zeta = 0.7$), then

$$
\alpha_c(s) = s^2 + s\sqrt{2} + 1.
\tag{7.158}
$$

The state feedback gain is found to be

$$
\mathbf{K} = \begin{bmatrix} 1 & \sqrt{2} \end{bmatrix}.
$$

If the estimator-error roots are at $\omega_n = 5$ rad/sec and $\zeta = 0.5$, then the desired estimator characteristic polynomial is

$$
\alpha_e(s) = s^2 + 5s + 25 = s + 2.5 \pm 4.3j,
\tag{7.159}
$$

and the estimator feedback-gain matrix is

$$
\mathbf{L} = \begin{bmatrix} 5 \\ 25 \end{bmatrix}.
$$

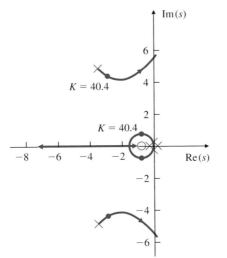

FIGURE 7.27
Root locus for the combined control and estimator, with process gain as the parameter

The compensator transfer function given by Eq. (7.153) is

$$D_c(s) = -\frac{40.4(s + 0.619)}{s + 3.21 \pm 4.77j},$$ (7.160)

which looks very much like a lead compensator in that it has a zero on the real axis to the right of its poles; however, rather than one real pole, Eq. (7.160) has two complex poles. The zero provides the derivative feedback with phase lead, and the two poles provide some smoothing of sensor noise.

The effect of the compensation on this system's closed-loop poles can be evaluated in exactly the same way we evaluated compensation in Chapters 5 and 6 using root-locus or frequency-response tools. The gain of 40.4 in Eq. (7.160) is a result of the pole selection inherent in Eqs. (7.158) and (7.159). If we replace this specific value of compensator gain with a variable gain K, then the characteristic equation for the closed-loop system of plant plus compensator becomes

$$1 + \frac{K(s+0.619)}{(s+3.21 \pm 4.77j)s^2} = 0.$$ (7.161)

The root-locus technique allows us to evaluate the roots of this equation with respect to K, as drawn in Fig. 7.27. Note that the locus goes through the roots selected for Eqs. (7.158) and (7.159), and, when $K = 40.4$, the four roots of the closed-loop system are equal to those specified.

Identical results of state space and frequency response design methods

The frequency-response plots given in Fig. 7.28 show that the compensation designed using state space accomplishes the same results that one would strive for using frequency-response design. Specifically, the uncompensated phase margin of $0°$ increases to $53°$ in the compensated case, and the gain $K = 40.4$ produces a crossover frequency $\omega_c = 1.35$ rad/sec. Both these values are roughly consistent with the controller closed-

FIGURE 7.28
Frequency response for $G(s) = 1/s^2$

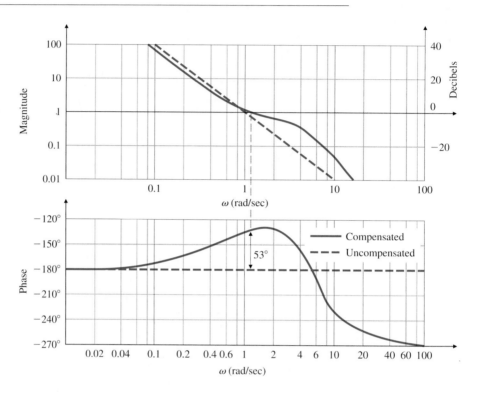

loop roots, with $\omega_n = 1$ rad/sec and $\zeta = 0.7$, as we would expect since these slow controller poles are dominant in the system response over the fast estimator poles.

Now we consider a reduced-order estimator for the same system.

◆ **EXAMPLE 7.20**   *Reduced-order Compensator Design for a Second-order System*

Repeat the design for the $1/s^2$ plant but use a reduced-order estimator. Place the one estimator pole at -5.

Solution.   From Eq. (7.133) we know that the estimator gain is

$$L = 5$$

FIGURE 7.29

Simplified block diagram of a reduced-order controller that is a lead network

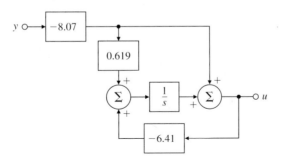

and the scalar compensator equations from Eq. (7.156) are

$$\dot{x}_c = -6.41x_c - 33.1y,$$
$$u = -1.41x_c - 8.07y,$$

where, from Eq. (7.135),

$$x_c = \hat{x}_2 - 5y.$$

The compensator has the transfer function calculated from Eq. (7.157) to be

$$D_{cr}(s) = -\frac{8.07(s+0.619)}{s+6.41}$$

and is shown in Fig. 7.29.

The reduced-order compensator here is precisely a lead network. This is a pleasant discovery, as it shows that transform and state-variable techniques can result in exactly the same type of compensation. The root locus of Fig. 7.30 shows that the closed-loop poles occur at the assigned locations. The frequency response of the compensated

FIGURE 7.30

Root locus of a reduced-order controller and $1/s^2$ process, root locations at $K = 8.07$ shown by the dots

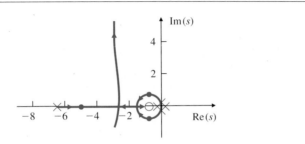

FIGURE 7.31
Frequency response for $G(s) = 1/s^2$ with a reduced-order estimator

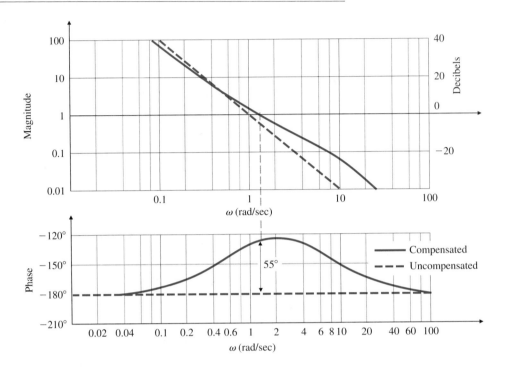

system seen in Fig. 7.31 shows a phase margin of about 55°. As with the full-order estimator, analysis by other methods confirms the selected root locations.

◆

More subtle properties of the pole-placement method can be illustrated by a third-order system.

◆ **EXAMPLE 7.21**   *A Third-order Compensator Design*

Use the state-space pole-placement method to design a compensator for the third-order system with the transfer function

$$G(s) = \frac{10}{s(s+2)(s+8)}.$$

Using a state description in observer canonical form, place the control poles at the third-order ITAE locations with $\omega_0 = 2$ and the full-order estimator poles at the third-order ITAE locations with $\omega_0 = 6$.

Solution. A block diagram of this system in observer canonical form is shown in Fig. 7.32. The corresponding state-space matrices are

$$\mathbf{F} = \begin{bmatrix} -10 & 1 & 0 \\ -16 & 0 & 1 \\ 0 & 0 & 0 \end{bmatrix}, \qquad \mathbf{G} = \begin{bmatrix} 0 \\ 0 \\ 10 \end{bmatrix},$$

$$\mathbf{H} = [1 \quad 0 \quad 0], \qquad J = 0.$$

We select the control poles from the third-order ITAE entry of Table 7.1 with s replaced by $s/2$. After clearing fractions we find the desired characteristic equation for the controller to be

$$\alpha_c(s) = (s + 1.42)(s + 1.04 \pm 2.14j)$$
$$= s^2 + 3.5s^2 + 8.6s + 8.$$

By comparing coefficients of α_c with the control characteristic equation

$$\det(s\mathbf{I} - \mathbf{F} + \mathbf{GK}) = \alpha_c(s),$$

we compute the state feedback gain to be

$$\mathbf{K} = [-46.4 \quad 5.76 \quad -0.65].$$

(The MATLAB functions acker or place can be used for this calculation.) We find the estimator poles also from Table 7.1, this time with $\omega_0 = 6$. In this case we compute the desired estimator characteristic equation to be

$$\alpha_e(s) = (s + 4.25)(s + 3.13 \pm 6.41j)$$
$$= s^3 + 10.5s^2 + 77.4s + 216.$$

By comparing coefficients of α_e in the estimator characteristic equation

$$\det(s\mathbf{I} - \mathbf{F} + \mathbf{LH}) = \alpha_e(s),$$

we calculate the estimator gain to be

$$\mathbf{L} = \begin{bmatrix} 0.5 \\ 61.4 \\ 216 \end{bmatrix}.$$

FIGURE 7.32
Third-order example in observer canonical form

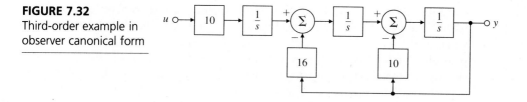

FIGURE 7.33
Root locus for ITAE pole assignment

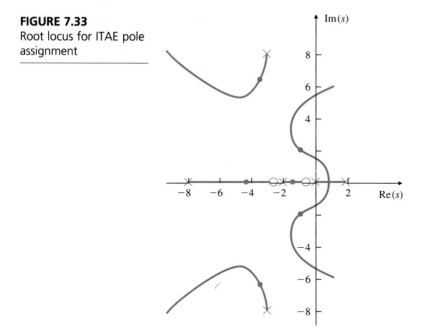

(The MATLAB functions acker or place can be used for this calculation also.) The compensator transfer function, as given by substituting into Eq. (7.153), is

$$D_c(s) = -190 \frac{(s+0.432)(s+2.10)}{(s-1.88)(s+2.94 \pm 8.32j)}.$$

Figure 7.33 shows the root locus of the system of compensator and plant in series plotted with the compensator gain as the parameter. It verifies that the roots are in the ITAE locations specified when the gain $K = 190$ in spite of the peculiar (unstable) compensation that has resulted. Note that this compensator has an unstable root at $s = +1.88$ but that all system closed-loop poles (controller and estimator) are stable.

◆

An unstable compensator is typically not acceptable because of the difficulty in testing either the compensator by itself or the system in open loop during a bench checkout. In some cases, however, better control can be achieved with an unstable compensator; then its inconvenience in checkout may be worthwhile.[5]

Figure 7.33 shows that a direct consequence of the unstable compensator is that the system becomes unstable as the gain is reduced from its nominal value. Such a system is called **conditionally stable** and should be avoided if possible. As we saw in Chapter 5, actuator saturation in response to large signals has the effect of lowering the effective gain, and in a conditionally stable

Conditionally stable compensator

5. There are even systems that cannot be stabilized with a stable compensator.

system instability can result. Also, if the electronics are such that the gain rises continuously from zero to the nominal value during startup, such a system would be expected to go unstable. These considerations lead us to consider alternative designs for this system.

◆ **EXAMPLE 7.22**   *Redesign of the Third-order System with a Reduced-order Esimator*

Design a compensator for the third-order system of Example 7.21 using the same control poles but with a reduced-order estimator. Place the estimator poles at the second-order ITAE positions with $\omega_0 = 6$.

Solution.   With reduced-order estimator poles at the second-order ITAE locations and with $\omega_0 = 6$, that is, at $-4.24 \pm 4.24j$, the desired estimator characteristic polynomial is

$$\alpha_e(s) = s^2 + 8.5s + 36.$$

After partitioning we have

$$\begin{bmatrix} \mathbf{F}_{aa} & \mathbf{F}_{ab} \\ \mathbf{F}_{ba} & \mathbf{F}_{bb} \end{bmatrix} = \begin{bmatrix} -10 & 1 & 0 \\ -16 & 0 & 1 \\ 0 & 0 & 0 \end{bmatrix}, \qquad \begin{bmatrix} \mathbf{G}_a \\ \mathbf{G}_b \end{bmatrix} = \begin{bmatrix} 0 \\ 0 \\ 10 \end{bmatrix}.$$

Solving for the estimator error characteristic polynomial,

$$\det(s\mathbf{I} - \mathbf{F}_{bb} + \mathbf{L}\mathbf{F}_{ab}) = \alpha_e(s),$$

we find that

$$\mathbf{L} = \begin{bmatrix} 8.5 \\ 36 \end{bmatrix}.$$

FIGURE 7.34
Root locus for an ITAE reduced-order controller

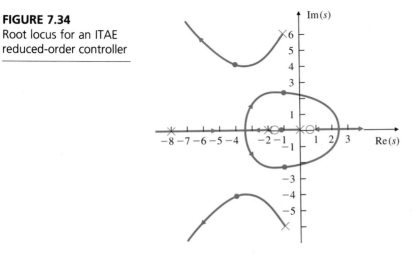

The compensator transfer function, given by Eq. (7.157), is computed to be

$$D_{cr}(s) = 20.93 \frac{(s - 0.735)(s + 1.871)}{s + 0.990 \pm 6.120j}.$$

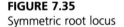 A nonminimum-phase compensator

The associated root locus for this system is shown in Fig. 7.34. Note that this time we have a stable but nonminimum-phase compensator and a zero-degree root locus.

◆

As a next pass at the design for this system, we attempt a design with the SRL.

◆ **EXAMPLE 7.23**   *Redesign of the Third-order Compensator Using the SRL*

Design a compensator for the third-order system of Example 7.21 using pole placement based on the SRL. For the control law, let the cost output z be the same as the plant output; for the estimator design, assume that the process noise enters at the same place as the system control signal. Select roots for a control bandwidth of about 2.5, and choose the estimator roots for a bandwidth of about three times faster than the control bandwidth.

Solution.   Since the problem has specified that $\mathbf{G}_1 = \mathbf{G}$ and $\mathbf{H}_1 = \mathbf{H}$, then the SRL is the same for the control as for the estimator, so we need to generate only one locus based on the plant transfer function. The SRL for the system is shown in Fig. 7.35. From the locus we select $-2 \pm 1.56j$ and -8.04 as the desired control poles and $-4 \pm 4.9j$ and -9.169 as the desired estimator poles. The state feedback gain is

$$\mathbf{K} = [-0.285 \quad 0.219 \quad 0.204],$$

FIGURE 7.35
Symmetric root locus

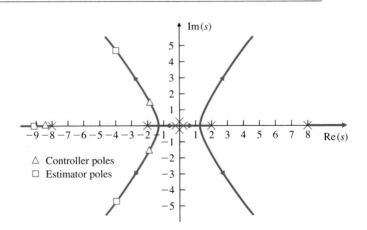

△ Controller poles
□ Estimator poles

FIGURE 7.36
Root locus for pole assignment from the SRL

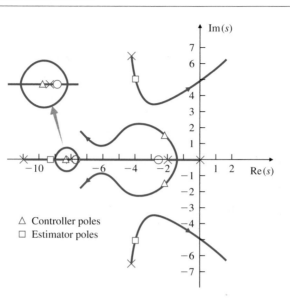

△ Controller poles
☐ Estimator poles

and the estimator gain is

$$\mathbf{L} = \begin{bmatrix} 7.17 \\ 97.4 \\ 367 \end{bmatrix}.$$

Notice that the feedback gains are much smaller than those required for ITAE placement. The resulting compensator transfer function is computed from Eq. (7.153) to be

$$D_c(s) = -\frac{94.5(s+7.98)(s+2.52)}{(s+4.28 \pm 6.42j)(s+10.6)}.$$

We now take this compensator, put it in series with the plant, and use the compensator gain as the parameter. The resulting ordinary root locus of the closed-loop system is shown in Fig. 7.36. When the root-locus gain equals the nominal gain of 94.5, the roots are at the closed-loop locations selected from the SRL, as they should be.

Note that the compensator is now stable and minimum-phase. This improved design comes about in large part because the plant pole at $s = -8$ is virtually unchanged by either controller or estimator. It does not need to be changed for good performance; in fact, the only feature in need of repair in the original $G(s)$ is the pole at $s = 0$. Using the SRL technique, we essentially discovered that the best use of control effort is to shift the two low-frequency poles at $s = 0$ and -2 and to leave the pole at $s = -8$ virtually unchanged. As a result, the control gains are much lower and the compensator design is less radical.

Armed with the knowledge gained from Example 7.23, let us go back, with a better selection of poles, to investigate the use of pole placement for this example. Initially we used the third-order ITAE locations, which produced three poles with a natural frequency of about 2 rad/sec. This design moved the pole at $s = -8$ to $s = -1.4$, thus violating the principle that open-loop poles should not be moved unless they are a problem. Now let us try it again, this time using the second-order ITAE to shift the slow poles and leaving the fast pole alone at $s = -8$.

◆ **EXAMPLE 7.24**   *Third-order System Redesign with Modified Dominant*
Second-order Pole Locations

Design a compensator for the third-order system of Example 7.21 using pole placement with control poles given by

$$\alpha_c(s) = (s + 1.41 \pm 1.41j)(s + 8)$$

and the estimator poles given by

$$\alpha_e(s) = (s + 4.24 \pm 4.24j)(s + 8).$$

Solution.   With these pole locations, we find that the required feedback gain is

$$\mathbf{K} = [-0.469 \quad 0.234 \quad 0.0828].$$

which has a smaller magnitude than the case where the pole at $s = -8$ was moved to the ITAE locations.
    We find the estimator gain to be

$$\mathbf{L} = \begin{bmatrix} 6.48 \\ 87.8 \\ 288 \end{bmatrix}.$$

The compensator transfer function is

$$D_c(s) = -\frac{414(s + 2.78)(s + 8)}{(s + 4.13 \pm 5.29j)(s + 9.05)},$$

which is stable and minimum-phase. This example illustrates the value of judicious pole selection and of the SRL technique.

◆

The poor pole selection inherent in the initial use of the third-order ITAE results in higher control effort and produces an unstable compensator. Both these undesirable features are eliminated by using the SRL or by improved pole selection.

7.7   Introduction of the Reference Input

The controller obtained by combining the control law studied in Section 7.3 with the estimator discussed in Section 7.6 is essentially a **regulator design**. This means that the characteristic equations of the control and the estimator are chosen for good disturbance rejection, that is, to give satisfactory transients to disturbances such as $w(t)$. However, this design approach does not consider a reference input, nor does it provide for **command following**, which is evidenced by a good transient response of the combined system to command changes. In general, good disturbance rejection and good command following both need to be taken into account in designing a control system. Good command following is done by properly introducing the reference input into the system equations.

Let us repeat the plant and controller equations for the full-order estimator; the reduced-order case is the same in concept, differing only in detail:

$$\text{Plant:} \qquad \dot{\mathbf{x}} = \mathbf{Fx} + \mathbf{G}u, \qquad\qquad (7.162a)$$

$$y = \mathbf{Hx}; \qquad\qquad (7.162b)$$

$$\text{Controller:} \qquad \dot{\hat{\mathbf{x}}} = (\mathbf{F} - \mathbf{GK} - \mathbf{LH})\hat{\mathbf{x}} + \mathbf{L}y, \qquad (7.163a)$$

$$u = -\mathbf{K}\hat{\mathbf{x}}. \qquad\qquad (7.163b)$$

Figure 7.37 shows two possibilities for introducing the command input r into the system. This figure illustrates the general issue of whether the compensation should be put in the feedback or feedforward path. The response of the system to command inputs is different, depending on the configuration, because the zeros of the transfer functions are different. The closed-loop poles are identical, however, as can be easily verified by letting $r = 0$ and noting that the systems are then identical.

The difference in the responses of the two configurations can be seen quite easily. Consider the effect of a step input in r. In Fig. 7.37(a) the step will excite the estimator in precisely the same way that it excites the plant; thus the estimator error will remain zero during and after the step. This means that the estimator dynamics are not excited by the command input so the transfer function from r to y must have zeros at the estimator pole locations that cancel those poles. As a result, a step command will excite system behavior that is consistent with the control poles alone, that is, with the roots of $\det(s\mathbf{I} - \mathbf{F} + \mathbf{GK}) = 0$.

In Fig. 7.37(b) a step command in r enters directly only into the estimator, thus causing an estimation error that decays with the estimator dynamic characteristics in addition to the response corresponding to the control poles. Therefore, a step command will excite system behavior consistent with both control roots and estimator roots; that is, the roots of

$$\det(s\mathbf{I} - \mathbf{F} + \mathbf{GK}) \cdot \det(s\mathbf{I} - \mathbf{F} + \mathbf{LH}) = 0.$$

FIGURE 7.37
Possible locations for
introducing the
command input:
(a) compensation in the
feedback path;
(b) compensation in the
feedforward path

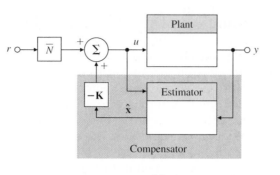

(a)

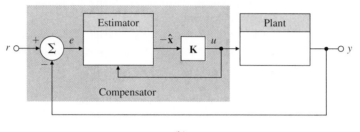

(b)

For this reason the configuration shown in Fig. 7.37(a) is typically the superior way to command the system where $\bar{\mathbf{N}}$ is found using Eqs. (7.79)–(7.81).

In Section 7.7.1 we will show a general structure for introducing the reference input with three choices of parameters that implement either the feedforward or the feedback case. We will analyze the three choices from the point of view of the system zeros and the implications the zeros have for the system transient response. Finally, in Section 7.7.2 we will show how to select the remaining parameter to eliminate constant errors.

■ 7.7.1 A General Structure for the Reference Input

Given a reference input $r(t)$, the most general linear way to introduce r into the system equations is to add terms proportional to it in the controller equations. We can do this by adding $\bar{N}r$ to Eq. (7.163b) and $\mathbf{M}r$ to Eq. (7.163a). Note that in this case, $\bar{N}$ is a scalar and $\mathbf{M}$ is an $n \times 1$ vector. With these additions the controller equations become

Controller equations

$$\dot{\hat{\mathbf{x}}} = (\mathbf{F} - \mathbf{GK} - \mathbf{LH})\hat{\mathbf{x}} + \mathbf{L}y + \mathbf{M}r,$$

$$u = -\mathbf{K}\hat{\mathbf{x}} + \bar{N}r.$$

(7.164)

The block diagram is shown in Fig. 7.38(a). The alternatives shown in Fig. 7.37 correspond to different choices of $\mathbf{M}$ and $\bar{N}$. Since $r(t)$ is an external signal, it is clear that neither $\mathbf{M}$ nor $\bar{N}$ affects the characteristic equation of the combined controller-estimator system. In transfer-function terms, the selection of **M** and $\bar{N}$ will only affect the zeros of transmission from r to y and as a consequence can significantly affect the transient response but not the stability. How can we choose **M** and $\bar{N}$ to obtain satisfactory transient response? We should point out that we assigned the poles of the system by feedback gains **K** and **L** and we are now going to assign zeros by feedforward gains **M** and $\bar{N}$. There are three strategies for choosing **M** and $\bar{N}$:

Three methods for selecting **M** and $\bar{N}$

1. *Autonomous estimator:* Select **M** and $\bar{N}$ so that the state estimator error equation is independent of r (Fig. 7.38b).
2. *Tracking-error estimator:* Select **M** and $\bar{N}$ so that only the tracking error, $e = (r - y)$, is used in the control (Fig. 7.38c).
3. *Zero-assignment estimator:* Select **M** and $\bar{N}$ so that n of the zeros of the overall transfer function are assigned at places of the designer's choice (Fig. 7.38a).

FIGURE 7.38
Alternative ways to introduce the reference input: (a) general case—zero assignment; (b) standard case—estimator not excited, zeros = $\alpha_e(s)$; (c) error-control case—classical compensation

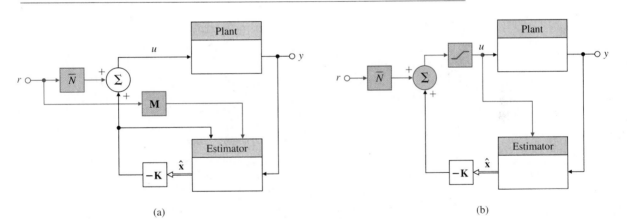

(a)   (b)

(c)

CASE 1: From the viewpoint of estimator performance, the first method is quite attractive and the most widely used of the alternatives. If $\hat{\mathbf{x}}$ is to generate a good estimate of $\mathbf{x}$, then surely $\tilde{\mathbf{x}}$ should be as free of external excitation as possible; that is, $\tilde{\mathbf{x}}$ should be uncontrollable from r. The computation of $\mathbf{M}$ and $\bar{N}$ to bring this about is quite easy. The estimator error equation is found by subtracting Eq. (7.164) from Eq. (7.162), with the plant output (Eq. 7.162b) substituted into the estimator (Eq. 7.163a) and the control (Eq. 7.163b) substituted into the plant (Eq. 7.162a):

$$\dot{\mathbf{x}} - \dot{\hat{\mathbf{x}}} = \mathbf{F}\mathbf{x} + \mathbf{G}(-\mathbf{K}\hat{\mathbf{x}} + \bar{N}r) - [(\mathbf{F} - \mathbf{G}\mathbf{K} - \mathbf{L}\mathbf{H})\hat{\mathbf{x}} + \mathbf{L}y + \mathbf{M}r], \quad (7.165a)$$

$$\dot{\tilde{\mathbf{x}}} = (\mathbf{F} - \mathbf{L}\mathbf{H})\tilde{\mathbf{x}} + \mathbf{G}\bar{N}r - \mathbf{M}r. \quad (7.165b)$$

If r is not to appear in Eqs. (7.165b), then we should choose

$$\mathbf{M} = \mathbf{G}\bar{N}. \quad (7.166)$$

Since $\bar{N}$ is a scalar, $\mathbf{M}$ is fixed to within a constant factor. Note that with this choice of $\mathbf{M}$ we can write the controller equations as

$$u = -\mathbf{K}\hat{\mathbf{x}} + \bar{N}r \quad (7.167a)$$

$$\dot{\hat{\mathbf{x}}} = (\mathbf{F} - \mathbf{L}\mathbf{H})\hat{\mathbf{x}} + \mathbf{G}u + \mathbf{L}y, \quad (7.167b)$$

which matches the configuration in Fig. 7.38(b). The net effect of this choice is that the control is computed from the feedback gain and the reference input *before* it is applied and then the same control is input to both the plant and the estimator. In this form, if the plant control is subject to saturation, the same control limits can be applied in Eq. (7.167) to the control entering the equation for the estimate $\hat{\mathbf{x}}$, and the nonlinearity cancels out of the $\tilde{\mathbf{x}}$ equation. This behavior is essential for proper estimator performance. The block diagram corresponding to this technique is shown in Fig. 7.38(b). We will return to the selection of the gain factor on the reference input, $\bar{N}$, in Section 7.7.3 after discussing the other two methods of selecting $\mathbf{M}$.

CASE 2: The second approach suggested earlier is to use the tracking error. This solution is sometimes forced on the control designer when the sensor measures only the output error. For example, in many thermostats the output is the difference between the temperature to be controlled and the setpoint temperature, and there is no absolute indication of the reference temperature available to the controller. Also, some radar tracking systems have a reading that is proportional to the pointing error, and this error signal alone must be used for feedback control. In these situations, we must select $\mathbf{M}$ and $\bar{N}$ so that Eqs. (7.164) are driven by the error only. This requirement is satisfied if we select

$$\bar{N} = 0 \quad \text{and} \quad \mathbf{M} = -\mathbf{L}. \quad (7.168)$$

Then the estimator equation is

$$\dot{\hat{x}} = (F - GK - LH)\hat{x} + L(y - r). \tag{7.169}$$

The compensator in this case, for low-order designs, is a lead compensator in the forward path. As we have seen in earlier chapters, this design can have a considerable amount of overshoot because of the zero of the compensator. This design corresponds exactly to the compensators designed by the transform methods given in Chapters 5 and 6.

CASE 3: The third method of selecting M and $\bar{N}$ is to choose the values so as to assign the systems zeros to arbitrary locations of the designers choice. This method provides the designer with the maximum flexibility in satisfying transient-response and steady-state gain constraints. The other two methods are special cases of this third method. All three methods depend on the zeros. As we saw in Section 7.3.2, when there is no estimator and the reference input is added to the control, the closed-loop system zeros remain fixed as the zeros of the open-loop plant. We now examine what happens to the zeros when an estimator is present. To do so, we reconsider the controller of Eqs. (7.164). If there is a zero of transmission from r to u, then there is necessarily a zero of transmission from r to y unless there is a pole at the same location as the zero. It is therefore sufficient to treat the controller alone to determine what effect the choices of M and $\bar{N}$ will have on the system zeros. The equations for a zero from r to u from Eqs. (7.164) are given by

$$\det \begin{bmatrix} sI - F + GK + LH & -M \\ -K & \bar{N} \end{bmatrix} = 0. \tag{7.170}$$

(We let $y = 0$ since we care only about the effect of r.) If we divide the last column by the (nonzero) scalar $\bar{N}$ and then add to the rest the product of K times the last column, we find the feedforward zeros are at the values of s such that

$$\det \begin{bmatrix} sI - F + GK + LH - \dfrac{M}{\bar{N}} K & -\dfrac{M}{\bar{N}} \\ 0 & 1 \end{bmatrix} = 0,$$

or

$$\det \left(sI - F + GK + LH - \dfrac{M}{\bar{N}} K \right) = \gamma(s) = 0. \tag{7.171}$$

Now Eq. (7.171) is exactly in the form of Eq. (7.111) for selecting L to yield desired locations for the estimator poles. Here we have to select $M/\bar{N}$ for a desired zero polynomial $\gamma(s)$ in the transfer function from the reference input to the control. Thus the selection of M provides a substantial amount of freedom to influence the transient response. We can add an arbitrary nth-order polynomial to the transfer function from r to u and hence from r to y; that is, we can assign n zeros in addition to all the poles that we assigned previously.

If the roots of $\gamma(s)$ are not canceled by the poles of the system, then they will be included in zeros of transmission from r to y.

Two considerations can guide us in the choice of $\mathbf{M}/\bar{N}$, that is, in the location of the zeros. The first is dynamic response. We have seen in Chapter 3 that the zeros influence the transient response significantly, and the heuristic guidelines given there may suggest useful locations for the available zeros. The second consideration, which will connect state-space design to another result from transform techniques, is steady-state error or velocity-constant control. In Chapter 4 we derived the relationship between the steady-state accuracy of a type I system and the closed-loop poles and zeros. If the system is type I, then the steady-state error to a step input will be zero and to a unit ramp input will be

$$e_\infty = \frac{1}{K_v}, \tag{7.172}$$

where K_v is the velocity constant. Furthermore, it was shown that if the *closed-loop* poles are at $\{p_i\}$ and the *closed-loop* zeros are at $\{z_i\}$, then (for a type I system) **Truxal's formula** gives

Truxal's formula

$$\frac{1}{K_v} = \sum \frac{1}{z_i} - \sum \frac{1}{p_i} \tag{7.173}$$

Equation (7.173) forms the basis for a partial selection of $\gamma(s)$, and hence of $\mathbf{M}$ and $\bar{N}$. The choice is based on two observations:

1. If $|z_i - p_i| \ll 1$, then the effect of this pole-zero pair on the dynamic response will be small, since the pole is almost canceled by the zero, and in any transient the residue of the pole at p_i will be very small.
2. Even though $z_i - p_i$ is small, it is possible for $1/z_i - 1/p_i$ to be substantial and thus to have a significant influence on K_v according to Eq. (7.173).

Application of these two guidelines to the selection of $\gamma(s)$, and hence of $\mathbf{M}$ and $\bar{N}$, results in a lag-network design. We illustrate this with an example.

◆ **EXAMPLE 7.25**   *Increasing the Velocity Constant through Zero Assignment*

Consider the second-order system described by

Lag compensation by a state space method

$$H(s) = \frac{1}{s(s+1)}$$

and with state description

$$\dot{x}_1 = x_2,$$

$$\dot{x}_2 = -x_2 + u.$$

Design a controller using pole placement so that it has control characteristic equation

$$\alpha_c(s) = (s + 2)^2 + 4 = s^2 + 4s + 8,$$

and a velocity constant $K_v = 10$.

Solution.   For this problem the state feedback gain

$$\mathbf{K} = [8 \quad 3]$$

results in the desired control poles. However, with this gain, $K_v = 2$, and we need $K_v = 10$. What effect will using estimators designed according to the three methods for $\mathbf{M}$ and $\bar{N}$ selection have on our design? Using the first strategy (the autonomous estimator), we find that the value of K_v does not change. If we use the second method (error control), we introduce a zero at a location unknown beforehand, and the effect on K_v will not be under direct design control. However, if we use the third option (zero placement) along with Truxal's formula (Eq. 7.173), we can satisfy both the dynamic response and the steady-state requirements.

First we must select the estimator pole p_3 and the zero z_3 to satisfy Eq. (7.173) for $K_v = 10$. We want to keep $z_3 - p_3$ small so that there is little effect on the dynamic response and yet have $1/z_3 - 1/p_3$ be large enough to increase the value of K_v. To do this, we arbitrarily set p_3 small compared with the control dynamics. For example, we let

$$p_3 = -0.1.$$

Notice that this approach is opposite to the usual philosophy of estimation design, where fast response is the requirement. Now using Eq. (7.173) to get

$$\frac{1}{K_v} = \frac{1}{z_3} - \frac{1}{p_1} - \frac{1}{p_2} - \frac{1}{p_3},$$

where $p_1 = -2 + 2j$, $p_2 = -2 - 2j$, and $p_3 = -0.1$. We solve for z_3 such that $K_v = 10$:

$$\frac{1}{K_v} = \frac{4}{8} + \frac{1}{0.1} + \frac{1}{z_3} = \frac{1}{10},$$

or

$$z_3 = -\frac{1}{10.4} = -0.096.$$

We thus design a reduced-order estimator to have a pole at -0.1 and choose $\mathbf{M}/\bar{N}$ such that $\gamma(s)$ has a zero at -0.096. A block diagram of the resulting system is shown in Fig. 7.39(a). You can readily verify that this system has the overall transfer function

$$\frac{Y(s)}{R(s)} = \frac{8.32(s + 0.096)}{(s^2 + 4s + 8)(s + 0.1)}, \tag{7.174}$$

for which $K_v = 10$, as specified.

The compensation shown in Fig. 7.39(a) is nonclassical in the sense that it has two inputs (e and y) and one output. If we resolve the equations to provide pure error compensation by finding the transfer function from e and u, which would give Eq. (7.174), we obtain the system shown in Fig. 7.39(b). This compensation is a classical

FIGURE 7.39
Servomechanism with assigned zeros (a lag network): (a) the two-input compensator; (b) equivalent unity feedback system

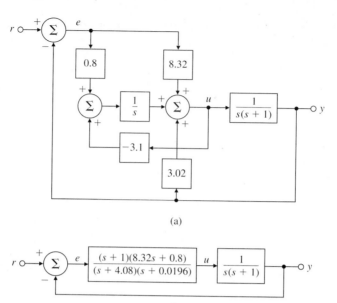

(a)

(b)

lag-lead network. The root locus of the system in Fig. 7.39(b) is shown in Fig. 7.40. Note the pole-zero pattern near the origin that is characteristic of a lag network. The Bode plot in Fig. 7.41 shows the phase lag at low frequencies and phase lead at high frequencies. The step response of the system is shown in Fig. 7.42.

FIGURE 7.40
Root locus of lag-lead compensation

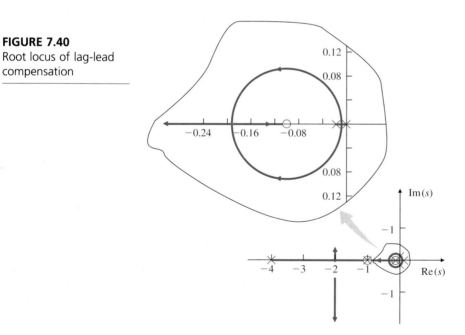

FIGURE 7.41
Frequency response of lag-lead compensation

FIGURE 7.42
Step response of the system with lag compensation

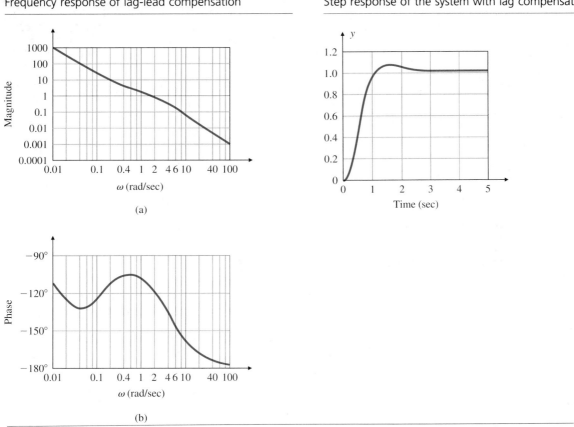

(a)

(b)

We now reconsider the first two methods for choosing $\mathbf{M}$ and $\bar{N}$, this time to examine their implications in terms of zeros. Under the first rule (for the autonomous estimator), we let $\mathbf{M} = \mathbf{G}\bar{N}$. Substituting this into Eq. (7.171) yields, for the controller feedforward zeros,

$$\det(s\mathbf{I} - \mathbf{F} + \mathbf{LH}) = 0. \qquad (7.175)$$

This is exactly the equation from which $\mathbf{L}$ was selected to make the characteristic polynomial of the estimator equation equal to $\alpha_e(s)$. Thus we have created n zeros in exactly the same locations as the n poles of the estimator. Because of this pole-zero cancellation (which causes "uncontrollability" of the estimator modes), the overall transfer function poles consist only of the state feedback controller poles.

The second rule (for a tracking-error estimator) selects $\mathbf{M} = -\mathbf{L}$ and $\bar{N} = 0$. If these are substituted into Eq. (7.170), then the feedforward zeros are given by

$$\det \begin{bmatrix} s\mathbf{I} - \mathbf{F} + \mathbf{GK} + \mathbf{LH} & \mathbf{L} \\ -\mathbf{K} & 0 \end{bmatrix} = 0. \tag{7.176}$$

If we postmultiply the last column by $\mathbf{H}$ and subtract the result from the first n columns, and then premultiply the last row by $\mathbf{G}$ and add it to the first n rows, Eq. (7.176) then reduces to

$$\det \begin{bmatrix} s\mathbf{I} - \mathbf{F} & \mathbf{L} \\ -\mathbf{K} & 0 \end{bmatrix} = 0. \tag{7.177}$$

If we compare Eq. (7.177) with the equations for the zeros of a system in a state description, Eq. (7.54), we see that the added zeros are those obtained by replacing the input matrix with $\mathbf{L}$ and the output with $\mathbf{K}$. Thus, if we wish to use error control, we have to accept the presence of these compensator zeros that depend on the choice of $\mathbf{K}$ and $\mathbf{L}$ and over which we have no direct control. For low-order cases this results, as we said before, in a lead compensator as part of a unity feedback topology.

Let us now summarize our findings on the effect of introducing the reference input. When the reference input signal is included in the controller, the overall transfer function of the closed-loop system is

$$T(s) = \frac{Y(s)}{R(s)} = \frac{K_s\gamma(s)b(s)}{\alpha_e(s)\alpha_c(s)}, \tag{7.178}$$

Transfer function for the closed-loop system when reference input is included in controller

where K_s is the total system gain and $\gamma(s)$ and $b(s)$ are monic polynomials. The polynomial $\alpha_c(s)$ results in a control gain $\mathbf{K}$ such that $\det[s\mathbf{I} - \mathbf{F} + \mathbf{GK}] = \alpha_c(s)$. The polynomial $\alpha_e(s)$ results in estimator gains $\mathbf{L}$ such that $\det[s\mathbf{I} - \mathbf{F} + \mathbf{LH}] = \alpha_e(s)$. Because as designers we get to choose $\alpha_c(s)$ and $\alpha_e(s)$, we have complete freedom in assigning the poles of the closed-loop system. There are three ways to handle the polynomial $\gamma(s)$: We can select it so that $\gamma(s) = \alpha_e(s)$ by using the implementation of Fig. 7.38(b), in which case $\mathbf{M}/\bar{N}$ is given by Eq. (7.166); we may accept $\gamma(s)$ as given by Eq. (7.177), so that error control is used; or we may give $\gamma(s)$ arbitrary coefficients by selecting $\mathbf{M}/\bar{N}$ from Eq. (7.171). It is important to point out that the plant zeros represented by $b(s)$ are not moved by this technique and remain as part of the closed-loop transfer function unless α_c or α_e are selected to cancel some of these zeros.

■ 7.7.2   Selecting the Gain

We now turn to the process of determining the gain $\bar{N}$ for the three methods of selecting $\mathbf{M}$. If we choose method 1, the control is given by Eq. (7.167a) and $\hat{x}_{ss} = x_{ss}$. Therefore, we can use either $\bar{N} = N_u + \mathbf{KN_x}$, as in Eq. (7.81), or

$u = N_u r - \mathbf{K}(\hat{\mathbf{x}} - \mathbf{N}_x r)$. *This is the most common choice.* If we use the second method, the result is trivial; recall that $\bar{N} = 0$ for error control. If we use the third method, we pick $\bar{N}$ such that the overall closed-loop DC gain is unity.[6] The overall system equations then are

$$
\begin{bmatrix} \dot{\mathbf{x}} \\ \dot{\tilde{\mathbf{x}}} \end{bmatrix} = \begin{bmatrix} \mathbf{F} - \mathbf{GK} & \mathbf{GK} \\ 0 & \mathbf{F} - \mathbf{LH} \end{bmatrix} \begin{bmatrix} \mathbf{x} \\ \tilde{\mathbf{x}} \end{bmatrix} + \begin{bmatrix} \mathbf{G} \\ \mathbf{G} - \bar{\mathbf{M}} \end{bmatrix} \bar{N}r,
$$

$$
y = [\mathbf{H} \quad 0] \begin{bmatrix} \mathbf{x} \\ \tilde{\mathbf{x}} \end{bmatrix},
$$

(7.179)

where $\bar{\mathbf{M}}$ is the outcome of selecting zero locations with either Eq. (7.171) or Eq. (7.166). The closed-loop system has unity DC gain if

$$
-[\mathbf{H} \quad 0] \begin{bmatrix} \mathbf{F} - \mathbf{GK} & \mathbf{GK} \\ 0 & \mathbf{F} - \mathbf{LH} \end{bmatrix}^{-1} \begin{bmatrix} \mathbf{G} \\ \mathbf{G} - \bar{\mathbf{M}} \end{bmatrix} \bar{N} = 1.
$$

(7.180)

If we solve Eq. (7.180) for $\bar{N}$, we get[7]

$$
\bar{N} = -\frac{1}{\mathbf{H}(\mathbf{F} - \mathbf{GK})^{-1}\mathbf{G}[1 - \mathbf{K}(\mathbf{F} - \mathbf{LH})^{-1}(\mathbf{G} - \bar{\mathbf{M}})]}.
$$

(7.181)

The techniques in this section can be readily extended to reduced-order estimators.

7.8   Integral Control and Robust Tracking

The choice of $\bar{N}$ will result in zero steady-state error to a step command, but the result is not robust because any change in the plant parameters will cause the error to be nonzero. We need to use integral control to obtain robust tracking.

In the state-space design methods discussed so far, no mention has been made of integral control, and no design examples have produced a compensation containing an integral term. In Section 7.8.1 we show how integral control can be introduced by a direct method of adding the integral of the system error to the equations of motion. Integral control is a special case of tracking a signal that does not go to zero in the steady state. We introduce (in Section 7.8.2) a general method for robust tracking that will present the internal model

6. A reasonable alternative is to select $\bar{N}$ such that, when r and y are both unchanging, the DC gain from r to u is the *negative* of the DC gain from y to u. The consequences of this choice are that our controller can be structured as a combination of error control and generalized derivative control, and if the system is capable of type I behavior, that capability will be realized.

7. We have used the fact that

$$
\begin{bmatrix} \mathbf{A} & \mathbf{C} \\ 0 & \mathbf{B} \end{bmatrix}^{-1} = \begin{bmatrix} \mathbf{A}^{-1} & -\mathbf{A}^{-1}\mathbf{CB}^{-1} \\ 0 & \mathbf{B}^{-1} \end{bmatrix}.
$$

principle, which solves an entire class of tracking problems and disturbance-rejection controls. Finally, in Section 7.8.3 we show that, if the system has an estimator and also needs to reject a disturbance of known structure, we can include a model of the disturbance in the estimator equations and use the computer estimate of the disturbance to cancel the effects of the real plant disturbance on the output.

7.8.1 Integral Control

We start with an ad hoc solution to integral control by augmenting the state vector with the desired dynamics. For the system

$$\dot{\mathbf{x}} = \mathbf{F}\mathbf{x} + \mathbf{G}u + \mathbf{G}_1 w,$$
$$y = \mathbf{H}\mathbf{x}, \tag{7.182}$$

we can feed back the integral of the error,[8] $e = y - r$, as well as the state of the plant, $\mathbf{x}$, by augmenting the plant state with the extra (integral) state x_I, which obeys the differential equation

$$\dot{x}_I = \mathbf{H}\mathbf{x} - r \qquad (= e).$$

Thus

$$x_I = \int^t e\, dt.$$

Augmented state equations with integral control

The augmented state equations become

$$\begin{bmatrix} \dot{x}_I \\ \dot{\mathbf{x}} \end{bmatrix} = \begin{bmatrix} 0 & \mathbf{H} \\ 0 & \mathbf{F} \end{bmatrix} \begin{bmatrix} x_I \\ \mathbf{x} \end{bmatrix} + \begin{bmatrix} 0 \\ \mathbf{G} \end{bmatrix} u - \begin{bmatrix} 1 \\ 0 \end{bmatrix} r + \begin{bmatrix} 0 \\ \mathbf{G}_1 \end{bmatrix} w, \tag{7.183}$$

and the feedback law is

$$u = -[K_1 \quad \mathbf{K}_0] \begin{bmatrix} x_I \\ \mathbf{x} \end{bmatrix},$$

Feedback law with integral control

or simply

$$u = -\mathbf{K} \begin{bmatrix} x_I \\ \mathbf{x} \end{bmatrix}.$$

With this revised definition of the system, we can apply the design techniques from Section 7.3 in a similar fashion; they will result in the control structure shown in Fig. 7.43.

---

8. Watch out for the sign here; we are using the negative of the usual convention.

FIGURE 7.43
Integral control structure

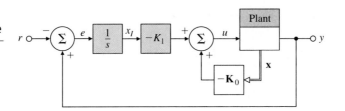

◆ **EXAMPLE 7.26** *Integral Control of a First-order System*

Consider the first-order system described by

$$\frac{Y(s)}{U(s)} = \frac{1}{s+3};$$

that is, $F = -3$, $G = 1$, and $H = 1$. Design the system to have integral control and two poles at $s = -5$. The disturbance enters at the same place as the control.

Solution. The pole-placement requirement is equivalent to asking for the following desired characteristic equation:

$$\alpha_c(s) = s^2 + 10s + 25 = 0.$$

The augmented system description including the disturbance w is

$$\begin{bmatrix} \dot{x}_I \\ \dot{x} \end{bmatrix} = \begin{bmatrix} 0 & 1 \\ 0 & -3 \end{bmatrix}\begin{bmatrix} x_I \\ x \end{bmatrix} + \begin{bmatrix} 0 \\ 1 \end{bmatrix}(u+w) - \begin{bmatrix} 1 \\ 0 \end{bmatrix}r.$$

Therefore, we can find **K** from

$$\det\left(s\mathbf{I} - \begin{bmatrix} 0 & 1 \\ 0 & -3 \end{bmatrix} + \begin{bmatrix} 0 \\ 1 \end{bmatrix}\mathbf{K} \right) = s^2 + 10s + 25,$$

or

$$s^2 + (3 + K_0)s + K_1 = s^2 + 10s + 25.$$

FIGURE 7.44
Integral control example

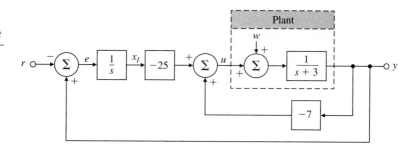

Therefore,

$$\mathbf{K} = [K_1 \quad K_0] = [25 \quad 7].$$

The system is shown with feedbacks in Fig. 7.44 along with a disturbance input w. This system will have dynamics according to the desired closed-loop poles at $s = -5$ and will exhibit integral control, having zero steady-state error to a step input r and zero steady-state error to a constant disturbance w.

◆

■7.8.2   Robust Tracking Control: The Error-space Approach

In Section 7.8.1 we introduced integral control in a direct way and selected the structure of the implementation so as to achieve integral action with respect to reference and disturbance inputs. We now present a more analytical approach to giving a control system the ability to track (with zero steady-state error) a nondecaying input and to reject (with zero steady-state error) a nondecaying disturbance such as a step, ramp, or sinusoidal input. The method is based on including the equations satisfied by these external signals as part of the problem formulation and solving the problem of control in an **error space** so we are assured that the error approaches zero even if the output is following a nondecaying, or even a growing, command (such as a ramp signal) and even if some parameters change (the robustness property). The method is illustrated in detail for signals that satisfy differential equations of order 2, but the extension to more complex signals is not difficult.

Suppose we have the system state equations

$$\dot{\mathbf{x}} = \mathbf{Fx} + \mathbf{G}u + \mathbf{G}_1 w,$$
$$y = \mathbf{Hx}, \tag{7.184}$$

and a reference signal that is known to satisfy a specific differential equation. The initial conditions on the equation generating the input are unknown. For example, the input could be a ramp whose slope and initial value are unknown. Plant disturbances of the same class may also be present. We wish to design a controller for this system so that the closed-loop system will have specified poles, and can also track input command signals, and reject disturbances of the type described without steady-state error. We will develop the results only for second-order differential equations. We define the reference input to satisfy the relation

$$\ddot{r} + \alpha_1 \dot{r} + \alpha_2 r = 0 \tag{7.185}$$

and the disturbance to satisfy exactly the same equation:

$$\ddot{w} + \alpha_1 \dot{w} + \alpha_2 w = 0. \tag{7.186}$$

The (tracking) error is defined as

$$e = y - r. \tag{7.187}$$

The problem of tracking r and rejecting w can be seen as an exercise in designing a control law to provide *regulation of the error*, which is to say that the error e tends to zero as time gets large. The control must also be **structurally stable** or **robust**, in the sense that regulation of e to zero in the steady state occurs even in the presence of "small" perturbations of the original system parameters. Note that in practice we never have a perfect model of the plant and the values of parameters are virtually always subject to some change, so robustness is always very important.

The meaning of robust control

We know that the command input satisfies Eq. (7.185), and we would like to eliminate the reference from the equations in favor of the error. We begin by replacing r in Eq. (7.185) with the error of Eq. (7.187). When we do this, the reference cancels because of Eq. (7.185), and we have the formula for the error in terms of the state

$$\ddot{e} + \alpha_1 \dot{e} + \alpha_2 e = \ddot{y} + \alpha_1 \dot{y} + \alpha_2 y$$
$$= \mathbf{H}\ddot{\mathbf{x}} + \alpha_1 \mathbf{H}\dot{\mathbf{x}} + \alpha_2 \mathbf{H}\mathbf{x}. \tag{7.188}$$

We now replace the plant state vector with the error-space state defined by

$$\boldsymbol{\xi} \triangleq \ddot{\mathbf{x}} + \alpha_1 \dot{\mathbf{x}} + \alpha_2 \mathbf{x}. \tag{7.189}$$

Similarly, we replace the control with the control in error space, defined as

$$\mu \triangleq \ddot{u} + \alpha_1 \dot{u} + \alpha_2 u. \tag{7.190}$$

With these definitions we can replace Eq. (7.188) with

$$\ddot{e} + \alpha_1 \dot{e} + \alpha_2 e = \mathbf{H}\boldsymbol{\xi}. \tag{7.191}$$

Robust control equations in the error space

The state equation for $\boldsymbol{\xi}$ is given by[9]

$$\dot{\boldsymbol{\xi}} = \ddot{\mathbf{x}} + \alpha_1 \ddot{\mathbf{x}} + \alpha_2 \dot{\mathbf{x}} = \mathbf{F}\boldsymbol{\xi} + \mathbf{G}\mu. \tag{7.192}$$

Notice that the disturbance as well as the reference cancels from Eq. (7.192). Equations (7.191) and (7.192) now describe the overall system in an error space. In standard state-variable form, the equations are

$$\dot{\mathbf{z}} = \mathbf{A}\mathbf{z} + \mathbf{B}\mu, \tag{7.193}$$

where $\mathbf{z} = [e \quad \dot{e} \quad \boldsymbol{\xi}^T]^T$ and

$$\mathbf{A} = \begin{bmatrix} 0 & 1 & \mathbf{0} \\ -\alpha_2 & -\alpha_1 & \mathbf{H} \\ \mathbf{0} & \mathbf{0} & \mathbf{F} \end{bmatrix}, \qquad \mathbf{B} = \begin{bmatrix} 0 \\ 0 \\ \mathbf{G} \end{bmatrix}. \tag{7.194}$$

---

9. Notice that this concept can be extended to more complex equations in r and to multivariable systems.

The error system $(\mathbf{A}, \mathbf{B})$ can be given arbitrary dynamics by state feedback if it is controllable. If the plant $(\mathbf{F}, \mathbf{G})$ is controllable and does not have a zero at any of the roots of the reference-signal characteristic equation

$$\alpha_r(s) = s^2 + \alpha_1 s + \alpha_2,$$

then the error system $(\mathbf{A}, \mathbf{B})$ is controllable.[10] We assume these conditions hold; therefore, there exists a control law of the form

$$\mu = -[K_2 \quad K_1 \quad \mathbf{K}_0] \begin{bmatrix} e \\ \dot{e} \\ \xi \end{bmatrix} = -\mathbf{K}\mathbf{z}, \qquad (7.195)$$

such that the error system has arbitrary dynamics by pole placement. We now need to express this control law in terms of the actual process state $\mathbf{x}$ and the actual control. We combine Eqs. (7.195), (7.189), and (7.190) to get the control law in terms of u and $\mathbf{x}$ (we write u^2 to mean d^2u/dt^2):

$$(u + \mathbf{K}_0\mathbf{x})^{(2)} + \sum_{i=1}^{2} \alpha_i(u + \mathbf{K}_0\mathbf{x})^{(2-i)} = -\sum_{i=1}^{2} K_i e^{(2-i)}. \qquad (7.196)$$

The structure for implementing Eq. (7.196) is very simple for tracking constant inputs. In that case the equation for the reference input is $\dot{r} = 0$. In terms of u and $\mathbf{x}$ the control law (Eq. 7.196) reduces to

$$\dot{u} + \mathbf{K}_0\dot{\mathbf{x}} = -K_1 e. \qquad (7.197)$$

Here we only need to integrate to reveal the control law and the action of integral control:

$$u = -K_1 \int^t e \, d\tau - \mathbf{K}_0\mathbf{x}. \qquad (7.198)$$

A block diagram of the system, shown in Fig. 7.45, clearly shows the presence of a pure integrator in the controller. In this case the only difference between the internal model method of Fig. 7.45 and the ad hoc method of Fig. 7.43 is the relative location of the integrator and the gain.

10. For example, it is not possible to add integral control to a plant that has a zero at the origin.

FIGURE 7.45

Integral control using the internal model approach

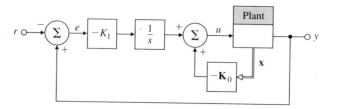

A more complex problem that clearly shows the power of the error-space approach to robust tracking is posed by requiring that a sinusoid be tracked with zero steady-state error. The problem arises, for instance, in the control of a mass-storage disk-head assembly.

◆ **EXAMPLE 7.27**   *Robust Control to Follow a Sinusoid*

A simple normalized model of a computer disk-drive servomechanism is given by the equations

$$\mathbf{F} = \begin{bmatrix} 0 & 1 \\ 0 & -1 \end{bmatrix}, \qquad \mathbf{G} = \begin{bmatrix} 0 \\ 1 \end{bmatrix};$$

$$\mathbf{G}_1 = \begin{bmatrix} 0 \\ 0 \end{bmatrix}; \qquad \mathbf{H} = [1 \quad 0]; \qquad J = 0.$$

Because the data on the disk is not exactly on a centered circle, the servo must follow a sinusoid of radian frequency ω_0 determined by the spindle speed. Give the structure of a controller for this system that will follow the given reference input with zero steady-state error.

Solution.   The reference input satisfies the differential equation $\ddot{r} = -\omega_0^2 r$ so that $\alpha_1 = 0$ and $\alpha_2 = \omega_0^2$. With these values the error-state matrices according to Eq. (7.194) are

$$\mathbf{A} = \begin{bmatrix} 0 & 1 & 0 & 0 \\ -\omega_0^2 & 0 & 1 & 0 \\ 0 & 0 & 0 & 1 \\ 0 & 0 & 0 & -1 \end{bmatrix}, \qquad \mathbf{B} = \begin{bmatrix} 0 \\ 0 \\ 0 \\ 1 \end{bmatrix}.$$

FIGURE 7.46
Structure of the compensator for the servomechanism to track exactly the sinusoid of frequency ω_0

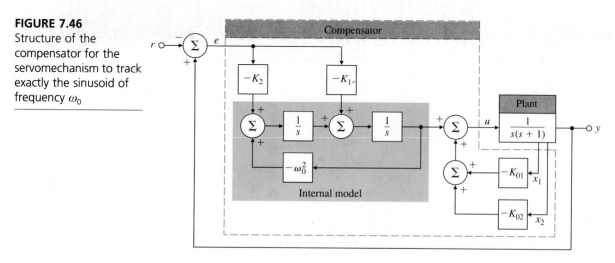

The characteristic equation of $\mathbf{A} - \mathbf{BK}$ is

$$s^4 + (1 + K_{02})s^3 + (\omega_0^2 + K_{01})s^2 + [K_1 + \omega_0^2(1 + K_{02})]s + K_{01}\omega_0^2 K_2 = 0,$$

from which the gain may be selected by pole assignment. The compensator implementation from Eq. (7.196) has the structure shown in Fig. 7.46, which clearly shows the

Internal model principle

presence of the oscillator with frequency ω_0 (known as the **internal model of the input generator**) in the controller.[11]

◆

From the nature of the pole-placement problem, the state $\mathbf{z}$ in Eq. (7.193) will tend toward zero for all perturbations in the system parameters as long as $\mathbf{A} - \mathbf{BK}$ remains stable. Notice that the signals that are rejected are those that satisfy the equations with the values of α_i actually implemented in the model of the external signals. The method assumes that these are known and implemented exactly. If the implemented values are in error, then a steady-state error will result.

Now let us repeat the example of Section 7.8.1 for integral control.

◆ **EXAMPLE 7.28**   *Integral Control Using the Error-space Design*

For the system

$$H(s) = \frac{1}{s+3}$$

with the state-variable description

$$F = -3, \qquad G = 1, \qquad H = 1,$$

construct a controller with poles at $s = -5$ to track an input that satisfies $\dot{r} = 0$.

Solution.   The error system is

$$\begin{bmatrix} \dot{e} \\ \dot{z} \end{bmatrix} = \begin{bmatrix} 0 & 1 \\ 0 & -3 \end{bmatrix} \begin{bmatrix} e \\ z \end{bmatrix} + \begin{bmatrix} 0 \\ 1 \end{bmatrix} \mu.$$

If we take the desired characteristic equation to be

$$\alpha_c(s) = s^2 + 10s + 25,$$

then the pole-placement equation for $\mathbf{K}$ is

$$\det[s\mathbf{I} - \mathbf{A} + \mathbf{BK}] = \alpha_c(s). \tag{7.199}$$

11. This is a particular case of the **internal model principle**, which requires that a model of the external or exogenous signal be in the controller for robust tracking and disturbance rejection.

FIGURE 7.47
Internal model as integral control with feedforward

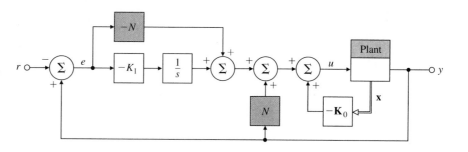

In detail Eq. (7.199) is

$$s^2 + (3 + K_1)s + K_0 = s^2 + 10s + 25,$$

which gives

$$\mathbf{K} = [25 \quad 7] = [K_1 \quad K_0],$$

and the system is implemented as shown in Fig. 7.47. The transfer function from r to e for this system,

$$\frac{E(s)}{R(s)} = -\frac{s(s + 5)}{s^2 + 10s + 25},$$

shows a blocking zero at $s = 0$, which prevents the constant input from affecting the error. The overall transfer function is

$$\frac{Y(s)}{R(s)} = \frac{25}{s^2 + 10s + 25}.$$

◆

The structure of Fig. 7.47 permits us to add a feedforward of the reference input, which provides one extra degree of freedom in zero assignment. If we add a term proportional to r in Eq. (7.198), then

$$u = -K_1 \int^t e(\tau)\, d\tau - \mathbf{K}_0\mathbf{x} + Nr. \tag{7.200}$$

This relationship has the effect of creating a zero at $-K_1/N$. The location of this zero can be chosen to improve the transient response of the system. For actual implementation we can rewrite Eq. (7.200) in terms of e to get

$$u = -K_1 \int^t e(\tau)\, d\tau - \mathbf{K}_0\mathbf{x} + N(y - e). \tag{7.201}$$

The block diagram for the system is shown in Fig. 7.48. For our example the overall transfer function now becomes

$$\frac{Y(s)}{R(s)} = \frac{Ns + 25}{s^2 + 10s + 25}.$$

FIGURE 7.48
Example of internal
model with feedforward

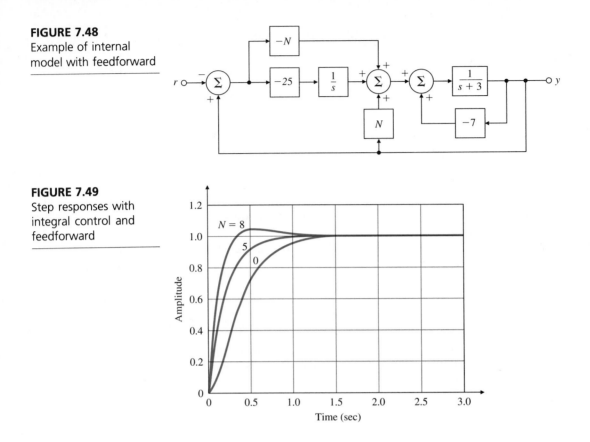

FIGURE 7.49
Step responses with
integral control and
feedforward

Notice that the DC gain is unity for any value of N and that through our choice of N we can place the zero at any real value to improve the dynamic response. A natural strategy for locating the zero is to have it cancel one of the system poles, in this case at $s = -5$. The step response of the system is shown in Fig. 7.49 for $N = 5$, as well as for $N = 0$ and 8. With the understanding that one pole can be canceled in integral control designs, we make sure to choose one of the desired control poles such that it is both real and able to be canceled through the proper choice of N.

■ 7.8.3 Disturbance Rejection by Disturbance Estimation

Our discussion of robust control so far has used a control based on full state feedback. If the state is not available, then as in the regular case, the component of the control $-\mathbf{Kx}$ can be replaced by the estimates $-\mathbf{K\hat{x}}$, where the estimator is built as before. As a final look at ways to design control with external inputs, in this section we develop a method for rejecting disturbances. The method is based on augmenting the estimator with the disturbance equations in a way that permits us to cancel out their effects on the output.

Suppose the plant is described by the equations

$$\dot{\mathbf{x}} = \mathbf{Fx} + \mathbf{G}u + \mathbf{G}_1 w_1,$$
$$y = \mathbf{Hx} + J_1 w_1. \tag{7.202}$$

Furthermore, assume that the disturbance w_1 is known to satisfy the equation[12]

$$\ddot{w}_1 + \alpha_1 \dot{w}_1 + \alpha_2 w_1 = 0. \tag{7.203}$$

The first step is to recognize that, as far as the steady-state response of the output is concerned, there is an input-equivalent disturbance w that satisfies the same equation as w_1 and enters the system at the same place as the control signal. As before, we must assume that the plant does not have a zero at any of the roots of Eq. (7.203). For our purposes here, we can replace Eqs. (7.202) with

$$\dot{\mathbf{x}} = \mathbf{Fx} + \mathbf{G}(u + w),$$
$$y = \mathbf{Hx}. \tag{7.204}$$

If we can estimate this equivalent disturbance, we can add to the control a term $-\hat{w}$ that will cancel out the effects of the real disturbance in the steady state. To do this we combine Eqs. (7.203) and (7.204) into a state description to get

$$\dot{\mathbf{z}} = \mathbf{Az} + \mathbf{B}u,$$
$$y = \mathbf{Cz}, \tag{7.205}$$

where $\mathbf{z} = [w \ \dot{w} \ \mathbf{x}^T]^T$. The matrices are

$$\mathbf{A} = \begin{bmatrix} 0 & 1 & \mathbf{0} \\ -\alpha_2 & -\alpha_1 & \mathbf{0} \\ \mathbf{G} & \mathbf{0} & \mathbf{F} \end{bmatrix}, \qquad \mathbf{B} = \begin{bmatrix} 0 \\ 0 \\ \mathbf{G} \end{bmatrix} \tag{7.206}$$

$$\mathbf{C} = [0 \ \ 0 \ \ \mathbf{H}]$$

The system given by Eqs. (7.206) is not controllable since we cannot influence w from u. However, if $\mathbf{F}$ and $\mathbf{G}$ are observable and if the system $(\mathbf{F}, \mathbf{G}, \mathbf{H})$ does not have a zero that is also a root of Eq. (7.203), then system (7.206) will be observable, and we can construct an observer that will compute estimates of both the state of the plant and of w. The estimator equations are standard, but the control is not:

$$\dot{\hat{\mathbf{z}}} = \mathbf{A}\hat{\mathbf{z}} + \mathbf{B}u + \mathbf{L}(y - \mathbf{C}\hat{\mathbf{z}}) \tag{7.207a}$$

$$u = -\mathbf{K}\hat{\mathbf{x}} + \bar{N}r - \hat{w} \tag{7.207b}$$

12. Again we develop the results for a second-order equation in the signal; the discussion can be extended to higher-order equations.

FIGURE 7.50
Block diagram of a system for bias cancellation: (a) standard form; (b) simplified form

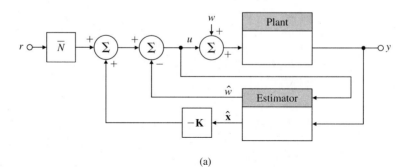

(a)

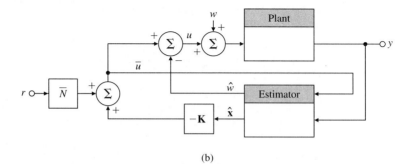

(b)

In terms of the original variables, the estimator equations are

$$\begin{bmatrix} \dot{\hat{w}} \\ \ddot{\hat{w}} \\ \dot{\hat{x}} \end{bmatrix} = \begin{bmatrix} 0 & 1 & \mathbf{0} \\ -\alpha_2 & -\alpha_1 & \mathbf{0} \\ \mathbf{G} & 0 & \mathbf{F} \end{bmatrix} \begin{bmatrix} \hat{w} \\ \dot{\hat{w}} \\ \hat{x} \end{bmatrix} + \begin{bmatrix} 0 \\ 0 \\ \mathbf{G} \end{bmatrix} u + \begin{bmatrix} l_1 \\ l_2 \\ \mathbf{L}_3 \end{bmatrix} [y - \mathbf{H}\hat{x}] \qquad (7.208)$$

The overall block diagram of the system is shown in Fig. 7.50(a). If we write out the last equation for $\hat{x}$ in Eq. (7.208) and substitute Eq. (7.207b), a simplification of sorts results because a term in $\hat{w}$ cancels out:

$$\dot{\hat{x}} = \mathbf{G}\hat{w} + \mathbf{F}\hat{x} + \mathbf{G}(-\mathbf{K}\hat{x} + \bar{N} - \hat{w}) + \mathbf{L}_3(y - \mathbf{H}\hat{x})$$
$$= \mathbf{F}\hat{x} + \mathbf{G}(-\mathbf{K}\hat{x} + \bar{N}r) + \mathbf{L}_3(y - \mathbf{H}\hat{x})$$
$$= \mathbf{F}\hat{x} + \mathbf{G}\bar{u} + \mathbf{L}_3(y - \mathbf{H}\hat{x}).$$

The simplified block diagram is drawn in Fig. 7.50(b). With the estimator of

Eq. (7.208) and the control of Eq. (7.207b), the state equation is

$$\dot{\mathbf{x}} = \mathbf{F}\mathbf{x} + \mathbf{G}(-\mathbf{K}\hat{\mathbf{x}} + \bar{N}r - \hat{w}) + \mathbf{G}w. \qquad (7.209)$$

In terms of the estimate errors, Eq. (7.209) can be rewritten as

$$\dot{\mathbf{x}} = (\mathbf{F} - \mathbf{G}\mathbf{K})\mathbf{x} + \mathbf{G}\bar{N}r + \mathbf{G}\mathbf{K}\tilde{\mathbf{x}} + \mathbf{G}\tilde{w}. \qquad (7.210)$$

Since we designed the estimator to be stable, the values of $\tilde{w}$ and $\tilde{\mathbf{x}}$ go to zero in the steady state, and the final value of the state is not affected by the disturbance input. A very simple example will illustrate the steps in this process.

◆ **EXAMPLE 7.29**   *Steady-state Disturbance Rejection by Estimating an Equivalent Disturbance*

Construct an estimator to control the state and cancel a constant bias at the output in the first-order system described by

$$\dot{x} = -3x + u,$$
$$y = x + w_1, \qquad (7.211)$$
$$\dot{w} = 0.$$

Place the control pole at $s = -5$ and the two estimator poles at $s = -15$.

Solution.   To begin, we design the control law by ignoring the disturbance. Rather, we notice by inspection that a gain of -2 will move the single pole from -3 to the desired -5. $K = 2$. Equations (7.79) and (7.80) can be solved to give the reference input gain $\bar{N} = 5$. The system with equivalent disturbance w, which replaces the actual disturbance w_1 at the control point, is given by

$$\dot{w} = 0,$$
$$\dot{x} = -3x + u + w,$$
$$y = x.$$

The estimator equations are

$$\dot{\hat{w}} = l_1(y - \hat{x})$$
$$\dot{\hat{x}} = -3\hat{x} + \hat{w} + u + l_2(y - \hat{x}).$$

The estimator error gain is found to be $\mathbf{L} = [225 \ \ 27]^T$ from the characteristic equation

$$\det \begin{bmatrix} s & l_1 \\ 1 & s + 3 + l_2 \end{bmatrix} = s^2 + 30s + 225.$$

A block diagram of the system is given in Fig. 7.51(a), and the step responses to input at the command r and at the disturbance w_1 are shown in Fig. 7.51(b,c).

FIGURE 7.51
First-order system with bias cancellation: (a) block diagram; (b) command step response;
(c) bias step response

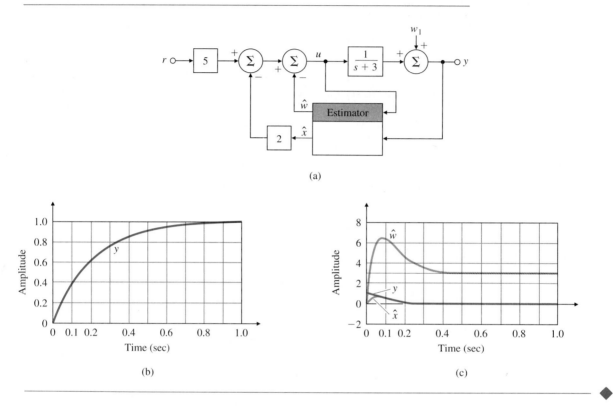

(a)

(b)

(c)

■ 7.9 Direct Design with Rational Transfer Functions

An alternative to the state-space methods discussed so far is to postulate a
general-structure dynamic controller with two inputs (r and y) and one output
(u) and to solve for the transfer function of the controller to give a specified
overall r-to-y transfer function. A block diagram of the situation is shown in
Fig. 7.52. We model the plant as the transfer function

$$\frac{Y(s)}{U(s)} = \frac{b(s)}{a(s)}, \tag{7.212}$$

rather than by state equations. The controller is also modeled by its transfer

FIGURE 7.52
Direct transfer-function
formulation

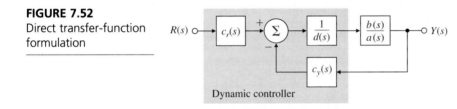

General controller in
polynomial form

function, in this case a transfer function with two inputs and one output:

$$U(s) = -\frac{c_y(s)}{d(s)} Y(s) + \frac{c_r(s)}{d(s)} R(s), \tag{7.213}$$

where $d(s)$, $c_y(s)$, and $c_r(s)$ are polynomials. In order for the controller of Fig. 7.52 and Eq. (7.213) to be implemented, the orders of the numerator polynomials $c_y(s)$ and $c_r(s)$ must not be higher than the order of the denominator polynomial $d(s)$.

To carry out the design, we require that the closed-loop transfer function defined by Eqs. (7.212) and (7.213) be matched to the desired transfer function

$$\frac{Y(s)}{R(s)} = \frac{c_r(s)b(s)}{\alpha_c(s)\alpha_e(s)}. \tag{7.214}$$

Equation (7.214) tells us that the zeros of the plant must be zeros of the overall system. The only way to change this is to have factors of $b(s)$ appear in either α_c or α_e. We continue Eqs. (7.212) and Eq. (7.213) to get

$$a(s)Y(s) = b(s)\left[-\frac{c_y(s)}{d(s)} Y(s) + \frac{c_r(s)}{d(s)} R(s) \right], \tag{7.215}$$

which can be rewritten as

$$[a(s)d(s) + b(s)c_y(s)] Y(s) = b(s)c_r(s)R(s). \tag{7.216}$$

Diophantine equation

Comparing Eq. (7.214) with Eq. (7.215) we immediately see that the design can be accomplished if we can solve the **Diophantine equation**

$$a(s)d(s) + b(s)c_y(s) = \alpha_c(s)\alpha_e(s) \tag{7.217}$$

for given arbitrary a, b, α_c, and α_e. Because each transfer function is a ratio of polynomials, we can assume that $a(s)$ and $d(s)$ are **monic polynomials**; that is, the coefficient of the highest power of s in each polynomial is unity. The question is, How many equations and how many unknowns are there if we match coefficients of equal powers of s in Eq. (7.217)? If $a(s)$ is of degree n (given) and $d(s)$ is of degree m (to be selected), then a direct count yields $2m + 1$ unknowns in $d(s)$ and $c_y(s)$ and $n + m$ equations from the coefficients of powers of s. Thus the requirement is that

$$2m + 1 \geqslant n + m$$

or

$$m \geqslant n - 1.$$

One possibility for a solution is to choose $d(s)$ of degree n and $c_y(s)$ of degree $n - 1$. In that case, which corresponds to the state-space design for a full-order estimator, there are $2n$ equations and $2n$ unknowns with $\alpha_c \alpha_e$ of degree $2n$. The resulting equations will then have a solution for arbitrary α_i if and only if $a(s)$ and $b(s)$ have no common factors.[13]

◆ **EXAMPLE 7.30**   *Pole Placement for Polynomial Transfer Functions*

Using the polynomial method, design a controller of order n for the third-order plant in Example 7.21. Note that if the polynomials $\alpha_c(s)$ and $\alpha_e(s)$ from Example 7.21 are multiplied, the result is the desired closed-loop characteristic equation:

$$\alpha_c(s)\alpha_e(s) = s^6 + 14s^5 + 122.75s^4 + 585.2s^3 + 1505.64s^2 + 2476.8s + 1728. \qquad (7.218)$$

Solution.   Using Eq. (7.217) with $b(s) = 10$, we find that

$$(d_0s^3 + d_1s^2 + d_2s + d_3)(s^3 + 10s^2 + 16s) + 10(c_0s^2 + c_1s + c_2) \equiv \alpha_c(s)\alpha_e(s). \quad (7.219)$$

We have expanded the polynomial $d(s)$ with coefficients d_i and the polynomial $c_y(s)$ with coefficients c_i.

Now we equate the coefficients of the like powers of s in Eq. (7.219) to find that the parameters must satisfy[14]

$$
\begin{bmatrix}
1 & 0 & 0 & 0 & 0 & 0 & 0 \\
10 & 1 & 0 & 0 & 0 & 0 & 0 \\
16 & 10 & 1 & 0 & 0 & 0 & 0 \\
0 & 16 & 10 & 1 & 0 & 0 & 0 \\
0 & 0 & 16 & 10 & 10 & 0 & 0 \\
0 & 0 & 0 & 16 & 0 & 10 & 0 \\
0 & 0 & 0 & 0 & 0 & 0 & 10
\end{bmatrix}
\begin{bmatrix}
d_0 \\ d_1 \\ d_2 \\ d_3 \\ c_0 \\ c_1 \\ c_2
\end{bmatrix}
=
\begin{bmatrix}
1 \\ 14 \\ 122.75 \\ 585.2 \\ 1505.64 \\ 2476.8 \\ 1728
\end{bmatrix}. \qquad (7.220)
$$

13. If they do have a common factor, it will show up on the left side of Eq. (7.217); for there to be a solution, the same factor must be on the right side of Eq. (7.217) and thus a factor of either α_c or α_e.

14. The matrix on the left side of Eq. (7.220) is called a **Sylvester matrix** and is nonsingular if and only if $a(s)$ and $b(s)$ have no common factor.

The solution to Eq. (7.220) is

$$
\begin{array}{ll}
d_0 = 1, & c_0 = 190.1, \\
d_1 = 4, & c_1 = 481.8, \\
d_2 = 66.75, & c_2 = 172.8. \\
d_3 = -146.3, &
\end{array}
$$

Hence the controller transfer function is

$$
\frac{c_y(s)}{d(s)} = \frac{190.1s^2 + 481.8s + 172.8}{s^3 + 4s^2 + 66.75s - 146.3}. \tag{7.221}
$$

Note that the coefficients of Eq. (7.221) are the same as those of the controller $D_c(s)$ (which we obtained using the state-variable techniques), once the factors in $D_c(s)$ are multiplied out.

◆

The reduced-order compensator can also be derived using a polynomial solution.

◆ **EXAMPLE 7.31**   *Reduced-order Design for a Polynomial Transfer Function Model*
Design a reduced-order controller for the third-order system in Example 7.21. The desired characteristic equation is

$$
\alpha_c(s)\alpha_e(s) = s^5 + 12s^4 + 74s^3 + 207s^2 + 378s + 288.
$$

Solution.   The equations needed to solve this problem are the same as those used to obtain Eq. (7.219), except that we take both $d(s)$ and $c_y(s)$ to be of degree $n - 1$. We need to solve

$$
(d_0 s^2 + d_1 s + d_2)(s^3 + 10s^2 + 16s) + 10(c_0 s^2 + c_1 s + c_2) \equiv \alpha_c(s)\alpha_e(s). \tag{7.222}
$$

Equating coefficients of like powers of s in Equation (7.222), we obtain

$$
\begin{bmatrix}
1 & 0 & 0 & 0 & 0 & 0 \\
10 & 1 & 0 & 0 & 0 & 0 \\
16 & 10 & 1 & 0 & 0 & 0 \\
0 & 16 & 10 & 1 & 0 & 0 \\
0 & 0 & 16 & 0 & 1 & 0 \\
0 & 0 & 0 & 0 & 0 & 1
\end{bmatrix}
\begin{bmatrix}
d_0 \\ d_1 \\ d_2 \\ c_0 \\ c_1 \\ c_2
\end{bmatrix}
=
\begin{bmatrix}
1 \\ 12 \\ 74 \\ 207 \\ 378 \\ 288
\end{bmatrix}. \tag{7.223}
$$

The solution is

$$
\begin{array}{ll}
d_0 = 1, & c_0 = -20.8, \\
d_1 = 2.0, & c_1 = -23.6, \\
d_2 = 38, & c_2 = 28.8,
\end{array}
$$

and the resulting controller is

$$\frac{c_y(s)}{d(s)} = \frac{-20.8s^2 - 23.6s + 28.8}{s^2 + 2.0s + 38}.\qquad(7.224)$$

Again, Eq. (7.224) is exactly the same as $D_{cr}(s)$ derived using the state-variable techniques in Example 7.22, once the polynomials of $D_{cr}(s)$ are multiplied out and minor numerical differences are considered.

◆

Notice that the reference input polynomial $c_r(s)$ does not enter into the analysis of Examples 7.30 and 7.31. We can select $c_r(s)$ so that it will assign zeros in the transfer function from $R(s)$ to $Y(s)$. This is the same role played by $\gamma(s)$ in Section 7.7. One choice is to select $c_r(s)$ to cancel $\alpha_e(s)$ so that the overall transfer function is

$$\frac{Y(s)}{R(s)} = \frac{K_s b(s)}{\alpha_c(s)}.$$

This corresponds to the first and most common choice of **M** and $\bar{N}$ for introducing the reference input described in Section 7.7.

Adding integral control to the polynomial solution

It is also possible to introduce integral control and, indeed, internal-model-based robust tracking control into the polynomial design method. What is required is that we have error control and that the controller have poles at the internal model locations. To get error control with the structure of Fig. 7.52, we need only let $c_r = c_y$. To get desired poles into the controller, we need to require that a specific factor be part of $d(s)$. For integral control—the most common case—this is almost trivial. The polynomial $d(s)$ will have a root at zero if we set the last term, d_m, to zero. The resulting equations can be solved if $m = n$. For a more general internal model we define $d(s)$ to be the product of a reduced-degree polynomial and a specified polynomial such as Eq. (7.203), and match coefficients in the Diophantine equation as before. The process is straightforward but tedious. Again we caution that, while the polynomial design method can be effective, the numerical problems of this method are often much worse than those associated with methods based on state equations. For higher-order systems the state-space methods are preferable.

■ 7.10   Design for Systems with Pure Time Delay

In any linear system consisting of lumped elements, the response of the system appears immediately after an excitation of the system. In some feedback systems—for example, process control systems, whether controlled by a human operator in the loop or by computer—there is a **pure time delay** (also called **transportation lag**) in the system. As a result of the distributed nature of these

FIGURE 7.53
A Smith regulator for systems with time delay

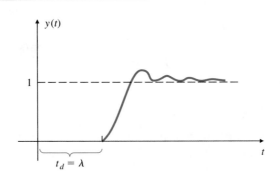

(a)

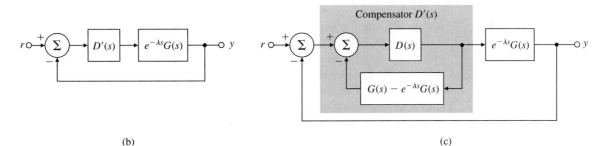

(b)   (c)

systems, the response remains identically zero until after a delay of λ seconds. A typical step response is shown in Fig. 7.53(a). The transfer function of a pure transportation lag is $e^{-\lambda s}$. We can represent an overall transfer function of a single input–single output (SISO) system with time delay as

$$G_I(s) = G(s)e^{-\lambda s}, \qquad (7.225)$$

where $G(s)$ has no pure time delay. Since $G_I(s)$ does not have a finite state description, standard use of state-variable methods is impossible. However, O. J. M. Smith (1958) showed how to construct a feedback structure that effectively takes the delay outside the loop and allows a feedback design based on $G(s)$ alone, which can be done with standard methods. The result of this method is a design having closed-loop transfer function with delay λ but otherwise showing the same response as the closed-loop design based on no delay. To see how the method works, let us consider the feedback structure shown in Fig. 7.53(b). The overall transfer function is

$$\frac{Y(s)}{R(s)} = \frac{D'(s)eG(s)^{-\lambda s}}{1 + D'(s)G(s)e^{-\lambda s}}. \qquad (7.226)$$

Smith suggested that we solve for $D'(s)$ by setting up a dummy overall transfer function in which the controller transfer function $D(s)$ is in a loop with $G(s)$ with *no loop* delay but with an overall delay of λ:

$$\frac{Y(s)}{R(s)} = \frac{D(s)G(s)}{1 + D(s)G(s)} e^{-\lambda s}. \tag{7.227}$$

We then equate Eqs. (7.226) and (7.227) to solve for $D'(s)$:

The Smith compensator

$$D'(s) = \frac{D(s)}{1 + D(s)[G(s) - G(s)e^{-\lambda s}]}. \tag{7.228}$$

If the plant transfer function and the delay are known, $D'(s)$ can be realized with real components by means of the block diagram shown in Fig. 7.53(c). With this knowledge we can design the compensator $D(s)$ in the usual way based on Eq. (7.227), as if there were no delay, and then implement it as shown in Fig. 7.53(c). The resulting closed-loop system would exhibit the behavior of a finite closed-loop system except for the time delay λ.

Notice that, conceptually, the Smith compensator is feeding back a simulated plant output to cancel the true plant output and then adding in a simulated plant output without the delay. It can be demonstrated that $D'(s)$ in Fig. 7.53(c) is equivalent to an ordinary regulator in line with a compensator that provides significant phase lead. To implement such compensators in analog systems it is usually necessary to approximate the delay required in $D'(s)$ by a Padé approximant; with digital compensators the delay can be implemented exactly (see Chapter 8). It is also a fact that the compensator D' is a strong function of $G(s)$, and a small error in the model of the plant used in the controller can lead to large errors in the closed loop, perhaps even to instability. This design is very sensitive.

◆ **EXAMPLE 7.32**   *Design with Pure Time Delay*

Figure 7.54 shows the heat exchanger from Example 2.16. The temperature of the product is controlled by controlling the flow rate of steam in the exchanger jacket. The temperature sensor is several meters downstream from the steam control valve, which introduces a transportation lag into the model. A suitable model is given by

$$G(s) = \frac{e^{-5s}}{(10s + 1)(60s + 1)}.$$

Design a controller for the heat exchanger using the Smith compensator and pole placement. The control poles are to be at

$$p_c = -0.05 \pm 0.087j,$$

and the estimator poles are to be at three times the control poles' natural frequency:

$$p_e = -0.15 \pm 0.26j.$$

FIGURE 7.54

A heat exchanger

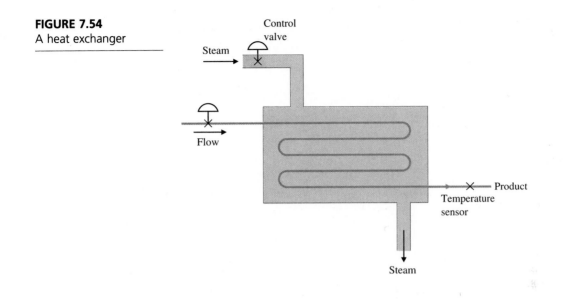

Control valve

Steam

Flow

Product

Temperature sensor

Steam

Solution.   A suitable set of state-space equations is

$$\dot{\mathbf{x}}(t) = \begin{bmatrix} -0.017 & 0.017 \\ 0 & -0.1 \end{bmatrix} \mathbf{x}(t) + \begin{bmatrix} 0 \\ 0.1 \end{bmatrix} u(t-5),$$

$$y = [1 \quad 0]\mathbf{x},$$

$$\lambda = 5.$$

For the specified control pole locations and for the moment ignoring the time delay we find that the state feedback gain is

$$\mathbf{K} = [5.2 \quad -0.17].$$

For the given estimator poles, the estimator gain matrix for a full-order estimator is

$$\mathbf{L} = \begin{bmatrix} 0.18 \\ 4.2 \end{bmatrix}.$$

If we choose to cancel the estimator poles with the feedforward zeros and adjust for unity DC gain, then

$$\bar{N} = 6.0 \quad \text{and} \quad \mathbf{M} = \mathbf{G}\bar{N}.$$

The resulting controller transfer function is

$$D(s) = \frac{U(s)}{Y(s)} = \frac{-0.25(s+1.8)}{s+0.14 \pm 0.27j}.$$

The open-loop and closed-loop step responses of the system and the control effort are shown in Figs. 7.55 and 7.56, and the root locus of the system (without the delay) is shown in Fig. 7.57. Note that the time delay of 5 sec in Figs. 7.55 and 7.56 is quite small compared with the response of the system, and is barely noticeable in this case.

FIGURE 7.55
Step response for a heat exchanger

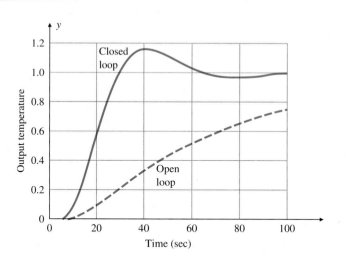

FIGURE 7.56
Control effort for a heat exchanger

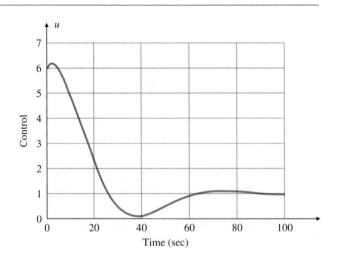

FIGURE 7.57
Root locus for a heat exchanger

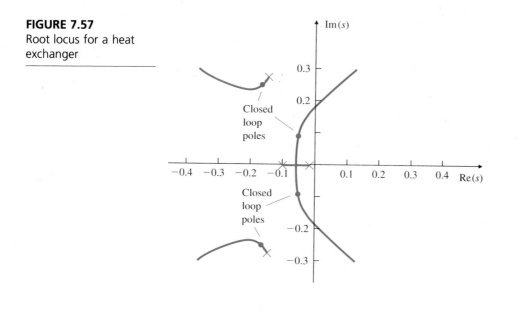

Closed loop poles

Closed loop poles

■ 7.11 Lyapunov Stability

It should be obvious by now that stability plays a major role in control systems design. We have showed that for linear constant systems, bounded input–bounded output (BIBO) stability requires that all system poles be in the LHP; also, we have given Routh's criterion based on the coefficients of the characteristic equation and Nyquist's criterion based on the frequency response to test for unstable poles. In this section we consider another important technique for analyzing the stability of nonlinear as well as linear dynamic systems based on the state-variable form of ordinary differential equations (ODEs).

A. M. Lyapunov[15] considered the stability of general systems described by ODEs in state-variable form. To illustrate a few points of his very useful theory, we consider first the general set of state equations

$$\dot{\mathbf{x}} = \mathbf{f}(\mathbf{x}). \tag{7.229}$$

If $\mathbf{f}(\mathbf{x}) = \mathbf{F}\mathbf{x}$, then we have the linear case considered throughout this chapter. We assume that the equations have been written so that $\mathbf{x} = \mathbf{0}$ is an equilibrium point, which is to say that $\mathbf{f}(\mathbf{0}) = \mathbf{0}$. This equilibrium point is said to be **Lyapunov-stable** if we are able to select a bound on initial conditions that will result in trajectories that remain within a chosen finite limit. The equilibrium is said to be **asymptotically stable** if it is Lyapunov-stable and if the state approaches zero as time approaches infinity. More formally, the system de-

Stability in the sense of Lyapunov

15. A. M. Lyapunov studied dynamic systems in Russia, and his fundamental work, *On the General Problem of Stability of Motion*, was published in 1892 by the Kharkov Mathematical Society.

FIGURE 7.58

Definition of Lyapunov
stability

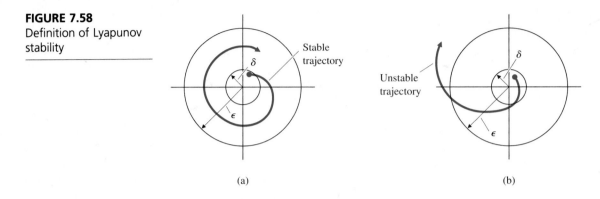

(a)   (b)

scribed by Eq. (7.229) has a stable equilibrium at $\mathbf{x} = \mathbf{0}$ if, for every ϵ there is
a δ such that if $\|\mathbf{x}(0)\| < \delta$ then $\|\mathbf{x}(t)\| < \varepsilon$ for all t.[16] Pictures of Lyapunov-
stable and -unstable responses are shown in Fig. 7.58.

For constant linear systems we see at once that $\dot{\mathbf{x}} = \mathbf{Fx}$ is Lyapunov-stable
if none of the eigenvalues of $\mathbf{F}$ are in the RHP and if any eigenvalues on the
imaginary axis are simple. (A multiple root on the imaginary axis would have
a response that grows in time and could not be stable.) Furthermore, the
response of a linear constant system is asymptotically stable if all the eigenvalues of $\mathbf{F}$ are inside the LHP. Comparing Lyapunov stability with the results
on BIBO stability (see Section 4.4), we see that linear constant systems that are
BIBO-stable are asymptotically stable in the sense of Lyapunov, but the
converse is not necessarily true: A system can be Lyapunov-stable but not
BIBO-stable. (The capacitor is a case of the latter.)

Let us consider a nonlinear system such as Eq. (7.229) and expand it into
the form

$$\dot{\mathbf{x}} = \mathbf{Fx} + \mathbf{g}(\mathbf{x}), \tag{7.230}$$

where $\mathbf{g}(\mathbf{x})$ contains all the higher powers of $\mathbf{x}$ to the extent that

$$\lim_{\|x\|\to 0} \frac{\|\mathbf{g}(\mathbf{x})\|}{\|\mathbf{x}\|} = 0$$

(that is, $\mathbf{g}(\mathbf{x})$ goes to zero faster than $\mathbf{x}$ does). Lyapunov showed that such a
system will be stable if all the eigenvalues of $\mathbf{F}$ are strictly inside the LHP and
unstable if at least one eigenvalue is in the RHP. For systems with eigenvalues
in the LHP and on the imaginary axis, the stability depends on the terms in $\mathbf{g}$.

In order to prove these results, Lyapunov introduced a function that has
many of the properties of energy. He then proved that if the "energy" never
increases, then the state must be confined to a volume bounded by a surface
of constant energy. This argument goes as follows: For a state $\mathbf{x}$ satisfying Eq.
(7.230), we consider the scalar function $V(\mathbf{x})$ with the following properties:

16. $\|\mathbf{x}\|^2 = \mathbf{x}^T\mathbf{x} = \Sigma_{i=1}^{n} x_i^2$ is a measure of the length of the vector $\mathbf{x}$.

Properties of the
Lyapunov function

1. $V(\mathbf{0}) = 0$;
2. $V(\mathbf{x}) > 0$, $\|\mathbf{x}\| \neq 0$;
3. V is continuous and has continuous derivatives with respect to all components of $\mathbf{x}$;
4. $\dot{V}(\mathbf{x}) \leqslant 0$ along trajectories of the equation.

We call V having these properties a **Lyapunov function** for the system. Properties 1 and 2 mean that, like energy, $V > 0$ if any state is different from zero, but $V = 0$ when the state is zero. Property 3 ensures that V is a smooth function and generally has the shape of a bowl near the equilibrium. Property 4 guarantees that any trajectory moves so as never to climb higher on the bowl than where it started out. If property 4 is made stronger so that $\dot{V} < 0$ for

$\|\mathbf{x}\| \neq 0$, then the trajectory must be drawn to the origin. The **Lyapunov stability theorem** states that, given the system of equations $\dot{\mathbf{x}} = \mathbf{f}(\mathbf{x})$ with $\mathbf{f}(\mathbf{0}) = \mathbf{0}$, if there exists a Lyapunov function for this equation, then the origin is a stable equilibrium point; in addition, if $\dot{V} < 0$, then the stability is asymptotic.

For the linear constant systems of main concern in this text, the quadratic function is adequate for demonstrating Lyapunov stability. Consider the function $V = \mathbf{x}^T \mathbf{P} \mathbf{x}$, where $\mathbf{P}$ is a symmetric positive matrix. For example, if $\mathbf{P} = \mathbf{I}$, then V is the sum of squares of the components of $\mathbf{x}$; that is, $V = \sum_{i=1}^{n} x_i^2$. In general, if $\mathbf{P} > \mathbf{0}$, it means we can find a matrix $\mathbf{T}$ such that $\mathbf{P} = \mathbf{T}^T \mathbf{T}$ and $V = \sum_{i=1}^{n} z_i^2$ where $\mathbf{z} = \mathbf{T}\mathbf{x}$. For such a $\mathbf{P}$ the quadratic function satisfies properties 1, 2, and 3 of a Lyapunov function. A well-known proof that a symmetric matrix is positive is if the determinants of the n principal minors of the matrix are all positive. We need to consider property 4, the derivative condition. To compute the derivative of V we use the chain rule to get

$$\dot{V} = \frac{d}{dt} \mathbf{x}^T \mathbf{P} \mathbf{x}$$

$$= \dot{\mathbf{x}}^T \mathbf{P} \mathbf{x} + \mathbf{x}^T \mathbf{P} \dot{\mathbf{x}}$$

$$= \mathbf{x}^T (\mathbf{F}^T \mathbf{P} + \mathbf{P}\mathbf{F}) \mathbf{x}$$

$$= -\mathbf{x}^T \mathbf{Q} \mathbf{x},$$

where $\mathbf{Q}$ is defined by

$$\mathbf{F}^T \mathbf{P} + \mathbf{P}\mathbf{F} = -\mathbf{Q}. \tag{7.231}$$

Lyapunov showed that, for any positive $\mathbf{Q}$, the solution $\mathbf{P}$ of the Lyapunov equation (7.231) is positive if and only if all the eigenvalues of $\mathbf{F}$ have negative real parts. In other words, given the system matrix $\mathbf{F}$, we can select a positive $\mathbf{Q}$ (such as the identity matrix $\mathbf{I}$), solve the Lyapunov equation (which is simply a system of linear equations in $n(n-1)/2$ unknowns), and test to see whether $\mathbf{P}$ is positive by looking at the determinants of the n principal minors. From this process we can determine the stability of the equations without either solving them or finding the eigenvalues, both much harder problems.

◆ **EXAMPLE 7.33**   *Lyapunov Stability for a Second-order System*

Use Lyapunov's method to find conditions for the stability of a linear system described by the state matrix

$$\mathbf{F} = \begin{bmatrix} -\alpha & \beta \\ -\beta & -\alpha \end{bmatrix}.$$ (7.232)

Solution.   For the linear case we can take any positive definite $\mathbf{Q}$ we like; the simplest is $\mathbf{Q} = \mathbf{I}$. The corresponding Lyapunov equation is

$$\begin{bmatrix} -\alpha & -\beta \\ \beta & -\alpha \end{bmatrix} \begin{bmatrix} p & q \\ q & r \end{bmatrix} + \begin{bmatrix} p & q \\ q & r \end{bmatrix} \begin{bmatrix} -\alpha & \beta \\ -\beta & -\alpha \end{bmatrix} = \begin{bmatrix} -1 & 0 \\ 0 & -1 \end{bmatrix}.$$ (7.233)

Multiplying out Eq. (7.233) and equating coefficients, we get

$$-\alpha p - \beta q - \alpha p - \beta q = -1,$$
$$-\alpha q - \beta r + \beta p - \alpha q = 0,$$ (7.234)
$$\beta q - \alpha r + \beta q - \alpha r = -1.$$

Equations (7.234) are readily solved to get $p = r = 1/2\alpha$, $q = 0$, so that

$$\mathbf{P} = \begin{bmatrix} \dfrac{1}{2\alpha} & 0 \\ 0 & \dfrac{1}{2\alpha} \end{bmatrix},$$

and the determinants are $1/2\alpha > 0$ and $1/4\alpha^2 > 0$.
Thus $\mathbf{P} > 0$, so we conclude that the system is stable if $\alpha > 0$.   ◆

For systems with many state variables, solution of the Lyapunov equation can be burdensome, but the result is an alternative to, for example, Routh's method for computing the conditions for stability in a system with literal parameters.

Testing for stability by considering the linear approximation to a differential equation is referred to as **Lyapunov's first, (or indirect) method**; using the idea of the Lyapunov function for a direct attack on the stability question is **Lyapunov's second, (or direct) method**. To illustrate the use of the direct method on a nonlinear system, we consider the position feedback system modeled in Fig. 7.59.

Lyapunov stability of
a nonlinear system

FIGURE 7.59
An elementary position feedback system with a nonlinear actuator

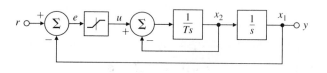

We assume that the actuator, which is perhaps only an amplifier in this case, has a significant nonlinearity which in the figure is shown as a saturation but is possibly more complex. We will assume only that $u = f(e)$ where the function lies in the first and third quadrants so that $\int_0^e f(e)\,de > 0$. We also assume that $f(e) = 0$ implies that $e = 0$ and we will assume $T > 0$, so the system is open-loop stable. The equations of motion are

$$\dot{e} = -x_2,$$
$$\dot{x}_2 = -\frac{1}{T}\,x_2 + \frac{f(e)}{T}. \tag{7.235}$$

For a Lyapunov function, consider something like kinetic plus potential energy:

$$V = \frac{T}{2}\,x_2^2 + \int_0^e f(\sigma)\,d\sigma. \tag{7.236}$$

Clearly, $V = 0$ if $x_2 = e = 0$ and, because of the assumptions about f, $V > 0$ if $x_2^2 + e^2 \neq 0$. To see whether the V in Eq. (7.236) is a Lyapunov function, we compute $\dot{V}$ as follows:

$$\dot{V} = Tx_2\dot{x}_2 + f(e)\dot{e}$$
$$= Tx_2\left[-\frac{1}{T}\,x_2 + \frac{f(e)}{T}\right] + f(e)(-x_2)$$
$$= -x_2^2.$$

Hence $\dot{V} \leqslant 0$ and the origin is Lyapunov-stable. Moreover, $\dot{V}$ is always decreasing if $x_2 \neq 0$, and Eq. (7.235) indicates that the system has no trajectory with $x_2 \equiv 0$, except $x_2 = 0$. Thus we can conclude that the system is asymptotically stable for every f that satisfies two conditions: (1) $\int f\,d\sigma > 0$ and (2) $f(e) = 0$ implies that $e = 0$.

As we mentioned earlier, the study of the stability of nonlinear systems is vast, so we have only touched here on some important points and methods. Further material for study can be found in LaSalle and Lefschetz (1961), Kalman and Bertram (1960), and Vidyasagar (1993).

Summary

- To every transfer function that has no more zeros than poles, there corresponds a differential equation in state-space form.

- State-space descriptions can be in several canonical forms. Among these are **control**, **observer** and **modal canonical forms**.

- Poles and zeros can be computed from the state description matrices (**F**, **G**, **H**, J).

$$\text{Poles:} \quad p = \text{eig}(\mathbf{F}), \quad \det(p\mathbf{I} - \mathbf{F}) = 0,$$

$$\text{Zeros:} \quad \det\begin{bmatrix} z\mathbf{I} - \mathbf{F} & -\mathbf{G} \\ \mathbf{H} & J \end{bmatrix} = 0.$$

- For any controllable system of order n, there exists a state feedback control law that will place the closed-loop poles at the roots of an arbitrary **control characteristic equation** of order n.

- The reference input can be introduced so as to result in zero steady-state error to a step command. This property is not expected to be robust to parameter changes.

- Good pole locations depend on the desired transient response, the robustness to parameter changes, and a balance between dynamic performance and control effort.

- Pole locations can be selected to result in a dominant second-order response, to match a prototype dynamic response, or to minimize a quadratic performance measure.

- For any observable system of order n, an estimator (or observer) can be constructed with only sensor inputs and a state that estimates the plant state. The poles of the estimator error system can be placed at the roots of an arbitrary **estimator characteristic equation** of order n.

- The control law and the estimator can be combined into a controller such that the poles of the closed-loop system are the sum of the control-law-only poles and the estimator-only poles.

- With the estimator-based controller the **reference input** can be introduced in such a way as to permit n arbitrary zeros to be assigned. The most common choice is to assign the zeros to cancel the estimator poles, thus not exciting an estimator error.

- **Integral control** can be introduced to obtain robust steady-state tracking of a step by augmenting the plant state.

- General **robust control** can be realized by combining the equations of the plant and the reference model into an **error space** and designing a control law for the extended system. Implementation of the robust design demonstrates the **internal model principle**. An estimator of the plant state can be added while retaining the robustness properties.

- Pole-placement designs, including integral control, can be computed using the polynomials of the plant transfer function in place of the state descriptions. Designs using polynomials frequently have problems with numerical accuracy.

- Controllers for plants that include a pure time delay can be designed as if there were no delay and then a controller implemented for the plant with the delay. The design can be expected to be sensitive to parameter changes.

- The stability of a nonlinear system in state-space description form can be studied by the methods of **Lyapunov**.

- Table 7.3 gives the important equations discussed in this chapter. The open squares indicate equations taken from optional sections in the text.

TABLE 7.3   Important Equations in Chapter 7

Name	Equation	Page
Control canonical form	$$\mathbf{A}_c = \begin{bmatrix} -a_1 & -a_2 & \cdots & -a_n \\ 1 & 0 & \cdots & 0 \\ \vdots & & & \vdots \\ 0 & & & \vdots \\ 0 & \cdots & 0 & 1 & 0 \end{bmatrix},$$ $$\mathbf{B}_c = \begin{bmatrix} 1 \\ 0 \\ \vdots \\ 0 \end{bmatrix}, \quad \mathbf{C}_c = [b_1 \quad b_2 \quad \cdots \quad b_n], \quad D_c = 0.$$	474
State description	$\dot{\mathbf{x}} = \mathbf{Fx} + \mathbf{G}u$	476
Output equation	$y = \mathbf{Hx} + Ju$	476
Transformation of state	$\mathbf{A} = \mathbf{T}^{-1}\mathbf{FT}$	477
	$\mathbf{B} = \mathbf{T}^{-1}\mathbf{G}$	
	$y = \mathbf{HTz} + Ju = \mathbf{Cz} + Du,$ where $\mathbf{C} = \mathbf{HT}, D = J$	
Controllability matrix	$\mathscr{C} = [\mathbf{G} \quad \mathbf{FG} \quad \cdots \quad \mathbf{F}^{n-1}\mathbf{G}]$	478
Transfer function from state equations	$G(s) = \dfrac{Y(s)}{U(s)} = \mathbf{H}(s\mathbf{I} - \mathbf{F})^{-1}\mathbf{G} + J$	486
Transfer-function poles	$\det(p_i\mathbf{I} - \mathbf{F}) = 0$	488
Transfer-function zeros	$\alpha_z(s) = \det\begin{bmatrix} z_i\mathbf{I} - \mathbf{F} & -\mathbf{G} \\ \mathbf{H} & J \end{bmatrix} = 0$	489
Control characteristic equation	$\det[s\mathbf{I} - (\mathbf{F} - \mathbf{GK})] = 0$	494
Ackermann's control formula for pole placement	$\mathbf{K} = [0 \quad \cdots \quad 0 \quad 1]\mathscr{C}^{-1}\alpha_c(\mathbf{F})$	497
Reference input gains	$\begin{bmatrix} \mathbf{F} & \mathbf{G} \\ \mathbf{H} & J \end{bmatrix}\begin{bmatrix} \mathbf{N}_x \\ N_u \end{bmatrix} = \begin{bmatrix} 0 \\ 1 \end{bmatrix}$	501

(Continued)

TABLE 7.3   Important Equations (*continued*)

Name	Equation	Page
Control equation with reference input	$u = N_u r - \mathbf{K}(\mathbf{x} - \mathbf{N}_x r)$ $= -\mathbf{Kx} + (N_u + \mathbf{KN}_x)r$ $= -\mathbf{Kx} + \bar{N}r$	501
Symmetric root locus	$1 + \rho G_0(-s)G_0(s) = 0$	511
Estimator-error characteristic equation	$\alpha_e(s) = \det[s\mathbf{I} - (\mathbf{F} - \mathbf{LH})] = 0$	516
Observer canonical form	$\dot{x} = \mathbf{F}_0 \mathbf{x}_0 + \mathbf{G}_0 u$ $y = \mathbf{H}_0 \mathbf{x}_0$ where $\mathbf{F}_o = \begin{bmatrix} -a_1 & 1 & 0 & 0 & \cdots & 0 \\ -a_2 & 0 & 1 & 0 & \cdots & \vdots \\ \vdots & \vdots & & \ddots & & 1 \\ -a_n & 0 & \cdots & 0 & 1 & 0 \end{bmatrix}$ $\mathbf{G}_o = \begin{bmatrix} b_1 \\ b_2 \\ \vdots \\ b_n \end{bmatrix} \quad \mathbf{H}_o = [1 \ \ 0 \ \ 0 \ \cdots \ 0]$	519
Observability matrix	$\mathcal{O} = \begin{bmatrix} \mathbf{H} \\ \mathbf{HF} \\ \vdots \\ \mathbf{HF}^{n-1} \end{bmatrix}$	520
Ackermann's estimator formula	$\mathbf{L} = \alpha_e(\mathbf{F})\mathcal{O}^{-1} \begin{bmatrix} 0 \\ 0 \\ \vdots \\ 1 \end{bmatrix}$	520
Compensator transfer function	$D_c(s) = \dfrac{U(s)}{Y(s)} = -\mathbf{K}(s\mathbf{I} - \mathbf{F} + \mathbf{GK} + \mathbf{LH})^{-1}\mathbf{L}$	529
Reduced-order compensator transfer function	$D_{cr}(s) = \dfrac{U(s)}{Y(s)} = \mathbf{C}_r(s\mathbf{I} - \mathbf{A}_r)^{-1}\mathbf{B}_r + D_r$	530
Controller equations	$\dot{\hat{\mathbf{x}}} = (\mathbf{F} - \mathbf{GK} - \mathbf{LH})\hat{\mathbf{x}} + \mathbf{L}y + \mathbf{M}r$ $u = -\mathbf{K}\hat{\mathbf{x}} + \bar{N}r$	542
Augmented state equations for integral control	$\begin{bmatrix} \dot{x}_I \\ \dot{x} \end{bmatrix} = \begin{bmatrix} 0 & \mathbf{H} \\ 0 & \mathbf{F} \end{bmatrix} \begin{bmatrix} x_I \\ x \end{bmatrix} + \begin{bmatrix} 0 \\ \mathbf{G} \end{bmatrix} u - \begin{bmatrix} 1 \\ 0 \end{bmatrix} r + \begin{bmatrix} 0 \\ \mathbf{G}_1 \end{bmatrix} w$	552
☐ General controller in polynomial form	$U(s) = -\dfrac{C_y(s)}{d(s)} Y(s) + \dfrac{c_r(s)}{d(s)} R(s)$	565
☐ Diophantine equation for closed-loop characteristic equation	$a(s)d(s) + b(s)c_y(s) = \alpha_c(s)\alpha_e(s)$	565
☐ Lyapunov's equation	$\mathbf{F}^T\mathbf{P} + \mathbf{PF} = -\mathbf{Q}$	575

Problems

7.1   Give the state description matrices in control-canonical form for the following transfer functions:

a) $\dfrac{1}{4s + 1}$

d) $\dfrac{s + 3}{s(s^2 + 2s + 2)}$

b) $\dfrac{5(s/2 + 1)}{(s/10 + 1)}$

e) $\dfrac{(s + 10)(s^2 + s + 25)}{s^2(s + 3)(s^2 + s + 36)}$

c) $\dfrac{2s + 1}{s^2 + 3s + 2}$

7.2   Use the MATLAB function tf2ss to obtain the state matrices called for in Problem 7.1.

7.3   Give the state description matrices in normal-mode form for the transfer functions of Problem 7.1. Make sure that all entries in the state matrices are real-valued by keeping any pairs of complex conjugate poles together, and realize them as a separate subblock in control canonical form.

7.4   A certain system with state $\mathbf{x}$ is described by the state matrices

$$\mathbf{F} = \begin{bmatrix} -2 & 1 \\ -2 & 0 \end{bmatrix}, \qquad \mathbf{G} = \begin{bmatrix} 1 \\ 3 \end{bmatrix},$$

$$\mathbf{H} = [1 \quad 0], \qquad J = 0.$$

Find the transformation $\mathbf{T}$ so that if $\mathbf{x} = \mathbf{Tz}$, the state matrices describing the dynamics of $\mathbf{z}$ are in control canonical form. Compute the new matrices $\mathbf{A}$, $\mathbf{B}$, $\mathbf{C}$, and D.

7.5   Use block-diagram reduction or Mason's rule to find the transfer function for the system in observer canonical form depicted by Fig. 7.23.

7.6   Suppose we are given a system with state matrices $\mathbf{F}$, $\mathbf{G}$, $\mathbf{H}$ ($J = 0$ in this case). Find the transformation $\mathbf{T}$ so that, under equations Eqs. (7.13) and (7.14), the new state description matrices will be in observer canonical form.

7.7   Use the transformation matrix in Eq. (7.30) to explicitly multiply out the equations at the end of Example 7.2.

7.8   Find the state transformation that takes the observer canonical form of Eq. (7.24) to the modal canonical form.

7.9   **a)** Find the transformation $\mathbf{T}$ that will keep the description of the tape-drive system of Example 7.3 in modal canonical form but will convert each element of the input matrix $\mathbf{B}_m$ to unity.

b) Use MATLAB to verify that your transformation does the job.

7.10   **a)** Find the state transformation that will keep the description of the tape-drive system of Example 7.3 in modal canonical form but will cause the poles to be displayed in $\mathbf{A}_m$ in order of increasing magnitude.

b) Use MATLAB to verify your result in part (a), and give the complete new set of state matrices as $\mathbf{A}$, $\mathbf{B}$, $\mathbf{C}$, and D.

7.11   Find the characteristic equation for the modal-form matrix $\mathbf{A}_m$ of Eq. (7.6) using Eq. (7.46).

7.12 Given the system

$$\dot{\mathbf{x}} = \begin{bmatrix} -4 & 1 \\ -2 & -1 \end{bmatrix} \mathbf{x} + \begin{bmatrix} 0 \\ 1 \end{bmatrix} u$$

with zero initial conditions, find the steady-state value of $\mathbf{x}$ for a step input u.

7.13 Consider the system shown in Fig. 7.60:
a) Find the transfer function from U to Y.
b) Write state equations for the system using the state variables indicated.

FIGURE 7.60
A block diagram for
Problem 7.13

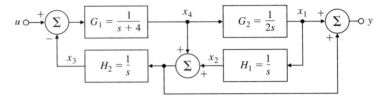

7.14 Using the indicated state variables, write the state equations for each of the systems shown in Fig. 7.61. Find the transfer function for each system using both block-diagram manipulation and matrix algebra (as in Eq. (7.36)).

FIGURE 7.61
Block diagrams for
Problem 7.14

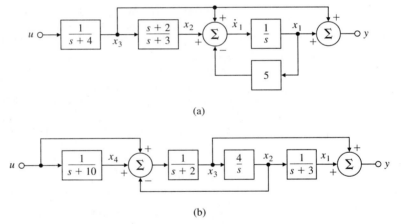

(a)

(b)

7.15 For each of the transfer functions below, write the state equations in both control and observer canonical form. In each case draw a block diagram and give the appropriate expressions for $\mathbf{F}$, $\mathbf{G}$, and $\mathbf{H}$.

a) $\dfrac{s^2 - 2}{s^2(s^2 - 1)}$ (control of an inverted pendulum by a force on the cart)

b) $\dfrac{3s + 4}{s^2 + 2s + 2}$

7.16 Consider the system

$$\mathbf{F} = \begin{bmatrix} -2 & 1 \\ 1 & 0 \end{bmatrix}, \qquad \mathbf{G} = \begin{bmatrix} 1 \\ 0 \end{bmatrix}, \qquad \mathbf{H} = [1 \quad 2],$$

and assume you are using feedback of the form $u = -\mathbf{Kx} + r$, where r is a reference input signal.

a) Show that $(\mathbf{F}, \mathbf{H})$ is observable.

b) Show that there exists a $\mathbf{K}$ such that $(\mathbf{F} - \mathbf{GK}, \mathbf{H})$ is unobservable.

c) Compute a $\mathbf{K}$ of the form $\mathbf{K} = [1, K_2]$ that will make the system unobservable as in part (b); that is, find K_2 so that the closed-loop system is not observable.

d) Compare the open-loop transfer function with the transfer function of the closed-loop system of part (c). What is the unobservability due to?

7.17 Consider the transfer function

$$G(s) = \frac{Y(s)}{U(s)} = \frac{s+1}{s^2 + 5s + 6}.$$ (7.237)

a) By rewriting Eq. (7.237) in the form

$$G(s) = \frac{1}{s+3}\left(\frac{s+1}{s+2}\right),$$

find a **series realization** of $G(s)$ as a cascade of two first-order systems.

b) Using a partial-fraction expansion of $G(s)$, find a **parallel realization** of $G(s)$.

c) Realize $G(s)$ in control canonical form.

7.18 A certain fifth-order system is found to have a characteristic equation with roots at $0, -1, -2$, and $-1 \pm 1j$. A decomposition into controllable and uncontrollable parts discloses that the controllable part has a characteristic equation with roots 0, and $-1 \pm 1j$. A decomposition into observable and nonobservable parts discloses that the observable modes are at $0, -1$, and -2.

a) Where are the zeros of $b(s) = \mathbf{H} \operatorname{adj}(s\mathbf{I} - \mathbf{F})\mathbf{G}$ for this system?

b) What are the poles of the reduced-order transfer function that includes only controllable and observable modes?

7.19 Explain how the controllability, observability, and stability properties of a linear system are related.

7.20 Consider the electric circuit shown in Fig. 7.62.

a) Write the internal (state) equations for the circuit. The input $u(t)$ is a current, and the output y is a voltage. Let $x_1 = i_L$ and $x_2 = v_c$.

b) What condition(s) on R, L, and C will guarantee that the system is controllable?

c) What condition(s) on R, L, and C will guarantee that the system is observable?

FIGURE 7.62
Electric circuit for
Problem 7.20

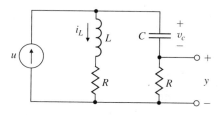

FIGURE 7.63

Block diagram for
Problem 7.21

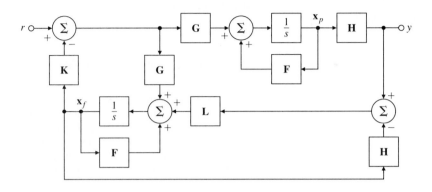

7.21 The block diagram of a feedback system is shown in Fig. 7.63. The system state is

$$x = \begin{bmatrix} x_p \\ x_f \end{bmatrix},$$

and the dimensions of the matrices are as follows:

$$\begin{array}{ll} F = n \times n, & L = n \times 1, \\ G = n \times 1, & x = 2n \times 1, \\ H = 1 \times n, & r = 1 \times 1, \\ K = 1 \times n, & y = 1 \times 1, \end{array}$$

a) Write state equations for the system.
b) Let $x = Tz$, where

$$T = \begin{bmatrix} I & 0 \\ I & -I \end{bmatrix}.$$

Show that the system is not controllable.
c) Find the transfer function of the system from r to y.

7.22 This problem is intended to give you more insight into controllability and observability. Consider the circuit in Fig. 7.64, with an input voltage source $u(t)$ and an output current $y(t)$.

a) Using the capacitor voltage and inductor current as state variables, write state and output equations for the system.
b) Find the conditions relating R_1, R_2, C, and L that render the system uncontrollable. Find a similar set of conditions that result in an unobservable system.

FIGURE 7.64

Electric circuit for
Problem 7.22

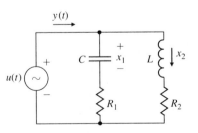

FIGURE 7.65

Coupled pendulums for Problem 7.23

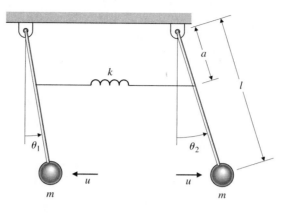

c) Interpret the conditions found in part (b) physically in terms of the time constants of the system.

d) Find the transfer function of the system. Show that there is a pole-zero cancellation for the conditions derived in part (b) (that is, when the system is uncontrollable or unobservable).

7.23 Two pendulums, coupled by a spring, are to be controlled by two equal and opposite forces u, which are applied to the pendulum bobs as shown in Fig. 7.65. The equations of motion are

$$ml^2\ddot{\theta}_1 = -ka^2(\theta_1 - \theta_2) - mgl\theta_1 - lu,$$
$$ml^2\ddot{\theta}_2 = -ka^2(\theta_2 - \theta_1) - mgl\theta_2 + lu.$$

a) Show that the system is uncontrollable. Can you associate a physical meaning with the controllable and uncontrollable modes?

b) Is there any way that the system can be made controllable?

7.24 Consider the system in Fig. 7.66.

a) Write the state-variable equations for the system, using $[\theta_1 \quad \theta_2 \quad \dot{\theta}_1 \quad \dot{\theta}_2]^T$ as the state vector and F as the single input.

b) Show that all the state variables are observable using measurements of θ_1 alone.

FIGURE 7.66

Coupled pendulums for Problem 7.24

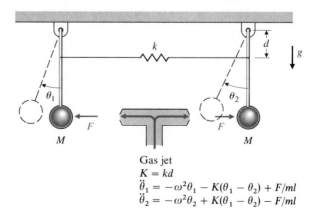

Gas jet
$K = kd$
$\ddot{\theta}_1 = -\omega^2\theta_1 - K(\theta_1 - \theta_2) + F/ml$
$\ddot{\theta}_2 = -\omega^2\theta_2 + K(\theta_1 - \theta_2) - F/ml$

c) Show that the characteristic polynomial for the system is the product of the polynomials for two oscillators. Do so by first writing a new set of system equations involving the state variables

$$\begin{bmatrix} y_1 \\ y_2 \\ \dot{y}_1 \\ \dot{y}_2 \end{bmatrix} = \begin{bmatrix} \theta_1 + \theta_2 \\ \theta_1 - \theta_2 \\ \dot{\theta}_1 + \dot{\theta}_2 \\ \dot{\theta}_1 - \dot{\theta}_2 \end{bmatrix}.$$

Hint: If **A** and **D** are invertible matrices, then

$$\begin{bmatrix} \mathbf{A} & \mathbf{0} \\ \mathbf{0} & \mathbf{D} \end{bmatrix}^{-1} = \begin{bmatrix} \mathbf{A}^{-1} & \mathbf{0} \\ \mathbf{0} & \mathbf{D}^{-1} \end{bmatrix}.$$

d) Deduce that the spring mode is controllable with F but the pendulum mode is not.

7.25 The linearized equations of motion for a satellite are

$$\dot{\mathbf{x}} = \mathbf{F}\mathbf{x} + \mathbf{G}\mathbf{u},$$
$$\mathbf{y} = \mathbf{H}\mathbf{x},$$

where

$$\mathbf{F} = \begin{bmatrix} 0 & 1 & 0 & 0 \\ 3\omega^2 & 0 & 0 & 2\omega \\ 0 & 0 & 0 & 1 \\ 0 & -2\omega & 0 & 0 \end{bmatrix}, \quad \mathbf{G} = \begin{bmatrix} 0 & 0 \\ 1 & 0 \\ 0 & 0 \\ 0 & 1 \end{bmatrix}, \quad \mathbf{H} = \begin{bmatrix} 1 & 0 & 0 & 0 \\ 0 & 0 & 1 & 0 \end{bmatrix},$$

$$\mathbf{u} = \begin{bmatrix} u_1 \\ u_2 \end{bmatrix}, \quad \mathbf{y} = \begin{bmatrix} y_1 \\ y_2 \end{bmatrix}.$$

The inputs u_1 and u_2 are the radial and tangential thrusts, the state variables x_1 and x_3 are the radial and angular deviations from the reference (circular) orbit, and the outputs y_1 and y_2 are the radial and angular measurements, respectively.

a) Show that the system is controllable using both control inputs.
b) Show that the system is controllable using only a single input. Which one is it?
c) Show that the system is observable using both measurements.
d) Show that the system is observable using only one measurement. Which one is it?

7.26 Consider the systems shown in Fig. 7.67, employing series, parallel, and feedback configurations.

a) Suppose we have controllable-observable realizations for each subsystem:

$$\dot{\mathbf{x}}_i = \mathbf{F}_i\mathbf{x}_i + \mathbf{G}_i u_i,$$
$$y_i = \mathbf{H}_i\mathbf{x}_i, \quad \text{where} \quad i = 1, 2.$$

Give a set of state equations for the combined systems in Fig. 7.67.
b) For each case, determine what condition(s) on the roots of the polynomials N_i and D_i is necessary for each system to be controllable and observable. Give a brief reason for your answer in terms of pole-zero cancellations.

7.27 *Output Controllability*: In many situations a control engineer may be interested in controlling the output y rather than the state **x**. A system is said to be **output**

FIGURE 7.67
Block diagrams for Problem 7.26: (a) series; (b) parallel; (c) feedback

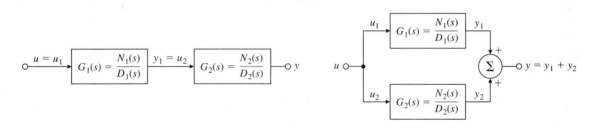

(a)   (b)

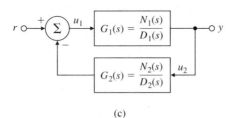

(c)

controllable if at any time you are able to transfer the output from zero to any desired output y^* in a finite time using an appropriate control signal u^*. Derive necessary and sufficient conditions for a continuous system $(\mathbf{F}, \mathbf{G}, \mathbf{H})$ to be output controllable. Are output and state controllability related? If so, how?

7.28 The state-space model for a certain application has been given to us with the following state description matrices:

$$\mathbf{F} = \begin{bmatrix} 0.174 & 0 & 0 & 0 & 0 \\ 0.157 & 0.645 & 0 & 0 & 0 \\ 0 & 1 & 0 & 0 & 0 \\ 0 & 0 & 1 & 0 & 0 \\ 0 & 0 & 0 & 1 & 0 \end{bmatrix}, \quad \mathbf{G} = \begin{bmatrix} -0.207 \\ -0.005 \\ 0 \\ 0 \\ 0 \end{bmatrix}, \quad \mathbf{H} = [1 \ 0 \ 0 \ 0 \ 0].$$

a) Draw a block diagram of the realization with an integrator for each state variable.

b) A student has computed $\det \mathscr{C} = 2.3 \times 10^{-7}$ and claims that the system is uncontrollable. Is the student right or wrong? Why?

c) Is the realization observable?

7.29 Consider the system $\ddot{y} + 3\dot{y} + 2y = \dot{u} + u$.

a) Find the state matrices $\mathbf{F}_c$, $\mathbf{G}_c$, and $\mathbf{H}_c$ in control canonical form that correspond to the given differential equation.

b) Sketch the eigenvectors of $\mathbf{F}_c$ in the (x_1, x_2) plane, and draw vectors that correspond to the completely observable $(\mathbf{x}_o)$ and the completely unobservable $(\mathbf{x}_{\bar{o}})$ state variables.

c) Express $\mathbf{x}_o$ and $\mathbf{x}_{\bar{o}}$ in terms of the observability matrix $\mathcal{O}$.

d) Give the state matrices in observer canonical form and repeat parts (b) and (c) in terms of controllability instead of observability.

7.30 *Staircase Algorithm (Van Dooren et al., 1978)*: Any realization $(\mathbf{F}, \mathbf{G}, \mathbf{H})$ can be transformed by an **orthogonal similarity transformation** to $(\bar{\mathbf{F}}, \bar{\mathbf{G}}, \bar{\mathbf{H}})$, where $\bar{\mathbf{F}}$ is an **upper Hessenberg matrix** (having one nonzero diagonal above the main diagonal):

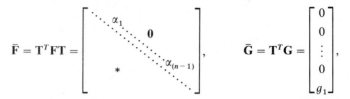

$$\bar{\mathbf{F}} = \mathbf{T}^T\mathbf{F}\mathbf{T} = \begin{bmatrix} \ddots & \alpha_1 & & & \\ & & \mathbf{0} & & \\ & & & \ddots & \\ & & & & \alpha_{(n-1)} \\ & * & & & \ddots \end{bmatrix}, \qquad \bar{\mathbf{G}} = \mathbf{T}^T\mathbf{G} = \begin{bmatrix} 0 \\ 0 \\ \vdots \\ 0 \\ g_1 \end{bmatrix},$$

where $g_1 \neq 0$, and

$$\bar{\mathbf{H}} = \mathbf{H}\mathbf{T} = [h_1 \quad \cdots \quad h_n], \qquad \mathbf{T}^{-1} = \mathbf{T}^T.$$

Orthogonal transformations correspond to a **rotation** of the vectors (represented by the matrix columns) being transformed with no change in length.

a) Prove that if $\alpha_i = 0$ and $\alpha_{i+1}, \ldots, \alpha_{n-1} \neq 0$ for some i, then the controllable and uncontrollable modes of the system can be identified after this transformation has been done.

b) How would you use this technique to identify the observable and unobservable modes of $(\mathbf{F}, \mathbf{G}, \mathbf{H})$?

c) What advantage does this approach for determining the controllable and uncontrollable modes have over transforming the system to any other form?

d) How can we use this approach to determine a basis for the controllable and uncontrollable subspaces, as in Problem 7.29?

This algorithm can be used to design a numerically stable algorithm for pole placement (see Minimis and Paige, 1982). The name of the algorithm comes from the multi-input version in which the α_i are the blocks that make $\bar{\mathbf{F}}$ resemble a staircase.

7.31 Consider the plant described by

$$\dot{\mathbf{x}} = \begin{bmatrix} 0 & 1 \\ 7 & -4 \end{bmatrix} \mathbf{x} + \begin{bmatrix} 1 \\ 2 \end{bmatrix} u,$$

$$y = [1 \quad 3]\mathbf{x}.$$

a) Draw a block diagram for the plant with one integrator for each state variable.

b) Find the transfer function using matrix algebra.

c) Find the closed-loop characteristic equation if the feedback is

(1) $u = -[K_1 \quad K_2]\mathbf{x}$; (2) $u = -K y$.

7.32 For the system

$$\dot{\mathbf{x}} = \begin{bmatrix} 0 & 1 \\ -6 & -5 \end{bmatrix} \mathbf{x} + \begin{bmatrix} 0 \\ 1 \end{bmatrix} u,$$

design a state feedback controller that satisfies the following specifications:

- Closed-loop poles have a damping coefficient $\zeta = 0.707$.
- Step-response peak time is under 3.14 sec.

7.33 a) Design a state feedback controller for the following system so that the closed-loop step response has an overshoot of less than 25% and a 1% settling time under 0.115 sec.:

$$\dot{\mathbf{x}} = \begin{bmatrix} 0 & 1 \\ 0 & -10 \end{bmatrix} \mathbf{x} + \begin{bmatrix} 0 \\ 1 \end{bmatrix} u.$$

b) Use the step command in MATLAB to verify that your design meets the specifications. If it does not, modify your feedback gains accordingly.

7.34 Consider the system

$$\dot{\mathbf{x}} = \begin{bmatrix} -1 & -2 & -2 \\ 0 & -1 & 1 \\ 1 & 0 & -1 \end{bmatrix} \mathbf{x} + \begin{bmatrix} 2 \\ 0 \\ 1 \end{bmatrix} u.$$

a) Design a state feedback controller for the following system so that the closed-loop step response has an overshoot of less than 5% and a 1% settling time under 4.6 sec.

b) Use the step command in MATLAB to verify that your design meets the specifications. If it does not, modify your feedback gains accordingly.

7.35 Consider the system in Fig. 7.68.

a) Write a set of equations that describes this system in the standard conanical control form as $\dot{\mathbf{x}} = \mathbf{F}\mathbf{x} + \mathbf{G}u$ and $y = \mathbf{H}\mathbf{x}$.

b) Design a control law of the form

$$u = -[K_1 \quad K_2]\begin{bmatrix} x_1 \\ x_2 \end{bmatrix}$$

that will place the closed-loop poles at $s = -2 \pm 2j$.

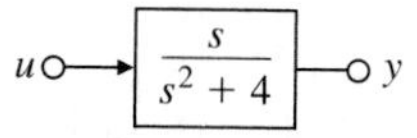

FIGURE 7.68
System for Problem 7.35

$u \circ\!\!\longrightarrow \boxed{\dfrac{s}{s^2 + 4}} \longrightarrow\!\circ\, y$

7.36 The normalized equations of motion for an inverted pendulum at angle θ on a cart are

$$\ddot{\theta} = \theta + u, \qquad \ddot{x} = -\beta\theta - u,$$

where x is the cart position, and the control input u is a force acting on the cart.

a) With the state defined as $\mathbf{x} = [\theta, \dot{\theta}, x, \dot{x}]^T$, find the feedback gain $\mathbf{K}$ that places the closed-loop poles at $s = -1, -1, -1 \pm 1j$.
For parts (b) through (d), assume that $\beta = -2$.

b) Use the symmetric root locus to select poles with a bandwidth as close as possible to those of part (a), and find the control law that will place the closed-loop poles at the points you selected.

c) Compare the responses of the closed-loop systems in parts (a) and (b) to an initial condition of $\theta = 10°$. You may wish to use the initial command in MATLAB.

d) Compute $\mathbf{N}_x$ and N_u for zero steady-state error to a constant command input on the cart position, and compare the step responses of each of the two closed-loop systems.

7.37 Consider the feedback system in Fig. 7.69. Find the relationship between K, T, and ξ such that the closed-loop transfer function minimizes the ITAE criterion for a step input.

FIGURE 7.69
Control system for
Problem 7.37

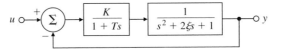

7.38 A certain process has the transfer function $G(s) = 4/(s^2 - 4)$.
  a) Find $\mathbf{F}$, $\mathbf{G}$, and $\mathbf{H}$ for this system in observer canonical form.
  b) If $u = -\mathbf{Kx}$, compute $\mathbf{K}$ so that the closed-loop control poles are located at $s = -2 \pm 2j$.
  c) Compute $\mathbf{L}$ so the estimator error poles are located at $s = -10 \pm 10j$.
  d) Give the transfer function of the resulting controller (for example, using Eq. (7.153)).
  e) What are the gain and phase margins of the controller and the given open-loop system?

7.39 An error analysis of an inertial navigator leads to the following set of normalized state equations:

$$\begin{bmatrix} \dot{x}_1 \\ \dot{x}_2 \\ \dot{x}_3 \end{bmatrix} = \begin{bmatrix} 0 & -1 & 0 \\ 1 & 0 & 1 \\ 0 & 0 & 0 \end{bmatrix} \begin{bmatrix} x_1 \\ x_2 \\ x_3 \end{bmatrix} + \begin{bmatrix} 0 \\ 0 \\ 1 \end{bmatrix} u,$$

where

   x_1 = east-velocity error,

   x_2 = platform tilt about the north axis,

   x_3 = north-gyro drift,

   u = gyro drift rate of change.

Design a reduced-order estimator with $y = x_1$ as the measurement, and place the observer error poles at -0.1 and -0.1. Be sure to provide all the relevant estimator equations.

7.40 The equations of motion for a station-keeping satellite (such as a weather satellite) are

$$\ddot{x} - 2\omega\dot{y} - 3\omega^2 x = 0, \qquad \ddot{y} + 2\omega\dot{x} = u,$$

where

   x = radial perturbation,

   y = longitudinal position perturbation,

   u = engine thrust in the y-direction.

as depicted in Fig. 7.70. If the orbit is synchronous with the earth's rotation, then $\omega = 2\pi/(3600 \times 24)$ rad/sec.

  a) Is the state $[x\dot{x}y\dot{y}]$ observable?

FIGURE 7.70

Diagram of a station-keeping satellite in orbit

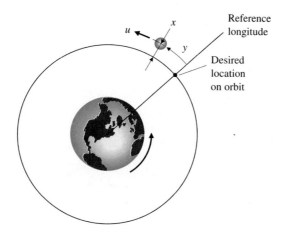

b) Choose $\mathbf{x} = [x \;\; \dot{x} \;\; y \;\; \dot{y}]^T$ as the state vector and y as the measurement, and design a full-order observer with poles placed at $s = -2\omega$, -3ω, and $-3\omega \pm 3j\omega$.

7.41 The linearized equations of motion of the simple pendulum in Fig. 7.71 are

$$\ddot{\theta} + \omega^2 \theta = u.$$

a) Write the equations of motion in state-space form.

b) Design an estimator (observer) that reconstructs the state of the pendulum given measurements of $\dot{\theta}$. Assume $\omega = 5$ rad/sec, and pick the estimator roots to be at $s = -10 \pm 10j$.

c) Write the transfer function of the estimator between the measured value of $\dot{\theta}$ and the estimated value of θ.

d) Design a controller (that is, determine the state feedback gain $\mathbf{K}$) so that the roots of the closed-loop characteristic equation are at $s = -4 \pm 4j$.

FIGURE 7.71

Pendulum diagram for Problem 7.41

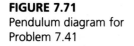

7.42 The linearized longitudinal motion of a helicopter near hover (Fig. 7.72) can be modeled by the normalized third-order system

$$\begin{bmatrix} \dot{q} \\ \dot{\theta} \\ \dot{u} \end{bmatrix} = \begin{bmatrix} -0.4 & 0 & -0.01 \\ 1 & 0 & 0 \\ -1.4 & 9.8 & -0.02 \end{bmatrix} \begin{bmatrix} q \\ \theta \\ u \end{bmatrix} + \begin{bmatrix} 6.3 \\ 0 \\ 9.8 \end{bmatrix} \delta,$$

FIGURE 7.72
Helicopter

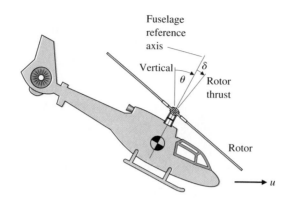

where

q = pitch rate,

θ = pitch angle of fuselage,

u = horizontal velocity (standard aircraft notation),

δ = rotor tilt angle (control variable).

Suppose our sensor measures the horizontal velocity u as the output; that is, $y = u$ and $s = -2$.

a) Find the open-loop pole locations.

b) Is the system controllable?

c) Find the feedback gain that places the poles of the system at $s = -1 \pm 1j$.

d) Design a full-order estimator for the system, and place the estimator poles at -8 and $-4 \pm 4j\sqrt{3}$.

e) Design a reduced-order estimator with both poles at -4. What are the advantages and disadvantages of the reduced-order estimator compared with the full-order case?

f) Compute the compensator transfer function using the control gain and the full-order estimator designed in part (d), and plot its frequency response. Draw a Bode plot for the closed-loop design, and indicate the corresponding gain and phase margins.

g) Repeat part (f) with the reduced-order estimator.

h) Draw the symmetrical root locus (SRL) and select roots for a control law that will give a control bandwidth matching the design of part (c), and select roots for a full-order estimator that will result in an estimator error bandwidth comparable to the design of part (d). Draw the corresponding Bode plot and compare the pole placement and SRL designs with respect to bandwidth, stability margins, step response, and control effort for a unit-step rotor-angle input.

7.43 Consider a system with the transfer function

$$G(s) = \frac{9}{s^2 - 9}.$$

a) Find $(\mathbf{F}_0, \mathbf{G}_0, \mathbf{H}_0)$ for this system in observer canonical form.

b) Is $(\mathbf{F}_0, \mathbf{G}_0)$ controllable?

c) Compute $\mathbf{K}$ so that the closed-loop poles are assigned to $s = -3 \pm 3j$.

d) Is the closed-loop system of part (c) observable?

e) Design   a   full-order   estimator   with   estimator-error   poles   at $s = -12 \pm 12j$.

f) Suppose the system is modified to have a zero:

$$G_1(s) = \frac{9(s + 1)}{s^2 - 9}.$$

Prove that if $u = -\mathbf{K}\mathbf{x} + r$, there is a feedback gain $\mathbf{K}$ that makes the closed-loop system unobservable. [Again assume an observer canonical realization for $G_1(s)$.]

7.44 Consider the system

$$\dot{\mathbf{x}} = \begin{bmatrix} 0 & 4 & 0 & 0 \\ -1 & -4 & 0 & 0 \\ 5 & 7 & 1 & 15 \\ 0 & 0 & 3 & -3 \end{bmatrix} \mathbf{x} + \begin{bmatrix} 0 \\ 0 \\ 1 \\ 0 \end{bmatrix} u.$$

a) Find the eigenvalues of this system. (*Hint:* Note the block-triangular structure.)

b) Find the controllable and uncontrollable modes of this system.

c) For each of the uncontrollable modes, find a vector $\mathbf{v}$ such that

$$\mathbf{v}^T \mathbf{G} = 0, \qquad \mathbf{v}^T \mathbf{F} = \lambda \mathbf{v}^T.$$

d) Show that there are an infinite number of feedback gains $\mathbf{K}$ that will relocate the modes of the system to -5, -3, -2, and -2.

e) Find the unique matrix $\mathbf{K}$ that achieves these pole locations and prevents initial conditions on the uncontrollable part of the system from ever affecting the controllable part.

7.45 Suppose a DC drive motor with motor current u is connected to the wheels of a cart in order to control the movement of an inverted pendulum mounted on the cart. The linearized and normalized equations of motion corresponding to this system can be put in the form

$$\ddot{\theta} = \theta + v + u,$$
$$\dot{v} = \theta - v - u,$$

where

θ = angle of the pendulum,

v = velocity of the cart.

a) We wish to control θ by feedback to u of the form

$$u = -K_1\theta - K_2\dot{\theta} - K_3 v.$$

Find the feedback gains so that the resulting closed-loop poles are located at -1, $-1 \pm j\sqrt{3}$.

b) Assume that θ and v are measured. Construct an estimator for θ and $\dot{\theta}$ of the form

$$\dot{\hat{\mathbf{x}}} = \mathbf{F}\hat{\mathbf{x}} + \mathbf{L}(y - \hat{y}),$$

where $\mathbf{x} = [\theta \ \dot{\theta}]^T$ and $y = \theta$. Treat both v and u as known. Select $\mathbf{L}$ so that the estimator poles are at -2 and -2.

c) Give the transfer function of the controller, and draw the Bode plot of the closed-loop system, indicating the corresponding gain and phase margins.

d) Plot the response of the system to an initial condition on θ, and give a physical explanation for the initial motion of the cart.

7.46 Consider the control of

$$G(s) = \frac{Y(s)}{U(s)} = \frac{10}{s(s+1)}.$$

a) Let $y = x_1$ and $\dot{x}_1 = x_2$, and write state equations for the system.

b) Find K_1 and K_2 so that $u = -K_1 x_1 - K_2 x_2$ yields closed-loop poles with a natural frequency $\omega_n = 3$ and a damping ratio $\zeta = 0.5$.

c) Design a state estimator for the system that yields estimator error poles with $\omega_{n1} = 15$ and $\zeta_1 = 0.5$.

d) What is the transfer function of the controller obtained by combining parts (a) through (c)?

e) Sketch the root locus of the resulting closed-loop system as plant gain (nominally 10) is varied.

7.47 Unstable equations of motion of the form

$$\ddot{x} = x + u$$

arise in situations where the motion of an upside-down pendulum (such as a rocket) must be controlled.

a) Let $u = -Kx$ (position feedback alone), and sketch the root locus with respect to the scalar gain K.

b) Consider a lead compensator of the form

$$U(s) = K \frac{s+a}{s+10} X(s).$$

Select a and K so that the system will display a rise time of about 2 sec and no more than 25% overshoot. Sketch the root locus with respect to K.

c) Sketch the Bode plot (both magnitude and phase) of the uncompensated plant.

d) Sketch the Bode plot of the compensated design, and estimate the phase margin.

e) Design state feedback so that the closed-loop poles are at the same locations as those of the design in part (b).

f) Design an estimator for x and $\dot{x}$ using the measurement of $x = y$, and select the observer gain $\mathbf{L}$ so that the equation for $\tilde{x}$ has characteristic roots with a damping ratio $\zeta = 0.5$ and a natural frequency $\omega_n = 8$.

g) Draw a block diagram of your combined estimator and control law, and indicate where $\hat{x}$ and $\dot{x}$ appear. Draw a Bode plot for the closed-loop system, and compare the resulting bandwidth and stability margins with those obtained using the design of part (b).

7.48 Consider the pendulum problem with control torque T_c and disturbance torque T_d:

$$\ddot{\theta} + 4\theta = T_c + T_d,$$

(here $g/l = 4$). Assume there is a potentiometer at the pin that measures the output angle θ, but with a constant unknown bias b. Thus the measurement equation is $y = \theta + b$.

a) Take the "augmented" state vector to be

$$\begin{bmatrix} \theta \\ \dot{\theta} \\ w \end{bmatrix},$$

where w is the input-equivalent bias. Write the system equations in state-space form. Give values for the matrices $\mathbf{F}$, $\mathbf{G}$, and $\mathbf{H}$.

b) Using state-variable methods, show that the characteristic equation of the model is $s(s^2 + 4) = 0$.

c) Show that w is observable if we assume $y = \theta$, and write the estimator equations for

$$\begin{bmatrix} \hat{\theta} \\ \dot{\hat{\theta}} \\ \hat{w} \end{bmatrix}.$$

Pick estimator gains $[l_1 \quad l_2 \quad l_3]^T$ to place all the roots of the estimator-error characteristic equation at -10.

d) Using full state feedback of the estimated (controllable) state-variables, derive a control law to place the closed-loop poles at $-2 \pm 2j$.

e) Draw a block diagram of the complete closed-loop system (estimator, plant, and controller) using integrator blocks.

f) Introduce the estimated bias into the control so as to yield zero steady-state error to the output bias b. Demonstrate the performance of your design by plotting the response of the system to a step change in b; that is, b changes from 0 to some constant value.

7.49 A simplified model for the control of a flexible robotic arm is shown in Fig. 7.73, where

$$k/M = 900 \text{ rad/sec}^2,$$

$$y = \text{output, the mass position;}$$

$$u = \text{input, the position of the end of the spring.}$$

FIGURE 7.73
Simple robotic arm
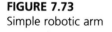

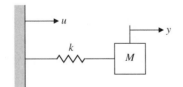

a) Write the equations of motion in state-space form.

b) Design an estimator with roots at $s = -100 \pm 100j$.

c) Could both state-variables of the system be estimated if only a measurement of $\dot{y}$ was available?

d) Design a full-state-feedback controller with roots at $s = -20 \pm 20j$.

e) Would it be reasonable to design a control law for the system with roots at $s = -200 \pm 200j$? State your reasons.

f) Write equations for the compensator, including a command input for y. Draw a Bode plot for the closed-loop system, and give the gain and phase margins for the design.

7.50 The linearized differential equations governing the fluid-flow dynamics for the two cascaded tanks in Fig. 7.74 are as follows:

$$\delta \dot{h}_1 + \sigma \delta h_1 = \delta u,$$

$$\delta \dot{h}_2 + \sigma \delta h_2 = \sigma \delta h_1,$$

where

$\delta h_1 =$ deviation of depth in tank 1 from the nominal level,

$\delta h_2 =$ deviation of depth in tank 2 from the nominal level,

$\delta u =$ deviation in fluid inflow rate to tank 1 (control).

a) *Level Controller for Two Cascaded Tanks:* Using state feedback of the form

$$\delta u = -K_1 \delta h_1 - K_2 \delta h_2,$$

choose values of K_1 and K_2 that will place the closed-loop eigenvalues at

$$s = -2\sigma(1 \pm j).$$

b) *Level Estimator for two Cascaded Tanks:* Suppose that only the deviation in the level of tank 2 is measured (that is, $y = \delta h_2$). Using this measurement, design an estimator that will give continuous, smooth estimates of the deviation in levels of tank 1 and tank 2, with estimator error poles at $-8\sigma(1 \pm j)$.

c) *Estimator/controller for Two Cascaded Tanks:* Sketch a block diagram (showing individual integrators) of the closed-loop system obtained by combining the estimator of part (b) with the controller of part (a).

d) Using MATLAB or equivalent software, compute and plot the response at y to an initial offset in δh_1. Assume $\sigma = 1$ for the plot.

FIGURE 7.74
Coupled tanks for
Problem 7.50

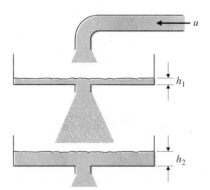

FIGURE 7.75
View of ship from above

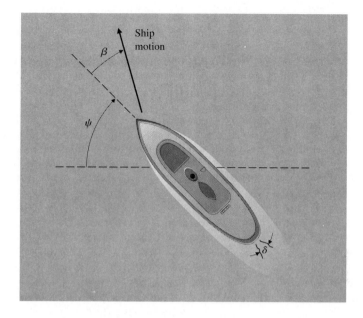

FIGURE 7.75
View of ship from above

7.51 The lateral motions of a ship that is 100 m long, moving at a constant velocity of 10 m/sec, are described by

$$\begin{bmatrix} \dot{\beta} \\ \dot{r} \\ \dot{\psi} \end{bmatrix} = \begin{bmatrix} -0.0895 & -0.286 & 0 \\ -0.0439 & -0.272 & 0 \\ 0 & 1 & 0 \end{bmatrix} \begin{bmatrix} \beta \\ r \\ \psi \end{bmatrix} + \begin{bmatrix} 0.0145 \\ -0.0122 \\ 0 \end{bmatrix} \delta,$$

where

β = sideslip angle, deg,

ψ = heading angles,

r = yaw rate. See Fig. 7.75.

a) Determine the transfer function from δ to ψ and the characteristic roots of the uncontrolled ship.

b) Using complete state feedback of the form

$$\delta = -K_1\beta - K_2r - K_3(\psi - \psi_d),$$

where ψ_d is the desired heading, determine values of K_1, K_2, and K_3 that will place the closed-loop roots at $s = -0.2$, $-0.2 \pm 0.2j$.

c) Design a state estimator based on the measurement of ψ (obtained from a gyrocompass, for example). Place the roots of the estimator error equation at $s = -0.8$ and $-0.8 \pm 0.8j$.

d) Give the state equations and transfer function for the compensator $D_c(s)$ in Fig. 7.76, and plot its frequency response.

e) Draw the Bode plot for the closed-loop system, and compute the corresponding gain and phase margins.

f) Compute the feedforward gains for a reference input, and plot the step response of the system to a change in heading of 5°.

FIGURE 7.76
Ship control block diagram

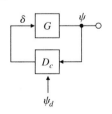

7.52 As mentioned in footnote 6 in section 7.7.2, a reasonable approach for selecting the feedforward gain in Eq. (7.181) is to choose $\bar{N}$ such that when r and y are both unchanging, the DC gain from r to u is the negative of the DC gain from y to u. Derive a formula for $\bar{N}$ based on this selection rule. Show that if the plant is type I, this choice is the same as that given by Eq. (7.181).

7.53 Assume the linearized and time-scaled equation of motion for the ball-bearing levitation device is $\ddot{x} - x = u + w$. Here w is a constant bias due to the power amplifier. Introduce integral error control, and select three control gains $\mathbf{K} = [K_1 \ \ K_2 \ \ K_3]$ so that the closed-loop poles are at -1 and $-1 \pm j$ and the steady-state error to w and to a (step) position command will be zero. Let $y = x$ and the reference input $r \triangleq y_{\text{ref}}$ be a constant. Draw a block diagram of your design showing the locations of the feedback gains K_i. Assume that both $\dot{x}$ and x can be measured. Plot the response of the closed-loop system to a step command input and the response to a step change in the bias input. Verify that the system is type I.

7.54 Consider a system with state matrices

$$\mathbf{F} = \begin{bmatrix} -2 & 1 \\ 0 & -3 \end{bmatrix}, \qquad \mathbf{G} = \begin{bmatrix} 1 \\ 1 \end{bmatrix}, \qquad \mathbf{H} = [1 \ \ 3].$$

a) Use feedback of the form $u(t) = -\mathbf{K}x(t) + \bar{N}r(t)$, where $\bar{N}$ is a nonzero scalar, to move the poles to $-3 \pm 3j$.

b) Choose $\bar{N}$ so that if r is a constant, the system has zero steady-state error; that is $y(\infty) = r$.

c) Show that if $\mathbf{F}$ changes to $\mathbf{F} + \delta\mathbf{F}$, where $\delta\mathbf{F}$ is an arbitrary 2×2 matrix, then your choice of $\bar{N}$ in part (b) will no longer make $y(\infty) = r$. Therefore, the system is not robust under changes to the system parameters in $\mathbf{F}$.

d) The system steady-state error performance can be made robust by augmenting the system with an integrator and using unity feedback; that is, by setting $\dot{x}_I = r - y$, where x_I is the state of the integrator. To see this, first use state feedback of the form $u = -\mathbf{K}x - K_1 x_I$ so that the poles of the augmented system are at $-3, -2 \pm j\sqrt{3}$.

e) Show that the resulting system will yield $y(\infty) = r$ no matter how the matrices $\mathbf{F}$ and $\mathbf{G}$ are changed, as long as the closed-loop system remains stable.

7.55 Consider a servomechanism for following the data track on a computer-disk memory system. Because of various unavoidable mechanical imperfections, the data track is not exactly a centered circle, and thus the radial servo must follow a sinusoidal input of radian frequency ω_0 (the spin rate of the disk). The state matrices for a linearized model of such a system are

$$\mathbf{F} = \begin{bmatrix} 0 & 1 \\ 0 & -1 \end{bmatrix}, \qquad \mathbf{G} = \begin{bmatrix} 0 \\ 1 \end{bmatrix}, \qquad \mathbf{H} = [1 \ \ 3].$$

The sinusoidal reference input satisfies $\ddot{r} = -\omega_0^2 r$.

a) Let $\omega_0 = 1$, and place the poles of the error system for an internal model design at

$$\alpha_c(s) = (s + 1 \pm j\sqrt{3})(s + \sqrt{3} \pm 1j)$$

and the pole of the reduced-order estimator at

$$\alpha_e(s) = s + 6.$$

b) Draw a block diagram of the system, and clearly show the presence of the oscillator with frequency ω_0 (the internal model) in the controller. Also verify the presence of the blocking zeros at $\pm j\omega_0$.

c) Use MATLAB or equivalent software to plot the time response of the system to a sinusoidal input at frequency $\omega_0 = 1$.

d) Draw a Bode plot to show how this system will respond to sinusoidal inputs at frequencies different from but near ω_0.

7.56 Consider the system with the transfer function $e^{-Ts}G(s)$, where

$$G(s) = \frac{1}{s(s+1)(s+2)}.$$

The Smith compensator for this system is given by

$$D'_c(s) = \frac{D_c}{1 + (1 - e^{-sT})G(s)D_c}.$$

Plot the frequency response of the compensator for $T = 5$ and $D_c = 1$, and draw a Bode plot that shows the gain and phase margins of the system.[17]

7.57 A first-order nonlinear system is described by the equation $\dot{x} = -f(x)$, where $f(x)$ is a continuous and differentiable nonlinear function that satisfies the following:

$$f(0) = 0,$$

$$f(x) > 0 \quad \text{for} \quad x > 0,$$

$$f(x) < 0 \quad \text{for} \quad x < 0.$$

Use the Lyapunov function $V(x) = x^2/2$ to show that the system is stable near the origin ($x = 0$).

7.58 Use the Lyapunov equation

$$\mathbf{F}^T\mathbf{P} + \mathbf{PF} = -\mathbf{Q} = -\mathbf{I}$$

to find the range of K for which the system in Fig. 7.77 will be stable. Compare your answer with the stable values for K obtained using Routh's stability criterion.

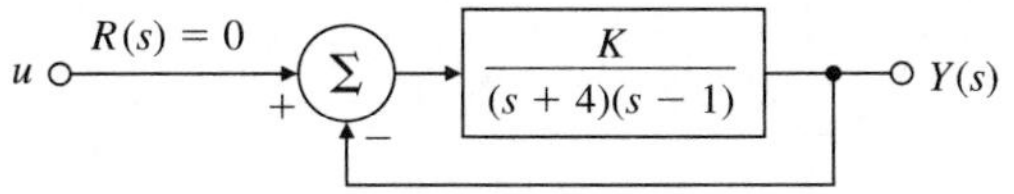

FIGURE 7.77
Control system for
Problem 7.58

7.59 Consider the system

$$\frac{d}{dt}\begin{bmatrix} x_1 \\ x_2 \end{bmatrix} = \begin{bmatrix} x_1 + x_2 u \\ x_2(x_2 + u) \end{bmatrix}, \qquad y = x_1.$$

Find all values of α and β for which the input $u(t) = \alpha y(t) + \beta$ will achieve the goal of maintaining the output $y(t)$ near 1.

17. This problem was given by Åström (1977).

7.60 Consider the nonlinear autonomous system

$$\frac{d}{dt}\begin{bmatrix} x_1 \\ x_2 \\ x_3 \end{bmatrix} = \begin{bmatrix} x_2(x_3 - x_1) \\ x_1^2 - 1 \\ -x_1 x_3 \end{bmatrix}.$$

a) Find the equilibrium point(s).

b) Find the linearized system about each equilibrium point.

c) For each case in part (b), what does Lyapunov theory tell us about the stability of the nonlinear system near the equilibrium point?

· 8 ·
Digital Control

A Perspective on Digital Control

All of the controllers we have studied so far can be built using analog electronics, such as those in Figs. 5.31 and 5.33. However, most control systems today use digital computers (usually microprocessors) to implement the controllers. The intent of this chapter is to show how to design control laws that will be implemented in a digital computer.

Unlike analog electronics, digital computers cannot integrate. Therefore, the differential equations describing compensation must be approximated by reducing them to algebraic equations involving addition and multiplication only, as developed in Section 3.6. This chapter expands on various ways to make these approximations. The resulting design can then be tuned up, if needed, using direct digital analysis and design.

You should be able to design and implement a digital control system from the material in this chapter. However, our treatment here is a limited version of a complex subject covered in more detail in *Digital Control of Dynamic Systems* by Franklin *et al.* (1990).

Chapter Overview

In Section 8.1 we describe how to approximate a continuous compensation $D(s)$ with a set of difference equations, a design method sometimes referred to as emulation. This method is sufficient for implementing a feedback control law in a digital control system, but not for analyzing how well the digital case matches the continuous case. The latter requires understanding the *z*-transform and its role in analyzing discrete systems, which we discuss in Section 8.2. Section 8.3 builds on this understanding of the *z*-transform to provide a better foundation for the emulation method.

In contrast with emulation, which is an approximate method, Section 8.4 explores direct digital design (also called discrete design), which provides an exact method independent of whether the sample rate is fast or not. Design using the state-variable form is introduced in Section 8.5. We end the chapter with a discussion of several topics related to digital control: hardware characteristics, word size, and sample rates, respectively, in Sections 8.6 to 8.8.

8.1   Digitization

Figure 8.1(a) shows the topology of the typical continuous system that we have been considering in previous chapters. The computation of the error signal *e* and the dynamic compensation $D(s)$ can all be accomplished in a digital

FIGURE 8.1
Block diagrams for a basic control system: (a) continuous system; (b) with a digital computer

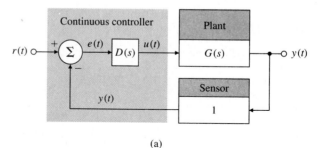

(a)

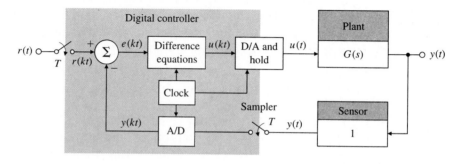

(b)

computer as shown in Fig. 8.1(b). The fundamental differences between the two implementations are that the digital system operates on **samples** of the sensed plant output rather than on the continuous signal and that the dynamics represented by $D(s)$ must be approximated by algebraic recursive equations.

We consider first the action of the **analog-to-digital (A/D) converter** on a signal. This device acts on a physical variable, most commonly an electrical voltage, and converts it into a binary number that usually consists of 10 or 12 bits. Conversion from the analog signal $y(t)$ occurs repeatedly at instants of time T seconds apart. T is the **sample period**, and $1/T$ is the **sample rate** in Hertz. The sampled signal is $y(kT)$, where k can take on any integer value. It is often written simply as $y(k)$. We call this type of variable a **discrete signal** to distinguish it from a continuous signal like $y(t)$, which changes continuously in time. A system having both discrete and continuous signals is called a **sampled data system**.

Sample period and sample rate

We make the assumption that the sample period is fixed. In practice, digital control systems sometimes have varying sample periods and/or different periods in different feedback paths. Usually, the computer logic includes a **clock** that supplies a pulse, or **interrupt**, every T seconds, and the A/D converter sends a number to the computer each time the interrupt arrives. An alternative implementation, often referred to as **free-running**, is to access the A/D converter after each cycle of executing the code has been completed. In the former case the sample period is precisely fixed; in the latter case the sample period is fixed essentially by the length of the code, providing that no logic branches are present, which could vary the amount of code executed.

There also may be a sampler and an A/D converter for the input command $r(t)$, which produce the discrete $r(kT)$ from which the sensed output $y(kT)$ will be subtracted to arrive at the discrete error signal $e(kT)$. As we shall see in Section 8.1.1, the continuous compensation is approximated by difference equations which are the discrete version of differential equations and can be made to duplicate the dynamic behavior of $D(s)$ if the sample period is short enough. The result of the difference equations is a discrete $u(kT)$ at each sample instant. This signal is converted to a continuous $u(t)$ by the **digital-to-analog (D/A) converter** and the hold: the D/A converter changes the binary number to an analog voltage, and a **zero-order hold** (ZOH) maintains that same voltage throughout the sample period. The resulting $u(t)$ is then applied to the actuator in precisely the same manner as the continuous implementation. There are two basic techniques for finding the difference equations for the digital controller. One technique, called **emulation**, consists of designing a continuous compensation $D(s)$ using methods described in the previous chapters, then approximating that $D(s)$ using one of the methods described in Section 8.1.1, 8.1.2, or 8.3. The other technique is **discrete design**, described in Sections 8.4 and 8.5. Here the difference equations are found directly without designing $D(s)$ first. Section 8.1.1 describes the use of Euler's method, and Section 8.1.2 describes the use of CACSD software as part of the emulation technique.

Zero-order hold (ZOH)

Emulation

FIGURE 8.2
The delay due to the
hold operation.

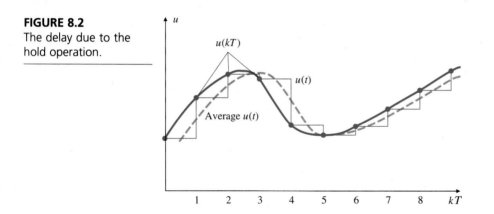

It is worth noting that the single most important impact of implementing a control system digitally is the delay associated with the hold. Because each value of $u(kT)$ in Fig. 8.1(b) is held constant until the next value is available from the computer, the continuous value of $u(t)$ consists of steps (see Fig. 8.2) that, on average, lag $u(kT)$ by $T/2$ as shown in the figure. If we simply incorporate this $T/2$ delay into a continuous analysis of a digital system, excellent agreement results for many reasonable sample rates. We shall explain this point further.

8.1.1　Euler's Method

One particularly simple way to make a digital computer approximate the real-time solution of differential equations is to use Euler's method (also called the **forward rectangular rule**). It resulted in the approximation (Eq. 3.50)

Approximation by
Euler's method

$$\dot{x} \cong \frac{x(k+1) - x(k)}{T}, \tag{8.1}$$

where

$$T = \text{the sample interval in seconds} = t_{k+1} - t_k,$$

$$t_k = kT,$$

$$k = \text{an integer},$$

$$x(k) = \text{value of } x \text{ at } t_k,$$

$$x(k+1) = \text{value of } x \text{ at } t_{k+1}.$$

This approximation can be used in place of all derivatives that appear in the controller differential equations to arrive at a set of equations that can be

Difference equations

solved by a digital computer. These **difference equations** are solved repetitively with time steps of length T. For systems having bandwidths of a few Hertz, sample rates are often on the order of 1 kHz, so that sample periods will be on the order of 1 msec and errors from the approximation can be quite small.

◆ **EXAMPLE 8.1**   *Difference Equations Using Euler's Method*

Using Euler's method, find the difference equations to be programmed into the control computer in Fig. 8.1(b) if $D(s)$ in Fig. 8.1(a) is given by

$$D(s) = \frac{U(s)}{E(s)} = K_o \frac{s + a}{s + b}. \tag{8.2}$$

Solution.   First we find the differential equation that corresponds to $D(s)$. After crossmultiplying Eq. (8.2) to obtain

$$(s + b)U(s) = K_o(s + a)E(s),$$

we can see by inspection that the corresponding differential equation is

$$\dot{u} + bu = K_o(\dot{e} + ae). \tag{8.3}$$

Using Euler's method to approximate Eq. (8.3) according to Eq. (8.1), we get the approximating difference equation

$$\frac{u(k + 1) - u(k)}{T} + bu(k) = K_o \left[\frac{e(k + 1) - e(k)}{T} + ae(k) \right]. \tag{8.4}$$

Rearranging Eq. (8.4) puts the difference equation in the desired form:

$$u(k + 1) = u(k) + T \left\{ -bu(k) + K_o \left[\frac{e(k + 1) - e(k)}{T} + ae(k) \right] \right\}. \tag{8.5}$$

Equation (8.5) computes the new value of the control, $u(k + 1)$, given the past value of the control, $u(k)$, and the new and past values of the error signal, $e(k + 1)$ and $e(k)$.

◆

In principle the difference equation is evaluated initially with $k = 0$, then $k = 1$, $2, 3, \ldots$. However, there is usually no requirement that values for all times be saved in memory. Therefore, the computer need only have variables defined for the current and past values. The instructions to the computer to implement the feedback loop in Fig. 8.1(b) with the difference equation from Eq. (8.5) would call for a continual looping through the following code:

READ y, r
$e = r - y$
$$u = u_p + T \left[-bu_p + K_o \left(\frac{e - e_p}{T} + ae \right) \right]$$
OUTPUT u
$u_p = u$ (where u_p will be the past value for the next loop through)
$e_p = e$
go back to READ when T seconds have elapsed since last READ

Design guideline
for sample rate

The sample rate required depends on the closed-loop bandwidth of the system. Generally, sample rates should be faster than 20 times the bandwidth in order to assure that the digital controller will match the performance of the continuous controller. Use of the discrete design method described in Section 8.4 allows for a much slower sample rate if that is required for the computer being used; however, best performance of a digital controller is obtained when the sample rate is greater than 20 times the bandwidth.

◆ **EXAMPLE 8.2**   *Lead Compensation Using a Digital Computer*

Find digital controllers to implement the lead compensation found in Example 5.13 for the plant

$$G(s) = \frac{1}{s(s + 1)}$$

using sample rates of 20 and 40 Hz. Implement the control equations on an experimental laboratory facility that includes a microprocessor for the control equations and analog electronics for the plant. Compute the step response of the continuous system, and compare that with the experimentally determined step response of the digitally controlled system.

Solution.   We concluded that the compensation transfer function for Example 5.13 was

$$D(s) = 70 \frac{s + 2}{s + 10}. \tag{8.6}$$

Therefore, the values of the parameters in Eq. (8.5) are $a = 2$, $b = 10$, and $K_o = 70$. For

FIGURE 8.3
Continuous and digital step response using Euler's method for the discretization:
(a) 20-Hz sample rate; (b) 40-Hz sample rate

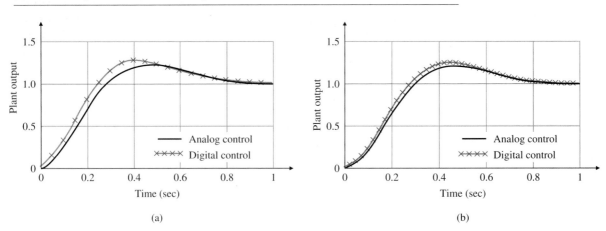

a sample rate of 20 Hz, $T = 0.05$ sec and Eq. (8.5) can be simplified to

$$u(k + 1) = 0.5u(k) + 70[e(k + 1) - 0.9e(k)].$$

For a sample rate of 40 Hz, $T = 0.025$ sec and Eq. (8.5) simplifies to

$$u(k + 1) = 0.75u(k) + 70[e(k + 1) - 0.95e(k)].$$

The code in MATLAB to compute the continuous step response is

```
num = 70 * [1  2]

den = conv([1  1  0],[1  10])

[numcl, dencl] = feedback(num, den, 1, 1)

step(numc,denc)
```

Figure 8.3 shows the step response of the two digital controllers compared with the continuous step response. Note that the 40-Hz sample rate (about 30 times bandwidth) behaves essentially like the continuous case, whereas the 20-Hz sample rate (about 15 times bandwidth) has a detectable increased overshoot signifying some degradation in the damping. The damping would degrade further if the sample rate were made any slower.

◆

8.1.2   Digitization Using CACSD Software

In addition to Euler's method, there are three other standard methods for obtaining the difference equations required for a digital computer; these will be described and analyzed in Section 8.3. In order to derive those methods and analyze their performance, it will be necessary to understand the underlying theory of discrete systems, which we will develop in Section 8.2.

None of the methods is exact for all types of input because no sampled system has access to the input time history between samples. In essence, each approximation makes a different assumption about what the continuous input is doing between samples. Most computer-aided control systems design (CACSD) software has a function that allows use of the various approximations. In MATLAB the function is called c2dm and can be used to compute the approximate difference equation in state-space form (Eq. 3.62) given a *differential* equation in state-space form (Eq. 3.56) using any of five different methods.

MATLAB c2dm

One popular approximation is the matched pole-zero method. We illustrate its use in the next example and discuss it in more detail in Section 8.3.1.

◆ **EXAMPLE 8.3**   *Matched Pole-zero versus Euler's Method for Difference Equations*

Determine the difference equations to implement the compensation in Eq. (8.6) at a sample rate of 15 Hz using MATLAB with the matched pole-zero approximation. Also use Euler's method by hand, and compare the resulting difference equations.

Solution. First we convert the continuous form of the compensation given by Eq. (8.6) to state-space form with the MATLAB statements

numD = 70 ∗ [1 2]

denD = [1 10]

[Fc,Gc,Hc,Jc] = tf2ss(numD,denD).

The resulting values of $\mathbf{F}_c$, $\mathbf{G}_c$, $\mathbf{H}_c$, and J_c define the continuous state-space description of the compensation where

$$\dot{\mathbf{x}}_c = \mathbf{F}_c\mathbf{x}_c + \mathbf{G}_c e,$$

$$u = \mathbf{H}_c\mathbf{x}_c + J_c e.$$

To find the discrete approximation using the matched pole-zero method, we use the MATLAB statements

T = 1/15

[Ac,Bc,Cc,Dc] = c2dm(Fc,Gc,Hc,Jc,T,'matched')

to get the parameter values

$$A_c = 0.5134, \qquad C_c = -2.672,$$
$$B_c = 7.387, \qquad D_c = 54.573.$$

We can now state the difference equations in state-variable form as

$$x_c(k+1) = 0.5134\,x_c(k) + 7.387\,e(k),$$
$$u(k) = -2.672\,x_c(k) + 54.573\,e(k). \tag{8.7}$$

Although Eqs. (8.7) could be implemented directly into the control computer, with some manipulation we can reduce them to the single equation

$$u(k+1) = 0.513u(k) + 54.6[e(k+1) - 0.875e(k)], \tag{8.8}$$

which is similar to the difference equations in Example 8.2.

FIGURE 8.4
Comparison of the Euler and matched pole-zero digitization methods at 15 Hz

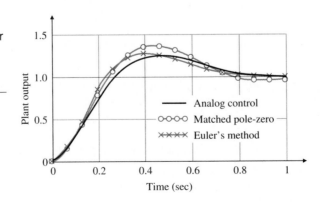

Euler's method is not an option for the c2dm function, but it can be computed by hand from Eq. (8.5) with $T = 1/15$. The result is

$$u(k + 1) = 0.333u(k) + 70[e(k + 1) - 0.867e(k)].\qquad(8.9)$$

A comparison of Eqs. (8.8) and (8.9) shows that they are similar in form, although there is a significant difference in some of the parameter values. This difference would diminish for faster sample rates. A comparison of the step responses of the two methods, along with the continuous response, is shown in Fig. 8.4. Note that both methods approximate the continuous closed-loop step response reasonably well, even though the sample rate of 15 Hz (approximately 12 times the bandwidth) is relatively slow.

◆

Digitization methods allow the designer to convert a compensation $D(s)$ into a set of difference equations that can be programmed directly into a control computer. As long as the sample rate is on the order of 20 times the bandwidth or faster, the digitally controlled system will behave similarly to its continuous counterpart, so the continuous analysis that has been the subject of the previous chapters will suffice. Analyzing the system to see how accurately the digital implementation matches the continuous analysis requires some understanding of z-transforms, which we cover in the next section.

8.2   Dynamic Analysis of Discrete Systems

The z-transform is the mathematical tool for the analysis of linear discrete systems. It plays the same role for discrete systems that the Laplace transform does for continuous systems. This section will give a short description of the z-transform, describe its use in analyzing discrete systems, and show how it relates to the Laplace transform.

8.2.1 z-Transform

In the analysis of continuous systems, we use the Laplace transform, which is defined by

$$\mathscr{L}\{f(t)\} = F(s) = \int_0^\infty f(t)e^{-st}\,dt,$$

which leads directly to the important property that (with zero initial conditions)

$$\mathscr{L}\{\dot{f}(t)\} = sF(s).\qquad(8.10)$$

Relation (8.10) enables us to find easily the transfer function of a linear continuous system, given the differential equation of that system.

For discrete systems a very similar procedure is available. The **z-transform**

FIGURE 8.5
A continuous, sampled
version of signal f

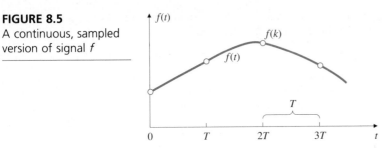

is defined by

$$\mathscr{Z}\{f(k)\} = F(z) = \sum_{k=0}^{\infty} f(k)z^{-k}, \tag{8.11}$$

where $f(k)$ is the sampled version of $f(t)$, as shown in Fig. 8.5, and $k = 1, 2, 3,\ldots$ refers to discrete sample times $t_1, t_2, t_3, \ldots$. This leads directly to a property analogous to Eq. (8.10), specifically, that

$$\mathscr{Z}\{f(k-1)\} = z^{-1}F(z). \tag{8.12}$$

This relation allows us to easily find the transfer function of a discrete system, given the difference equations of that system. For example, the general second-order difference equation

$$y(k) = -a_1y(k-1) - a_2y(k-2) + b_0u(k) + b_1u(k-1) + b_2u(k-2)$$

can be converted from this form to the z-transform of the variables $y(k)$, $u(k)$, $\ldots$ by invoking Eq. (8.12) once or twice to arrive at

$$Y(z) = (-a_1z^{-1} - a_2z^{-2})Y(z) + (b_0 + b_1z^{-1} + b_2z^{-2})U(z). \tag{8.13}$$

Eq. (8.13) then results in the transfer function

Discrete transfer function

$$\frac{Y(z)}{U(z)} = \frac{b_0 + b_1z^{-1} + b_2z^{-2}}{1 + a_1z^{-1} + a_2z^{-2}}.$$

8.2.2 z-Transform Inversion

Table 8.1 relates simple discrete-time functions to their z-transforms and gives the Laplace transforms for the same time functions.

Given a general z-transform, we could expand it into a sum of elementary terms using partial-fraction expansion and find the resulting time series from the table. These procedures are exactly the same as those used for continuous systems; like the continuous case, most designers would use a numerical evaluation of the discrete equations to obtain a time history rather than inverting the z-transform.

TABLE 8.1   Laplace Transforms and z-transforms of Simple Discrete Time Functions

$F(s)$ is the Laplace transform of $f(t)$, and $F(z)$ is the z-transform of $f(kT)$. Note: $f(t) = 0$ for $t = 0$.

Number	$\mathscr{F}(s)$	$f(kT)$	$F(z)$
1		$1,\ k = 0;\ 0,\ k \neq 0$	1
2		$1,\ k = k_o;\ 0,\ k \neq k_o$	z^{-k_o}
3	$\dfrac{1}{s}$	$1(kT)$	$\dfrac{z}{z - 1}$
4	$\dfrac{1}{s^2}$	kT	$\dfrac{Tz}{(z - 1)^2}$
5	$\dfrac{1}{s^3}$	$\dfrac{1}{2!}(kT)^2$	$\dfrac{T^2}{2}\left[\dfrac{z(z + 1)}{(z - 1)^3}\right]$
6	$\dfrac{1}{s^4}$	$\dfrac{1}{3!}(kT)^3$	$\dfrac{T^3}{6}\left[\dfrac{z(z^2 + 4z + 1)}{(z - 1)^4}\right]$
7	$\dfrac{1}{s^m}$	$\lim\limits_{a \to 0} \dfrac{(-1)^{m-1}}{(m - 1)!}\left(\dfrac{\partial^{m-1}}{\partial a^{m-1}} e^{-akT}\right)$	$\lim\limits_{a \to 0} \dfrac{(-1)^{m-1}}{(m - 1)!}\left(\dfrac{\partial^{m-1}}{\partial a^{m-1}} \dfrac{z}{z - e^{-aT}}\right)$
8	$\dfrac{1}{s + a}$	e^{-akT}	$\dfrac{z}{z - e^{-aT}}$
9	$\dfrac{1}{(s + a)^2}$	kTe^{-akT}	$\dfrac{Tze^{-aT}}{(z - e^{-aT})^2}$
10	$\dfrac{1}{(s + a)^3}$	$\dfrac{1}{2}(kT)^2 e^{-akT}$	$\dfrac{T^2}{2} e^{-aT} z \dfrac{(z + e^{-aT})}{(z - e^{-aT})^3}$
11	$\dfrac{1}{(s + a)^m}$	$\dfrac{(-1)^{m-1}}{(m - 1)!}\left(\dfrac{\partial^{m-1}}{\partial a^{m-1}} e^{-akT}\right)$	$\dfrac{(-1)^{m-1}}{(m - 1)!}\left(\dfrac{\partial^{m-1}}{\partial a^{m-1}} \dfrac{z}{z - e^{-aT}}\right)$
12	$\dfrac{a}{s(s + a)}$	$1 - e^{-akT}$	$\dfrac{z(1 - e^{-aT})}{(z - 1)(z - e^{-aT})}$
13	$\dfrac{a}{s^2(s + a)}$	$\dfrac{1}{a}(akT - 1 + e^{-akT})$	$\dfrac{z[(aT - 1 + e^{-aT})z + (1 - e^{-aT} - aTe^{-aT})]}{a(z - 1)^2(z - e^{-aT})}$
14	$\dfrac{b - a}{(s + a)(s + b)}$	$e^{-akT} - e^{-bkT}$	$\dfrac{(e^{-aT} - e^{-bT})z}{(z - e^{-aT})(z - e^{-bT})}$
15	$\dfrac{s}{(s + a)^2}$	$(1 - akT)e^{-akT}$	$\dfrac{z[z - e^{-aT}(1 + aT)]}{(z - e^{-aT})^2}$
16	$\dfrac{a^2}{s(s + a)^2}$	$1 - e^{-akT}(1 + akT)$	$\dfrac{z[z(1 - e^{-aT} - aTe^{-aT}) + e^{-2aT} - e^{-aT} + aTe^{-aT}]}{(z - 1)(z - e^{-aT})^2}$
17	$\dfrac{(b - a)s}{(s + a)(s + b)}$	$be^{-bkT} - ae^{-akT}$	$\dfrac{z[z(b - a) - (be^{-aT} - ae^{-bT})]}{(z - e^{-aT})(z - e^{-bT})}$
18	$\dfrac{a}{s^2 + a^2}$	$\sin akT$	$\dfrac{z \sin aT}{z^2 - (2 \cos aT)z + 1}$
19	$\dfrac{s}{s^2 + a^2}$	$\cos akT$	$\dfrac{z(z - \cos aT)}{z^2 - (2 \cos aT)z + 1}$
20	$\dfrac{s + a}{(s + a)^2 + b^2}$	$e^{-akT} \cos bkT$	$\dfrac{z(z - e^{-aT} \cos bT)}{z^2 - 2e^{-aT}(\cos bT)z + e^{-2aT}}$
21	$\dfrac{b}{(s + a)^2 + b^2}$	$e^{-akT} \sin bkT$	$\dfrac{ze^{-aT} \sin bT}{z^2 - 2e^{-aT}(\cos bT)z + e^{-2aT}}$
22	$\dfrac{a^2 + b^2}{s[(s + a)^2 + b^2]}$	$1 - e^{-akT}\left(\cos bkT + \dfrac{a}{b}\sin bkT\right)$	$\dfrac{z(Az + B)}{(z - 1)[z2^2 - 2e^{-aT}(\cos bT)z + e^{-2aT}]}$

$$A = 1 - e^{-aT}\cos bT - \dfrac{a}{b}e^{-aT}\sin bT$$

$$B = e^{-2aT} + \dfrac{a}{b}e^{-aT}\sin bT - e^{-aT}\cos bT$$

A z-transform inversion technique that has no continuous counterpart is called **long division**. Given the z-transform

$$Y(z) = \frac{N(z)}{D(z)}, \tag{8.14}$$

we simply divide the denominator into the numerator using long division. The result is a series (perhaps with an infinite number of terms) in z, from which the time series can be found by using Eq. (8.11).

For example, a first-order system described by the difference equation

z-transform inversion: long division

$$y(k) = \alpha y(k-1) + u(k)$$

yields

$$\frac{Y(z)}{U(z)} = \frac{1}{1 - \alpha z^{-1}}.$$

For a unit pulse input defined by

$$u(0) = 1,$$
$$u(k) = 0 \qquad k \neq 0,$$

the z-transform is then

$$U(z) = 1, \tag{8.15}$$

so

$$Y(z) = \frac{1}{1 - \alpha z^{-1}}. \tag{8.16}$$

Therefore, to find the time series, we divide the numerator of Eq. (8.16) by its denominator using long division,

$$
\begin{array}{r}
1 + \alpha z^{-1} + \alpha^2 z^{-2} + \alpha^3 z^{-3} + \cdots \\
1 - \alpha z^{-1} \overline{)\, 1 } \\
\underline{1 - \alpha z^{-1}} \\
\alpha z^{-1} + 0 \\
\underline{\alpha z^{-1} - \alpha^2 z^{-2}} \\
\alpha^2 z^{-2} + 0 \\
\underline{\alpha^2 z^{-2} - \alpha^3 z^{-3}} \\
\alpha^3 z^{-3} \\
\ddots
\end{array}
$$

to yield the infinite series

$$Y(z) = 1 + \alpha z^{-1} + \alpha^2 z^{-2} + \alpha^3 z^{-3} + \cdots. \tag{8.17}$$

From Eqs. (8.17) and (8.11) we see that the sampled time history of y is

$$y(0) = 1,$$
$$y(1) = \alpha,$$
$$y(2) = \alpha^2,$$
$$\vdots \qquad \vdots$$
$$y(k) = \alpha^k.$$

8.2.3 Relationship between s and z

For continuous systems we saw in Chapter 3 that certain behaviors result from different pole locations in the s-plane: oscillatory behavior for poles near the imaginary axis, exponential decay for poles on the negative real axis, and unstable behavior for poles with a positive real part. A similar kind of association would also be useful to know when designing discrete systems. Consider the continuous signal

$$f(t) = e^{-at}, \qquad t > 0,$$

which has the Laplace transform

$$F(s) = \frac{1}{s + a}$$

and corresponds to a pole at $s = -a$. The z-transform of $f(kT)$ is

$$F(z) = \mathscr{L}\{e^{-akT}\}. \tag{8.18}$$

From Table 8.1 we can see Eq. (8.18) is equivalent to

$$F(z) = \frac{z}{z - e^{-aT}},$$

which corresponds to a pole at $z = e^{-aT}$. This means that a pole at $s = -a$ in the s-plane corresponds to a pole at $z = e^{-aT}$ in the discrete domain. This is true in general:

Relationship between z-plane and s-plane characteristics

The equivalent characteristics in the z-plane are related to those in the s-plane by the expression

$$z = e^{sT}, \tag{8.19}$$

where T is the sample period.

Table 8.1 also includes the Laplace transforms, which demonstrates the $z = e^{sT}$ relationship for the roots of the denominators of the table entries for $F(s)$ and $F(z)$.

Figure 8.6 shows the mapping of lines of constant damping ζ and natural frequency ω_n from the s-plane (Fig. 3.11) to the upper half of the z-plane, using Eq. (8.19). The mapping has several important features (See Problem 8.20):

1. The stability boundary is the unit circle $|z| = 1$.
2. The small vicinity around $z = +1$ in the z-plane is essentially identical to the vicinity around $s = 0$ in the s-plane.
3. The z-plane locations give response information normalized to the sample rate, rather than to time as in the s-plane.
4. The negative real z-axis always represents a frequency of $\omega_s/2$, where $\omega_s = 2\pi/T =$ sample rate in radians per second.
5. Vertical lines in the left half of the s-plane (the constant real part or time constant) map into circles within the unit circle of the z-plane.
6. Horizontal lines in the s-plane (the constant imaginary part of the frequency) map into radial lines in the z-plane.

Nyquist rate $= \omega_s/2$

7. Frequencies greater than $\omega_s/2$, called the **Nyquist rate**, appear in the

FIGURE 8.6
Natural frequency (solid color) and damping loci (light color) in the z-plane; the portion below the Re(z)-axis (not shown) is the mirror image of the upper half shown.

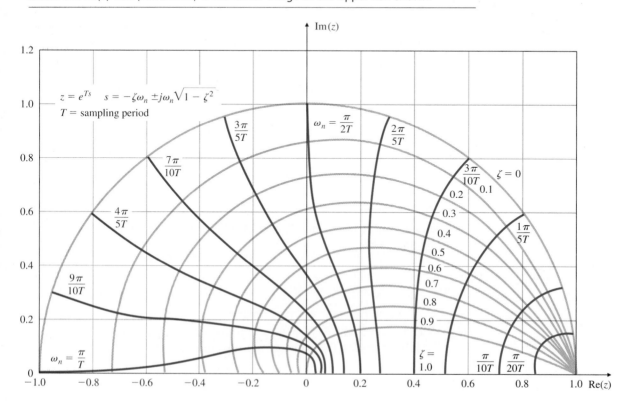

FIGURE 8.7
Time sequences associated with points in the z-plane

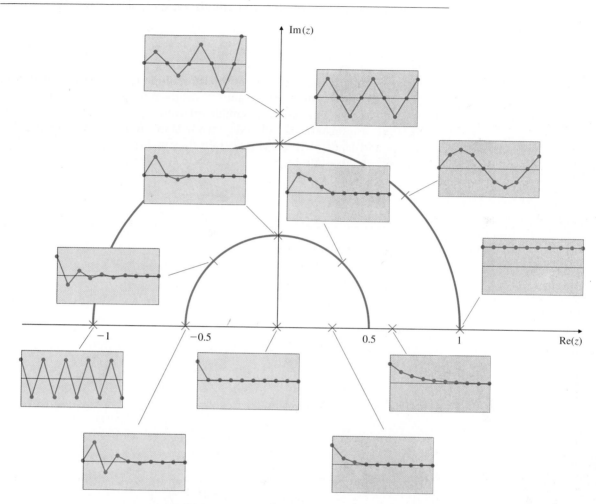

z-plane on top of corresponding lower frequencies because of the circular character of the trigonometric functions imbedded in Eq. (8.19). This overlap is called **aliasing** or **folding**. As a result it is necessary to sample at least twice as fast as a signal's highest frequency component in order to represent that signal with the samples. (We will discuss aliasing in greater detail in Section 8.6.3.)

To provide insight into the correspondence between z-plane locations and the resulting time sequence, Fig. 8.7 sketches time responses that would result from poles at the indicated locations. This figure is the discrete companion of Fig. 3.9.

8.2.4   Final Value Theorem

The Final Value Theorem for continuous systems, which we discussed in Section 3.1.6, states that

$$\lim_{t \to \infty} x(t) = x_{ss} = \lim_{s \to 0} sX(s), \tag{8.20}$$

as long as all the poles of $sX(s)$ are in the left half-plane (LHP). It is often used to find steady-state system errors and/or steady-state gains of portions of a control system. We can obtain a similar relationship for discrete systems by noting that a constant continuous steady-state response is denoted by $X(s) = A/s$ and leads to the multiplication by s in Eq. (8.20). Therefore, since the constant steady-state response for discrete systems is

$$X(z) = \frac{A}{1 - z^{-1}},$$

Final Value Theorem for discrete systems

the discrete Final Value Theorem is

$$\lim_{k \to \infty} x(k) = x_{ss} = \lim_{z \to 1} (1 - z^{-1})X(z), \tag{8.21}$$

if all the poles of $(1 - z^{-1})X(z)$ are inside the unit circle.

For example, to find the DC gain of the transfer function

$$G(z) = \frac{X(z)}{U(z)} = \frac{0.58(1 + z)}{z + 0.16},$$

we let $u(k) = 1$ for $k \geqslant 0$, so that

$$U(z) = \frac{1}{1 - z^{-1}}$$

and

$$X(z) = \frac{0.58(1 + z)}{(1 - z^{-1})(z + 0.16)}.$$

Applying the Final Value Theorem yields

$$x_{ss} = \lim_{z \to 1} \left[\frac{0.58(1 + z)}{z + 0.16} \right] = 1,$$

DC gain

so the DC gain of $G(z)$ is unity. To find the DC gain of any stable transfer function, we simply substitute $z = 1$ and compute the resulting gain. Since the DC gain of a system should not change whether represented continuously or discretely, this calculation is an excellent aid in finding a discrete controller that best matches a continuous controller. It is also a good check on the calculations associated with determining the discrete model of a system.

8.3 Design by Emulation

Design by emulation, partially described in Section 8.1, proceeds through the following stages:

Stages in emulation design

1. Designing a continuous compensation as described in Chapters 1 through 7.
2. Digitizing the continuous compensation.
3. Using discrete analysis, simulation, or experimentation to verify the design.

In Section 8.1.1 we discussed Euler's method for performing the digitization. Armed with an understanding of the z-transform from Section 8.2, we now develop more digitization procedures and analyze the performance of the digitally controlled system.

8.3.1 Digitization Procedures

Discrete approximations to a continuous transfer function

Assume we are given a continuous compensation $D(s)$ as shown in Fig. 8.1(a). We wish to find a set of difference equations (or $D(z)$) for the digital implementation of that compensation in Fig. 8.1(b). First we rephrase the problem as one of finding the best $D(z)$ in the digital implementation shown in Fig. 8.8(a) to match the continuous system represented by $D(s)$ in Fig. 8.8(b). In this section we examine and compare three methods for solving this problem.

It is important to remember, as stated earlier, that these methods are approximations; there is no exact solution for all possible inputs because $D(s)$ responds to the complete time history of $e(t)$, whereas $D(z)$ only has access to the sample $e(kT)$. In a sense the various digitization techniques simply make different assumptions about what happens to $e(t)$ between the sample points.

Tustin's Method

One digitization technique is to approach the problem as one of numerical integration. Suppose

$$\frac{U(s)}{E(s)} = D(s) = \frac{1}{s},$$

FIGURE 8.8
Comparison of (a) digital and (b) continuous implementation

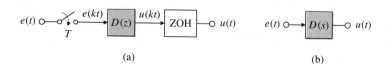

(a)

(b)

which is integration. Therefore,

$$u(kT) = \int_0^{kT-T} e(t)\,dt + \int_{kT-T}^{kT} e(t)\,dt, \tag{8.22}$$

which can be rewritten as

$$u(kT) = u(kT - T) + \text{area under } e(t) \text{ over last } T, \tag{8.23}$$

where T is the sample period.

For Tustin's method, the task at each step is to use trapezoidal integration, that is, to approximate $e(t)$ by a straight line between the two samples (Fig. 8.9). Writing $u(kT)$ as $u(k)$ and $u(kT - T)$ as $u(k - 1)$ for short, we convert Eq. (8.23) to

$$u(k) = u(k - 1) + \frac{T}{2}\,[e(k - 1) + e(k)], \tag{8.24}$$

or, taking the z-transform,

$$\frac{U(z)}{E(z)} = \frac{T}{2}\left(\frac{1 + z^{-1}}{1 - z^{-1}}\right) = \frac{1}{\dfrac{2}{T}\left(\dfrac{1 - z^{-1}}{1 + z^{-1}}\right)}. \tag{8.25}$$

For $D(s) = a/(s + a)$, applying the same integration approximation yields

$$D(z) = \frac{a}{\dfrac{2}{T}\left(\dfrac{1 - z^{-1}}{1 + z^{-1}}\right) + a}.$$

In fact, substituting

$$s = \frac{2}{T}\left(\frac{1 - z^{-1}}{1 + z^{-1}}\right)$$

for every occurrence of s in any $D(s)$ yields a $D(z)$ based on the trapezoidal integration formula. This is called **Tustin's** or the **bilinear approximation**. Finding Tustin's approximation by hand for even a simple transfer function requires fairly extensive algebraic manipulations. MATLAB's c2dm function expedites the process, as shown in the following example.

FIGURE 8.9
Trapezoidal integration

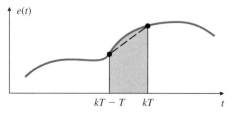

◆ **EXAMPLE 8.4**   *Difference Equations Using Tustin's Approximation*

Determine the difference equations to implement the compensation in Eq. (8.6) at a sample rate of 15 Hz using MATLAB with Tustin's approximation.

Solution.   The continuous form of the compensation is given by Eq. (8.6), or

$$D(s) = 70 \, \frac{s + 2}{s + 10}.$$

The MATLAB statements

    numDs = 70 * [1 2]

    denDs = [1 10]

    T = 1/15

    [numDz,denDz] = c2dm(numDs,denDs,T,'tustin').

produce values for numDz and denDz that result in the discrete transfer function

$$D(z) = \frac{56 - 49 \, z^{-1}}{1 - 0.5 \, z^{-1}}. \tag{8.26}$$

We can then write the difference equation by inspecting Eq. (8.26) to get

$$u(k) = 0.223 \, u(k - 1) + 70e(k) - 59.12 \, e(k - 1)$$

or, equivalently

$$u(k + 1) = 0.5u(k) + 56[e(k + 1) - 0.875e(k)]. \tag{8.27}$$

Note that Eq. (8.27) is quite similar to that obtained using Euler's method in Example 8.3, or Eq. (8.9).

◆

Matched Pole-zero (MPZ) Method

Another digitization method, called the **matched pole-zero** method, is found by extrapolating from the relationship between the s- and z-planes stated in Eq. (8.19). If we take the z-transform of a sampled function $x(k)$, then the poles of $X(z)$ are related to the poles of $X(s)$ according to the relation $z = e^{sT}$. The MPZ technique applies the relation $z = e^{sT}$ to the poles and zeros of a transfer function, even though, strictly speaking, this relation applies neither to transfer functions nor even to the zeros of a time sequence. Like all transfer-function digitization methods, the MPZ method is an approximation; here the approximation is motivated partly by the fact that $z = e^{sT}$ is the correct s-to-z transformation for the poles of the transform of a time sequence, and partly by the minimal amount of algebra required to determine the digitized transfer function if performing the calculations by hand.

Since physical systems often have more poles than zeros, it is useful to arbitrarily add zeros at $z = -1$, resulting in a $1 + z^{-1}$ term in $D(z)$. This causes an averaging of the current and past input values, as in Tustin's method. We select the low-frequency gain of $D(z)$ so that it equals that of $D(s)$.

Summary of the MPZ Method

1. Map poles and zeros according to the relation $z = e^{sT}$.
2. If the numerator is of lower order than the denominator, add powers of $1 + z^{-1}$ to the numerator until numerator and denominator are of equal order.
3. Set the DC or low-frequency gain of $D(z)$ equal to that of $D(s)$.

The MPZ approximation of

$$D(s) = K_c \frac{s + a}{s + b} \tag{8.28}$$

is

$$D(z) = K_d \frac{z - e^{-aT}}{z - e^{-bT}}. \tag{8.29}$$

where K_d is found by causing the DC gain of $D(z)$ to equal the DC gain of $D(s)$ using the continuous and discrete versions of the Final Value Theorem. The result is

$$K_c \frac{a}{b} = K_d \frac{1 - e^{-aT}}{1 - e^{-bT}},$$

or

$$K_d = K_c \frac{a}{b} \left(\frac{1 - e^{-bT}}{1 - e^{-aT}} \right).$$

For a $D(s)$ with a higher-order denominator, Step 2 in the method calls for adding the $1 + z^{-1}$ term. For example,

$$D(s) = K_c \frac{s + a}{s(s + b)} \Rightarrow D(z) = K_d \frac{(z + 1)(z - e^{-aT})}{(z - 1)(z - e^{-bT})}, \tag{8.30}$$

where, after dropping the poles at $s = 0$ and $z = 1$,

$$K_d = K_c \frac{a}{2b} \left(\frac{1 - e^{-bT}}{1 - e^{-aT}} \right).$$

In the digitization methods described so far, the same power of z appears in the numerator and denominator of $D(z)$. This implies that the difference equation output at time k will require a sample of the input at time k. For example, the $D(z)$ in Eq. (8.29) can be written

$$\frac{U(z)}{E(z)} = D(z) = K \frac{1 - \alpha z^{-1}}{1 - \beta z^{-1}}, \tag{8.31}$$

where $\alpha = e^{-aT}$ and $\beta = e^{-bT}$. By inspection we can see that Eq. (8.31) results in the difference equation

$$u(k + 1) = \beta u(k) + K[e(k + 1) - \alpha e(k)]. \tag{8.32}$$

It is impossible to sample $e(k + 1)$, compute $u(k + 1)$, and then output $u(k + 1)$ all in zero elapsed time; therefore, Eq. (8.32) is technically impossible to implement. However, if the equation is simple enough and/or the computer is fast enough, a slight computation delay between the $e(k + 1)$ sample and the $u(k + 1)$ output will have a negligible effect on the actual response of the system compared with that expected from the original design. A rule of thumb would

Design guideline for computation delay

be to keep the computation delay on the order of $1/10$ of T. The real-time code and hardware can be structured so that this delay is minimized by making sure that computations between read A/D and write D/A are minimized and that $u(k + 1)$ is sent to the ZOH immediately after its calculation.

Modified Matched Pole-zero (MMPZ) Method

The $D(z)$ in Eq. (8.30) would also result in $u(k)$ being dependent on $e(k)$, the input at the same time point. If the structure of the computer hardware prohibits this relation or if the computations are particularly lengthy, it may be desirable to derive a $D(z)$ that has one less power of z in the numerator than in the denominator; hence the computer output $u(k)$ would only require input from the previous time, that is, $e(k - 1)$. To do this, we simply modify Step 2 in the matched pole-zero procedure so that the numerator is of lower order than the denominator by 1. For example, if

$$D(s) = K_c \frac{s + a}{s(s + b)},$$

we skip Step 2 to get

$$D(z) = K_d \frac{z - e^{-aT}}{(z - 1)(z - e^{-bT})}, \tag{8.33}$$

where

$$K_d = K_c \frac{a}{b} \left(\frac{1 - e^{-bT}}{1 - e^{-aT}} \right).$$

To find the difference equation, we multiply the top and bottom of Eq. (8.33) by z^{-2} to find

$$D(z) = K_d \frac{z^{-1}(1 - e^{-aT}z^{-1})}{1 - z^{-1}(1 + e^{-bT}) + z^{-2}e^{-bT}}. \tag{8.34}$$

By inspecting Eq. (8.34) we can see that the difference equation is

$$u(k) = (1 + e^{-bT})u(k - 1) - e^{-bT}u(k - 2) + K_d[e(k - 1) - e^{-aT}e(k - 2)].$$

In this equation an entire sample period is available to perform the calculation and to output $u(k)$ because it depends only on $e(k - 1)$. A discrete analysis of this controller would therefore more accurately explain the behavior of the actual system. However, because this controller is using data that are one cycle old, it will typically not perform as well as the MPZ controller in terms of the deviations of the desired system output in the presence of random disturbances.

Comparison of Digital Approximation Methods

A numerical comparison of the magnitude of the frequency response for a first-order lag,

$$D(s) = \frac{5}{s + 5}$$

is made in Fig. 8.10 for the three approximation techniques at two different sample rates. The results of the $D(z)$ computations used in Fig. 8.10 are shown in Table 8.2.

Figure 8.10 shows that all the approximations are quite good at frequencies below about 1/4 the sample rate, or $\omega_s/4$. If $\omega_s/4$ is sufficiently larger than the filter break-point frequency—that is, if the sampling is fast enough—the break-point characteristics of the lag will be accurately reproduced. Tustin's technique and the two MPZ methods show a notch at $\omega_s/2$ because of their zero at $z = -1$ from the $z + 1$ term. Other than the large difference at $\omega_s/2$, which is typically outside the range of interest, the three methods have similar

FIGURE 8.10
A comparison of the frequency response of three discrete approximations

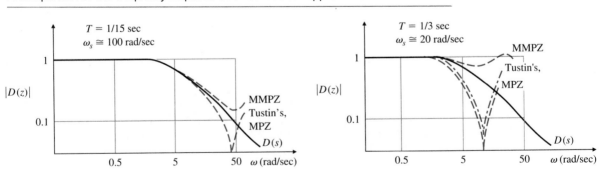

TABLE 8.2   Comparing Digital Approximations of $D(z)$ for $D(s) = 5/(s + 5)$

| | ω_s | |
Method	100 rad/sec	20 rad/sec
Matched pole-zero (MPZ)	$0.143 \dfrac{z + 1}{z - 0.715}$	$0.405 \dfrac{z + 1}{z - 0.189}$
Modified MPZ (MMPZ)	$0.285 \dfrac{1}{z - 0.715}$	$0.811 \dfrac{1}{z - 0.189}$
Tustin's	$0.143 \dfrac{z + 1}{z - 0.713}$	$0.454 \dfrac{z + 1}{z - 0.0914}$

accuracies. Since the MPZ techniques require less algebra than Tustin's method, they are preferred when the designer is performing the calculations by hand.

8.3.2   Design Example

To recapitulate, the process of designing by emulation consists of the following steps:

1. Assume that the system and compensation are continuous, and use the design methods discussed in Chapters 1 to 7 to meet all specifications.
2. Use a digitization procedure to approximate $D(s)$ with $D(z)$.
3. Implement $D(z)$ directly with a difference equation in the control computer.

In this section we follow this design process through with an example.

◆ **EXAMPLE 8.5**   *Emulation Design of a Digital Controller*

For a system with the plant transfer function

$$G(s) = \frac{1}{s^2},$$

design a digital controller to have a closed-loop natural frequency $\omega_n \cong 0.3$ rad/sec and a damping ratio $\zeta = 0.7$.

Solution.   The first step is to find the proper $D(s)$ for the system defined in Fig. 8.11. After some trial and error, we find the specifications can be met by the lead compensation

$$D(s) = K_c \frac{s + a}{s + b}, \tag{8.35}$$

FIGURE 8.11
Continuous-design definition for Example 8.5

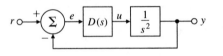

where

$$a = 0.2, \quad b = 2.0, \quad K_c = 0.81.$$

The root locus in Fig. 8.12 verifies the appropriateness of using Eq. (8.35).

To digitize this $D(s)$, we first need to select a sample rate. For a system with $\omega_n = 0.3$ rad/sec, the bandwidth will also be about 0.3 rad/sec, and a very safe sample rate would be faster than ω_n by a factor of 20. Thus

$$\omega_s = 0.3 \times 20 = 6 \text{ rad/sec.}$$

A sample rate of 6 rad/sec is about 1 Hz; therefore, the sample period should be $T = 1$ sec. The MPZ digitization of Eq. (8.35), given by Eq. (8.29), yields

$$D(z) = 0.389 \, \frac{z - 0.82}{z - 0.135}$$

$$= \frac{0.389 - 0.319z^{-1}}{1 - 0.135z^{-1}}. \tag{8.36}$$

Inspection of Eq. (8.36) gives us the difference equation

$$u(k) = 0.135u(k - 1) + 0.389e(k) - 0.319e(k - 1), \tag{8.37}$$

where

$$e(k) = r(k) - y(k),$$

FIGURE 8.12
s-plane locus with respect to K

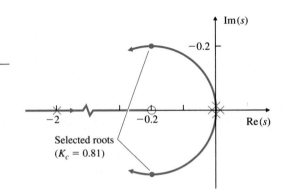

FIGURE 8.13

A digital control system for emulating Fig. 8.11

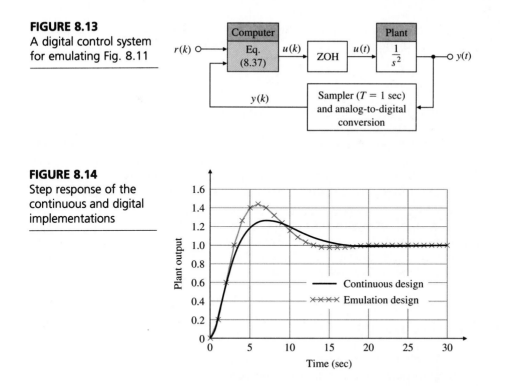

FIGURE 8.14

Step response of the continuous and digital implementations

and this completes the digital algorithm design. The complete digital system is shown in Fig. 8.13.

The last step in emulation design is to verify the design by implementing it on the computer. Figure 8.14 compares the step response of the digital system with the step response of the continuous compensation. Note that there is greater overshoot and a longer settling time in the digital system, which suggests a decrease in the damping. The delay shown in Fig. 8.2 is the cause of the reduced damping.

◆

8.3.3 Applicability Limits of the Emulation Design Method

If we performed an exact discrete analysis or a simulation of a system and determined the digitization for a wide range of sample rates, the system would typically be unstable for rates slower than approximately $5\omega_n$, and the damping would be substantially degraded for rates slower than $10\omega_n$. At sample rates on the order of $20\omega_n$ (or 20 times the bandwidth for more complex systems), design by emulation yields reasonable results and can be used with confidence for sample rates of 30 times the bandwidth or higher.

As shown by Fig. 8.2, the errors come about because the technique ignores the lagging effect of the ZOH which, on the average, is $T/2$ sec. A method to account for this is to approximate the $T/2$ delay with Eq. (5.54) by including a transfer function approximation for the ZOH:

ZOH transfer function

$$G_{ZOH}(s) = \frac{2/T}{s + 2/T}.$$ (8.38)

We could therefore improve on our original discrete design by inserting Eq. (8.38) into the original plant model and finding the $D(s)$ that yields a satisfactory response. However, one of the advantages of using the emulation design method is that the sample rate need not be selected until *after* the basic feedback design is completed. Use of Eq. (8.38) eliminates this advantage, although it does partially alleviate the approximate nature of the method.

Rather than using an approximation for the delay and a continuous analysis, a better approach is to analyze the entire system using an exact discrete analysis. If a discrete analysis shows an unacceptable degradation of performance due to the sampling, the design can then be refined using exact discrete methods. We cover this approach next.

8.4   Discrete Design

It is possible to obtain an exact discrete model that relates the samples of the continuous plant $y(k)$ to the input control sequence $u(k)$. This plant model can be used as part of a discrete model of the feedback system including the compensation $D(z)$. Analysis and design using this discrete model is called **discrete design** or, alternatively, **direct digital design**. The following sections will describe how to find the discrete plant model (Section 8.4.1), what the feedback compensation looks like when designing with a discrete model (Sections 8.4.2 and 8.4.3), and how the design process is carried out (Section 8.4.4).

8.4.1   Analysis Tools

The first step in performing a discrete analysis of a system with some discrete elements is to find the discrete transfer function of the continuous portion. For a system similar to that shown in Fig. 8.1(b), we wish to find the transfer function between $u(kT)$ and $y(kT)$. Unlike the cases discussed in the previous sections, there is an *exact* discrete equivalent for this system because the ZOH precisely describes what happens between samples of $u(kT)$, and the output $y(kT)$ is dependent only on the input at the sample times $u(kT)$.

The exact discrete equivalent

For a plant described by $G(s)$ and preceded by a ZOH, the discrete transfer function is

$$G(z) = (1 - z^{-1})\mathscr{Z}\left\{\frac{G(s)}{s}\right\},$$ (8.39)

FIGURE 8.15

Comparison of a (a) mixed control system and (b) its pure discrete equivalent

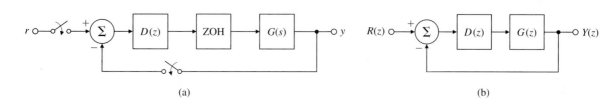

(a)   (b)

where $\mathscr{Z}\{F(s)\}$ is the z-transform of the sampled time series whose Laplace transform is the expression for $F(s)$, given on the same line in Table 8.1. Equation (8.39) has the term $G(s)/s$ because the control comes in as a step input during each sample period. The term $1 - z^{-1}$ reflects the fact that a one-sample duration step can be thought of as an infinite duration step followed by a negative step one cycle delayed. For a more complete derivation, see Franklin *et al.* (1990). Equation 8.39 allows us to replace the mixed (continuous and discrete) system shown in Fig. 8.15(a) with the equivalent pure discrete system shown in Fig. 8.15(b).

The analysis and design of discrete systems is very similar to the analysis and design of continuous systems; in fact, all the same rules apply. The closed-loop transfer function of Fig. 8.15(b) is obtained using the same rules of block-diagram reduction; that is,

$$\frac{Y(z)}{R(z)} = \frac{D(z)G(z)}{1 + D(z)G(z)}. \tag{8.40}$$

To find the characteristic behavior of the closed-loop system, we need to find the factors in the denominator of Eq. (8.40), that is, the roots of the discrete characteristic equation

$$1 + D(z)G(z) = 0.$$

The root-locus techniques used in continuous systems to find roots of a polynomial in s apply equally well and without modification to the polynomial in z; however, the interpretation of the results is quite different, as we saw in Fig. 8.6. A major difference is that the stability boundary is now the unit circle instead of the imaginary axis.

◆ **EXAMPLE 8.6** *Discrete Root Locus*

For the case where $G(s)$ in Fig. 8.15(a) is

$$G(s) = \frac{a}{s + a}$$

FIGURE 8.16
Root loci for (a) the
z-plane and (b) the
s-plane

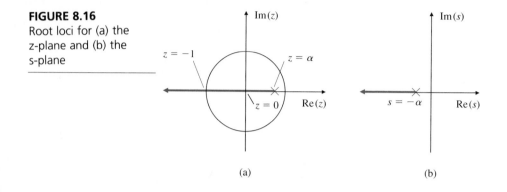

(a)

(b)

and $D(z) = K$, draw the root locus with respect to K, and compare your results with a root locus of a continuous version of the system. Discuss the implications of your loci.

Solution. It follows from Eq. (8.39) that

$$G(z) = (1 - z^{-1})\mathscr{Z}\left[\frac{a}{s(s + a)}\right]$$

$$= (1 - z^{-1})\left[\frac{(1 - e^{-aT})z^{-1}}{(1 - z^{-1})(1 - e^{-aT}z^{-1})}\right]$$

$$= \frac{1 - \alpha}{z - \alpha}$$

where

$$\alpha = e^{-aT}.$$

To analyze the performance of the closed-loop system, standard root-locus rules apply. The result is shown in Fig. 8.16(a) for the discrete case and in Fig. 8.16(b) for the continuous case. In contrast to the continuous case, in which the system remains stable for all values of K, in the discrete case the system becomes oscillatory with decreasing damping ratio as z goes from 0 to -1 and eventually becomes unstable. This instability is due to the lagging effect of the ZOH, which is properly accounted for in the discrete analysis.

◆

8.4.2 Feedback Properties

In continuous systems we typically start the design process by using the following basic design elements: proportional, derivative, or integral control laws or some combination of these, sometimes with a lag included. The same ideas can be used in discrete design. Alternatively, the $D(z)$ resulting from the digitization of a continuously designed $D(s)$ will produce these basic design elements, which will then be used as a starting point in a discrete design. The discrete control laws are as follows.

Discrete control laws

Proportional

$$u(k) = Ke(k) \Rightarrow D(z) = K. \tag{8.41}$$

Derivative

$$u(k) = K_p T_D[e(k) - e(k - 1)], \tag{8.42}$$

for which the transfer function is

$$D(z) = K_p T_D(1 - z^{-1}) = K_p T_D \frac{z - 1}{z}. \tag{8.43}$$

Integral

$$u(k) = u(k - 1) + \frac{K_p}{T_I} e(k), \tag{8.44}$$

for which the transfer function is

$$D(z) = \frac{K_p}{T_I} \left(\frac{1}{1 - z^{-1}} \right) = \frac{K_p}{T_I} \left(\frac{z}{z - 1} \right). \tag{8.45}$$

Lead Compensation   The examples in Section 8.3 showed that a continuous lead compensation leads to difference equations of the form

$$u(k + 1) = \beta u(k) + K[e(k + 1) - \alpha e(k)], \tag{8.46}$$

for which the transfer function is

$$D(z) = K \frac{1 - \alpha z^{-1}}{1 - \beta z^{-1}}. \tag{8.47}$$

8.4.3   Discrete Design Example

Digital control design consists of using the basic feedback elements of Eqs. (8.41) to (8.47) and iterating on the unknown parameters until all specifications are met.

◆ **EXAMPLE 8.7**   *Direct Discrete Design of a Digital Controller*

Design a digital controller to meet the same specifications as in Example 8.5 using discrete design.

Solution.   The discrete model of the $1/s^2$ plant, preceded by a ZOH, is found through

FIGURE 8.17
z-plane root locus for a
$1/s^2$ plant with
proportional feedback

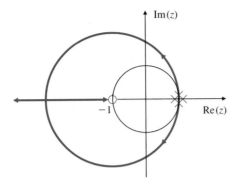

Eq. (8.39) to be

$$G(z) = \frac{T^2}{2}\left[\frac{z+1}{(z-1)^2}\right],$$

which, with $T = 1$ sec, becomes

$$G(z) = \frac{1}{2}\left[\frac{z+1}{(z-1)^2}\right].$$

Proportional feedback in the continuous case yields pure oscillatory motion, so in the discrete case we should expect even worse results. The root locus in Fig. 8.17 verifies this. For very low values of K (where the locus represents roots at very low frequencies compared to the sample rate) the locus is tangent to the unit circle ($\zeta \cong 0$ indicating pure oscillatory motion), thus matching the proportional continuous design.

For higher values of K, Fig. 8.17 shows that the locus diverges into the unstable region because of the effect of the ZOH and sampling. To compensate for this, let us add a derivative term to the proportional term so that the control law is

$$U(z) = K[1 + T_D(1 - z^{-1})]E(z), \tag{8.48}$$

which yields compensation of the form

$$D(z) = K\,\frac{z - \alpha}{z} \tag{8.49}$$

where the new K and α replace the K and T_D in Eq. (8.48). Now the task is to find the values of α and K that yield good performance. The specifications for the design are that $\omega_n = 0.3\,\text{rad/sec}$ and $\zeta = 0.7$. Figure 8.6 indicates that this s-plane root location maps into a desired z-plane location of

$$z = 0.78 \pm 0.18j.$$

Figure 8.18 is the locus with respect to K for $\alpha = 0.85$ and shows that the desired values of z lie on the locus. The value of the gain when the locus passes through $z = 0.78 \pm 0.18j$ is $K = 0.374$. Equation (8.49) now becomes

$$D(z) = 0.374\,\frac{z - 0.85}{z}. \tag{8.50}$$

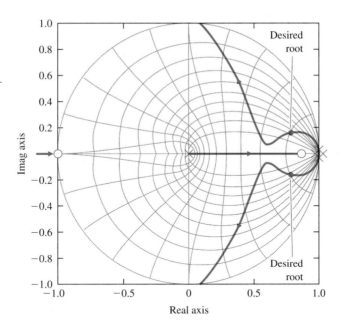

FIGURE 8.18

z-plane locus for the $1/s^2$ plant with $D(z) = K(z - 0.85)/z$

Normally, it is not particularly advantageous to match specific z-plane root locations; rather it is only necessary to pick K and α (or T_D) to obtain acceptable z-plane roots, a much easier task. In this example, we want to match a specific location only so that we can compare the result with the design in Example 8.5.

The control law that results is

$$U(z) = 0.374(1 - 0.85z^{-1})E(z),$$

or

$$u(k) = 0.374e(k) - 0.318e(k - 1),\tag{8.51}$$

which is similar to the control equation (8.37) obtained previously.

◆

Comparison of discrete and continuous controllers

The controller in Eq. (8.51) basically differs from the continuously designed controller (Eq. 8.37) only in the absence of the $u(k - 1)$ term. The $u(k - 1)$ term in Eq. (8.37) results from the lag term $s + b$ in the compensation (Eq. 8.35). The lag term is typically included in analog controllers both because it supplies noise attenuation and because pure analog differentiators are difficult to build. Some equivalent lag in discrete design naturally appears as a pole at $z = 0$ (see Fig. 8.18) and represents the one sample delay in computing the derivative by a first difference. For more noise attenuation we could move the pole to the right of $z = 0$, thus resulting in less derivative action and more smoothing, the same tradeoff that exists in continuous control design.

8.4.4   Discrete Analysis of Designs

Any digital controller, whether designed by emulation or directly in the z-plane, can be analyzed using discrete analysis, which consists of the following steps:

1. Find the discrete model of the plant and ZOH using Eq. (8.39).
2. Form the feedback system including $D(z)$.
3. Analyze the resulting discrete system.

We can determine the roots of the system using a root locus as described in Section 8.4.3, or we can determine the time history (at the sample instants) of the discrete system. Time histories are tedious to do by hand, but CACSD software alleviates that difficulty.

◆ **EXAMPLE 8.8**   *Damping and Step Response in Digital versus Continuous Design*

Use discrete analysis to determine the equivalent *s*-plane damping and the step responses of the digital designs in Examples 8.5 and 8.7, and compare your results with the damping and step response of the continuous case in Example 8.5.

Solution.   The MATLAB statements to evaluate the damping and step response of the continuous case in Example 8.5 are

> numGs = 1
>
> denGs = [1 0 0]
>
> numDs = .81 * [1 .2]
>
> denDs = [1 2]
>
> [numCL,denCL] = feedback(conv(numGs,numDs),conv(denGs,denDs),1,1)
>
> step(numCL,denCL)
>
> damp(denCL)

To analyze the digital control cases, the model of the plant preceded by the ZOH is found using the statements

> T = 1
>
> [numGz,denGz] = c2dm(numGs,denGs,T,'zoh')

Analysis of the digital control designed by emulation (Eq. 8.37) in Example 8.5 is performed by the statements

> numDz = [.389 −.319]
>
> denDz = [1 −.135]
>
> [numCLz,denCLz] = feedback(conv(numGz,numDz),conv(denGz,denDz),1,1)

FIGURE 8.19
Step response of the continuous and digital systems in Examples 8.5 and 8.7

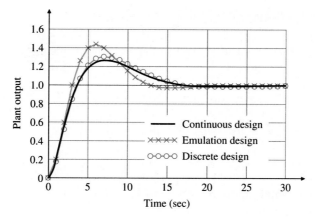

dstep(numCLz,denCLz)
ddamp(denCLz,T)

Likewise, the discrete design from Eq. (8.50) can be analyzed by the same sequence.

The resulting step responses are shown in Fig. 8.19. The calculated damping ζ and complex root natural frequencies ω_n of the closed-loop systems are

Continuous case: $\zeta = 0.705$, $\omega_n = 0.324$;
Emulation design: $\zeta = 0.645$, $\omega_n = 0.441$;
Discrete design: $\xi = 0.733$, $\omega_n = 0.306$.

The figure shows increased overshoot for the emulation method that occurred because of the decreased damping of that case. Very little increased overshoot occurred in the discrete design because that compensation was adjusted specifically so that the equivalent s-plane damping of the discrete system was approximately at the desired damping value of $\zeta = 0.7$.

◆

Although the analysis showed some difference between the performance of the digital controllers designed by the two methods, neither the performance nor the control equations (Eqs. 8.37 and 8.51) are very different. This similarity results because the sample rate is fairly fast compared with ω_n; that is, $\omega_s \cong 20\omega_n$. If we were to decrease the sample rate, the numerical values in the compensations would become increasingly different and the performance would degrade considerably for the emulation case.

As a general rule, discrete design should be used if the sampling frequency is slower than $10\omega_n$. At the very least, an emulation design with slow sampling ($\omega_s < 10\omega_n$) should be verified by a discrete analysis or by simulation as described in Section 3.6, and the compensation adjusted if needed. A simulation of a digital control system is a good idea in any case. If it properly accounts for all delays and possibly asynchronous behavior of different modules, it may expose instabilities that are impossible to detect using continuous or discrete linear analysis.

8.5   State-space Design Methods

We have seen in previous chapters that a linear, constant-coefficient continuous system can be represented by a set of first-order matrix differential equations of the form

$$\dot{\mathbf{x}} = \mathbf{F}\mathbf{x} + \mathbf{G}u, \tag{8.52}$$

where u is the control input to the system. The output equation can be expressed as

$$y = \mathbf{H}\mathbf{x} + Ju. \tag{8.53}$$

The solution to these equations, discussed in Section 3.6.2, resulted in Eq. (3.59), which we repeat here:

$$\mathbf{x}(t) = e^{\mathbf{F}(t-t_0)}\mathbf{x}(t_0) + \int_{t_0}^{t} e^{\mathbf{F}(t-\tau)}\mathbf{G}u(\tau)\,d\tau. \tag{8.54}$$

It is possible to use Eq. (8.54) to obtain a discrete state-space representation of the system. Because the solution over one sample period results in a difference equation, we can alter the notation a bit (letting $t = kT + T$ and $t_0 = kT$) to arrive at a particularly useful version of Eq. (8.54):

$$\mathbf{x}(kT + T) = e^{\mathbf{F}T}\mathbf{x}(kT) + \int_{kT}^{kT+T} e^{\mathbf{F}(kT+T-\tau)}\mathbf{G}u(\tau)\,d\tau. \tag{8.55}$$

This result is not dependent on the type of hold, since u is specified in terms of its continuous time history $u(\tau)$ over the sample interval. To find the discrete model of a continuous system where the input $u(t)$ is the output of a ZOH, we let $u(\tau)$ be a constant throughout the sample interval; that is,

$$u(\tau) = u(kT), \qquad kT \leqslant \tau < kT + T.$$

To facilitate the solution of Eq. (8.55) for a ZOH, we let

$$\eta = kT + T - \tau,$$

which converts Eq. (8.55) to

$$\mathbf{x}(kT + T) = e^{\mathbf{F}T}\mathbf{x}(kT) + \left(\int_{0}^{T} e^{\mathbf{F}\eta}\,d\eta \right) \mathbf{G}u(kT).$$

If we let

$$\mathbf{\Phi} = e^{\mathbf{F}T}$$

and

$$\mathbf{\Gamma} = \left(\int_{0}^{T} e^{\mathbf{F}\eta}\,d\eta \right) \mathbf{G}, \tag{8.56}$$

Eqs. (8.55) and (8.53) reduce to difference equations in standard form:

Difference equations in standard form

$$\mathbf{x}(k+1) = \mathbf{\Phi x}(k) + \mathbf{\Gamma} u(k), \tag{8.57a}$$

$$y(k) = \mathbf{Hx}(k) + Ju(k), \tag{8.57b}$$

where $\mathbf{x}(k+1)$ is a shorthand notation for $\mathbf{x}(kT+T)$, $\mathbf{x}(k)$ for $\mathbf{x}(kT)$, and $u(k)$ for $u(kT)$. The series expansion

$$\mathbf{\Phi} = e^{\mathbf{F}T} = \mathbf{I} + \mathbf{F}T + \frac{\mathbf{F}^2 T^2}{2!} + \frac{\mathbf{F}^3 T^3}{3!} + \cdots$$

can also be written

$$\mathbf{\Phi} = \mathbf{I} + \mathbf{F}T\mathbf{\Psi}, \tag{8.58}$$

where

$$\mathbf{\Psi} = \mathbf{I} + \frac{\mathbf{F}T}{2!} + \frac{\mathbf{F}^2 T^2}{3!} + \cdots.$$

The $\mathbf{\Gamma}$ integral in Eq. (8.56) can be evaluated term by term to give

$$\mathbf{\Gamma} = \sum_{k=0}^{\infty} \frac{\mathbf{F}^k T^{k+1}}{(k+1)!}\mathbf{G}$$

$$= \sum_{k=0}^{\infty} \frac{\mathbf{F}^k T^k}{(k+1)!} T\mathbf{G}$$

$$= \mathbf{\Psi} T\mathbf{G}. \tag{8.59}$$

We evaluate $\mathbf{\Psi}$ by a series in the form

$$\mathbf{\Psi} \cong \mathbf{I} + \frac{\mathbf{F}T}{2}\left\{\mathbf{I} + \frac{\mathbf{F}T}{3}\left[\mathbf{I} + \cdots \frac{\mathbf{F}T}{N-1}\left(\mathbf{I} + \frac{\mathbf{F}T}{N}\right)\right]\right\},$$

which has better numerical properties than the direct series. We then find $\mathbf{\Gamma}$ from Eq. (8.59) and $\mathbf{\Phi}$ from Eq. (8.58). For a discussion of various methods of numerical determination of $\mathbf{\Phi}$ and $\mathbf{\Gamma}$, see Franklin *et al.* (1990) and Moler and van Loan (1978). The evaluation of the $\mathbf{\Phi}$ and $\mathbf{\Gamma}$ matrices in practice is carried out by the c2d or c2dm functions in MATLAB.

MATLAB c2d

To compare this method of representing the plant with the discrete transfer function, we can take the z-transform of Eqs. (8.57) with $J = 0$ to obtain

$$(z\mathbf{I} - \mathbf{\Phi})\mathbf{X}(z) = \mathbf{\Gamma} U(z), \tag{8.60a}$$

$$Y(z) = \mathbf{HX}(z). \tag{8.60b}$$

Therefore,

$$\frac{Y(z)}{U(z)} = G(z) = \mathbf{H}(z\mathbf{I} - \mathbf{\Phi})^{-1}\mathbf{\Gamma}. \tag{8.61}$$

◆ **EXAMPLE 8.9**   *Discrete State-space Representation of $1/s^2$ Plant*

Use the relation in this section to verify that the discrete model of the $1/s^2$ plant preceded by a ZOH is that given in the solution to Example 8.7.

Solution.   The $\mathbf{\Phi}$ and $\mathbf{\Gamma}$ matrices can be calculated using Eqs. (8.58) and (8.59). Example 2.7 (with $I = 1$) showed that the values for $\mathbf{F}$ and $\mathbf{G}$ are

$$\mathbf{F} = \begin{bmatrix} 0 & 1 \\ 0 & 0 \end{bmatrix}, \qquad \mathbf{G} = \begin{bmatrix} 0 \\ 1 \end{bmatrix}.$$

Since $\mathbf{F}^2 = \mathbf{0}$ in this case, we have

$$\mathbf{\Phi} = \mathbf{I} + \mathbf{F}T + \frac{\mathbf{F}^2 T^2}{2!} + \cdots$$

$$= \begin{bmatrix} 1 & 0 \\ 0 & 1 \end{bmatrix} + \begin{bmatrix} 0 & 1 \\ 0 & 0 \end{bmatrix} T = \begin{bmatrix} 1 & T \\ 0 & 1 \end{bmatrix},$$

$$\mathbf{\Gamma} = \left(\mathbf{I} + \mathbf{F}\frac{T}{2!} \right) T\mathbf{G}$$

$$= \left(\begin{bmatrix} T & 0 \\ 0 & T \end{bmatrix} + \begin{bmatrix} 0 & 1 \\ 0 & 0 \end{bmatrix} \frac{T^2}{2} \right) \begin{bmatrix} 0 \\ 1 \end{bmatrix} = \begin{bmatrix} T^2/2 \\ T \end{bmatrix}.$$

Hence, using Eq. (8.61), we obtain

$$G(z) = \frac{Y(z)}{U(z)} = \begin{bmatrix} 1 & 0 \end{bmatrix} \left(z \begin{bmatrix} 1 & 0 \\ 0 & 1 \end{bmatrix} - \begin{bmatrix} 1 & T \\ 0 & 1 \end{bmatrix} \right)^{-1} \begin{bmatrix} T^2/2 \\ T \end{bmatrix}$$

$$= \frac{T^2}{2} \left[\frac{z + 1}{(z - 1)^2} \right]. \tag{8.62}$$

This is the same result we obtained using Eq. (8.39) and the z-transform tables in Example 8.7.

Note that to compute Y/U we find that the denominator of Eq. (8.62) is $\det(z\mathbf{I} - \mathbf{\Phi})$, which was created by the matrix inverse in Eq. (8.61). This determinant is the characteristic polynomial of the transfer function, and the zeros of the determinant are the poles of the plant. We have two poles at $z = 1$ in this case, corresponding to two integrations in this plant's equations of motion.

◆

We can further explore the question of poles and zeros and the state-space description by considering again the transform formulas (Eqs. 8.60). One way to interpret transfer-function poles from the perspective of the corresponding difference equation is that a pole is a value of z such that the equation has a nontrivial solution when the forcing input is zero. From Eq. (8.60a), this interpretation implies that the linear equations

$$(z\mathbf{I} - \mathbf{\Phi})\mathbf{X}(z) \doteq \mathbf{0}$$

have a nontrivial solution. From matrix algebra the well-known requirement for a nontrivial solution is that $\det(z\mathbf{I} - \mathbf{\Phi}) = 0$. Using the system in Example 8.9, we get

System poles from the state-space description

$$\det(z\mathbf{I} - \mathbf{\Phi}) = \det\left(\begin{bmatrix} z & 0 \\ 0 & z \end{bmatrix} - \begin{bmatrix} 1 & T \\ 0 & 1 \end{bmatrix}\right)$$

$$= \det\begin{bmatrix} z - 1 & -T \\ 0 & z - 1 \end{bmatrix}$$

$$= (z - 1)^2 = 0,$$

which is the characteristic equation, as we have seen.

Along the same line of reasoning, a system zero is a value of z such that the system output is zero even with a nonzero state-and-input combination. Thus, if we are able to find a nontrivial solution for $\mathbf{X}(z_0)$ and $U(z_0)$ such that $Y(z_0)$ is identically zero, then z_0 is a zero of the system. In combining Eqs. (8.60a) and (8.60b), we must satisfy the requirement that

$$\begin{bmatrix} z\mathbf{I} - \mathbf{\Phi} & -\mathbf{\Gamma} \\ \mathbf{H} & 0 \end{bmatrix}\begin{bmatrix} \mathbf{X}(z) \\ U(z) \end{bmatrix} = \mathbf{0}.$$

Once more the condition for the existence of nontrivial solutions is that the determinant of the square coefficient system matrix be zero. For Example 8.9, the calculation is

$$\det\begin{bmatrix} z - 1 & -T & -T^2/2 \\ 0 & z - 1 & -T \\ 1 & 0 & 0 \end{bmatrix} = \det\begin{bmatrix} -T & -T^2/2 \\ z - 1 & -T \end{bmatrix}$$

$$= T^2 + \frac{T^2}{2}(z - 1)$$

$$= \frac{T^2}{2}z + \frac{T^2}{2}$$

$$= \frac{T^2}{2}(z + 1).$$

Thus we have a single zero at $z = -1$, as we have seen from the transfer function.

Much of the algebra for discrete state-space control design is the same as for the continuous time case discussed in Chapter 7. The poles of a discrete system can be moved to desirable locations by linear state-variable feedback:

$$u = -\mathbf{K}\mathbf{x}$$

such that

$$\det(z\mathbf{I} - \mathbf{\Phi} + \mathbf{\Gamma}\mathbf{K}) = \alpha_c(z), \tag{8.63}$$

provided the system is controllable. The system is controllable if the controllability matrix

$$\mathscr{C} = [\boldsymbol{\Gamma} \quad \boldsymbol{\Phi}\boldsymbol{\Gamma} \quad \boldsymbol{\Phi}^2\boldsymbol{\Gamma} \quad \cdots \quad \boldsymbol{\Phi}^{n-1}\boldsymbol{\Gamma}]$$

is full-rank.

A discrete full-order estimator has the form

$$\bar{\mathbf{x}}(k + 1) = \boldsymbol{\Phi}\bar{\mathbf{x}}(k) + \boldsymbol{\Gamma}u(k) + \mathbf{L}[y(k) - \mathbf{H}\bar{\mathbf{x}}(k)],$$

where $\bar{\mathbf{x}}$ is the state estimate. The error equation

$$\tilde{\mathbf{x}}(k + 1) = (\boldsymbol{\Phi} - \mathbf{L}\mathbf{H})\tilde{\mathbf{x}}(k)$$

can be given arbitrary dynamics $\alpha_e(z)$, provided that the system is observable, which requires that the observability matrix

$$\mathcal{O} = \begin{bmatrix} \mathbf{H} \\ \mathbf{H}\boldsymbol{\Phi} \\ \mathbf{H}\boldsymbol{\Phi}^2 \\ \vdots \\ \mathbf{H}\boldsymbol{\Phi}^{n-1} \end{bmatrix}$$

be full-rank.

As was true for the continuous-time case, if the open-loop transfer function is

$$G(z) = \frac{Y(z)}{U(z)} = \frac{b(z)}{a(z)},$$

then a state-space compensator can be designed such that

$$\frac{Y(z)}{R(z)} = \frac{K_s\gamma(z)b(z)}{\alpha_c(z)\alpha_e(z)},$$

where r is the reference input. The polynomials $\alpha_c(z)$ and $\alpha_e(z)$ are selected by the designer using exactly the same methods discussed in Chapter 7 for continuous systems. $\alpha_c(z)$ results in a control gain $\mathbf{K}$ such that $\det(z\mathbf{I} - \boldsymbol{\Phi} + \boldsymbol{\Gamma}\mathbf{K}) = \alpha_c(z)$ and $\alpha_e(z)$ results in an estimator gain $\mathbf{L}$ such that $\det(z\mathbf{I} - \boldsymbol{\Phi} + \mathbf{L}\mathbf{H}) = \alpha_e(z)$. If the estimator is structured according to Fig. 7.37(a), the system zeros $\gamma(z)$ will be identical to the estimator poles $\alpha_e(z)$, thus removing the estimator response from the closed-loop system response. However, if desired, we can arbitrarily select the polynomial $\gamma(z)$ by providing suitable feedforward from the reference input. Refer to Franklin *et al.* (1990) for details.

◆ **EXAMPLE 8.10**   *State-space Design of a Digital Controller*

Design a digital controller for a $1/s^2$ plant to meet the specifications given in Example 8.5. Use state-space design methods including use of an estimator, and structure the reference input in two ways: (a) Use the error command of case 2 in Section 7.7.1, and (b) use the state command described in Section 7.3.2 and case 1 in Section 7.7.1.

Solution. We find the state-space model of the $1/s^2$ plant preceded by a ZOH using the MATLAB statements

> F = [0 1;0 0]
>
> G = [0; 1]
>
> H = [1 0]
>
> J = 0
>
> T = 1
>
> [Phi,Gam] = c2d(F, G, T)

Using discrete analysis for Example 8.7, we find that the desired z-plane roots are at $z = 0.78 \pm 0.18j$. Solving the discrete pole-placement problem involves placing the eigenvalues of $\mathbf{\Phi} - \mathbf{\Gamma K}$ as indicated by Eq. (8.63). Likewise, the solution of the continuous pole-placement problem involves placing the eigenvalues of $\mathbf{F} - \mathbf{GK}$ as indicated by Eq. (7.60). Since these two tasks are identical, we use the same function in MATLAB for the continuous and discrete cases. Therefore, the control feedback matrix $\mathbf{K}$ is found by

> j = sqrt(−1)
>
> pc = [0.78 + 0.18∗j; 0.78 − 0.18∗j]
>
> K = acker(Phi,Gam,pc)

which yields,

$$\mathbf{K} = [0.0808 \quad 0.3996].$$

To ensure that the estimator roots are substantially faster than the control roots (so that the estimator roots will have little effect on the output), we choose them to be at $z = 0.2 \pm 0.2j$. Therefore, the estimator feedback matrix $\mathbf{L}$ is found by

> pe = [0.2 + 0.2∗j; 0.2 − 0.2∗j]
>
> L = acker(Phi′,H′,pe)′

which yields

$$\mathbf{L} = \begin{bmatrix} 1.6 \\ 0.68 \end{bmatrix}.$$

The equations of the compensation for $r = 0$ (regulation to $\mathbf{x}^T = [0 \ 0]$) are then

$$\bar{\mathbf{x}}(k + 1) = \mathbf{\Phi}\bar{\mathbf{x}}(k) + \mathbf{\Gamma}u(k) + \mathbf{L}[y(k) - \mathbf{H}\bar{\mathbf{x}}(k)], \tag{8.64a}$$

$$u(k) = -\mathbf{K}\bar{\mathbf{x}}(k). \tag{8.64b}$$

a) For the error command structure where the compensator is placed in the feedforward path, as shown in Fig. 7.37(b), $y(k)$ from Eq. (8.64a) is replaced with $y(k) - r$, so the state description of the plant plus the estimator (a fourth-order system whose

FIGURE 8.20
Step response of Example 8.10

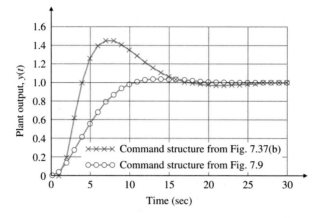

state vector is $[\mathbf{x} \ \bar{\mathbf{x}}]^T$) is

A = [Phi − Gam∗K; L∗H   Phi − Gam∗K − L∗H]

B = [0; 0; −L]

C = [1 0 0 0]

D = 0

dstep(A,B,C,D).

The resulting step response in Fig. 8.20 shows a response similar to that of the previous step responses in Fig. 8.19.

b) For the state command structure described in Section 7.3.2, we wish to command the position element of the state vector so that

$$\mathbf{N}_x = \begin{bmatrix} 1 \\ 0 \end{bmatrix},$$

and the $1/s^2$ plant requires no steady control input for a constant output y, therefore $N_u = 0$. To analyze a system with this command structure, we need to modify matrix B from the MATLAB statement above to properly introduce the reference input r according to Fig. 7.9. The MATLAB statement

B = [Gam∗K∗Nx; Gam∗K∗Nx]

channels r into both the plant and estimator equally, thus not exciting the estimator dynamics. The resulting step response in Fig. 8.20 shows a substantial reduction in the overshoot with this structure. In fact, the overshoot is now about 5%, which is expected for a second-order system with $\zeta \cong 0.7$. The previous designs all had considerably greater overshoot because of the effect of the extra zero and pole. ◆

8.6 Hardware Characteristics

A digital control system includes several unique components not found in continuous control systems: an analog-to-digital converter is a device to sample the continuous signal voltage from the sensor and to convert that signal to a digital word; a digital-to-analog converter is a device to convert the digital word from the computer to an analog voltage, an analog prefilter is a device designed to reduce the effects of aliasing, and the computer is the device where the compensation $D(z)$ is programmed and the calculations are carried out. This section provides a brief description of each of these.

8.6.1 Analog-to-Digital (A/D) Converters

As discussed in Section 8.1, A/D converters are devices that convert a voltage level from a sensor to a digital word usable by the computer. At the most basic level, all digital words are binary numbers consisting of many bits that are set to either 1 or 0. Therefore, the task of the A/D converter at each sample time is to convert a voltage level to the correct bit pattern and often to hold that pattern until the next sample time.

Common A/D conversion schemes

Of the many A/D conversion techniques that exist, the most common are based on counting schemes or a successive-approximation technique. In counting methods the input voltage may be converted to a train of pulses whose frequency is proportional to the voltage level. The pulses are then counted over a fixed period using a binary counter, thus resulting in a binary representation of the voltage level. A variation on this scheme is to start the count simultaneously with a voltage that is linear in time and to stop the count when the voltage reaches the magnitude of the input voltage to be converted.

The successive-approximation technique tends to be much faster than the counting methods. It is based on successively comparing the input voltage to reference levels representing the various bits in the digital word. The input voltage is first compared with a reference value that is half the maximum. If the input voltage is greater, the most significant bit is set, and the signal is then compared with a reference level that is 3/4 the maximum to determine the next bit, and so on. One clock cycle is required to set each bit, so an n-bit converter would require n cycles. At the same clock rate a counter-based converter might require as many as 2^n cycles, which would usually be much slower.

With either technique, the greater the number of bits, the longer it will take to perform the conversion. Not surprisingly, the price of A/D converters goes up with both speed and bit size. For example, in 1993 an 8-bit (resolution of 0.4%)converter with the minimal performance capability of a 100-μsec conversion time sold for a modest \$3, while faster models with a 50-nsec conversion time sold for approximately \$40.

If more than one channel of data needs to be sampled and converted to digital words, it is usually accomplished by use of a multiplexer rather than by multiple A/D converters. The multiplexer sequentially connects the converter into the channel being sampled.

8.6.2   Digital-to-Analog (D/A) Converters

D/A converters, as mentioned in Section 8.1, are used to convert the digital words from the computer to a voltage level for driving actuators or perhaps a recording device such as an oscilloscope or strip-chart recorder. The basic idea behind their operation is that the binary bits cause switches (electronic gates) to open or close, thus routing the electric current through an appropriate network of resistors to generate the correct voltage level. Since no counting or iteration is required for such converters, they tend to be much faster than A/D converters. In fact, A/D converters that use the successive-approximation method of conversion include D/A converters as components.

8.6.3   Analog Prefilters

An **analog prefilter** is often placed between the sensor and the A/D converter. Its function is to reduce the higher-frequency noise components in the analog signal in order to prevent aliasing, that is, having the noise be modulated to a lower frequency by the sampling process.

Analog prefilters reduce aliasing

An example of aliasing is shown in Fig. 8.21, where a 60-Hz oscillatory signal is being sampled at 50 Hz. The figure shows the result from the samples as a 10-Hz signal and also shows the mechanism by which the frequency of the signal is aliased from 60 to 10 Hz. Aliasing will occur any time the sample rate is not at least twice as fast as any of the frequencies in the signal being sampled. Therefore, to prevent aliasing of a 60-Hz signal, the sample rate would have to be faster than 120 Hz, clearly much higher than the 50-Hz rate in the figure.

Nyquist–Shannon sampling theorem

Aliasing is one of the consequences of the **sampling theorem of Nyquist and Shannon**. Their theorem basically states that, for the signal to be accurately reconstructed from the samples, it must have no frequency component greater than half the sample rate ($\omega_s/2$). Another consequence of their theorem is that

FIGURE 8.21
An example of aliasing

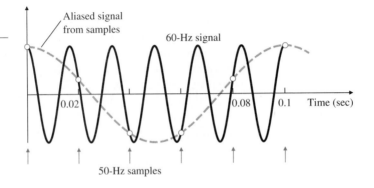

the highest frequency that can be unambiguously represented by discrete samples is the Nyquist rate of $\omega_s/2$, an idea we discussed in Section 8.2.3.

The consequence of aliasing on a digital control system can be substantial. In a continuous system, noise components with a frequency much higher than the control-system bandwidth normally have a small effect because the system will not respond at the high frequency. However, in a digital system, the frequency of the noise can potentially be aliased down in frequency to the vicinity of the system bandwidth so that the closed-loop system would respond to the noise. Thus, the noise in a poorly designed digitally controlled system could have a substantially greater effect than if the control had been implemented using analog electronics.

The solution is to place an analog prefilter before the sampler. In many cases a simple first-order low-pass filter will do, that is,

$$H_p(s) = \frac{a}{s + a},$$

where the break point a is selected to be lower than $\omega_s/2$ so that any noise present with frequencies greater than $\omega_s/2$ is attenuated by the prefilter. The lower the break-point frequency selected, the more the noise above $\omega_s/2$ is attenuated. However, too low a break point may force the designer to reduce the control system's bandwidth. The prefilter does not completely eliminate the aliasing; however, through judicious choice of the prefilter break point and the sample rate, the designer has the ability to reduce the magnitude of the aliased noise to some acceptable level.

8.6.4 The Computer

The computer is the unit that does all the computations. Most digital controllers used today are built around a microcontroller that contains both a microprocessor and most of the other functions needed, including the A/D and D/A conversion. For development purposes in a laboratory, a digital controller could be a desktop-sized workstation or a PC. The relatively low cost of micropocessor technology has accounted for the large increase in the use of digital control systems, which started in the 1980s and continues in the 1990s.

The computer consists of a central processor unit (CPU), which does the computations and provides the system logic; a clock to synchronize the system; memory modules for data and instruction storage; and a power supply to provide the various required voltages. The memory modules come in three basic varieties:

1. **Read-only memory (ROM)** is the least expensive, but after its manufacture its contents cannot be changed. Most of the memory in products manufactured in quantity is ROM. It retains its stored values when power is removed.

2. **Random-access memory (RAM)** is the most expensive, but its values can be changed by the CPU. It is only required to store the values that will be changed during the control process and typically represents only a small fraction of the total memory of a developed product. It loses the values in memory when power is removed.

3. **Programmable read-only memory (PROM)** is a ROM whose values can be changed by a technician using a special device. It is typically used during product development to enable the designer to try different algorithms and parameter values. It retains its stored values when power is removed.

Microprocessors for control applications generally come with a digital word size of 8, 16, or 32 bits, although some have been available with 12 bits. Larger word sizes give better accuracy, but at an increase in cost. The most economical solution is often to use an 8-bit microprocessor, but to use two digital words to store one value (**double precision**) in the areas of the controller that are critical to the system accuracy. Many digital control systems use computers originally designed for digital signal-processing applications, so-called DSP chips.

8.7 Word-size Effects

A numerical value can be represented with only limited precision in a digital computer. For fixed-point arithmetic, the resolution is 0.4% of full range for 8 bits and 0.1% for 10 bits; resolution drops by a factor of 2 for each additional bit. The effect of this quantization shows up in A/D conversion, multiplication truncation, and parameter storage errors. If the computer uses **floating-point arithmetic**, the resolution of the multiplication and parameter storage changes with the magnitude of the number being stored; thus the resolution affects only the mantissa, while the exponent continually adjusts scale factors relating the bit size to the physical variable.

8.7.1 Random Effects

As long as a system has varying inputs or disturbances, A/D errors and multiplication errors act in a random manner on the system producing noise at the output of the system. The output noise due to a particular noise source (multiplication or A/D conversion) has a mean value of

$$\bar{n}_o = H_{\mathrm{DC}} \bar{n}_I,$$

where

H_{DC} = DC gain of transfer function between noise source and output,

$\bar{n}_I$ = mean value of noise source.

The mean value of the noise will be zero for a roundoff process, but otherwise it has the value

$$\bar{n}_I = \frac{q}{2},$$

where q is the resolution level for a truncation process. Although most A/D converters round off, producing no mean error, some truncate and thus do produce an error. The total noise effect is the sum of all noise sources.

The variance of the output noise, σ_o, is always nonzero, irrespective of whether the process truncates or rounds off. The value of the output variance is most easily found by first solving the discrete **Lyapunov equation** for the covariance of the state,

Lyapunov equation

$$\mathbf{R}_x = \mathbf{\Phi}\mathbf{R}_x\mathbf{\Phi}^T + \mathbf{\Gamma}\mathbf{\Gamma}^T\sigma_I^2, \qquad (8.65)$$

and then solving

$$\sigma_o^2 = \mathbf{H}\mathbf{R}_x\mathbf{H}^T, \qquad (8.66)$$

where

σ_o = output noise root-mean-square (rms),

$\mathbf{R}_x$ = state covariance matrix,

σ_I^2 = input noise variance,

$\mathbf{\Phi}, \mathbf{H}$ = system description matrices from Section 8.5,

$\mathbf{\Gamma}$ = noise input matrix.

The magnitude of the input noise variance is

$$\sigma_I^2 = \frac{q^2}{12}$$

for either roundoff or truncation.

The evaluation of Eq. (8.65) is usually done using computer tools, as described in Franklin *et al.* (1990). Carrying out the calculations for a system with a small-word-size computer (8 bits or less) will generally show that the noise response of the system becomes more sensitive as the sampling rate increases and can be a design issue for very fast sample rates. However, for 16- or 32-bit computers, the onset of significant noise errors occurs at much higher sample rates, and quantization noise is typically not a factor in design.

Noise response vs. sampling rate

The sensitivity of a system to A/D errors can be alleviated by sampling faster or by adding more bits to the A/D converter. Different structures of the digital controller have no effect.

In contrast, multiplication errors can be reduced substantially for high-order (third-order and above) controllers by properly structuring a given control transfer function. For example, a second-order transfer function with

FIGURE 8.22
An example of parallel
implementation

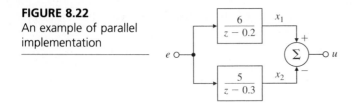

real roots,

$$D(z) = \frac{U(z)}{E(z)} = \frac{z - 0.8}{(z - 0.2)(z - 0.3)}, \qquad (8.67)$$

Direct implementation   can be implemented in a direct manner, yielding

$$u(k) = 0.5u(k - 1) - 0.06u(k - 2) + e(k) - 0.8e(k - 1). \qquad (8.68)$$

Equation (8.67) can also be implemented in a parallel manner by performing a partial-fraction expansion on the equation. The result, shown in Fig. 8.22, yields

Parallel implementation

$$x_1(k) = 0.2x_1(k - 1) + 6e(k), \qquad (8.69a)$$

$$x_2(k) = 0.3x_2(k - 1) + 5e(k), \qquad (8.69b)$$

$$u(k) = x_1(k) - x_2(k). \qquad (8.69c)$$

Note that the transfer functions to the output from the multiplications in Eq. (8.68) are substiantially different from those in Eqs. (8.69). It is also possible to
Cascade implementation   implement $D(z)$ with a cascade factorization, which, for this example, would consist of two first-order blocks arranged serially.

   For digital controllers with a small word size, either cascade or parallel implementation is preferable to the direct implementation because of the substantially reduced response of such controllers to quantization errors.

8.7.2 Systematic Effects

Parameters such as the numerical values in Eqs. (8.68) and (8.69), if in error, will change the dynamic behavior of a system. In a high-order controller with a small word size and a direct implementation, a very small percentage error in a stored parameter can result in substantial root-location changes and can sometimes cause instability. In the Apollo command module, a 14-bit word size for parameter storage would have resulted in instability if a direct implementation had been used in the sixth-order compensator. These effects are amplified when there are two compensator poles close together (or repeated), or when there are fast sample rates, in which case all poles tend to clump around $z = +1$ and are close to one another. These effects can be reduced by using larger word sizes, parallel or cascade implementations, double-precision parameter storage, and/or slower sample rates.

Under conditions of constant disturbances and input commands, multiplication errors can also cause systematic errors. Typically the result is a steady-state error or possibly a limit cycle. The steady-state error results from the dead band created by the quantization. These effects are also reduced by larger word sizes, parallel or cascade implementations, and slower sample rates. For more detail, see Franklin *et al.* (1990).

8.8   Sample-rate Selection

The selection of the best sample rate for a digital control system is the result of a compromise of many factors. The basic motivation for lowering the sample rate ω_s is cost. A decrease in sample rate means more time is available for the control calculations; hence slower computers can be used for a given control function or more control capability can be achieved from a given computer. Either way, the cost per function is lowered. For systems with A/D converters, less demand on conversion speed will also lower cost. These economic arguments indicate that the best engineering choice is the slowest-possible sample rate that still meets all performance specifications.

Design guidelines for sample rate

There are several factors that could provide a lower limit on the acceptable sample rate:

1. tracking effectiveness as measured by closed-loop bandwidth or by time-response requirements, such as rise time and settling time;
2. regulation effectiveness as measured by the error response to random plant disturbances;
3. error due to measurement noise and the associated prefilter design methods.

A fictitious limit occurs when using emulation design techniques. The inherent approximation in the method may give rise to system instabilities as the sample rate is lowered. This can lead the designer to conclude that a lower limit on ω_s has been reached when, in fact, the proper conclusion is that the approximations are invalid; the solution is not to sample faster but to refine the design with a direct digital-design method.

8.8.1   Tracking Effectiveness

An absolute lower bound on the sample rate is set by a specification to track a command input with a certain frequency (the system bandwidth). The sampling theorem (see Section 8.6.3 and Franklin *et al.*, 1990) states that in order to reconstruct an unknown, band-limited, continuous signal from samples of that signal, we must sample at least twice as fast as the highest frequency contained in the signal. Therefore, in order for a closed-loop system to track an input at a certain frequency, it must have a sample rate twice as fast; that is, ω_s must be at least twice the system bandwidth ($\omega_s \Rightarrow 2\omega_{BW}$). We also saw from the results of mapping the s-plane into the z-plane ($z = e^{sT}$) that the highest frequency that can be represented by a discrete system is $\omega_s/2$, which supports the conclusion of the theorem.

It is important to note the distinction between the closed-loop bandwidth ω_{BW} and the highest frequency in the open-loop plant dynamics, since the two frequencies can be quite different. For example, closed-loop bandwidths can be an order of magnitude *less* than open-loop modes of resonances for some control problems. Information concerning the state of the plant resonances for purposes of control can be extracted from sampling the output without satisfying the sampling theorem because some a priori knowledge concerning these dynamics (albeit imprecise) is available, and the system is not required to track these frequencies. Thus a priori knowledge of the dynamic model of the plant can be included in the compensation in the form of a notch filter.

The closed-loop-bandwidth limitation provides the fundamental lower bound on the sample rate. In practice, however, the theoretical lower bound of sampling at twice the bandwidth of the reference input signal would not be judged sufficient in terms of the quality of the desired time responses. For a system with a rise time on the order of 1 sec (thus yielding a closed-loop bandwidth on the order of 0.5 Hz), it is reasonable to insist on a sampling rate of 10 to 20 Hz, which is a factor of 20 to 40 times ω_{BW}. The purposes of choosing a sample rate much greater than the bandwidth are to reduce the delay between a command and the system response to the command and also to smooth the system output to the control steps coming out of the ZOH.

8.8.2   Disturbance Rejection

Disturbance rejection is an important—if not the most important—aspect of any control system. Disturbances enter a system with various frequency characteristics ranging from steps to white noise. For the purpose of sample-rate selection, the higher-frequency random disturbances are the most influential.

The ability of the control system to reject disturbances with a good continuous controller represents a lower bound on the error response that we can hope for when implementing the controller digitally. In fact, some degradation relative to the continuous design must occur because the sampled values are slightly out of date at all times except precisely at the sampling instants. However, if the sample rate is very fast compared with the frequencies contained in the noisy disturbance, we should expect no appreciable loss from the digital system as compared with the continuous controller. At the other extreme, if the sample time is very long compared with the characteristic frequencies of the noise, the response of the system because of noise is essentially the same as the response we would get if the system had no control at all. The selection of a sample rate will place the response somewhere in between these two extremes. Thus the impact of the sample rate on the ability of the system to reject disturbances may be very important to consider when choosing the sample rate.

Although the best choice of sample rate in terms of the ω_{BW} multiple is dependent on the frequency characteristics of the noise and the degree to which random disturbance rejection is important to the quality of the controller, sample rates on the order of 20 times ω_{BW} or higher are typical.

8.8.3 Control Systems Design Methodology

Prefilter Design

Digital control systems with analog sensors typically include an analog prefilter between the sensor and the sampler as an antialiasing device as described in Section 8.6.3. The prefilters are low-pass, and the simplest transfer function is

$$H_p(s) = \frac{a}{s + a}$$

so that the noise above the prefilter break point a is attenuated. The goal is to provide enough attenuation at half the sample rate $(\omega_s/2)$ so that the noise above $\omega_s/2$, when aliased into lower frequencies by the sampler, will not be detrimental to control system performance.

Design guideline for selecting prefilter break point and ω_s

A conservative design procedure is to select ω_s and the break point to be sufficiently higher than the system bandwidth so that the phase lag from the prefilter does not significantly alter the system stability. This would allow the prefilter to be ignored in the basic control system design. Furthermore, for a good reduction in the high-frequency noise at $\omega_s/2$, we choose a sample rate that is about 5 or 10 times higher than the prefilter break point. The implication of this prefilter design procedure is that sample rates need to be on the order of 30 to 100 times faster than the system bandwidth. If done this way, the prefilter design procedure is likely to provide the lower bound on the selection of the sample rate.

An alternative design procedure is to allow significant phase lag from the prefilter at the system bandwidth. This requires us to include the analog prefilter characteristics when carrying out the control design. This procedure allows us to use very low sample rates, but at the expense of increased complexity in the original design, since the prefilter must be included in the plant transfer function. If this procedure is used and low prefilter break points are allowed, the effect of sample rate on sensor noise is small and essentially places no limits on the sample rate.

Asynchronous Modules

As noted in the previous paragraphs, divorcing the prefilter design from the control-law design may require using a faster sample rate than otherwise. This same result may show up in other types of architecture. For example, a smart sensor with its own computer running asynchronously relative to the primary control computer will not be amenable to direct digital design because the overall system transfer function depends on the phasing between the smart sensor and the primary digital controller. This situation is similar to that of the digitization errors discussed in Section 8.4. Therefore, if asynchronous digital subsystems are present, sample rates on the order of $20\omega_{BW}$ or slower in any module should be used with caution and the system performance checked through simulation or experiment.

Summary

- There are two basic design techniques for finding compensation equations for implementation in a digital computer: emulation and discrete design.

- **Emulation design** entails: (a) finding the continuous compensation $D(s)$ using the ideas in Chapters 1 to 7, and (b) approximating $D(s)$ with difference equations using Euler's method, Tustin's method, or the matched-pole-zero method.

- **Discrete design** entails: (a) finding the discrete model of the plant $G(s)$, and (b) using the discrete model to design the compensation directly in its discrete form.

- The **z-transform** is the primary tool used to determine the behavior of discrete linear systems. The z-transform of a time sequence $f(k)$ is given by

$$\mathscr{Z}\{f(k)\} = F(z) = \sum_{k=0}^{\infty} f(k)z^{-k}$$

and has the key property that

$$\mathscr{Z}\{f(k-1)\} = z^{-1}F(z).$$

This property allows us to find the discrete transfer function of a difference equation. Analysis using z-transforms closely parallels that using Laplace transforms.

- Normally z-transforms are found using Table 8.1 or by computer.

- The discrete Final Value Theorem is

$$\lim_{k \to \infty} x(k) = \lim_{z \to 1} (1 - z^{-1})X(z)$$

provided all poles of $(1 - z^{-1})X(z)$ are inside the unit circle.

- For a continuous signal $f(t)$ whose samples are $f(k)$, the poles of $F(s)$ are related to the poles of $F(z)$ by

$$z = e^{sT}.$$

- The most common emulation methods are:

a) **Tustin's approximation**:

$$D(z) = D(s)\Big|_{s = \frac{2}{T}\left(\frac{z-1}{z+1}\right)}$$

b) the **matched-pole-zero approximation**:

- Map poles and zeros by $z = e^{sT}$.
- Add powers of $z + 1$ to the numerator until numerator and denominator are of equal order.
- Set the low-frequency gain of $D(z)$ equal to that of $D(s)$.

- The discrete model of the continuous plant $G(s)$ preceded by a ZOH is

$$G(z) = (1 - z^{-1}) \mathscr{Z} \left\{ \frac{G(s)}{s} \right\}.$$

- Discrete design using $G(z)$ closely parallels continuous design, but the stability boundary and interpretation of z-plane root locations is different. Figure 8.7 summarizes the response characteristics.

- The continuous state-space form of differential equation,

$$\dot{\mathbf{x}} = \mathbf{Fx} + \mathbf{G}u,$$
$$y = \mathbf{Hx} + Ju,$$

has a discrete counterpart in the difference equations

$$\mathbf{x}(k + 1) = \mathbf{\Phi x}(k) + \mathbf{\Gamma}u(k),$$
$$y(k) = \mathbf{H}x(k) + Ju(k),$$

where

$$\mathbf{\Phi} = e^{\mathbf{F}T}$$

$$\mathbf{\Gamma} = \left(\int_0^T e^{\mathbf{F}\eta} \, d\eta \right) \mathbf{G}$$

- The pole placement and estimation ideas are identical in the continuous and discrete domains.

- If designing by emulation, a sample rate of 20 times the bandwidth is recommended. If using discrete design, system stability can be assured when sampling at a rate as slow as 2 times the bandwidth. However, to reject random disturbances, best results are obtained by sampling at 20 times the closed-loop bandwidth or faster.

- Analog **prefilters** are commonly placed before the **sampler** in order to attenuate the effects of high-frequency measurement noise. A sampler **aliases** all frequencies in the signal that are greater than half the sample frequency to lower frequencies; therefore, prefilter break points should be selected so that no significant frequency content remains above half the sample rate.

Problems

8.1 The z-transform of a discrete-time filter $h(k)$ at a $1\,Hz$ sample rate is

$$H(z) = \frac{1 + (1/2)z^{-1}}{[1 - (1/2)z^{-1}][1 + (1/3)z^{-1}]}.$$

a) Let $u(k)$ and $y(k)$ be the discrete input and output of this filter. Find a difference equation relating $u(k)$ and $y(k)$.
b) Find the natural frequency and damping coefficient of the filter's poles.
c) Is the filter stable?

8.2 Use the z-transform to solve the difference equation

$$y(k) - 3y(k - 1) + 2y(k - 2) = 2u(k - 1) - 2u(k - 2),$$

where

$$u(k) = \begin{cases} k, & k \geqslant 0, \\ 0, & k < 0, \end{cases}$$

$$y(k) = 0, \qquad k > 0.$$

8.3 The one-sided z-transform is defined as

$$F(z) = \sum_0^\infty f(k)z^{-k}.$$

a) Show that the one-sided transform of $f(k + 1)$ is $\mathscr{Z}\{f(k + 1)\} = zF(z) - zf(0)$.
b) Use the one-sided transform to solve for the transforms of the Fibonacci numbers generated by the difference equation $u(k + 2) = u(k + 1) + u(k)$. Let $u(0) = u(1) = 1$. [*Hint:* You will need to find a general expression for the transform of $f(k + 2)$ in terms of the transform of $f(k)$].
c) Compute the pole locations of the transform of the Fibonacci numbers.
d) Compute the inverse transform of the Fibonacci numbers.
e) Show that, if $u(k)$ represents the kth Fibonacci number, then the ratio $u(k + 1)/u(k)$ will approach $(1 + \sqrt{5})/2$. This is the golden ratio valued so highly by the Greeks.

8.4 A unity feedback system has an open-loop transfer function given by

$$G(s) = \frac{250}{s[(s/10) + 1]}.$$

The following lag compensator added in series with the plant yields a phase margin of 50°:

$$D_c(s) = \frac{s/1.25 + 1}{50s + 1}.$$

Using the matched pole-zero approximation, determine an equivalent digital realization of this compensator.

8.5 a) The following transfer function is a lead network designed to add about 60° of phase at $\omega_1 = 3$ rad/sec:

$$H(s) = \frac{s+1}{0.1s+1}.$$

Assume a sampling period of $T = 0.25$ sec, and compute and plot in the z-plane the pole and zero locations of the digital implementations of $H(s)$ obtained using (1) Tustin's method and (2) pole-zero mapping. For each case, compute the amount of phase lead provided by the network at $z_1 = e^{j\omega_1 T}$.

b) Using a log-log scale for the frequency range $\omega = 0.1$ to $\omega = 100$ rad/sec, plot the magnitude Bode plots for each of the equivalent digital systems you found in part (a), and compare with $H(s)$. (*Hint:* Magnitude Bode plots are given by $|H(z)| = |H(e^{j\omega T})|$.)

8.6 a) The following transfer function is a lag network designed to introduce a gain attenuation of 10 (-20 dB) at $\omega = 3$ rad/sec:

$$H(s) = \frac{10s+1}{100s+1}.$$

Assume a sampling period of $T = 0.25$ sec, and compute and plot in the z-plane the pole and zero locations of the digital implementations of $H(s)$ obtained using (1) Tustin's method and (2) pole-zero mapping. For each case, compute the amount of gain attenuation provided by the network at $z_1 = e^{j\omega_1 T}$.

b) For each of the equivalent digital systems in part (a), plot the Bode magnitude curves over the frequency range $\omega = 0.01$ to 10 rad/sec.

8.7 Consider the linear equation $\mathbf{Ax = b}$, where $\mathbf{A}$ is an $n \times n$ matrix. When $\mathbf{b}$ is given, one way of solving for $\mathbf{x}$ is to use the discrete-time recursion

$$\mathbf{x}(k+1) = (\mathbf{I} + c\mathbf{A})\mathbf{x}(k) - c\mathbf{b},$$

where c is a scalar to be chosen.

a) Show that the solution of $\mathbf{Ax = b}$ is the equilibrium point $\mathbf{x}^*$ of the discrete-time system. An equilibrium point $\mathbf{x}^*$ of a discrete-time system $\mathbf{x}(k+1) = \mathbf{f}(\mathbf{x}(k))$ satisfies the relation $\mathbf{x}^* = \mathbf{f}(\mathbf{x}^*)$.

b) Consider the error $\mathbf{e}(k) = \mathbf{x}(k) - \mathbf{x}^*$. Write the linear equation that relates the error $\mathbf{e}(k+1)$ to $\mathbf{e}(k)$.

c) Suppose $|1 + c\lambda_i(\mathbf{A})| < 1$, $i = 1, \ldots, n$, where $\lambda_i(\mathbf{A})$ denotes the ith eigenvalue of $\mathbf{A}$. Show that starting from any initial guess $\mathbf{x}_0$, the algorithm converges to $\mathbf{x}^*$. [*Hint:* For any matrix $\mathbf{B}$, $\lambda_i(\mathbf{I} + \mathbf{B}) = 1 + \lambda_i(\mathbf{B})$.]

8.8 The open-loop plant of a unity feedback system has the transfer function

$$G(s) = \frac{1}{s(s+2)}.$$

Determine the transfer function of the equivalent digital plant using a sampling period of $T = 1$ sec, and design a proportional controller for the discrete-time system that yields dominant closed-loop poles with a damping ratio ζ of 0.7.

FIGURE 8.23
Control system for
Problem 8.9

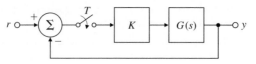

8.9 Consider the system configuration shown in Fig. 8.23, where

$$G(s) = \frac{40(s + 2)}{(s + 10)(s^2 - 1.4)}.$$

a) Find the transfer function $G(z)$ for $T = 1$ assuming the system is preceded by a ZOH.
b) Find the range of K for which the closed-loop system is stable.
c) Draw the root locus of the system with respect to K.
d) Compare your results of part (b) to the case where an analog controller is used (that is, where the sampling switch is always closed). Which system has a larger allowable value of K?
e) Use CACSD software to compute the step response of both the continuous and discrete systems with K chosen to yield a damping factor of $\zeta = 0.5$ for the continuous case. If you did parts (a) to (d) by hand, verify those results using the computer.

8.10 Write a computer program to compute Φ and Γ from F, G, and the sample period T. Use your program to compute Φ and Γ when

a) $F = \begin{bmatrix} -1 & 0 \\ 0 & -2 \end{bmatrix}$, $G = \begin{bmatrix} 1 \\ 1 \end{bmatrix}$, $T = 0.2$ sec.

b) $F = \begin{bmatrix} -3 & -2 \\ 1 & 0 \end{bmatrix}$, $G = \begin{bmatrix} 1 \\ 0 \end{bmatrix}$, $T = 0.2$ sec.

8.11 Consider the following discrete-time system in state-space form:

$$\begin{bmatrix} x_1(k + 1) \\ x_2(k + 1) \end{bmatrix} = \begin{bmatrix} 0 & 1 \\ 0 & -1 \end{bmatrix} \begin{bmatrix} x_1(k) \\ x_2(k) \end{bmatrix} + \begin{bmatrix} 0 \\ 10 \end{bmatrix} u(k).$$

Use state feedback to relocate all of the system's poles to 0.5.

8.12 a) For

$$\Phi = \begin{bmatrix} 1 & T \\ 0 & 1 \end{bmatrix} \quad \text{and} \quad \Gamma = \begin{bmatrix} T^2/2 \\ T \end{bmatrix},$$

find a transformation matrix T so that, if $x = Tw$, the state equations for w will be in control canonical form.
b) Compute the gain K_w so that if $u = -K_w w$, the characteristic equation will be $\alpha_c(z) = z^2 - 1.6z + 0.7$.
c) Use T from part (a) to compute K_x, the feedback gain required by the state equations in x to achieve the desired characteristic polynomial.

8.13 Consider a system whose plant transfer function is $1/s^2$ and which has a piecewise constant input of the form

$$u(t) = u(kT), \qquad kT \leqslant t < (k + 1)T.$$

a) Show that if we restrict attention to the time instants kT, $k = 0, 1, 2,...$, the resulting sampled-data system can be described by the equations

$$\begin{bmatrix} x_1(k+1) \\ x_2(k+1) \end{bmatrix} = \begin{bmatrix} 1 & 0 \\ T & 1 \end{bmatrix} \begin{bmatrix} x_1(k) \\ x_2(k) \end{bmatrix} + \begin{bmatrix} T \\ T^2/2 \end{bmatrix} u(k),$$

$$y(k) = [0 \quad 1][x_1(k) \quad x_2(k)]^T.$$

b) Design a second-order estimator that will always drive the error in the estimate of the initial state vector to zero in time $2T$ or less.

c) Is it possible to estimate the initial state exactly with a first-order estimator? Justify your answer.

8.14 *Single-axis Satellite Attitude Control:* Satellites often require attitude control for proper orientation of antennas and sensors with respect to Earth. Figure 2.6 shows a communication satellite with a three-axis attitude-control system. To gain insight into the three-axis problem we often consider one axis at a time. Figure 8.24 depicts this case where motion is only allowed about an axis perpendicular to the page. The equations of motion of the system are given by

$$I\ddot{\theta} = M_C + M_D,$$

where

I = moment of inertia of the satellite about its mass center,

M_C = control torque applied by the thrusters,

M_D = disturbance torques,

θ = angle of the satellite axis with respect to an inertial reference with no angular acceleration.

We normalize the equations of motion by defining

$$u = \frac{M_C}{I}, \qquad w_d = \frac{M_D}{I},$$

and obtain

$$\ddot{\theta} = u + w_d.$$

Taking the Laplace transform yields

$$\theta(s) = \frac{1}{s^2}[u(s) + w_d(s)],$$

FIGURE 8.24

Satellite control schematic for Problem 8.14

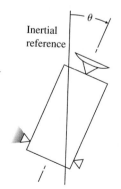

Inertial reference

which with no disturbance becomes

$$\frac{\theta(s)}{u(s)} = \frac{1}{s^2} = G_1(s).$$

In the discrete case where u is applied through a ZOH, we can use the methods described in this chapter to obtain the discrete transfer function

$$G_1(z) = \frac{\theta(z)}{u(z)} = \frac{T^2}{2}\left[\frac{z+1}{(z-1)^2}\right].$$

a) Sketch the root locus of this system assuming proportional control.
b) Add a lead network to your controller so that the dominant poles correspond to $\zeta = 0.5$ and $\omega_n = 3\pi/(10T)$.
c) What is the feedback gain if $T = 1$ sec? If $T = 2$ sec.
d) Plot the closed-loop step response for $T = 1$ sec.

8.15 In this problem you will show how to compute Φ by changing states so that the system matrix is diagonal.

a) Using an infinite series expansion, compute e^{AT} for

$$\mathbf{A} = \begin{bmatrix} -1 & 0 \\ 0 & -2 \end{bmatrix}.$$

b) Show that if $\mathbf{F} = \mathbf{TAT}^{-1}$ for some nonsingular transformation matrix $\mathbf{T}$, then

$$e^{\mathbf{F}T} = \mathbf{T}e^{\mathbf{A}T}\mathbf{T}^{-1}.$$

c) Show that if

$$\mathbf{F} = \begin{bmatrix} -3 & 1 \\ -2 & 0 \end{bmatrix},$$

there exists a $\mathbf{T}$ such that $\mathbf{TAT}^{-1} = \mathbf{F}$. (*Hint*: Write $\mathbf{AT} = \mathbf{TF}$, assume four unknowns for the elements of $\mathbf{T}$, and solve. Next show that the columns of $\mathbf{T}$ are the eigenvectors of $\mathbf{F}$.)
d) Compute $e^{\mathbf{F}T}$.

8.16 It is possible to suspend a mass of magnetic material by means of an electromagnet whose current is controlled by the position of the mass (Woodson and Melcher, 1968). The schematic of a possible setup is shown in Fig. 8.25, and a photo of a working system at Stanford University is shown in Fig. 2.34. The equations of motion are

$$m\ddot{x} = -mg + f(x, I),$$

where the force on the ball due to the electromagnet is given by $f(x, I)$. At equilibrium the magnet force balances the gravity force. Suppose we let I_0 represent the current at equilibrium. If we write $I = I_0 + i$, expand f about $x = 0$ and $I = I_0$, and neglect higher-order terms, we obtain the linearized equation

$$m\ddot{x} = k_1 x + k_2 i. \tag{8.70}$$

Reasonable values for the constants in Eq. (8.70) are $m = 0.02$ kg, $k_1 = 20$ N/m, and $k_2 = 0.4$ N/A.

a) Compute the transfer function from i to x, and draw the (continuous) root locus for the simple feedback $i = -Kx$.

FIGURE 8.25
Schematic of magnetic levitation device for Problems 8.16 and 8.17

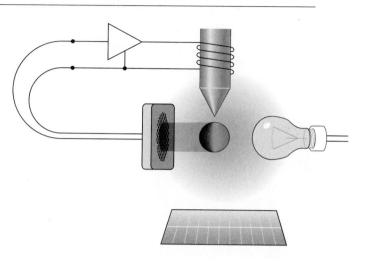

b) Assume the input is passed through a ZOH, and let the sampling period be 0.02 sec. Compute the transfer function of the equivalent discrete-time plant.

c) Design a digital control for the magnetic levitation device so that the closed-loop system meets the following specifications: $t_r \leqslant 0.1$ sec, $t_s \leqslant 0.4$ sec, and overshoot $\leqslant 20\%$.

d) Plot a root locus with respect to k_1 for your design, and discuss the possibility of using your closed-loop system to balance balls of various masses.

e) Plot the step response of your design to an initial disturbance displacement on the ball, and show both x and the control current i. If the sensor can measure x only over a range of $\pm 1/4$ cm and the amplifier can only provide a current of 1 A, what is the *maximum* displacement possible for control, neglecting the nonlinear terms in $f(x, I)$?

8.17 In Problem 8.16 we described an experiment in magnetic levitation described by Eq. (8.70) which reduces to

$$\ddot{x} = 1000x + 20i.$$

Let the sampling time be 0.01 sec.

a) Use pole placement to design a controller for the magnetic levitator so that the closed-loop system meets the following specifications: $t_s \leqslant 0.25$ sec, and overshoot to an initial offset in x that is less than 20%.

b) Plot the step response of x, $\tilde{x}$, and i to an initial displacement in x.

c) Plot the root locus for changes in the plant gain, and mark the pole locations of your design.

d) Introduce a command reference input r (as discussed in Section 7.7) that does not excite the estimate of x. Measure or compute the frequency response from r to the system error $r - x$ and give the highest frequency for which the error amplitude is less than 20% of the command amplitude.

FIGURE 8.26
Satellite-tracking
antenna *(courtesy Space
Systems/Loral)*

FIGURE 8.27
Schematic diagram of
satellite-tracking
antenna

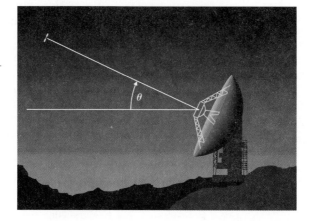

8.18 *Servomechanism for Antenna Elevation Control:* Suppose it is desired to control the
elevation of an antenna designed to track a satellite. A photo of such a system is
shown in Fig. 8.26 and a schematic diagram is depicted in Fig. 8.27. The antenna
and drive parts have a moment of inertia J and damping B, arising to some extent
from bearing and aerodynamic friction, but mostly from the back emf of the DC
drive motor. The equation of motion is

$$J\ddot{\theta} + B\dot{\theta} = T_c + T_d,$$

where

T_c = net torque from the drive motor,

T_d = disturbance torque due to wind.

If we define

$$\frac{B}{J} = a, \qquad u = \frac{T_c}{B}, \qquad w_d = \frac{T_d}{B},$$

the equation simplifies to

$$\frac{1}{a}\ddot{\theta} + \dot{\theta} = u + w_d.$$

After Laplace transformation, we obtain

$$\theta(s) = \frac{1}{s(s/a + 1)} [u(s) + w_d(s)],$$

or, with no disturbance,

$$\frac{\theta(s)}{u(s)} = \frac{1}{s(s/a + 1)} = G_2(s).$$

With $u(k)$ applied through a ZOH, the transfer function for an equivalent discrete-time system is

$$G_2(z) = \frac{\theta(z)}{u(z)} = K \frac{z + b}{(z - 1)(z - e^{-aT})},$$

where

$$K = \frac{aT - 1 + e^{-aT}}{a}, \qquad b = \frac{1 - e^{-aT} - aTe^{-aT}}{aT - 1 + e^{-aT}}.$$

a) Let $a = 0.1$ and $x_1 = \dot{\theta}$, and write the continuous-time state equations for the system.

b) Let $T = 1$ sec, and find a state feedback gain $\mathbf{K}$ for the equivalent discrete-time system that yields closed-loop poles corresponding to the following points in the s-plane: $s = -1/2 \pm j(\sqrt{\frac{3}{2}})$. Plot the step response of the resulting design.

c) Design an estimator: Select $\mathbf{L}$ so that $\alpha_e(z) = z^2$.

d) Using the values for $\mathbf{K}$ and $\mathbf{L}$ computed in parts (b) and (c) as the gains for a combined estimator/controller, introduce a reference input that will leave the state estimate undisturbed. Plot the response of the closed-loop system due to a step change in the reference input. Also plot the system response to a step wind-gust disturbance.

e) Plot the root locus of the closed-loop system with respect to the plant gain, and mark the locations of the closed-loop poles.

8.19 *Tank Fluid Temperature Control:* The temperature of a tank of fluid with a constant inflow and outflow rate is to be controlled by adjusting the temperature of the incoming fluid. The temperature of the incoming fluid is controlled by a mixing valve that adjusts the relative amounts of hot and cold supplies of the fluid (see Fig. 8.28). The distance between the valve and the point of discharge into the tank creates a time delay between the application of a temperature change at the mixing valve and the discharge of the flow with the changed temperature into the tank. The differential equation governing the tank temperature is

$$\dot{T}_e = \frac{1}{cM} (q_{in} - q_{out}),$$

FIGURE 8.28
Tank temperature
control

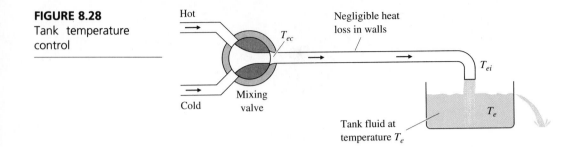

where

$$T_e = \text{tank temperature,}$$

$$c = \text{specific heat of the fluid,}$$

$$M = \text{fluid mass contained in the tank,}$$

$$q_{in} = c\dot{m}_{in}T_{ei},$$

$$q_{out} = c\dot{m}_{out}T_e,$$

$$\dot{m} = \text{mass flow rate } (\dot{m}_{in} = \dot{m}_{out}),$$

$$T_{ei} = \text{temperature of fluid entering tank.}$$

However, the temperature at the input to the tank at time t is equal to the control temperature τ_d seconds in the past. This relationship may be expressed as

$$T_{ei}(t) = T_{ec}(t - \tau_d),$$

where

$$\tau_d = \text{delay time,}$$

$$T_{ec} = \text{temperature of fluid immediately after the control valve and directly controllable by the valve.}$$

Combining constants, we obtain

$$\dot{T}_e(t) + aT_e(t) = aT_{ec}(t - \tau_d),$$

where

$$a = \frac{\dot{m}}{M}.$$

The transfer function of the system is thus

$$\frac{T_e(s)}{T_{ec}(s)} = \frac{e^{-\tau_d s}}{s/a + 1} = G_3(s).$$

To form a discrete transfer function equivalent to G_3 preceded by a ZOH, we must compute

$$G_3(z) = \mathscr{Z}\left\{ \left(\frac{1 - e^{-sT}}{s} \right) \left(\frac{e^{-\tau_d s}}{s/a + 1} \right) \right\}.$$

We assume that for some integer ℓ, $\tau_d = \ell T - mT$, where $0 \leqslant m < 1$. Then

$$G_3(z) = \mathscr{L}\left\{\left(\frac{1 - e^{-sT}}{s}\right)\left(\frac{e^{-\ell sT}e^{msT}}{s/a + 1}\right)\right\}$$

$$= (1 - z^{-1})z^{-\ell}\mathscr{L}\left\{\frac{e^{msT}}{s(s/a + 1)}\right\}$$

$$= (1 - z^{-1})z^{-\ell}\mathscr{L}\left\{\frac{e^{msT}}{s} - \frac{e^{msT}}{s + a}\right\}$$

$$= \frac{z - 1}{z}\left(\frac{1}{z^{\ell}}\right)\mathscr{L}\{1(t + mT) - e^{-a(t + mT)}1(t + mT)\}$$

$$= \frac{z - 1}{z}\left(\frac{1}{z^{\ell}}\right)\left(\frac{z}{z - 1} - \frac{e^{-amT}z}{z - e^{-aT}}\right)$$

$$= \frac{1}{z^{\ell}}\left[\frac{(1 - e^{-amT})z + e^{-amT} - e^{-aT}}{z - e^{-aT}}\right]$$

$$= \left(\frac{1 - e^{-amT}}{z^{\ell}}\right)\left(\frac{z + \alpha}{z - e^{-aT}}\right);$$

and

$$\alpha = \frac{e^{-amT} - e^{-aT}}{1 - e^{-amT}}.$$

The zero location $-\alpha$ varies from $\alpha = \infty$ at $m = 0$ to $\alpha = 0$ as $m \to 1$. Note also that $G_3(1) = 1.0$ for all a, m, and ℓ. For the specific values $\tau_d = 1.5$, $T = 1$, and $a = 1$, $\ell = 2$, and $m = \frac{1}{2}$, the transfer function reduces to

$$G_3(z) = 0.3935\,\frac{z + 0.6065}{z^2(z - 0.3679)}.$$

a) Write the discrete-time system equations in state-space form.
b) Design a state feedback gain that yields $\alpha_c(z) = z^3$.
c) Design a state estimator with $\alpha_e(z) = z^3$.
d) Plot the root locus of the system with respect to the plant gain.
e) Plot the step response of the system.

8.20 Prove the seven properties of the s-plane-to-z-plane mapping listed in Section 8.2.3.

8.21 For the system shown in Fig. 8.29, find values for K_P, T_D, and T_I so that the closed-loop poles satisfy $\zeta > 0.5$ and $\omega_n > 1$ rad/sec. Discretize the PID controller using:

a) Euler's method   c) matched pole-zero method
b) Tustin's method

FIGURE 8.29
Control system for
Problem 8.21

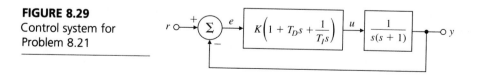

Use MATLAB or equivalent software to simulate the step response of each of these digital implementations for sample times of $T = 1, 0.1$, and 0.01 sec.

8.22 Repeat Problem 5.37 by constructing discrete root loci and performing the designs directly in the z-plane. Assume that the output y is sampled, the input u is passed through a ZOH as it enters the plant, and the sample rate is 50 Hz.

8.23 Design a digital controller for the antenna servo system shown in Figs. 3.50 and 3.51 and described in Problem 3.40. The design should provide a step response with an overshoot of less than 10% and a rise time of less than 80 sec.
a) What should the sample rate be?
b) Use emulation design with the matched pole-zero method.
c) Use discrete design and the z-plane root locus.

8.24 The system

$$G(s) = \frac{1}{(s + 0.1)(s + 3)}$$

is to be controlled with a digital controller having a sampling period of $T = 0.1$ sec. Using a z-plane root locus, design compensation that will respond to a step with a rise time $t_r \leqslant 1$ sec and an overshoot $M_p \leqslant 5\%$. What can be done to reduce the steady-state error?

8.25 The transfer function for pure derivative control is

$$D(z) = K T_D \frac{z - 1}{Tz},$$

where the pole at $z = 0$ adds some destabilizing phase lag. Can this phase lag be removed by using derivative control of the form

$$D(z) = K T_D \frac{(z - 1)}{T}?$$

Support your answer with the difference equation that would be required, and discuss the requirements to implement it.

· 9 ·

Control-system Design:
Principles and Case Studies

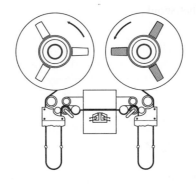

A Perspective on Design Principles

In Chapters 5, 6, and 7 we presented techniques for analyzing and designing feedback systems based on the root-locus, frequency-response, and state-variable methods. Thus far we have had to consider somewhat isolated, idealized aspects of larger systems and to focus on applying one analysis method at a time. In this chapter we return to the theme of Chapter 4—the advantages of feedback control—to reconsider the overall problem of control systems design with the sophisticated tools developed in Chapters 5 to 7 in hand. We will apply these tools to several complex, real-world applications in a case-study–type format.

Having an overarching, step-by-step design approach serves two purposes: It provides a useful starting point for any real-world controls problem, and it provides meaningful checkpoints once the design process is underway. This chapter develops just such a general approach, which will be applied in the case studies.

Chapter Overview

Section 9.1 opens the chapter with a step-by-step design process that is sufficiently general to apply to any control design process, but which also provides useful definition and direction. We then apply the design process to four practical, complex applications: design of an attitude control system for a satellite (Section 9.2), lateral and longitudinal control of a Boeing 747 (Section 9.3); control of the fuel-air ratio in an automotive engine (Section 9.4), and control of a digital tape transport (Section 9.5).

9.1 An Outline of Control Systems Design

Control engineering is an important part of the design process of many dynamic systems. As suggested in Chapter 4, the deliberate use of feedback can stabilize an otherwise unstable system, reduce the error due to disturbance inputs, reduce the tracking error while following a command input, and reduce the sensitivity of a closed-loop transfer function to small variations in internal system parameters. In those situations where feedback control is required, it is possible to outline an approach to control systems design that often leads to a satisfactory solution.

Before describing this approach, we wish to emphasize that the purpose of control is to aid the product or process—the mechanism, the robot, the chemical plant, the aircraft, or whatever—to do its job. Engineers engaged in other areas of the design process are increasingly taking the contribution of control into account early in their plans. As a result, more and more systems have been designed so that they will not work at all without feedback. This is especially significant in the design of high-performance aircraft, where control has taken its place along with structures and aerodynamics as essential to assuring that the craft will even fly at all. It is impossible to give a description of such overall design in this book, but recognizing the existence of such cases places in perspective not only the specific task of control system design but also the central role this task can play in an enterprise.

Control system design begins with a proposed product or process whose satisfactory dynamic performance depends on feedback for stability, disturbance regulation, tracking accuracy, or reduction of the effects of parameter variations. We will give an outline of the design process that is general enough to be useful whether the product is an electronic amplifier or a large structure to be placed in earth orbit. Obviously, to be so widely applicable, our outline has to be vague with respect to physical details and specific only with respect to the feedback-control problem. To present our results, we will divide the control design problem into a sequence of characteristic steps.

Specifications

STEP 1. *Understand the process and translate dynamic performance requirements into time, frequency, or pole-zero specifications.* The importance of understanding the process, what it is intended to do, how much system error is permissible, how to describe the class of command and disturbance signals to be expected, and what the physical capabilities and limitations are can

FIGURE 9.1
Examples of (a) time-response, (ʙ) frequency-response, and (c) pole-zero specifications resulting from Step 1

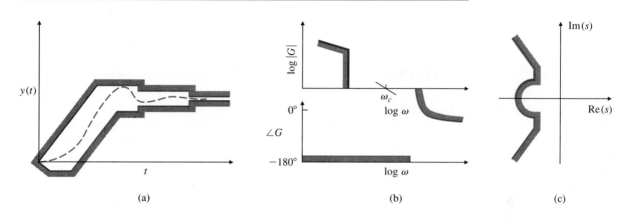

(a)   (b)   (c)

hardly be overemphasized. Regrettably, in a book such as this, it is easy to view the process as a linear, time-invariant transfer function capable of responding to inputs of arbitrary size, and we tend to overlook the fact that the linear model is a *very* limited representation of the real system, valid only for small signals, short times, and particular environmental conditions. Do not confuse the approximation with the reality. You must be able to use the simplified model for its intended purpose, and to return to an accurate model or the actual physical system to really verify the design performance.

Typical results of this step are specifications that the system have a step response inside some constraint boundaries (as shown in Fig. 9.1a), an open-loop frequency response satisfying certain constraints (Fig. 9.1b), or closed-loop poles to the left of some constraint boundary (Fig. 9.1c).

Sensors

STEP 2.   *Select a sensor.* In sensor selection, consider which variables are important to control and which variables can physically be measured. For example, in a jet engine there are critical internal temperatures that must be controlled but that cannot be measured directly in an operational engine. Select sensors that indirectly allow a good estimate to be made of these critical variables. It is important to consider sensors for the disturbance. Sometimes, especially in chemical processes, it is beneficial to sense a load disturbance directly because improved performance can be obtained if this information is fed forward to the controller.

Factors that influence sensor selection are:

Technology	Electric or magnetic, mechanical, electro-mechanical, electrooptical, piezoelectric
Functional performance	Linearity, bias, accuracy, dynamic range, noise

Physical properties	Weight, size, strength
Quality factors	Reliability, durability, maintainability
Cost	Expense, availability, facilities for testing and maintenance

Actuators

STEP 3. *Select an actuator.* In order to control a dynamic system, obviously you must be able to influence the response. The device that does this is the **actuator**. Before choosing a specific actuator, consider which variables can be influenced. For example, in a flight vehicle many configurations of movable surfaces are possible, and the influence these have on the performance and controllability of the craft can be profound. The locations of jets or other torque devices are also a major part of the control design of spacecraft.

Having selected a particular variable to control, you may need to consider other factors:

Technology	Electric, hydraulic, pneumatic, thermal, other
Functional performance	Maximum force possible, extent of the linear range, maximum speed possible, power, efficiency
Physical properties	Weight, size, strength
Quality factors	Reliability, durability, maintainability
Cost	Expense, availability, facilities for testing and maintenance

Linearization

STEP 4. *Make a linear model.* Here you take the best choice for process, actuator, and sensor; identify the equilibrium point of interest; and construct a small-signal dynamic model valid over the range of frequencies included in the specifications of Step 1. You should also validate the model with experimental data where possible. To be able to make use of all the available tools, express the model in state-variable and pole-zero form as well as in frequency-response form. As we have seen, MATLAB and other computer-aided control systems design (CACSD) software packages have the means to perform the transformations among these forms.

Simple compensation

PID/lead-lag design

STEP 5. *Try a simple proportional-integral-derivative (PID) or lead-lag design.* To form an initial estimate of the complexity of the design problem, sketch a frequency response (Bode plot) and a root locus with respect to plant gain. If the plant-actuator-sensor model is stable and minimum-phase, the Bode plot will probably be the most useful; otherwise, the root locus shows very important information with respect to behavior in the right half-plane (RHP). In any case, try to meet the specifications with a simple controller of the lead-lag variety, including integral control if steady-state error response requires it. Do not overlook feedforward of the disturbances if the necessary sensor information is available. Consider the effect of sensor noise, and compare a lead network to a direct sensor of velocity to see which gives a better design.

Reevaluate the specifications, the physical configuration of the process, and the actuator and sensor selections in the light of the preliminary design, and return to Step 1 if improvement seems necessary or feasible. For example, in many motion-control problems, after testing the first-pass design, you might find vibrational modes that prevent the design from meeting the initial specifications of the problem. It may be much easier to meet the specifications by altering the structure of the plant through the addition of stiffening members or by passive damping than to meet them by control strategies alone. An alternative solution may be to move a sensor so it is at a node of a vibration mode, thus providing no feedback of the motion. Also, some actuator technologies (such as hydraulic) have many more low-frequency vibrations than others (such as electric) do. It is important to consider all parts of the design, not only the control logic, to meet the specifications in the most cost-effective way. If the design now seems satisfactory, go to Step 7; otherwise try Step 6.

Optimal design

STEP 6.   *Try an optimal design.* If the trial-and-error compensators do not give entirely satisfactory performance, consider a design based on optimal control. The symmetric root locus will show possible root locations from which to select locations for the control poles that meet the response specifications; you can select locations for the estimator poles that represent a compromise between sensor and process noise. Plot the corresponding open-loop frequency response and the root locus to evaluate the stability margins of this design and its robustness to parameter changes. You can modify the pole locations until a best compromise results. Returning to the symmetric root locus with different cost measures is often a part of this step.

Compare the optimal design yielding the most satisfactory frequency response with the transform-method design you derived in Step 5. Select the better of the two before proceeding to Step 7.

Simulation

STEP 7.   *Build a computer model, and compute (simulate) the performance of the design.*[1] After reaching the best compromise among process modification, actuator and sensor selection, and controller design choice, run a computer model of the system. This model should include important nonlinearities such as actuator saturation, realistic noise sources, and parameter variations you expect to find during operation of the system. The simulation will confirm stability and robustness and allow you to predict the true performance you can expect from the system. As part of this simulation you can often include parameter optimization, in which the computer tunes the free parameters for best performance. In the early stages of design the model you simulate will be relatively simple; as the design progresses, you will study more complete and detailed models. At this step it is also possible to compute a digital equivalent

1. Extensive computer software is available to assist in this difficult but critical step. SIMULINK (from The Mathworks), SystemBuild (Integrated Systems, Inc.), Model-C (Systems Control Technology), SIMNON (Lund Institute of Technology, Lund, Sweden), ACSL (Mitchell and Gauthier), DSL (IBM), and EASY5 (Boeing Computer Services), are just a few of these packages.

of the analog controller as described in Chapter 8. This allows the final design to be implemented with digital processor logic.

If the results of the simulation prove the design satisfactory, go to Step 8; otherwise return to Step 1.

Prototype
Prototype testing

STEP 8.   *Build a prototype.* As the final test before production, it is common to build and test a prototype. At this point you verify the quality of the model, discover unsuspected vibration and other modes, and consider ways to improve the design. After these tests, you may want to reconsider the sensor, actuator, and process and return to Step 1—unless time, money, or ideas have run out.

This outline is an approximation of good practice; other engineers will have variations on these themes. In some cases you way wish to carry out the steps in a different order, to omit a step, or to add one. The stages of simulation and prototype construction vary widely, depending on the nature of the system. For systems where a prototype is difficult to test and rework (for example, a satellite) or where a failure is dangerous (for example, active stabilization of a high-speed centrifuge or landing a human on the moon), most of the design verification is done through simulation of some sort. It may take the form of a digital numerical simulation, a laboratory-scale model, or a full-size laboratory model with a simulated environment. For systems that are easy to build and modify (for example, feedback control for an automotive fuel system), the simulation step is often skipped entirely; design verification and refinement are instead accomplished by working with prototypes.

Implicit in the process of design is the well-known fact that designs within a given category often draw on experience gained from earlier models. Thus good designs evolve rather than appear in their best form after the first pass. We will illustrate the method with a few cases (Sections 9.2 to 9.5). For easy reference, we summarize the steps here.

Summary of Control Design Steps

1. Understand the process and its performance requirements.
2. Select the best sensor considering location, technology, and noise.
3. Select the best actuator considering location, technology, and power.
4. Make a linear model of the process, actuator, and sensor.
5. Make a simple trial design based on the concepts of lead-lag compensation or PID control. If satisfied, go to Step 7.
6. Make a trial pole-placement design based on optimal control or other criteria.
7. Simulate the design, including the effects of nonlinearities, noise,

and parameter variations. If the performance is not satisfactory, return to Step 1 and repeat.

8. Build a prototype and test it. If not satisfied, return to Step 1 and repeat.

9.2   Design of a Satellite's Attitude Control

Our first example, taken from the space program, is suggested by the need to control the pointing direction, or attitude, of a satellite in orbit about the earth. We will go through each step in our design outline and touch on some of the factors that might be considered for the control of such a system.

STEP 1.   *Understand the process and its performance specifications.* A satellite is sketched in Fig. 9.2. We imagine that the vehicle has an astronomical survey mission requiring accurate pointing of a scientific sensor package. This package must be maintained in the quietest possible environment, which entails isolating it from the vibrations and electrical noise of the main service body and from its power supplies, thrusters, and communication gear. We model the resulting structure as two masses connected by a flexible boom. Disturbance torques due to solar pressure, micrometeorites, and orbit perturbations are computed to be negligible. The pointing requirement arises when it is necessary to point the unit in another direction, it can be met by dynamics with a transient settling time of 20 sec and an overshoot of no more than 15%. The

FIGURE 9.2
Diagram of a satellite and its two-body model

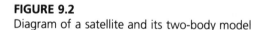

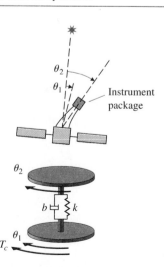

dynamics of the satellite include parameters that can vary. The control must be satisfactory for any parameter values in a prespecified range to be given when the equations are written.

STEP 2. *Select a sensor.* In order to orient the scientific package, it is necessary to measure the attitude angles of the vehicle. For this purpose we propose to use a **star tracker**, a system based on gathering an image of a specific star and keeping it centered on the focal plane of a telescope. This sensor gives a relatively noisy but very accurate (on the average) reading proportional to θ_2, the angle of deviation of the instrument package from the desired angle. To stabilize the control, we include a rate gyro to give a clean reading of $\dot\theta_2$, since a lead network on the star-tracker signal would amplify the noise too much. Furthermore, the rate gyro can stabilize large motions before the star tracker has acquired the target star image.

STEP 3. *Select an actuator.* Major considerations in selecting the actuator are precision, reliability, weight, power requirements, and lifetime. Alternatives for applying torque are cold-gas jets, reaction wheels or gyros, magnetic torquers, and a gravity gradient. The jets have the most power and are the least accurate. Reaction wheels are precise but can only transfer momentum, so jets or magnetic torquers are required to "dump" momentum from time to time. Magnetic torquers provide relatively low levels of torque and are only suitable for some low-altitude satellite missions. A gravity gradient also provides a very small torque that limits the speed of response and places severe restrictions on the shape of the satellite. For purposes of this mission, we select cold-gas jets as being fast and adequately accurate.

STEP 4. *Make a linear model.* For the satellite we assume two masses connected by a spring with torque constant k and viscous-damping constant b as shown in Fig. 9.2. The equations of motion are

$$J_1\ddot\theta_1 + b(\dot\theta_1 - \dot\theta_2) + k(\theta_1 - \theta_2) = T_c, \tag{9.1a}$$

$$J_2\ddot\theta_2 + b(\dot\theta_2 - \dot\theta_1) + k(\theta_2 - \theta_1) = 0, \tag{9.1b}$$

where T_c is the control torque on the main body. With inertias $J_1 = 1$ and $J_2 = 0.1$, the transfer function is

$$G(s) = \frac{10bs + 10k}{s^2(s^2 + 11bs + 11k)}. \tag{9.2}$$

If we choose

$$\mathbf{x} = [\theta_2 \quad \dot\theta_2 \quad \theta_1 \quad \dot\theta_1]^T$$

as the state vector, then, using Eq. (9.1) and assuming $T_c \equiv u$, the equations of motion in state-variable form are

$$\dot{\mathbf{x}} = \begin{bmatrix} 0 & 1 & 0 & 0 \\ -\dfrac{k}{J_2} & -\dfrac{b}{J_2} & \dfrac{k}{J_2} & \dfrac{b}{J_2} \\ 0 & 0 & 0 & 1 \\ \dfrac{k}{J_1} & \dfrac{b}{J_1} & -\dfrac{k}{J_1} & -\dfrac{b}{J_1} \end{bmatrix} \mathbf{x} + \begin{bmatrix} 0 \\ 0 \\ 0 \\ \dfrac{1}{J_1} \end{bmatrix} u, \tag{9.3a}$$

$$y = [1 \quad 0 \quad 0 \quad 0]\mathbf{x}. \tag{9.3b}$$

Physical analysis of the boom leads us to assume that the parameters k and b vary as a result of temperature fluctuations but are bounded by

$$0.09 \leqslant k \leqslant 0.4 \tag{9.4a}$$

$$0.038\sqrt{\frac{k}{10}} \leqslant b \leqslant 0.2\sqrt{\frac{k}{10}}. \tag{9.4b}$$

As a result, the vehicle's natural resonance frequency ω_n can vary between 1 and 2 rad/sec, and the damping ratio ζ varies between 0.02 and 0.1.

One approach to control design when parameters are subject to variation is to select nominal values for the parameters, contruct the design for this model, and then test the controller performance with other parameter values. In the present case we choose nominal values of $\omega_n = 1$ and $\zeta = 0.02$. The choice is somewhat arbitrary, being based on experience and heuristic analysis. However, note that these are the lowest values in their respective ranges and thus correspond to the plant that is probably the most difficult to control so as to meet the specifications. We assume that a design for this model has a good chance to meet the specifications for other parameter values as well. (Another choice would be to select a model with average values for each parameter). The selected parameter values are $k = 0.091$ and $b = 0.0036$; with $J_1 = 1$ and $J_2 = 0.1$, the nominal equations become

Selecting nominal values for varying parameters

$$\dot{\mathbf{x}} = \begin{bmatrix} 0 & 1 & 0 & 0 \\ -0.91 & -0.036 & 0.91 & 0.036 \\ 0 & 0 & 0 & 1 \\ 0.091 & 0.0036 & -0.091 & -0.0036 \end{bmatrix} \mathbf{x} + \begin{bmatrix} 0 \\ 0 \\ 0 \\ 1 \end{bmatrix} u, \tag{9.5a}$$

$$y = [1 \quad 0 \quad 0 \quad 0]\mathbf{x}. \tag{9.5b}$$

The corresponding transfer function is then

$$G(s) = \frac{0.036(s + 25)}{s^2(s^2 + 0.04s + 1)}. \tag{9.6}$$

When a trial design is completed, the computer simulation should be run with a range of possible parameter values to ensure that the design has

FIGURE 9.3
Root locus of $KG(s)$

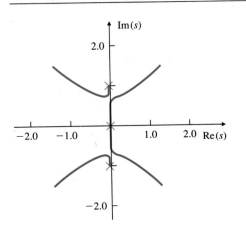

FIGURE 9.4
Open-loop Bode plot of $KG(s)$ for $K = 0.5$

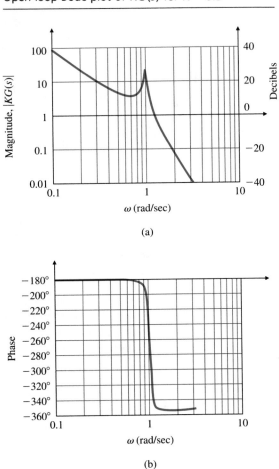

(a)

(b)

adequate robustness to tolerate these changes. Equations (3.42) tell us that the dynamic performance specifications will be met if the closed-loop poles have a natural frequency of 0.5 rad/sec and a closed-loop damping ratio of 0.5; these coprrespond to an open-loop crossover frequency of $\omega_c \cong 0.5$ rad/sec and a phase margin of about PM = 50°. We will try to meet these design criteria.

STEP 5.   *Try a lead-lag or PID controller.* The proportional-gain root locus for the nominal plant is drawn in Fig. 9.3, and the Bode plot is given in Fig. 9.4. We can see from Fig. 9.4 that this may be a difficult design problem because the frequency of the lightly damped resonance is greater than the crossover-frequency design point by only a factor of 2. This situation will require that the compensation can correct for the phase lag of the plant at the

FIGURE 9.5
Root locus of $KD_1(s)G(s)$

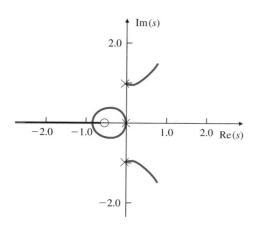

FIGURE 9.6
Bode plot of $KD_1(s)G(s)$

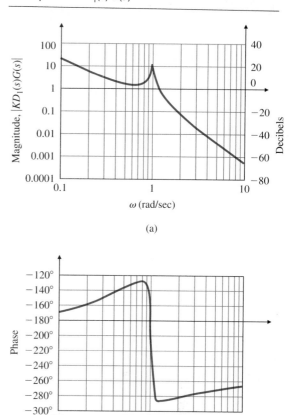

(a)

(b)

resonance frequency. Such correction is very dependent on knowing the resonance frequency, which is subject to change in this case. There may be trouble ahead.

In order to illustrate some important aspects of compensation design, we will at first ignore the resonance and generate a design for the rigid body alone. We take the process transfer function to be $1/s^2$, the feedback to be position plus derivative (star tracker plus rate gyro) or PD control with the transfer function $D(s) = K(sT + 1)$, and the response objective to be $\omega_n = 0.5\,\text{rad/sec}$ and $\zeta = 0.5$. A suitable controller is thus

$$D_1(s) = 0.25(2s + 1). \tag{9.7}$$

The root locus is shown in Fig. 9.5 and the Bode plot in Fig. 9.6. From these

FIGURE 9.7
Root locus of $KD_2(s)G(s)$

FIGURE 9.8
Bode plot of $D_2(s)G(s)$

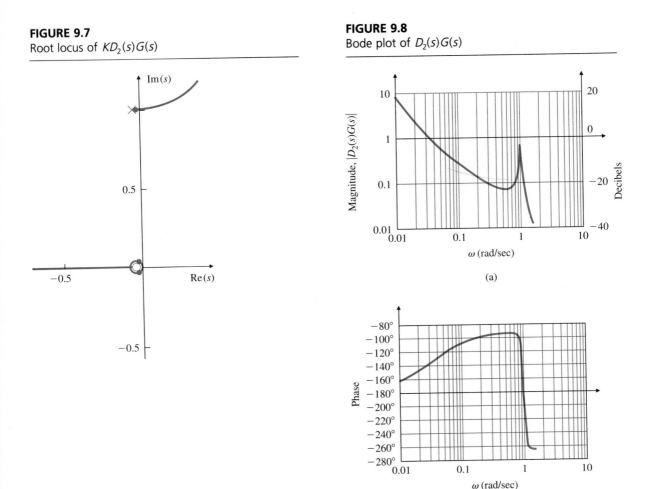

plots we can see that the low-frequency poles are reasonable but that the system will be unstable because of the resonance.[2] At this point we take the simple actions of reducing our expectation with respect to bandwidth and of slowing the system down by lowering the gain until the system is stable. With so little damping, we must really go slowly. A bit of experimentation leads to

$$D_2(s) = 0.001(30s + 1), \tag{9.8}$$

2. If this system were built, the actuator jets would saturate as the response grew. We could analyze the response using the method described in Section 5.7.2 for nonlinear systems. From the analysis we would expect the signal to grow and the equivalent gain of the actuator to fall until the roots return to the imaginary axis near ω_n. The resulting limit cycle would rapidly deplete the control gas supply.

for which the root locus is drawn in Fig. 9.7 and the Bode plot given in Fig. 9.8. The Bode plot shows we have a phase margin of 50° but a crossover frequency of only $\omega_c = 0.04$ rad/sec. While this is too low to meet the settling-time specification, a low crossover frequency is unavoidable if we expect to keep the gain at the resonance frequency below unity.

An alternative approach to the problem is to place zeros near the lightly damped poles and use them to hold these poles back from the RHP. Such a compensation has a frequency response with a very low gain near the frequency of the offending poles and a reasonable gain elsewhere. Because the frequency response seems to have a dent or notch in it, the device is called a **notch filter**. (It is also called a **band reject filter** in electric network theory.) An RC circuit with a notch characteristic is shown in Fig. 9.9, its pole-zero pattern in Fig. 9.10, and its frequency response in Fig. 9.11. The $+180°$ phase lead of the

FIGURE 9.9
Twin-tee realization of a notch filter

FIGURE 9.11
Bode plot of a notch filter

FIGURE 9.10
Notch filter pole-zero patter

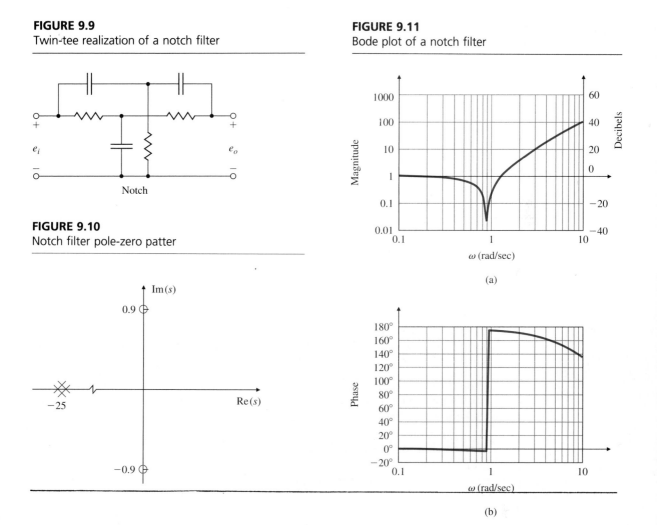

FIGURE 9.12

Bode plot of $KD_3(s)G(s)$

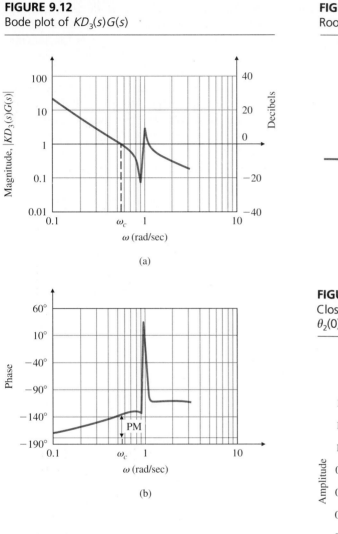

(a)

(b)

FIGURE 9.13

Root locus of $KD_3(s)G(s)$

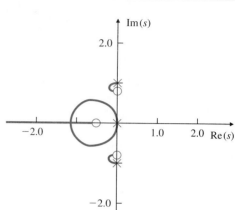

FIGURE 9.14

Closed-loop step response of $D_3(s)G(s)$ where $\theta_2(0) = 0.2$ rad

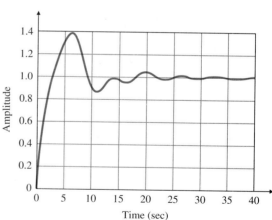

notch can be used to correct for the 180° phase lag of the resonance; if the notch frequency is *lower* than the plant's resonance frequency, the system phase is kept above 180° near resonance.

With this idea we return to the compensation given by Eq. (9.7) and add the notch, producing the revised compensator transfer function,

$$D_3(s) = 0.25(2s + 1) \frac{(s/0.9)^2 + 1}{[(s/25) + 1]^2}. \tag{9.9}$$

The Bode plot for this case is shown in Fig. 9.12, the root locus in Fig. 9.13, and the unit step response in Fig. 9.14. The settling time of the design is too long for the specification and the overshoot is too large, but this design approach seems promising; with iteration it could lead to a satisfactory compensator.

We now recall that the compensator is expected to provide adequate performance as the parameters vary over the ranges given by Eqs. (9.4). An examination of the robustness of the design can be made by looking at the root locus shown in Fig. 9.15, which is drawn using the compensator of Eq. (9.9) and the plant with $\omega_n = 2$ rather than 1, such that

$$\hat{G}(s) = \frac{s/50 + 1}{s^2(s^2/4 + 0.02s + 1)}. \tag{9.10}$$

FIGURE 9.15
Root locus of $KD_3(s)\hat{G}(s)$

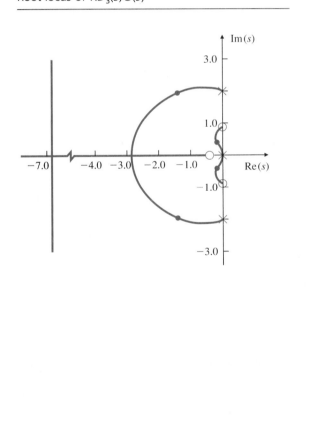

FIGURE 9.16
Bode plot of $KD_3(s)\hat{G}(s)$

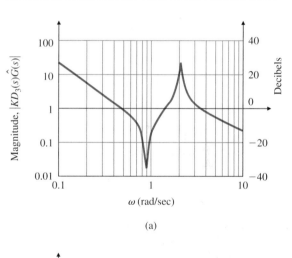

(a)

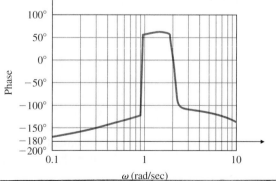

(b)

FIGURE 9.17

Closed-loop step
response of $D_3(s)\hat{G}(s)$

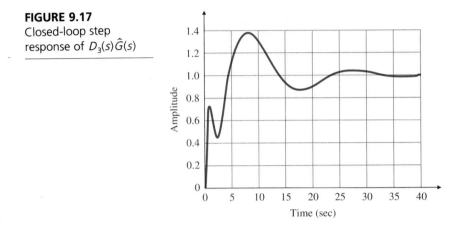

This assumes that the boom is as stiff as possible. Notice that now the low-frequency poles have a damping ratio of only 0.25. Combining the various parameter values we get the frequency response and transient response shown in Figs. 9.16 and 9.17. We could make a few more trial-and-error iterations with the notch filter and rate feedback, but the system is complex enough that a look at state-space designs now seems reasonable. We go to Step 6.

STEP 6.   *Try an optimal design using pole placement.* Using the state-variable formulation of the equations of motion in Eq. (9.5), we devise a controller that will place the closed-loop poles in arbitrary locations. Of course, used without thought, the method of pole placement can also result in a design that requires unreasonable levels of control effort or is very sensitive to changes in the plant transfer function. Guidelines for pole placement are given in Chapter 7; an often successful approach is to derive optimal pole locations using the symmetric root locus (SRL). Figure 9.18 shows the SRL for the problem at hand. To obtain a bandwidth of about 0.5 rad/sec, we select closed-loop control poles from this locus at $-0.45 \pm 0.34j$ and $-0.15 \pm 1.05j$.

If we select $\alpha_c(s)$, as discussed earlier, from the SRL, the control gain is

$$\mathbf{K} = [-0.2788 \quad 0.0546 \quad 0.6814 \quad 1.1655]. \tag{9.11}$$

Figure 9.19 shows the step responses for the nominal-plant-parameters and stiff-spring-plant models. The Bode plot of the SRL controller design with the nominal plant parameters (Fig. 9.20) shows a phase margin of about 60°. While the speed of response of the design meets the specifications with the nominal plant, the settling time when the plant has the stiff spring is a bit longer than the specifications call for. We might be able to get a better compromise between the nominal and the stiff-spring cases by selecting another point on

FIGURE 9.18
Symmetric root locus of the satellite

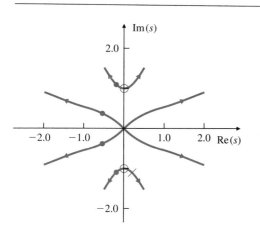

FIGURE 9.20
Frequency response of the SRL design from u to $K\mathbf{x}$

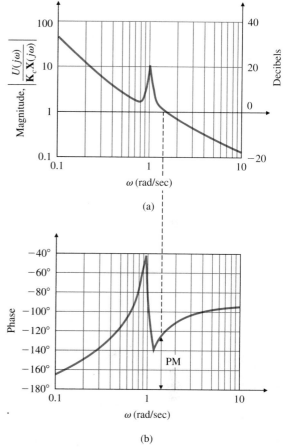

(a)

(b)

FIGURE 9.19
Closed-loop step response of the SRL design

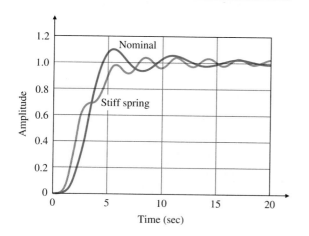

Time (sec)

the SRL; at this point we do not know. The designer must face alternatives such as these and select the best compromise for the problem at hand.

The design of Fig. 9.19 is based on full state feedback. To complete the optimal design we need an estimator. We select the closed-loop estimator error poles to be about eight times faster than the control poles. The reason for this is to keep the error-poles from reducing the robustness of the design; a fast estimator will have almost the same effect on the response as no estimator at all. We choose the error poles from the SRL at $-7.7 \pm 3.12j$ and $-3.32 \pm 7.85j$.

FIGURE 9.21
Bode plot of optimal
compensator $D_4(s)$

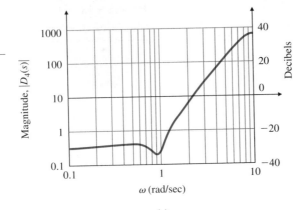

(a)

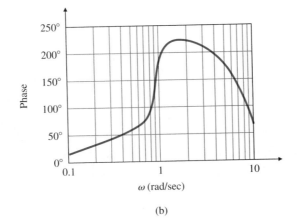

(b)

Pole placement with these values leads to an estimator (filter) gain of

$$\mathbf{L} = \begin{bmatrix} 22 \\ 242.3 \\ 1515.4 \\ 5503.9 \end{bmatrix}. \tag{9.12}$$

After we combine the control gain and estimator as described in Section 7.6, the compensator transfer function that results is

$$D_4(s) = \frac{-745(s + 0.3217)(s + 0.0996 \pm 0.9137j)}{(s + 3.1195 \pm 8.3438j)(s + 8.4905 \pm 3.6333j)}. \tag{9.13}$$

The frequency response of this compensator (Fig. 9.21) shows that pole placement has introduced a notch directly. The frequency response and the root locus of the combined system $D_4(s)G(s)$ are given in Figs. 9.22 and 9.23,

FIGURE 9.22
Bode plot of the compensated system $D_4(s)G(s)$

FIGURE 9.23
Root locus of $D_4(s)G(s)$

FIGURE 9.24
Closed-loop step response of $D_4(s)G(s)$

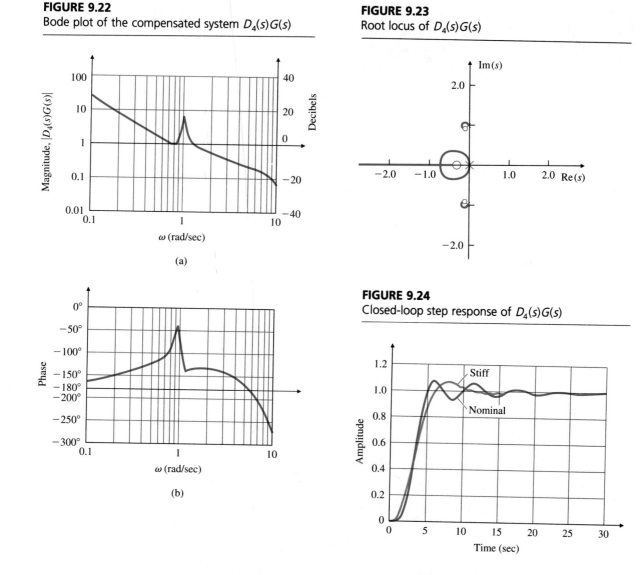

(a)

(b)

while Fig. 9.24 shows the step response for both the nominal and the stiff-spring plants. Notice that the design almost meets the specifications.

STEP 7.   *Simulate the design, and compare the alternatives.* At this point we have two designs, with differing complexities and different robustness properties. The notch-filter design might be improved with further iterations or by starting with a different nominal case. The SRL design meets the specifications for the nominal plant but is too slow for the stiff-spring case, although alternative selections for the pole locations might lead to a better design. In either case, much more extensive studies need to be made to explore the robustness and noise-response properties. Rather than follow any of these paths, we consider some aspects of the physical system.

Both designs are strongly influenced by the presence of the lightly damped resonant mode caused by the coupled masses. However, the transfer function of this system is strongly dependent on the fact that the actuator is on one body and the sensor is on the other (that is, not collocated). Suppose that, rather than considering pointing the star tracker on the small mass, we have the mission of pointing the main mass, perhaps toward a Earth station for communications purposes. For this purpose we can put the sensor on the *same* mass that holds the actuator—to give control with a collocated actuator and sensor. Due to the physics of the situation, the system's transfer function now has zeros close to the flexible modes, so control can be achieved by using PD feedback alone since the plant already has the effect of a notch compensator. Consider the transfer function of the satellite with collocated actuator and sensor (to measure θ_1) for which the state matrices are

$$\mathbf{F} = \begin{bmatrix} 0 & 1 & 0 & 0 \\ -0.91 & -0.036 & 0.91 & 0.036 \\ 0 & 0 & 0 & 1 \\ 0.091 & 0.0036 & -0.091 & -0.0036 \end{bmatrix}, \quad \mathbf{G} = \begin{bmatrix} 0 \\ 0 \\ 0 \\ 1 \end{bmatrix},$$

$$\mathbf{H} = [0 \quad 0 \quad 1 \quad 0].$$

The transfer function of the system is

$$G_{co}(s) = \mathbf{H}(s\mathbf{I} - \mathbf{F})^{-1}\mathbf{G} = \frac{s + 0.018 \pm 0.954j}{s^2(s + 0.02 \pm j)}. \tag{9.14}$$

Notice the presence of the zeros in the vicinity of the complex conjugate poles. If we now use the same PD feedback as before,

$$D_5(s) = 0.25(2s + 1), \tag{9.15}$$

then the system will not only be stabilized, but will also have a satisfactory response (if we consider θ_1 as the output), since the resonant poles tend to be cancelled by the complex conjugate zeros. Figures 9.25 to 9.27 show the frequency response, the root locus and the step response, respectively, for this system. Note from Fig. 9.27 that the step response has the excess overshoot associated with the zero of the compensator in the forward path of the transfer function.

The result is a very simple robust design achieved by moving the sensor from a noncollocated position to one collocated with the actuator. The result illustrates that, to achieve good feedback control, it is very important to consider sensor location and other features of the physical problem. However, this last control design will *not* do for pointing the star tracker. This is evident

FIGURE 9.25
Bode plot of $D_5(s)G_{co}(s)$

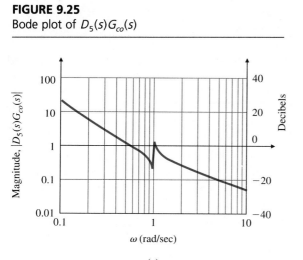

(a)

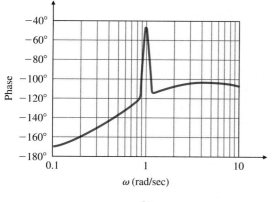

(b)

FIGURE 9.26
Root locus for $D_5(s)G_{co}(s)$

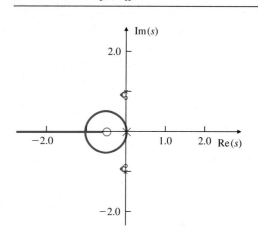

FIGURE 9.27
Closed-loop step response of the system with collocated control, $D_5(s)G_{co}(s)$ and $D_5(s)\hat{G}_{co}(s)$

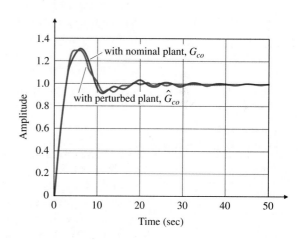

FIGURE 9.28
Response at θ_2 of the collocated design

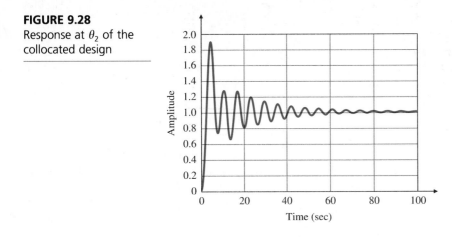

from plotting the output θ_2 corresponding to the nice step response of Fig. 9.27. The result is shown in Fig. 9.28.

9.3   Lateral and Longitudinal Control of a Boeing 747

The Boeing 747 (Fig. 9.29) is a large wide-body transport jet. A schematic with the relevant coordinates that move with the airplane is shown in Fig. 9.30. The linearized equations of (rigid-body) motion for the Boeing 747 are of eighth order but are separated into two fourth-order sets representing the perturbations in longitudinal (u, w, θ, and q in Fig. 9.30) and lateral motion (ϕ, β, r, and p). The longitudinal motion consists of axial (x), vertical (z), and pitching (θ, q)

FIGURE 9.29
Boeing 747 (*courtesy Boeing Commercial Airplane Co.*)

FIGURE 9.30

Definition of aircraft coordinates

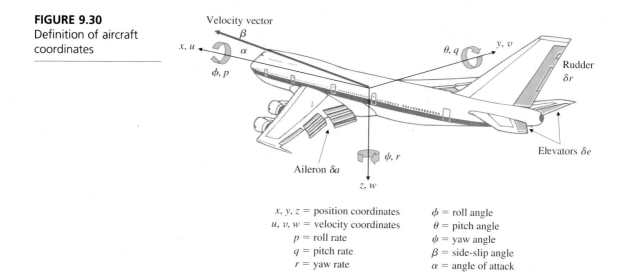

x, y, z = position coordinates ϕ = roll angle
u, v, w = velocity coordinates θ = pitch angle
p = roll rate ψ = yaw angle
q = pitch rate β = side-slip angle
r = yaw rate α = angle of attack

motion, while the lateral motion consists of rolling (ϕ, p), yawing (r), and lateral (y) movement. The side-slip angle β is a measure of the direction of forward velocity relative to the direction of the nose of the airplane. The elevator control surfaces and the throttle affect the longitudinal motion, whereas the aileron and rudder primarily affect lateral motion. Although there is a small amount of couping of lateral motion into longitudinal motion, this is usually ignored, so the equations of motion are treated as two decoupled fourth-order sets for designing the control, or **stability augmentation**, for the aircraft. We will discuss the design of a stability-augmentation system for the lateral dynamics, called a **yaw damper**, and the autopilot affecting the longitudinal behavior.

9.3.1   Yaw Damper

STEP 1.   *Understand the process and its performance specifications.* Swept-wing aircraft have a natural tendency to be lightly damped in one of the lateral modes of motion. At typical commercial-aircraft cruising speeds and altitudes, this dynamic mode is sufficiently difficult to control that virtually every swept-wing aircraft has a feedback system to help the pilot. Therefore, the goal of our control system is to modify the natural dynamics so that the plane is pleasant for the pilot to fly. Studies have shown that pilots like natural frequencies $\omega_n \lesssim 0.5$ and damping ratio of $\rho \geqslant 0.5$. Aircraft with dynamics that violate these guidelines are generally considered fatiguing to fly and highly undesirable. Thus our system specifications are to achieve lateral dynamics that meet these constraints.

STEP 2.   *Select a sensor.* The easiest measurement of aircraft motion to take is the angular rate. Velocities can also be measured using pitot tubes and wind-vane devices, but these are noisier and less reliable for stabilization. Two angular rates—roll and yaw—partake in the lateral motion. Study of the lightly damped lateral mode indicates that it is primarily a yawing phenomenon, so measurement of the yaw rate is a logical starting point for the design. Until the early 1980s the measurement was made with a **gyroscope**, essentially a small, fast-spinning rotor that can yield an electric output proportional to the angular yaw rate of the aircraft. Since the early 1980s most new aircraft systems have relied on a laser device (called a **ring-laser gyroscope**) for the measurement. Here two laser beams traverse a closed path (often a triangle) in opposite directions. As the triangular device rotates, the detected frequencies of the two beams shift according to the Doppler effect, and this frequency shift is measured, producing a measure of rotational rate. These devices have fewer moving parts and are more reliable at less cost than the spinning-rotor variety of gyroscope.

STEP 3.   *Select an actuator.* Two aerodynamic surfaces typically influence the lateral aircraft motion: the rudder and the ailerons (see Fig. 9.30). The lightly damped yaw mode that will be stabilized by the yaw damper is most affected by the rudder. Therefore, use of that single control input is a logical starting point for the design. Hydraulic devices are universally employed in large aircraft to provide the force that moves the aerodynamic surfaces. No other kind of device has been developed to provide the combination of high force, high speed, and light weight desirable for the actuation of the controlling aerodyamic surfaces. On the other hand, the low-speed flaps, which are extended slowly prior to landing, are typically actuated by an electric motor with a worm gear. For small aircraft with no stabilization system, no actuator is required at all; the pilot wheel is directly connected to the aerodynamic surface by means of wire cables, and all the force required to move the surfaces is provided by the pilot.

STEP 4.   *Make a linear model.* The lateral-perturbation equations of motion in horizontal flight at 40,000 ft and nominal forward speed $U_0 = 774\,\text{ft/sec}$ (Mach 0.8) are (Heffley and Jewell, 1972)

$$\begin{bmatrix} \dot{\beta} \\ \dot{r} \\ \dot{p} \\ \dot{\phi} \end{bmatrix} = \begin{bmatrix} -0.0558 & -0.9968 & 0.0802 & 0.0415 \\ 0.598 & -0.115 & -0.0318 & 0 \\ -3.05 & 0.388 & -0.4650 & 0 \\ 0 & 0.0805 & 1 & 0 \end{bmatrix} \begin{bmatrix} \beta \\ r \\ p \\ \phi \end{bmatrix} + \begin{bmatrix} 0.00729 \\ -0.475 \\ 0.153 \\ 0 \end{bmatrix} \delta r,$$

$$y = \begin{bmatrix} 0 & 1 & 0 & 0 \end{bmatrix} \begin{bmatrix} \beta \\ r \\ p \\ \phi \end{bmatrix},$$

where β and ϕ are in radians and r and p are in radians per second. The transfer function is

$$G(s) = \frac{r(s)}{\delta r(s)} = \frac{-0.475(s + 0.498)(s + 0.012 \pm 0.488j)}{(s + 0.0073)(s + 0.563)(s + 0.033 \pm 0.947j)} \tag{9.16}$$

so that the system has two stable real poles and a pair of stable complex poles. Notice first that the low-frequency gain is negative, corresponding to the simple physical fact that a positive or clockwise rudder motion causes a negative or counterclockwise yaw rate. In other words, turning the rudder left (clockwise) causes the front of the aircraft to rotate left (counterclockwise). The natural motion corresponding to the complex poles is referred to as the **Dutch roll**; the name comes from the motions of a person skating on the frozen canals of Holland. The motion corresponding to the stable real poles is referred to as the **spiral mode** ($s_1 = -0.0073$) and the **roll mode** ($s_2 = -0.563$). From looking at the system poles, we see that the offending mode that needs repair for good pilot handling is the Dutch roll, with the poles at $s = -0.033 \pm 0.95j$. The roots have an acceptable frequency, but their damping ratio $\zeta \cong 0.03$ is far short of the desired value $\zeta \cong 0.5$.

STEP 5.   *Try a lead-lag or PID design.* As a first try at the design, we will consider proportional feedback of the yaw rate to the rudder. The root locus with respect to the gain of this feedback is shown in Fig. 9.31, and its frequency response is shown in Fig. 9.32. The figures show that $\zeta \cong 0.45$ is achievable and can be computed to occur at a gain of about 3.0.

In practice, however, we find that this simple feedback changes the overall gain from the pilot to the roll rate at low frequencies from about 15 to one over the feedback gain or about 0.33. This change in DC or zero-frequency gain creates an objectionable situation during a steady turn: Because the feedback produces a steady rudder input opposite the pilot's input, the pilot must introduce a much larger steady command for the same yaw rate than is

FIGURE 9.31

Root locus for yaw damper with proportional feedback

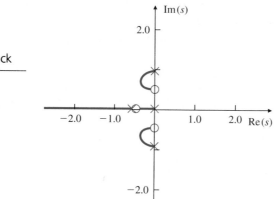

FIGURE 9.32
Bode plot of yaw
damper with
proportional feedback

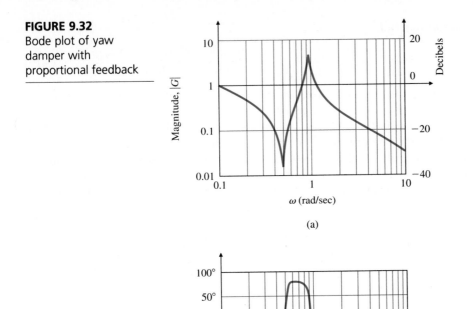

(a)

(b)

necessary in the open-loop case. This dilemma is solved by feeding back not the yaw rate but the *derivative* of the yaw rate, since the derivative-feedback term will be zero in the steady state. In practice the derivative is approximated by including in the feedback path a lead compensation with its zero at the origin (called a **washout circuit**). The result is a feedback that passes transients and provides the desired damping but washes out steady and low-frequency signals. Figure 9.33 shows a block diagram of the yaw damper with the washout.

The washout circuit

The transfer function of the washout circuit is

$$H(s) = \frac{e(s)}{e_{yrg}(s)} = \frac{s}{s + 1/\tau}.$$

FIGURE 9.33
Yaw damper:
(a) Functional block
diagram; (b) block
diagram for analysis

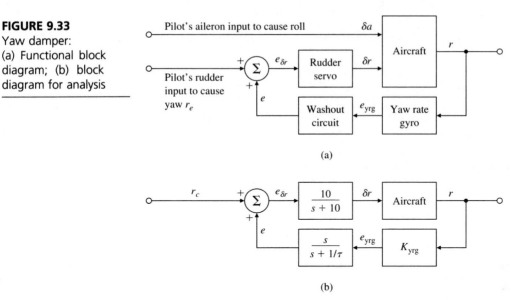

(a)

(b)

For a more complete model, we include the rudder servo, which represents the actuator dynamics and has the transfer function

$$A(s) = \frac{\delta r(s)}{e_{\delta r}(s)} = \frac{10}{s + 10},$$

which is fast compared with the dynamics of the rest of the system and is not expected to change the response very much. The root locus including actuator dynamics and a washout circuit with $\tau = 3$ is shown in Fig. 9.34. As seen from the root locus, the addition of the washout allows the damping ratio to be

FIGURE 9.34
Root locus with washout
circuit, $\tau = 3$

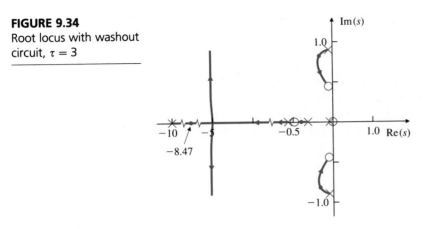

FIGURE 9.35
Bode plot of yaw
damper, including
washout and actuator

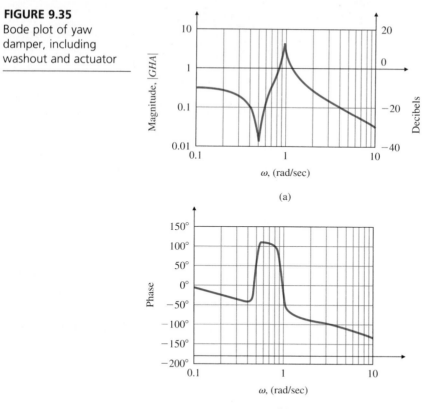

(a)

(b)

FIGURE 9.36
Initial-condition response
with yaw damper and
washout for $\beta_0 = 1°$

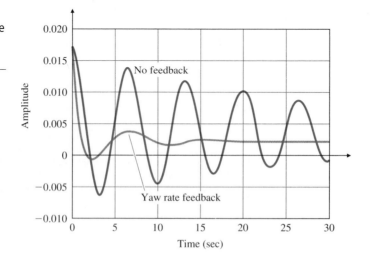

increased from 0.03 to about 0.35. The associated frequency response of the system is shown in Fig. 9.35. The response of the closed-loop system to an initial condition of $\beta_0 = 1°$ is shown in Fig. 9.36 for a root-locus gain of 2.6. For reference the response of yaw rate with no feedback is also given. Although feedback of yaw rate through the washout circuit results in a considerable improvement over the original aircraft control, the response is not as good as originally specified. Further interactions, not included here, could include other gain values or more complex compensations.

STEP 6. *Try an optimal design using pole placement.* If we augment the dynamic model of the system by adding the actuator and washout, we obtain the following state-variable model:

$$\begin{bmatrix} \dot{x}_A \\ \dot{\beta} \\ \dot{r} \\ \dot{p} \\ \dot{\phi} \\ \dot{x}_{co} \end{bmatrix} = \begin{bmatrix} -10 & 0 & 0 & 0 & 0 & 0 \\ 0.0729 & -0.0558 & -0.997 & 0.0802 & 0.0415 & 0 \\ -4.75 & 0.598 & -0.1150 & -0.0318 & 0 & 0 \\ 1.53 & -3.05 & 0.388 & -0.465 & 0 & 0 \\ 0 & 0 & 0.0805 & 1 & 0 & 0 \\ 0 & 0 & 1 & 0 & 0 & -0.333 \end{bmatrix} \begin{bmatrix} x_A \\ \beta \\ r \\ p \\ \phi \\ x_{wo} \end{bmatrix} + \begin{bmatrix} 10 \\ 0 \\ 0 \\ 0 \\ 0 \\ 0 \end{bmatrix} e_{\delta r},$$

$$e = \begin{bmatrix} 0 & 0 & 1 & 0 & 0 & -0.333 \end{bmatrix} \begin{bmatrix} x_A \\ \beta \\ r \\ p \\ \phi \\ x_{\omega o} \end{bmatrix},$$

where $e_{\delta r}$ is the input to the actuator and e is the output of the washout. The SRL for the augmented system is as shown in Fig. 9.37. If we select the state-feedback poles from the SRL so that the complex roots have maximum damping ($\zeta = 0.4$), we find that

$$\alpha_c(s) = (s + 0.0051)(s + 0.468)(s + 0.279 \pm 0.628j)(s + 1.106)(s + 9.89).$$

Then we can compute the state-feedback gain to be

$$\mathbf{K} = \begin{bmatrix} 1.059 & -0.191 & -2.32 & 0.0992 & 0.0370 & 0.486 \end{bmatrix}.$$

Note that the third entry in **K** is larger than the others, so the feedback of all six state variables is essentially the same as proportional feedback of r. This is also evident from the similarity of the root locus in Fig. 9.31 and the SRL of Fig. 9.37. If we select the estimator poles to be five times faster than the controller poles, then

$$\alpha_e(s) = (s + 0.0253)(s + 2.34)(s + 1.39 \pm 3.14j)(s + 5.53)(s + 49.5),$$

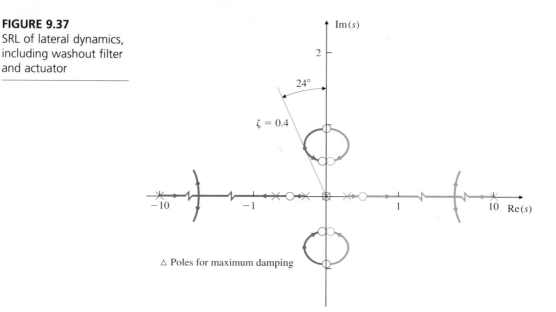

and the estimator gain is found to be

$$
\mathbf{L} = \begin{bmatrix}
25.0 \\
-2{,}044 \\
-5{,}158 \\
-24{,}843 \\
-40{,}113 \\
-15{,}624
\end{bmatrix}.
$$

The compensator transfer function from Eq. (7.153) is

$$
D_c(s) = \frac{-844(s + 10.0)(s - 1.04)(s + 0.974 \pm 0.559j)(s + 0.0230)}{(s + 0.0272)(s + 0.837 \pm 0.671j)(s + 4.07 \pm 10.1j)(s + 51.3)} \quad (9.17)
$$

Figure 9.38 shows the response of the system to an initial condition of $\beta_0 = 1°$.
It is clear from the root locus that the damping can be improved by the SRL
approach, and this is borne out by the reduced oscillatory behavior in the
transient response of the system. However, this improvement has come at a
considerable price. Note that the order of the compensator has increased from
1 in the original design (Fig. 9.33) to 6 in the design obtained using the
controller-estimator-SRL approach.

Design tradeoff:
system response vs.
system complexity

Aircraft yaw dampers in use today generally employ a proportional
feedback of yaw rate to rudder through a washout or through minor modifications to this design. The improved performance achievable with an optimal
design approach utilizing full state feedback and estimation is not judged to be
worth the increase in complexity.

FIGURE 9.38

Initial condition response for $\beta_0 = 1°$, $\dot{x}_0 = 0$

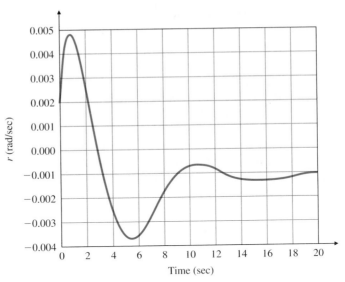

Perhaps a more fruitful approach to improving the design would be to add the aileron surface as a control variable along with the rudder. This approach also has not been considered worth its complexity.

STEPS 7 and 8.   *Verify the design.* Linear models of aircraft motion are reasonably accurate as long as the motion is small enough that the actuators and surfaces do not saturate. Since actuators are sized for safety in order to handle large transients, such saturation is very rare. Therefore, the linear-analysis-based design is reasonably accurate, and we will not pursue a nonlinear simulation or further design verification.

9.3.2   Altitude-hold Autopilot

STEP 1.   *Understand the process and its performance specifications.* One of the pilot's many tasks is to hold a specific altitude. As an aid to keeping aircraft from colliding, those craft on an easterly path are required to be on an odd multiple of 1000 ft and those on a westerly path on an even multiple of 1000 ft. Therefore, the pilot needs to be able to hold the altitude to within a few hundred feet. A well-trained, attentive pilot can easily accomplish this task manually to within ± 50 ft, and air-traffic controllers expect pilots to maintain this kind of tolerance. However, since this task requires the pilot to be fairly diligent, sophisticated aircraft often have an altitude-hold autopilot to lessen the pilot's work. This system differs fundamentally from the yaw damper because its role is to replace the pilot for certain periods of time, while the yaw damper's role is to help the pilot fly. Dynamic specifications, therefore, need not require that pilots like the craft's "feel" (how it responds to their handling of the controls); instead, the design should provide the kind of ride that pilots and passengers like. The damping ratio should still be in the vicinity of $\zeta \cong 0.5$, but for a smooth ride the natural frequency should be much slower than $\omega_n = 1 \, \text{rad/sec}$.

STEP 2. *Select a sensor.* Clearly needed is a device to measure altitude, a task most easily done by measuring the atmospheric pressure. Almost from the time of the first Wright brothers' flight, this basic idea has been used in a device called a **barometric altimeter**. Before autopilots the device consisted of a bellows whose free end was connected to a needle that directly indicated altitude on a dial. The same bellows concept is used today, but the free end motion that is proportional to pressure is sensed electrically.

Because the transfer function from the controlling elevator input to the altitude control consists of essentially four poles, stabilization of the feedback loop cannot be accomplished by simple proportional feedback. Therefore pitch rate q is also used as a stabilizing feedback; it is measured by a gyroscope or ring-laser gyro identical to that used for yaw-rate measurement. Further stabilization using pitch-angle feedback is also helpful. It is obtained either from an inertial reference system based on a ring-laser gyro or from a rate-integrating gyro. The latter is a device similar to the rate gyro but structured differently so that its outputs are proportional to the angles of the aircraft's pitch θ and roll ϕ.

STEP 3. *Select an actuator.* The only aerodynamic surface typically used for pitch control on most aircraft is the elevator δe. It is located on the horizontal tail, well removed from the aircraft's center of gravity so that its force produces an angular pitch rate and thus a pitch angle, which acts to change the lift from the wing. In some high-performance aircraft there are direct-lift control devices on the wing or perhaps small canard surfaces, which are like tiny wings forward of the main wing, which produce vertical forces on the aircraft that are much faster than elevators on the tail are able to generate. However, for purposes of our altitude hold, we will only consider the typical case of an elevator surface on the tail.

As for the rudder, hydraulic actuators are the preferred devices to move the elevator surface mainly because of their favorable force-to-weight ratio.

STEP 4. *Make a linear model.* The longitudinal perturbation equations of motion for the Boeing 747 in horizontal flight at a nominal speed $U_0 = 830 \, \text{ft/sec}$ at 20,000 ft (Mach 0.8) with a weight of 637,000 lb, are

$$\dot{\mathbf{x}} = \mathbf{Fx} + \mathbf{G}\delta e$$

$$\begin{bmatrix} \dot{u} \\ \dot{w} \\ \dot{q} \\ \dot{\theta} \\ \dot{h} \end{bmatrix} = \begin{bmatrix} -0.00643 & 0.0263 & 0 & -32.20 \\ -0.0941 & -0.624 & 820 & 0 & 0 \\ -0.000222 & -0.00153 & -0.668 & 0 & 0 \\ 0 & 0 & 1 & 0 & 0 \\ 0 & -1 & 0 & 830 & 0 \end{bmatrix} \begin{bmatrix} u \\ w \\ q \\ \theta \\ h \end{bmatrix} + \begin{bmatrix} 0 \\ -32.7 \\ -2.08 \\ 0 \\ 0 \end{bmatrix} \delta e$$

$$(9.18)$$

FIGURE 9.39
Altitude-hold feedback system

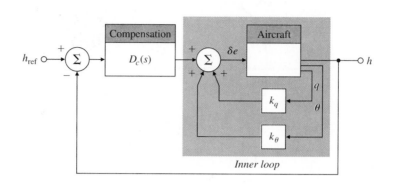

where the desired output for an altitude-hold autopilot is

$$h = \mathbf{Hx}$$

$$h = [0 \quad 0 \quad 0 \quad 0 \quad 1] \begin{bmatrix} u \\ w \\ q \\ \theta \\ h \end{bmatrix}. \tag{9.19}$$

The system has two pairs of stable complex poles and a pole at $s = 0$. The complex pair at $-0.003 \pm 0.0098j$ are referred to as the **phugoid mode**,[3] and the poles at -0.6463 ± 1.1211 are the **short-period modes**.

STEP 5.   *Try a lead-lag or PID controller.* As a first step in the design, it is

Inner-loop design

typically helpful to use an inner-loop feedback of pitch rate q to δe so as to improve the damping of the short-period mode of the aircraft (see Fig. 9.39). The transfer function from δe to q is

$$\frac{q(s)}{\delta e(s)} = -\frac{2.08s(s + 0.0105)(s + 0.596)}{(s + 0.003 \pm 0.0098j)(s + 0.646 \pm 1.21j)}. \tag{9.20}$$

The inner loop root locus for q feedback using Eq. (9.20) is as shown in Fig. 9.40. Since k_q is the root-locus parameter, the system matrix (Eq. 9.18) is now modified as follows:

$$\mathbf{F}_q = \mathbf{F} + k_q \mathbf{GH}_q, \tag{9.21}$$

where $\mathbf{F}$ and $\mathbf{G}$ are defined in Eq. (9.18) and $\mathbf{H}_q = [0 \ 0 \ 1 \ 0 \ 0]$. The process of picking a suitable gain k_q is an iterative one. The selection procedure is the

3. The name was adopted by F. W. Lanchester (1908), who was the first to study the dynamic stability of aircraft analytically. It is apparently an incorrect version of a Greek word.

FIGURE 9.40
Inner-loop root locus for
altitude-hold dynamics
with q feedback

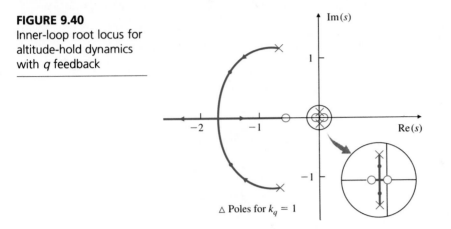

same one discussed in Chapter 5 (recall the tachometer feedback example in Section 5.4.1). If we choose $k_q = 1$, then the closed-loop poles will be located at $-0.0039 \pm 0.0067j$, $-1.683 \pm 0.277j$ on the root locus and

$$\mathbf{F}_q = \begin{bmatrix} -0.00643 & 0.0263 & 0 & -32.20 \\ -0.0941 & -0.624 & 787.3 & 0 & 0 \\ -0.000222 & -0.00153 & -2.75 & 0 & 0 \\ 0 & 0 & 1 & 0 & 0 \\ 0 & -1 & 0 & 830 & 0 \end{bmatrix}. \tag{9.22}$$

Note that only the third column of $\mathbf{F}_q$ is different from $\mathbf{F}$. To further improve the damping, the pitch angle of the aircraft must be controlled. We achieve this by also feeding back the pitch angle θ to elevator control. After we select

$$\mathbf{K}_{\theta q} = \begin{bmatrix} 0 & 0 & -0.8 & -6 & 0 \end{bmatrix}$$

in order to feed back θ and q, the system matrix becomes

$$\mathbf{F}_{\theta q} = \mathbf{F}_q - \mathbf{G}\mathbf{K}_{\theta q}$$

$$= \begin{bmatrix} -0.0064 & 0.0263 & 0 & -32.2 & 0 \\ -0.0941 & -0.624 & 761 & -196.20 \\ -0.0002 & -0.0015 & -4.41 & -12.480 \\ 0 & 0 & 1 & 0 & 0 \\ 0 & -1 & 0 & 830 & 0 \end{bmatrix},$$

with poles at $s = 0$, $-2.25 \pm 2.99j$, -0.531, -0.0105.

So far, the inner loop of the aircraft has been stabilized significantly. The uncontrolled aircraft has a natural tendency to return to equilibrium in level flight as evidenced by the open-loop roots in the LHP. The inner-loop stabilization is necessary to enable an outer-loop feedback of h and $\dot{h}$ to be

FIGURE 9.41
Response of altitude-
hold autopilot to a
command in θ

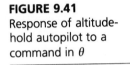

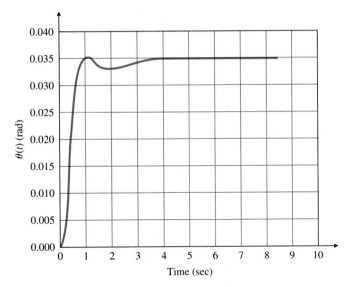

Time (sec)

successful; furthermore, the feedbacks of θ and q can be used by themselves in an attitude-hold mode of the autopilot, where a pilot wishes to control θ directly through input command. Fig. 9.41 shows the response of the inner loop to a 2° (0.035-rad) step command in θ. With the inner loop the transfer function of the system from elevator angle to altitude is now

$$\frac{h(s)}{\delta e(s)} = \frac{32.7(s + 0.0045)(s + 5.645)(s - 5.61)}{s(s + 2.25 \pm 2.99j)(s + 0.0105)(s + 0.0531)}. \tag{9.23}$$

The root locus for this system, given in Fig. 9.42, shows that proportional feedback of altitude by itself does not yield an acceptable design. For stabiliz-

FIGURE 9.42
0° root locus with
feedback of h only

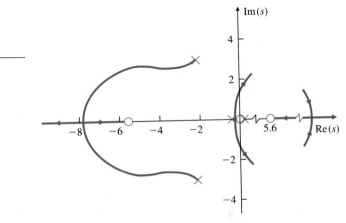

FIGURE 9.43
0° root locus with
feedback of h and $\dot{h}$

FIGURE 9.44
SRL for altitude-hold
design

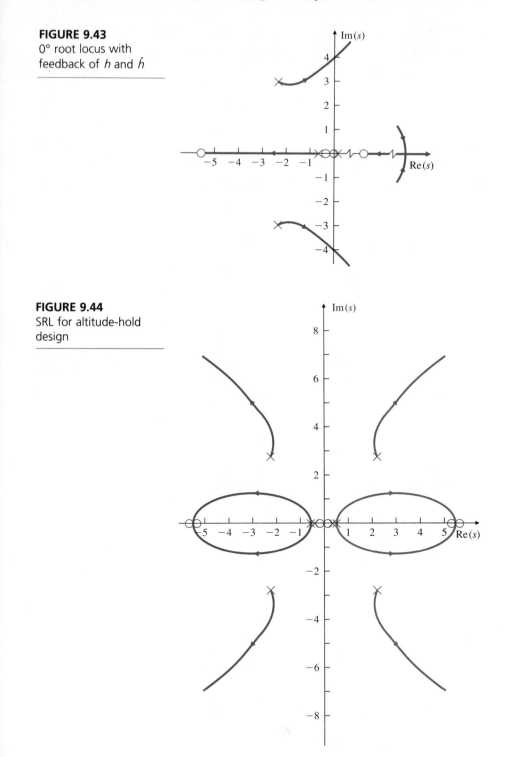

ation we may also feed back the rate of change in the altitude in a PD controller. The root locus of the system with feedback of both h and $\dot{h}$ is shown in Fig. 9.43. After some iteration we find that the best ratio of $\dot{h}$ to h is 10:1, that is,

$$D_e(s) = K_h(s + 0.1).$$

The final design is the result of iterations between the q, θ, $\dot{h}$, and h feedback gains, obviously a lengthy process. Although this trial design was successful, use of the SRL approach promises to expedite the process.

STEP 6.   *Do an optimal design.* The symmetric root locus of the system is show in Fig. 9.44. If we choose the closed-loop poles at

$$\alpha_c(s) = (s + 0.0045)(s + 0.145)(s + 0.513)(s + 2.25 \pm 2.98j),$$

then the required feedback gain is

$$\mathbf{K} = [-0.0009 \quad 0.0016 \quad -1.883 \quad -7.603 \quad -.001].$$

The step response of the system to a 100-ft step command in h is shown in Fig. 9.45, and the associated control effort is shown in Fig. 9.46.

This design has been carried out with the assumption that the linear model is valid for the altitude changes under consideration. We should perform simulations to verify this or to determine the range of validity of the linear model. In order to achieve large altitude changes, such as the one shown in Fig. 9.45, we may need to increase engine thrust.

FIGURE 9.45
Step response of altitude-hold autopilot to a 100-ft step command

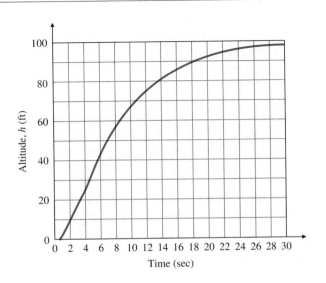

FIGURE 9.46

Control effort for 100-ft
step command in
altitude

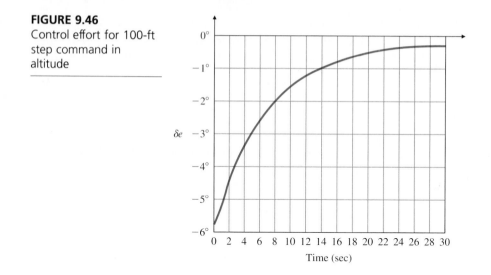

STEPS 7 AND 8.   *Verify the design.* The comments in Steps 7 and 8 of Section
9.3.1 apply to this design as well.

9.4   Control of the Fuel-air Ratio in an Automotive Engine

Until the 1980s most automobile engines had a carburetor to meter the fuel so
that the ratio of the gasoline-mass flow to air-mass flow, or fuel-to-air ratio
(F/A), remained in the vicinity of 1 : 15. This device metered the fuel by relying
on a pressure drop produced by the air flowing through a venturi. The device
has performed adequately for the last 100 years in terms of keeping the engine
running satisfactorily, but it has historically allowed excursions of up to 20%
in the F/A. After the implementation of federal exhaust-pollution regulations,
this level of inaccuracy in the F/A was unacceptable since neither excess
hydrocarbons (HCs) nor excess oxygen could be accepted. During the 1970s,
automobile companies improved the design and manufacturing process of the
carburetors so that they became more accurate and delivered a F/A accuracy
in the vicinity of 3 to 5%. Through a combination of factors, this improved F/A
accuracy helped lower the exhaust pollution levels. However, the carburetors
were still open-loop devices since the system did not measure the F/A of the
mixture entering the engine for subsequent feedback into the carburetor.
During the 1980s almost all manufacturers have turned to feedback control
systems to provide a much-improved level of F/A accuracy, an action made
necessary by the decreasing levels of allowable exhaust pollutants.

We now turn to the design of a typical feedback system for engine control,
again using the step-by-step design outline given in Section 9.1.

FIGURE 9.47
F/A feedback control system

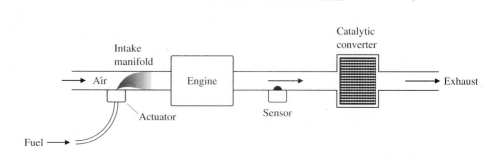

STEP 1.   *Understand the process and its performance.* The method chosen to meet the exhaust-pollution standards has been to use a catalytic converter that simultaneously oxidizes excess levels of exhaust carbon monoxide (CO) and unburned HCs and reduces excess levels of the oxides of nitrogen (NO and NO_2, or NO_x). This device is usually referred to as a three-way catalyst because of its effect on all three pollutants. This catalyst is ineffective when the F/A is more than 1% different from the stoichiometric level of $1:14.7$; therefore, a feedback control system is required to maintain the F/A within $\pm 1\%$ of that desired level. The system is depicted in Fig. 9.47.

The dynamic phenomena that affect the relationship between the sensed F/A output from the exhaust and the fuel-metering command in the intake manifold are (1) intake fuel and air mixing, (2) cycle delays due to the piston strokes in the engine, and (3) the time required for the exhaust to travel from the engine to the sensor. All these effects are strongly dependent on the speed and load of the engine. For example, engine speeds will vary from 600 to 6000 rpm. The result of these variations is that the time delays in the system that will affect the feedback control-system behavior will also vary by at least $10:1$, depending on the operating condition. The system undergoes transients as the driver demands more or less power through changes in the accelerator pedal, with the changes taking place over fractions of a second. Ideally, the feedback control system should be able to keep up with these transients.

STEP 2.   *Select a sensor.* The discovery and development of the exhaust sensor was the key technological step that made possible this concept of exhaust-emission reduction by feedback control. The active element in the device, zirconium oxide, is placed in the exhaust stream where it yields a voltage that is a monotonic function of the oxygen content of the exhaust gas. The F/A is uniquely related to the oxygen level. The voltage of the sensor is highly nonlinear with respect to F/A (Fig. 9.48); almost all the change in voltage occurs precisely at the F/A value at which the feedback system must operate for effective performance of the catalyst. Therefore, the gain of the sensor will be very high when the F/A is at the desired point ($1:14.7$) but will fall off considerably for F/A excursions away from $1:14.7$.

FIGURE 9.48

Exhaust sensor output

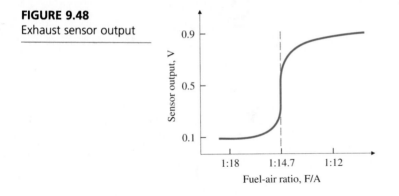

Although other sensors are under development for possible use in F/A feedback control, no other cost-effective sensor has so far demonstrated the capability to perform adequately. All manufacturers of production-line automobiles currently use zirconium oxide sensors in their feedback control systems.

STEP 3. *Select an actuator.* Fuel metering can be accomplished by a carburetor or by fuel injection. Implementing a feedback F/A system requires the capability of adjusting the fuel metering electrically, since the sensor used provides an electric output. Carburetors have been designed to provide this capability by including adjustable orifices that modify the primary fuel flow in reponse to the electric error signal. Most designs accomplish this by the use of on-off solenoid valves that modulate the pressure applied to a bellows in the carburetor. The bellows moves a needle to adjust the orifice size.

Since fuel-injection systems are typically electrical by nature, they can be used to perform the fuel adjustment for F/A feedback simply by including the capability of using the feedback signal from the sensor. In some cases, fuel injectors are placed at the inlet to every cylinder (called **multipoint injection**); in others there is one large injector upstream from all the cylinders (called **single-point** or **throttle body injection**). With the growing popularity of microprocessor engine control, multipoint fuel injection is becoming much more popular. Differences in performance between the carburetor and single-point injection are not major, since they both introduce fuel at approximately the same location. Multipoint injection does offer improved performance because the fuel is introduced much closer to the engine, with better distribution to the cylinders. Being closer reduces the time delays and thus yields better engine response.

STEP 4. *Make a linear model.* The sensor nonlinearity shown in Fig. 9.48 is severe enough that any design effort based on a linearized model of it should be used with caution. Figure 9.49 shows a block diagram of the system, with the sensor shown to have a gain K_s. The time constants τ_1 and τ_2 indicated for the inlet manifold dynamics represent, respectively, fast fuel flow in the form

FIGURE 9.49
Block diagram of an F/A control system

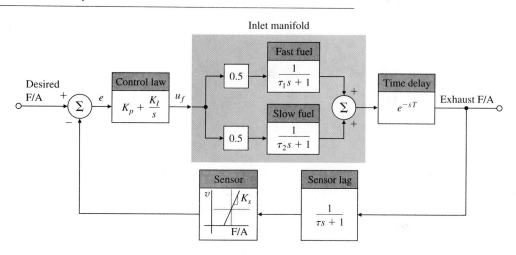

of vapor or droplets and slow fuel flow in the form of a liquid film on the manifold walls. The time delay is the sum of (1) the time it takes the pistons to move through the four strokes from the intake process until the exhaust process and (2) the time required for the exhaust to travel from the engine to the sensor located roughly 1 ft away. A sensor lag with time constant τ is also included in the process to account for the mixing that occurs in the exhaust manifold. Although the time constants and the delay time change considerably, primarily as a function of engine load and speed, we will examine the design at a specific point where the values are

$$\tau_1 = 0.02 \, \text{sec}, \qquad T = 0.2 \, \text{sec},$$
$$\tau_2 = 1 \, \text{sec}, \qquad \tau = 0.1 \, \text{sec}.$$

STEP 5. *Try a lead-lag or PID controller.* Given the tight error specifications and the wide variations in the required fuel command u_f due to varying engine-operating conditions, an integral control term is mandatory. With integral control, any required steady state u_f can be provided when the error signal $e = 0$. The addition of a proportional term, although not often used, allows for an increase (doubling) in the bandwidth without degrading steady-state characteristics. In this example we use a control law that is proportional plus integral (PI). The output from the control law is a voltage that drives the injector's pulse former to give a fuel pulse whose duration is proportional to the voltage. The controller transfer function can be written as

$$D_c(s) = K_p + \frac{K_I}{s} = \frac{K_p}{s}(s + z), \tag{9.24}$$

FIGURE 9.50
Bode plot of a PI F/A controller

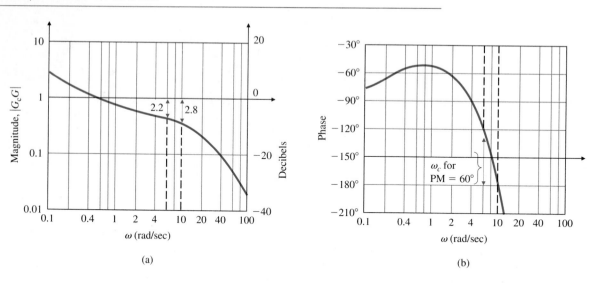

(a)   (b)

where

$$z = \frac{K_I}{K_p}$$

and z can be chosen as desired.

First, let us assume that the sensor is linear and can be represented by a gain K_s. Then we can choose z for good stability and good response of the system. Figure 9.50 shows the frequency response of the system for $K_s K_p = 1.0$ and $z = 0.3$, while Fig. 9.51 shows a root locus of the system with respect to $K_s K_p$ with $z = 0.3$. Both analyses show that the system becomes unstable for $K_s K_p \cong 2.8$. Fig. 9.50 shows that to achieve a phase margin of approximately 60° the gain $K_s K_p$ should be ~ 2.2. Figure 9.50 also shows that this produces a crossover frequency of 6.0 rad/sec (~ 1 Hz). The root locus in Fig. 9.51 verifies that this candidate design will achieve acceptable damping ($\zeta \cong 0.5$).

Although this linear analysis shows that acceptable stability at a reasonable bandwidth (~ 1 Hz) can be achieved with a proportional-integral (PI) controller, a look at the nonlinear sensor characteristics (Fig. 9.48) shows that this indeed may not be achievable. Note that the slope of the sensor output is extremely high near the desired setpoint, thus producing a very high value of K_s. Therefore, lower values of the controller gain K_p need to be used to maintain the overall $K_s K_p$ value of 2.2 when including the effect of the high sensor gain. On the other hand, a value of K_p low enough to yield a stable system at F/A = 1 : 14.7 will yield a sluggish response to transient errors that

Complications of nonlinearity

FIGURE 9.51
Root locus of a PI F/A
controller

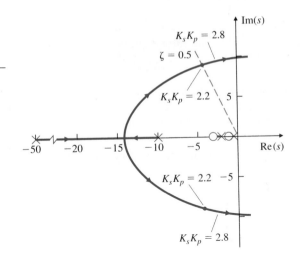

deviate much from the set point, because the effective sensor gain will be reduced substantially. It is therefore necessary to account for the sensor nonlinearity in order to obtain satisfactory response characteristics of the system for anything other than minute disturbances about the setpoint. A first approximation to the sensor is shown in Fig. 9.52. Since the actual sensor gain at the setpoint is still quite different from its approximation, this approximation will yield erroneous conclusions regarding stability about the set point; however, it will be useful in a simulation to determine the response to initial conditions.

STEP 6.   *Try an optimal controller.* The response of this system is dominated by the sensor nonlinearity, and any fine tuning of the control needs to account for that feature. Furthermore, the system dynamics are relatively simple, and it is unlikely that an optimal design approach will yield any improvement over the PI controller used. We will thus omit this step.

FIGURE 9.52
Sensor approximation

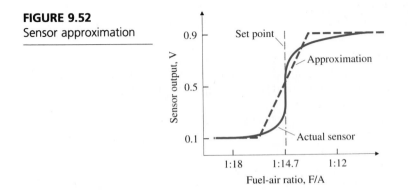

STEP 7. *Simulate design with nonlinearities.* Figure 9.53(a) is a plot of the system error using the approximate sensor of Fig. 9.52 and $K_p K_s = 2.0$. The slow response is apparent with 12.5 sec before the error comes out of saturation and a time constant of almost 5 sec once the linear region is reached. In real automobiles these systems are operated with much higher gains. To show these effects, a simulation with $K_p K_s = 6.0$ is plotted in Fig. 9.53(b, c). At this gain the linear system is unstable and up until about 5 sec the signals grow. The growth halts after 5 sec due to the fact that, as the input to the sensor nonlinearity gets large, the *effective* gain of the sensor decreases due to the

FIGURE 9.53

System response with nonlinear sensor approximation

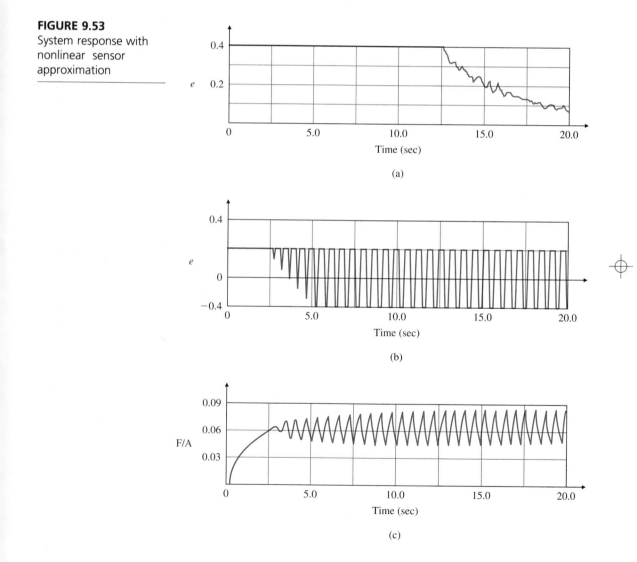

saturation, and eventually, a limit cycle is reached. The frequency of this limit cycle corresponds to the point where the root locus crosses the imaginary axis and has an amplitude such that the total effective gain of $K_p K_{s,eq} = 2.8$. As described in Section 5.7.2, the effective gain of a saturation for moderately large inputs can be computed and is approximately $4N/\pi A$, where N is the saturation level and A is the amplitude of the input signal. Here $N = 0.4$, and, if $K_p = 0.1$, then $K_{s,eq} = 28$. Thus we predict an input signal amplitude of $A = 4(0.4)/28\pi = 0.018$. This value is closely verified by the plot of Fig. 9.53(c), the input to the nonlinearity in this case. The frequency of oscillation is also nearly 10.1 rad/sec, as predicted by the root locus in Fig. 9.51.

In the actual implementation of F/A feedback controllers in automobile engines, sensor degradation over thousands of miles of use is of primary concern since the federal government mandates that the engines meet the exhaust-pollution standards for the first 50,000 mi. In order to reduce the sensitivity of the average set point to changes in the sensor output characteristics, manufacturers typically modify the design discussed here. One approach is to feed the sensor output into a relay function (see Fig. 5.36b), thus completely eliminating any dependency on the sensor gain at the set point. The frequency of the limit cycle is then solely determined by controller constants and engine characteristics. Average steady-state F/A accuracy is also improved.

9.5 Control of a Digital Tape Transport

STEP 1. *Understand the process and its performance specifications.* Some high-performance tape transports are designed with a small capstan to pull the tape past the read/write heads with the take-up reels turned by DC motors. The capstan motion is isolated from the reels by vacuum columns that provide nearly constant tension to the tape. This structure permits separate design of capstan, vacuum, and reel controls. A sketch of the system is shown in Fig. 9.54, and a schematic of the capstan control is drawn in Fig. 9.55. The objective of the capstan control system is to control the speed and the tension of the tape at the read/write head. The tape is to be controlled at speeds of up to 200 in/sec and start-up is to be as fast as possible but the tension must remain below 12 N at all times to prevent permanent distortion of the tape. For the purposes of this design exercise, we will consider the speed of response to be satisfactory if the settling time of the response to a step input is less than 12 msec and the overshoot of the response to a step input is less than 5%.

STEP 2. *Select a sensor.* Selecting the sensors is easy in this case: We will use a DC tachometer to measure idler speed and a shaft encoder to measure idler position. An alternative to the tachometer would be an AC generator, which gives a sinusoidal voltage signal whose amplitude is proportional to speed. We chose not to use it because the output voltage needs additional electronic signal processing (such as rectifying and filtering) to provide a suitable signal for the

FIGURE 9.54
Front view of a typical
tape-drive mechanism
with reference track 1
nearest the front
(observer) edge of the
tape

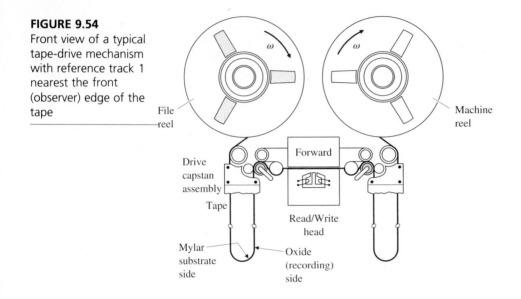

FIGURE 9.55
Model of a tape drive

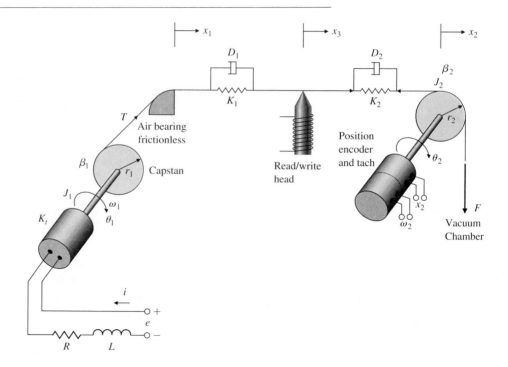

controller. As for the shaft recorder, models that provide a digital reading of shaft position have low inertia and low friction and thus are often used in digital control, which is increasingly attractive for motion-control problems. In the present case, however, the output of the encoder will be processed to provide an analog measure of position, which can be used in the feedback to the drive motor.

STEP 3.   *Select an actuator.* There are many choices for motion-control actuators, including hydraulic, pneumatic, AC electric, DC electric, and stepping motors. Hydraulic and pneumatic actuators require an auxiliary pump but can provide high forces and torques in a lightweight package. They are common in aircraft control-surface servo mechanisms and chemical-process valve controls. However, for our purpose the electric motors are the best choice, since they are reliable, inexpensive, and easy to maintain. Of these, the DC motor is the machine of choice here because it has the best low-speed acceleration and low-ripple characteristics.

STEP 4.   *Make a linear model.* The system is in static equilibrium when the tape tension equals the vacuum forces, $T_0 = F$ and when the torque from the motor equals the torque on the capstan, $K_t i_0 = r_1 T_0$ where

T_0 = tape tension at the read/write head at equilibrium, N,

F = constant force (tape tension from vacuum column), $6\,N$,

K = motor torque constant,

i_0 = equilibrium motor current,

r_1 = radius of the capstan take-up wheel.

We define the variables as deviations from this equilibrium. Thus the linearized system has torque deviations limited to $6\,N$. The equations of motion of the system are given by the laws of dynamics as follows:

$$J_1 \frac{d\omega_1}{dt} + \beta_1 \omega_1 - r_1 T = K_t i, \tag{9.25}$$

$$\dot{x}_1 = r_1 \omega_1, \tag{9.26}$$

$$L \frac{di}{dt} + Ri + K_e \omega_1 = e, \tag{9.27}$$

$$\dot{x}_2 = r_2 \omega_2, \tag{9.28}$$

$$J_2 \frac{d\omega_2}{dt} + \beta_2 \omega_2 + r_2 T = 0, \tag{9.29}$$

$$T = K_1(x_3 - x_1) + D_1(\dot{x}_3 - \dot{x}_1) \tag{9.30}$$

$$T = K_2(x_2 - x_3) + D_2(\dot{x}_2 - \dot{x}_3), \tag{9.31}$$

$$x_1 = r_1\theta_1, \tag{9.32}$$

$$x_2 = r_2\theta_2, \tag{9.33}$$

$$x_3 = \frac{x_1 + x_2}{2}, \tag{9.34}$$

where

$D_{1,2}$ = damping in the tape-stretch motion = $20\,\text{N/m}\cdot\text{sec}$,

e = applied voltage, V,

i = current into capstan motor

J_1 = combined inertia of the wheel and take-up motor, $4 \times 10^{-5}\,\text{kg}\cdot\text{m}^2$,

J_2 = inertia of the idler = $10^{-5}\,\text{kg}\cdot\text{m}^2$,

$K_{1,2}$ = spring constant in the tape-stretch motion, $4 \times 10^4\,\text{N/m}$,

K_e = electric constant of the motor = $0.03\text{V}\cdot\text{sec}$,

K_t = torque constant of the motor = $0.03\,\text{V}\cdot\text{sec}$,

L = armature inductance = $10^{-3}\,\text{H}$,

R = armature resistance = $1\,\Omega$,

r_1 = radius of take-up wheel, 0.02 m,

r_2 = radius of the tape on the idler = 0.02 m,

T = tape tension at the read/write head, N,

x_3 = position of the tape at the head, m,

$\dot{x}_3$ = velocity of the tape at the head, m/sec.

β_1 = viscous friction at the take-up wheel, $0.01\,\text{N}\cdot\text{m}\cdot\text{sec}$,

β_2 = viscous friction at the wheel = $0.01\,\text{N}\cdot\text{m}\cdot\text{sec}$,

θ_1 = angular displacement of the capstan,

θ_2 = tachometer shaft angle,

ω_1 = speed of the drive wheel = $\dot{\theta}_1$, rad/sec,

ω_2 = output speed measured by the tachometer output $\dot{\theta}_2$, rad/sec,

The equations of the system are fifth-order and are already almost in the state form

$$\dot{\mathbf{x}} = \mathbf{Fx} + \mathbf{G}u, \tag{9.35a}$$

$$y = \mathbf{Hx}, \tag{9.35b}$$

where

$$\mathbf{x} = [x_1 \quad \omega_1 \quad x_2 \quad \omega_2 \quad i]^T,$$

$$u = e.$$

There are two outputs of interest, tape position at the head, x_3, and tape tension T. For the tape position we have Eq. (9.34). Thus, for the position at the head,

$$\mathbf{H}_3 = [0.5 \quad 0 \quad 0.5 \quad 0 \quad 0].$$

The tension is given by Eq. (9.30), which we can write as

$$T = K_1 \left(\frac{x_1 + x_2}{2} - x_1 \right) + D_1 \left(\frac{r_1\omega_1 + r_2\omega_2}{2} - r_1\omega_1 \right)$$

$$= \frac{K_1}{2} (x_2 - x_1) + \frac{D_1}{2} (r_2\omega_2 - r_1\omega_1). \tag{9.36}$$

Using time and amplitude scaling

If we substitute the numbers into these equations, we find very large and very small values, indicating the need for time and amplitude scaling. Ideally, we would like the elements in $\mathbf{F}$, $\mathbf{G}$, and $\mathbf{H}$ to be in the range 0.1 to 10 if possible. We write the equations in state form with the numbers and use x_1' and x_2' for the unscaled positions:

$$\dot{x}_1' = 0.02\omega_1,$$

$$4 \times 10^{-5}\dot{\omega}_1 = -400x_1' - 0.014\omega_1 + 400x_2' + 0.004\omega_2 + 0.03i,$$

$$\dot{x}_2' = 0.02\omega_2,$$

$$10^{-5}\dot{\omega}_2 = 400x_1' + 0.004\omega_1 - 400x_2' - 0.014\omega_2$$

$$10^{-3}i = -0.03\omega_1 - i + e. \tag{9.37}$$

After some experimentation, we decide to measure time in milliseconds, so $\tau = 10^3 t$, and to measure position in units of 10^{-5} m, so we take $x_1 = 10^5 x_1'$ and $x_2 = 10^5 x_2'$. With these conversions Eqs. (9.37) become

$$\frac{dx_1}{d\tau} = 2\omega_1,$$

$$\frac{d\omega_1}{d\tau} = -0.1x_1 - 0.35\omega_1 + 0.1x_2 + 0.1\omega_2 + 0.75i,$$

$$\frac{dx_2}{d\tau} = 2\omega_2,$$

$$\frac{d\omega_2}{d\tau} = 0.4x_1 + 0.4\omega_1 - 0.4x_2 - 1.4\omega_2,$$

$$\frac{di}{d\tau} = -0.03\omega_1 - i + e. \tag{9.38}$$

These equations are described by the matrices:

$$\mathbf{F} = \begin{bmatrix} 0 & 2 & 0 & 0 & 0 \\ -0.1 & -0.35 & 0.1 & 0.1 & 0.75 \\ 0 & 0 & 0 & 2 & 0 \\ 0.4 & 0.4 & -0.4 & -1.4 & 0 \\ 0 & -0.03 & 0 & 0 & -1 \end{bmatrix}, \qquad \mathbf{G} = \begin{bmatrix} 0 \\ 0 \\ 0 \\ 0 \\ 1 \end{bmatrix},$$

$\mathbf{H}_2 = \begin{bmatrix} 0 & 0 & 1 & 0 & 0 \end{bmatrix}$ x_2 output,

$\mathbf{H}_3 = \begin{bmatrix} 0.5 & 0 & 0.5 & 0 & 0 \end{bmatrix}$ x_3 output,

$\mathbf{H}_T = \begin{bmatrix} -0.2 & -0.2 & 0.2 & 0.2 & 0 \end{bmatrix}$ tension output.

The transfer function to measured position x_2 is

$$G(s) = \frac{X_2(s)}{E(s)} = \frac{0.6(s + 2)}{s^5 + 2.75s^4 + 3.22s^3 + 1.88s^2 + 0.418s}$$

$$= \frac{0.6(s + 2)}{s(s + 0.507)(s + 0.968)(s + 0.637 \pm 0.667j)}. \tag{9.39}$$

STEP 5.   *Try a lead-lag or PID controller.* The root locus with respect to K for proportional feedback only and the corresponding open-loop frequency response of the system are shown in Figs. 9.56 and 9.57. The problem states that we are to achieve a settling time of about 12 msec and low overshoot in response to a step command.

As a start on this design, we will aim for a natural frequency ω_n of about 1 rad/msec at the "dominant" second-order poles, and check the response of the higher-order system at that time to see which direction the design modifications should go. With a pole-zero excess (that is, the difference between number of poles and zeros) of 4, we expect the compensator to be required to provide substantial lead. One way to obtain this is to design a compensator with *both* derivative feedback (from the tachometer) and lead-network feedback. If we place both the lead-network zero and the zero due to the

FIGURE 9.56

Uncompensated root locus of a tape drive

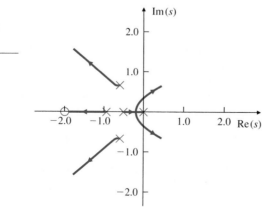

FIGURE 9.57
Uncompensated Bode plot of tape transport, $K_v = 1$

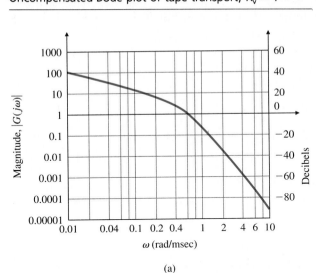

(a)

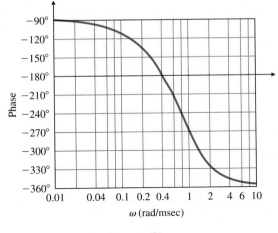

(b)

FIGURE 9.58
Compensated root locus of the tape drive

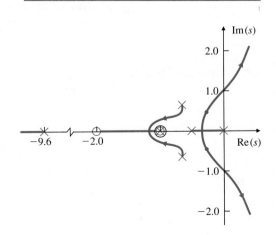

tachometer[4] at -0.968 and place the lead network pole a factor of 10 away at -9.68, then the compensator is

$$D_c(s) = K_c \frac{(s + 0.968)^2}{s + 9.68}. \tag{9.40}$$

4. Pole-zero cancellation like this is usually avoided because of sensitivity problems, but it may be aceptable if, as in this case, the canceled pole is sufficiently stable. Cancellation of poles in the unstable region is, of course, absolutely forbidden.

The root locus of the system is as shown in Fig. 9.58. The ramp response of the system and the associated tension are shown in Fig. 9.59. The frequency response of the compensated system is given in Fig. 9.60.

FIGURE 9.59
Ramp response with $D_c(s)$ compensation: (a) position; (b) tension

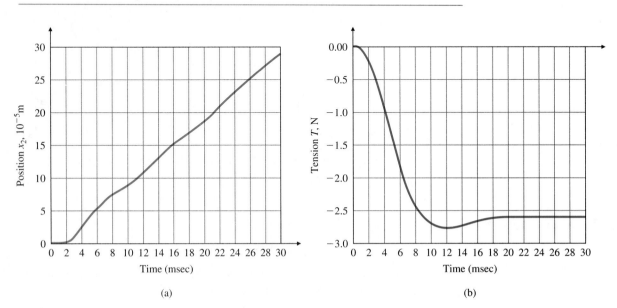

(a)

(b)

FIGURE 9.60
Bode plot of the compensated system

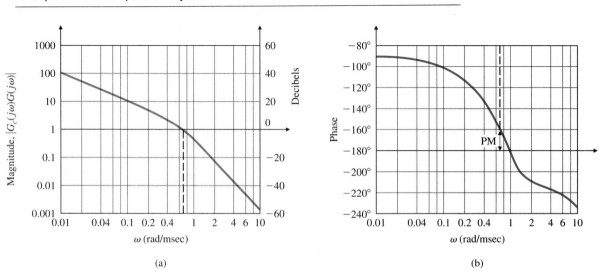

(a)

(b)

FIGURE 9.61
SRL of the tape drive

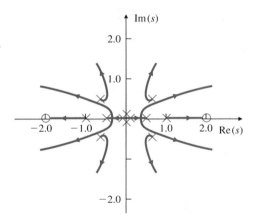

STEP 6.   *An optimal design using pole-placement.* We now consider designing a compensator using the state-variable approach. The SRL of the system is shown in Fig. 9.61. If we select the closed-loop poles of the system to be on the SRL at a bandwidth of about 1 rad/msec,

$$\alpha_c(s) = (s + 0.451 \pm 0.937j)(s + 0.947 \pm 0.581j)(s + 1.16),$$

then the required state feedback gain is

$$\mathbf{K} = [0.802 \quad 2.58 \quad 0.489 \quad 0.964 \quad 1.21].$$

For the step response of the system shown in Fig. 9.62, the reference input has been multiplied by the inverse of the closed-loop DC gain to get

$$u = -\mathbf{K}\mathbf{x} + 1.29r,$$

FIGURE 9.62
Step response of full state feedback

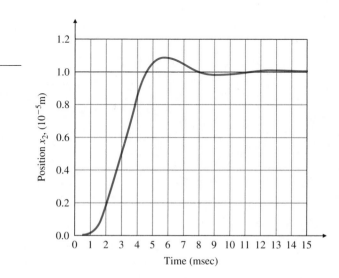

FIGURE 9.63
Full state feedback: (a) control effort; (b) tension

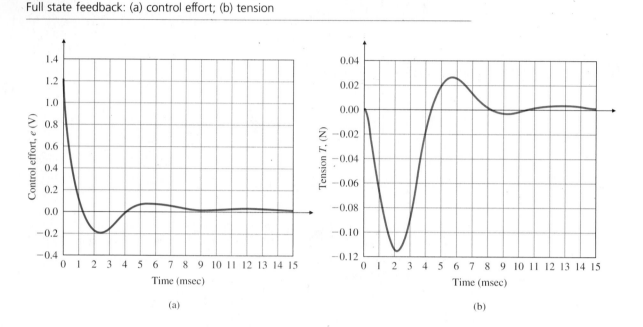

(a)

(b)

FIGURE 9.64
Ramp response with full-state feedback: (a) position; (b) control effort

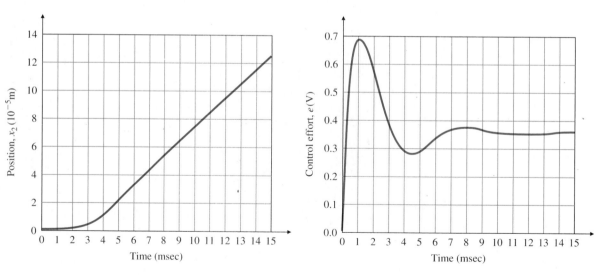

(a)

(b)

which results in zero steady-state error to a unit step input. The associated control effort and tension are given in Fig. 9.63, while Fig. 9.64 shows the ramp response of the system and its associated control effort. If measurement of all the state is not available and we are to design an estimator based on position measurement alone, such that its poles are five times faster than the control, then

$$\alpha_e(s) = (s + 1.80 \pm 3.75j)(s + 3.79 \pm 2.32j)(s + 4.64),$$

and the estimator gain is

$$\mathbf{L} = \begin{bmatrix} 403.9 \\ 50.6 \\ 13.1 \\ 38.6 \\ 1166.2 \end{bmatrix}.$$

The compensator transfer function is then given by

$$D_c(s) = -\frac{1905(s + 0.640 \pm 0.797j)(s + 0.984 \pm 0.247j)}{(s + 2.38)(s + 1.60 \pm 4.73j)(s + 5.73 \pm 2.64j)}. \tag{9.41}$$

It has the frequency response shown in Fig. 9.65, which is essentially a lead network.

FIGURE 9.65
Bode plot of the optimal compensator

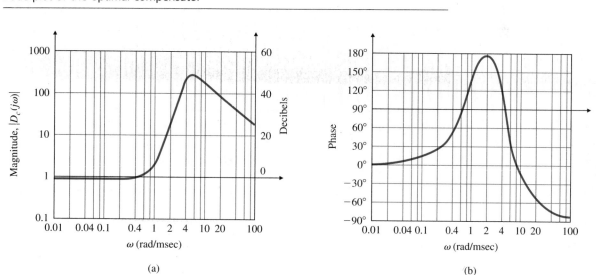

(a)

(b)

FIGURE 9.66
Bode plot of the compensated system

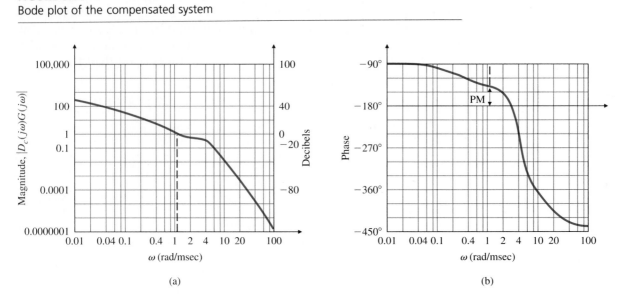

(a)

(b)

FIGURE 9.67
Root locus of the
compensated tape drive

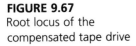

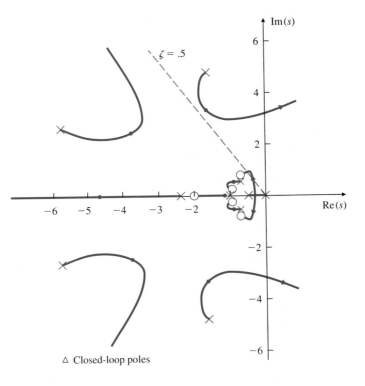

△ Closed-loop poles

FIGURE 9.68
Initial conditions $x_3(0) = 1$, $\hat{\mathbf{x}}_0 = \mathbf{0}$

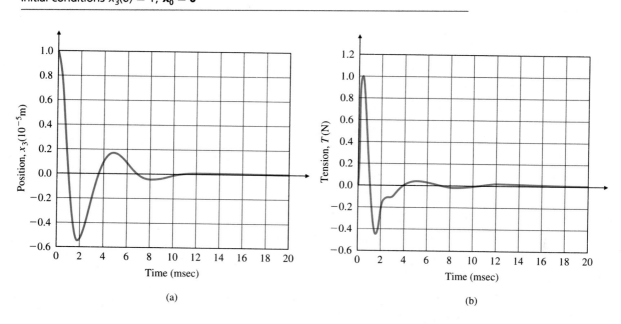

(a)

(b)

For the complete compensated system, Fig. 9.66 shows the frequency response, and Fig. 9.67 the root locus. Figure 9.68(a) shows a transient response, including the effect of the estimator, and Fig. 9.68(b) shows the associated tension.

STEPS 7 and 8.   Verify the design. At this point the system should be simulated with all known non-linear effects such as friction and the best design selected. A prototype system is typically built and tested with the proposed conrol hardware.

Summary

- In this chapter we have laid out a basic outline of control systems design and applied it to four typical case studies. The design outline calls for a number of explicit steps.

1. *Make a system model and determine the required performance specifications.* The purpose of this step is to answer the question, What is the system, and what is it supposed to do?

2. *Select a sensor.* A basic rule of control is that if you can't observe it,

you can't control it. Factors to consider in the selection of sensors are:

technology to be used;

performance of the sensor, such as its accuracy;

physical size and weight;

quality of the sensor such as life time and robustness to environment changes;

cost.

3. *Select an actuator.* The actuator must be capable of driving the system so as to meet the required performance specifications. The selection is governed by the same factors that apply to sensor selection.

4. *Make a linear model.* All our design methods are based on linear models. Both small-signal perturbation models and feedback-linearization methods can be used.

5. *Try a simple PID controller.* An effort to meet the specifications with a PID or its cousin, the lead-lag compensator, may succeed; in any case such an effort will expose the nature of the control problem.

6. *Try an optimal design.* The SRL method for control-law selection and estimator design based on state equations is guaranteed to produce a stable control system and can be structured to show a tradeoff between error reduction and control effort. A related alternative is arbitrary pole placement, which gives the designer direct control over the dynamic response. Both the SRL and the pole-placement methods may result in designs that are not robust to parameter changes.

7. *Simulate the design, and verify its performance.* All the tools of analysis should be used here, including the root locus, the frequency response, GM and PM measurements, and transient responses. Also, the performance of the design can be tested in simulation against changes in model parameters and the effects of approximating the compensator with a discrete model if digital control is to be used.

8. *Build a prototype, and measure the performance with typical input signals.* The proof of the pudding is in the eating, and no control design is acceptable until it has been tested. No model can include all the features of a real physical device, so the final step before fixing the design is to try it out on a physical prototype if time and budget permit.

- The satellite case study illustrated particularly the use of a notch compensation for a system with lightly damped resonance. Also shown was that collocated actuator and sensor systems are much easier to control than the noncollocated case.

- The Boeing 747 lateral-stabilization case study illustrated the use of feedback as an inner-loop designed to aid the pilot, who provides the primary outer-loop control.

- The Boeing 747 altitude control showed how to combine inner-loop feedback with outer-loop compensation to design a complete control system.

- The automobile fuel-air ratio control illustrated the use of the Bode plot to design a system that includes time delay. Simulation of the design with the nonlinear sensor verified our heuristic analysis of limit cycles using the concept of equivalent gain with a root locus.

- The magnetic-tape case study illustrated a design where two output responses are important: position and tension.

- In all cases the designer needs to be able to use multiple tools including the root locus, the frequency response, pole placement by state feedback, and simulation of time responses to get a good design. We promised an understanding of these tools at the beginning of the text, and we trust you are now ready to practice the art of control engineering.

Problems

9.1 Of the three types of PID control (proportional, integral, or derivative), which one is the most effective in reducing the error resulting from a constant disturbance? Explain.

9.2 Is there a greater chance of instability when the sensor in a feedback control system for a mechanical structure is not collocated with the actuator? Explain.

9.3 Consider the plant $G(s) = 1/s^3$. Determine whether or not it is possible to stabilize this plant by adding the lead compensator

$$D_c(s) = K\frac{s+a}{s+b}, \qquad (a < b).$$

a) What is the maximum phase margin of the resulting feedback system?
b) Can a system with this plant, together with any number of lead compensators, be made unconditionally stable? Explain why or why not.

9.4 Consider the closed-loop system shown in Fig. 9.69.

a) What is the phase margin if $K = 70,000$?
b) What is the gain margin if $K = 70,000$?
c) What value of K will yield a phase margin of $\sim 70°$?
d) What value of K will yield a phase margin of $\sim 0°$?

FIGURE 9.69
Control system for
Problem 9.4

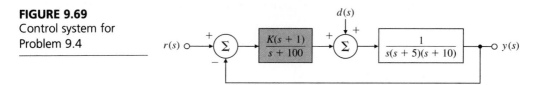

e) Sketch the root locus with respect to K for the system, and determine what value of K causes the system to be on the verge of instability.

f) If the disturbance w is a constant and $K = 10,000$, what is the maximum allowable value for w if $y(\infty)$ is to remain less than 0.1? (Assume $r = 0$.)

g) Suppose the specifications require you to allow larger values of w than the value you obtained in part (f) but with the same error constraint $[|y(\infty)| < 0.1]$. Discuss what steps you could take to alleviate the problem.

9.5 Consider the system shown in Fig. 9.70, which represents the attitude rate control for a certain aircraft.

a) Design a compensator so that the dominant poles are at $-2 \pm 2j$.

b) Sketch the Bode plot for your design, and select the compensation so that the crossover frequency is at least $2\sqrt{2}$ rad/sec and PM $\geqslant 50°$.

c) Sketch the root locus for your design, and find the velocity constant when $\omega_n > 2\sqrt{2}$ and $\zeta \geqslant 0.5$.

FIGURE 9.70
Block diagram for
aircraft-attitude rate
control

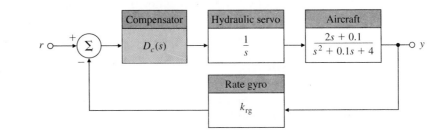

9.6 Consider the block diagram for the servomechanism drawn in Fig. 9.71. Which of the following claims are true?

a) The actuator dynamics (the pole at 1000 rad/sec) must be included in an analysis to evaluate a usable maximum gain for which the control system is stable.

b) The gain K must be negative for the system to be stable.

c) There exists a value of K for which the control system will oscillate at a frequency between 4 and 6 rad/sec.

FIGURE 9.71
Servomechanism for Problem 9.6

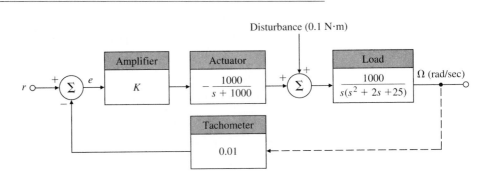

d) The system is unstable if $|K| > 10$.

e) If K must be negative for stability, the control system cannot counteract a positive disturbance.

f) A positive constant disturbance will speed up the load, thereby making the final value of e negative.

g) With only a positive constant command input r, the error signal e must have a final value greater than zero.

h) For $K = -1$ the closed-loop system is stable, and the disturbance results in a speed error whose steady-state magnitude is less than 5 rad/sec.

9.7 A stick balancer and its corresponding control block diagram are shown in Fig. 9.72. The control is a torque applied about the pivot.

   a) Using root-locus techniques, design a compensator $D(s)$ that will place the dominant roots at $s = -5 \pm 5j$ (corresponding to $\omega_n = 7$ rad/sec, $\zeta = 0.707$).

   b) Use Bode plotting techniques to design a compensator $D(s)$ to meet the following specifications:

   ● steady state θ-displacement of less than 0.001 for a constant input torque $T_d = 1$,

   ● Phase Margin $\geqslant 50°$,

   ● Closed-loop bandwidth $\cong 7$ rad/sec.

FIGURE 9.72
Stick balancer

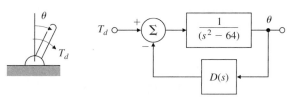

9.8 Consider the standard feedback system drawn in Fig. 9.73.

   a) Suppose

   $$G(s) = \frac{2500K}{s(s + 25)}.$$

   Design a lead compensator so that the phase margin of the system is more than 45°; the steady-state error due to a ramp should be less than or equal to 0.01.

   b) Using the plant transfer function from part (a), design a lead compensator so that the overshoot is less than 25% and the 1% settling time is less than 0.1 sec.

   c) Suppose

   $$G(s) = \frac{K}{s(1 + 0.1s)(1 + 0.2s)}$$

FIGURE 9.73
Block diagram of a standard feedback control system

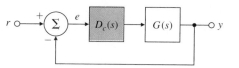

and the performance specifications are now $K_v = 100$, and PM $\geqslant 40°$. Is the lead compensation effective for this system? Find a lag compensator, and plot the root locus of the compensated system.

d) Using $G(s)$ from part (c), design a lag compensator such that the peak overshoot is less than 20% and $K_v = 100$.

e) Repeat part (c) using a lead-lag compensator.

f) Find the root locus of the compensated system in part (e), and compare your findings with those from part (c).

9.9 Consider the system in Fig. 9.73 with the plant transfer function

$$G(s) = \frac{10}{s(1 + s/10)}.$$

We wish to find a compensator $D_c(s)$ that satisfies the following design specifications:

- $K_v = 100$,
- PM $= 45°$,
- sinusoidal inputs of up to 1 rad/sec to be reproduced with $\leqslant 2\%$ error,
- sinusoidal inputs with a frequency of greater than 100 rad/sec to be attenuated at the output to $\leqslant 5\%$ of their input value.

a) Sketch the Bode plot of $G(s)$, choosing the open-loop gain so that $K_v = 100$.

b) Show that a *sufficient* condition for meeting the specification on sinusoidal inputs is that the magnitude plot lies outside the shaded regions in Fig. 9.74. Recall that

$$\frac{Y}{R} = \frac{KG}{1 + KG} \quad \text{and} \quad \frac{E}{R} = \frac{1}{1 + KG}.$$

c) Explain why introducing a lead network alone cannot meet the design specifications.

d) Explain why a lag network alone cannot meet the design specifications.

e) Develop a full design using a lead-lag compensator that meets all the design specifications, without altering the previously chosen open-loop gain.

FIGURE 9.74
Control system for
Problem 9.9

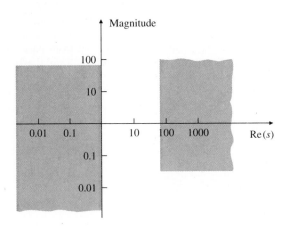

9.10 Consider the system in Fig. 9.73, where

$$G(s) = \frac{300}{s(s + 0.225)(s + 4)(s + 180)}.$$

The compensator $D_c(s)$ is to be designed so that the closed-loop system satisfies the following specifications:

- zero steady-state error for step inputs,
- $PM = 55°$, $GM \geqslant 6\,dB$,
- gain crossover frequency is not smaller than that of the uncompensated plant.

a) What kind of compensation should be used and why?

b) Design a suitable compensator $D_c(s)$ to meet the specifications.

9.11 We have discussed three design methods: the root-locus method of Evans, the frequency-response method of Bode, and the state-variable pole-assignment method. Explain which of these methods is *best* described by the following statements. If you feel more than one method fits a given statement equally well, say so and explain why.

a) This method is the one most commonly used when the plant description must be obtained from experimental data.

b) This method provides the most direct control over dynamic response characteristics such as rise time, percent overshoot, and settling time.

c) This method lends itself most easily to an automated (computer) implementation.

d) This method provides the most direct control over the steady-state error constants K_p and K_v.

e) This method is most likely to lead to the *least complex* controller capable of meeting the dynamic and static accuracy specifications.

f) This method allows the designer to guarantee that the final design will be unconditionally stable.

g) This method can be used without modification for plants that include transportation lag terms, for example,

$$G(s) = \frac{e^{-2s}}{(s + 3)^2}.$$

9.12 Lead and lag networks are typically employed in designs based on frequency response (Bode) methods. Assuming a type I system, indicate the effect of these compensation networks on each of the following performance specifications. In each case, indicate the effect as "an increase," "substantially unchanged," or "a decrease." Use the second-order plant $G(s) = K/[s(s + 1)]$ to illustrate your conclusions.

a) K_v

b) Phase margin

c) Closed-loop bandwidth

d) Percent oveshoot

e) Settling time

FIGURE 9.75

Hot-air balloon

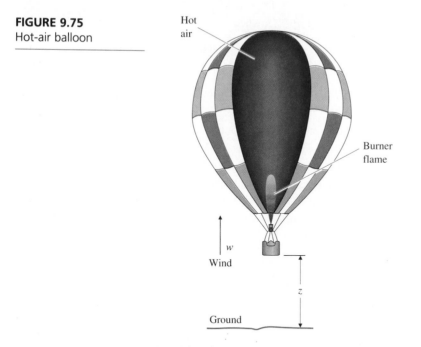

Hot air

Burner flame

w

Wind

z

Ground

9.13 *Altitude Control of a Hot-air Balloon*: The equations of vertical motion for a hot-air balloon (Fig. 9.75) linearized about vertical equilibrium are

$$\delta\dot{T} + \frac{1}{\tau_1}\,\delta T = \delta q,$$

$$\tau_2\ddot{z} + \dot{z} = a\delta T + w,$$

where

δT = deviation of the hot-air temperature from the equilibrium temperature where buoyant force = weight,

z = altitude of the balloon,

δq = deviation in the burner heating rate from the equilibrium rate (normalized by the thermal capacity of the hot air),

w = vertical component of wind-velocity

τ_1, τ_2, a = parameters of the equations

An altitude-hold autopilot is to be designed for a balloon whose parameters are

$$\tau_1 = 250\,\mathrm{sec} \qquad \tau_2 = 25\ \mathrm{sec} \qquad a = 0.3\,\mathrm{m/(sec \cdot {}^\circ C)}.$$

Only altitude is sensed, so a control law of the form

$$\delta q(s) = D(s)[z_d(s) - z(s)]$$

will be used, where z_d is the desired (commanded) altitude.

a) Sketch a root locus of the closed-loop eigenvalues with respect to the gain K for a proportional feedback controller, $\delta q = -K(z - z_d)$. Use Routh's criterion (or let $s = j\omega$ and find the roots of the characteristic polynomial) to determine the value of the gain and the associated frequency at which the system is marginally stable.

b) Our intuition and the results of part (a) indicate that a relatively large amount of lead compensation is required to produce a satisfactory autopilot. Sketch a root locus of the closed-loop eigenvalues with respect to the gain K for a double-lead compensator, $\delta q = D(s)(z_d - z)$, where

$$D(s) = K \left(\frac{s + 0.03}{s + 0.12} \right)^2 .$$

c) Sketch the magnitude portions of the Bode plots (straight-line asymptotes only) for the open-loop transfer functions of the proportional feedback and lead-compensated systems.

d) Select a gain K for the lead-compensated system to give a crossover frequency of 0.06 rad/sec.

e) With the gain selected in part (d), what is the steady-state error in altitude for a steady vertical wind of 1 m/sec? (Be careful: First find the closed-loop transfer function from w to the error.)

f) If the error in part (e) is too large, how would you modify the compensation to give higher low-frequency gain? (Give a qualitative answer only.)

9.14 Satellite-attitude control systems often use a reaction wheel to provide angular motion. The equations of motion for such a system are

$$\text{Satellite:} \quad I\ddot{\phi} = T_c + T_{ex},$$

$$\text{Wheel:} \quad J\dot{r} = -T_c,$$

$$\text{Measurement:} \quad \dot{Z} = \dot{\phi} - aZ,$$

$$\text{Control:} \quad T_c = -D(s)(Z - Z_d).$$

where

T_c = control torque,

T_{ex} = disturbance torque,

ϕ = angle to be controlled,

Z = measurement from the sensor,

Z_d = reference angle,

I = satellite inertia (1000 kg/m^2),

a = sensor constant (1 rad/sec),

$D(s)$ = compensation.

a) Suppose $D(s) = K_0$, a constant. Draw the root locus with respect to K_0 for the resulting closed-loop system.

b) For what range of K_0 is the closed-loop system stable?

c) Add a lead network with a pole at $s = -1$ so that the closed-loop system has a bandwidth $\omega_{BW} = 0.04$ rad/sec and a damping ratio $\zeta = 0.5$ and the compensation is given by

$$D(s) = K_1 \frac{s+z}{s+1}.$$

Where should the zero of the lead network be located? Draw the root locus of the compensated system, and give the value of K_1 that allows the specifications to be met.

d) For what range of K_1 is the system stable?

e) What is the steady-state error (the difference between Z and some reference input Z_d) to a constant disturbance torque T_{ex} for the design of part (c)?

f) What is the type of this system with respect to rejection of T_{ex}?

g) Draw the Bode plot asymptotes of the *open-loop* system, with the gain adjusted for the value of K_1 computed in part (c). Add the compensation of part (c), and compute the phase margin of the closed-loop system.

h) Write state equations for the open-loop system, using the state variables ϕ, $\dot{\phi}$, and Z. Select the gains of a state-feedback controller $T_c = -K_\phi \phi - K_{\dot\phi} \dot\phi$ to locate the closed-loop poles at $s = -0.02 \pm 0.02j\sqrt{3}$.

9.15 Three alternative designs are sketched in Fig. 9.76 for the closed-loop control of a system with the plant transfer function $G(s) = 1/s(s+1)$. The signal w is the plant noise and may be analyzed as if it were a step; the signal v is the sensor noise and may be analyzed as if it contained power to very high frequencies.

a) Compute values for the parameters K_1, a, K_2, K_T, K_3, d, and K_D so that in each case (assuming $w = 0$ and $v = 0$),

$$\frac{Y}{R} = \frac{16}{s^2 + 4s + 16}.$$

Note that in system III, a pole is to be placed at $s = -4$.

b) Complete the following table. Express the last entries as A/s^k to show how fast noise from v is attenuated at high frequencies.

System	K_v	$\left.\dfrac{y}{w}\right\|_{s=0}$	$\left.\dfrac{y}{v}\right\|_{s\to\infty}$
I			
II			
III			

c) Rank the three designs according to the following characteristics (the best as "1," the poorest as "3"):

	I	II	III
Tracking			
Plant-noise rejection			
Sensor-noise rejection			

FIGURE 9.76
Alternative feedback
structures for
Problem 9.15

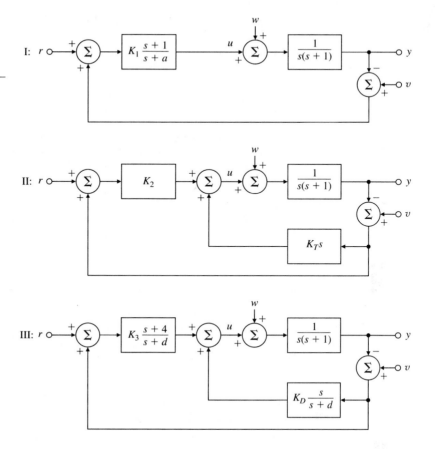

9.16 The equations of motion for a cart-stick balancer with state variables of stick angle, stick angular velocity, and cart velocity are

$$\dot{x} = \begin{bmatrix} 0 & 1 & 0 \\ 31.33 & 0 & 0.016 \\ -31.33 & 0 & -0.216 \end{bmatrix} x + \begin{bmatrix} 0 \\ -0.649 \\ 8.649 \end{bmatrix} u,$$

$$y = [10 \quad 0 \quad 0] x,$$

where the output is stick angle, and the control input is voltage on the motor that drives the cart wheels.

a) Compute the transfer function from u to y, and determine the poles and zeros.
b) Determine the feedback gain K necessary to move the poles of the system to the third-order ITAE locations -2.832 and $-0.521 \pm 1.068j$, with $\omega_n = 4$ rad/sec.
c) Determine the estimator gain L needed to place the three estimator poles at -10.
d) Determine the transfer function of the estimated-state-feedback compensator defined by the gains computed in parts (b) and (c).

e) Suppose we use a reduced-order estimator with poles at -10, and -10. What is the required estimator gain?

f) Repeat part (d) using the reduced-order estimator.

g) Compute the frequency response of the two compensators.

9.17 A 282-ton Boeing 747 is approaching land at sea level. If we use the state given in the case study (Section 9.3) and assume a velocity of 221 ft/sec (Mach 0.198), then the lateral-direction perturbation equations are

$$\begin{bmatrix} \dot{\beta} \\ \dot{r} \\ \dot{p} \\ \dot{\phi} \end{bmatrix} = \begin{bmatrix} -0.0890 & -0.989 & 0.1478 & 0.1441 \\ 0.168 & -0.217 & -0.166 & 0 \\ -1.33 & 0.327 & -0.975 & 0 \\ 0 & 0.149 & 1 & 0 \end{bmatrix} \begin{bmatrix} \beta \\ r \\ p \\ \phi \end{bmatrix} + \begin{bmatrix} 0.0148 \\ -151 \\ 0.0636 \\ 0 \end{bmatrix} \delta r,$$

$$y = \begin{bmatrix} 0 & 1 & 0 & 0 \end{bmatrix} \begin{bmatrix} \beta \\ r \\ p \\ \phi \end{bmatrix}.$$

The corresponding transfer function is

$$G(s) = \frac{r(s)}{\delta r(s)} = \frac{-0.151(s + 1.05)(s + 0.0328 \pm 0.414j)}{(s + 1.109)(s + 0.0425)(s + 0.646 \pm 0.731j)}.$$

a) Draw the uncompensated root loocus [for $1 + KG(s)$] and the frequency response of the system. What type of classical controller could be used for this system?

b) Try a state-variable design approach by drawing a symmetric root locus for the system. Choose the closed-loop poles of the system on the SRL to be

$$\alpha_c(s) = (s + 1.12)(s + 0.165)(s + 0.162 \pm 0.681j),$$

and choose the estimator poles to be five times faster at

$$\alpha_e(s) = (s + 5.58)(s + 0.825)(s + 0.812 \pm 3.40j).$$

c) Compute the transfer function of the SRL compensator.

d) Discuss the robustness properties of the system with respect to parameter variations and unmodeled dynamics.

e) Note the similarity of this design with the one developed for different flight conditions earlier in the chapter. What does this suggest about providing a continuous (nonlinear) control throughout the operating envelope?

9.18 (Contributed by Prof. L. Swindlehurst) The feedback control system shown in Fig. 9.77 is proposed as a position control system. A key component of this system is an armature-controlled DC motor. The input potentiometer produces a voltage E_i that is proportional to the desired shaft position: $E_i = K_p \theta_i$. Similarly, the output potentiometer produces a voltage E_o that is proportional to the actual shaft position: $E_o = K_p \theta_o$. Note we have assumed that both potentiometers have the same proportionality constant. The error signal $E_i - E_o$ drives a compensator, which in turn produces an armature voltage that drives the motor. The motor has an armature resistance R_a, an armature inductance L_a, a torque constant K_t, and

FIGURE 9.77

A servomechanism with gears on the motor shaft and potentiometer sensors

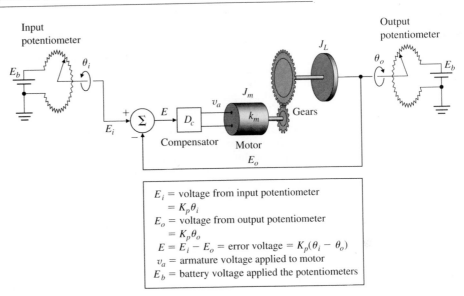

E_i = voltage from input potentiometer
   = $K_p \theta_i$
E_o = voltage from output potentiometer
   = $K_p \theta_o$
$E = E_i - E_o$ = error voltage = $K_p(\theta_i - \theta_o)$
v_a = armature voltage applied to motor
E_b = battery voltage applied the potentiometers

a back-emf constant K_e. The moment of inertia of the motor shaft is J_m, and the rotational damping due to bearing friction is B_m. Finally, the gear ratio is $N:1$, the moment of inertia of the load is J_L, and the load damping is B_L.

a) Write the differential equations that describe the operation of this feedback system.

b) Find the transfer function relating $\theta_o(s)$ and $\theta_i(s)$ for a general compensator $D_c(s)$.

c) The open-loop frequency-response data shown in Table 9.1 was taken using the armature voltage v_a of the motor as an input and the output potentiometer

TABLE 9.1   Frequency-response Data for Problem 9.18

Frequency (rad/sec)	$\left\|\dfrac{E_o(s)}{V_a(s)}\right\|$ (dB)	Frequency (rad/sec)	$\left\|\dfrac{E_o(s)}{V_a(s)}\right\|$ (dB)
0.1	60.0	10.0	14.0
0.2	54.0	20.0	2.0
0.3	50.0	40.0	−10.0
0.5	46.0	60.0	−20.0
0.8	42.0	65.0	−21.0
1.0	40.0	80.0	−24.0
2.0	34.0	100.0	−30.0
3.0	30.5	200.0	−48.0
4.0	27.0	300.0	−59.0
5.0	23.0	500.0	−72.0
7.0	19.5		

voltage E_o as the output. Assuming the motor is linear and minimum-phase, make an estimate of the transfer function of the motor:

$$G(s) = \frac{\theta_m(s)}{V_a(s)},$$

where θ_m is the angular position of the motor shaft.

d) Determine a set of performance specifications that are appropriate for a position control system and will yield good performance. Design $D_c(s)$ to meet these specifications.

e) Verify your design through analysis and simulation.

9.19. Design and construct a device to keep a ball centered on a freely swinging beam. An example of such a device is shown in Fig. 9.78. It uses coils surrounding permanent magnets as the actuator to move the beam, solar cells to sense the ball position, and a hall-effect device to sense the beam position. Reseach other possible actuators and sensors as part of your design effort. Compare the quality of the control achievable for ball-position-feedback only with that of multiple-loop feedback of both ball and beam position.

FIGURE 9.78
Ball-balancer design example

◆ Appendix A ◆
Laplace Transforms

Laplace transforms can be used to study the complete response characteristics of feedback systems, including the transient response. This is in contrast to Fourier transforms, in which the steady-state response is the main concern. The Laplace transform of $f(t)$, denoted by $\mathcal{L}\{f(t)\} = F(s)$, is a function of the complex variable $s = \sigma + j\omega$, where

$$F(s) \triangleq \int_{0^-}^{\infty} f(t)e^{-st}\, dt \tag{A.1}$$

and is referred to as the **unilateral** (or **one-sided**) **Laplace transform.**[1] A function $f(t)$ will have a Laplace transform if it is of **exponential order**, which means there exists a real number σ such that

$$\lim_{t \to \infty} |f(t)e^{-\sigma t}| = 0. \tag{A.2}$$

For example, ae^{bt} is of exponential order, whereas e^{t^2} is not. If $F(s)$ exists for some $s_0 = \sigma_0 + j\omega_0$, then it exists for all values of s such that

$$\mathrm{Re}(s) \geqslant \sigma_0. \tag{A.3}$$

The smallest value of σ_0 for which $F(s)$ exists is called the **abscissa of convergence**, and the region to the right of $\mathrm{Re}(s) \geqslant \sigma_0$ is called the **region of convergence**. Typically, two-sided Laplace transforms exist for a specified range,

$$\alpha < \mathrm{Re}(s) < \beta, \tag{A.4}$$

which defines the strip of convergence. Table A.1 lists some properties of Laplace transforms, Table A.2 gives some Laplace transform pairs. For a thorough study of Laplace transforms and extensive tables, see Churchill (1972) and Campbell and Foster (1948); for the two-sided transform, see Van der Pol and Bremmer (1955).

---

1. **Bilateral** (or **two-sided**) **Laplace transforms** and the so-called $\mathcal{L}_+$ transforms, in which the value of integral is 0^+, also arise elsewhere.

TABLE A.1   Properties of Laplace Transforms

Number	Laplace Transform	Time Function	Comment
—	$F(s)$	$f(t)$	Transform pair
1	$\alpha F_1(s) + \beta F_2(s)$	$\alpha f_1(t) + \beta f_2(t)$	Superposition
2	$F(s)e^{-s\lambda}$	$f(t - \lambda)$	Time delay $(\lambda \geqslant 0)$
3	$\dfrac{1}{\lvert a \rvert} F\left(\dfrac{s}{a}\right)$	$f(at)$	Time scaling
4	$F(s + a)$	$e^{-at}f(t)$	Shift in frequency
5	$s^m F(s) - s^{m-1}f(0)$ $\quad - s^{m-2}\dot{f}(0) - \cdots - f^{(m-1)}(0)$	$f^{(m)}(t)$	Differentiation
6	$\dfrac{1}{s} F(s)$	$\displaystyle\int f(\zeta)\,d\zeta$	Integration
7	$F_1(s)F_2(s)$	$f_1(t) \star f_2(t)$	Convolution
8	$\displaystyle\lim_{s \to \infty} sF(s)$	$f(0^+)$	Initial Value Theorem
9	$\displaystyle\lim_{s \to 0} sF(s)$	$\displaystyle\lim_{t \to \infty} f(t)$	Final Value Theorem
10	$\dfrac{1}{2\pi j} \displaystyle\int_{c-j\infty}^{c+j\infty} F_1(\zeta)F_2(s - \zeta)\,d\zeta$	$f_1(t)f_2(t)$	Time product
11	$-\dfrac{d}{ds} F(s)$	$tf(t)$	Multiplication by time

TABLE A.2   Table of Laplace Transforms

Number	$F(s)$	$f(t), \quad t \geqslant 0$
1	1	$\delta(t)$
2	$1/s$	$1(t)$
3	$1/s^2$	t
4	$2!/s^3$	t^2
5	$3!/s^4$	t^3
6	$m!/s^{m+1}$	t^m
7	$\dfrac{1}{s+a}$	e^{-at}
8	$\dfrac{1}{(s+a)^2}$	te^{-at}
9	$\dfrac{1}{(s+a)^3}$	$\dfrac{1}{2!}t^2 e^{-at}$
10	$\dfrac{1}{(s+a)^m}$	$\dfrac{1}{(m-1)!}t^{m-1}e^{-at}$
11	$\dfrac{a}{s(s+a)}$	$1 - e^{-at}$
12	$\dfrac{a}{s^2(s+a)}$	$\dfrac{1}{a}(at - 1 + e^{-at})$
13	$\dfrac{b-a}{(s+a)(s+b)}$	$e^{-at} - e^{-bt}$
14	$\dfrac{s}{(s+a)^2}$	$(1 - at)e^{-at}$
15	$\dfrac{a^2}{s(s+a)^2}$	$1 - e^{-at}(1 + at)$
16	$\dfrac{(b-a)s}{(s+a)(s+b)}$	$be^{-bt} - ae^{-at}$
17	$\dfrac{a}{s^2 + a^2}$	$\sin at$
18	$\dfrac{s}{s^2 + a^2}$	$\cos at$
19	$\dfrac{s+a}{(s+a)^2 + b^2}$	$e^{-at}\cos bt$
20	$\dfrac{b}{(s+a)^2 + b^2}$	$e^{-at}\sin bt$
21	$\dfrac{a^2 + b^2}{s[(s+a)^2 + b^2]}$	$1 - e^{-at}\left(\cos bt + \dfrac{a}{b}\sin bt\right)$

◆ Appendix B ◆
A Review of Complex Variables

This appendix is a brief summary of some results on complex variables theory with emphasis on the facts needed in control theory. For a comprehensive study of basic complex variables theory, see standard textbooks such as Churchill *et al.* (1976).

B.1 Definition of a Complex Number

The complex numbers are distinguished from purely real numbers in that they also contain the **imaginary operator**, which we shall denote j. By definition,

$$j^2 = -1 \quad \text{or} \quad j = \sqrt{-1}. \tag{B.1}$$

A **complex number** may be defined as

$$A = \sigma + j\omega, \tag{B.2}$$

where σ is the real part and ω is the imaginary part, denoted respectively as

$$\sigma = \text{Re}(A), \qquad \omega = \text{Im}(A). \tag{B.3}$$

Note that the imaginary part of A is itself a real number.

Graphically, we may represent the complex number A in two ways: in the Cartesian coordinate system (Fig. B.1a), A is represented by a single point in the complex plane. In the polar coordinate system, A is represented by a vector with length r and an angle θ; the angle is measured in radians counterclockwise from the positive real axis (Fig. B.1b). In polar form the complex number A is denoted by

$$A = |A| \cdot \angle \arg A = r \cdot \angle \theta = re^{j\theta}, \qquad 0 \leqslant \theta \leqslant 2\pi, \tag{B.4}$$

where r—called the **magnitude**, **modulus**, or **absolute value** of A—is the length of the vector representing A,

$$r = |A| = \sqrt{\sigma^2 + \omega^2}, \tag{B.5}$$

736

FIGURE B.1
The complex number A
represented in
(a) Cartesian and
(b) polar coordinates

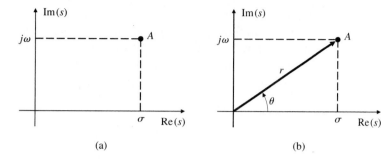

(a)   (b)

and where θ is given by

$$\tan \theta = \frac{\omega}{\sigma}. \tag{B.6}$$

The **conjugate** of A is defined as

$$A^* = \sigma - j\omega. \tag{B.7}$$

Therefore,

$$(A^*)^* = A, \tag{B.8}$$

$$(A_1 \pm A_2)^* = A_1^* \pm A_2^*, \tag{B.9}$$

$$\left(\frac{A_1}{A_2}\right)^* = \frac{A_1^*}{A_2^*}, \tag{B.10}$$

$$\operatorname{Re}(A) = \frac{A + A^*}{2}, \qquad \operatorname{Im}(A) = \frac{A - A^*}{2j}, \tag{B.11}$$

$$AA^* = (|A|)^2. \tag{B.12}$$

B.2   Algebraic Manipulations

The rules for addition, multiplication, and division are as would be expected.

Complex Addition

If we let

$$A_1 = \sigma_1 + j\omega_1 \qquad \text{and} \qquad A_2 = \sigma_2 + j\omega_2, \tag{B.13}$$

then

$$A_1 + A_2 = (\sigma_1 + j\omega_1) + (\sigma_2 + j\omega_2) = (\sigma_1 + \sigma_2) + j(\omega_1 + \omega_2). \tag{B.14}$$

Because each complex number is represented by a vector extending from the origin, we can add or subtract complex numbers graphically. The sum is obtained by adding the two vectors. This we do by constructing a parallelogram and finding its diagonal, as shown in Fig. B.2(a). Alternatively, we could start at the tail of one vector, draw a vector parallel to the other vector, and then connect the origin to the new arrowhead.

Complex subtraction is very similar to complex addition.

Complex Multiplication

For two complex numbers defined according to Eq. (B.13),

$$
\begin{aligned}
A_1 A_2 &= (\sigma_1 + j\omega_1)(\sigma_2 + j\omega_2) \\
&= (\sigma_1 \sigma_2 - \omega_1 \omega_2) + j(\omega_1 \sigma_2 + \sigma_1 \omega_2).
\end{aligned} \tag{B.15}
$$

The product of two complex numbers may be done graphically using polar representations as shown in Fig. B.2(b).

Complex Division

The division of two complex numbers is carried out by **rationalization**. This means that both the numerator and denominator in the ratio are multiplied by the conjugate of the denominator:

$$
\begin{aligned}
\frac{A_1}{A_2} &= \frac{A_1 A_2^*}{A_2 A_2^*} \\
&= \frac{(\sigma_1 \sigma_2 + \omega_1 \omega_2) + j(\omega_1 \sigma_2 - \sigma_1 \omega_2)}{\sigma_2^2 + \omega_2^2}.
\end{aligned} \tag{B.16}
$$

FIGURE B.2
Arithmetic of complex numbers: (a) addition; (b) multiplication; (c) division

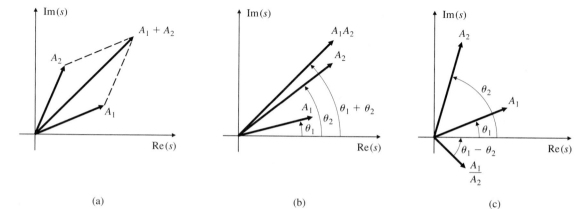

From Eq. (B.4) it follows that

$$A^{-1} = \frac{1}{r} e^{-j\theta}, \qquad r \neq 0. \tag{B.17}$$

Also, if $A_1 = r_1 e^{j\theta_1}$ and $A_2 = r_2 e^{j\theta_2}$, then

$$A_1 A_2 = r_1 r_2 e^{j(\theta_1 + \theta_2)} \tag{B.18}$$

and

$$\frac{A_1}{A_2} = \frac{r_1}{r_2} e^{j(\theta_1 - \theta_2)}, \qquad r_2 \neq 0. \tag{B.19}$$

The division of complex numbers may be carried out graphically in polar coordinates as shown in Fig. B.2(c).

B.3   Graphical Evaluation of Magnitude and Phase

Consider the transfer function

$$G(s) = \frac{\displaystyle\prod_{i=1}^{m} (s + z_i)}{\displaystyle\prod_{i=1}^{n} (s + p_i)}. \tag{B.20}$$

The value of the transfer function for sinusoidal inputs is found by replacing s with $j\omega$. The gain and phase are given by $G(j\omega)$ and may be determined analytically or by a graphical procedure. Consider the pole-zero configuration for such a $G(s)$ and a point $s_0 = j\omega_0$ on the imaginary axis, as shown in Fig. B.3. Also consider the vectors drawn from the poles and the zero to s_0. The magnitude of the transfer function evaluated at $s_0 = j\omega_0$ is simply the ratio of

FIGURE B.3
Graphical determination
of magnitude and phase

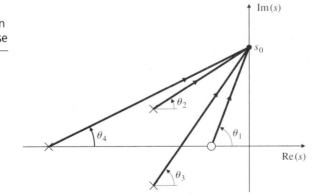

FIGURE B.4
Illustration of graphical
computation of $s + z_1$

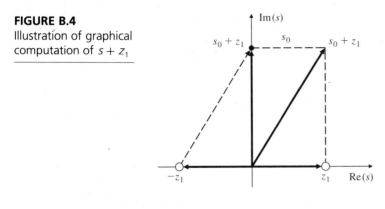

the distance from the zero to the product of all the distances from the poles:

$$|G(j\omega_0)| = \frac{r_1}{r_2 r_3 r_4}. \tag{B.21}$$

The phase is given by the sum of the angles from the zero minus the sum of the angles from the poles:

$$\arg G(j\omega_0) = \angle\, G(j\omega_0) = \theta_1 - (\theta_2 + \theta_3 + \theta_4). \tag{B.22}$$

This may be explained as follows. The term $s + z_1$ is a vector addition of its two components. We may determine this equivalently as $s - (-z_1)$, which amounts to translation of the vector $s + z_1$ starting at $-z_1$, as shown in Fig. B.4. This means that a vector drawn from the zero location to s_0 is equivalent to $s + z_1$. The same reasoning applies to the poles. We reflect p_1, p_2 and p_3 about the origin to obtain the pole locations. Then the vectors drawn from $-p_1$, $-p_2$ and $-p_3$ to s_0 are the same as the vectors in the denominator represented in polar coordinates. Note that this method may also be used to evaluate s_0 at places in the complex plane besides the imaginary axis.

B.4 Differentiation and Integration

The usual rules apply to complex differentiation. Let $G(s)$ be differentiable with respect to s. Then the derivative at s_0 is defined as

$$G'(s_0) = \lim_{s \to s_0} \frac{G(s) - G(s_0)}{s - s_0}, \tag{B.23}$$

provided the limit exists.

The standard rules also apply to integration, except that the constant of integration c is a complex constant:

$$\int G(s)\, ds = \int \mathrm{Re}[G(s)]\, ds + j \int \mathrm{Im}[G(s)]\, ds + c. \tag{B.24}$$

B.5 Euler's Relations

Let us now derive an important relationship involving the complex exponential. If we define

$$A = \cos\theta + j\sin\theta, \tag{B.25}$$

where θ is in radians, then

$$\frac{dA}{d\theta} = -\sin\theta + j\cos\theta = j^2\sin\theta + j\cos\theta$$

$$= j(\cos\theta + j\sin\theta) = jA. \tag{B.26}$$

We collect the terms involving A to obtain

$$\frac{dA}{A} = j\,d\theta. \tag{B.27}$$

Integrating both sides of Eq. (B.27) yields

$$\ln A = j\theta + c, \tag{B.28}$$

where c is a constant of integration. If we let $\theta = 0$ in Eq. (B.28), we find that $c = 0$ or

$$A = e^{j\theta} = \cos\theta + j\sin\theta. \tag{B.29}$$

Similarly,

$$A^* = e^{-j\theta} = \cos\theta - j\sin\theta. \tag{B.30}$$

From Eqs. (B.29) and (B.30) it follows that

Euler's relations

$$\cos\theta = \frac{e^{j\theta} + e^{-j\theta}}{2}, \tag{B.31}$$

$$\sin\theta = \frac{e^{j\theta} - e^{-j\theta}}{2j}. \tag{B.32}$$

B.6 Analytic Functions

Let us assume that G is a complex-valued function defined in the complex plane. Let s_0 be in the domain of G, which is assumed to be finite within some disk centered at s_0. Thus $G(s)$ is defined not only at s_0 but also at all points in the disk centered at s_0. The function G is said to be **analytic** if its derivative exists at s_0 and at each point in the neighborhood of s_0.

FIGURE B.5
Contours in the s-plane:
(a) A closed contour; (b)
two different paths
between A_1 and A_2

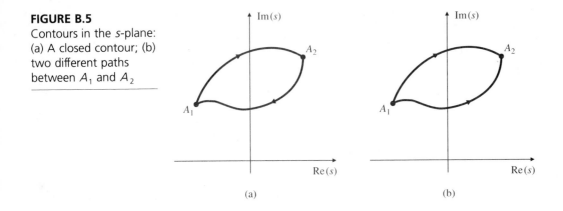

(a)

(b)

B.7   Cauchy's Theorem

A **contour** is a piecewise-smooth arc that consists of a number of smooth arcs joined together. A **simple closed contour** is a contour that does not intersect itself and ends on itself. Let C be a closed contour as shown in Fig. B.5(a) and let G be analytic inside and on C. Cauchy's theorem states that

$$\oint_C G(s)\,ds = 0. \tag{B.33}$$

A corollary to this theorem is the following. Let C_1 and C_2 be two paths connecting the points A_1 and A_2 as in Fig. B.5(b), then

$$\int_{C_1} G(s)\,ds = \int_{C_2} G(s)\,ds. \tag{B.34}$$

B.8   Singularities and Residues

If a function $G(s)$ is not analytic at s_0 but is analytic at some point in every neighborhood of s_0, it is said to be a **singularity**. A singular point is said to be an **isolated singularity** if $G(s)$ is analytic everywhere else in the neighborhood of s_0 except at s_0. Let $G(s)$ be a **rational function** (that is, a ratio of polynomials). If the numerator and denominator are both analytic, then $G(s)$ will be analytic except at the locations of the poles (that is, at roots of the denominator). All singularities of rational algebraic functions are the pole locations.

Let $G(s)$ be analytic except at s_0. Then we may write $G(s)$ in its Laurent series-expansion form

$$G(s) = \frac{A_{-n}}{(s-s_0)^n} + \cdots + \frac{A_{-1}}{s-s_0} + B_0 + B_1(s-s_0) + \cdots. \tag{B.35}$$

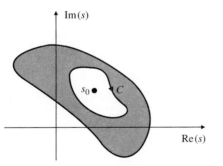

The coefficient A_{-1} is called the **residue** of $G(s)$ at s_0 and may be evaluated as

$$A_{-1} = \text{Res}[G(s); s_0] = \frac{1}{2\pi j} \oint_C G(s)\,ds, \tag{B.36}$$

where C denotes a closed arc within an analytic region centered at s_0 that contains no other singularity as shown in Fig. B.6. When s_0 is not repeated with $n = 1$, then

$$A_{-1} = \text{Res}[G(s); s_0] = (s - s_0)G(s)|_{s=s_0}. \tag{B.37}$$

This is the familiar cover-up method of computing residues.

B.9 Residue Theorem

If the contour C contains l singularities, then Eq. (B.37) may be generalized to yield **Cauchy's residue theorem**:

$$\frac{1}{2\pi j} \oint G(s)\,ds = \sum_{i=1}^{l} \text{Res}[G(s); s_i]. \tag{B.38}$$

B.10 The Argument Principle

Before stating the argument principle, we need a preliminary result from which the principle follows readily.

Number of Poles and Zeros

Let $G(s)$ be an analytic function inside and on a closed contour C except for a finite number of poles inside C. Then, for C described in the positive sense (clockwise direction),

$$\frac{1}{2\pi j} \oint \frac{G'(s)}{G(s)}\,ds = N - P \tag{B.39}$$

or

$$\frac{1}{2\pi j} \oint d(\ln G) = N - P, \tag{B.40}$$

where N and P are the total number of zeros and poles of G inside C, respectively. A pole or zero of multiplicity k is counted k times.

Proof   Let s_0 be a zero of G with multiplicity k. Then in some neighborhood of that point we may write $G(s)$ as

$$G(s) = (s - s_0)^k f(s), \tag{B.41}$$

where $f(s)$ is analytic and $f(s_0) \neq 0$. If we differentiate Eq. (B.41) we obtain

$$G'(s) = k(s - s_0)^{k-1} f(s) + (s - s_0)^k f'(s). \tag{B.42}$$

Equation (B.39) may be rewritten as

$$\frac{G'(s)}{G(s)} = \frac{K}{s - s_0} + \frac{f'(s)}{f(s)}. \tag{B.43}$$

Therefore, $G'(s)/G(s)$ has a pole at $s = s_0$ with residue k. This analysis may be repeated for every zero. Hence, the sum of the residues of $G'(s)/G(s)$ is the number of zeros of $G(s)$ inside C. If s_0 is a pole with multiplicity l, we may write

$$h(s) = (s - s_0)^l G(s), \tag{B.44}$$

where $h(s)$ is analytic and $h(s_0) \neq 0$. Then Eq. (B.44) may be rewritten as

$$G(s) = \frac{h(s)}{(s - s_0)^l}. \tag{B.45}$$

Differentiating Eq. (B.45) we obtain,

$$G'(s) = \frac{h'(s)}{(s - s_0)^l} - \frac{lh(s)}{(s - s_0)^{l+1}} \tag{B.46}$$

so that

$$\frac{G'(s)}{G(s)} = \frac{-l}{s - s_0} + \frac{h'(s)}{h(s)}. \tag{B.47}$$

This analysis may be repeated for every pole. The result is that the sum of the residues of $G'(s)/G(s)$ at all the poles of $G(s)$ is $-P$.

The Argument Principle

Using Eq. (B.36) we get

$$\frac{1}{2\pi j} \oint_C d[\ln G(s)] = N - P, \tag{B.48}$$

where $d[\ln G(s)]$ was substituted for $G'(s)/G(s)$. If we write $G(s)$ in polar form, then

$$\oint_\Gamma d[\ln G(s)] = \oint_\Gamma d\{\ln|G(s)| + j\arg[\ln G(s)]\}$$

$$= \ln|G(s)|\Big|_{s=s_1}^{s=s_2} + j\arg G(s)\Big|_{s=s_1}^{s=s_2}. \tag{B.49}$$

Since Γ is a closed contour, the first term is zero, but the second term is 2π times the net encirclements of the origin:

Number of poles and zeros

$$\frac{1}{2\pi j}\oint_\Gamma d[\ln G(s)] = N - P. \tag{B.50}$$

Intuitively, the argument principle may be stated as follows. We let $G(s)$ be a rational function that is analytic except at possibly a finite number of points. We select an arbitrary contour in the s-plane so that $G(s)$ is analytic at every point on the contour (the contour does not pass through any of the singularities). The corresponding mapping into the $G(s)$-plane may encircle the origin. The number of times it does so is determined by the difference between the number of zeros and the number of poles of $G(s)$ encircled by the s-plane contour. The direction of this encirclement is determined by which is greater, N (clockwise) or P (counterclockwise). For example, if the contour encircles a single zero, the mapping will encircle the origin once in the clockwise direction. Similarly, if the contour encloses only a single pole, the mapping will encircle the origin, this time in the counterclockwise direction. If the contour encircles no singularities, or if the contour encloses an equal number of poles and zeros, there will be no encirclement of the origin. A contour evaluation of $G(s)$ will encircle the origin if there is a nonzero net difference between the encircled singularities. The mapping is **conformal** as well, which means that the magnitude and sense of the angles between smooth arcs is preserved. Chapter 6 provides a more detailed intuitive treatment of the argument principle and its application to feedback control in the form of the Nyquist stability theorem.

◆ Appendix C ◆
Summary of Matrix Theory

In the text, we assume you are already somewhat familiar with matrix theory and with the solution of linear systems of equations. However, for the purposes of review we present here a brief summary of matrix theory with an emphasis on the results needed in control theory. For further study, see Strang (1988) and Gantmacher (1959).

C.1 Matrix Definitions

An array of numbers arranged in rows and columns is referred to as a **matrix**. If $\mathbf{A}$ is a matrix with m rows and n columns, an $m \times n$ (read "m by n") matrix, it is denoted by

$$\mathbf{A} = \begin{bmatrix} a_{11} & a_{12} & \cdots & a_{1n} \\ a_{21} & a_{22} & \cdots & a_{2n} \\ \vdots & \vdots & & \vdots \\ a_{m1} & a_{m2} & \cdots & a_{mn} \end{bmatrix}, \tag{C.1}$$

where the entries a_{ij} are its elements. If $m = n$, then the matrix is **square**; otherwise it is **rectangular**. Sometimes a matrix is simply denoted by $\mathbf{A} = [a_{ij}]$. If $m = 1$ or $n = 1$, then the matrix reduces to a **row vector** or a **column vector**, respectively. A **submatrix** of $\mathbf{A}$ is the matrix with certain rows and columns removed.

C.2 Elementary Operations on Matrices

If $\mathbf{A}$ and $\mathbf{B}$ are matrices of the same dimension, then their sum is defined by

$$\mathbf{C} = \mathbf{A} + \mathbf{B}, \tag{C.2}$$

where

$$c_{ij} = a_{ij} + b_{ij}. \tag{C.3}$$

That is, the addition is done element by element. It is easy to verify the following properties of matrices:

Commutative law for addition	$$\mathbf{A} + \mathbf{B} = \mathbf{B} + \mathbf{A},$$	(C.4)
Associative law for addition	$$(\mathbf{A} + \mathbf{B}) + \mathbf{C} = \mathbf{A} + (\mathbf{B} + \mathbf{C}).$$	(C.5)

Two matrices can be multiplied if they are compatible. Let $\mathbf{A} = m \times n$ and $\mathbf{B} = n \times p$. Then the $m \times p$ matrix

$$\mathbf{C} = \mathbf{AB} \tag{C.6}$$

is the product of the two matrices, where

$$c_{ij} = \sum_{k=1}^{n} a_{ik} b_{kj}. \tag{C.7}$$

Matrix multiplication satisfies the associative law

Associative law for multiplication

$$\mathbf{A(BC)} = \mathbf{(AB)C}, \tag{C.8}$$

but not the commutative law; that is, in general,

$$\mathbf{AB} \neq \mathbf{BA}. \tag{C.9}$$

C.3 Trace

The **trace** of a square matrix is the sum of its diagonal elements:

$$\text{trace}\,\mathbf{A} = \sum_{i=1}^{n} a_{ii}. \tag{C.10}$$

C.4 Transpose

The $n \times m$ matrix obtained by interchanging the rows and columns of $\mathbf{A}$ is called the **transpose of matrix $\mathbf{A}$**:

$$\mathbf{A}^T = \begin{bmatrix} a_{11} & a_{21} & \cdots & a_{m1} \\ a_{12} & a_{22} & \cdots & a_{m2} \\ \vdots & \vdots & & \vdots \\ a_{1n} & a_{2n} & \cdots & a_{mn} \end{bmatrix}. \tag{C.11}$$

A matrix is said to be **symmetric** if

$$\mathbf{A}^T = \mathbf{A}. \tag{C.12}$$

It is easy to show that

$$(\mathbf{AB})^T = \mathbf{B}^T \mathbf{A}^T \tag{C.13}$$

Transposition

$$(\mathbf{ABC})^T = \mathbf{C}^T \mathbf{B}^T \mathbf{A}^T \tag{C.14}$$

$$(\mathbf{A} + \mathbf{B})^T = \mathbf{A}^T + \mathbf{B}^T. \tag{C.15}$$

C.5   Determinant and Matrix Inverse

The **determinant** of a square matrix is defined by Laplace's expansion:

$$\det A = \sum_{j=1}^{n} a_{ij}\gamma_{ij} \qquad \text{for any } i = 1, 2, \ldots, n, \tag{C.16}$$

where γ_{ij} is called the **cofactor** and

$$\gamma_{ij} = (-1)^{i+j}\det M_{ij}, \tag{C.17}$$

where $\det M_{ij}$ is called a **minor**. M_{ij} is the same as the matrix A except that its ith row and jth column have been removed. Note that M_{ij} is always an $(n-1) \times (n-1)$ matrix, and that the minors and cofactors are identical except possibly for a sign.

The **adjugate** of a matrix is the transpose of the matrix of its cofactors:

$$\text{adj } A = [\gamma_{ij}]^{T}. \tag{C.18}$$

It can be shown that

$$A \text{ adj } A = (\det A)I, \tag{C.19}$$

where I is called the **identity matrix**:

Identity matrix

$$I = \begin{bmatrix} 1 & 0 & \cdots & \cdots & 0 \\ 0 & 1 & 0 & \cdots & 0 \\ \vdots & \vdots & \ddots & & \vdots \\ 0 & \cdots & \cdots & 0 & 1 \end{bmatrix}, \tag{C.20}$$

that is, with ones along the diagonal and zeros elsewhere.

If $\det A \neq 0$, then the **inverse** of a matrix A is defined by

Inverse matrix

$$A^{-1} = \frac{\text{adj } A}{\det A} \tag{C.21}$$

and has the property that

$$AA^{-1} = A^{-1}A = I. \tag{C.22}$$

Note that a matrix has an inverse—that is, it is **nonsingular**—if its determinant is nonzero.

The inverse of the product of two matrices is the product of the inverse of the matrices in reverse order:

$$(AB)^{-1} = B^{-1}A^{-1} \tag{C.23}$$

Inversion

and

$$(ABC)^{-1} = C^{-1}B^{-1}A^{-1}. \tag{C.24}$$

C.6 Properties of the Determinant

When dealing with determinants of matrices, the following elementary (row or column) operations are useful:

1. If any row (or column) of $\mathbf{A}$ is multiplied by a scalar α, the resulting matrix $\bar{\mathbf{A}}$ has the determinant

$$\det \bar{\mathbf{A}} = \alpha \det \mathbf{A}. \qquad (C.25)$$

Hence

$$\det(\alpha \mathbf{A}) = \alpha^n \det \mathbf{A}. \qquad (C.26)$$

2. If any two rows (or columns) of $\mathbf{A}$ are interchanged to obtain $\bar{\mathbf{A}}$, then

$$\det \bar{\mathbf{A}} = -\det \mathbf{A}. \qquad (C.27)$$

3. If a multiple of a row (or column) of $\mathbf{A}$ is added to another to obtain $\bar{\mathbf{A}}$, then

$$\det \bar{\mathbf{A}} = \det \mathbf{A}. \qquad (C.28)$$

4. It is also easy to show that

$$\det \mathbf{A} = \det \mathbf{A}^T \qquad (C.29)$$

and

$$\det \mathbf{AB} = \det \mathbf{A} \det \mathbf{B}. \qquad (C.30)$$

Applying Eq. (C.30) to Eq. (C.22) we have that

$$\det \mathbf{A} \det \mathbf{A}^{-1} = 1. \qquad (C.31)$$

If $\mathbf{A}$ and $\mathbf{B}$ are square matrices, then the determinant of the block triangular matrix

$$\det \begin{bmatrix} \mathbf{A} & \mathbf{C} \\ \mathbf{0} & \mathbf{B} \end{bmatrix} = \det \mathbf{A} \det \mathbf{B} \qquad (C.32)$$

is the product of the determinants of the diagonal blocks. If $\mathbf{A}$ is nonsingular, then

$$\det \begin{bmatrix} \mathbf{A} & \mathbf{B} \\ \mathbf{C} & \mathbf{D} \end{bmatrix} = \det \mathbf{A} \det(\mathbf{D} - \mathbf{C}\mathbf{A}^{-1}\mathbf{B}). \qquad (C.33)$$

Using this identity, the transfer function of a scalar system can be written in a compact form:

$$G(s) = \mathbf{H}(s\mathbf{I} - \mathbf{F})^{-1}\mathbf{G} + J = \frac{\det \begin{bmatrix} s\mathbf{I} - \mathbf{F} & \mathbf{G} \\ -\mathbf{H} & J \end{bmatrix}}{\det(s\mathbf{I} - \mathbf{F})}. \qquad (C.34)$$

C.7   Inverse of Block Triangular Matrices

If $\mathbf{A}$ and $\mathbf{B}$ are square invertible matrices, then

$$\begin{bmatrix} \mathbf{A} & \mathbf{C} \\ \mathbf{0} & \mathbf{B} \end{bmatrix}^{-1} = \begin{bmatrix} \mathbf{A}^{-1} & -\mathbf{A}^{-1}\mathbf{C}\mathbf{B}^{-1} \\ \mathbf{0} & \mathbf{B}^{-1} \end{bmatrix}. \tag{C.35}$$

C.8   Special Matrices

Some matrices have special structures and are given names. We have already defined the identity matrix, which has a special form. A **diagonal matrix** has (possibly) nonzero elements along the main diagonal and zeros elsewhere:

Diagonal matrix

$$\mathbf{A} = \begin{bmatrix} a_{11} & & & & \mathbf{0} \\ & a_{22} & & & \\ & & a_{33} & & \\ & & & \ddots & \\ \mathbf{0} & & & & a_{nn} \end{bmatrix}. \tag{C.36}$$

A matrix is said to be **(upper) triangular** if all the elements below the main diagonal are zeros:

Upper triangular matrix

$$\mathbf{A} = \begin{bmatrix} a_{11} & a_{12} & & \cdots & a_{1n} \\ 0 & a_{22} & & & \\ \vdots & 0 & & & \vdots \\ 0 & \vdots & \ddots & \ddots & \\ 0 & 0 & \cdots & 0 & a_{nn} \end{bmatrix}. \tag{C.37}$$

The determinant of a diagonal or triangular matrix is simply the product of its diagonal elements.

A matrix is said to be in the **(upper) companion form** if it has the structure

$$\mathbf{A}_c = \begin{bmatrix} -a_1 & -a_2 & & \cdots & -a_n \\ 1 & 0 & & \cdots & 0 \\ 0 & 1 & 0 & \cdots & 0 \\ \vdots & & \ddots & & \vdots \\ 0 & \cdots & \cdots & 1 & 0 \end{bmatrix}. \tag{C.38}$$

Note that all the information is contained in the first row. Variants of this form are the lower, left, or right companion matrices. A **Vandermonde matrix** has the following structure:

$$\mathbf{A} = \begin{bmatrix} 1 & a_1 & a_1^2 & \cdots & a_1^{n-1} \\ 1 & a_2 & a_2^2 & \cdots & a_2^{n-1} \\ \vdots & \vdots & \vdots & & \vdots \\ 1 & a_n & a_n^2 & \cdots & a_n^{n-1} \end{bmatrix}. \tag{C.39}$$

C.9   Rank

The **rank** of a matrix is the number of its linearly independent rows or columns. If the rank of $\mathbf{A}$ is r, then all $(r + 1) \times (r + 1)$ submatrices of $\mathbf{A}$ are singular, and there is at least one $r \times r$ submatrix that is nonsingular. It is also true that

$$\text{row rank of } \mathbf{A} = \text{column rank of } \mathbf{A}. \tag{C.40}$$

C.10   Characteristic Polynomial

The **characteristic polynomial** of a matrix $\mathbf{A}$ is defined by

$$a(s) \triangleq \det(s\mathbf{I} - \mathbf{A})$$
$$= s^n + a_1 s^{n-1} + \cdots + a_{n-1}s + a_n, \tag{C.41}$$

where the roots of the polynomial are referred to as **eigenvalues** of $\mathbf{A}$. We can write

$$a(s) = (s - \lambda_1)(s - \lambda_2)\cdots(s - \lambda_n), \tag{C.42}$$

where $\{\lambda_i\}$ are the eigenvalues of $\mathbf{A}$. The characteristic polynomial of a companion matrix (e.g., Eq. C.38) is

$$a(s) = \det(s\mathbf{I} - \mathbf{A}_c)$$
$$= s^n + a_1 s^{n-1} + \cdots + a_{n-1}s + a_n. \tag{C.43}$$

C.11   Cayley–Hamilton Theorem

The Cayley–Hamilton theorem states that every square matrix $\mathbf{A}$ satisfies its characteristic polynomial. This means that if $\mathbf{A}$ is an $n \times n$ matrix with characteristic equation $a(s)$, then

$$a(\mathbf{A}) \triangleq \mathbf{A}^n + a_1\mathbf{A}^{n-1} + \cdots + a_{n-1}\mathbf{A} + a_n\mathbf{I} = 0. \tag{C.44}$$

C.12   Eigenvalues and Eigenvectors

Any scalar λ and nonzero vector $\mathbf{v}$, that satisfy

$$\mathbf{Av} = \lambda\mathbf{v}, \tag{C.45}$$

are referred to as the eigenvalue and the associated (**right**) **eigenvector** of the matrix $\mathbf{A}$ (since $\mathbf{v}$ appears to the right of $\mathbf{A}$ in Eq. C.45). By rearranging terms in Eq. (C.45) we get

$$(\lambda\mathbf{I} - \mathbf{A})\mathbf{v} = 0. \tag{C.46}$$

Since $\mathbf{v}$ is nonzero, then

$$\det(\lambda\mathbf{I} - \mathbf{A}) = 0, \tag{C.47}$$

so λ is an eigenvalue of the matrix $\mathbf{A}$ as defined in Eq. (C.45). The normalization of the eigenvectors is arbitrary; that is, if $\mathbf{v}$ is an eigenvector, so is $\alpha\mathbf{v}$. The eigenvectors are usually normalized to have unit length; that is, $\|\mathbf{v}\|^2 = \mathbf{v}^T\mathbf{v} = 1$.
    If $\mathbf{w}^T$ is a nonzero row vector such that

$$\mathbf{w}^T\mathbf{A} = \lambda\mathbf{w}^T, \tag{C.48}$$

then $\mathbf{w}$ is called a **left eigenvector** of $\mathbf{A}$ (since $\mathbf{w}^T$ appears to the left of $\mathbf{A}$ in Eq. C.48). Note that we can write

$$\mathbf{A}^T\mathbf{w} = \lambda\mathbf{w} \tag{C.49}$$

so that $\mathbf{w}$ is simply a right eigenvector of $\mathbf{A}^T$.

C.13   Similarity Transformations

Consider the arbitrary nonsingular matrix $\mathbf{T}$ such that

$$\bar{\mathbf{A}} = \mathbf{T}^{-1}\mathbf{AT}. \tag{C.50}$$

The matrix operation shown in Eq. (C.50) is referred to as a **similarity transformation**. If $\mathbf{A}$ has a full set of eigenvectors, then we can choose $\mathbf{T}$ to be the set of eigenvectors and $\bar{\mathbf{A}}$ will be diagonal.
    Consider the set of equations in state-variable form:

$$\dot{\mathbf{x}} = \mathbf{Fx} + \mathbf{G}u. \tag{C.51}$$

If we let

$$\mathbf{T}\boldsymbol{\xi} = \mathbf{x}, \tag{C.52}$$

then Eq. (C.51) becomes

$$\mathbf{T}\dot{\boldsymbol{\xi}} = \mathbf{FT}\boldsymbol{\xi} + \mathbf{G}u, \tag{C.53}$$

and premultiplying both sides by $\mathbf{T}^{-1}$, we get

$$\begin{aligned}\dot{\xi} &= \mathbf{T}^{-1}\mathbf{F}\mathbf{T}\xi + \mathbf{T}^{-1}\mathbf{G}u \\ &= \bar{\mathbf{F}}\xi + \bar{\mathbf{G}}u,\end{aligned} \tag{C.54}$$

where

$$\begin{aligned}\bar{\mathbf{F}} &= \mathbf{T}^{-1}\mathbf{F}\mathbf{T}, \\ \bar{\mathbf{G}} &= \mathbf{T}^{-1}\mathbf{G}.\end{aligned} \tag{C.55}$$

The characteristic polynomial of $\bar{\mathbf{F}}$ is

$$\begin{aligned}\det(s\mathbf{I} - \bar{\mathbf{F}}) &= \det(s\mathbf{I} - \mathbf{T}^{-1}\mathbf{F}\mathbf{T}) \\ &= \det(s\mathbf{T}^{-1}\mathbf{T} - \mathbf{T}^{-1}\mathbf{F}\mathbf{T}) \\ &= \det[\mathbf{T}^{-1}(s\mathbf{I} - \mathbf{F})\mathbf{T}] \\ &= \det \mathbf{T}^{-1} \det(s\mathbf{I} - \mathbf{F}) \det \mathbf{T}.\end{aligned} \tag{C.56}$$

Using Eq. (C.31), Eq. (C.56) becomes

$$\det(s\mathbf{I} - \bar{\mathbf{F}}) = \det(s\mathbf{I} - \mathbf{F}). \tag{C.57}$$

From Eq. (C.57) we can see that $\bar{\mathbf{F}}$ and $\mathbf{F}$ both have the same characteristic polynomial, giving us the important result that a similarity transformation does not change the eigenvalues of a matrix. From Eq. (C.52) a new state made up of a linear combination of old state has the same eigenvalues as the old set.

C.14   Matrix Exponential

Let $\mathbf{A}$ be a square matrix. The **matrix exponential** of $\mathbf{A}$ is defined as the series

$$e^{\mathbf{A}t} = \mathbf{I} + \mathbf{A}t + \frac{1}{2!}\mathbf{A}^2 t^2 + \frac{\mathbf{A}^3 t^3}{3!} + \cdots. \tag{C.58}$$

It can be shown that the series converges. If $\mathbf{A}$ is an $n \times n$ matrix, then $e^{\mathbf{A}t}$ is also an $n \times n$ matrix and can be differentiated:

$$\frac{d}{dt} e^{\mathbf{A}t} = \mathbf{A}e^{\mathbf{A}t}. \tag{C.59}$$

Other properties of the matrix exponential are

$$e^{\mathbf{A}t_1} e^{\mathbf{A}t_2} = e^{\mathbf{A}(t_1 + t_2)} \tag{C.60}$$

and, in general,

$$e^{\mathbf{A}} e^{\mathbf{B}} \neq e^{\mathbf{B}} e^{\mathbf{A}}. \tag{C.61}$$

(In the exceptional case where $\mathbf{A}$ and $\mathbf{B}$ commute—that is, $\mathbf{AB} = \mathbf{BA}$—then $e^{\mathbf{A}} e^{\mathbf{B}} = e^{\mathbf{B}} e^{\mathbf{A}}$.)

C.15   Fundamental Subspaces

The **range space** of $\mathbf{A}$, denoted by $\mathscr{R}(\mathbf{A})$ and also called the **column space** of $\mathbf{A}$, is defined by the set of vectors $\mathbf{x}$ where

$$\mathbf{x} = \mathbf{A}\mathbf{y} \tag{C.62}$$

for some vector $\mathbf{y}$. The **null space** of $\mathbf{A}$, denoted by $\mathscr{N}(\mathbf{A})$, is defined by the set of vectors $\mathbf{x}$ such that

$$\mathbf{A}\mathbf{x} = \mathbf{0}. \tag{C.63}$$

If $\mathbf{x} \in \mathscr{N}(\mathbf{A})$ and $\mathbf{y} \in \mathscr{R}(\mathbf{A}^T)$, then $\mathbf{y}^T\mathbf{x} = 0$; that is, every vector in the null space of $\mathbf{A}$ is **orthogonal** to every vector in the range space of $\mathbf{A}^T$.

C.16   Singular-value Decomposition

The **singular-value decomposition (SVD)** is one of the most useful tools in linear algebra and has been widely used in control theory during the last decade. Let $\mathbf{A}$ be an $m \times n$ matrix. Then there always exist matrices $\mathbf{U}$, $\mathbf{S}$, and $\mathbf{V}$ such that

$$\mathbf{A} = \mathbf{U}\mathbf{S}\mathbf{V}^T. \tag{C.64}$$

Here $\mathbf{U}$ and $\mathbf{V}$ are **orthogonal matrices**; that is

$$\mathbf{U}\mathbf{U}^T = \mathbf{I}, \qquad \mathbf{V}\mathbf{V}^T = \mathbf{I}, \tag{C.65}$$

$\mathbf{S}$ is a **quasidiagonal matrix** with singular values as its diagonal elements; that is,

$$\mathbf{S} = \begin{bmatrix} \Sigma & 0 \\ 0 & 0 \end{bmatrix}, \tag{C.66}$$

where Σ is a diagonal matrix of nonzero singular values in descending order:

$$\sigma_1 \geqslant \sigma_2 \geqslant \cdots \geqslant \sigma_r > 0. \tag{C.67}$$

The unique diagonal elements of $\mathbf{S}$ are called the **singular values**. The maximum singular value is denoted by $\bar{\sigma}(\mathbf{A})$, and the minimum singular value is denoted by $\underline{\sigma}(\mathbf{A})$. The rank of the matrix is the same as the number of nonzero singular values. The columns of $\mathbf{U}$ and $\mathbf{V}$,

$$\begin{aligned} \mathbf{U} &= [u_1 \quad u_2 \quad \cdots \quad u_m], \\ \mathbf{V} &= [v_1 \quad v_2 \quad \cdots \quad v_n], \end{aligned} \tag{C.68}$$

are called the left and right **singular vectors**, respectively. SVD provides complete information about the fundamental subspaces associated with a matrix:

$$\mathcal{N}(\mathbf{A}) = \text{span}[v_{r+1} \quad v_{r+2} \quad \cdots \quad v_n]$$

$$\mathcal{R}(\mathbf{A}) = \text{span}[u_1 \quad u_2 \quad \cdots \quad u_r]$$

$$\mathcal{R}(\mathbf{A}^T) = \text{span}[v_1 \quad v_2 \quad \cdots \quad v_r]$$

$$\mathcal{N}(\mathbf{A}^T) = \text{span}[u_{r+1} \quad u_{r+2} \quad \cdots \quad u_m]. \tag{C.69}$$

The **norm** of the matrix $\mathbf{A}$, denoted by $\|\mathbf{A}\|_2$, is given by

$$\|\mathbf{A}\|_2 = \bar{\sigma}(\mathbf{A}). \tag{C.70}$$

If $\mathbf{A}$ is a function of ω, then the infinity norm of $\mathbf{A}$, $\|\mathbf{A}\|_\infty$, is given by

$$\|\mathbf{A}(j\omega)\|_\infty = \max_\omega \bar{\sigma}(\mathbf{A}). \tag{C.71}$$

C.17   Positive Definite Matrices

A matrix $\mathbf{A}$ is said to be **positive semidefinite** if

$$\mathbf{x}^T \mathbf{A}\mathbf{x} \geq 0 \qquad \text{for all } \mathbf{x}. \tag{C.72}$$

The matrix is said to be **positive definite** if equality holds in Eq. (C.72) only for $\mathbf{x} = 0$. A symmetric matrix is positive definite if and only if all of its eigenvalues are positive. It is positive semidefinite if and only if all of its eigenvalues are nonnegative.

An alternate method for determining positive definiteness is to test the minors of the matrix. A matrix is positive definite if all the leading principal minors are positive, and positive semidefinite if they are all nonnegative.

◆ Appendix D ◆
Controllability and Observability

Controllability and observability are important structural properties of a dynamic system. First identified and studied by Kalman (1960) and later by Kalman *et al.* (1961), these properties have continued to be examined during the last three decades. We will discuss only a few of the known results for linear constant systems with one input and one output. In the text we discuss these concepts in connection with control law and estimator designs. For example, in Section 7.2 we suggest that if the square matrix given by

$$\mathscr{C} = [\mathbf{G} \quad \mathbf{FG} \quad \mathbf{F^2G} \quad \cdots \quad \mathbf{F}^{n-1}\mathbf{G}] \tag{D.1}$$

is nonsingular, then by transformation of the state we can convert the given description into control canonical form. We can then construct a control law that will give the closed-loop system an arbitrary characteristic equation.

Controllability

We begin our formal discussion of controllability with the first of three definitions:

> *Definition I:* The system $(\mathbf{F}, \mathbf{G})$ is **controllable** if for any given nth-order polynomial $\alpha_c(s)$ there exists a (unique) control law $u = -\mathbf{Kx}$ such that the characteristic polynomial of $\mathbf{F} - \mathbf{GK}$ is $\alpha_c(s)$.

From the results of Ackermann's formula (see Appendix E), we have the following mathematical test for controllability: $(\mathbf{F}, \mathbf{G})$ is a controllable pair if and only if the rank of $\mathscr{C}$ is n. Definition I based on pole placement is a frequency-domain concept. Controllability can be equivalently defined in the time domain.

> *Definition II:* The system $(\mathbf{F}, \mathbf{G})$ is **controllable** if there exists a (piecewise continuous) control signal $u(t)$ that will take the state of the system from any initial state $\mathbf{x}_0$ to any desired final state $\mathbf{x}_f$ in a finite time interval.

We will now show that the system is controllable by this definition if and only if $\mathscr{C}$ is full-rank. We first assume that the system is controllable but that

$$\operatorname{rank}[\mathbf{G} \quad \mathbf{FG} \quad \mathbf{F}^2\mathbf{G} \quad \cdots \quad \mathbf{F}^{n-1}\mathbf{G}] < n. \tag{D.2}$$

We can then find a vector $\mathbf{v}$ such that

$$\mathbf{v}[\mathbf{G} \quad \mathbf{FG} \quad \mathbf{F}^2\mathbf{G} \quad \cdots \quad \mathbf{F}^{n-1}\mathbf{G}] = \mathbf{0}, \tag{D.3}$$

or

$$\mathbf{vG} = \mathbf{vFG} = \mathbf{vF}^2\mathbf{G} = \cdots = \mathbf{vF}^{n-1}\mathbf{G} = 0. \tag{D.4}$$

The Cayley–Hamilton theorem states that $\mathbf{F}$ satisfies its own characteristic equation, namely,

$$-\mathbf{F}^n = a_1\mathbf{F}^{n-1} + a_2\mathbf{F}^{n-2} + \cdots + a_n\mathbf{I}. \tag{D.5}$$

Therefore,

$$-\mathbf{vF}^n\mathbf{G} = a_1\mathbf{vF}^{n-1}\mathbf{G} + a_2\mathbf{vF}^{n-2}\mathbf{G} + \cdots + a_n\mathbf{vG} = 0. \tag{D.6}$$

By induction, $\mathbf{vF}^{n+k}\mathbf{G} = 0$ for $k = 0, 1, 2, \ldots$, or $\mathbf{vF}^m\mathbf{G} = 0$ for $m = 0, 1, 2, \ldots$, and thus

$$\mathbf{v}e^{\mathbf{F}t}\mathbf{G} = \mathbf{v}\left(\mathbf{I} + \mathbf{F}t + \frac{1}{2!}\mathbf{F}^2t^2 + \cdots\right)\mathbf{G} = 0 \tag{D.7}$$

for all t. However, the zero initial-condition response $(\mathbf{x}_0 = \mathbf{0})$ is

$$\mathbf{x}(t) = \int_0^t e^{\mathbf{F}(t-\tau)}\mathbf{G}u(\tau)\,d\tau$$

$$= e^{\mathbf{F}t}\int_0^t e^{-\mathbf{F}\tau}\mathbf{G}u(\tau)\,d\tau. \tag{D.8}$$

Using Eq. (D.7), Eq. (D.8) becomes

$$\mathbf{vx}(t) = \int_0^t \mathbf{v}e^{\mathbf{F}(t-\tau)}\mathbf{G}u(\tau)\,d\tau = 0 \tag{D.9}$$

for all $u(t)$ and $t > 0$. This implies that all points reachable from the origin are orthogonal to $\mathbf{v}$. This restricts the reachable space and therefore contradicts the second definition of controllability. Thus if $\mathscr{C}$ is singular, $(\mathbf{F}, \mathbf{G})$ is not controllable by Definition II.

Next we assume that $\mathscr{C}$ is full-rank but $(\mathbf{F}, \mathbf{G})$ is uncontrollable by Definition II. This means that there exists a nonzero vector $\mathbf{v}$ such that

$$\mathbf{v}\int_0^{t_f} e^{\mathbf{F}(t_f-\tau)}\mathbf{G}u(\tau)\,d\tau = 0, \tag{D.10}$$

since the whole state space is not reachable. But Eq. (D.10) implies that

$$\mathbf{v}e^{\mathbf{F}(t_f-\tau)}\mathbf{G} = 0, \qquad 0 \leqslant \tau \leqslant t_f. \tag{D.11}$$

FIGURE D.1

Block diagram of a
system with a diagonal
matrix

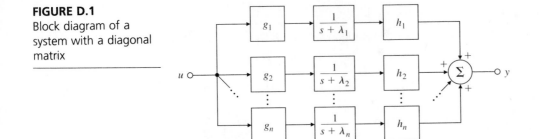

If we set $\tau = t_f$, we see that $\mathbf{vG} = 0$. Also, differentiating Eq. (D.11) and letting $\tau = t_f$ gives $\mathbf{vFG} = 0$. Continuing this process, we find that

$$\mathbf{vG} = \mathbf{vFG} = \mathbf{vF^2G} = \cdots = \mathbf{vF}^{n-1}\mathbf{G} = 0, \tag{D.12}$$

which contradicts the assumption that $\mathscr{C}$ is full-rank.

We have now shown that the system is controllable by Definition II if and only if the rank of $\mathscr{C}$ is n, exactly the same condition we found for pole assignment.

Our final definition comes closest to the structural character of controllability:

Definition III: The system $(\mathbf{F}, \mathbf{G})$ is **controllable** if every mode of $\mathbf{F}$ is connected to the control input.

Because of the generality of the modal structure of systems, we will only treat the case of systems for which $\mathbf{F}$ can be transformed to diagonal form. (The double-integration plant does *not* qualify.) Suppose we have a diagonal matrix $\mathbf{F}_d$ and its corresponding input matrix $\mathbf{G}_d$ with elements g_i. The structure of such a system is shown in Fig. D.1. By definition, for a controllable system the input must be connected to each mode so that the g_i are all nonzero. However, this is not enough if the poles (λ_i) are not distinct. Suppose, for instance, that $\lambda_1 = \lambda_2$. The first two state equations are then

$$\dot{x}_{1d} = \lambda_1 x_{1d} + \tilde{g}_1 u,$$
$$\dot{x}_{2d} = \lambda_1 x_{2d} + g_2 u. \tag{D.13}$$

If we define a new state, $\xi = g_2 x_{1d} - g_1 x_{2d}$, the equation for ξ is

$$\dot{\xi} = g_2 \dot{x}_{1d} - g_1 \dot{x}_{2d} = g_2 \lambda_1 x_{1d} + g_2 g_1 u - g_1 \lambda_1 x_{2d} - g_1 g_2 u = \lambda_1 \xi, \tag{D.14}$$

which does not include the control u; hence ξ is not controllable. The point is that if any two poles are equal *in a diagonal $\mathbf{F}_d$ system with only one input*, we effectively have a hidden mode that is not connected to the control, and the system is not controllable (Fig. D.2a). This is because the two state variables move together exactly, so we cannot *independently* control x_{1d} and x_{2d}.